GAMMA-RAY BURSTS: 30 YEARS OF DISCOVERY

Other Proceedings on Gamma-Ray Bursts

Year	Huntsville Symposia	Publisher	ISBN
1999	Fifth	AIP Conf. Proceedings Vol. 526	1-56396-947-5
1997	Fourth	AIP Conf. Proceedings Vol. 428	1-56396-766-9
1995	Third	AIP Conf. Proceedings Vol. 384	1-56396-685-9
1993	Second	AIP Conf. Proceedings Vol. 307	1-56396-336-1
1991	First	AIP Conf. Proceedings Vol. 265	1-56396-018-4

Other Related Titles from AIP Conference Proceedings

662 Gamma-Ray Burst and Afterglow Astronomy 2001: A Workshop Celebrating the First Year of the HETE Mission
Edited by G. R. Ricker and R. K. Vanderspek, April 2003, 0-7354-0122-5

599 X-Ray Astronomy: Stellar Endpoints, AGN, and the Diffuse X-Ray Background
Edited by Nicholas E. White, December 2001, 0-7354-0043-1

587 Gamma 2001: Gamma-Ray Astrophysics 2001
Edited by Steven Ritz, Neil Gehrels, and Chris R. Shrader, October 2001, 0-7354-0027-X
CD-ROM: 0-7354-0030-X

586 Relativistic Astrophysics: 20th Texas Symposium
Edited by J. Craig Wheeler and Hugo Martel, October 2001, 0-7354-0026-1

515 GeV-TeV Gamma Ray Astrophysics Workshop: Towards a Major Atmospheric Cherenkov Detector VI
Edited by Brenda L. Dingus, Michael H. Salamon, and David B. Kieda, May 2000, 1-56396-938-6

510 The Fifth Compton Symposium
Edited by Mark L. McConnell and James M. Ryan, March 2000, 1-56396-932-7

499 Small Missions for Energetic Astrophysics: Ultraviolet to Gamma-Ray
Edited by Steven P. Brumby, December 1999, 1-56396-912-2

To learn more about these titles, or the AIP Conference Proceedings Series, please visit the webpage **http://proceedings.aip.org**

GAMMA-RAY BURSTS: 30 YEARS OF DISCOVERY

Gamma-Ray Burst Symposium

Santa Fe, New Mexico   8 – 12 September 2003

EDITORS
E. E. Fenimore
M. Galassi
Los Alamos National Laboratory

SPONSORING ORGANIZATIONS
Los Alamos National Laboratory
Center for Space Science and Exploration, Los Alamos
Swift Satellite Mission
Spectrum Astro, Inc.

Melville, New York, 2004
AIP CONFERENCE PROCEEDINGS ■ VOLUME 727

Editors:

E. E. Fenimore
M. Galassi

Los Alamos National Laboratory
MS B244
P. O. Box 1663
Los Alamos, NM 87545
USA

E-mail: efenimore@lanl.gov
mgalassi@lanl.gov

The articles on pp. 613–617, 637–641, and 671–674 were authored by U. S. Government employees and are not covered by the below mentioned copyright.

The article on pp. 3–6 is reprinted with permission from The Astrophysical Journal, 182:L85-L88, 1973 June 1 © 1973. The Astronomical Society.

Authorization to photocopy items for internal or personal use, beyond the free copying permitted under the 1978 U.S. Copyright Law (see statement below), is granted by the American Institute of Physics for users registered with the Copyright Clearance Center (CCC) Transactional Reporting Service, provided that the base fee of $22.00 per copy is paid directly to CCC, 222 Rosewood Drive, Danvers, MA 01923. For those organizations that have been granted a photocopy license by CCC, a separate system of payment has been arranged. The fee code for users of the Transactional Reporting Service is: 0-7354-0208-6/04/$22.00.

© 2004 American Institute of Physics

Individual readers of this volume and nonprofit libraries, acting for them, are permitted to make fair use of the material in it, such as copying an article for use in teaching or research. Permission is granted to quote from this volume in scientific work with the customary acknowledgment of the source. To reprint a figure, table, or other excerpt requires the consent of one of the original authors and notification to AIP. Republication or systematic or multiple reproduction of any material in this volume is permitted only under license from AIP. Address inquiries to Office of Rights and Permissions, Suite 1NO1, 2 Huntington Quadrangle, Melville, N.Y. 11747-4502; phone: 516-576-2268; fax: 516-576-2450; e-mail: rights@aip.org.

L.C. Catalog Card No. 2004112016
ISBN 0-7354-0208-6
ISSN 0094-243X
Printed in the United States of America

CONTENTS

Preface xix
Committees and Sponsors xxi

HISTORY

Observations of Gamma-Ray Bursts of Cosmic Origin 3
    R. W. Klebesadel, I. B. Strong, and R. A. Olson
Gamma-Ray Bursts in Their Historic Context 7
    V. Trimble

GLOBAL PROPERTIES OF GRBs

A Unified Jet Model of X-Ray Flashes and Gamma-Ray Bursts 19
    D. Q. Lamb, T. Q. Donaghy, and C. Graziani
Radiation Processes in GRBs. Prompt Emission 25
    M. V. Medvedev
Broad Band (2-700 keV) Properties of the GRBs Observed
with BeppoSAX 31
    F. Frontera
Observation and Implications of the $E_{peak}-E_{iso}$ Correlation in
Gamma-Ray Bursts 37
    J.-L. Atteia, G. R. Ricker, D. Q. Lamb, T. Sakamoto, C. Graziani,
    T. Donaghy, C. Barraud, and the HETE-2 Science Team
Evidence from HETE-2 for GRB Evolution with Redshift 42
    C. Graziani, D. Q. Lamb, T. Sakamoto, T. Donaghy, J.-L. Atteia, and the
    HETE-2 Science Team
GRB Redshift Evolution within the Unified Jet Model 47
    T. Q. Donaghy, D. Q. Lamb, and C. Graziani
Quiescent Burst Evidence for Two Distinct GRB
Emission Components 52
    J. Hakkila and T. W. Giblin
The HETE-2 Burst Catalog 57
    R. Vanderspek, A. Dullighan, N. Butler, G. B. Crew, J. N. Villasenor,
    G. R. Ricker, T. Tamagawa, T. Sakamoto, M. Suzuki, Y. Shirasaki,
    T. Yamazaki, K. Hurley, C. Graziani, T. Donaghy, D. Q. Lamb, C. Barraud,
    and J.-L. Atteia
Similarities in the Temporal Properties of Gamma-Ray Bursts and
Soft Gamma-Ray Repeaters 61
    S. McBreen, L. Moran, B. McBreen, L. Hanlon, J. French, and M. Conway
Burst Statistics Using the Lag-Luminosity Relationship 65
    D. L. Band, J. P. Norris, and J. T. Bonnell
Short-Bright GRBs: Spectral Properties 69
    G. Ghirlanda, G. Ghisellini, and A. Celotti

The Internal Luminosity Function and GRB Properties 73
    J. Hakkila, T. W. Giblin, S. P. Fuller, K. C. Young, A. D. Stallworth,
    and A. J. Sprague
Prompt Comparison of Data for Optical Transients of
Gamma-Ray Bursts 77
    G. Pizzichini, P. Ferrero, C. Bartolini, A. Guarnieri, and A. Piccioni
Spectral Analysis of 50 GRBs Detected by HETE-2 81
    C. Barraud, J.-L. Atteia, J. F. Olive, K. Hurley, G. Ricker, D. Q. Lamb,
    N. Kawai, R. Vanderspek, T. Sakamoto, and the HETE-2 Science Team
The Cosmological Evolution Trends of GRB Features 86
    D. M. Wei and W. H. Gao
Evidence for Different Spectral Behaviours for Long and Short GRBs 90
    B. M. Belli
Gamma-Ray Bursts in Wavelet Space 94
    Z. Bagoly, I. Horváth, A. Mészáros, and L. G. Balázs

X-RAY FLASHES

Origin of XRFs: Low $\dot{E}$, Low Contrast of Γ or Large Viewing Angle? 101
    C. Barraud, F. Daigne, R. Mochkovitch, and J.-L. Atteia
HETE-2 Observation of the Extremely Soft X-Ray Flashes,
XRF010213, and XRF020903 106
    T. Sakamoto, M. Suzuki, N. Kawai, Y. Nakagawa, A. Yoshida, Y. Shirasaki,
    T. Tamagawa, K. Torii, M. Matsuoka, E. E. Fenimore, M. Galassi,
    D. Q. Lamb, C. Graziani, T. Q. Donaghy, J.-L. Atteia, C. Barraud, M. Boer,
    J.-P. Dezalay, J.-F. Olive, G. Ricker, J. Doty, R. Vanderspek, G. B. Crew,
    J. Villasenor, N. Butler, J. G. Jernigan, K. Hurley, S. E. Woosley,
    G. Pizzichini, and the HETE-2 Science Team
Optical and X-Ray Observations of the Afterglow to XRF 030723 111
    N. Butler, A. Dullighan, P. Ford, G. Ricker, R. Vanderspek, K. Hurley,
    J. Jernigan, D. Lamb, and C. Graziani
Cosmological X-Ray Flashes from Off-Axis Jets 115
    R. Yamazaki, K. Ioka, and T. Nakamura
Comparing Prompt Emission from X-Ray Flashes and
Gamma-Ray Bursts 119
    R. M. Kippen, J. J. M. in 't Zand, P. M. Woods, J. Heise, R. D. Preece,
    and M. S. Briggs

ULTRA-HIGH ENERGY GAMMA-RAYS, NEUTRINOS, GRAVITY WAVES

Ultra-High Energy Gamma-Rays, Neutrinos, and Gravitational Waves
from GRBs... 125
    P. Mészáros, S. Kobayashi, S. Razzaque, and B. Zhang
Milagro—A TeV Observatory for Gamma Ray Bursts 131
    B. L. Dingus and the Milagro Collaboration

Gravitational Radiation from Gamma-Ray Burst Progenitors 136
  S. Kobayashi and P. Mészáros
**High-Energy Cosmic Rays from Galactic and Extragalactic
Gamma-Ray Bursts** 141
  S. D. Wick, C. D. Dermer, and A. Atoyan
**Method for Detecting Neutrinos from Internal Shocks in GRB
Fireballs with AMANDA** 146
  M. Stamatikos and the AMANDA Collaboration
The ARGO-YBJ Sensitivity to GRBs 150
  T. Di Girolamo, G. Di Sciascio, and S. Vernetto
Neutrino Oscillation in Gamma-Ray Burst Fireball 154
  J. C. D'Olivo, J. F. Nieves, and S. Sahu
**The AMANDA Search for High Energy Neutrinos from
Gamma-Ray Bursts** 158
  R. Hardtke and the AMANDA Collaboration
**Limits on Very High Energy Emission from Gamma-Ray Bursts with
the Milagro Observatory** 162
  M. F. Morales for the Milagro Collaboration
GRB Observations around 100 GeV with STACEE 166
  D. A. Williams, L. M. Boone, D. Bramel, J. Carson, C. E. Covault,
  P. Fortin, D. M. Gingrich, D. Hanna, A. Jarvis, J. Kildea, T. Lindner,
  C. Mueller, R. Mukherjee, R. A. Ong, K. Ragan, R. A. Scalzo,
  and J. Zweerink
**Neutrinos and Gamma Rays from Photomeson Processes in
Gamma-Ray Bursts** 170
  A. Atoyan and C. D. Dermer
**Gravitational Waves and GRBs from Tidal Disruption of Stars in the
Center of Galaxies** 174
  P. Fortini and A. Ortolan

PROMPT EMISSION AND EARLY AFTERGLOWS

Some Recent Peculiarities of the Early Afterglow 181
  T. Piran, E. Nakar, and J. Granot
Early Stages of the GRB Explosion 187
  A. M. Beloborodov
**Broad-Band (2-400 keV) Spectra of Gamma-Ray Bursts and
X-Ray Flashes Based on HETE-2 Observations** 192
  N. Kawai, T. Sakamoto, M. Suzuki, M. Matsuoka, A. Yoshida, Y. Shirasaki,
  T. Tamagawa, Y. Nakagawa, Y. Yamazaki, R. Sato, K. Torii,
  E. E. Fenimore, M. Galassi, D. Q. Lamb, C. Graziani, T. Q. Donaghy,
  G. Ricker, J. Doty, R. Vanderspek, G. B. Crew, J. Villasenor, N. Butler,
  J.-L. Atteia, C. Barraud, M. Boer, J.-P. Dezalay, J.-F. Olive, J. G. Jernigan,
  K. Hurley, S. E. Woosley, G. Pizzichini, and the HETE-2 Science Team
Heating and Deceleration of GRB Fireballs by Neutron Decay 198
  E. M. Rossi, A. M. Beloborodov, and M. J. Rees

**Discovery of a Distinct Higher Energy Spectral Component
in GRB941017** 203
    M. M. González, B.-L. Dingus, Y. Kaneko, R. D. Preece, C. D. Dermer,
    and M. S. Briggs

**Early Afterglow, Magnetized Central Engine, and a Quasi-Universal
Jet Configuration for Long GRBs** 208
    B. Zhang, S. Kobayashi, P. Mészáros, N. M. Lloyd-Ronning, and X. Dai

Further Analysis of GRB 030501 213
    M. Topinka

**Durations of Gamma-Ray Bursts and X-Ray Flashes in X-Ray and
Gamma-Ray Bands Observed with HETE-2** 217
    M. Suzuki, N. Kawai, A. Yoshida, Y. Shirasaki, M. Matsuoka,
    T. Tamagawa, K. Torii, T. Sakamoto, C. Graziani, D. Q. Lamb, J.-L. Atteia,
    E. E. Fenimore, M. Galassi, T. Donaghy, G. Ricker, J. Doty, R. Vanderspek,
    G. B. Crew, J. Villasenor, N. Butler, J. G. Jernigan, C. Barraud, M. Boer,
    J.-P. Dezalay, J.-F. Olive, K. Hurley, S. E. Woosley, and the HETE-2
    Science Team

**Crude Limits on Prior and Prompt Optical Emission from GRBs
from CONCAMs of the Night Sky Live Global Network** 221
    R. J. Nemiroff, D. Perez-Ramirez, and D. Cordell

INTEGRAL Spectrometer Analysis of GRB030227 and GRB030131 225
    L. Moran, L. Hanlon, B. McBreen, R. Preece, Y. Kaneko, O. R. Williams,
    K. Bennett, R. M. Kippen, A. Von Kienlin, V. Beckmann, S. McBreen,
    and J. French

XMM-Newton Observations of Gamma-Ray Burst Afterglows 229
    N. Schartel

**Particle Acceleration via Relativistic Magnetic-Dominated Expansion
and GRBs** 233
    E. Liang

**Spectral Time Evolution for GRBs Observed by BATSE
and EGRET-TASC** 236
    M. M. González, B. L. Dingus, Y. Kaneko, R. D. Preece, and M. S. Briggs

GRB Optical Prompt Emission: The Role of Monitors 240
    R. Hudec

**COMPTEL Observation of GRB941017 with Distinct
High-Energy Component** 244
    Y. Kaneko, L. Hanlon, R. D. Preece, M. M. González, B. L. Dingus,
    M. S. Briggs, O. R. Williams, K. Bennett, and C. Winkler

RELATIVISTIC JETS AND POLARIZATION

**Linear Polarization on Gamma-Ray Bursts: From the Prompt to the
Late Afterglow** 251
    D. Lazzati

Magnetic Acceleration and Collimation of Gamma-Ray Burst Jets 257
    A. Königl

Polarization Measurements of GRBs with RHESSI 262
   W. Coburn and S. E. Boggs
The Polarization Evolution of the Optical Afterglow of GRB 030329 269
   J. Greiner, S. Klose, K. Reinsch, H. M. Schmid, R. Sari, D. H. Hartmann,
   C. Kouveliotou, A. Rau, E. Palazzi, C. Straubmeier, B. Stecklum,
   S. Zharikov, G. Tovmassian, O. Bärnbantner, C. Ries, E. Jehin, A. Henden,
   A. A. Kaas, T. Grav, J. Hjorth, H. Pedersen, R. A. M. J. Wijers, A. Kaufer,
   H.-S. Park, G. Williams, and O. Reimer
Comparison of Three Afterglow Morphologies 274
   J. D. Salmonson, E. Rossi, and D. Lazzati
Collapsar Jet Stability at Breakout 278
   E. A. Gómez and P. E. Hardee
**Large-Scale Magnetic Fields in GRB Outflows: Acceleration,
Collimation, and Neutron Decoupling** 282
   N. Vlahakis and A. Königl
Computational Relativistic Fluids and Jet Formation 286
   G. Richardson, K.-I. Nishikawa, S. Koide, and K. Shibata
**Particle Acceleration and Radiation Associated with Magnetic Field
Generation from Relativistic Collisionless Shocks** 290
   K.-I. Nishikawa, P. Hardee, G. Richardson, R. Preece, H. Sol,
   and G. J. Fishman
**The "Supercritical Pile" Model of GRB: Thresholds, Polarization,
Time Lags** 294
   D. Kazanas, M. Georganopoulos, and A. Mastichiadis

GRB030329

The GRB-SN Connection: GRB 030329 and XRF 030723 301
   J. P. U. Fynbo, J. Hjorth, J. Sollerman, P. Møller, J. Gorosabel, F. Grundahl,
   B. L. Jensen, M. I. Andersen, P. Vreeswijk, A. Castro-Tirado, and the
   GRACE Collaboration
**Earliest Detection of the Optical Afterglow of GRB 030329
and its Variability** 307
   R. Sato, N. Kawai, M. Suzuki, Y. Yatsu, J. Kataoka, R. Takagi,
   K. Yanagisawa, and H. Yamaoka
A New Astrophysical "Triptych": GRB030329/SN2003dh/URCA-2 312
   M. G. Bernardini, C.-L. Bianco, P. Chardonnet, F. Fraschetti, R. Ruffini,
   and S.-S. Xue
The X-Ray Afterglow of GRB030329 at Early and Late Times 316
   A. Tiengo, S. Mereghetti, G. Ghisellini, E. Rossi, G. Ghirlanda,
   and N. Schartel
**Structure in Early Afterglow Light Curves: GRB021004
and GRB030329** 320
   M. Uemura, R. Ishioka, T. Kato, D. Nogami, and H. Yamaoka
High Resolution Observations of GRB 030329 324
   G. Taylor, D. Frail, E. Berger, and S. Kulkarni

The Low-Luminosity Tail of the GRB Distribution:
The Case of GRB 980425 328
    F. Daigne and R. Mochkovitch
GRB 030329 with SARA and TLS 333
    K. Lindsay, A. Zeh, D. H. Hartmann, S. Klose, S. Shaw, M. Leake,
    J. Webb, B. Stecklum, M. Williams, and E. Howard
A Search for Short Time-Scale Optical Variability in the
GRB 030329 Afterglow 337
    N. Mirabal, J. P. Halpern, M. Bureau, and K. Fathi
Colors of the Optical Afterglow of GRB030329/SN 2003dh 339
    V. Šimon, R. Hudec, and G. Pizzichini

GRB PROGENITORS

The Collapsar Model for Gamma-Ray Bursts 343
    S. E. Woosley, W. Zhang, and A. Heger
A Field Guide to Collapsars 349
    E. Ramirez-Ruiz
Circumburst Environments of Gamma-Ray Bursts....................... 355
    R. A. Chevalier
Dynamos, Super-Pulsars and Gamma-Ray Bursts....................... 361
    S. Rosswog and E. Ramirez-Ruiz
GRB 021004: A Possible Shell Nebula around a Wolf-Rayet Star
Gamma-Ray Burst Progenitor 366
    N. Mirabal, J. P. Halpern, R. Chornock, A. V. Filippenko, and
    D. M. Terndrup
Stellar Collapse and the Formation of Black Holes....................... 371
    C. L. Fryer and R. Dupuis
Numerical Simulations of Relativistic Jets in Collapsars 376
    W. Zhang, S. E. Woosley, and A. Heger
The First Steps in the Life of a Short GRB 380
    M.-A. Aloy, H.-T. Janka, and E. Müller
MHD Simulations of the Collapsar Model for GRBs 384
    D. Proga, A. I. MacFadyen, P. J. Armitage, and M. C. Begelman
Searching for GRB Remnants in Nearby Galaxies 388
    S. G. Bhargavi, J. Rhoads, R. Perna, J. Feldmeier, and J. Greiner
General Relativistic MHD Simulations of the Gravitational
Collapse of a Rotating Star with Magnetic Field as a Model
of Gamma-Ray Bursts 392
    Y. Mizuno, S. Yamada, S. Koide, and K. Shibata

GRB CONNECTION TO SUPERNOVAE

Previously Claimed(/Unclaimed) X-Ray Emission Lines in High Resolution Afterglow Spectra 399
    N. Butler, A. Dullighan, P. Ford, G. Ricker, R. Vanderspek, K. Hurley,
    J. Jernigan, and D. Lamb

SN 2002lt and GRB 021211: A SN/GRB Connection at $z=1$ 403
    M. Della Valle, D. Malesani, S. Benetti, V. Testa, M. Hamuy,
    L. A. Antonelli, G. Chincarini, G. Cocozza, S. Covino, P. D'Avanzo,
    D. Fugazza, G. Ghisellini, R. Gilmozzi, D. Lazzati, E. Mason, P. Mazzali,
    and L. Stella

Search for Correlations between BATSE Gamma-Ray Bursts and Supernovae 408
    J. Polcar, M. Topinka, R. Hudec, V. Hudcová, N. Masetti, G. Pizzichini,
    and E. Palazzi

How Can the SN-GRB Time Delay be Measured? 412
    J. P. Norris and J. T. Bonnell

GRB 980425 in the Off-Axis Jet Model of the Standard GRBs............ 416
    R. Yamazaki, D. Yonetoku, and T. Nakamura

Color Superconductivity in Compact Stars and Gamma-Ray Bursts........ 420
    A. Drago, A. Lavagno, and G. Pagliara

The GRB 980425-SN1998bw Association in the EMBH Model 424
    F. Fraschetti, M. G. Bernardini, C. L. Bianco, P. Chardonnet, R. Ruffini,
    and S.-S. Xue

GRB 970228 within the EMBH Model 428
    A. Corsi, M. G. Bernardini, C. L. Bianco, P. Chardonnet, F. Fraschetti,
    R. Ruffini, and S.-S. Xue

DARK VERSUS BRIGHT GRBs

Chandra Observations of the Optically Dark GRB 030528 435
    N. Butler, A. Dullighan, P. Ford, G. Ricker, R. Vanderspek, K. Hurley,
    J. Jernigan, and D. Lamb

Discovery of the Faint Near-IR Afterglow of GRB 030528 439
    A. Rau, J. Greiner, S. Klose, J. Castro Cerón, A. Fruchter,
    A. Küpcü Yoldaş, J. Gorosabel, A. Levan, J. Rhoads, and N. Tanvir

Dust and Gamma-Ray Bursts: Mutual Implications 443
    S. D. Vergani, E. Molinari, F. M. Zerbi, and G. Chincarini

Four Years of Observations of GRB Localizations with TAROT 447
    M. Boër, A. Klotz, C. Thiébaud, J.-L. Atteia, R. Malina,
    J. de Freitas Pacheco, and H. Pedersen

LATE AFTERGLOWS

Damped Lyα Systems in GRB Afterglows 453
    P. Vreeswijk, S. Ellison, C. Ledoux, R. Wijers, J. Hjorth, J. Fynbo,
    A. Fruchter, and the GRACE Collaboration
On the Shallow Decay of Some GRB Afterglows 458
    A. Panaitescu and P. Kumar
Relativistic Wind Bubbles 463
    Z. G. Dai
IR and Optical Observations of GRB 030115 467
    A. Dullighan, G. Ricker, N. Butler, and R. Vanderspek
**Observations of Optical Afterglows of Gamma-Ray Bursts
from Loiano** 471
    C. Bartolini, A. Guarnieri, A. Piccioni, G. Pizzichini, and P. Ferrero
GRB Afterglows in the Deep Newtonian Phase 475
    Y. F. Huang, K. S. Cheng, Z. G. Dai, and T. Lu
Optical Orphan Afterglows: Observational Aspects 479
    R. Hudec
The Optical Afterglow of GRB 030226 483
    S. Klose, J. Greiner, A. Zeh, A. Rau, A. A. Henden, D. H. Hartmann,
    N. Masetti, A. J. Castro-Tirado, J. Hjorth, E. Pian, N. R. Tanvir,
    R. A. M. J. Wijers, and E. van den Heuvel
Colors of Optical Afterglows of GRBs and Their Time Evolution 487
    V. Šimon, R. Hudec, G. Pizzichini, and N. Masetti
**The GRB 030227 Detected by INTEGRAL: Another Sign of
Compton Scattering in X-Rays** 491
    A. J. Castro-Tirado, J. Gorosabel, S. Guziy, D. Reverte, J. M. Castro Cerón,
    A. de Ugarte Postigo, N. Tanvir, S. Mereghetti, A. Tiengo, S. B. Pandey,
    N. Masetti, H. Pedersen, M. D. Pérez Ramírez, and the
    GRACE Collaboration

GRBs AND COSMOLOGY

Energetics and the GRB Hubble Diagram 497
    J. S. Bloom
Towards Measuring the Cosmic Gamma-Ray Burst Rate 503
    P. A. Price and B. P. Schmidt
SCUBA Observations of the Host Galaxies of Gamma-Ray Bursts 508
    V. E. Barnard, N. R. Tanvir, A. W. Blain, A. Fruchter, C. Kouveliotou,
    P. Natarajan, E. Ramirez-Ruiz, E. Rol, I. A. Smith, R. P. J. Tilanus, and
    R. A. M. J. Wijers
**Near-Infrared Colors of Gamma-Ray Burst Afterglows and
Cosmic Reionization History** 514
    A. K. Inoue, R. Yamazaki, and T. Nakamura
Detectability of Long GRB Afterglows from Very High Redshifts 518
    L. Gou, P. Mészáros, T. Abel, and B. Zhang

Probing Cosmological Parameters with GRBs..............................522
    T. Di Girolamo, M. Vietri, and G. Di Sciascio

GENERAL OBSERVATIONS

GRBlog: A Database for Gamma-Ray Bursts..............................529
    R. Quimby, E. McMahon, and J. Murphy
Was the X-Ray Afterglow of GRB 970815 Detected? 533
    N. Mirabal, J. P. Halpern, E. V. Gotthelf, and R. Mukherjee
The Optical Afterglow of GRB 020305 537
    J. Gorosabel, J. P. U. Fynbo, A. S. Fruchter, P. Nugent, J. M. Castro Cerón,
    A. Levan, J. Rhoads, D. Bersier, I. Burud, A. J. Castro-Tirado, and
    J. Hjorth
DMSP 14 Observations of GRB011121 and the Giant SGR1900+14 Flare
of 98/08/27 541
    J. Terrell and R. W. Klebesadel

GENERAL THEORY

An Integrated Universal Collapsar Gamma-Ray Burst Model 547
    J. D. Salmonson
Electromagnetic (versus Fireball) Model of GRBs 552
    M. Lyutikov
On Hadronic Models for the Anomalous γ-Ray Emission Component
in GRB 941017 557
    C. D. Dermer and A. Atoyan
Evidence for GRB Induced Extinctions in the Fossil Record?...............562
    T. G. Kaye
Observations of X-Ray Bursts by HETE-2.................................566
    Y. E. Nakagawa, T. Yamazaki, M. Suzuki, A. Yoshida, N. Kawai,
    D. Takahashi, M. Matsuoka, Y. Shirasaki, T. Tamagawa, K. Torii,
    T. Sakamoto, Y. Urata, R. Sato, Y. Yamamoto, E. E. Fenimore, M. Galassi,
    D. Q. Lamb, C. Graziani, and G. Ricker
Magnetic Field Generation in Relativistic Shocks 570
    J. Wiersma and A. Achterberg
Very Short Gamma-Ray Bursts: New Physics? 574
    D. B. Cline
Firework Model: Time Dependent Spectral Evolution of GRB.............578
    G. Barbiellini, F. Longo, G. Ghirlanda, A. Celotti, and Z. Bosnjak

ANALYSIS AND OBSERVATION TECHNIQUES

The Gamma-Ray Burst ToolSHED is Open for Business....................585
    T. W. Giblin, J. Hakkila, D. J. Haglin, and R. J. Roiger
Future Prospects for High Energy Polarimetry of Gamma-Ray Bursts........589
    M. L. McConnell

The KLENOT Telescope and GRBs..593
    M. Tichý, J. Tichá, M. Kočer, R. Hudec, and V. Šimon
Burst Populations and Detector Sensitivity..............................597
    D. L. Band
A Brief Comment on One of the "Amati Relationships" for
Gamma-Ray Bursts 601
    G. Pizzichini

PRESENT SATELLITES

Gamma-Ray Bursts Observed by INTEGRAL 607
    S. Mereghetti
The Past, Present, and Future of the Third Interplanetary Network..........613
    K. Hurley and T. Cline
In-Flight Calibration of the HETE-2 WXM Detector Response 618
    T. Sakamoto, Y. Nakagawa, K. Torii, Y. Shirasaki, T. Tamagawa, N. Kawai,
    A. Yoshida, M. Matsuoka, E. E. Fenimore, M. Galassi, D. Q. Lamb,
    C. Graziani, T. Q. Donaghy, J.-L. Atteia, C. Barraud, M. Boer,
    J.-P. Dezalay, J.-F. Olive, and the HETE-2 Science Team
Gamma-Ray Bursts Observed with the Spectrometer SPI
Onboard INTEGRAL 622
    A. von Kienlin, A. Rau, V. Beckmann, and S. Deluit
The Growing SXC Burst Catalog: A Transient for Each Detection 626
    J. Villasenor, J. G. Jernigan, G. Crew, R. Vanderspek, A. Dullighan,
    N. Butler, G. Prigozhin, J. Doty, and G. Ricker
Current Status of HETE-2 Operations 630
    R. Vanderspek, N. Butler, G. B. Crew, A. Dullighan, G. Prigozhin,
    J. P. Doty, J. N. Villasenor, G. R. Ricker, T. Tamagawa, K. Torii, N. Kawai,
    T. Sakamoto, R. Sato, M. Suzuki, Y. Urata, Y. Yamamoto, A. Yoshida,
    Y. E. Nakagawa, T. Yamazaki, Y. Shirasaki, C. Graziani, T. Donaghy,
    D. Q. Lamb, J. G. Jernigan, K. Hurley, J.-L. Atteia, E. E. Fenimore,
    and M. Galassi

SWIFT SATELLITE

The Swift Gamma-Ray Burst Mission....................................637
    N. Gehrels
The X-ray Telescope for the SWIFT Gamma-Ray Burst Mission.............642
    A. Wells, D. N. Burrows, J. E. Hill, J. A. Nousek, G. Chincarini,
    A. F. Abbey, A. Beardmore, J. Bosworth, H. W. Brauninger, W. Burkert,
    S. Campana, M. Capalbi, W. Chang, O. Citterio, M. J. Freyberg,
    P. Giommi, G. D. Hartner, R. Killough, B. Kittle, R. Klar, C. Mangels,
    M. McMeekin, B. J. Miles, A. Moretti, K. Mori, D. C. Morris,
    K. Mukerjee, J. P. Osborne, G. Tagliaferri, F. Tamburelli, D. J. Watson,
    R. Willingale, and M. Zugger

Flight Calibration and Operations of the Swift X-Ray
Telescope (XRT) 647
    D. N. Burrows, J. E. Hill, J. A. Nousek, A. A. Wells, J. P. Osborne,
    K. Mukerjee, G. Chincarini, G. Tagliaferri, and S. Campana

The Swift Ultra-Violet/Optical Telescope (UVOT) 651
    P. W. A. Roming, S. D. Hunsberger, J. A. Nousek, M. Ivanushkina,
    K. O. Mason, and A. A. Breeveld

Swift Burst Alert Telescope Hard X-Ray Monitor and Survey 655
    H. A. Krimm, P. Banat, S. D. Barthelmy, T. Belloni, J. R. Cummings,
    A. Dean, E. E. Fenimore, N. Gehrels, C. B. Markwardt, D. M. Palmer,
    A. M. Parsons, J. Tueller, and D. Willis

Swift Burst Alert Telescope Data Products and Analysis Software 659
    H. A. Krimm, L. M. Barbier, S. D. Barthelmy, J. R. Cummings,
    E. E. Fenimore, N. Gehrels, D. D. Hullinger, C. B. Markwardt,
    D. M. Palmer, A. M. Parsons, and J. Tueller

The BAT-Swift Science Software 663
    D. M. Palmer, E. E. Fenimore, M. Galassi, K. Mclean, T. Tavenner,
    S. Barthelmy, M. Blau, J. Cummings, N. Gehrels, D. Hullinger, H. Krimm,
    C. Markwardt, R. Mason, J. Ong, J. Polk, A. Parsons, L. Shackelford,
    J. Tueller, S. Walling, Y. Okada, H. Takahashi, M. Toshiro, M. Suzuki,
    G. Sato, T. Takahashi, and S. Watanabe

Setting the Triggering Thresholds on Swift............................... 667
    K. M. McLean, E. E. Fenimore, D. Palmer, S. Barthelmy, N. Gehrels,
    H. Krimm, C. B. Markwardt, and A. Parsons

Swift Burst Alert Telescope (BAT) Instrument Response 671
    A. Parsons, S. Barthelmy, J. Cummings, N. Gehrels, D. Hullinger,
    H. Krimm, C. Markwardt, J. Tueller, E. E. Fenimore, D. Palmer, G. Sato,
    T. Takahashi, K. Nakazawa, Y. Okada, H. Takahashi, M. Suzuki,
    and M. Tashiro

FUTURE SATELLITES

Observing GRBs with EXIST 677
    D. H. Hartmann, J. Grindlay, J. Hong, A. Loeb, R. Blandford, W. Craig,
    J. Fishman, C. Kouveliotou, N. Gehrels, D. Band, F. Harrison, and
    S. E. Woosley

GLAST and Gamma-Ray Bursts: Probing Photon Propagation over
Cosmological Distances 681
    N. Omodei, J. Cohen-Tanugi, and F. Longo

The GLAST Burst Monitor... 684
    P. N. Bhat, C. A. Meegan, G. G. Lichti, M. S. Briggs, V. Connaughton,
    R. Diehl, G. J. Fishman, J. Greiner, R. M. Kippen, C. Kouveliotou,
    W. S. Paciesas, R. D. Preece, V. Schönfelder, R. B. Wilson, and A. von
    Kienlin

GLAST's GBM Burst Trigger.. 688
    D. Band, M. Briggs, V. Connaughton, R. M. Kippen, and R. Preece

Analysis of Burst Observations by GLAST's LAT Detector 692
  D. L. Band, S. W. Digel, the GLAST LAT Collaboration, and the GLAST
  Science Support Center
SuperAGILE: The Hard X-Ray Imager of AGILE......................... 696
  M. Feroci, E. Costa, L. Barbanera, E. Del Monte, G. Di Persio, M. Frutti,
  I. Lapshov, F. Lazzarotto, M. Mastropietro, E. Morelli, L. Pacciani,
  G. Porrovecchio, B. Preger, M. Rapisarda, A. Rubini, P. Soffitta, M. Tavani,
  A. Argan, G. Ghirlanda, S. Mereghetti, A. Pellizzoni, S. Vercellone,
  G. Barbiellini, F. Longo, M. Prest, and E. Vallazza
The Test Equipment of the AGILE Minicalorimeter Prototype.............. 700
  M. Trifoglio, A. Bulgarelli, F. Gianotti, E. Celesti, G. Di Cocco, C. Labanti,
  A. Mauri, M. Prest, E. Vallazza, and T. Froysland
AGILE Sensitivity and GRB Spectral Properties 704
  G. Ghirlanda, M. Galli, F. Longo, B. Preger, A. Argan, G. Barbiellini,
  S. Mereghetti, A. Pellizzoni, M. Tavani, and S. Vercellone
Scaling and GRB Mission Optimization 708
  J. Doty
X-Ray Monitoring of GRBs with Lobster Eye Telescopes 712
  L. Svéda, R. Hudec, A. Inneman, L. Pína, and G. Pizzichini

ROBOTIC OBSERVING SYSTEMS

**Exploring the First Minute: New Technology for Measuring Color
and Polarization Variations in Prompt Optical Emission**................... 719
  W. T. Vestrand, D. J. Casperson, C. Ho, E. Raby, R. Shirey, D. Thompson,
  R. R. White, and J. Wren
**The Search for Optical and Near-Infrared Counterparts of GRBs
with the Super-LOTIS Telescope** 723
  G. G. Williams, H. S. Park, S. D. Barthelmy, D. H. Hartmann,
  K. C. Hurley, P. A. Milne, K. J. Lindsay, M. Bradshaw, R. E. Wurtz, and
  J. Wickersham
**Mining the Sky for Explosive Optical Transients with
Both Eyes Open** 728
  W. T. Vestrand, K. Borozdin, D. J. Casperson, S. Davidoff, H. Davis,
  E. E. Fenimore, M. Galassi, K. McGowan, D. Starr, R. R. White,
  P. Wozniak, and J. Wren
**RAPTOR-Scan: Identifying and Tracking Objects through
Thousands of Sky Images** 733
  S. Davidoff and P. Wozniak
**A Rapid-Response Gamma-Ray Burst Afterglow Observing Program
at Etelman Observatory in the US Virgin Islands**......................... 737
  T. W. Giblin, J. E. Neff, J. Hakkila, D. Hartmann, N. Andresian-Thomas,
  and D. M. Drost
**Watcher: A Telescope for Rapid Gamma-Ray Burst
Follow-Up Observations**.. 741
  J. French, L. Hanlon, B. McBreen, S. McBreen, L. Moran, N. Smith,
  A. Giltinan, P. Meintjes, and M. Hoffman

The University of Wyoming GRB Afterglow Follow-Up Program 745
    S. L. Savage, J. P. Norris, A. S. Kutyrev, M. Pierce, and R. Canterna

Scout or Cavalry? Optimal Discovery Strategies for GRBs 749
    R. J. Nemiroff

RTS2—Remote Telescope System, 2nd Version 753
    P. Kubánek, M. Jelínek, M. Nekola, M. Topinka, J. Štrobl, R. Hudec,
    T. de J. Mateo Sanguino, A. de Ugarte Postigo, and A. J. Castro-Tirado

**Wide Field Optical Camera for Search and Investigation of
Fast Cosmic Transients** 757
    A. Pozanenko, G. Beskin, S. Bondar, A. Biryukov, K. Hurley, E. Ivanov,
    S. Karpov, V. Loznikov, V. Rumyantsev, and Y. Zolotukhin

**BOOTES: Technological Developments and Scientific Results by a
Stereoscopic System with Two Stations Spaced by 240 km** 761
    T. de J. Mateo Sanguino, A. J. Castro-Tirado,
    A. de Ugarte Postigo, M. T. Fernández Palomo, J. M. Castro Cerón,
    J. A. Berná Galiano, P. Páta, J. Soldán, M. Bernas, R. Hudec, M. Jelínek,
    S. Vítek, P. Kubánek, S. McBreen, J. Gorosabel, C. E. García Dabó,
    T. Soria, B. A. de la Morena Carretero, and J. Torres Riera

REM. Rapid Eye Mount 765
    E. Molinari, S. D. Vergani, F. M. Zerbi, S. Covino, and G. Chincarini

GRB 2003 Conference Participants 769
Author Index 777

PREFACE

In 1969 Ray Klebesadel set out to show that there were no natural events that could spoof the "Vela" instruments recently built by Los Alamos to detect violation of the 1963 treaty banning nuclear weapons in space. At the time the universe was considered mostly static, just as Aristotle had said it should be. This was before the discoveries of x-ray bursts, active galactic nuclei, cataclysmic variables, blazars, jets, and other denizens that we now know inhabit the sky, making it a wonderous place to see physics at work. To his surprise, Ray found near simultaneous events in multiple satellites. Others had seen strange glitches in their instruments, but no one had multiple satellite measurements to allow them to eliminate particle disturbances. Still, the discovery could not be asserted because the satellites did not have high enough time resolution to eliminate the sun as a possible source. By 1973 new Vela satellites with better time resolution had been launched, sixteen non-solar events found, and the landmark paper "Observations of Gamma-Ray Bursts of Cosmic Origin" published.

Thirty years later, the scientific community gathered in Santa Fe, New Mexico, in the shadow of Los Alamos to commemorate the recent progress in this ever-expanding field. Gamma-Ray Bursts are now known to come from explosions literally on the other side of the universe. The Burst and Transient Source Experiment (BATSE) discovered that gamma-ray bursts prefer no direction, implying they are farther than all nearby galaxies. The BeppoSax satellite located gamma-ray bursts and within hours slewed its x-ray telescope to discover fading afterglows which could be identified with galaxies so far away that the universe has doubled in size since the burst. The High Energy Transient Explorer (HETE) autonomously located bursts on-board the satellite within seconds and fed the information to waiting telescopes on the ground. Stanek et al. observed the March 29 2003 HETE burst within hours to discover the signature of a special type of supernova, solidifying that some gamma-ray bursts are the birth pains of black holes when massive stars in the early universe exhaust their fuel and collapse.

Over four days in September 2003, 220 participants discussed the latest results and their hopes for the future. Much progress has been made in understanding what makes Gamma-ray bursts tick and how to use their brilliance to backlight and study ever deeper regions of the universe. Soon we will be in the *Swift* era when bursts are not only located on-board the satellite within seconds, but the satellite will re-point within minutes to observe the afterglows with x-ray and ultra-violet telescopes. The field founded by Ray Klebesadel 30 years ago will continue to expand our knowledge of unique physics and the evolution of the universe.

Santa Fe, New Mexico, USA
June 2004

Ed Fenimore
Mark Galassi

SCIENTIFIC ORGANIZING COMMITTEE

Brenda Dingus
Enrico Costa
Ed Fenimore (chair)
Jerry Fishman
Dale Frail
Filippo Frontera
Andy Fruchter
Chris Fryer
Neil Gehrels
Kevin Hurley
Nobu Kawai
Shri Kularni

Chyryssa Kouveliotou
Don Lamb
Nicole Lloyd-Ronning
Luigi Piro
Dan Reichart
George Ricker
Re'em Sari
Marco Tavani
Tom Vestrand
Ralph Wijers
Stan Woosley

LOCAL ORGANIZING COMMITTEE

Brenda Dingus
Mary Dugan
Richard Epstein
Chris Fryer
Ed Fenimore (chair)
Herb Funsten

Mark Galassi
Marc Kippen
David Palmer
Bill Priedhorsky
Tom Vestrand

SPONSORS

Los Alamos National Laboratory
Center for Space Science and Exploration (Los Alamos)
Swift Satellite Mission
Spectrum Astro, Inc.

HISTORY

OBSERVATIONS OF GAMMA-RAY BURSTS OF COSMIC ORIGIN

Ray W. Klebesadel, Ian B. Strong, and Roy A. Olson

University of California, Los Alamos Scientific Laboratory, Los Alamos, New Mexico
Received 1973 March 16; revised 1973 April 2

ABSTRACT

Sixteen short bursts of photons in the energy range 0.2–1.5 MeV have been observed between 1969 July and 1972 July using widely separated spacecraft. Burst durations ranged from less than 0.1 s to $\sim$30 s, and time-integrated flux densities from $\sim 10^{-5}$ ergs cm^{-2} to $\sim 2 \times 10^{-4}$ ergs cm^{-2} in the energy range given. Significant time structure within bursts was observed. Directional information eliminates the Earth and Sun as sources.

Subject headings: gamma rays — X-rays — variable stars

I. INTRODUCTION

On several occasions in the past we have searched the records of data from early *Vela* spacecraft for indications of gamma-ray fluxes near the times of appearance of supernovae. These searches proved uniformly fruitless. Specific predictions of gamma-ray emission during the initial stages of the development of supernovae have since been made by Colgate (1968). Also, more recent *Vela* spacecraft are equipped with much improved instrumentation. This encouraged a more general search, not restricted to specific time periods. The search covered data acquired with almost continuous coverage between 1969 July and 1972 July, yielding records of 16 gamma-ray bursts distributed throughout that period. Search criteria and some characteristics of the bursts are given below.

II. INSTRUMENTATION

The observations were made by detectors on the four *Vela* spacecraft, *Vela 5A*, *5B*, *6A*, and *6B*, which are arranged almost equally spaced in a circular orbit with a geocentric radius of $\sim 1.2 \times 10^5$ km.

On each spacecraft six 10 cm^3 CsI scintillation counters are so distributed as to achieve a nearly isotropic sensitivity. Individual detectors respond to energy depositions of 0.2–1.0 MeV for *Vela 5* spacecraft and 0.3–1.5 MeV for *Vela 6* spacecraft, with a detection efficiency ranging between 17 and 50 percent. The scintillators are shielded against direct penetration by electrons below ~ 0.75 MeV and protons below ~ 20 MeV. A high-Z shield attenuates photons with energy below that of the counting threshold. No active anticoincidence shielding is provided.

Normalized output pulses from the six detectors are summed into the counting and logics circuitry. Logical sensing of a rapid, statistically significant rise in count rate initiates the recording of discrete counts in a series of quasi-logarithmically increasing time intervals. This capability provides continuous coverage in time which, coupled with isotropic response, is unique in observatonal astronomy. A time measurement is also associated with each record.

The data accumulations include a background component due to cosmic particles and their secondary effects. The observed background rate, which is a function of the energy threshold, is ~ 150 counts per second for the *Vela 5* spacecraft and ~ 20 counts per second for the *Vela 6* spacecraft.

III. OBSERVATIONS

Since these detectors are susceptible to stimulation by energetic particles, the following evidence is offered in support of the interpretation that the signals reported here are due to fluxes of photons within the quoted energy range. Other Vela detectors with high sensitivity to energetic charged particles and neutrons recorded no deviation from the steady counting rate induced by cosmic particle fluxes at the time of any of the observed bursts. It has been noted, furthermore, that the detailed time structure of each burst is reproduced at all spacecraft recording the event, even though the radiation must, in most cases, have traversed an appreciable portion of the geomagnetic field. Simple calculations show that electron energies of many GeV and proton energies of many MeV would be required to produce this degree of rigidity, and fluxes of such particles would create observable effects in the other instruments on the spacecraft. Additionally, no difference in the time of arrival of the stimulating signals at two different spacecraft has been found which exceeds 0.8 s, the maximum transit time for light, even though the search allowed a deviation from simultaneity as great as 4 s.

A count-rate record is generated only in response to a rapid rise in count rate to a level significantly above background. The frequency with which individual records are generated is relatively high for *Vela 5* spacecraft. Modifications to *Vela 6* detectors reduced this frequency, at some cost in sensitivity, to an insignificant level. Only 47 such records have been generated by both *Vela 6* spacecraft over a 2-year period, 22 of which are responses, in coincidence, to the bursts reported here. Present processing requires that at least two spacecraft record the burst with a deviation from simultaneity of 4 s or less. Sixteen events have been observed to meet these criteria, two of which were recorded by all four spacecraft. Absence of consistent response from all four spacecraft can be attributed in most cases to an inappropriate mode of operation or to marginal signal levels.

These bursts display a wide variety of characteristics. Time durations range from less than a second to about 30 s. Some count-rate records have a number of clearly resolved peaks while others do not appear to display any significant structure. The time-integrated flux density in the measured energy interval ranges from the minimum identifiable level of $\sim 10^{-5}$ ergs cm^{-2} to more than 2×10^{-4} ergs cm^{-2}. Instantaneous flux densities have exceeded 4×10^{-4} ergs cm^{-2} s^{-1}. An indication of the spectral distribution of the incident flux may be derived from the ratio of the response in the two energy intervals in those cases where both *Vela 5* and *Vela 6* spacecraft recorded the burst.

Allowing for differing energy thresholds and statistical fluctuations, the integrated flux for a particular event is independent of the recording spacecraft. Differences in the time of arrival of the signals at the various spacecraft imply that the spacecraft are not equidistant from any given source. Inverse-square law considerations thereby place the sources at a distance of at least 10 orbit diameters, or several million kilometers.

Arrival-time differences have been derived approximately in all cases, and fairly accurate (± 0.05 s) for a number of cases. For a two-spacecraft coincidence the transit delay defines a circle on the celestial sphere on which the source position must lie. For three spacecraft we can define intersecting circles, whose points of intersection represent the source position and its mirror image in the orbital plane of the spacecraft, a presently unresolved ambiguity. Nevertheless, it has been possible by this technique to rule out the sun as a source. Also, in none of the 16 cases was there found any close correlation with any recorded indications of solar activity.

One event has been observed which almost certainly was associated with a solar outburst. It differs distinctly from the 16 bursts reported here, and will be described in detail at a later date.

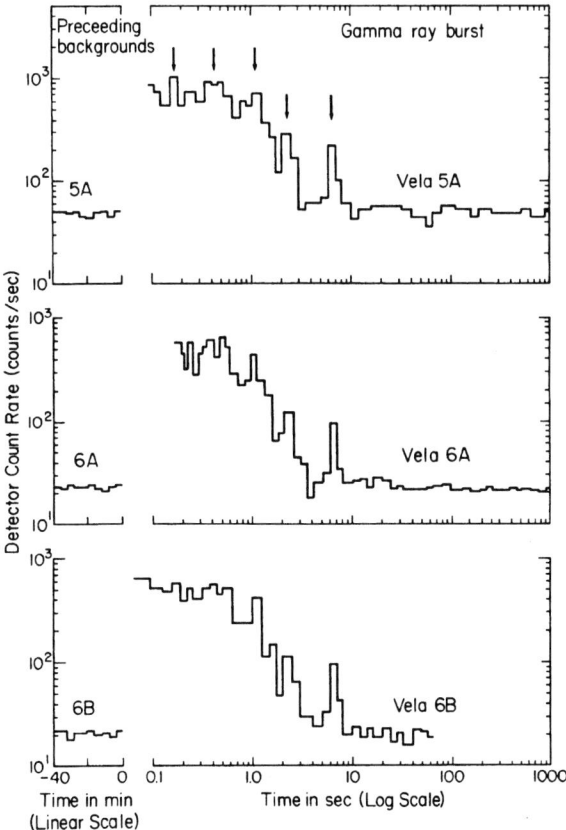

Fig. 1.—Count rate as a function of time for the gamma-ray burst of 1970 August 22 as recorded at three Vela spacecraft. Arrows indicate some of the common structure. Background count rates immediately preceding the burst are also shown. *Vela 5A* count rates have been reduced by 100 counts per second (a major fraction of the background) to emphasize structure.

A burst observed on 1970 August 22 is presented as an example. Figure 1 shows the count rate as a function of time. Each plot is presented in two parts. On the left, on a linear time scale, are plotted 10 measurements of count rate made at 4-minute intervals for the time immediately preceding the burst. These establish a background count rate. The record of the burst is plotted on the right on a logarithmic time scale. All the *Vela 5A* data have had a uniform 100 counts per second (a major fraction of the background) subtracted before plotting in order to facilitate comparison of time structure.

The initial part of the burst (extending to ~ 4 s) has an integrated flux density of $\sim 8 \times 10^{-5}$ ergs cm^{-2} in the range 0.2–1.0 MeV, and $\sim 6 \times 10^{-5}$ ergs cm^{-2} in the range 0.3–1.5 MeV. Within these 4 s there appears structure common to the records of all three spacecraft. Although the exact statistical significance of this structure has not yet been firmly established, it has been used to adjust these three records in time, relative to the initiation of the recordings. Exclusion of the Sun as the source, based on directional resolution, is unaffected by this correction.

In addition to the initial structure, all three records show a distinct peak centered around 6.5 s. For each record this peak is statistically significant to about 6 standard deviations. It represents integrated flux densities of 10^{-5} ergs cm^{-2} and 4×10^{-6} ergs cm^{-2} in the lower and higher energy ranges, respectively. The spectrum is clearly softer than that of the initial part of the burst.

IV. DISCUSSION

A search was made for reports of a nova or supernova within a reasonable time ($\sim$ several weeks) of each gamma-ray burst. No reported novae were related in time or direction to any of the bursts. Only two reported supernovae reached maximum apparent magnitude within a few days of an observed burst. In both cases, however, reports of prediscovery observations were later made which preceded the gamma-ray burst by at least several days. In addition, the source positions derived from preliminary timing data are inconsistent with the locations of the supernovae.

The lack of correlation between gamma-ray bursts and reported supernovae does not conclusively argue against such an association, since it is possible that there are supernovae, not necessarily bright in the optical region ("theoreticians' supernovae"), whose rate of occurrence may exceed those which are optically visible (see, e.g., Thorne 1969). A source at a distance of 1 Mpc would need to emit $\sim 10^{46}$ ergs in the form of electromagnetic radiation between 0.2 and 1.5 MeV in order to produce the level of response observed here. Since this represents only a small fraction ($<10^{-3}$) of the energy usually associated with supernovae, the energy observed is not inconsistent with a supernova as a source.

The authors wish to acknowledge the interest shown in the past by Edward Teller, Stirling Colgate, and A. G. W. Cameron who have on a number of occasions encouraged us to look for bursts of energetic photons.

We also wish to thank J. H. Coon and all of our colleagues in the Space Science Group at Los Alamos who have helped with this work. The detector electronics were the responsibility of the Space Electronics section at Los Alamos, under the direction of J. P. Glore. Logics were developed by the Satellite Systems Division at Sandia Laboratories; in particular we wish to mention R. E. Spalding, G. J. Dodrill, and J. G. Mitchell.

This research was performed as part of the Vela Satellite Program, which is jointly sponsored by the U.S. Department of Defense and the U.S. Atomic Energy Commission. The program is managed by the U.S. Air Force, and satellite operation activities are under the jurisdiction of the Air Force Satellite Control Facility, Sunnyvale, California.

REFERENCES

Colgate, S. A. 1968, *Canadian J. Phys.*, **46**, S476.
Thorne, K. S. 1969, in *Supernovae and Their Remnants*, ed. Peter J. Brancazio and A. G. W. Cameron (New York: Gordon & Breach).

Gamma Ray Bursts In Their Historic Context

Virginia Trimble

Department of Physics and Astronomy, University of California, Irvine CA 92697 USA, and Astronomy Department, University of Maryland, College Park MD 20742 USA

Abstract. Gamma ray bursts remained essentially non-understood or misunderstood from their 1973 discovery (not, I will claim, "serendipitous") to the first, 1997, redshift. This is by no means a record. The poster explored some of the examples of longer-standing puzzles and the after-dinner talk some of the details of the GRB case. The most striking feature of the GRB history is probably the unanimity with which "all we, like sheep, went astray," which followed the epoch of "we have turned everyone to his own way." Some of the reasons for this, the range of hypotheses, and how GRBs were presented to the astronomical and larger communities are discussed.

INTRODUCTION: THE PATTERN OF PUZZLING PHENOMENA

Some astronomical discoveries are understood as soon as they are made (pulsars as rapidly rotating, magnetized neutron stars, for instance) or even before (this is called prediction, black holes between 1800 and 1972, are a classic case). For those where understanding is considerably delayed, there seems to be a characteristic pattern, which also applies to the gamma ray bursts.

First, an observation is perceived as puzzling, sometimes by the person who made it or his immediate contemporaries; sometimes only considerably later (Tycho's nova stella, for example). Second, one or more explanations are forthcoming, at most one of which can be right. Eventually the issue is resolved, and the solution holds down to the present. Then, with expert hindsight, it becomes possible to recognize what additional input was needed for the resolution. (Very occasionally the need is spotted in advance and the items sought.) The time between perception of a puzzle and basic resolution was, for the GRBs, 24 years (1973 discovery to 1997 redshift), and this seemed like a very long stretch to those of us who lived through it. The record is, however, considerably longer.

THE LONGEST-LASTING PUZZLES

Pulsating variable stars

Boulliau discovered the periodicity of Mira in 1667 and attributed it to rotation at that period and spots more numerous and darker than those on the sun. When, in 1782-84, Piggott and Goodricke found periods for Algol and Delta Cephei, they explained both as eclipses by a dark companion. And there the situation remained for more than 100 years, until it became clear that, at least for Delta Cephei, the period was short enough that one star would have to be inside the other. Plummer in 1911 and Shapley in 1914 suggested pulsation, with $P \simeq (G\rho)^{-1/2}$, and Eddington by 1926 [4] had worked out the kappa (opacity) mechanism, persuasively for the Cepheids and tentatively for the Miras. What was needed was simply the idea of a dynamical time scale for stars, which would have been possible for Kelvin and Helmholtz (or even Newton, but he did not have enough information about stellar densities), and the duration of the puzzle from Boulliau to Eddington was 259 years. This is the longest I am aware of.

Sources of Stellar Energy

Newton remarked that the sun had been bright for a long time and that there was probably a reason for it. John Herschel put forward friction or electrical discharges in 1833 (arguably the recognition of a puzzle). Gravitational potential energy, in the form of infall of meteoric material or gravitational contraction came from Julius Mayer in 1841 and James Waterston in 1843-53 respectively. Their papers were both rejected, and the core idea habitually therefore called the Kelvin-Helmholtz time scale. It is only about 30,000,000 years, too short for some geological and biological considerations of the earth. Perrine, Eddington, and others advocated "subatomic" energy early in the 20th century. Atkinson and Hourtermans put forward catalyzed cyclic nuclear fusion in 1929-31 (before the discovery of the neutron, so that their helium atom when it spun off had four protons and two electrons in its nucleus). The details came from Hans Bethe in 1939, and had required all sorts of advances in fundamental physics, including special relativity, mass spectroscopy, the Saha equation (to get the composition of the sun), and quantum mechanical tunneling. Duration: 106 years.

Coronal lines in Solar Eclipse Spectra

Charles A. Young found the 5303 Angstrom line in 1869. The explanation was, of course, the element coronium, by analogy with helium (for a solar photospheric feature) and nebulium (for the 3727 line and others in planetary nebulae). As the periodic table filled out, it became clear there was no room for the latter, and Ira Bowen tied the nebular lines to forbidden transitions of oxyen and nitrogen in 1927. It was not, however, until 1939-42 that Grotrian and Edlen were able to recognize 5303 as a transition of Fe XIV, for a duration of 70 years. Both a new idea (that the corona is hot, rather than that the gas responsible is lighter than hydrogen) and a great deal of 20th century physics were needed.

Advance of the Perihelion of Mercury

LeVerrier reported a value of 0.5 arcmin/century in 1859. The intra-Mercurial planet responsible, Vulcan, was supposedly seen the same year and many times thereafter.

The alternative of gravity deviating from $1/r^2$ was largely rejected as conflicting with lunar and Venusian data, but tentatively supported around 1900 by Simon Newcomb (the first president of the American Astronomical Society, who generally gets fairly bad press). The solution was, of course, to be found in exceedingly new physics, as noted immediately by Einstein in his 1915 publication on the general theory of relativity. Duration, 56 years.

The Solar Neutrinos

The construction of Ray Davis's chlorine detector was carried out in close collaboration with theorists, who said he would see a good many events per month. By 1968 it was clear that the observed rate was, at most, one third of the prediction (3 SNU vs. 9 SNU). Explanations blossomed, if not quite 1000. These included modest adjustments in nuclear physics, much odder physics (quark catalysis, multiplicative creation, Maxwell tail depletion, WIMP energy transport), inefficient extraction of Ar^{37}, and large or small changes in solar physics (complete mixing, a metal poor interior, rapid rotation or magnetic support, and central black hole or an iron core). Curiously, the right answer, neutrino oscillation (propounded by Bruno Pontecorvo) was in the inventory from the beginning. But it took several new experiments and several new neutrino detectors to make this clear. SAGE and GALLEX were supposed to look at the problem, but the key discoveries were really the detection of oscillation of atmospheric neutrinos (SuperKamiokande 1999) and detection of the rotated flavors in the solar flux (SNO 2001). The duration of the puzzle was a personal matter, with opinion gradually swinging in the direction of new weak interaction physics between about 1985 and 2001. I think I opted for oscillation in 1994, about the same time Hans Bethe did, which was not a coincidence.

The Also-Rans

These are the puzzles for which I suspect that the correct answer is not yet on our plates or anyhow not entirely recognized. Some examples are (a) stability of spiral arms (winding-up problem recognized in 1896, density waves from Lin and Shu in 1964, and on to the present), (b) acceleration of cosmic rays (recognized as extrasolar in 1911, Fermi mechanism 1949, GRB component 2003...), (c) nature of the dark matter (cluster velocities, Zwicky 1933, LSPs, axions, Wimpzillas and all 1974 to present), (d) heating of solar and stellar coronae (recognition of Fe XIV in 1939, heating by acoustic waves, MHD waves, microflares, electric currents, and so forth down to the present, for a duration of at least 64 years). And there are probably others.

Credits

Most of these topics are discussed in one or more of Hoskins [7, 8], Bahcall [1], Hufbauer [9], Eddington [4], and Russell, Dugan, & Stewart [13].

THE DISCOVERY OF "GAMMA-RAY BURSTS OF COSMIC ORIGIN": SERENDIPITY AND MISCONCEPTIONS

Casual reviews and textbooks quite often announce that GRBs were discovered in 1969 or even 1967. In so far as science is shared knowledge, both are clearly wrong. The discovery must be placed in 1973, when Klebesadel, Strong, and Olson (1973) [11] submitted and published their *Astrophysical Journal Letter*. The 1969 date is that of the first burst reported in that letter and the 1967 one a burst recognized slightly later, for which, however, there is a little directional information, so that it could conceivably have come from the sun.

Others have argued that the events should have been announced earlier. The same statement was made in 1968 about pulsars, and I think both again are clearly wrong. The time to announce is when you are as sure as you can be, short of building a new array or launching a new satellite, that you have seen something interesting. Thus post-mature publication is rather rare, but premature publication fairly common. It is easy to see the difference by tracking the rates of citations to the "discovery papers" over the years. For GRBs and pulsars (and also the classic 1963 Schmidt paper on quasars, the 1965 announcement of the 3K microwave background, and many others), a dozen or so citations in the first year rise to many tens or hundreds in the next couple of years and then jaggedly taper off, with occasional secondary peaks when something new and exciting happens in the field. The GRB spike with optical identifications in 1997 is one example; another is a sudden recrudescence of citations to a 1965 paper by Gunn and Ostriker when the effect they had considered finally became detectable.

In contrast, a premature publication has an equally clear, but different signal. There is a similar enormous peak in the first couple of years, sometimes as high as that for appropriate publication, but the taper is almost as fast as the rise, and continues right on down to zero, apart from occasional mentions in comprehensive review papers. Examples include the announcements of a pulsar in the remnant of SN 1987A (whose period was an alias of the video camera scan rate) and of a planet orbiting the pulsar 1829-10 (with a six month period that was incorrect allowance for the eccentricity of our orbit).

Were GRB's predicted? It is easy to say, and I have said earlier, that two kinds of events were predicted and two have been seen. Unfortunately, they were not the same kinds. The predictions were for shock break out in core collapse supernovae (Colgate [3]) and evaporation of primordial black holes (Hawking [6]). The latter counts as a prediction despite its date because the author does not seem to have been aware of the data (compare Einstein and the Michelson-Morley experiment, perhaps!). The two sorts seen were the classical GRBs and the soft gamma repeaters. I think that the quip about the two kinds not being the same is no longer entirely true. Indeed, there is still no observational evidence for Hawking radiation from evaporating PBHs, but the current understanding, at least of events like GRB 030329 is not really so very different (apart from extreme asymmetry) from what Colgate suggested.

Was the discovery "serendipitous"? The answer here is, I conclude, yet another firm no. The fictional princes of Serendip (Sri Lanka) went about making fortunate discoveries by accident. A much better model is the remark of Louis Pasteur, in French, of course, that chance favors (only) the prepared mind. The discovery paper, and other discussions from the 1970s say firmly that the Los Alamos group were aware of the 1968 report by Cline, Holt, and Hones [2] of a burst of gamma rays from the solar flare of 1966, near the beginning of a new solar cycle. Thus the discoverers immediately recognize some solar flares as the cause of the near-simultaneous triggering of detectors on widely-separated satellites, and set out to make sure that there were no other such non-terrestrial triggers. Except that there were. Unexpected, but not serendipitous.

Two "alternative universes" come to mind. First, could GRBs have been discovered any earlier? That is, was anything up before the Vela and Kosmos series satellites carrying gamma ray sensors with very good time-tagging of photon arrivals? The answer is yes, and one (possible) GRB was recognized retrospectively in the data stream from a balloon that had been looking for positrons connected with aurorae on 26 September 1966. Gamma ray emission at 511 keV just after a positively charged particle passes through your outer layer is, of course, a positron signal. It is, however, a little hard to imagine this sort of record being recognized as anything cosmically interesting unless there had been simultaneous balloon flights more or less on opposite sides of the earth.

Second, would GRBs have been discovered fairly soon anyhow? To this it is a bit harder to say no, categorically. Confirmations of specific events came very quickly from a large fraction of the devices above the atmosphere at the time and soon thereafter, including OSO-5, 6, and 7, OGO 3 and 5, SAS-1, IMP-6, Apollo-16, Uhuru, TD-1, IMP-7, and 1972-076B, as well as the Kosmos series. None of these, of course, had been planned with any such entities as GRBs in mind. In particular, event 1972-2 appeared in detectors, anti-coincidence shields, and so forth in IMP-6, SAS-1, OGO-5, and OSO-6 as well as three Vela satellites. One is led to suspect that sooner or later people associated with these projects might by chance have found out about each others' excess noise, in somewhat the same way that Princeton found out about the Penzias and Wilson detection of excess antenna temperature.

WE HAVE TURNED EVERY ONE TO HIS OWN WAY

I do not know whether GRBs resulted in more "prompt" models than any other astrophysics phenomenon in history. What is true is that the set is better documented. As early as the 1974 "Texas" symposium on relativistic astrophysics, Mal Ruderman remarked, first, that the only model not taken seemed to be anti-matter comets hitting white holes, second, that the only theorist who had not published a model was J.P. Ostriker (who still has not; I asked him, in the only original research done for this presentation; but he and R.D. Blandford have discussed an electromagnetic model), and, third, that he personally was betting on Black Hole ridden by Accretion to win, with Glitch to place.

The first 118 models, up to 1992, were tabulated by Nemiroff [12] and indeed include a remarkable range of compact objects, large and small, innovative physics, known and unknown, and so forth. Nemiroff's list is not quite complete even for its period. Colgate's prediction is there, but not Hawking's. Also missing are Jelley (1974) [10] who put forward a Leblanc-Wilson type collapse of a rotating magnetized star (that is, something rather similar to the current best buy) and Harris [5] who contemplated exhaust trails from interstellar space craft (accounting for the possible positron annihilation line in some spectra), looked for events in straight lines across the sky, and didn't find any. Slightly later came magnetar models and collisions of comets in the Kuiper belt.

Among my favorites within the list are directed stellar flares (the first beamed model), proposed by Brecher and Morrison, and Goblins (chunks of neutron star material released from gravitational confinement and exploding to white dwarf densities), the last published paper of Fritz Zwicky. Should one disparage this wild blooming of scenarios? It is easy to do so, in the form, for instance of a theorem said to be due to R.O. Redman, "A competent theorist can explain any set of observations using any theory," and a corollary actually due to Malcolm Longair, "In many cases he need not to even be competent." But the reality is (another quote, this time from Joe Weber) that theorists are cheap and telescopes are expensive, and so the fullest possible consideration of any datum is only just good sense.

Several curious sidelights appear in any extended list of models. One is the "incident of the gamma ray in the nighttime," when, in 1974, at that same Texas symposium, Stirling Colgate said, "I do not believe that hard X-ray pulses are not created in supernova events. We have just not been fortunate enough to observe the phenomenon." Most GRB2003 participants would probably say that, with GRB 0303029 = SN2003dh, we have now observed it. Another oddity is the distinction between "premature" considerations and "just in time" delivery. The Brecher and Morrison beaming was perhaps too early to make a mark, as was a 1993 paper by Rhoads and Paczynski that included a forecast of radio afterglows. In contrast, the 1997 prediction by Meszaros and Rees of optical afterglows slipped into print just before the photons were recorded. Only Stanton Peale's conclusion that the interior of Io should be very hot (because of tidal stressing by Jupiter) published days before the discovery of the volcanos strikes us as better timed.

ALL WE, LIKE SHEEP, WENT ASTRAY

The earliest models included a mix of galactic and extragalactic events, but, throughout the 1980s and lingering even into the BATSE era, there was truly remarkable concensus in favor of a picture where the bursts were something which happened on the surfaces of old, fairly nearby, neutron stars that had managed (somehow) to maintain magnetic fields near 10^{12} G and which the neutron stars would survive. How did this happen? There were, I think, something like seven contributing factors.

First was the 1976 discovery of X-ray bursts by Grindlay and his colleagues seen largely in the direction of the galactic center and correctly interpreted by Joss and by Woosley as nuclear explosions on neutron star surfaces. Indeed Woosley's model had the status of a prediction, because it was originally intended to apply to GRBs.

Next was the discovery of cyclotron resonance features in the X-ray spectrum of the intermediate mass X-ray binary Her X-1, recorded in a 1976 balloon flight coordinated by Truemper. The implied field was about the 10^{12} G expected from pulsar data. Third came an excessive distrust of any deviation of the N(S) relation from a -3/2 power law. The 3/2 slope is an indicator of a uniform distribution in space, while flattening would mean we are seeing the edge (and steepening that the sources are commoner far away and/or long ago). Indeed a couple of early announcements of flattening were over-interpretations of data with awkward selection effects and such, but I believe that the White et al. (1983) balloon results [14], which actually included several events down to a fluence of 10^{-7} erg/cm^2 should have received wider credence that it did.

Fourth were spectral features, of two sorts, reported in the GRBs themselves. According to the Soviet detectors, the March 5th, 1979 burst (now a part of a long trail of a soft gamma repeater) had a redshifted positron annihilation feature near 400 keV (which a real detection of the line in the region of Sgr A* had conditioned us to believe). Less soft events in the next couple of years also seemed to have positron annihilation redshifted just about as much as you would expect for photons that had climbed out of a neutron star potential. These were gradually discounted beginning in 1984, when SMM did not record them.

The other sort of spectral feature was cyclotron resonance, like that of Her X-1, reported by the Soviet group, soon after from SMM and HEAO-1, and especially from Ginga data. The implied magnetic fields for the 10-70 keV resonances were again in the 10^{12} G range expected for neutron stars. This was apparently a defining datum for a large fraction of the GRB community. Indeed even I, in a 1990 review, listed as the best buy model "mergers of binary neutron stars with strong magnetic fields at cosmological distances." BATSE never really saw any such spectral features, and the number of GRBers who would still defend them is small, but not zero.

Fifth "distractor" was the reported optical flashes at GRB positions but at earlier times, for instance a 1928 blip on a Harvard plate where a 1978 GRB happened. Schaefer and Hudec each found some of these, and they meant that the source had to survive a burst and do it again, something like once per century. In retrospect, the inventory included plate flaws, flare stars, and possibly a few SNe in the GRB host galaxies.

Sixth was the "no host" problem, the absence of bright galaxies in the error boxes of some of the best localized, brightest GRBs. This also came from examination of archival plates, again by Schaefer, and has to be ascribed to a combination of the hosts (at least of long duration events) being genuinely rather faint and bad luck. Seventh and last were the neutron star runaways. Some pulsars are indeed relatively high velocity objects, probably kicked by asymmetrical supernovae and binary star disruption during their formation. For the Crab pulsar, the velocity is somewhat less than 200 km/sec, which won't have carried it far in its lifetime, but there were claims of 1000 km/sec and more for pulsars once in (but now on the edge or outside of) much older remnants. That sort of velocity would easily take a 1-10 Gyr neutron start into an extended galactic halo. And an extended galactic halo of sources could, just barely, account for the early CGRO burst census, in which they remained very isotropic on the sky but, at long last, we definitely saw the edge of the distribution.

Even at the time of the 1995 staging of the 75th anniversary celebration of the Great (Curtis-Shapley) Debate, organized by Robert Nemiroff, a vote of the participants showed about equal numbers for "galactic" and "extragalactic." The only effect of the debate itself (between D.Q. Lamb and B Paczynski) was a large increase in the faction of the audience who declined to vote. Not all minds were made up until (or perhaps after) the 1997 redshift measurement, though in September 1991, Martin Rees offered Bohdan Paczynski a 100:1 bet in favor of "galactic," and then, after the fall of 1992 announcement of BATSE isotropy and turn-over in N(S), opined, "We were both fools. I for offering the bet; Bohdan for not accepting it."

SHARING GRBS WITH THE WORLD

I looked two sorts of places to see how this played out, first the proceedings of the biennial Texas Symposia, and, second semi-popular journals (Sky and Telescope and Scientific American). At Texas, GRBs did not exist in 1972, in 1974 got both a data review and the Ruderman theoretical presentation already mentioned. In 1976 and 1978 they again did not exist. The talks in 1980 and 1982 relied heavily on the spectral features, and 1984 saw another eclipse. Local neutron stars with strong fields were the only respectable models in 1986 and 1988, and Hartmann in 1990 again emphasized the cyclotron (Ginga) features as our definitive line of evidence.

The 1992 Texas GRB talk was given by Paczynski. Not that he had ever really deviated from extragalactic models, but it marked the sea change introduced by early CGRO data that he was the person asked to review the topic. Rees voted 50% for extragalactic, 30% for extended halo, and 20% for galactic disk and the N(S) turnover still not being real.

On the popular front, both journals recognized the existence of the phenomenon (with articles written by or drawing directly on the expertise of the discoverers). Optical counterparts yielded renewed attention, and then the early BATSE data. S&T also reported a range of false alarms (correlations with QSOs, repeaters, etc), but both settled into the extragalactic camp by the end of 1997. Most striking to this reader, going back to the four Scientific American articles andthe 16 S&T reports and reading them all at once, was how much more informative the early articles, written by active GRB astronomers, were than the later ones written by staff members. In one of the worst (S&T 1996) the issue of the cyclotron features is not even mentioned.

How have GRBs done in (non-technical) astronomy education? Not very well. Of 15 standard texts in multiple editions that I examined, only Jay Pasachoff's regards them as interesting enough to mention the galactic/extragalactic issue before 1997 or to address how one might resolve it by optical counterparts with spectra. Since 1997, about half of the textbooks mention GRBs and say that the optical data settled a previously existing (but apparently not known to them!) controversy. Most of the books also do not score high on two other items from the history of astronomy (big bang vs. steady state and heliocentric vs. geocentric) where we have the opportunity to use the idea of falsifiability to indicate how science differs from "other ways of knowing."

ACKNOWLEDGEMENTS

I am indebted to Ed Fenimore for the invitation to deliver the after dinner talk at GRB2003 and to Mark Galassi for efforts above and beyond the call of duty to figure out the logistics of dinner in two separated rooms followed by a talk with images. Special thanks go to an anonymous colleague who generously downloaded and printed out pictures of the astronomers and physicists who appear in Sect. 2, and some others, for display with poster, but who very sensibly declined any further involvement with the presentation or, indeed, with me. It is not possible to leave this topic without quoting Philip Morrison, who said (in another after dinner talk) that "it is hard to waste 10^8 dollars." He had in mind both the GRB discovery and evidence for plate tectonics that came from a seismic array also intended to detect illicit tests for nuclear weapons. Only inflation and LIGO make me suspect this should now be 10^9 dollars.

REFERENCES

1. Bahcall, J.N., Neutrino Astrophysics. Cambridge University Press (1989).
2. Cline, T.R., Holt, S.S & Hones, E.W., JGR **73**, 434 (1968).
3. Colgate, S.A., Canadian J Phys. **46**, S476 (1968).
4. Eddington, A.S., The Internal Constitution of the Stars. Cambridge University Press (1926).
5. Harris, M.J., J. British Interplanetary Assoc. **43**, 551 (1990).
6. Hawking, S.W., Nature **248**, 30 (1974).
7. Hoskin, M.A. (Ed.) The Cambridge Illustrated History of Astronomy, Cambridge University Press (1997).
8. Hoskin, M.A. (Ed.) 1 The Cambridge Concise History of Astronomy, Cambridge University Press (1999).
9. Hufbauer, K., Exploring the Sun: Solar Science Since Galileo. Johns Hopkins University Press (1991).
10. Jelley, J.V., Nature **249**, 747 (1974).
11. Klebesadel, R.W., Strong, I.B., & Olsen, R.A., ApJ **182**, L85 (1973).
12. Nemiroff, R.J., Comments on Astrophysics **17**, 189 (1994).
13. Russell, H.N., Dugan, R.S. & Stewart, J.Q., Astronomy. Boston: Ginn & Co. (1926).
14. White, R.S. et al.,Nature **271**, 635 (1983).

GLOBAL PROPERTIES OF GRBs

A Unified Jet Model of X-Ray Flashes and Gamma-Ray Bursts

D. Q. Lamb[*†], T. Q. Donaghy[*] and C. Graziani[*]

[*]*Department of Astronomy & Astrophysics, University of Chicago, Chicago, IL 60637*
[†]*d-lamb@uchicago.edu*

Abstract. HETE-2 has provided strong evidence that the properties of X-Ray Flashes (XRFs) and GRBs form a continuum, and therefore that these two types of bursts are the same phenomenon. We show that both the structured jet and the uniform jet models can explain the observed properties of GRBs reasonably well. However, if one tries to account for the properties of both XRFs and GRBs in a unified picture, the uniform jet model works reasonably well while the structured jet model fails utterly. The uniform jet model of XRFs and GRBs implies that most GRBs have very small jet opening angles ($\sim$ half a degree). This suggests that magnetic fields play a crucial role in GRB jets. The model also implies that the energy radiated in gamma rays is $\sim$ 100 times smaller than has been thought. Most importantly, the model implies that there are $\sim 10^4 - 10^5$ more bursts with very small jet opening angles for every such burst we see. Thus the rate of GRBs could be comparable to the rate of Type Ic core collapse supernovae. Accurate, rapid localizations of many XRFs, leading to identification of their X-ray and optical afterglows and the determination of their redshifts, will be required in order to confirm or rule out these profound implications.

INTRODUCTION

Two-thirds of all HETE-2–localized bursts are either "X-ray-rich" or X-Ray Flashes (XRFs); of these, one-third are XRFs [1][1]. These events have received increasing attention in the past several years [2, 3], but their nature remains unknown.

XRFs have t_{90} durations between 10 and 200 sec and their sky distribution is consistent with isotropy. In these respects, XRFs are similar to "classical" GRBs. A joint analysis of WFC/BATSE spectral data showed that the low-energy and high-energy photon indices of XRFs are -1 and ~ -2.5, respectively, which are similar to those of GRBs, but that the XRFs had spectral peak energies $E^{\rm obs}_{\rm peak}$ that were much lower than those of GRBs [3]. The only difference between XRFs and GRBs therefore appears to be that XRFs have lower $E^{\rm obs}_{\rm peak}$ values. It has therefore been suggested that XRFs might represent an extension of the GRB population to bursts with low peak energies.

Clarifying the nature of XRFs and X-ray-rich GRBs, and their connection to GRBs, could provide a breakthrough in our understanding of the prompt emission of GRBs. Analyzing 42 X-ray-rich GRBs and XRFs seen by FREGATE and/or the WXM instruments on HETE-2, [1] find that the XRFs, the X-ray-rich GRBs, and GRBs form a continuum

---

[1] We define "X-ray-rich" GRBs and XRFs as those events for which $\log[S_X(2-30 \text{ kev})/S_\gamma(30-400 \text{ kev})] > -0.5$ and 0.0, respectively.

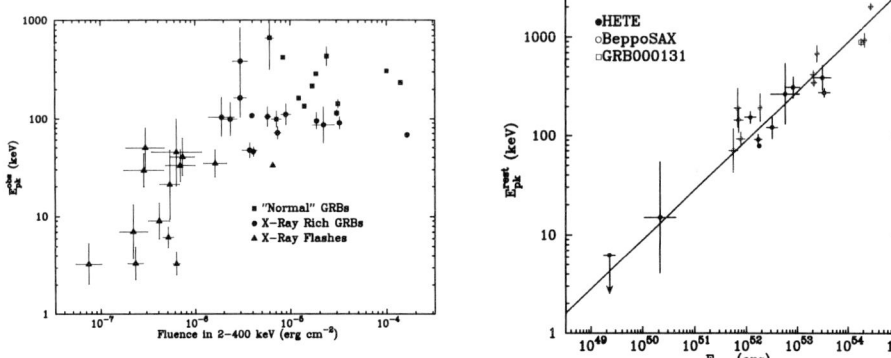

FIGURE 1. Distribution of HETE-2 bursts in the $[S(2-400 \text{ keV}), E_{\text{peak}}^{\text{obs}}]$-plane, showing XRFs, X-ray-rich GRBs, and GRBs (left panel). From [1]. Distribution of HETE-2 and BeppoSAX bursts in the $(E_{\text{iso}}, E_{\text{peak}})$-plane, where E_{iso} and E_{peak} are the isotropic-equivalent GRB energy and the peak of the GRB spectrum in the source frame (right panel). The HETE-2 bursts confirm the relation between E_{iso} and E_{peak} found by Amati et al. (2002), and extend it by a factor ~ 300 in E_{iso}. The bursts with the lowest and second-lowest values of E_{iso} are XRFs 020903 and 030723. From [4].

in the $[S_\gamma(2-400 \text{ kev}), E_{\text{peak}}^{\text{obs}}]$-plane (see Figure 1, left-hand panel). This result strongly suggests that all of these events are the same phenomenon.

Furthermore, [4] have placed 9 HETE-2 GRBs with known redshifts and 2 XRFs with known redshifts or strong redshift constraints in the $(E_{\text{iso}}, E_{\text{peak}})$-plane (see Figure 1, right-hand panel). Here E_{iso} is the isotropic-equivalent burst energy and E_{peak} is the energy of the peak of the burst spectrum, measured in the source frame. The HETE-2 bursts confirm the relation between E_{iso} and E_{peak} found by [5] (see also [6]) for GRBs and extend it down in E_{iso} by a factor of 300. The fact that XRF 020903, one of the softest events localized by HETE-2 to date, and XRF 030723, the most recent XRF localized by HETE-2, lie squarely on this relation [7, 4] provides strong evidence that XRFs and GRBs are the same phenomenon. However, additional redshift determinations are clearly needed for XRFs with 1 keV $< E_{\text{peak}} <$ 30 keV energy in order to confirm these results.

Figure 2 shows a simulation of the expected distribution of bursts in the $(E_{\text{iso}}, E_{\text{peak}})$-plane (left panel) and in the $(F_N^{\text{peak}}, E_{\text{peak}})$-plane (right panel), assuming that the [5] relation holds for XRFs as well as for GRBs [8], as is strongly suggested by the HETE-2 results. The SXC, WXM, and FREGATE instruments on HETE-2 have thresholds of $1-6$ keV and considerable effective areas in the X-ray energy range. Thus HETE-2 is ideally suited for detecting and studying XRFs. In contrast, BAT on *Swift* has a nominal threshold of 20 keV. This simulation shows that the WXM and SXC instruments on HETE-2 detect many times more bursts with $E_{\text{peak}} < 10$ keV than will BAT on *Swift*.

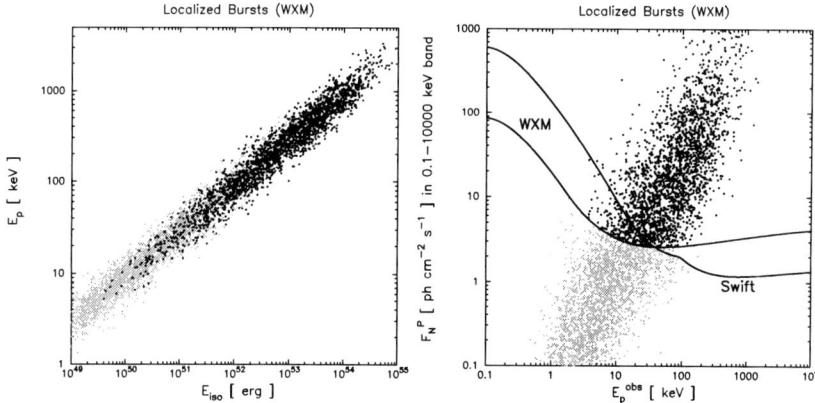

FIGURE 2. Expected distribution of bursts in the ($E_{\rm iso}, E_{\rm peak}$)-plane (left panel) and in the ($F_N^{\rm peak}, E_{\rm peak}$)-plane (right panel), assuming that the Amati et al. (2002) relation holds for XRFs as well as for GRBs, as strongly suggested by the HETE-2 results. Black dots are simulated bursts that the WXM on HETE-2 detects; gray dots are simulated bursts that it does not detect. The curved lines in the right-hand panel show the threshold sensitivities of the WXM on HETE-2 and BAT on Swift. From [8].

XRFS AS A PROBE OF GRB JET STRUCTURE, GRB RATE, AND CORE COLLAPSE SUPERNOVAE

Most GRBs have a "standard" energy [9, 10, 11]; i.e, if their isotropic equivalent energy is corrected for the jet opening angle inferred from the jet break time, most GRBs have the same radiated energy, $E_\gamma = 1.3 \times 10^{51}$ ergs, to within a factor of $\sim$ 2-3.

Two models of GRB jets have received widespread attention:

- The "structured jet" model (see the left-hand panel of Figure 3). In this model, all GRBs produce jets with the same structure [12, 13, 14, 15]. The isotropic-equivalent energy and luminosity is assumed to decrease as the viewing angle θ_v as measured from the jet axis increases. The wide range in values of $E_{\rm iso}$ is attributed to differences in the viewing angle θ_v. In order to recover the "standard energy" result [9], $E_{\rm iso}(\theta_v) \sim \theta_v^{-2}$ is required [14].

- The "uniform jet" model (see the right-hand panel of Figure 3). In this model GRBs produce jets with very different jet opening angles $\theta_{\rm jet}$. For $\theta < \theta_{\rm jet}$, $E_{\rm iso}(\theta_v) = $ constant while for $\theta > \theta_{\rm jet}$, $E_{\rm iso}(\theta_v) = 0$.

As we have seen, HETE-2 has provided strong evidence that the properties of XRFs, X-ray-rich GRBs, and GRBs form a continuum, and that these bursts are therefore the same phenomenon. If this is true, it immediately implies that the E_γ inferred by [9] is too large by a factor of at least 100 [8]. The reason is that the values of $E_{\rm iso}$ for XRF 020903 [7] and XRF 030723 [4] are $\sim$ 100 times smaller than the value of E_γ inferred by Frail et al. – an impossibility.

HETE-2 has also provided strong evidence that, in going from XRFs to GRBs, $E_{\rm iso}$ changes by a factor $\sim 10^5$ (see Figure 1, right-hand panel). If one tries to explain only

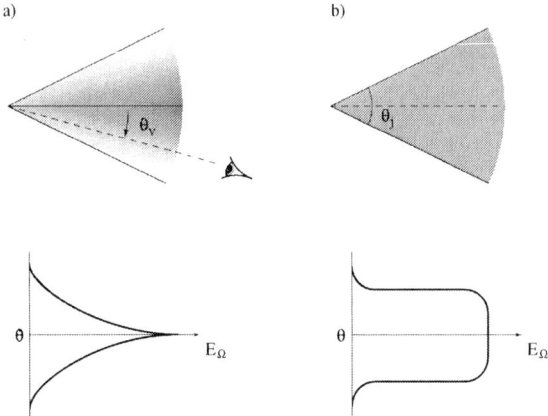

FIGURE 3. Schematic diagrams of universal jet model and jet model of GRBs [16]. In the universal jet model, the isotropic-equivalent energy and luminosity is assumed to decrease as the viewing angle θ_v as measured from the jet axis increases. In order to recover the "standard energy" result [9], $E_{\text{iso}}(\theta_v) \sim \theta_v^{-2}$ is required. In the uniform jet model, GRBs produce jets with a large range of jet opening angles θ_{jet}. For $\theta < \theta_{\text{jet}}$, $E_{\text{iso}}(\theta_v) =$ constant while for $\theta > \theta_{\text{jet}}$, $E_{\text{iso}}(\theta_v) = 0$.

the range in E_{iso} corresponding to GRBs, both the uniform jet model and the structured jet model work reasonably well. However, if one tries to explain the range in E_{iso} of a factor $\sim 10^5$ that is required in order to accommodate both XRFs and GRBs in a unified description, the uniform jet works reasonably well while the structured jet model does not.

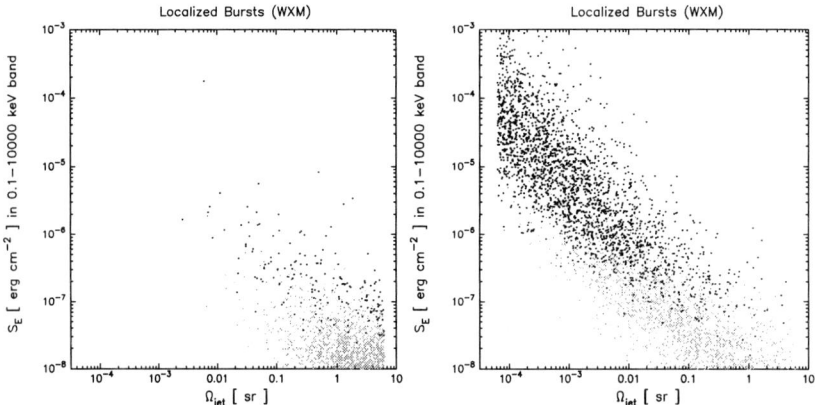

FIGURE 4. Expected distribution of bursts in the $(\Omega_{\text{jet}}, S_E)$-plane for the universal jet model (left panel) and uniform jet model (right panel), assuming that the Amati et al. (2002) relation holds for XRFs as well as for GRBs, as the HETE-2 results strongly suggest. From [8].

The reason is the following: the observational implications of the structured jet model and the uniform jet model differ dramatically if they are required to explain XRFs and GRBs in a unified picture. In the structured jet model, most viewing angles θ_v are $\approx 90°$.

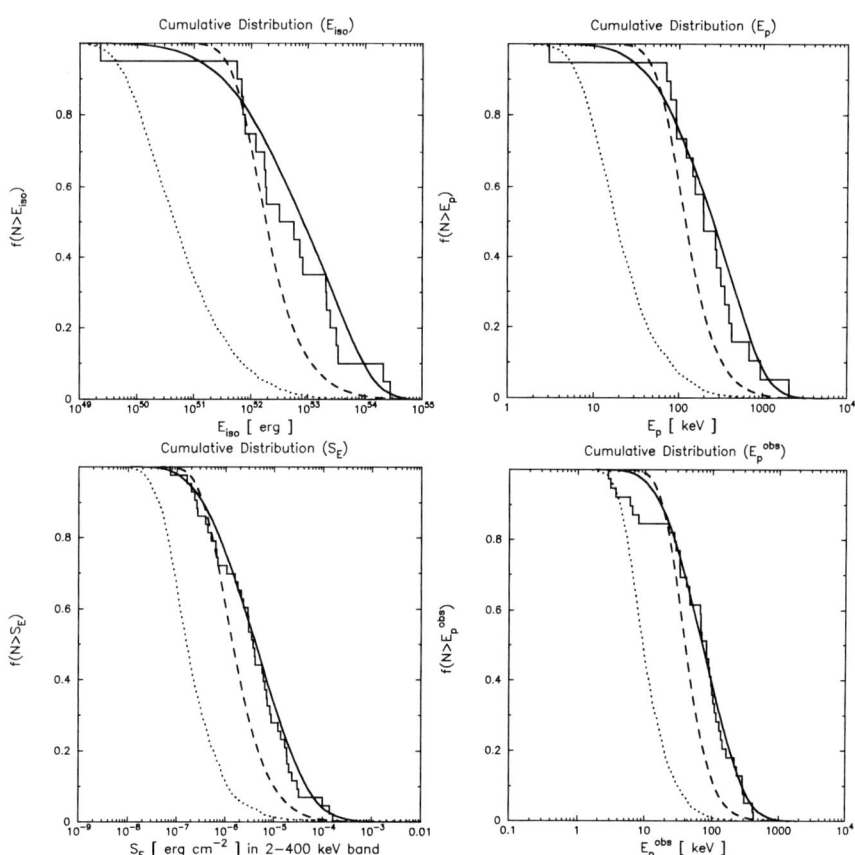

FIGURE 5. Top row: cumulative distributions of E_{iso} (left panel) and E_{peak} (right panel) predicted by various models, compared to the observed cumulative distributions of these quantities. Bottom row: cumulative distributions of $S(2-400\text{keV})$ (left panel) and E_{peak}^{obs} (right panel) predicted by various models, compared to the observed cumulative distributions of these quantities. The uniform jet model is shown as a solid line. The cumulative distributions corresponding to the best-fit structured jet model that explains XRFs and GRBs are shown as dotted lines; the cumulative distributions corresponding to the best-fit structured jet model that explains GRBs alone are shown as dashed lines. The structured jet model provides a reasonable fit to GRBs alone but cannot provide a unified picture of both XRFs and GRBs, whereas the uniform jet model can. From [8].

This implies that the number of XRFs should exceed the number of GRBs by many orders of magnitude, something that HETE-2 does not observe (see Figures 1, 2, 4, and 5). On the other hand, by choosing $N(\Omega_{jet}) \sim \Omega_{jet}^{-2}$, the uniform jet model predicts equal numbers of bursts per logarithmic decade in E_{iso} (and S_E), which is exactly what HETE-2 sees (again, see Figures 1, 2, 4, and 5) [8]. Thus, if E_{iso} spans a range $\sim 10^5$, as the HETE-2 results strongly suggest, the uniform jet model can provide a unified picture of both XRFs and GRBs, whereas the structured jet model cannot. This means that XRFs provide a powerful probe of GRB jet structure.

A range in $E_{\rm iso}$ of 10^5, which is what the HETE-2 results strongly suggest, requires a *minimum* range in $\Delta\Omega_{\rm jet}$ of $10^4 - 10^5$ in the uniform jet model. Thus the unified picture of XRFs and GRBs in the uniform jet model implies that there are $\sim 10^4 - 10^5$ more bursts with very small $\Omega_{\rm jet}$'s for every such burst we see; i.e., the rate of GRBs may be ~ 100 times greater than has been thought.

In addition, since the observed ratio of the rate of Type Ic SNe to the rate of GRBs in the observable universe is $R_{\rm Type\ Ic}/R_{\rm GRB} \sim 10^5$ [17], a unified picture of XRFs and GRBs in the uniform jet model implies that the GRB rate is comparable to that of Type Ic SNe [8]. More spherically symmetric jets yield XRFs and narrow jets produce GRBs. Thus XRFs and GRBs provide a combination of GRB/SN samples that would enable astronomers to study the relationship between the degree of jet-like behavior of the GRB and the properties of the supernova (brightness, polarization ⇔ asphericity of the explosion, velocity of the explosion ⇔ kinetic energy of the explosion, etc.). GRBs may therefore provide a unique laboratory for understanding Type Ic core collapse supernovae.

A unified picture of XRFs and GRBs in the uniform jet model also implies that many Type Ic SNe produce narrow jets, which may suggest that the collapsing cores of many Type Ic supernovae are rapidly rotating. Finally, such a unified picture implies that the total radiated energy in gamma rays E_γ is ~ 100 times smaller than has been thought [8].

REFERENCES

1. Sakamoto, T. et al., ApJ, to be submitted (2003).
2. Heise, J., in't Zand, J., Kippen, R. M., & Woods, P. M., in Proc. 2nd Rome Workshop: Gamma-Ray Bursts in the Afterglow Era, eds. E. Costa, F. Frontera, J. Hjorth (Berlin: Springer-Verlag), 16 (2000).
3. Kippen, R. M., Woods, P. M., Heise, J., in't Zand, J., Briggs, M.S., & Preece, R. D., in Gamma-Ray Burst and Afterglow Astronomy, AIP Conf. Proc. 662, ed. G. R. Ricker & R. K. Vanderspek (New York: AIP), 244 (2002).
4. Lamb, D. Q., et al., ApJ, submitted (2003).
5. Amati, L., et al., A & A **390**, 81 (2002).
6. Lloyd, N. M., Petrosian, V. & Mallozzi, R. S., ApJ **534**, 227 (2000).
7. Sakamoto, T. et al., ApJ, in press (2003).
8. Lamb, D. Q., Donaghy, T. Q., & Graziani, C., ApJ, submitted (2003).
9. Frail, D. et al., ApJ **562**, L55 (2001).
10. Panaitescu, A. & Kumar, P., ApJ **556**, 1002 (2001).
11. Bloom, J., Frail, D. A. & Kulkarni, S. R., ApJ **588**, 945 (2003).
12. Rossi, E., Lazzati, D., & Rees, M. J., MNRAS **332**, 945 (2002).
13. Woosley, S. E., Zhang, W. & Heger, A., ApJ, in press (2003).
14. Zhang, B. & Mészáros, P., ApJ **571**, 876 (2002).
15. Mészáros, P., Ramirez-Ruiz, E., Rees, M. J., & Zhang, B., ApJ **578**, 812 (2002).
16. Ramirez-Ruiz, E. & Lloyd-Ronning, N., New Astronomy **7**, 197 (2002).
17. Lamb, D. Q., A&A **138**, 607 (1999).

Radiation Processes in GRBs. Prompt Emission

Mikhail V. Medvedev

Department of Physics and Astronomy, University of Kansas, Lawrence, KS 66045

Abstract. A substantial fraction of prompt GRB spectra have soft spectral indexes exceeding the maximum allowed by the synchrotron model $\alpha_{max} = -2/3$. Some spectra also exhibit a very sharp break at E_p, inconsistent with the smooth synchrotron spectra. These facts pose a serious problem for the "optically thin synchrotron" interpretation of the prompt emission. We review various models suggested in order to resolve this puzzle.

INTRODUCTION

Time-resolved spectral analyses of *BATSE* and *BeppoSAX* [1, 2] clearly demonstrate that 30-50% of spectra violate the so-called "synchrotron line of death" (LoD), i.e., they have the soft photon indexes α greater than $-2/3$ (note, $F_\nu \propto \nu^{\alpha+1}$). In addition, a significant number of the spectra are better fit with the sharply broken power-law (BPL) model than with the smooth Band function. These facts make the simplest synchrotron interpretation of the prompt GRB emission at least questionable. Some attention has been paid to this problem and here we review alternative models suggested by several authors.

SELF-ABSORBED SYNCHROTRON MODEL

The simplest model which can produce a hard spectrum at low energies suggests that synchrotron radiation may be self-absorbed. This possibility has been considered by several authors; for more discussion and references, see Ref. [3]. The low-energy power-law index depends on the relative values of the self-absorption frequency ν_a and the peak synchrotron frequency ν_m:

$$F_\nu \propto \begin{cases} \nu^{5/2}, & \text{for } \nu_m < \nu \ll \nu_a; \\ \nu^{1/3}, & \text{for } \nu_a < \nu < \nu_m; \\ \nu^2, & \text{for } \nu \ll \min(\nu_m, \nu_a). \end{cases} \quad (1)$$

Note that the second case corresponds to the optically thin regime. A typical self-absorbed spectrum with $\nu_m < \nu_a$ is shown in Fig. 1(a) by the curve labeled SAS. In addition to a large spectral index, a self-absorbed spectrum has also a much narrower peak than the optically thin spectrum. Both these properties often lead to improved spectral fits of LoD-violating and BPL bursts.

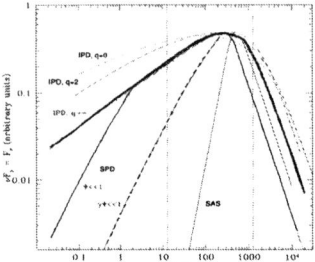 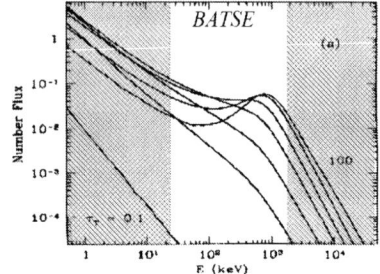

FIGURE 1. *(a)* The synchrotron self-absorbed spectrum (SAS) and the synchrotron spectrum in the small pitch-angle regime (from [3]), the thick solid line shows a standard synchrotron spectrum for comparison. *(b)* The spectrum produced by saturated Comptonization (from[4]).

What conditions of a fireball are needed to have the self-absorption frequency in the *BATSE*'s spectral window? The optical depth to synchrotron self-absorption

$$\tau \sim \left(\frac{l}{10^{13}\text{ cm}}\right)\left(\frac{n}{10^8\text{ cm}^{-3}}\right)\left(\frac{B}{10^8\text{ G}}\right)^{2/3}\left(\frac{\gamma_m}{50}\right)^{-8/3}\left(\frac{\Gamma}{10^3}\right)^3\left(\frac{\nu_{obs}}{10^{19}\text{ Hz}}\right)^{-5/3} \quad (2)$$

must be of order unity for the observed frequency ν_{obs} to be in the *BATSE* range. Here l and n are the line-of-sight path length and particle density in the co-moving frame, γ_m and Γ are the minimum Lorentz factor of power-law electrons and the bulk Lorentz factor of the ejecta, and B is the co-moving magnetic field strength. Apparently, the values of the parameters are rather extreme, e.g., the magnetic field strength is (much) greater than the equipartition field of $\sim 10^5...10^6$ G, typically assumed within the standard synchrotron shock model (SSM). Another problem of this model is a very low efficiency of the fireball shock because the peak synchrotron frequency (where most of the energy is emitted) is deeply in the optically thick range.

SATURATED COMPTONIZATION MODEL

Another model that may be of interest to us is the so-called *saturated synchrotron self-Compton*, proposed in a series of papers (see e.g., Ref. [4]) primarily in an attempt to explain the spectral peak energy – fluence anti-correlation observed in several long, bright, smooth GRBs. The model proposes that impulsively accelerated, non-thermal, relativistic electrons (and, perhaps, pairs) repeatedly Compton up-scatter self-emitted radio/infrared synchrotron photons into gamma-ray energies. The Thompson optical depth is initially large, $\tau_T \gg 1$, so that the emerging gamma-rays are in thermal equilibrium with electrons and α approaches the Wien limit, $\alpha = +2$, whereas the synchrotron soft-photon source is strongly self-absorbed by internal free-free and synchrotron opacities. As time goes on, the Thompson opacity decreases and for $\tau_T \ll 1$ the spectrum reduces to a single-scattering Compton spectrum with the slope $\alpha = -(p+1)/2$ (for the electron distribution $N(E) \propto E^{-p}$). Thus, this model can naturally explain the hard-to-soft evolution in prompt GRB spectra.

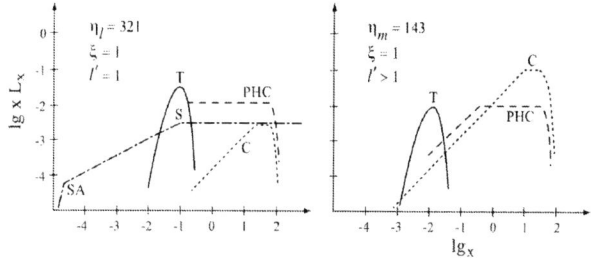

FIGURE 2. Spectra predicted by the photospheric model (from [5]).

Typical spectra are shown in Fig. 1(b) for various values of τ_T. It is quite clear that the emerging spectrum differs dramatically from the Band spectrum (even within a narrow *BATSE* window) which nicely fits the majority of GRBs. The required values of the fireball parameters are also not very likely: the comptonizing electrons ought to be "warm" with $\gamma_e \sim$ few (in contrast to the SSM, in which electrons are in near equipartition, hence $\gamma_e \sim 1000$) and the required magnetic fields are also too week, $B \sim 0.1...10$ G.

PHOTOSPHERIC MODEL

A synthetic model incorporating a standard synchrotron internal shock model and an extended photosphere can also explain steep low-energy spectra [5]. In this model synchrotron photons are re-processed in the photosphere. The single-scattering Comptonized and photosphetic components naturally have low-energy spectra with $\alpha = 0$. Fig. 2 shows some examples; here T: thermal photosphere, PHC: photospheric comptonized component, S: shock synchrotron, C: shock paid-dominated comptonized component. For a detailed discussion the reader is referred to Ref. [5]. The weakness of this model is that it requires moderate Thompson opacities $\tau_T \sim 1$ which, in turn, requires either fine tuning of plasma parameters or, alternatively, some sort of self-regulated pair opacity which produces the column density which self-adjusts itself to a column density of few g cm^{-2}. This model also requires very low baryonic load, that is, large bulk Lorentz factors $\Gamma > 1000$.

SMALL PITCH-ANGLE RADIATION MODEL

Optically thin synchrotron radiation has a low-energy asymptotic power-law index $\alpha = -2/3$ only if the the emitting electrons have an isotropic distribution. For electrons having anisotropic velocity distribution this may not be the case. Let us consider a beam of mono-energetic highly-relativistic electrons ($\gamma_e \gg 1$ is their Lorentz factor) propagating almost along a homogeneous magnetic field, so that the parallel velocity is

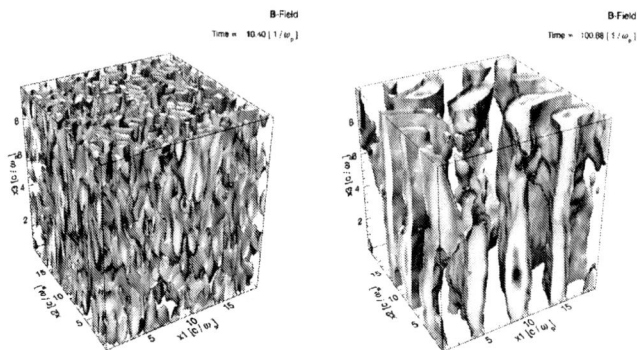

FIGURE 3. The magnetic field 3D structure (surfaces of constant B^2) at a relativistic shock (from [7]).

much larger than the transverse velocity:

$$v_\perp/v_\parallel \sim \Psi \ll 1/\gamma_e, \qquad (3)$$

where Ψ is the pitch-angle. In this case, from $v^2 = v_\parallel^2 + v_\perp^2$ and $1 - v^2 = 1/\gamma_e^2$, it follows that the transverse motion of such electrons is non-relativistic, $v_\perp \ll 1$. Radiation emitted by these electrons will be *cyclotron* (not synchrotron), relativistically boosted with the Lorentz factor $\sim \gamma_e$ along the magnetic field. The low-energy asymptotic spectrum of cyclotron radiation is steeper, $\alpha = 0$. This is called the "small pitch-angle" regime.

Small pitch-angle radiation has been suggested as yet another way of producing steep low-energy spectra, see Ref. [3] and references therein. The typical spectra for $1/\gamma_e < \Psi \ll 1$ and $\Psi \ll 1/\gamma_e$ are shown in Fig. 1(a) by the curves labeled SPD. Note that in the latter case, the spectrum has a very sharp break. Thus, this model can naturally explain both LoD-violating and BPL bursts. The small pitch-angle radiation model relies, however, on a crucial assumption: a highly anisotropic electron distribution is somehow created and maintained at the shock. Moreover, it is a well-known fact that (highly) anisotropic particle distributions are always unstable with respect to a number of plasma instabilities. Finally, in the small pitch-angle regime only the transverse energy $\sim m_e v_\perp^2/2$ can be converted into radiation. Because of the condition $v_\perp/v_\parallel \ll 1/\gamma_e$ with $\gamma_e \sim 1000$ or more, the radiation efficiency will be enormously low.

JITTER RADIATION MODEL

It is now becoming a widely accepted fact that magnetic fields of sub-equipartition strength are generated at the front of a relativistic shock via the two-stream (or Weibel) instability. The magnetic field generation has been predicted theoretically [6] and then confirmed via 3D PIC kinetic simulations [7, 8]. The produced magnetic fields have rather unusual properties. The field is predominantly generated in the direction, perpendicular to the shock propagation direction. In the plane of the shock, the field is highly

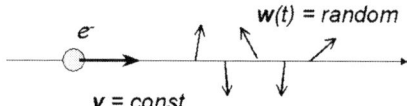

FIGURE 4. An electron motion in the jitter regime.

chaotic with the correlation length being of order the relativistic skin depth

$$\lambda_s \simeq \frac{c\sqrt{\gamma_s}}{\omega_{ps}} \sim (3 \text{ cm}) \gamma_e^{1/2} \left(\frac{m_s}{m_e}\right)^{1/2} \left(\frac{n_s}{10^{10} \text{ cm}^{-3}}\right)^{-1}, \qquad (4)$$

where $\omega_{ps} = (4\pi e^2 n_s/m_s)^{1/2}$ is the non-relativistic co-moving plasma frequency of species $s = e^-, p$ (both species, electrons and protons, generate the field). A typical strength of the field is $\varepsilon_B = B^2/(8\pi \Gamma n m_p c^2) \sim 10^{-3}$ (here n is the density downstream). The correlation scale λ is not constant, rather it increases with time, i.e., with the distance from the shock front. Fig. 3 represents three-dimensional contours of constant B^2 close to the shock front and far downstream.

It can straightforwardly be evaluated that the correlation length of the electron-produced field, λ_e, is much smaller than the Larmor radius of a relativistic radiating electron, ρ_e. The electron trajectory is not helical, so the standard synchrotron theory is not applicable. Quantitatively, jitter regime (for details, see Ref. [9]) occurs when the deflection angle of the electron is smaller than the relativistic beaming angle $\sim 1/\gamma_e$, i.e.:

$$\delta \sim \lambda_e/(\rho_e/\gamma_e) \sim (eB\lambda_e)/(m_e c^2) < 1. \qquad (5)$$

In the case $\delta \ll 1$, the particle motion may safely be approximated as straight. As the electron moves at a constant velocity, it experiences short accelerations in random directions, perpendicular to the direction of motion, as represented in Fig. 4. The power spectrum of radiation is obtained from the Lienard-Wichert potentials:

$$\frac{dW}{d\omega} = \frac{e^2 \omega}{2\pi c^3} \int_{\omega/2\gamma_e^2}^{\infty} \frac{|\mathbf{w}_{\omega'}|^2}{\omega'^2} \left(1 - \frac{\omega}{\omega'\gamma_e^2} + \frac{\omega^2}{2\omega'^2 \gamma_e^4}\right) d\omega', \qquad (6)$$

where $\mathbf{w}_{\omega'} = \int \mathbf{w} e^{i\omega' t} dt$ is the Fourier component of the particle's acceleration, which is related to the spectrum of the magnetic field as $w_{\omega'} = (eB_{\omega'})/(\gamma_e m_e) = (eB_{k'})/(\gamma_e m_e c)$. For a standard energy distribution of electrons (power-law with a cutoff at low energies, $\gamma_{e,min}$), the resultant spectrum is shown in Fig. 5(a). It is well described by a BPL model with $\alpha = 0$, the high energy exponent $\beta = -(p+1)/2$, and the jitter break frequency:

$$\nu_j \simeq (c/\lambda_e) \gamma_{e,min}^2 \Gamma \sim (10^{10} \text{ Hz}) \gamma_{e,min}^{3/2} \Gamma. \qquad (7)$$

Note that this frequency is independent of the magnetic field strength.

Unlike the electron-produced fields, the proton-produced magnetic field has a larger spatial correlation scale, λ_p, for which $\delta > 1$. An electron radiates synchrotron radiation

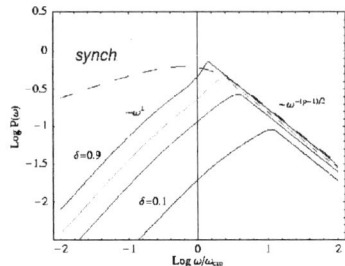 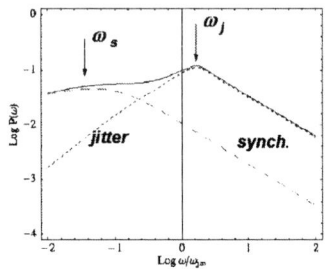

FIGURE 5. *(a)* The spectrum of jitter radiation for the power-law distributed electrons (from [9]). *(b)* A generalized jitter+synchrotron spectral model (from [9]).

in such a field. Also, a large-scale magnetic field may be ejected from a magnetized progenitor. Therefore, in general, the spectrum may consist of two components, a jitter component (due to small-scale fields, B_{SS}, with $\delta < 1$) and a synchrotron component (due to large-scale fields, B_{LS}, with $\delta > 1$), as in Fig. 5(b). The jitter-to-synchrotron peak frequency ratio and the ratio of the *photon* fluxes at these peak frequencies uniquely determine two free parameters, δ and B_{LS}/B_{SS}:

$$\frac{v_j}{v_m} \simeq \frac{3}{2}\frac{B_{LS}}{B_{SS}}\delta, \qquad \frac{F(v_j)}{F(v_m)} \simeq \delta^2, \qquad (8)$$

which offers a unique diagnostic of GRB shocks.

CONCLUSIONS

We reviewed several models which have been proposed in order to resolve the puzzle of LoD-violating and BPL gamma-ray bursts. It seems that the jitter model is the most promising one, because it readily follows from the collisionless shock physics and results in minimal changes in the standard optically thin synchrotron shock model.

REFERENCES

1. Preece, R. B., *et al.*, ApJS, **126**, 19 (2000).
2. Frontera, *et al.*, ApJS, **127**, 59 (2000).
3. Lloyd-Ronning, N. M. and Petrosian, V., ApJ, **565**, 182 (2002).
4. Liang, E., Kukunose, M., Smith, I. A., and Crider, A., ApJL, **479**, L35 (1997).
5. Mészáros, P. and M. J. Rees, ApJ, **530**, 292 (1997).
6. Medvedev, M. V., and Loeb, A., ApJ, **526**, 697 (1999).
7. Silva, L. O., Fonseca, R. A., Tonge, J. W., Dawson, J. M., Mori, W. B., and Medvedev, M. V., ApJL, **596**, L121 (2003)
8. Frederiksen, J. T., Hederal, C. D., Haugbølle, T., Nordlund, Å., astro-ph/0308104 (2003).
9. M.V. Medvedev, M. V., ApJ, **540**, 704 (2000).

Broad Band (2-700 keV) Properties of the GRBs Observed With BeppoSAX

Filippo Frontera

Physics Department, University of Ferrara, Ferrara, Italy
and
IASF, CNR, Bologna, Italy

Abstract. In this paper I review some significant results obtained with the wide field instruments (2 Wide Field Cameras and a Gamma Ray Burst Monitor) aboard the *BeppoSAX* satellite, and their implications for the current models concerning GRB progentors and their environments.

INTRODUCTION

It is universally recognized that the *BeppoSAX* mission [1] has been crucial for the great step forward in Gamma Ray Burst (GRB) astronomy. The satellite, switched off on April 29 2002, not only has allowed the settling of the distance scale of GRBs, but has provided and/or allowed most of the exciting results of the last 7 years on GRBs. The high performance of *BeppoSAX* for GRB studies was due to a particularly well-matched configuration of its payload, with both wide field instruments (Wide Field Cameras and Gamma Ray Burst Monitor) covering a broad energy band (2–700 keV), and narrow field telescopes (NFTs) with focusing optics in the 0.1 to 10 keV energy interval.

After the accurate GRB localizations and X–ray afterglow discoveries of 1997, many other GRB events were accurately localized and their afterglow sources detected during the operational life of *BeppoSAX*. The summary of GRB detections with the wide field instruments and follow–ups with the *BeppoSAX* NFTs is the following: 1082 events were detected with the GRBM (catalog in preparation), 168 of which (corresponding to $\sim 16\%$) are short (<2 s). 51 long (>2 s) GRBs were simultaneously detected with the GRBM and WFCs, 37 of which were followed-up with the NFTs. The 40–700 keV fluence of these GRBs ranges from 1.9×10^{-4} erg cm^{-2} down to 2.5×10^{-7} erg cm^{-2}. X–ray afterglow emission was discovered in $\sim 90\%$ of the followed-on GRBs [2]. However only $\sim 50\%$ of the followed-up GRBs were detected in the optical band (and $\sim 40\%$ in the radio band). These lower detection rates have raised the question of the origin of the so called 'dark' GRBs (i.e., GRBs with no optical counterparts). It is likely that many of them have origin in stellar formation regions with high mass densities, which likely absorb the ultraviolet radiation (in the rest frame of the GRB source). However, some dark GRBs could have origin in galaxies at very high redshifts (>5). The discovery of these high redshifts would have great cosmological implications.

In this paper I concentrate on some relevant results obtained so far from a systematic analysis, now in progress, of the prompt emission properties of GRBs detected with either the *BeppoSAX* WFCs plus GRBM or with the GRBM alone.

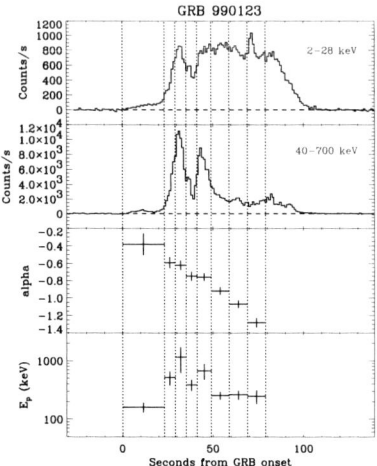

FIGURE 1. Spectral evolution of GRB990123. From top to bottom: X–ray light curve, gamma–ray light curve, low energy photon index of the Band model [9]; peak energy of the $EF(E)$ spectrum.

SPECTRAL EVOLUTION OF THE PROMPT EMISSION

In addition to the study of the time averaged spectral properties of the GRB prompt emission in the broad 2–700 keV energy band [3, 4], in order to better constrain the GRB emission mechanism we have systematically investigated the spectral evolution of all GRBs detected with the *BeppoSAX* GRBM and WFCs. Results of this analysis will be found in Frontera et al. [5], while the results obtained from the analysis of a sample of 8 GRBs were already reported [6]. Our investigation has shown that there is a general evolution of the spectra, from hard to soft, except for the most intense events in gamma rays ($\sim 10^{-4}$ erg cm^{-2}), whose spectral hardness either mimics the gamma–ray time profile (e.g., GRB990123, see Figure 1) or does not show any significant evolution with time [5]. The fit with an optically thin synchrotron shock model [7] is found to be consistent with most of the time resolved spectra, except at early times. As discussed by Frontera et al. [6], at these times some other emission mechanism (likely Inverse Compton) is at work. The important role of the synchrotron radiation has been confirmed by the recent polarization measurement for GRB021206 [8].

GRB X–RAY RICHNESS

Most of the GRBs observed with the *BeppoSAX* WFCs and GRBM emit most of their energy in the γ–ray band. Defining as X-ray richness of a GRB the softness ratio SR between its fluence S_X in the 2–10 keV band and the fluence, S_γ, in the 40–700 keV band, the number of *BeppoSAX* GRBs vs. their S_X/S_γ is shown in Figure 2. As can be seen, most of the GRBs have $SR < 0.3$. The $\log N/\log SR$ distribution is found to be consistent with a Gaussian profile plus a separated component which includes the X-

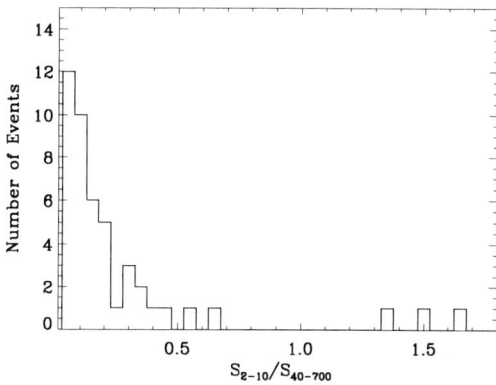

FIGURE 2. Distribution of the *BeppoSAX* GRBs detected with both WFC and GRBM according to their softness ratio (see text).

ray richest events shown in Figure 2. X-ray Flashes (XRF) cover part of the Gaussian, showing that they are a subclass of the classical GRBs. Typically *SR* is much higher for GRBs with lowest S_γ, confirming that their non–detection in gamma–rays is related to the limited sensitivity of the gamma–ray instruments.

The three X-ray richest GRBs detected with *BeppoSAX* (GRB981226 [10], GRB990704 [11] and GRB000615 [12, 5]), show the usual hard-to-soft spectral evolution, but with a maximum value of the peak energy E_p of their $EF(E)$ spectra which is much lower than that measured in normal GRBs.

GRBs which are probably associated with type Ib/Ic supernovae, in particular GRB990425 / SN1998bw [13], show spectral properties which are very similar to those of the other GRBs. However GRB011121 ($S_\gamma = 1.0 \times 10^{-4}$ erg cm^{-2}, $SR = 0.059$), associated with SN2001ke [14], like other strong GRBs, shows a peculiar spectral behavior: the peak energy E_p remains above the GRBM passband for the entire duration of the event [5].

THE E_P VS. E_{RAD} RELATIONSHIP

An initiative to search for correlations between parameters derived from the redshift–corrected energy spectra of GRBs with known redshift has permitted us to discover [4] a power–law relation between intrinsic peak energy E_p of the $\nu F(\nu)$ spectra and isotropic electromagnetic energy E_{rad} released in the GRB event:

$$E_p \propto E_{rad}^{0.52\pm0.06} \qquad (1)$$

The relation is now confirmed by more *BeppoSAX* and HETE-2 results [15] . It puts strong constraints to the GRB emission models: independently of the radiation pattern geometry, the E_p vs. E_{rad} relation has to be satisfied. The optically thin synchrotron shock model yields a similar relation, but with assumptions that are too simplistic, like requiring the same duration for all GRBs [16]. A discussion of the possible interpretations of the above relation is given by Zhang and Mészáros [17] within the internal and external shock scenarios.

GRB ENVIRONMENT

Low energy cutoffs and absorption features in the GRB spectra are a key tool to get information on the circumburst environment and, more importantly, to unveil the nature of GRB progenitors. Indeed, a dense star-forming medium is expected in the case of collapse of a massive star (hypernova model, Woosley [18]), and absorption features in the prompt emission are expected in the case of a supernova explosion preceding the occurrence of a GRB [19]. We have evidence of both these observational features.

High hydrogen column densities N_H, two orders of magnitude higher than the Galactic N_H^G along the line of sight, with decreasing behavior in time, have been discovered in the prompt emission of two GRBs: GRB980329 [6] and GRB000528 [20]. Variable N_H has also been found for GRB010214 [21], GRB010222 [22] and GRB990705 [23]. The time behaviour of N_H in the case of GRB980329 was investigated by Lazzati & Perna [24]. According to their model, the N_H time profile can be explained if the GRB occurs in overdense regions similar to the cocoon of star formation within molecular clouds.

Evidence of transient absorption features in the prompt emission has been found for two events: GRB990705 [23, 25] and GRB011211 [20]. In both cases the features (at 3.8 keV in the case of GRB990705, at 6.9 keV in the case of GRB011211) are apparent only during the rise time of the burst.

The absorption feature from GRB990705 was interpreted by Amati et al. [23] as a cosmologically redshifted K edge due to neutral Fe around the GRB location. The implied redshift (0.86 ± 0.17) was later confirmed by the optical redshift ($z_{opt} = 0.84$) of the associated host galaxy [26]. Lazzati et al. [27] assumed that the feature is an absorption line due to resonant scattering of GRB photons on H-like Iron (transition 1s-2p, $E_{rest} = 6.927$ keV). Also in this case the redshift derived is consistent with that of the host galaxy, and the line width is interpreted as being due to the outflow velocity dispersion (up to $\sim 0.1c$) of the material. In both scenarios, the observed feature points to the presence of an iron-rich environment, like that left by a supernova which explodes before the GRB event.

In the case of the transient line feature from GRB011211, the scenario is much more complex. Given that the redshift of the GRB optical counterpart is known ($z = 2.14$), if the line feature is interpreted as due to resonant scattering of GRB photons off H–like Ni XXVIII (which give the highest energy, 8.1 keV in the environment rest frame), the measured line energy implies a very high blue-shift (by $0.75c$) of the absorbing material. A possible intepretation of the origin of this feature is discussed by Frontera et al. [28]. But, also in this case, a prior supernova explosion appears to be the only way to accommodate the observed feature properties.

Peak Luminosity Versus Short Time Variability

Using the BATSE data, Reichart et al. [29] found a correlation between isotropic equivalent peak luminosity L_p of GRBs with known redshift and a measure V of their time variability on 64 ms timescale (the shortest achievable with BATSE). The constructed variability measure gives the non Poissonian variance of the light curve with respect to a smoothed light curve. Reichart et al. [29] found that only smoothing timescales proportional to the burst duration lead to a significant correlation between L_p and V: $L_p \sim V^{3.3^{+1.1}_{-0.9}}$. Smoothing timescales of a fixed duration in the source rest frame were found not to lead to significant correlations.

We performed the same analysis using the *BeppoSAX* GRBM data, which have a much better time resolution (7.8 ms for the entire GRB duration) [30]. We substantially confirm with a larger sample of GRBs the Reichart et al. [29] results. We have also searched for a correlation between L_p and V at shorter timescales (7.8 ms), and only when we select smoothing timescales of fixed duration in the source rest frame. After having corrected for the instrument dead time and a small non Poissonian component found in the GRBM background, the preliminary results show that, for smoothing timescales from 31.25 ms to 4 s, no statistically significant correlation is apparent.

CONCLUSIONS

Some significant conclusions can be drawn from the systematic analysis of the prompt X–/gamma–ray emission from GRBs detected with the *BeppoSAX* satellite.

The hard-to-soft evolution is typical for all GRBs, except the strong events. For example, in the case of GRB011121, the peak energy E_p of its $EF(E)$ spectrum stays high for the entire GRB duration, while in the case of GRB990123 it mimics the GRB time profile.

X-ray rich events are related only to GRBs with low gamma-ray fluence. This shows that XRFs are completely similar to classical GRBs with low E_p. Their spectral evolution is hard-to-soft like classic GRBs.

A decreasing N_H, first observed in the prompt emission of GRB980329, is now found in the prompt emission of another event, GRB000528. It shows a dense environment which is quickly ionized by the GRB event.

We have evidence of another transient absorption line, from GRB011211. It raises new problems, but points to the supranova model [19] as a likely scenario for the production of this event.

We confirm the results obtained by Reichart et al. (2001, [29]) with a larger sample of GRBs. We do not find a statistically significant correlation between peak luminosity and 7.8 ms variability for a smoothing timescale from 31.25 ms to 4 s.

ACKNOWLEDGMENTS

I wish to thank Lorenzo Amati and Cristiano Guidorzi for their important contribution to get the results discussed in this paper. This research is supported by the Italian Space Agency ASI and the Ministry of Education, University, and Research (COFIN funds 2001).

REFERENCES

1. Boella, G., et al., A&AS, **122**, 299 (1997).
2. Frontera, F., in *Supernovae and Gamma Ray Bursters*, ed. K. Weiler (Berlin, New York: Springer), Lecture Notes in Physics, Vol. 598, p. 317 (2003).
3. Amati, L., et al., in: *Gamma–Ray Bursts in the Afterglow Era*, ed. by E. Costa, F. Frontera, J. Hjorth (Springer, Berlin, Heidelberg), pp. 34–36 (2001).
4. Amati, L., et al., A&A, **390**, 81 (2002).
5. Frontera, F., et al., in preparation (2003).
6. Frontera, F., et al., ApJS, **127**, 59 (2000).
7. Tavani, M., ApJ, **466**, 768 (1996).
8. Coburn, W. & Boggs, S.E.,, Nature, **423**, 415 (2003).
9. Band, D., et al., ApJ, **413**, 281 (1993).
10. Frontera, F., et al., ApJ, **540**, 697 (2000).
11. Feroci, M., et al., A&A, **378**, 441 (2001).
12. Maiorano, E. et al., in *Gamma Ray Bursts in the Afterglow Era III*, ed.s M. Feroci, F. Frontera, N. Masetti, and L. Piro (ASP), in press (2003).
13. Galama, T. et al., Nature, **395**, 670 (1998).
14. Bloom, J.S., et al., ApJ, **572**, L45 (2002).
15. Amati, L., Chinese J. of Astr. & Ap, in press (2004).
16. Lloyd, N.M. et al., ApJ, **534**, 227 (2000).
17. Zhang, B. & Mészáros, P., ApJ, **581**, 1236 (2002).
18. Woosley, S.E., ApJ, **405**, 273 (1993).
19. Vietri, M, & Stella, L., ApJ, **507**, L45 (1998).
20. Frontera, F., et al., in preparation (2003).
21. Guidorzi, C. et al., A&A, bf 401, 491 (2003).
22. in 't Zand, J.J.M. et al., ApJ, **559**, 710 (2001).
23. Amati, L., et al., Science, **290**, 953 (2000).
24. Lazzati, D. & Perna, R., MNRAS, **330**, 383 (2002).
25. F. Frontera, et al., in *Gamma Ray Bursts in the Afterglow Era*, eds E. Costa, F. Frontera and J. Hjorth (Springer, Berlin, Heidelberg), p. 106 (2001).
26. Le Floc'h, E. et al., ApJ, **581**, L81 (2002).
27. Lazzati, D., et al., ApJ, 556, 471 (2001).
28. Frontera, F., et al., ApJ submitted (2003).
29. Reichart, D. et al., ApJ, **552**, 57 (2001).
30. Guidorzi, C. et al., in preparation (2003).

Observation and implications of the E_{peak}- E_{iso} correlation in Gamma-Ray Bursts

J-L. Atteia[*], G. R. Ricker[†], D. Q. Lamb[**], T. Sakamoto[‡§], C. Graziani[**], T. Donaghy[**], C. Barraud[*] and The HETE-2 Science Team[¶]

[*]*Laboratoire d'Astrophysique, Observatoire Midi-Pyrénées, Toulouse, France*
[†]*Center for Space Research, Massachusetts Institute of Technology, Cambridge, MA 02139, USA*
[**]*Department of Astronomy & Astrophysics, University of Chicago, Chicago, IL 60637, USA*
[‡]*Tokyo Institute of Technology, 2-12-1 Ookayama, Meguro-ku, Tokyo 152-8551, Japan*
[§]*RIKEN (The Institute of Physical and Chemical Research), Saitama 351-0198, Japan*
[¶]*An international collaboration of institutions including MIT, LANL, U. Chicago, U.C. Berkeley, U.C. Santa Cruz (USA), CESR, CNES, Sup'Aero (France), RIKEN, NASDA (Japan), IASF/CNR (Italy), INPE (Brazil), TIFR (India)*

Abstract. The availability of a few dozen GRB redshifts now allows studies of the intrinsic properties of these high energy transients. Amati et al. recently discovered a correlation between E_{peak}, the intrinsic peak energy of the $\nu f \nu$ spectrum, and E_{iso}, the isotropic equivalent energy radiated by the source. Lamb et al. have shown that HETE-2 data confirm and extend this correlation. We discuss here one of the consequences of this correlation: the existence of a 'spectral standard candle', which can be used to construct a simple redshift indicator for GRBs.

THE E_{PEAK} – E_{ISO} RELATION FOR GAMMA-RAY BURSTS

The growing sample of GRBs with spectroscopic redshifts allows the measure of some *intrinsic* properties of these explosions, like the energy radiated at various wavelengths or the energy at which most of the power is emitted. In 2002, Amati et al. performed a systematic analysis of 12 GRBs with known redshifts in order to derive their intrinsic spectral parameters (the spectral parameters at the source). In their paper, they report a strong correlation between E_{peak}, the intrinsic peak energy of the $\nu f \nu$ spectrum, and E_{iso}, the isotropic equivalent energy radiated by the source. For many years this correlation was suspected because it is the best way to explain the well known Hardness-Intensity correlation observed in GRBs (e.g. Mallozzi et al. 1995 [1], Dezalay et al. 1997 [2], Lloyd et al. 2000 [3], Lloyd-Ronning & Ramirez-Ruiz 2002 [4], and ref. therein). However the lack of distance measurements kept the work on the hardness-luminosity correlation qualitative. Recently Lamb et al. (2004 [5]) pointed out that HETE observations not only confirm this correlation, but also suggest its extension to the population of X-Ray Flashes (Fig. 1a). According to Fig. 1a, the E_{peak}- E_{iso} correlation can be approximated by the following relation: $E_{peak}/(100\,\mathrm{keV}) = \sqrt{E_{iso}/(10^{52}\mathrm{erg})}$. The origin of this correlation (which is reminiscent of the Temperature-Luminosity correlation for clusters of galaxies) is not discussed here. The aim of this paper is to discuss one of its consequences: the existence of a "spectral standard candle", which can be used to construct a redshift indicator for GRBs.

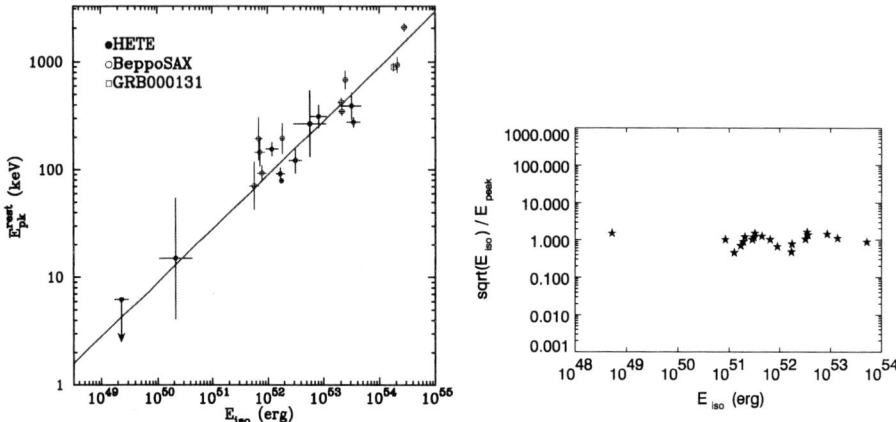

FIGURE 1. *Left Panel.* The E_{peak}-E_{iso} correlation measured at the end of 2003 with 21 GRBs detected by BeppoSAX (Amati et al. 2002 [6]), HETE-2 (Sakamoto et al. 2004 [7], Lamb et al. 2004 [5]), and the IPN (Andersen et al. 2000 [8]). Note the extent of the correlation in E_{iso}. *Right Panel.* Illustration of the fact that the ratio $\sqrt{E_{iso}}$ / E_{peak} is close to a standard candle. This ratio appears almost constant over 4-5 orders of magnitude in E_{iso}. The ratio $\sqrt{E_{iso}}$ / E_{peak} is plotted here for 20 GRBs with known redshift detected with BeppoSAX, HETE-2, and the IPN.

BUILDING A REDSHIFT INDICATOR FOR GRBS

The good correlation between E_{peak} and E_{iso} suggests that the ratio $\sqrt{E_{iso}}$ / E_{peak} is close to a standard candle. This is illustrated in Fig. 1b which shows this ratio as a function of E_{iso} for 20 gamma-ray bursts with known redshifts. Assuming that $\sqrt{E_{iso}}$ / E_{peak} is constant at the source, it is easy to compute its evolution with redshift. This is the dotted curve in the lower right panel of Fig. 2. This curve shows that unfortunately the *observed* ratio has a very small dependence on redshift beyond z=1. This illustrates the fact that when one wants to find a redshift indicator, it must be a quantity which has not only a small intrinsic dispersion, but also a strong dependence on redshift over a large range of redshifts. We performed an empirical search for such a quantity, starting from the fact that $\sqrt{E_{iso}}$ / E_{peak} is close to a standard candle. This work led us to conclude that the quantity $X_0 = N_\gamma/(E_{peak}*\sqrt{T_{90}})$ has the right properties for a redshift indicator (we do not claim however that it is the best redshift indicator which can be constructed from gamma-ray data only). Fig. 2 shows the intrinsic dispersion of X_0 (lower left panel), and its dependence on redshift (solid curve in the lower right panel). In the definition of X_0, E_{peak} is the peak energy of the $\nu f \nu$ spectrum, N_γ is the number of photons emitted by the GRB between ($E_{peak}/100$) and ($E_{peak}/2$), and T_{90} is the duration of the burst (see Atteia 2003 [9] for additional discussion on X_0). All these parameters are measured at the source.

Using X_0 as a redshift indicator, we can compute pseudo-redshifts by assuming that X_0 at the source is constant and that the observed value $X=n_\gamma/(e_{peak}*\sqrt{t_{90}})$ (the

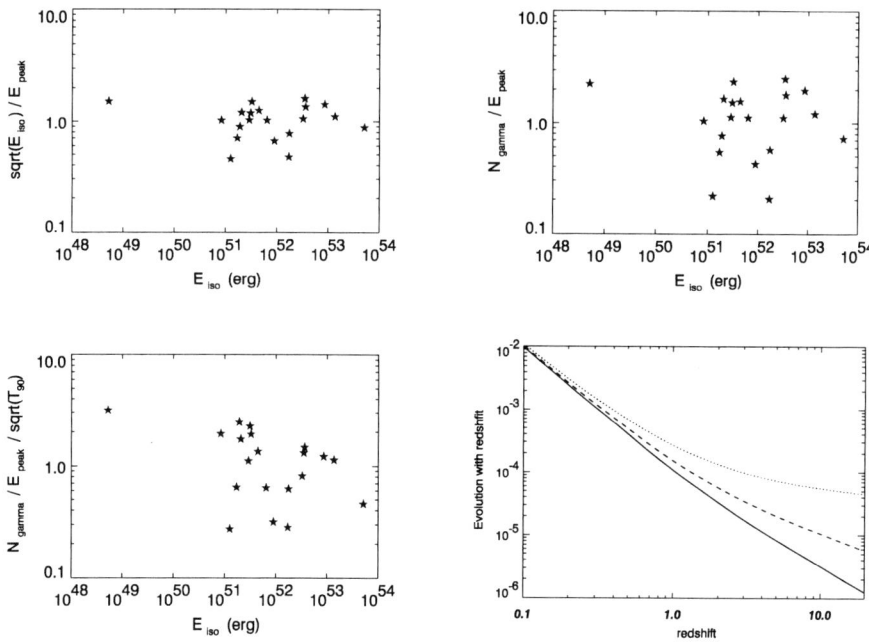

FIGURE 2. Comparison of various redshift indicators. The lower right panel shows the theoretical dependence on redshift of $\sqrt{E_{iso}}/E_{peak}$ (dotted line), N_γ/E_{peak} (dashed line), and $N_\gamma/(E_{peak}*\sqrt{T_{90}})$ (solid line), where N_γ is the number of photons emitted by the GRB between ($E_{peak}/100$) and ($E_{peak}/2$), and T_{90} is the duration of the burst. The other three panels show the intrinsic dispersion of these ratios.

lower case letters indicating that the parameters are now measured in the observer's framework) differs from X_0 only for the effect of the redshift. The validity of these pseudo-redshifts can be assessed from Fig. 3 which compares pseudo-redshifts and spectroscopic redshifts of 20 GRBs detected and localized by BeppoSAX, HETE-2, and the IPN. Possible applications of these pseudo-redshifts are discussed in Atteia (2003 [9]).

PSEUDO-REDSHIFTS OF HETE-2 GRBS

Table 1 presents the pseudo-redshifts of 42 long GRBs detected by HETE-2, with comments on their spectral properties, and on the detection of an afterglow. Pseudo-redshifts range from 0.20 (GRB 030824) to 14.0 (GRB 031026). Nine of the GRBs in Table 1 have spectroscopic redshifts (in the range 0.25 to 3.2). The pseudo-redshifts of these bursts are all within a factor of two of the spectroscopic redshifts, which leds us to conclude that pseudo-redshifts provide a robust redshift indicator in the range z=0.2-3. Table 1 contains three GRBs (in boldface) which have pseudo-redshifts

TABLE 1. Pseudo-redshifts of 42 long GRBs detected by HETE. The pseudo-redshifts of a few HETE GRBs could not be calculated because their e_{peak} is outside the energy range of HETE (e.g. GRB 021004). The 'Comment' column indicates the spectral hardness of the burst, and the detection of an afterglow when appropriate. OA, XA, RA, and IRA respectively stand for Optical Afterglow, X-ray Afterglow, Radio Afterglow, and Infra-Red Afterglow. Three GRBs with pseudo-redshifts greater than 5 are indicated in bold.

Name	redshift	pseudo-redshift	Comment	Name	redshift	pseudo-redshift	Comment
grb001225		0.69	Bright GRB	grb021104		0.88	X-Ray Flash
grb010126		1.52		grb021211	1.01	0.86	OA
grb010213		0.23	X-Ray Flash	grb030115		1.44	X-Ray Rich
grb010326		3.43		grb030226	1.98	3.56	XA, OA
grb010612		**9.50**		grb030324		3.93	dark, X-Ray Rich
grb010613		0.70		grb030328	1.52	1.39	OA, XA
grb010629		0.57	X-Ray Rich	grb030329	0.17	0.24	Bright, OA, XA, RA, SN 2003dh, X-Ray Rich
grb010921	0.45	0.62	OA, RA	grb030418		1.10	X-Ray Flash
grb010928		3.22		grb030429	2.65	1.44	OA
grb020124	3.20	2.28	OA	grb030519		2.53	
grb020127		2.67	XA, RA, X-Ray Rich	grb030528		0.36	IRA, XA
grb020305		**5.88**	OA	grb030723		0.59	OA, XA, SN, X-Ray Flash
grb020317		1.86	X-Ray Flash	grb030725		1.21	OA
grb020331		2.90	OA	grb030821		2.36	X-Ray Rich
grb020418		1.92		grb030823		0.64	X-Ray Flash
grb020801		0.95	X-Ray Rich	grb030824		0.20	X-Ray Flash
grb020812		3.03		**grb031026**		**14.0**	
grb020813	1.25	1.37	OA, XA, RA	grb031109a		1.29	
grb020819		1.52	RA, X-Ray Rich	grb031109b		1.39	X-Ray Flash
grb020903	0.25	0.31	OA, RA	grb031111a		4.20	
grb021016		1.45		grb031111b		0.56	X-Ray Rich

40

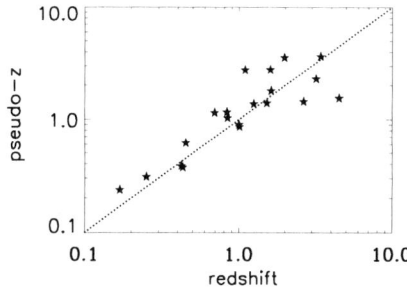

FIGURE 3. Pseudo-redshifts of 20 GRBs detected by BeppoSAX and HETE, compared with their spectroscopic redshifts.

larger than 5, suggesting that they could be GRBs at high redshifts. Unfortunately the spectroscopic redshifts of these three GRBs have not been measured, leaving open the issue of whether pseudo-redshifts are reliable at large redshifts. If pseudo-redshifts are reliable beyond z=5-6, they may become a useful tool to quickly identify high-z GRBs, and trigger the follow-up actions which are appropriate for these bursts (e.g. X-ray and IR observations).

Finally, we would like to mention that after discussions with the participants at the grb2003 Conference in Santa Fe, the HETE-2 Science Team and Operation Team now routinely provide the spectral parameters and the pseudo-redshifts of GRBs localized by HETE-2. These parameters are made available to the community on a web page just a few minutes after the determination of the burst localization (see GCNs 2421 and 2444).

ACKNOWLEDGMENTS

The authors acknowledge the wonderful work of the HETE operation team.

REFERENCES

1. Mallozzi, R. S., Paciesas, W. S., and Pendleton G.N. et al., *ApJ*, **454**, 597–+ (1995).
2. Dezalay, J.-P., Atteia, J.-L., and Barat C. et al., *ApJ Letters*, **490**, L17+ (1997).
3. Lloyd, N. M., Petrosian, V., and Mallozzi, R. S., *ApJ*, **534**, 227–238 (2000).
4. Lloyd-Ronning, N. M., and Ramirez-Ruiz, E., *ApJ*, **576**, 101–106 (2002).
5. Lamb D.Q. et al., *ApJ*, p. submitted (2004).
6. Amati, L., Frontera, F., and Tavani M. et al., *A&A*, **390**, 81–89 (2002).
7. Sakamoto, T., Lamb, D. Q., and Graziani C. et al., *ApJ*, **602**, 875–885 (2004).
8. Andersen, M. I., Hjorth, J., and Pedersen H. et al., *A&A Letters*, **364**, L54–L61 (2000).
9. Atteia, J.-L., *A&A Letters*, **407**, L1–L4 (2003).

Evidence From HETE-2 For GRB Evolution With Redshift

Carlo Graziani*, Donald Q. Lamb*, Takanori Sakamoto[†][**], Timothy Donaghy*, Jean-Luc Atteia[‡] and The HETE-2 Science Team[§]

*Department of Astronomy & Astrophysics, University of Chicago
[†]Department of Physics, Tokyo Institute of Technology
[**]RIKEN (Institute of Physical and Chemical Research)
[‡]Centre D'Etude Spatiale des Rayonnements, France
[§] An international collaboration of institutions including MIT, LANL, U. Chicago, U.C. Berkeley, U.C. Santa Cruz (USA), CESR, CNES, Sup'Aero (France), RIKEN, NASDA (Japan), IASF/CNR (Italy), INPE (Brazil), TIFR (India)

Abstract. After taking into account threshold effects, we find that the isotropic-equivalent energies E_{iso} and luminosities L_{iso} of gamma-ray bursts (GRBs) are correlated with redshift at the 5% and 0.9% signficance levels, respectively. Our results are based on 10 *Beppo*SAX GRBs and 11 HETE-2 GRBs with known redshifts. Our results suggest that the isotropic-equivalent energies and luminosities of GRBs increase with redshift. They strengthen earlier clues to this effect from analyses of the BATSE catalog of GRBs, using the variability of burst time histories as an estimator of burst luminosities (and therefore redshifts), and from an analysis of *Beppo*SAX bursts only. If the isotropic-equivalent energies and luminosities of GRBs really do increase with redshift, it suggests that GRB jets at high redshifts may be narrower and thus the cores of GRB progenitor stars at high redshifts may be rotating more rapidly. It also suggests that GRBs at very high redshifts may be more luminous – and therefore easier to detect – than has been thought, which would make GRBs a more powerful probe of cosmology and the early universe than has been thought.

INTRODUCTION

GRB sources are a cosmologically-distributed population. Like all other such populations, their properties presumably evolve with redshift. In this paper, we address two key questions: what evidence exists bearing on the evolution of GRB energetics as a function of redshift, and what is the nature and magnitude of that evolution?

PREVIOUS INDICATIONS OF GRB EVOLUTION

Using a Cepheid-like variability-based redshift measure developed together with Fenimore & Ramirez-Ruiz [1], Lamb & Reichart [2] found evidence for a positive correlation between isotropic-equivalent peak luminosity L_{iso} and redshift, using BATSE data. The left panel of Figure 1 shows their variability-based luminosity estimator, plotted as a function of redshift. The diagonal lines represent the BATSE 10% and 90% detection thresholds.

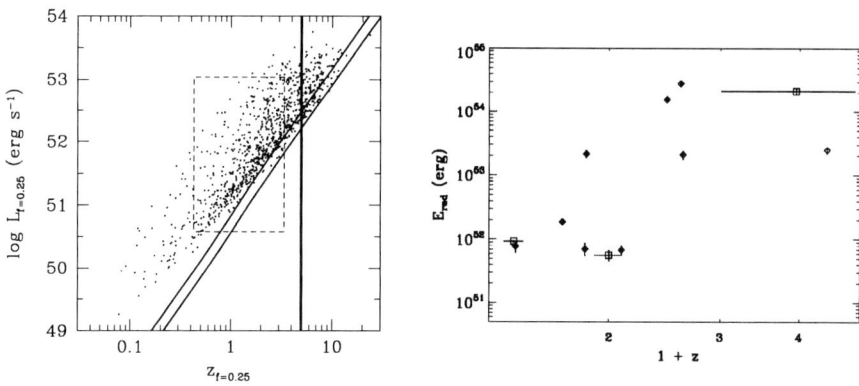

FIGURE 1. Left Panel: Variability Measure of Luminosity versus redshift. Lamb & Reichart [2]. Right panel: $E_{\rm iso}$ versus redshift for *Beppo*SAX GRB afterglows with redshift determinations. Amati et al. [4].

As can be seen in the figure, there is a dearth of high-luminosity bursts at estimated redshifts below $z \sim 1$ and the estimated luminosities also "peel off" the threshold at estimated redshifts below $z \sim 1$, hinting at a correlation of burst luminosity with redshift. Nevertheless, threshold effects, together with the possibility of unknown systematic effects intrinsic to the variability measure, are causes for concern.

Lloyd-Ronning, Fryer, & Ramirez-Ruiz [3] addressed threshold truncation effects in a study of 220 BATSE GRBs. They found that after applying the correction, there was significant evidence for a correlation slope $L_{\rm iso} \sim (1+z)^{1.4}$.

The number of GRBs with known redshifts has now grown to the point that it is possible to do meaningful studies of the distributions of GRBs with spectroscopic redshifts, as opposed to the less certain redshifts derived from the GRB variability measure. Amati et al. [4], using a sample of 12 *Beppo*SAX GRBs with afterglows and redshift estimates, found a positive correlation between the isotropic-equivalent prompt energy $E_{\rm iso}$ and redshift, with a reported significance of 7%. The right panel of Figure 1 shows the data used by Amati et al. [4]. While the result is promising, threshold effects were not quantified, and the distribution of $L_{\rm iso}$ was not investigated.

The HETE-2 bursts increase the size of the sample of GRBs with known redshifts by a factor of 2, and therefore offer the opportunity to confirm (or possibly refute) the Amati et al. results. In addition, it is interesting to extend the Amati et al. results to isotropic-equivalent peak luminosities ($L_{\rm iso}$), to make contact with the results of Lamb & Reichart [2] and of Lloyd-Ronning, Fryer, & Ramirez-Ruiz [3].

THE HETE-2-AUGMENTED GRB SAMPLE

The HETE-2 events considered in this study are:

- 9 HETE-2-localized classical GRBs with afterglows and redshifts — GRB010921,

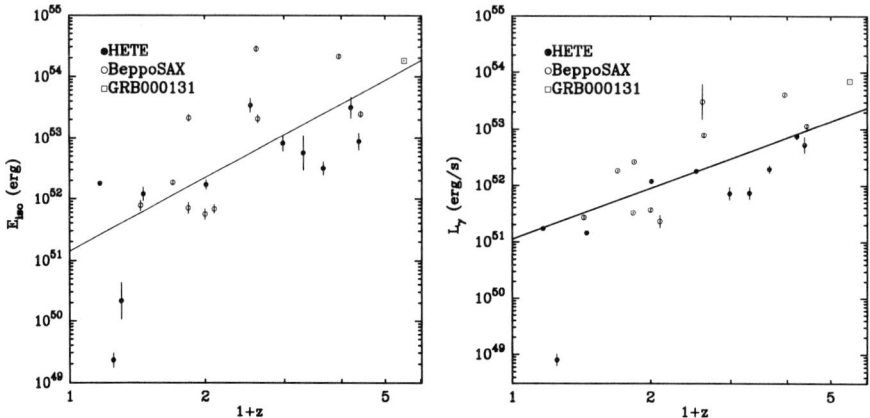

FIGURE 2. Left Panel: Isotropic-equivalent energy versus redshift. Right Panel: Isotropic-equivalent Luminosity versus redshift.

GRB020124, GRB021004, GRB021211, GRB030226, GRB030328, GRB030329, GRB030429, and GRB030323.

- 2 HETE-2-localized XRFs with afterglows and redshifts — XRF020903 ($z = 0.25$), and XRF030723 ($z \lesssim 0.3$).

In addition to these eleven events, we consider GRB000131 ($z = 4.511$), an IPN-located, BATSE-detected GRB. We further consider 10 events with well-determined redshifts from the Amati et al. [4] sample. In treating these events, we extract values of $L_{\rm iso}$ (not otherwise reported in [4]) using the values of $E_{\rm iso}$, S, and $F_{\rm peak}$ provided in their paper, with due attention to cosmological corrections.

The data are shown in Figure 2. The straight lines show fits obtained by excluding the two HETE-2-located XRFs, which would otherwise obviously bias the fits toward very high slopes, whose value for extrapolation to high-z would be dubious.

ANALYSIS

We perform a linear-regression analysis to the isotropic-equivalent prompt energy and peak luminosity data, obtaining best-fit slopes and intercepts. We also calculate correlation coefficients, and evaluate the formal correlation significance using the standard t-test.

This formal significance does not account in any way for threshold effects. We address the question of threshold-related systematic distortions by calculating detection thresholds using the methods described in Band [5]. The thresholds corresponding to WXM and to *Beppo*SAX are shown by the dashed lines in Figure 3.

We calculate threshold-corrected correlation significance by generating simulated samples of *Beppo*SAX- and HETE-2-observed data, which are truncated by the thresholds. The key assumption we make in the simulations is that the burst luminosity func-

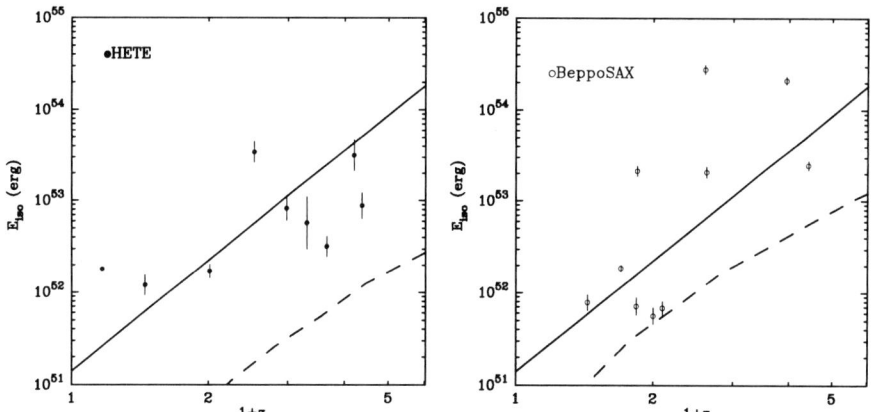

FIGURE 3. Isotropic-equivalent energy versus redshift. The dashed line shows the estimated detection threshold, calculated using the results in [5]. Left Panel: HETE. Right Panel: *Beppo*SAX

tion $f(L_{\mathrm{iso}}) \propto L_{\mathrm{iso}}^{-1}$ [6] (see also [7]). The significance is the fraction of the simulated samples with correlation coefficients that exceed the observed value.

RESULTS

The results of the analysis of the correlation of L_{iso} with $1+z$ are reported in table 1, those of the analysis of the correlation of E_{iso} with $1+z$ are reported in table 2.

Threshold effects are clearly playing an important role, and reduce the significance of the evidence for evolution substantially. Nonetheless, after accounting for them, there remains modest (5% significance) evidence for evolution of E_{iso}, and encouraging (9.5×10^{-3} significance) evidence for evolution of L_{iso}.

The slopes that we find are typically 3-4. These values are considerably higher than that found by Lloyd-Ronning, Fryer, & Ramirez-Ruiz [3]. However, our slopes are uncertain, and while our significances are corrected for threshold effects, our slopes are not. The magnitude of this effect remains to be estimated.

DISCUSSION

If the isotropic-equivalent energies and luminosities of GRBs really do increase with redshift, as the evidence presented here appears to suggest, the consequences for GRB models are profound. Models that rely on variations of the characteristic size of the jet opening angle to produce the observed variations in the isotropic-equivalent energies and luminosities of GRBs (e.g., the uniform jet model [6]) interpret the correlation as saying that higher-z GRBs have narrower jets. This suggests that the cores of the high-z progenitor stars may rotate more rapidly than those at low z. On the other hand, models that rely exclusively on variations of viewing angle to produce the observed variations

TABLE 1. Correlation Results: $\log(L_{\mathrm{iso}})$ vs. $\log(1+z)$

Quantity	HETE-2 + GRB000131	*Beppo*SAX + GRB000131	Combined
Sample Size	10	11	20
Correlation Coefficient	0.87	0.83	0.76
Formal Significance	9.0×10^{-4}	1.7×10^{-3}	8.9×10^{-5}
Threshold-Corrected Significance	1.2×10^{-2}	4.9×10^{-2}	9.5×10^{-3}
Slope*	3.2 ± 0.5	4.1 ± 0.8	3.3 ± 0.6

TABLE 2. Correlation Results: $\log(E_{\mathrm{iso}})$ vs. $\log(1+z)$

Quantity	HETE-2 + GRB000131	*Beppo*SAX + GRB000131	Combined
Sample Size	10	11	20
Correlation Coefficient	0.74	0.75	0.66
Formal Significance	1.3×10^{-2}	8.1×10^{-3}	5.3×10^{-4}
Threshold-Corrected Significance	7.8×10^{-2}	1.1×10^{-1}	5.1×10^{-2}
Slope*	2.0 ± 1.0	4.3 ± 1.1	3.0 ± 0.8

* Uncertainties in the slope are 68% confidence levels.

in the isotropic-equivalent energies and luminosities of GRBs (e.g., the universal jet model) run into difficulty explaining why the viewing angle distribution should change as a function of redshift.

If the isotropic-equivalent energies and luminosities of GRBs really do increase with redshift, GRBs at very high redshifts may be more luminous – and therefore easier to detect – than has been thought, which would make GRBs a more powerful probe of cosmology and the early universe than has been thought.

REFERENCES

1. Fenimore, E. E., and Ramirez-Ruiz, E., ApJ, submitted (astro-ph/0004176) (2000).
2. Lamb, D. Q. & Reichart, D. E., ApJ, **535**, 1 (2000).
3. Lloyd-Ronning, N. M., Fryer, C. L., and Ramirez-Ruiz, E., ApJ, **574**, 554 (2000).
4. Amati, L., et al., A&A, **390**, 81 (2002).
5. Band, D. L., ApJ, **588**, 945 (2003).
6. Donaghy, T., Lamb, D. Q., and Graziani, C., these proceedings (2004).
7. Schmidt, M., ApJ, **552**, 36 (2001).

GRB Redshift Evolution Within the Unified Jet Model

T. Q. Donaghy*[†], D. Q. Lamb* and C. Graziani*

Department of Astronomy & Astrophysics, University of Chicago, Chicago, IL 60637
[†]*quinn@oddjob.uchicago.edu*

Abstract. HETE-2 has provided new evidence that gamma-ray bursts may evolve with redshift [1]. We investigate the consequences of this possibility for the unified jet model of XRFs and GRBs [2]. We find that burst evolution with redshift can be naturally explained within the unified jet model, and the resulting model provides excellent agreement with existing HETE-2 and *Beppo*SAX data sets. In addition, this evolution model produces reasonable fits to the BATSE peak photon number flux distribution – something that cannot be easily done without redshift evolution.

INTRODUCTION

Most objects at cosmological distances (stars, galaxies, AGN) display evolution of their observable properties with redshift. Since gamma-ray bursts (GRBs) are thought to originate in the core-collapse of massive stars and are observed over a wide range in redshift, it is reasonable to suppose that they might also evolve.

Since the advent of rapid GRB localizations with *Beppo*SAX and HETE-2, and the consequent follow-up observations, over 30 GRBs have reported redshift measurements. Using 9 *Beppo*SAX bursts from [3], plus an additional 11 bursts localized with HETE-2, [1] have been able to strengthen earlier indications that GRBs are brighter at higher redshifts.

A uniform-jet model proposed by [2] has been shown to provide a unified picture of GRBs and XRFs. Considering all bursts to have a "standard energy" but a range of jet opening solid-angles that spans five orders of magnitude, this unified jet model can account for the full range of observed burst properties seen by HETE-2 . Here we extend this unified jet model to account for redshift evolution and show that it can also explain the observed properties of BATSE bursts.

OBSERVATIONS OF GRB EVOLUTION

Analysis of the BATSE catalog has revealed evidence that GRBs may evolve strongly with redshift. The use of redshift estimators based on burst variability has provided evidence that bursts are intrinsically brighter at larger redshifts than at smaller [4, 9]. Analyzing a set of 9 GRBs with spectroscopically determined redshifts observed by the BeppoSAX satellite, [3] also claim evidence for an increase in the isotropic-equivalent

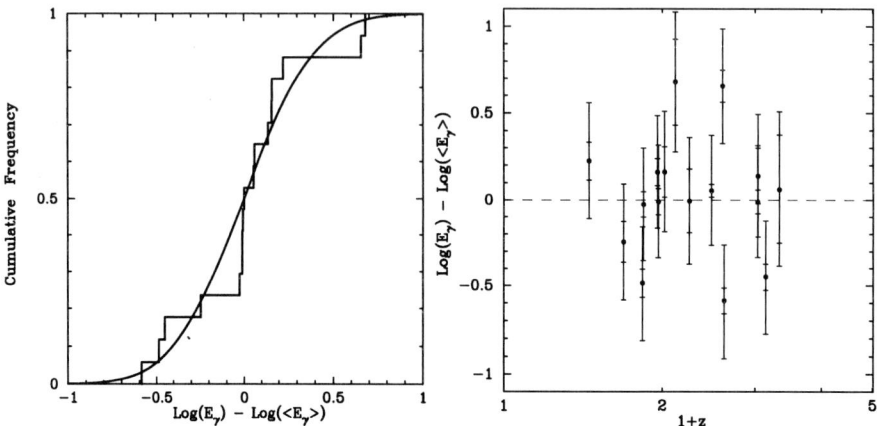

FIGURE 1. *Left:* Cumulative distribution of E_γ values reported by [6], shown with the best-fit Gaussian. *Right:* Displacement of E_γ values from the best-fit central value as a function of redshift. From [2].

energy (E_{iso}) with redshift, but they did not include a discussion of possible threshold selection effects.

Recent results from HETE-2 strengthen the evidence for this relationship. Figure 2 from [1] shows the isotropic-equivalent energies E_{iso} and luminosities L_{iso} as a function of redshift for the HETE-2 and BeppoSAX events. After correcting for threshold effects, [1] find that E_{iso} is correlated with redshift at the 5.1% confidence level, and L_{iso} is correlated with redshift at the 0.9% confidence level. The observed relationship goes roughly as $E_{\text{iso}} \sim (1+z)^3$.

UNIFIED JET MODEL SIMULATIONS

The unified jet model of GRBs and XRFs [2] provides a natural explanation for redshift evolution in GRBs. Namely, each burst exhibits a "standard energy" [5, 6], but the possible range of jet opening angles varies from fairly large values at low redshift, to very small values at high redshift, according to the relationship given above. That is, evolution in E_{iso} is explained by an evolution of the jet opening solid-angle, Ω_{jet}.

The burst simulations that we have implemented to test the unified jet model also provide a powerful way to explore models including burst evolution with redshift. For each burst, we obtain: (1) A redshift z by drawing from a model of the star-formation rate [7], and (2) a jet-opening solid angle Ω_{jet} by drawing from specific distribution range in Ω_{jet} that is fixed at $z = 0$ and shifts to smaller values at higher redshifts. We also introduce three Gaussian smearing functions to generate: (1) A spread in jet energy (E_γ, see Figure 1), (2) a spread in E_{peak} around the E_{iso}-E_{peak} relation, and (3) a spread in the timescale T that converts fluence to flux. Using these five quantities, we calculate various rest-frame quantities (E_{iso}, E_{peak}, etc.), and finally, we construct a Band function for each burst and transform it to the observer frame, which allows us to calculate fluences and

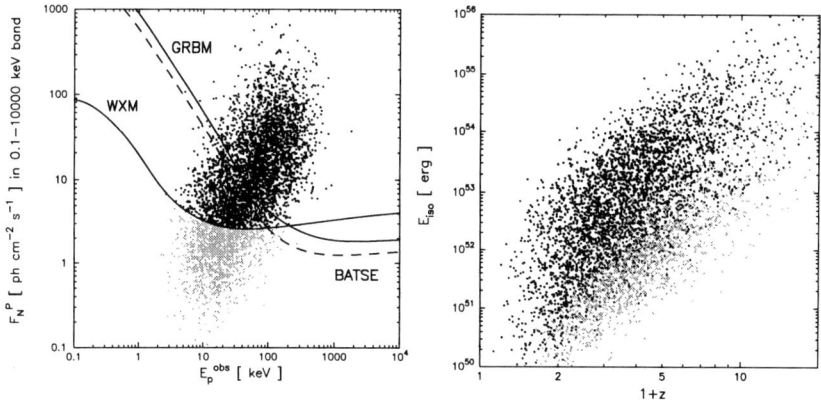

FIGURE 2. *Left:* Distribution of bursts in the $(E_{\text{peak}}^{\text{obs}}, F_N^P)$-plane, showing the threshold curves we use to determine if a burst is detected by various instruments. Black points are bursts detected by the WXM, while gray points are not detected. *Right:* Distribution of bursts in the $(1+z, E_{\text{iso}})$-plane for our best uniform jet with redshift evolution model.

peak fluxes and determine if the burst would be detected by various instruments (see Figure 2a).

To obtain the model presented here, we sought to roughly match the observed distribution in the $(1+z, E_{\text{iso}})$-plane (compare Figure 2b with Figure 2 of [1]), which displays a range in E_{iso} of 3×10^{49} to 1.5×10^{52} ergs at $z = 0$ and 6.4×10^{51} to 3.2×10^{54} ergs at $z = 5$. Assuming the faintest burst at $z = 0$ corresponds to $\Omega_{\text{jet}} = 2\pi$, this translates into a range of Ω_{jet} of 2π to 0.0125 steradians at $z = 0$ and 0.0291 to 5.79×10^{-5} steradians at $z = 5$.

The observed values of E_γ are taken from [6], and their distribution is well-fit by narrow Gaussian [5, 6, 8] (see Figure 1a). Figure 1b plots the displacement of these values from the central value as a function of redshift and shows no evidence for evolution of E_γ with redshift [2]. This rules out the possibility that redshift evolution might be explained by an evolution of the "standard energy".

Since it relies on the random distribution of burst jet axes with respect to the viewing angle, the universal structured-jet model cannot easily accommodate redshift evolution. The only solution would be for the "jet energy", E_γ, to evolve with redshift, but that would make it difficult to explain Figure 1 and the results of [5, 6].

RESULTS

Figure 4 compares the cumulative distributions of four observable burst quantities against several possible models. We find that the uniform jet model with burst evolution can adequately describe the observed distributions of localized bursts.

In addition, adding evolution to the uniform jet model replicates the observed distribution of peak photon number fluxes as observed by BATSE. Models without redshift evolution (Figure 3a, dotted curve) tend to overpredict the number of high peak flux bursts. However, models with strong evolution (solid curve) provide excellent agreement with the BATSE distribution. Figure 3b shows the differential distribution of redshifts (for bursts detected by WXM) for the models with and without redshift evolution.

This model makes several predictions. Most observed XRFs are predicted to be at $z < 1$. Bursts at low z have $\Omega_{jet} > 10^{-2}$ or $\theta_{jet} \sim$ a few degrees. We require a value of the "standard energy" to be $E_\gamma \sim 5 \times 10^{49}$ ergs, or about 50 times less than the value reported by [6]. Thus, the fraction of Type Ic supernovae producing GRBs increases from $\sim 0.1\%$ at $z = 0$ to $\sim 10\%$ at $z \sim 5$. Finally, 70% of bursts with $z > 5$ are detected by the WXM.

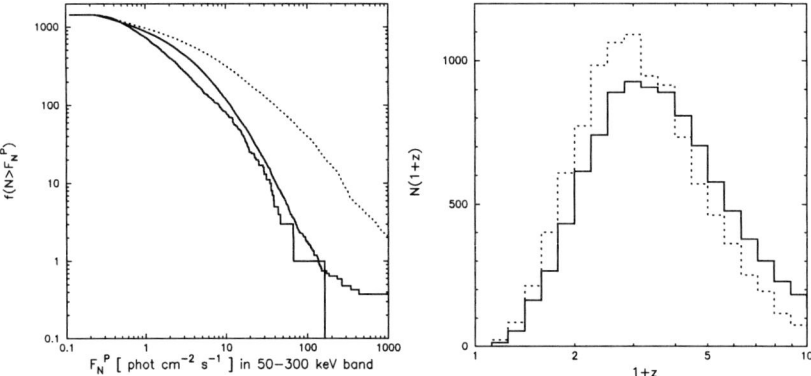

FIGURE 3. *Left:* Comparison of uniform jet models with redshift evolution (solid) and without (dotted) with the BATSE peak photon number flux cumulative distribution. *Right:* Predicted observed distribution of redshifts for the uniform jet model with redshift evolution (solid) and without (dotted).

CONCLUSIONS

HETE-2 has strengthened the evidence that GRBs evolve with z. The uniform jet model can describe XRFs and GRBs and can accommodate evolution whereas the universal jet model cannot.

Confirmation of this model will require the localization of many more XRFs, the determination of E_{peak} and E_{iso} for many more XRFs and GRBs, and the identification of optical afterglows and the measurement of redshifts for these bursts.

HETE-2 is ideally suited to localize XRFs and study their spectra, but this will be difficult for *Swift*, which has a nominal threshold of $E_{min} \sim 15$ keV and a narrow energy band of 15 keV $< E <$ 150 keV. However, *Swift* is optimized for pinpointing X-ray and optical afterglows, and facilitating spectroscopic redshift measurements. Therefore, it is very important that the HETE-2 mission continue, even after *Swift* is fully operational. A partnership between HETE-2 and *Swift* can confirm or rule out GRB evolution with redshift.

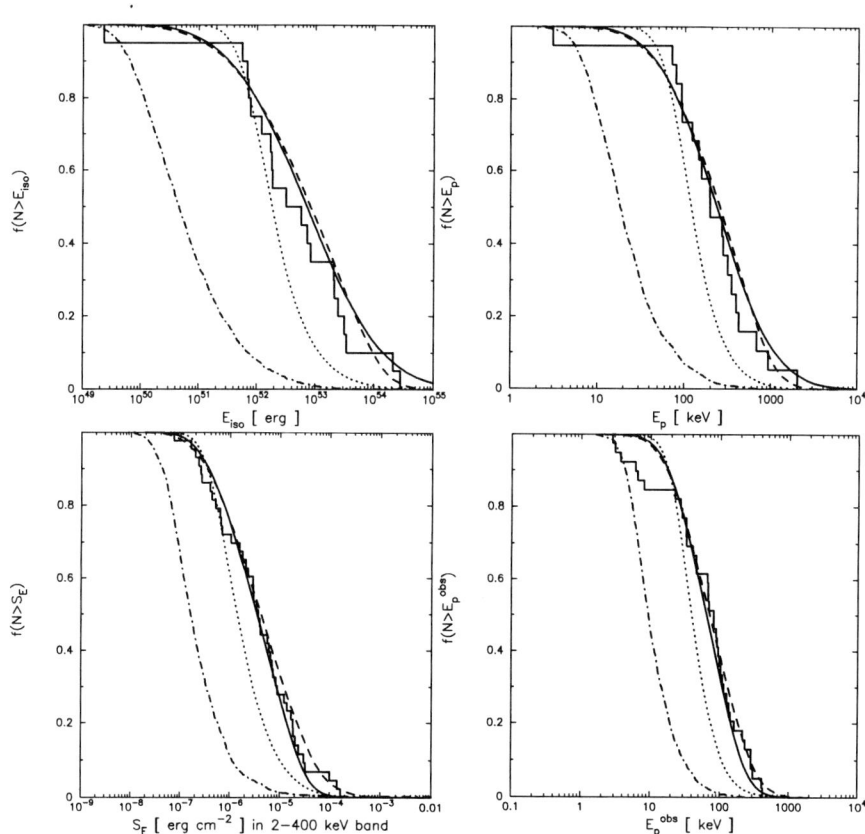

FIGURE 4. Comparisons of different models with the observed cumulative distributions for E_{iso} (upper left), E_{peak} (upper right), $S_E(2-400\text{ keV})$ (lower left) and E_{peak}^{obs} (lower right). The solid curve is the uniform jet model with redshift evolution and the dashed curve is the uniform jet model without redshift evolution. The dotted and dash-dotted curves are two variants of the universal or structured jet model.

REFERENCES

1. Graziani, C., et al., in these proceedings (2003).
2. Lamb, D. Q., Donaghy, T. Q. & Graziani, C., submitted to ApJ, (astro-ph/0312634) (2003).
3. Amati, L., et al., A & A, **390**, 81 (2002).
4. Lloyd-Ronning, N. M., Fryer, C. L. & Ramirez-Ruiz, E., ApJ, **574**, 565 (2002).
5. Frail, D. A., et al., ApJ, **562**, L55 (2001).
6. Bloom, J. S., Frail, D. A. & Kulkarni, S. R., ApJ, **594**, 674 (2003).
7. Rowan-Robinson, M., ApJ, **549**, 745 (2001).
8. Lamb, D. Q., et al., submitted to ApJ (2004).
9. Reichart, D. E. & Lamb, D. Q. (2001).
10. Sakamoto, T., et al., submitted to ApJ (2004).

Quiescent Burst Evidence for Two Distinct GRB Emission Components

Jon Hakkila* and Timothy W. Giblin*

Department of Physics and Astronomy, College of Charleston, Charleston, SC 29424-0001

Abstract. We have identified two GRBs in which early emission is very different than late emission. Post-quiescent pulses of BATSE GRBs 960530 and 980125 have long lags, smooth morphologies, and soft spectral evolution. The early emission satisfies the standard internal shock paradigm, while the late emission is consistent with external shocks. The peak luminosity ratio between early and late episodes is not in agreement with that predicated by the lag vs. peak luminosity relationship. We briefly discuss these observations in terms of a current hypernova jet model.

INTRODUCTION

Quiescent GRBs release their prompt energy in more than one distinct emission episode [1, 2]; e.g. they have at least one extended period during which emission is absent. Normally, post quiescent emission is similar to prequiescent emission, but not for GRBs 960530 and 980125 (Figure 1). These components have different morphologies, spectral lags, Color-Color Diagrams (CCDs), and Internal Luminosity Functions (ILFs). Despite these differences, localization measurements verify that the multiple emission episodes are associated with the same sources for both GRB 960530 and GRB 980125.

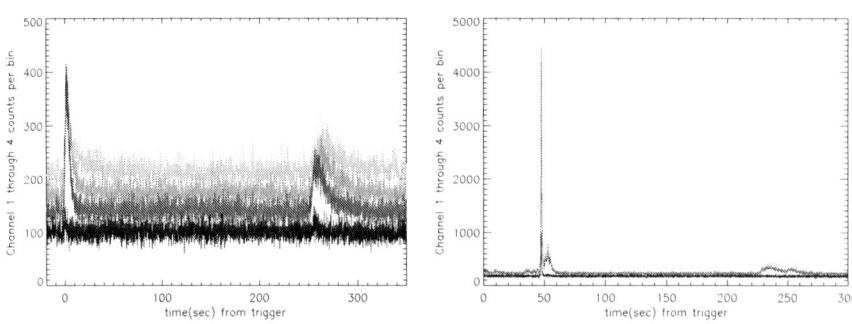

FIGURE 1. (a) GRB 960530 (BATSE trigger 5478) and (b) GRB 980125 (BATSE trigger 6581).

GRB 960530 (BATSE trigger 5478) has two distinct emission episodes separated by 233 s. Episode (a) is a single-pulse FRED (Fast Rise Exponential Decay) of duration 23 s. Episode (b) is a smooth single-pulse episode of 37 s. GRB 980125 (BATSE trigger 6581) has two distinct emission episodes preceded by a faint preburst episode of 13

TABLE 1. Properties of Quiescent GRB 960530 (BATSE trigger 5478).

Time Period	morphology	ILF α	Lag$_{21}$ (s)	Lag$_{31}$ (s)	Lag$_{32}$ (s)
before 200 sec	spiky FRED	0.9 ± 0.4	0.1 ± 0.11	0.58 ± 0.12	0.58 ± 0.04
after 200 sec	smooth gradual	-7.5 ± 1.2	1.17 ± 0.26	5.04 ± 0.06	2.77 ± 0.06

TABLE 2. Properties of Quiescent GRB 980125 (BATSE Trigger 6581).

Time Period	morphology	ILF α	Lag$_{21}$ (s)	Lag$_{31}$ (s)	Lag$_{32}$ (s)
before 150 sec	complex + FRED	-1.3 ± 0.5	0.04 ± 0.02	0.08 ± 0.01	0.10 ± 0.01
150 to 250 sec	smooth gradual	-3.1 ± 0.5	0.43 ± 0.32	1.66 ± 0.19	0.90 ± 0.40
after 250 sec	smooth FRED	-3.9 ± 0.4	0.07 ± 0.03	0.13 ± 0.08	0.05 ± 0.08

s; the properties of the preburst episode are unfortunately too faint to be accurately measured. Complex emission episode (a) lasts 31 s and occurs 18 s after the preburst episode. Episode (b) is smooth with two distinct pulses and a duration of roughly 45 s; it is preceded by a quiescent period of 161 s. The first of these two distinct pulses is a smooth broad pulse similar in temporal structure to GRB 960530's emission episode (b), while the second pulse (which overlaps the first) is more indicative of a standard FRED.

ANALYSIS

Spectral lags represent the maximum value of the Cross-Correlation Function (CCF) between two energy channels [3]. Identification of the CCF peak is complicated by Poisson background variations, requiring a fitting function. The Norris et al. pulse model [4] is a good fitting function because it describes a normalized comparison between episodes whose primary structures are pulses. Spectral lags appear to correlate with burst parameters such as beaming angles and peak luminosities (measured on the 256 ms timescale) [5]. We find that the broad smooth pulses of GRBs 960530 and 980125 have significantly longer lags than those found elsewhere in the bursts (Tables 1 and 2).

The Internal Luminosity Function (ILF) $\psi(L)$ is the distribution of luminosity within a GRB; $\psi(L)\Delta L$ represents the fraction of time during which a burst's luminosity lies between L and $L + \Delta L$ [6, 7]. We have calculated the ILF in the 50 to 300 keV energy range using 4-channel 64-ms data; The ILF is normalized to the number of Poisson background-subtracted time intervals during which emission is observed 2σ or more above the background and by $\Sigma\psi(L)\Delta L = 1$. The ILF is best fit with a quasi power-law form for most bursts such that $\log \psi(L) \propto \alpha \log L + \beta (\log L)^2$. The power-law index α and the curvature index β are indicators of burst temporal morphology; FRED-like bursts typically have large α while spiky (complex) bursts have small α. The index α is also a strong indicator of β. The ILFs of the different components of quiescent GRBs 960530 and 980125 are indicated in Table 1 and Table 2. The late-appearing, smooth, long-lag components of both bursts are fit by noticeably steeper ILF values than are the other episodes. The second (FRED) pulse (at 255 s) in GRB 980125b is smoother than a typical FRED and has a steep α, but has a short lag consistent with the burst's episode (a). It is possible that overlap with the broad, long-lag pulse has affected its ILF.

Color-color diagrams (CCDs) [8] describe GRB spectral evolution (which typically evolves from hard to soft). The hard color index HC is the ratio of BATSE channel 3 to channel 2 photon fluxes, while the soft color index SC is the ratio of channel 2 to channel 1 photon fluxes. The color indices are reliable for spectral evolution studies when a constant signal-to-noise is maintained (by varying the timescale on which the color indices are measured). The predicted colors of fast-and slow-cooling external (synchrotron) shocks tend to be soft in a particular range of values. GRB 960530 has previously been considered to be an afterglow candidate because it exhibits a soft gamma-ray component that was coincident with the onset of the afterglow phase [8]. The CCD of this soft gamma-ray component was in agreement with the theoretical prediction for synchrotron cooling. The late emission of GRB 980125 is consistent with both fast- and slow-cooling synchrotron models, and the second episode is in better agreement with synchrotron afterglow emission than the first [9].

There are other reasons why the long-lag pulses are consistent with external shocks. For one thing, external shocks are expected to produce broad pulses [10, 11, 12, 13]. Also, internal shocks can explain the one-to-one ratio of the post-quiescent emission duration to the quiescent time observed in most GRBs [1, 2] (interpreted as a metastable outflow that reserves a significant fireball energy fraction for liberation through later internal shocks); but it cannot explain the ratios found in these quiescent GRBs (it is 0.16 in GRB 960530 while it is 0.28 in GRB 980125).

It is possible that some bursts exhibit *only* the smooth, long lag (external shock) component. The variable, short lag (internal shock) component would have to be either suppressed or undetected in these bursts. There is evidence that the long-lag emission component is common in many GRBs, even though it has thus far been identified in only two quiescent GRBs. For example, the lag calculations of [14] use an apodization technique to remove low-intensity emission from the lag calculation. This is done because there is a long-lag signature hidden in the low-intensity emission capable of distorting the lag measurement (e.g. GRB 990123). The long-lag component might be present prior to the end of many bursts (n GRB 980125 the long-lag pulse occurs prior to a FRED pulse); it would be hard to isolate because it is smooth, broad, and faint. Note that many GRBs can be characterized by combinations of overlapping broad and narrow pulses.

The long-lag emission pulses of GRBs 960530 and 980125 potentially provide a dilemma for the lag vs. peak luminosity relation, since the relation implicitly assumes that each GRB can be characterized by a single lag. The lag vs. peak luminosity relation [14] is $L_{53} \approx 1.3 \times \tau^{-1.14}$ where $\tau = (\text{lag}_{31}/0.01\text{s})$, lag_{31} is the lag between the 100 to 300 keV channel and the 25 to 50 keV channel, and L_{53} is the isotropic peak luminosity on the 256 ms timescale in units of 10^{53} erg s^{-1}. The relation was originally calibrated for a half-dozen GRBs of known redshift. However, the anomalously-underluminous long-lag GRB 980425 was not included in the calibration. Since each quiescent episode in GRBs 960530 and 980125 exhibits a distinct lag, one would expect each component to be independently consistent with the lag vs. peak luminosity relation. Thus, the post-quiescent to pre-quiescent luminosity ratios of these GRBs as obtained from lags should be equal to their peak flux ratios. For GRB 960530, the lag vs. peak luminosity relation

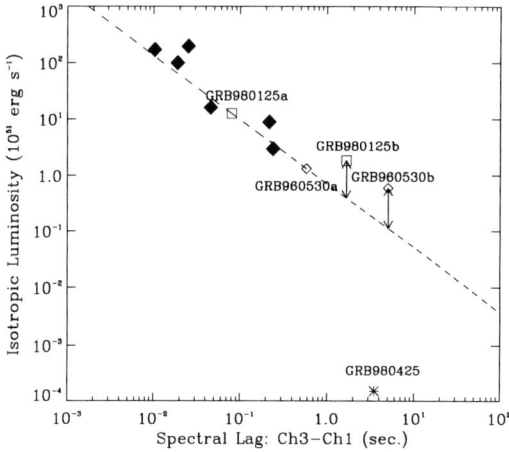

FIGURE 2. Placing the separate emission episodes of GRBs 960530 and 980125 on the peak luminosity versus lag diagram.

predicts $L_b/L_a = 0.065$, while the ratio of 256 ms peak fluxes yields $P_b/P_a = L_b/L_a = 0.46$. For GRB 980125, the lag vs. peak luminosity relation predicts $L_b/L_a = 0.024$, while the ratio of 256 ms peak fluxes yields $P_b/P_a = L_b/L_a = 0.15$. There is a distinct possibility that [5] has applied the lag vs. peak luminosity relation to GRBs dominated by short-lag components as well as to potentially different bursts dominated by long-lag components. According to Figure 2 of [5], roughly 300 individual GRBs have $\mathrm{lag}_{21} > 0.43$ s, where lag_{21} is the lag between the 50 to 100 keV channel and the 25 to 50 keV channel ($\mathrm{lag}_{21} = 1.17$s for GRB 960530 and $\mathrm{lag}_{21} = 0.42$s for GRB 980125).

The peak luminosities of the short-lag (a) and long-lag (b) components are plotted in Figure 2 together with the GRBs (filled diamonds) used to calibrate the lag vs. peak luminosity relation [14]. The luminosities of the long-lag components are predicted using both the lag vs. peak luminosity relation and peak flux ratios; the values predicted by the lag vs. peak luminosity relation lie on the theoretical (dashed) line while those obtained from peak flux ratios are shifted vertically. The luminosities calculated from peak fluxes are larger than those calculated from the lag vs. peak luminosity relation, suggesting that this relation underestimates long-lag burst luminosities. This result is difficult to reconcile with long-lag GRB 980425, which is underluminous relative to the lag vs. peak luminosity relation (see Figure 2). Even though GRB 980425 and the post-quiescent emission of GRBs 960530 and 980125 all have long lags of similar durations, it appears that these lags are not good predictors of peak luminosity. The long-lag GRBs do not obey the same lag vs. peak luminosity relation as short-lag GRBs, and it is possible that they do not obey a lag vs. peak luminosity relation at all. It is also surprising that the two emission components appear to have different opening angles [15, 16, 17].

The collapsar jet model of [18] could naturally explain the two emission components observed here. This model naturally produces complex relativistic jets having two distinct components. The first is a high-Γ, highly-beamed component containing the internally-shocked material, and the second is a lower-Γ component with a larger beaming angle that results from the jet's Lorentz factor and its internal energy at breakout. The beaming angle of the high-Γ is limited to $\sim 5°$, whereas the low-Γ component can have a beaming angle exceeding $15°$. There is no obvious reason why the relationship between lags and peak luminosities produced by a highly-relativistic beamed jet containing internal shocks should be similar to that produced by mildly-relativistic broad jets presumably exhibiting external shocks.

CONCLUSIONS

At least two GRBs exhibit multiple lags. The short-lag component is consistent with internal shocks while an external shock explanation is preferred for the smooth long-lag component. The long-lag component is broad and faint, and is possibly present but unrecognizable in many bursts. This result is not entirely surprising; two types of GRB shocks have been predicted and expected [19, 20]. The lag vs. peak luminosity relation apparently requires re-calibration. This work is supported by NSF grant AST00-98499.

REFERENCES

1. Ramirez-Ruiz, E. & Merloni, A., MNRAS, **320**, L25 (2001).
2. Ramirez-Ruiz, E., Merloni, A., & Rees, M. J. MNRAS, **324**, 1147 (2001).
3. Band, D. L. ApJ, **486**, 928 (1997).
4. Norris, J. P., Nemiroff, R. J., Bonnell, J. T., Scargle, J. D., Kouveliotou, C., Paciesas, W. S., Meegan, C. A., & Fishman, G. J., ApJ, 459, 393 (1996).
5. Norris, J. P., ApJ, **579**, 386 (2002).
6. Horack, J. M. & Hakkila, J., ApJ, **479**, 371 (1997).
7. Hakkila, J., Giblin, T. W., Young, K. C., Fuller, S. P., Stallworth, A. D., & Sprague, A. P., in proceedings of the 2003 Santa Fe Gamma-Ray Burst Conference (ed. E. E. Fenimore, M. Galassi), submitted (2004).
8. Giblin, T. W., Connaughton, V., van Paradijs, J., Preece, R. D., Briggs, M. S., Kouveliotou, C., Wijers, R. A. M. J., & Fishman, G. J., ApJ, **570**, 573 (2002).
9. Freismuth, T. M., Giblin, T., & Hakkila, J., American Astronomical Society Meeting, 202 (2003).
10. Fenimore, E. E., Madras, C. D., & Nayakshin, S., ApJ, **473**, 998 (1996).
11. Sari, R. & Piran, T. ApJ, **485**, 270 (1997).
12. Fenimore, E. E., Cooper, C., Ramirez-Ruiz, E., Sumner, M. C., Yoshida, A., & Namiki, M. ApJ, **512**, 683 (1999).
13. Fenimore, E. E. & Ramirez-Ruiz, E., ASP Conf. Ser. 190: Gamma-Ray Bursts: The First Three Minutes, 67 (1999).
14. Norris, J. P., Marani, G. F., & Bonnell, J. T., ApJ, **534**, 248 (2000).
15. Sari, R., ApJL, **524**, L43 (1999).
16. Frail, D. A. et al. ApJL, **562**, L55 (2001).
17. Panaitescu, A. & Kumar, P., ApJ, **571**, 779 (2002).
18. Zhang, W., Woosley, S. E., & MacFadyen, A. I., ApJ, **586**, 356 (2003).
19. Rees, M. J. & Meszaros, P., ApJL, **430**, L93 (1994).
20. Sari, R. & Piran, T. ApJ, **520**, 641 (1999).

The HETE-2 Burst Catalog

R. Vanderspek*, A. Dullighan*, N. Butler*, G. B. Crew*, J. N. Villasenor*,
G. R. Ricker*, T. Tamagawa†, T. Sakamoto**†, M. Suzuki**, Y. Shirasaki‡†,
T. Yamazaki§, K. Hurley¶, C. Graziani‖, T. Donaghy‖, D. Q. Lamb‖, C.
Barraud†† and J-L. Atteia††

*Center for Space Research, Massachussetts Institute of Technology, Cambridge, MA 02139 USA
†RIKEN (The Institute of Physical and Chemical Research), 2-1 Hirosawa, Wako, Saitama 351-0198, Japan
**Tokyo Institute of Technology, 2-12-1 Ookayama, Meguro-ku, Tokyo 152-8551, Japan
‡National Astronomical Observatory of Japan, 2-21-1 Osawa, Mitaka, Tokyo 181-8588, Japan
§Department of Physics, College of Science and Engineering, Aoyama Gakuin University, 5-10-1 Fuchinobe, Sagamihara, Kanagawa 229-8558, Japan
¶Space Sciences Laboratory, UC Berkeley, Berkeley, CA 94720-7450
‖Department of Astronomy & Astrophysics, University of Chicago, 5640 South Ellis Avenue, Chicago, IL 60637
††Laboratoire d'Astrophysique, Observatoire Midi-Pyrénées, 14, Avenue E. Belin 31400 Toulouse France

Abstract. During its >2.5 years of operation, the High Energy Transient Explorer (HETE-2) has detected nearly 300 GRBs, over 800 XRBs, and ~50 bursts from SGRs in on-orbit operations and sophisticated ground searches of archived data. In addition to these bursts, there have been over 1500 other triggers detected using on-board or ground algorithms, hundreds of which are as yet unidentified and may come from cosmic sources. We have developed the HETE Burst Catalog as a means of organizing and analyzing the data from these nearly 3000 triggers in a systematic way. We present preliminary results from the HETE burst catalog.

INTRODUCTION

During its >2.5 years of operation, the High Energy Transient Explorer (HETE-2) has detected over 170 GRBs in on-orbit operations; over 100 additional bursts have been detected in more sensitive ground searches of survey data. In this same period, over 800 XRBs and ~50 bursts from SGRs have been detected. In addition to these ~1200 bursts, there have been over 1500 additional triggers detected by on-board software (hereafter "triggered" bursts) or ground algorithms (hereafter "untriggered" bursts), hundreds of which are as yet unidentified and may come from cosmic sources.

We have developed the HETE Burst Catalog as a means of organizing and analyzing the data from these nearly 3000 triggers in a systematic way. We apply the same analyses to all triggers, allowing us to identify patterns in the data that identify particular types of triggers. The output of these analyses are text files in a "keyword=value" format; these text files are reprocessed and put into the master HETE trigger database, aka the HETE Burst Catalog. Simple scripts can be used to extract data from the catalog to create tables and illuminative plots, such as the ones in Figure 1 and Table 2; in addition,

TABLE 1. Distribution of HETE Trigger Types

Trigger Type	Triggered	Untriggered	Description
GRB	104	101	Bursts with significant band C emission
XRB	281	561	Bursts localized to known XRB sources. The initial classification of XRB is also given to unlocalized bursts detected near the Galactic Center if the burst is bright and FRED-like.
Collimator	58	0	Real GRBs which entered Fregate through the graded-shield collimator (incident angle $>70°$). The shield absorbs everything <100 keV, so collimator bursts have no band A or B emission.
XRF	10	8	Bursts with significant low-energy fluence, yet not consistent with an XRB source
SGR	40	6	Bursts from known SGR; also assigned to very short, soft bursts when the Galactic Center is in the Fregate FOV.
Particles	298	4	Particle events from either the Ecuador or South Atlantic Anomaly.
Emersion	28	1	Emersion of a bright X-ray source from behind the Earth's limb.
Immersion	3	0	Immersion of a bright X-ray source behind the Earth's limb.
Corruption	13	0	Data corruption
Error	61	0	Operational error
Poisson	92	2	False trigger due to statistical fluctuations
Sco X-1	56	0	Trigger due to Sco X-1 fluctuation
Solar	46	1	Reflected solar X-rays
Spin	180	0	Trigger due to spacecraft spin
Other	65	763	Unclassified or unknown trigger sources

intermediate data products can be extracted and further processed (e.g., color-color plots, T_{50} distributions, etc.).

As more sophisticated analyses are developed, they are incorporated into the suite of analysis routines and modified to create standard output files; new database keywords are created as needed. The new analysis routines are run over the entire collection of triggers, and a new Burst Catalog is created.

One caveat to the Burst Catalog in general and to the results presented here: because the new analysis routines are developed and perfected using subsets of HETE data before they are added to the suite of routine analyses, the catalog sometimes does not have the most up-to-date results. At this point in the development of the catalog, these differences are typically small; some of the results from detailed spectral analyses of certain bursts, presented elsewhere in these proceedings, may differ slightly from what is in the catalog and presented here.

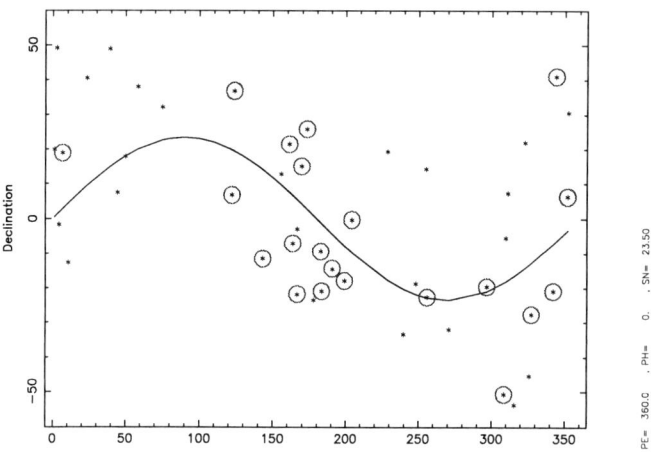

FIGURE 1. Plotted are the celestial coordinates of all GRBs localized by HETE; the sinusoid is the Ecliptic, the nominal (antisolar) pointing direction of the HETE instrument complement during the year. The circled localizations are those with detected afterglows, either optical, radio, infrared, or X-ray. The localized GRBs are distributed relatively randomly in a band $<50°$ from the Ecliptic, as expected. The apparent clustering of the detected afterglows is curious, but likely not statistically significant.

TRIGGER CLASSIFICATIONS

Table 1 gives a list of the major classifications of HETE triggers and their frequency in triggered and untriggered searches. The classification is assigned usually within 1–2 hours of the burst trigger; when the burst is reviewed later on, the classification can change.

Some classifications require significant analysis to confirm. For example, "collimator" events are those which propagated through the Fregate graded shield (which absorbs the lower-energy photons) and are typically hard and quite often short: The primary way to distinguish "collimator" events from "short, hard GRBs" is to examine the distribution of counts in the four Fregate detectors: any significant asymmetry is a sign that the burst came in through the collimator. The precise definition of the "classical GRB" category vs. "X-ray Flash" or the recently-added "X-ray Rich" categories depends on spectral analyses of the burst data; as more sophisticated analyses are performed, the designation can change.

The "other" category contains hundreds of weak triggers of the ground trigger software. Many of these may be weak bursts; further analyses are required to confirm them as cosmic events.

A tabular summary of those bursts localized by the WXM and/or SXC by 2003 September 1 is shown in Table 2. A summary of their locations on the celestial sphere is shown in Figure 1.

TABLE 2. Table of WXM/SXC localized bursts before 2003 September 1.

GRB	Class*	Loc†	AG**	Z	α	β	Error‡	IPN§
010110	GRB	Y/n						K,U
010213	XRF	Y/n						
010225	XRF	Y/n						
010326B	XRF	Y/n			$11^h\ 50^m\ 59^s$	$-23°\ 32'\ 44.0''$	10.0	K,U
010612	GRB	Y/n			$18^h\ 03^m\ 18^s$	$-32°\ 08'\ 02.0''$	36.0	K,S,U
010613	GRB	Y/n			$17^h\ 00^m\ 40^s$	$+14°\ 16'\ 05.0''$	36.0	K,U
010629B	GRB	Y/n			$16^h\ 32^m\ 38^s$	$-18°\ 43'\ 23.0''$	15.0	K,S,U
010921	GRB	1D/n	O	0.451	$23^h\ 00^m\ 00^s$	$+44°\ 00'\ 00.0''$	300.0	K,S,U
010928	GRB	1D/n			$23^h\ 29^m\ 00^s$	$+30°\ 39'\ 90.0''$	480.0	D
011019	XRF	Y/n			$00^h\ 43^m\ 11^s$	$-12°\ 38'\ 56.0''$	12.0	
011103¶	XRF	Y/n			$03^h\ 19^m\ 49^s$	$+17°\ 52'\ 05.0''$	840.0	
011130	XRF	Y/n			$02^h\ 58^m\ 09^s$	$+07°\ 24'\ 40.0''$	60.0	
011212	XRF	Y/n			$05^h\ 00^m\ 23^s$	$+32°\ 09'\ 58.0''$	10.0	
020124	GRB	Y/n	O	3.20	$09^h\ 32^m\ 49^s$	$-11°\ 27'\ 35.0''$	12.0	K,U
020127	GRB	Y/n	X		$8^h\ 15^m\ 06^s$	$+36°\ 44'\ 31.0''$	7.8	K,U
020305	GRB	Y/n	O		$12^h\ 43^m\ 03^s$	$-14°\ 33'\ 06.0''$	25.0	M,S,U
020317	XRF	Y/n			$10^h\ 23^m\ 21^s$	$+12°\ 44'\ 38.0''$	18.0	
020331	GRB	Y/n	O		$13^h\ 16^m\ 23^s$	$-17°\ 55'\ 23.0''$	7.8	K,U
020531	GRB	Y/n			$15^h\ 14^m\ 45^s$	$+19°\ 21'\ 30.0''$	36.0	M,U
020625	XRF	Y/n			$20^h\ 44^m\ 03^s$	$+07°\ 14'\ 28.0''$	14.0	
020801	GRB	Y/n			$21^h\ 2^m\ 14^s$	$-53°\ 46'\ 14.0''$	13.9	U
020812	GRB	Y/n			$20^h\ 38^m\ 48^s$	$-05°\ 23'\ 34.0''$	12.6	
020813	GRB	Y/Y	OX	1.254	$19^h\ 46^m\ 38^s$	$-19°\ 35'\ 16.0''$	1.0	K,M,U
020819	GRB	Y/Y	R		$23^h\ 27^m\ 25^s$	$+06°\ 16'\ 46.0''$	1.1	H,K,U
020903	XRF	Y/1D	OR	0.25	$22^h\ 49^m\ 01^s$	$-20°\ 55'\ 47.0''$	15.0	
021004	GRB	Y/Y	OIRX	2.328	$00^h\ 26^m\ 57^s$	$+18°\ 55'\ 44.0''$	2.0	K
021016	GRB	1D/n			$00^h\ 11^m\ 04^s$	$+49°\ 08'\ 20.0''$	330.0	K,M,U
021021	XRF	Y/n			$00^h\ 17^m\ 23^s$	$-01°\ 37'\ 01.0''$	20.0	
021104	XRF	Y/n			$3^h\ 53^m\ 48^s$	$+37°\ 57'\ 12.0''$	25.9	U
021112	XRF	Y/n			$02^h\ 36^m\ 52^s$	$+48°\ 50'\ 56.0''$	20.0	U
021113	GRB	Y/n			$01^h\ 33^m\ 53^s$	$+40°\ 27'\ 45.0''$	13.6	U,K,I
021211	GRB	Y/Y	O	1.006	$08^h\ 09^m\ 00^s$	$+06°\ 44'\ 20.0''$	2.0	U,K
030115	GRB	Y/Y	OIR		$11^h\ 18^m\ 30^s$	$+15°\ 2'\ 17.0''$	2.0	
030226	GRB	Y/Y	OX	> 1.986	$11^h\ 33^m\ 01^s$	$+25°\ 53'\ 56.0''$	2.0	
030323	GRB	Y/1D	O	3.3	$11^h\ 06^m\ 06^s$	$-21°\ 54'\ 20.0''$	12.0	
030324	GRB	Y/n	O		$13^h\ 37^m\ 11^s$	$+00°\ 19'\ 22.0''$	7.2	
030328	GRB	Y/Y	OX	1.52	$12^h\ 10^m\ 51^s$	$-09°\ 21'\ 05.0''$	0.8	
030329	GRB	1D/Y	ORX	0.1685	$10^h\ 44^m\ 50^s$	$+21°\ 30'\ 54.0''$	1.8	
030416	XRF	Y/n			$11^h\ 06^m\ 51^s$	$-02°\ 52'\ 58.0''$	7.2	
030418	GRB	Y/n	O		$10^h\ 54^m\ 53^s$	$-06°\ 59'\ 22.0''$	9.0	
030429	XRF	Y/Y	O	2.656	$12^h\ 13^m\ 06^s$	$-20°\ 56'\ 00.0''$	2.0	
030519	GRB	1D/n			$15^h\ 59^m\ 02^s$	$-33°\ 29'\ 19.0''$	136.6	
030528	XRF	Y/n	IX		$17^h\ 04^m\ 02^s$	$-22°\ 39'\ 00.0''$	2.0	
030723	XRF	Y/Y	OX		$21^h\ 49^m\ 30^s$	$-27°\ 42'\ 07.0''$	2.0	
030725	GRB	Y/1D	O		$20^h\ 33^m\ 47^s$	$-50°\ 45'\ 50.0''$	14.4	
030821	GRB	1D/n			$21^h\ 44^m\ 08^s$	$-45°\ 21'\ 25.0''$	49.8	K,I,M
030823	GRB	Y/n			$21^h\ 30^m\ 47^s$	$+21°\ 59'\ 46.0''$	5.4	
030824	XRF	Y/n			$00^h\ 05^m\ 02^s$	$+19°\ 55'\ 37.0''$	11.2	

* Classification at the time of the conference; some designations may have changed (see text)
† WXM/SXC Localization
** Afterglow: O=optical, I=infrared, R=radio, X=X-ray
‡ radius of circumscribing circle in arcminutes
§ M=Mars Observer; H=RHESSI; U=Ulysses; I=INTEGRAL; S=SAX; D=DMS
¶ The correct burst localization was discovered in reprocessing archival burst data: no GCN Notice was issued for this burst

Similarities in the Temporal Properties of Gamma-Ray Bursts and Soft Gamma-Ray Repeaters

S. McBreen*, L. Moran*, B. McBreen*, L. Hanlon*, J. French* and M. Conway*

*Department of Experimental Physics, University College Dublin, Dublin 4, Ireland.

Abstract. Magnetars are modelled as sources that derive their output from magnetic energy that substantially exceeds their rotational energy [1]. An implication of the recent polarization measurement of GRB 021206 is that the emission mechanism may be dominated by a magnetic field that originates in the central engine [2]. Similarities in the temporal properties of SGRs and GRBs are considered in light of the fact that the central engine in GRBs may be magnetically dominated. The results show that 1) the time intervals between outbursts in SRG 1806 − 20 and pulses in GRBs are consistent with lognormal distributions and 2) the cumulative outputs of SGRs and GRBs increase linearly with time. This behaviour can be successfully modelled by a relaxation system that maintains a steady state situation.

INTRODUCTION

Over the past decade the evidence for neutron stars with ultra-strong magnetic fields or magnetars has become convincing. Soft gamma-ray repeaters (SGRs) are associated with supernova remnants and have multiple bursts of gamma-rays which distinguish them from gamma-ray bursts (GRBs) e.g. [3, 4]. SGRs have intensely active periods which can last for weeks or months that are separated by quiescent phases lasting for years or decades [5]. The most intense outburst recorded to date has $\sim 10^{44}$ ergs in γ-rays. Several SGRs and anomalous X-ray pulsars have been found to be X-ray pulsars that have unusually high spin down rates. The rapid reduction in spin is usually attributed to magnetic breaking caused by the super-strong magnetic fields with values above 10^{14} G. In the magnetar model the magnetic field provides the burst energy [1]. A common scenario is that stresses build up in the magnetic field and cause a quake in the crust of the neutron star which ejects plasma Alfvèn waves through the magnetosphere.

GRBs are non-recurrent catastrophic events that radiate $\sim 10^{51}$ ergs in γ-rays. There is substantial evidence that GRBs are linked to supernova explosions e.g. [6]. They are often modelled as accretion onto a newly formed black hole e.g. [7]. Recent measurements suggest that the γ-rays are highly polarised by the magnetic field which may originate in the central engine with values of B above 10^{17} G [2]. It is therefore interesting to compare the pulse properties of SGRs and GRBs. It has been noted that the lognormal distributions apply to the pulse properties of SGRs [8, 9, 10] and GRBs [11, 12, 13]. Furthermore the radio afterglow from the giant flare from SGR 1900 + 14 is similar to radio afterglows from GRBs [14].

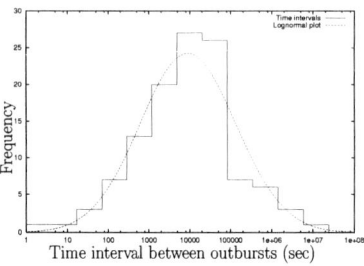

FIGURE 1. Histogram of the number of bursts from SGR 1806-20 versus time interval between the bursts. The dashed curve is a lognormal fit to the data. The SGR was observed by the International Cometary Explorer (ICE) and the data are taken from Ulmer et al [16].

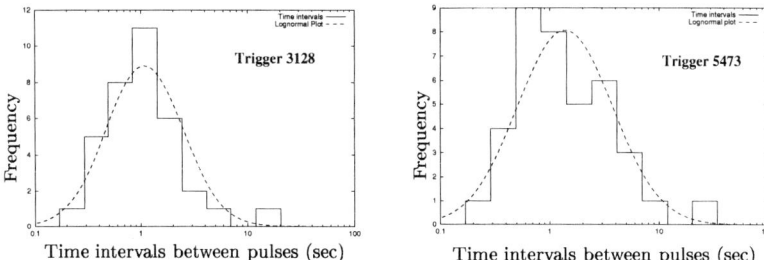

FIGURE 2. Histograms of the time interval between the pulses and lognormal representations for BATSE triggers GRB 940817 and GRB 960524 with 37 and 39 pulses respectively.

TIME INTERVALS BETWEEN OUTBURSTS IN SGRS AND GRBS

SGR 1806 −20 had a very active phase during 1983 when more than 100 outbursts were recorded [15, 16, 8]. The distribution of the time intervals between the outbursts is given in Figure 1 along with a lognormal fit to the data. Long time intervals between active phases (> 10 years) can easily be accomodated by the long tail of the lognormal distribution [17]. The time intervals between pulses in two BATSE GRBs with large numbers of pulses are given in Figure 2 along with lognormal plots. The pulse properties and time intervals between the pulses in short and long GRBs have been found to be lognormally distributed [11, 12, 18, 13, 19, 20]. It is interesting to note that the time intervals between radio glitches in the Vela pulsar are lognormally distributed [8].

In Fig. 1 the median time interval between SGR outbursts is $\sim 10^4$ s and in Fig. 2 the time interval between pulses in the GRBs is ~ 1 s.

CUMULATIVE LIGHT CURVES FOR SGRS AND GRBS

The cumulative light curve has been shown to be an interesting parameter for SGRs [21] and GRBs [22]. The running and cumulative light curves for the outbursts from SGR

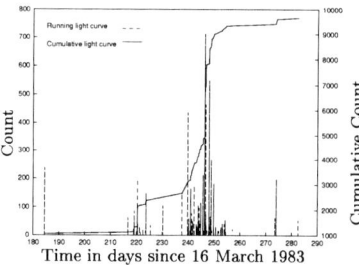

FIGURE 3. The running (dashed line) and cumulative (solid line) lightcurves of the outbursts from SGR 1806 − 20 in 1983. The presentation is adapted from Palmer [21]. The cumulative output approximates to straight lines of different slopes in the active region in late 1983.

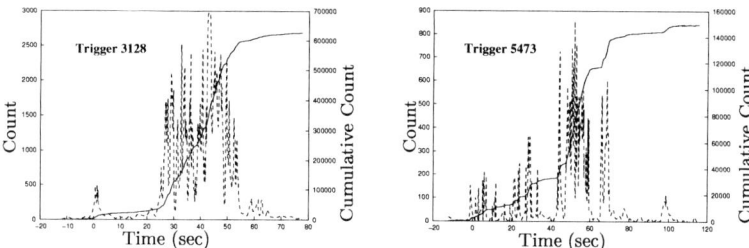

FIGURE 4. The running light curves (dashed line) and cumulative light curves (solid line) of GRB 940817 and GRB 960524 The cumulative light curves are approximately linear over the majority of the burst. The cumulative light curve of trigger 5473 can be fit by two linear sections. Trigger 3128 can be fit by one straight line despite the temporal complexity.

1806 − 20 are given in Figure 3. The cumulative light curves of a large majority of short and long GRBs has been found to be approximately linear with time. The running and cumulative light curves of the two GRBs in Figure 2 are given in Figure 4. The cumulative output increases linearly with time over the most active part of the GRB.

The cumulative output of the large majority of long and short GRBs was found to increase linearly with time [22].

RELAXATION SYSTEMS

The behaviour of bursting sources can be usefully compared with relaxation systems [21, 22]. A relaxation system is taken to be one that continuously accumulates energy from some process and discontinuously releases it. The energy in the reservoir at any time t is

$$E(t) = E_o + \int_o^t R(t)dt - \Sigma S_i \quad (1)$$

where E_o is the energy stored in the reservoir that accumulates energy at a rate $R(t)$ and discontinuously releases events of size S_i.

The simplest system is referred to as a relaxation oscillator where there is a fixed level or trip-point that triggers a release of the energy when $E = E_{max}$. More complicated behaviour occurs when the accumulation rate, trigger rate or release strength are not constant [23]. If the system starts from a minimum level $E = E_{min}$, accumulates energy at a constant rate $R = r$, the sum of the releases is approximately a linear function of time i.e. $\Sigma S_i \propto rt$. This model can account for the approximately linear increase in cumulative counts. The output from GRBs and SGRs has a tendency to keep the cumulative count close to a linear function and maintain a steady state situation

In a minority of bright GRBs with many pulses it was found that the cumulative output increases with time as t^2 in $\sim 10\%$ of bursts and as $(1-t)^2$ in $\sim 5\%$. This behaviour was attributed to a change in spin of the newly formed black hole [24].

ACKNOWLEDGEMENT

SMcB acknowledges IRCSET grant number RS/2002/820-8M for support. LH, BMcB and JF thank IRCSET Basic Research Grant number SC-2002-377 for support.

REFERENCES

1. Duncan, R., and Thompson, C., *ApJ*, **392** (1992).
2. Coburn, W., and Boggs, S. E., *Nature*, **423**, 415–417 (2003).
3. Zhang, B., and Meszaros, P. (2003), [astro-ph/0311321].
4. Dar, A., and De Rujula, A. (2003), [astro-ph/0308248].
5. Mazets, E. P., et al., *Nature*, **282**, 587–589 (1979).
6. Hjorth, J., et al., *Nature*, **423**, 847–850 (2003).
7. MacFadyen, A. I., and Woosley, S. E., *ApJ*, **524**, 262–289 (1999).
8. Hurley, K. J., McBreen, B., Rabbette, M., and Steel, S., *A&A*, **288**, L49–L52 (1994).
9. Göğüş, E., et al., *ApJ*, **526**, L93–L96 (1999).
10. Göğüş, E., Woods, P. M., Kouveliotou, C., van Paradijs, J., Briggs, M. S., Duncan, R. C., and Thompson, C., *ApJ*, **532**, L121–L124 (2000).
11. McBreen, B., Hurley, K. J., Long, R., and Metcalfe, L., *MNRAS*, **271**, 662 (1994).
12. Li, H., and Fenimore, E. E., *ApJ*, **469**, L115 (1996).
13. McBreen, S., Quilligan, F., McBreen, B., Hanlon, L., and Watson, D., *A&A*, **380**, L31 (2001).
14. Cheng, K. S., and Wang, X. Y., *ApJ*, **593**, L85–L88 (2003).
15. Laros, J. G., et al., *ApJ*, **320**, L111 (1987).
16. Ulmer, A., Fenimore, E. E., Epstein, R. I., Ho, C., Klebesadel, R. W., Laros, J. G., and Delgado, F., *ApJ*, **418**, 395 (1993).
17. Hurley, K. J., McBreen, B., Delaney, M., and Britton, A., *Ap&SS*, **231**, 81–84 (1995).
18. Gupta, V., Das Gupta, P., and Bhat, P. N., "AIP Conf. Proc. 526: Gamma-ray Bursts, 5th Huntsville Symposium. Edited by R.M. Kippen et al.," 2000, p. 215.
19. Quilligan, F., et al., *A&A*, **385**, 377–398 (2002).
20. Nakar, E., and Piran, T., *MNRAS*, **331**, 40–44 (2002).
21. Palmer, D. M., *ApJ*, **512**, L113–L116 (1999).
22. McBreen, S., McBreen, B., Hanlon, L., and Quilligan, F., *A&A*, **393**, L29–L32 (2002).
23. Ramirez-Ruiz, E., and Merloni, A., *MNRAS*, **320**, L25 (2001).
24. McBreen, S., McBreen, B., Hanlon, L., and Quilligan, F., *A&A*, **393**, L15–L19 (2002).

Burst Statistics Using the Lag-Luminosity Relationship

D. L. Band*, J. P. Norris* and J. T. Bonnell*

*Code 661, NASA/Goddard Space Flight Center, Greenbelt, MD 20771

Abstract. Using the lag-luminosity relation and various BATSE catalogs we create a large catalog of burst redshifts, peak luminosities and emitted energies. These catalogs permit us to evaluate the lag-luminosity relation, and to study the burst energy distribution. We find that this distribution can be described as a power law with an index of $\alpha = 1.76 \pm 0.05$ (95% confidence), close to the $\alpha = 2$ predicted by the original quasi-universal jet model.

INTRODUCTION

Jet models predict the distribution of the isotropic-equivalent energy E_{iso}: quasi-universal jet profile models predict an approximate power law distribution with index $\alpha = 2$, where $N(E_{iso}) \propto E_{iso}^{-\alpha}$[1]. The isotropic-equivalent energy E_{iso} is the total energy radiated if the observed flux were radiated isotropically. To study the distribution of burst intensities, we used the lag-luminosity relationship to create a burst database with redshifts, peak luminosities, and burst energies, and then we fit energy distributions to the burst database. Of course, this database can be used for other studies.

In the lag-luminosity relation[2] the peak bolometric luminosity L_B is a function of the lag τ_B between two energy bands in the burst's frame—$L_B = Q(\tau_B)$. But τ_0 is measured in our frame. We model $\tau_B = (1+z)^c \tau_0$: time dilation contributes -1 to c, while the redshifting of temporal structure with a smaller lag from higher energy contributes $\sim 1/3$ (pulses are narrower at high energy). The peak bolometric luminosity is related to the peak bolometric energy flux $F_B = L_B/[4\pi D_L^2]$, where D_L is the luminosity distance. The peak bolometric energy flux is related to the peak photon flux P (integrated over an energy band, e.g., 50–300 keV for BATSE data): $F_B = \langle E \rangle P$. The result is an implicit equation that must be solved for each burst:

$$P = Q((1+z)^c \tau_0) / \left[\langle E \rangle 4\pi D_L^2 \right] \tag{1}$$

After solving eq. 1 for the redshift, L_B and E_{iso} can be calculated from F_B and the energy fluence, respectively.

The original lag-luminosity relation was a single power law, e.g., $L_B \propto \tau_B^{-1.15}$. But this power law over-predicts the luminosity of GRB980425 (assuming this burst was SN1998bw). Consequently Salmonson[3] and Norris[4] suggested breaking the single power law; for $\tau_B > 0.35$ s the power law index is -4.7. A population of nearby, long lag bursts resulted.

With a database of bursts with E_{iso} we can now calculate the energy distribution. The methodology presented in [5] can also be applied to the luminosity function. The probability of detecting a given energy is truncated by the detection threshold: $p(E_{iso}|E_{iso,th}M(\vec{a}))$ where $E_{iso,th}$ is the threshold value of E_{iso} for that burst and $M(\vec{a})$ is the model (e.g., the functional form of the energy distribution) with parameters $\vec{a}$. For the ensemble of bursts the probability of detecting bursts with the observed energies is

$$\Lambda = \prod_i p\left(E_{iso,i}|E_{iso,th,i}M(\vec{a})\right) \tag{2}$$

where the product is over each burst. This probability is the likelihood for the model $M(\vec{a})$. In frequentist statistics, we maximize Λ with respect to the parameters $\vec{a}$ to get a best fit value. In Bayesian statistics the likelihood is a factor in the "posterior," which can be used for confidence ranges and best fit values; the Bayesian approach allows the use of "priors" reflecting our expectations for $\vec{a}$.

Note that to study the energy distribution we do NOT need a complete sample in terms of observed fluences, only a sample that has no bias on the intrinsic E_{iso}. There can be gaps in the distribution of peak fluxes, but E_{iso} has to be drawn uniformly from $p(E_{iso}|E_{iso,th}M(\vec{a}))$ in our sample. On the other hand, if we want the burst rate per comoving volume as a function of redshift, then we do need a complete sample.

Is the resulting energy distribution a good representation of the data? The likelihood (frequentist approach) or posterior (Bayesian approach) can be used to compare models (functional forms), but do not indicate "goodness-of-fit." However, our methodology assumes the energies are drawn uniformly from $p(E_{iso}|E_{iso,th}M(\vec{a}))$. The cumulative distribution of $p(E_{iso}|E_{iso,th}M(\vec{a}))$ should therefore be a straight line, and the average value should be 1/2, with a statistical uncertainty of $[12N]^{-1/2}$ for N bursts.

RESULTS

We started with 1438 BATSE bursts for which we calculated lags. Of these, 1218 have positive lags. These bursts also have hardness ratios, peak fluxes and durations. To calculate the average energy $\langle E \rangle$ we used the "GRB" spectral fits of Mallozzi et al.[6] to the peaks of 580 of these bursts. For the 858 bursts without fits we assumed average spectral indices $\alpha = -0.8$ and $\beta = -2.3$. Plotting HR_{32} (the 100–300 keV to 50–100 keV hardness ratio) vs. E_p shows a clear correlation which can be approximated by E_p=240 HR_{32}^2 keV; we used this relation for the bursts without spectral fits.

Redshifts were calculated for this database for both the original simple power law lag-luminosity relation and the broken power law Salmonson[3] and Norris[4] introduced to incorporate GRB980425. As expected, the difference in the lag-luminosity relations is apparent at low redshifts: the broken power law results in a population of nearby bursts. There were few physically implausible high z bursts (e.g., $z > 20$) and thus no additional cutoffs on the lag-luminosity relation are required. In the absence of additional information, the choice between the two lag-luminosity relations depends on whether GRB980425 is considered to be a typical low luminosity burst. For the remainder of this analysis we use a single power law lag-luminosity relation.

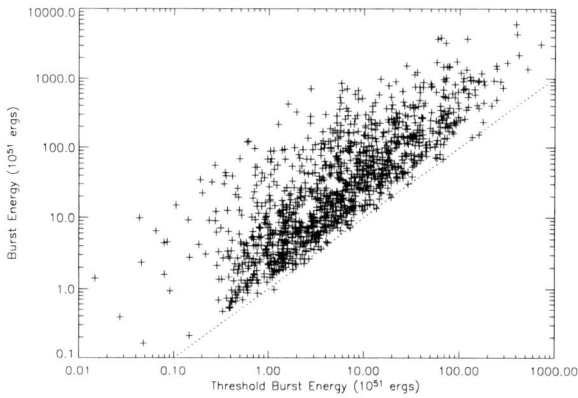

FIGURE 1. Scatter plot of the isotropic equivalent energy E_{iso} vs. the detection threshold.

As an aside, we found that the redshift calculation is sensitive to the value of $\langle E \rangle$. Calculating this quantity inconsistently can introduce errors into the resulting database.

We calculated the energy E_{iso} for each burst from the redshift and energy fluence. The results were reasonable (see Fig. 1): few bursts had $E_{iso} > 10^{54}$ erg or $E_{iso} < 10^{51}$ erg.

The energy detection threshold $E_{iso,th}$ can be calculated by scaling E_{iso} by the ratio of the threshold peak photon flux to the observed peak flux. BATSE's threshold peak flux was $P_{min} \sim 0.3$ ph cm^{-2} s^{-1}; however, the number of bursts in our sample with E_{iso} just above the threshold is suspiciously low (see Fig. 1), suggesting that the sample's true threshold was greater than 0.3 ph cm^{-2} s^{-1}. Consequently we used $P_{min} \sim 0.5$ ph cm^{-2} s^{-1} as the threshold, deleting bursts with $P < 0.5$ ph cm^{-2} s^{-1}.

The left hand side of Fig. 2 shows the likelihood surface for our sample assuming a power law functional form, where the two parameters are the low energy cutoff E_2 and the power law index α (i.e., $N(E_{iso}) \propto E_{iso}^{-\alpha}$ for $E_{iso} \geq E_2$). The likelihood is maximized by E_2 equal to the lowest observed value E_{iso}, although lower values are not ruled out. The best fit spectral index is $\alpha = 1.76 \pm 0.05$ (95% confidence). Although $\langle P(>E_{iso}) \rangle$ = 0.4642±0.0089 (N=1054, assuming $P_{min} \sim 0.5$ ph cm^{-2} s^{-1}) deviates from 1/2 by 4σ, considering the possible systematic errors (e.g., in the estimation of E_p from the hardness ratio), this value of $\langle P(>E_{iso}) \rangle$ indicates that a power law energy distribution is a fairly good characterization of the data.

We also tried a lognormal energy distribution (right hand side of Fig. 2). The maximum likelihood occurs at $E_{iso,cen} = 3 \times 10^{51}$ ergs and $\sigma_E = 2.7$. The surface's shape indicates that the data permit a high central value of E_{iso} and a narrow distribution, or a low central value of E_{iso} and a broad distribution. The observational cutoff truncates the true energy distribution, and the low energy extent is relatively unknown. We find $\langle P(>E) \rangle$=0.4821±0.0089 (N=1054, assuming $P_{min} \sim 0.5$ ph cm^{-2} s^{-1}), consistent with 1/2 at the 2σ level.

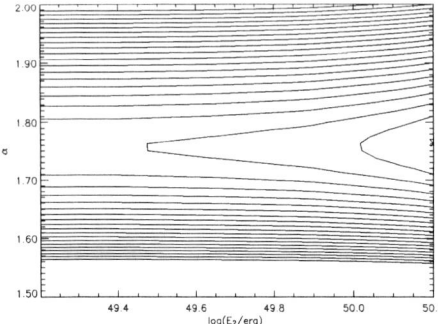 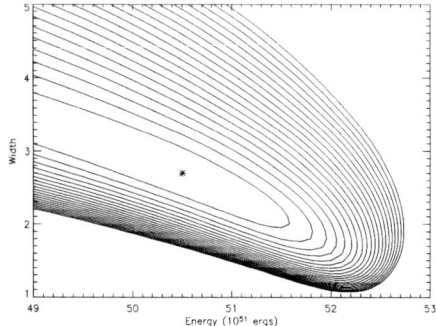

FIGURE 2. Contour plots of the likelihood surface for a power law energy distribution (left) and lognormal energy distribution (right). The power law has a low energy cutoff E_2 and power law index α; the contours are spaced by $\Delta\log(\text{likelihood})=1$. The lognormal distribution has a central value $E_{\text{iso,cen}}$ and a logarithmic width σ_E; the contours are spaced by $\Delta\log(\text{likelihood})=10$.

IMPLICATIONS

A quasi-universal jet profile that is a power law in the off-axis angle θ—the energy per solid angle $\varepsilon(\theta) \propto \theta^k$—results in a power law energy distribution (or luminosity function) with index $\alpha = 1 - 2/k$ (hence $\alpha = 2$ for $k = -2$), while a Gaussian profile results in $\alpha = 1$. Lloyd-Ronning et al.[1] found that if the profile parameters are distributions, the luminosity functions could be approximated by power laws with $\alpha \sim 2$ for power law profiles and $\alpha \sim 1$ for Gaussian profiles, but with curvature. The additional degrees of freedom introduced by varying the parameters give the jet models the freedom to fit a wide variety of energy distribution shapes. We find that our burst data can be fit by a power law energy distribution with $\alpha = 1.76 \pm 0.05$ (95% confidence); considering only the statistical uncertainty the power law distribution is formally not a good fit, but with the likely systematic uncertainties the power law distribution is probably a good description of the data. While our power law fit is inconsistent with the original jet profile model ($k = -2$ and therefore $\alpha = 2$), it is consistent with the jet profile models where parameters are permitted to vary.

A log-normal energy distribution also describes the data; the data permit a smaller average energy if the distribution is wider.

REFERENCES

1. Lloyd-Ronning, N., Dai, X., & Zhang, B., ApJ, submitted [astro-ph/0310431] (2003).
2. Norris, J. P., Marani, G. F., & Bonnell, J. T., ApJ, **534**, 248, (2000).
3. Salmonson, J., ApJ, **546**, 29, (2001).
4. Norris, J. P., ApJ, **579**, 386, (2002).
5. Band, D., ApJ, **563**, 582, (2001).
6. Mallozzi, R., et al., in Gamma-Ray Bursts, 4th Huntsville Symposium, AIP Conference Proceedings 428, eds. C. Meegan, R. Preece and T. Koshut (AIP: Woodbury, NY), 273, (1998).

Short-Bright GRBs: Spectral Properties

G. Ghirlanda*, G. Ghisellini† and A. Celotti**

*IASF - Via Bassini 15, I-20133 Milano, Italy.
†OAB - Via Bianchi 46, I-23807 Merate, Italy.
**SISSA - Via Beirut 2, I-34014 Trieste, Italy.

Abstract. We study the spectra of short–bright GRBs detected by BATSE and compare them with the average and time resolved spectral properties of long–bright bursts. We confirm that short events are harder than long bursts, as already found from the comparison of their (fluence) hardness ratio, but we find that this difference is mainly due to a harder low energy spectral component present in short bursts, rather than to a (marginally) different peak energy. Moreover, we find that short GRBs are similar to the first 1 sec emission of long bursts. The comparison of the energetic of short and long bursts also suggests that short GRBs do not obey the peak energy–equivalent isotropic energy correlation recently proposed for long events, implying that short GRBs emit lower energy than long ones. Nonetheless, short bursts seem to emit a luminosity similar to long GRBs and under such hypothesis their redshift distribution appears consistent with that observed for long events. These findings might suggest the presence of a common mechanism at the beginning of short and long bursts which operates on different timescales in the these two classes.

INTRODUCTION

Evidences supporting the possible different nature of short and long GRBs are: (1) their bimodal duration distribution [1] with mean duration of $\sim$ 20 sec and $\sim$ 0.3 sec for long and short events, respectively; (2) their different temporal properties (e.g. number and width of pulses in the light curve, [2], [3], [4]); (3) the larger hardness ratio of short bursts [1], [5], [6]. Moreover, theoretical models for the progenitors of GRBs associate short bursts to the merger of compact objects in a binary system [7], [8] while long GRBs seem to be connected to the core–collapse of massive stars [9], [10].

Despite the increasing understanding of the nature of long duration γ–ray bursts, the population ($\sim$ 1/3) of sub-second short GRBs is still largely not understood. This is mainly due to (i) the low signal-to-noise ratio which strongly limits the analysis of the prompt (temporal and spectral) emission of short events, and (ii) the lack of any firm afterglow measurement for short GRBs, which indeed represented a major advance in unveiling the mystery of long bursts.

SHORT VS. LONG: SPECTRAL PROPERTIES

The emission properties of long GRBs have been studied in details [11], [12] by fitting their time average [11] and time resolved [12], [13] broad-band high-resolution spectra. Nonetheless, the comparison with the spectral properties of short bursts has been based mainly on the analysis of their fluence hardness ratio (e.g. [14], [15]) which is marginally

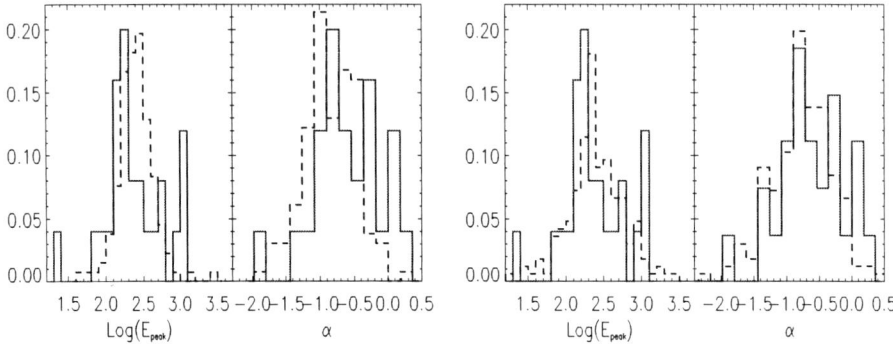

FIGURE 1. Distribution of the spectral parameters for short (*solid line*) and long (*dotted line*) GRBs. *Left*: peak energy (E_{peak}) of the EF_E spectrum and low energy photon spectral index (α) for the average spectra of long and short bursts. *Right*: E_{peak} and α of the first 1 second of long bursts compared to the average spectra of short events.

representative of the effective spectral shape. For this reason we analyzed [16] a sample of bright–short BATSE bursts (selected with peak flux ≥ 10 phot/cm^2 sec between 50-300 keV) by fitting their γ–ray spectra ($\sim 30 - 1800$ keV) with the standard spectral functions (e.g. [12]). The set of spectral parameters (namely the low energy photon spectral index α and the EF_E peak energy E_{peak}) were compared with those of long bright bursts [12] both considering time integrated and time resolved spectra.

As shown in Figure 1 (*Left*) short and long GRBs present different α distributions and only marginally different E_{peak} distributions (with a small Kolmogorov-Smirnov test probability – i.e. 4% and 10%, respectively – for the two population to be similar). The average values are $\langle\alpha\rangle = -0.84\pm0.15$, $\langle E_{peak}\rangle = 305\pm22$ keV and $\langle\alpha\rangle = -0.58\pm0.10$, $\langle E_{peak}\rangle = 355\pm30$ keV for long and short GRBs, respectively. This suggests that short bursts have (average) harder spectra than long events, as also indicated by the analysis of their hardness ratio, *but*, intriguingly, this difference is due to a considerably *harder low energy spectral component* present in short bursts rather than to a different peak energy. The comparison of short GRBs with the time resolved spectral properties of long bursts shows that short events have spectra similar to the first 1 sec of long GRBs, as shown in Figure 1 (*Right*), with a high KS probability of 23% and 80% for α and E_{peak} to be similar. This might suggest that a similar mechanism operates for the complete duration of short bursts and in the first 1 sec of long GRBs.

SHORT VS. LONG: ENERGY AND LUMINOSITY

The analysis of the intrinsic properties of (still few) long bursts with known redshift [17] highlighted a possible correlation ($E_{iso} \propto E_{peak}^{1.93}$) between is the equivalent isotropic burst energy E_{iso} and the spectral peak energy E_{peak}. This correlation has been also confirmed by Hete–II [18] and a similar relation has been found between E_{peak} and the burst isotropic luminosity L_{iso} [19]. If short bursts were similar to long events they

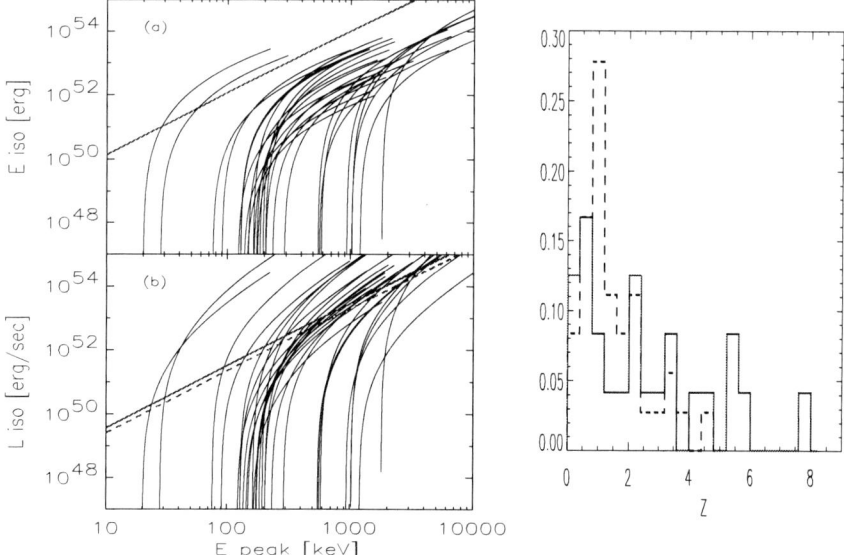

FIGURE 2. *Left*: Equivalent isotropic energy E_{iso} (a) and equivalent isotropic luminosity L_{iso} (b) vs. E_{peak}. Solid lines represent the correlations between these intrinsic properties found for long bursts [17] and [19]. Curves represent the intrinsic E_{iso} (L_{iso}) and E_{peak} for short bursts assuming a variable redshift between 0.1 and 10. *Right*: Redshift distribution of short bursts (*solid line*) assuming that they indeed satisfy the L_{iso} vs. E_{peak} relation. For comparison also the observed distribution of long bursts (*dotted line*) is reported.

might satisfy these correlations between E_{iso} (L_{iso}) and E_{peak}. However, no redshift of short GRBs has been measured so far. Nonetheless, we can still verify ([16]) the above hypothesis by assuming a variable redshift (between 0.1 and 10) and scaling the observed spectral properties of the sample of bright short bursts in the source rest frame. These bursts, due to the varibale redshift assumed, describe a line (solid curves) in the E_{peak} - $E_{iso}(L_{iso})$ plane. Figure 2-*a* shows that for any assumed z short bursts (solid lines) populate a region below the E_{peak} - E_{iso} correlation of long bursts ([17] - solid line). On the other hand (Figure 2-*b*) the luminosity of short events is consistent with the proposed relation for long bursts ([19] - solid line). In conclusion short and long bursts seem to emit a *similar equivalent isotropic luminosity* but different (lower in short bursts) energy due to their different duration. Nonetheless, short and long bursts might still have similar emitted energies if short bursts are less collimated than long events although, in this case, short bursts would have a much larger luminosity. Under the first hypothesis we can further extract from the L_{iso}-E_{peak} relation the possible redshift distribution of short bursts (Figure 2-*right*, *solid line*). This is consistent with the observed z distribution of long events (Figure 2-*right*, *dotted line*) and again supports a possible similarity of short and long GRBs.

DISCUSSION

The comparison of the spectral properties of short and long bursts pointed out that short bursts are harder than long events due to a (average) harder low energy spectral component (rather than to a different peak energy). The spectra of short bursts are similar to the first 1 sec of the emission of long bursts. Intriguingly, short bursts present a similar intrinsic luminosity but a lower (isotropic) energy than long bursts if the recently found correlations between these quantities were true also for short GRBs. Under such hypothesis the implied redshift distribution of short bursts results similar to that observed in long GRBs, and this prediction will be tested in the forthcoming Swift era. Nonetheless, short GRBs might still have a similar energy to that of long bursts if their collimation angle is much larger than that of long events, but in this case their intrinsic luminosity should be much larger than that of long bursts. These results suggest the presence of a common mechanism operating at the beginning of short and long bursts which could explain their similar spectral properties and luminosity. If this is the case a possible difference in the burst dynamical evolution (e.g. fallback of the pre-GRB ejected material) might play a crucial role in distinguishing these two classes.

ACKNOWLEDGMENTS

G. Ghirlanda would like to thank the organizing committee for the full grant support for the participation to this conference.

REFERENCES

1. Kouveliotou, C., *ApJL*, **413**, L101–L104 (1993).
2. Norris, J. P., Scargle, J. D., and Bonnell, J. T., *Bull. of the Amer. Ast. Soc.*, **32**, 1244 (2000).
3. Nakar, E., and Piran, T., *MNRAS*, **330**, 920–926 (2002).
4. McBreen, S., Quilligan, F., McBreen, B., Hanlon, L., and Watson, D., *ApJL*, **380**, L31–L34 (2001).
5. Tavani, M., *ApJL*, **497**, L21 (1998).
6. Paciesas, W. S., Preece, R. D., Briggs, M. S., and Mallozzi, R. S., "Gamma-ray Bursts in the Afterglow Era," 2001, p. 13.
7. Goodman, J., *ApJL*, **308**, L47–L50 (1986).
8. Meszaros, P., and Rees, M. J., *ApJ*, **397**, 570–575 (1992).
9. Woosley, S. E., *ApJ*, **405**, 273–277 (1993).
10. Vietri, M., and Stella, L., *ApJL*, **507**, L45–L48 (1998).
11. Band, D. et al., *ApJ*, **413**, 281–292 (1993).
12. Preece, et al., *ApJS*, **126**, 19–36 (2000).
13. Ghirlanda, G., Celotti, A., and Ghisellini, G., *A&A*, **393**, 409–423 (2002).
14. Cline, D. B., Matthey, C., and Otwinowski, S., *ApJ*, **527**, 827–834 (1999).
15. Qin, Y., Xie, G., Liang, E., and Zheng, X., *A&A*, **369**, 537–543 (2001).
16. Ghirlanda, G., Ghisellini, G., and Celotti, A., *astro-ph/0310861* (2003).
17. Amati, L. et al., *A&A*, **390**, 81–89 (2002).
18. Lamb, D. Q. et al., *astro-ph/0309462* (2003).
19. Yonetoku, et al., *astro-ph/0309217* (2003).

The Internal Luminosity Function and GRB Properties

Jon Hakkila*, Timothy W. Giblin*, Stephen P. Fuller*, Kevin C. Young*, Andrew D. Stallworth* and Amanda J. Sprague*

*Department of Physics and Astronomy,
College of Charleston, Charleston, SC 29424-0001*

Abstract. The Internal Luminosity Function (ILF) is the distribution of luminosity within a gamma-ray burst and is thus an intrinsic burst property. The ILF of most bursts can be described accurately using a power-law index modified by a power-law curvature parameter. A flat power-law index indicates a large amount of high luminosity emission relative to low luminosity emission, while a steep power-law index indicates a large amount of low luminosity emission relative to high luminosity emission. Bursts with steep power-law indices tend to show large ILF curvatures, indicating that they are also deficient in lower-luminosity emission. The ILF anticorrelates with duration and fluence, indicating that bursts with large ILF curvatures are longer and brighter (as measured by fluence) than bursts emitting a broad range of luminosities. The ILF might be an indicator of Long vs. Short GRB class.

INTRODUCTION

The internal luminosity function (or ILF) $\psi(L)$ is the distribution of luminosity within a GRB (Gamma-Ray Burst). The ILF can be calculated from a GRB flux distribution with the temporal resolution limited by the detector's temporal window. The quantity $\psi(L)\Delta L$ represents the fraction of time that a GRB's luminosity lies between L and $L+\Delta L$ [1].

The ILF of Long GRBs (e.g. those with T90 > 1.4 s; [2]) can be calculated using BATSE 64-ms data for numerous four- broadband energy channel combinations (25 to 50 keV, 50 to 100 keV, 100 to 300 keV, and 300 keV to 1 MeV). Each energy channel combination produces a slightly different ILF due to spectral evolution.

The following procedure has been used to calculate the ILF from 64-ms data. First, Poisson background rate variations are used to obtain an estimated constant background rate (this corrects the background to account for prior HEASARC background subtractions). Second, Monte Carlo models of Poisson variations are used to noisify time intervals with poor (e.g. 1024 ms) resolution data. Next, a distribution function is constructed by binning count rates relative to a defined minimum (we have chosen this minimum to be 2σ above background). Expected Poisson background rates are then subtracted from each bin so that only counts statistically associated with an ILF signal remain. Finally, the ILF is normalized to the peak luminosity by the requirement that

$$\Sigma \psi(L)\Delta L = 1. \tag{1}$$

ILF PROPERTIES

We have calculated the ILF for 2069 BATSE bursts in the online Current BATSE Catalog using 64 ms data. It was previously [1, 3] found that the ILF could be fit by a power-law distribution. However, with a larger dataset and a careful study of systematics, we find that the ILF fit can be improved using the quasi power-law form:

$$\psi(L) = AL^\alpha 10^{\beta[\log(L)]^2} \qquad (2)$$

Here, α is the power-law index, β is the curvature index, and the fit spans the ILF range from 2σ above background to the 64 ms peak flux. We limit our statistical results to the 478 highest-quality measurements for which the ILF calculated in the 50 to 300 keV range has at least 5 degrees of freedom and for which $\sigma_\alpha \leq 0.5$ and $\sigma_\beta \leq 0.5$ GRBs in this sample have $-7 < \alpha < 3$ and $-6 < \beta < 4$.

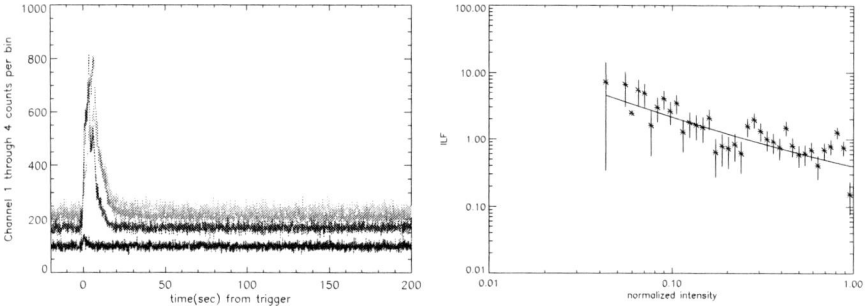

FIGURE 1. Four channel time history (a) and ILF (b) of BATSE trigger 829.

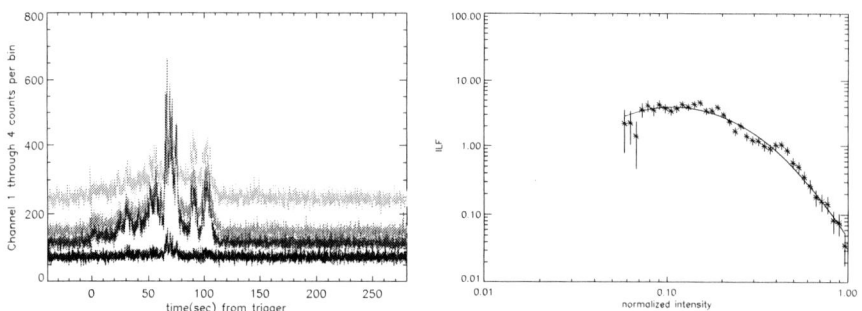

FIGURE 2. Four channel time history (a) and ILF (b) of BATSE trigger 3035.

We demonstrate the best fits for two sample bursts; BATSE triggers 829 and 3035. Trigger 829 (Figure 1) is a FRED that produces flat ILF indices ($\alpha = -0.5, \beta = 0.1$)

indicating that there is a moderate amount of high-luminosity emission relative to low-luminosity emission. Trigger 3035 (Figure 2) is a complex, spiky burst that has steep ILF indices ($\alpha = -4.0, \beta = -1.8$) indicating a relative depletion of both high- and low-luminosity emission.

Although α and β represent orthogonal attributes, we find that β is a strong predictor of α (Figure 3a). Bursts with lots of high luminosity emission (large α) have negligible or positive ILF curvature (large β), while those with small amounts of high luminosity emission (small α) also tend to be lacking in low luminosity emission (small β). The probability that this correlation is random is 2.0×10^{-231}, and the best fit linear relation between the two yields $\alpha \approx (4/3)(\beta - 1)$.

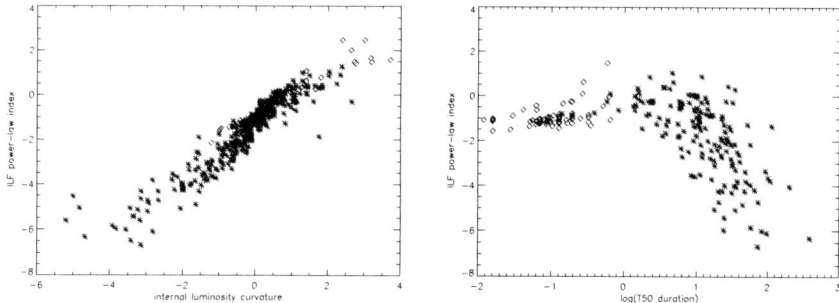

FIGURE 3. Correlations between power-law index α and (a) curvature index β, and (b) T50 duration.

It has previously been reported [1] that the power-law index correlates with a variety of GRB attributes. We present the results of a correlation analysis in Table 1. These results verify strong anticorrelations between α (and thus β) and both log(duration) and log(fluence), with weaker correlations between α and log(internal luminosity range) and log(hardness). The correlation between fluence and duration is expected: principal component analysis [4] indicates that these are strongly dependent attributes. The relationship between α and log(T50) is demonstrated in Figure 3b.

TABLE 1. Correlation between 50 to 300 keV ILF power-law index and other GRB attributes

Correlation between α and...	Type of correlation found	Prob. of random correlation
log(internal luminosity range)	correlation	7.9×10^{-4}
log(1024 ms peak flux)	none	0.75
log(fluence)	anticorrelation	2.9×10^{-15}
log(dual timescale peak flux)[2]	none	0.75
log($T90$)	anticorrelation	1.5×10^{-24}
log($T50$)	anticorrelation	1.3×10^{-29}
log($HR32$)	correlation	2.0×10^{-3}
duty cycle[5]	none	0.75

Short, low fluence GRBs tend to have a much narrower range of α values than Long GRBs. This indicates that the anticorrelation between α and duration is even more pronounced if the discussion is limited to Long GRBs. It also suggests that Short GRBs have inherently different ILFs than Long ones.

It is possible that the Short GRB ILFs suffer from a systematic error, since their durations are often of the order of the 64 ms temporal sampling window. We have tested this by calculating the ILF for 30 Short GRBs using both 64 ms data and TTE (Time Tagged Event) data. Although the 64 ms data produces a narrow range of α values, the TTE data provides a wide range of α values *for the same bursts*. There is essentially no correlation between the α values obtained from the separate data types.

On the other hand, the anticorrelation between α and fluence for Long bursts does not appear to be attributable to sampling bias. It instead appears to indicate different morphologies of bright and faint Long GRBs: a large percentage of bright bursts appear to be complex and are composed of bright narrow pulses combined with faint broad ones, whereas a large percentage of faint bursts appear to be smooth and are composed of broad emission *only*. More of the bright, complex Long bursts have correspondingly steep α values while more of the faint, smooth bursts have flat α values.

CONCLUSIONS

We note what appears to be a relationship between α and Long GRB morphology: simple, smooth bursts have flat power-law indices, while complex bursts with large pulse intensity variations have steep power-law indices and large curvature indices. Short GRBs might have different ILFs than Long ones, but further analysis is needed to remove systematic biases. The ILFs of Long GRBs anticorrelate with duration and fluence, and weakly correlate with internal luminosity range and hardness. Correlations of α with other Long burst attributes suggest that these bursts are composed of broad pulses combined in differing relative amounts with narrow pulses.

ACKNOWLEDGMENTS

This work is sponsored by NASA grant NRA-98-OSS-03 and NSF grant AST00-98499.

REFERENCES

1. Horack, J. M. & Hakkila, J., ApJ, **479**, 371 (1997).
2. Hakkila, J., Giblin, T. W., Roiger, R. J., Haglin, D. J., Paciesas, W. S., & Meegan, C. A. ApJ, **582**, 320 (2003).
3. Hakkila, J. et al. American Institute of Physics Conference Series, **662**, 147 (2003).
4. Bagoly, Z., Meszaros, A., Horvath, I., Balazs, L. G., & Meszaros, P., ApJ, **498**, 342 (1998).
5. Hakkila, J., Preece, R. D., & Pendleton, G. N., AIP Conf. Proc. 526: Gamma-ray Bursts, 5th Huntsville Symposium, **526**, 83 (2000).

Prompt Comparison of Data for Optical Transients of Gamma-Ray Bursts

G. Pizzichini[*], P. Ferrero[*], C. Bartolini[†], A. Guarnieri[†] and A. Piccioni[†]

[*]*IASF/CNR, Sezione di Bologna, via Gobetti 101, 40129 Bologna, Italy*
[†]*Astronomy Department, University of Bologna, via Ranzani 1, 40127, Bologna, Italy*

Abstract. The prompt information given by the GRB Coordinates Network (GCN) differs very much for different GRBs. We compare the optical magnitudes quoted in GCNs for some recent events. Our compilation might help in planning coordinated observations in the future.

INTRODUCTION

When a new Gamma-Ray Burst is observed, messages by the GRB coordinates Network (GCN), available either via e-mail or at: http://gcn.gsfc.nasa.gov/gcn/gcn_main.html
are extremely useful in first alerting the "GRB community" of the new event and, later, giving details of observations of the possible afterglow. Thus, for example, many optical observatories around the world shall either try to detect a new Optical Transient from the GRB or to continue its observations during the subsequent days, by taking spectra or following the light curve behaviour in various colors. We are especially interested in the latter, because of our observations at the 152 cm telescope in Loiano (see Bartolini et al., these proceedings).

DATA COLLECTION

Prompt knowledge of the magnitude of an OT at recent epochs is quite helpful in planning for more (possibly coordinated) observations. For this reason we compiled all the data acquired via GCN during the first days for the OTs detected for several recent GRBs, in the hope that showing what happened for them might help in future observation plans.

We limited our compilation to the first hours after the event. For some events, that is GRBs 021004 (HETE 2380), 030329 (HETE 2651), and, to some extent, also 030226 (HETE U10893), the GCNs reported the magnitudes in several colors. In other cases, for example GRB 030328 (HETE 2650), practically only Rc magnitudes were promptly reported. This fact can be partially, but not entirely, attributed to the observed magnitude of the transient, which evidently encourages observations in different filters only for the most intense OTs, as was the case for GRB 030329. For lack of space we cannot quote separately the GCN used.

In order to estimate the magnitudes in other colors we can resort to the findings of [1], also reported and updated in these proceedings, which show that, at least during the first ten days, the color indices of most OTs from GRBs fall into narrow intervals which are fairly typical of these phenomena and independent of the OT magnitude, at least during the first ten days after the event, a property which we hope to test carefully for as many OTs as possible.

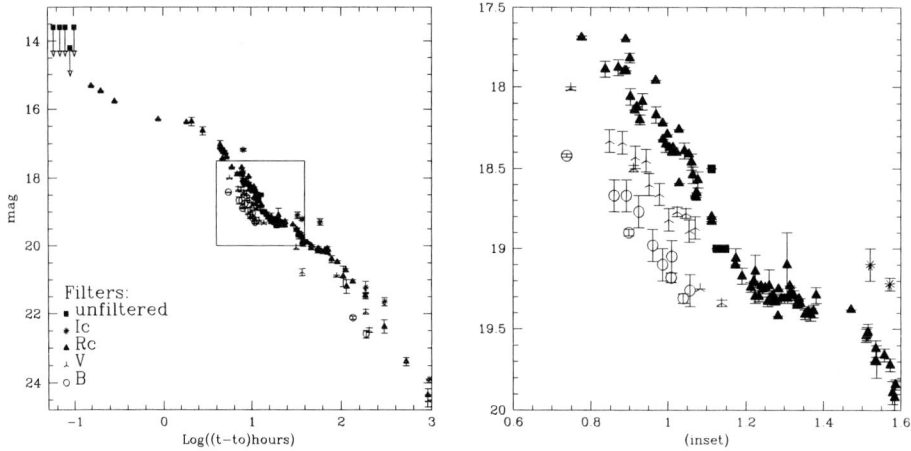

FIGURE 1. Light curve in different filters for GRB021004/HETE2380, data from GCN.

CONCLUSIONS

Our conclusions are fairly simple and could be easily anticipated, but we wanted to stress them by showing what actually happened for some observations of recent Optical Transients from Gamma-Ray Bursts: it would indeed be very helpful and productive, in order to obtain the best set of data, if at least early estimates of OT magnitudes in several filters were to be distributed as soon as possible via GCN, which is, as we all know, a very useful and fast way of sharing information between observers. Possibly, also setting up a chat-line might help in planning and sharing the task of following the OTs with multifilter photometry between different observatories. It is also evident from Figure 2 that it would be desirable to converge as soon as possible on using the same field photometry in all the magnitude estimates.

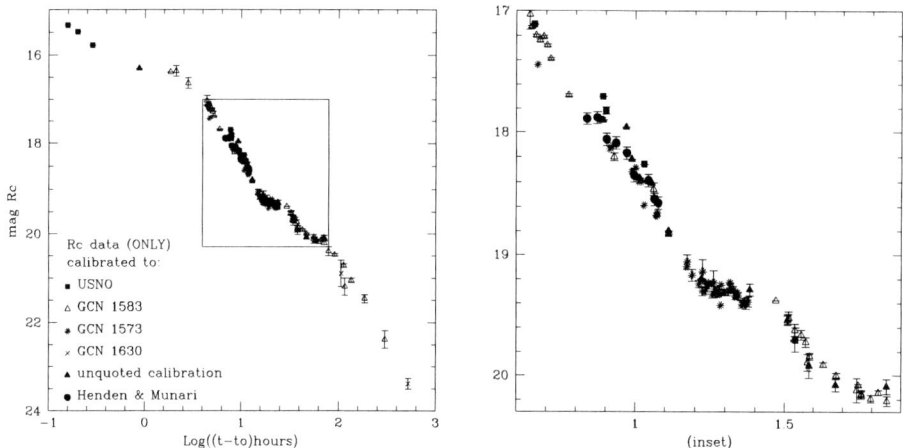

FIGURE 2. Light curve for GRB021004/HETE 2380, Rc data only from GCN.

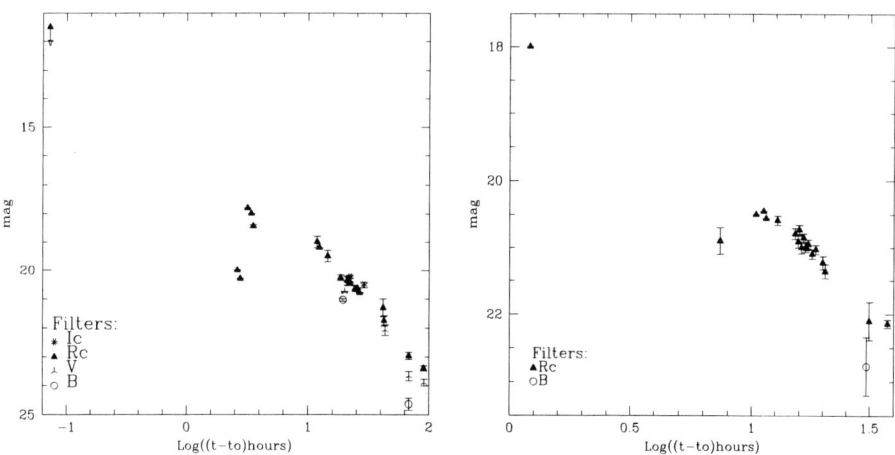

FIGURE 3. Light curves for GRBs 030226/HETE-U10893 and 030328/HETE 2650, data from GCN.

ACKNOWLEDGMENTS

We acknowledge the extremely useful work done by Dr. Scott Barthelmy at NASA/GSFC in setting up and maintaining the GRB Coordinates Network (GCN).

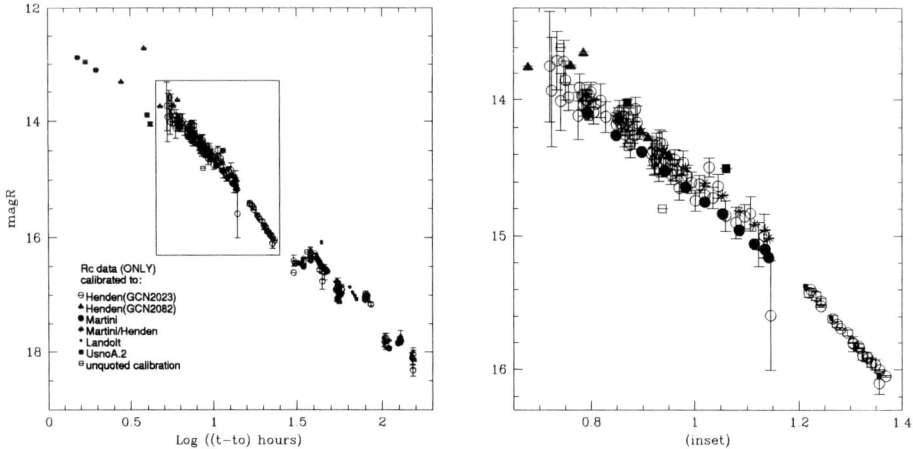

FIGURE 4. Light curve for GRB030329/HETE 2651, data from GCN.

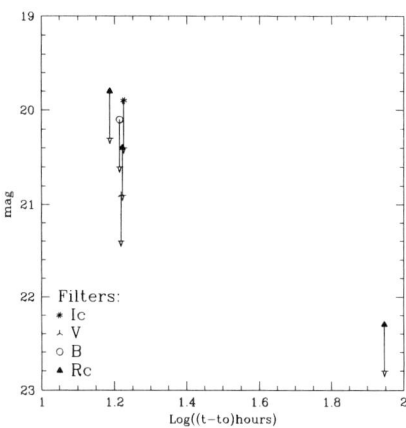

FIGURE 5. Upper limits for GRB030416/HETE-U10897, data from GCN.

REFERENCES

1. Simon, V., Hudec, R., Pizzichini, G., & Masetti, N., Astron. & Astrophys. **377**, 450, (2001).

Spectral Analysis of 50 GRBs Detected by HETE-2

C. Barraud*, J. L. Atteia*, J. F. Olive[†], K.Hurley**, G. Ricker[‡], D. Q. Lamb[§], N. Kawai[∥], R. Vanderspek[‡], T. Sakamoto[††] and The HETE-2 Science Team[‡‡]

*Laboratoire d'Astrophysique, Observatoire Midi-Pyrénées, Toulouse, France
[†]C.E.S.R., Observatoire Midi-Pyrénée, Toulouse, France
**UC Berkeley Space Sciences Laboratory, Berkeley CA 94720-7450, USA
[‡]Center for Space Research, MIT, Cambridge, MA, USA
[§]Department of Astronomy and Astrophysics, University of Chicago, Chicago, IL 60637.
[¶]Tokyo Institute of Technology, 2-12-1 Ookayama, Meguro-ku, Tokyo 152-8551, Japan.
[∥]RIKEN, 2-1 Hirosawa, Wako, Saitama 351-0198, Japan.
[††]Los Alamos National Laboratory, P.O. Box 1663, Los Alamos, NM, 87545.
[‡‡]An international collaboration of institutions including, MIT, LANL, U. Chicago, U.C. Berkeley, U.C. Santa Cruz (USA), CESR, CNES, Sup'Aero (France), RIKEN, NASDA (Japan), IASF/CNR (Italy), INPE (Brazil), TIFR (India).

Abstract. FREGATE, the gamma-ray detector of HETE-2 is entirely dedicated to the study of GRBs. Its main characteristic is its broad energy range, from 7 keV to 400 keV. This energy range can be further extended down to 2 keV using the data from the WXM, the X-ray detector of HETE-2. Such a large energy range allows studies of the prompt emission of GRBs, determining with a high precision their spectral parameters. Moreover, because this energy range is at low energies, the sample of GRBs detected by both FREGATE and WXM contains a significant fraction of X-Ray Rich GRBs and X-Ray Flashes.

We present here the distributions of the spectral parameters mesured for the time integrated spectra of 50 GRBs. We put emphasis on the distribution of the low energy spectral index α. Because FREGATE and WXM detected all classes of GRBs, we also discuss the connection between GRBs, X-Ray Rich GRBs and X-Ray Flashes.

INTRODUCTION

FREGATE is the gamma-ray detector of HETE-2 (see [1] for a description of FREGATE). Its broad-energy range 7–400 keV which can be extended down to 2 keV using the WXM instrument (see [2] for a description of WXM) allows us to determine with high precision the spectral parameters of the prompt emission of the GRBs seen by both instruments. The two instruments also detected an important fraction of X-Ray Rich GRBs and X-Ray Flashes (see [3] for a description of these new classes) and we are now able to discuss the differences and the similarities between these three populations.

We present in this paper an update of the results presented in Barraud et al. [4]: 'Spectral analysis of 35 GRBs/XRFs observed with HETE-2/FREGATE' . This paper

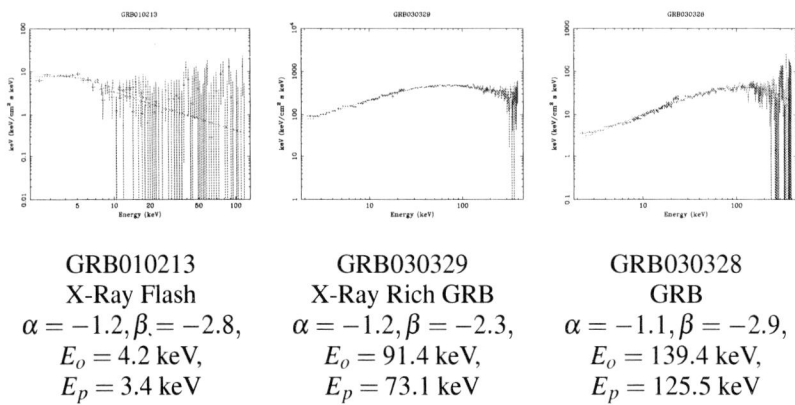

GRB010213	GRB030329	GRB030328
X-Ray Flash	X-Ray Rich GRB	GRB
$\alpha = -1.2, \beta = -2.8$,	$\alpha = -1.2, \beta = -2.3$,	$\alpha = -1.1, \beta = -2.9$,
$E_o = 4.2$ keV,	$E_o = 91.4$ keV,	$E_o = 139.4$ keV,
$E_p = 3.4$ keV	$E_p = 73.1$ keV	$E_p = 125.5$ keV

FIGURE 1. Spectra of the different classes of GRBs: X-Ray Flashe, X-Ray Rich GRB and GRB using both FREGATE and WXM data.

presented a first spectral analysis of 35 GRBs detected by HETE-2/FREGATE since its launch in October 2000 and which were well localized by either the instruments on-board HETE-2 (WXM or SXC, see [6] for a description of the Soft X-Ray Camera), or by the GRB Interplanetary Network.

The update of the paper [4] corresponds to an increase of the number of GRBs seen by both FREGATE and WXM which now reaches 50, all in the class of long GRBs. We did not include the two short/hard bursts GRB020113 and GRB020531 detected by HETE-2. Another improvement is that the spectral parameters are now obtained from a joint fit of WXM and FREGATE data. We thus obtain spectra ranging from 2 keV to 400 keV.

We focus here on the distribution of the spectral parameters: we show that the distribution of the low energy spectral index α is compatible with the predictions of the synchrotron shock model and we show that a significant fraction of bursts have their peak energy E_p lower than 50 keV. We also put emphasis on the hardness-intensity correlation. This correlation shows that the three populations, GRBs, X-Ray-Rich GRBs and X-Ray-Flashes form a continuum which strongly suggest that they are all produced by the same phenomenon.

THE SPECTRAL ANALYSIS

Our sample is made of 50 GRBs localized either with the HETE-instrument or with the GRB Interplanetary Network and which were within 60° of the FREGATE line of sight. GRB spectra are usually fit with the BAND function (Band et al. [5]), which is two power laws smoothly connected:
$$N(E) = AE^\alpha exp\left(\frac{-E}{E_o}\right) \quad \text{for } E > (\alpha - \beta)E_o,$$
$$N(E) = BE^\beta \quad \text{otherwise.}$$

In this equation α and β are the photon indices of respectively the low and the high energy power laws, E_o is the energy break and the peak energy E_p of the νf_ν spectrum

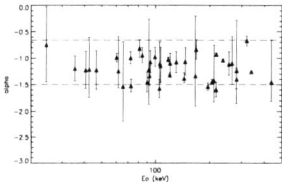
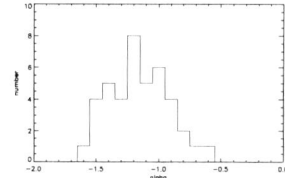

FIGURE 2. left: The photon index of the low energy power law α versus the break energy E_o. right: Distribution of the spectral index α

is defined by : $Ep = Eo * (2 + \alpha)$. We have to notice that in the case of GRBs detected by HETE-2, the energy range is often not broad enough to determine accurately all the parameters of the spectra especially the index of the high energy power law β. In these cases, and in order to not neglect the flux at high energies, we fix the value of β to an arbitrary value which is -2.3. The combination of WXM and FREGATE data allows us to study spectra down to 2 keV and determine accurately the parameter E_p, even at low energies for the X-Ray-Rich GRBs and X-Ray-Flashes.

Figure 1 shows the νf_ν spectrum of one GRB in each of the three classes derived from joint fits of the WXM and FREGATE data. The left panel shows the first X-Ray-Flash detected by HETE-2, GRB010213. The addition of the WXM data allowed to determine the E_p which has a value of 3.4 keV. This is the weakest GRB detected by FREGATE. The middle panel is GRB030329 the "monster burst", an X-Ray Rich with $E_p = 73.1$ keV, and the right panel is GRB030328 a "standard" GRB with $E_p = 125.5$ keV.

THE DISTRIBUTION OF THE SPECTRAL PARAMETERS

Figure 2 displays α, the photon index of the low energy power law versus E_o, the energy break for 41 GRBs for which we were able to mesure these parameters. For clarity of the figure, the 90% error bars are shown for α only. The dotted lines represent the limits predicted by the classical synchrotron shock model which are $-3/2$ and $-2/3$. The values used in this plot result from a fit of the time-intagreted spectra with a cutoff power law model. This model is similar to the Band model but using only the low energy part and the spectral break of the band function. The definition of E_p is not affected by the choice of this model. We use this procedure because in most cases the energy range of HETE–2 (2–400 keV) is not broad enough to determine good values of β and the values of α and E_o are less constrained if we use the Band function .

This figure also shows that whatever the value of Eo, all values of α are compatible with the values expected from the synchrotron shock model. In this model, the emission comes from synchrotron radiation emitted by a population of shock accelerated electrons ([7, 8, 9]). We can also notice that there is a significant fraction of GRBs with E_o lower than 50 keV.

The histogram 2 displays the distribution of the photon index of the low energy power law α. This distribution peaks at -1.2 and has a full width at half maximum of approx-

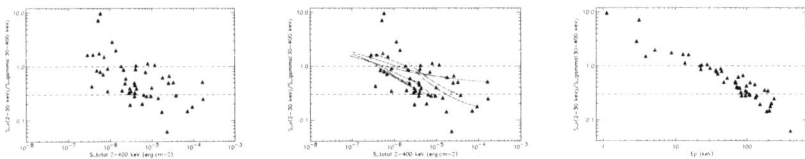

FIGURE 3. left: softness (S_x/S_γ) versus total fluence for GRBs observed by HETE-2. center: softness (S_x/S_γ) versus total fluence for GRBs observed by HETE-2. The lines indicates how GRBs with known redshift would evolve on this diagram if their redshifts were increased to z = 10. right: softness (S_x/S_γ) versus E_p for GRBs observed by HETE-2.

imately 0.5.

THE HARDNESS-INTENSITY CORRELATION AND THE CONNECTION BETWEEN GRBS, X-RAY-RICH GRBS, AND X-RAY-FLASHES

Figure 3 shows the hardness-intensity correlation observed by HETE-2. The y-axis shows the inverse of the hardness or the *softness* (S_x/S_γ) which is the ratio between the fluence in 2–30 keV (S_x) and the fluence in 30–400 keV (S_γ). The x-axis shows the intensity, the fluence in 2-400 keV. The first point highlighted by this figure is the strong correlation between these two quantities over 3 orders of magnitude in fluence, it shows that the weaker a burst is, the softer it is.

The second point is that this figure does not clearly seperate X-Ray-Rich GRBs and X-Ray-Flashes from GRBs. We define here X-Ray-Rich GRB as GRBs which have a softness in the range 0.3–1, and X-Ray-Flashes as GRBs which have a softness greater than 1. The two dotted lines represent the limits (in terms of S_x/S_γ) of the 3 populations. It is clear that there is no gap between these populations and the continuum strongly suggest that these three types of bursts are all produced from the same phenomenon.

We now discuss if X-Ray-Rich GRBs and X-Ray-Flashes can be highly redshifted GRBs, which was one of the first hypothesis to explain such weak and soft bursts. To this end we looked how GRBs with known redshift would evolve on this diagram (3) if their redshifts were increased to z = 10. We added lines on figure 3 in the center which indicate the evolution of these GRBs (ie GRB010921 z = .45, GRB020124 z = 3.2, GRB020813 z = 1.25, GRB021004 z = 2.31, GRB021211 z = 1.01, GRB030226 z = 1.98, GRB030323 z = 3.37, GRB030328 z = 1.52, GRB030329 z = .17, GRB030429 z = 2.65). Redshifts 1 and 5 are marked with crosses and redshifts 2 and 10 with empty squares. What we notice here is that these GRBs have their total fluence which decreases while their softness increases with the redshift. We notice that the higher value of softness we reach at z = 10 with this method is $S_x/S_\gamma = 2$. This value is very small compared to the $S_x/S_\gamma = 10$ found for two bursts. This mechanism, putting GRBs at high redshift, can apparently produce X-Ray Rich GRBs and X-Ray-Flashes but it seems to reach an upper limit and

can't produce the very high values of the softness observed for X-Ray-Flashes.
Figure 3 on the right shows the softness versus the value of the peak energy E_p in keV for all the GRBs detected by HETE-2. The horizontal dashed lines represent the limits of the three classes. This figure first indicates that the softness is very well representative of the value of the E_p. This is very important for spectral analysis indeed the E_p is often hard to calculate because the value of α and E_o are sometimes not well constrained, especially for soft bursts which have their value of E_o near the lower limit of the energy range of HETE-2 (for example GRB010213 has $E_o = 4.2$ keV and it is clear that the value of α can't be well determined and so the value of E_p) whereas the fluence in all energy ranges can always be calculated. This plot also shows that the distribution of E_p covers a broad energy range similar to that covered by HETE-2 from few keV to several hundred keV. In addition of the distribution found by BATSE which peaks at 200 keV HETE-2 had allowed the detection of many bursts which have their E_p lower than 50 keV. This makes this distribution very broad. GRBs which have a low E_p are associated with X-Ray-Rich GRBs (the middle part of the plot) and X-Ray-Flashes (the upper part of the plot).

CONCLUSION

In this paper, we update the results presented in Barraud et al. [4]. The update consists of an increase of the number of GRBs to 50, and an analysis which is now based on joint spectra with both WXM and FREGATE data. Joint spectra allow to study spectra from 2 keV to 400 keV and thus provide more accurate values of α, E_o and fluence ratios.

The first important result of this study comes from the distribution of the spectral parameter α which is fully in agreement with the predictions of the synchrotron shock model. We have also shown that the new class of "soft" GRBs cannot apparently be explained as high redshift GRBs. But we have confirmed and extended the hardness-intensity correlation which strongly suggests that the three classes of GRBs, X-Ray-Rich GRBs and X-Ray-Flashes, which distinguish themselves by the values of their E_p and their softness are all from the same phenomenon. More GRBs and more broad energy coverage of GRB missions will allow to refine these results and constrain models of the prompt emission of GRBs.

REFERENCES

1. Atteia et al., AIP Conf. Proc. **662**, 3-16 (2003).
2. Shirasaki,Y., PASJ **55**, 1033 (2003).
3. Heise, J. et al: GRB Conf pg. 16 (2001)
4. Barraud, C. et al., A&A **400**, 1021 (2003).
5. Band, D. et al., ApJ **413**, 281 (1993).
6. Villasenor, J.N., et al. AIP Conf.Proc. **662** 3-33 (2003).
7. Katz, J.I., ApJ **432**, L107 (1994).
8. Cohen, E., Katz, J.I., Piran, T., Sari, R., Preece, R.D., & Band,D.L., ApJ **488**, 330 (1997)
9. Llyod,N.M. & Petrosian, V., ApJ **543**, 722 (2000).

The Cosmological Evolution Trends of GRB Features

D.M. Wei* and W.H. Gao*

Purple Mountain Observatory, Nanjing, 210008, China

Abstract. Amati et al. presented the results of spectral and energetic properties of several GRBs with known redshifts. Here we analyse the properties of two groups of GRBs, one group with known redshift from afterglow observation, and another group with redshift derived from the luminosity - variability relation. We study the redshift dependence of various GRBs features in their cosmological rest frames, and find that, for these two group GRB groups, their properties are all redshift dependent, i.e. their intrinsic duration, luminosity, radiated energy and peak energy E_p, are all correlated with the redshift, which means that there are cosmological evolution effects on gamma-ray bursts features, and this can provide an interesting clue to the nature of GRBs.

INTRODUCTION

Two important correlations have been discovered, i.e. between the degree of variability of the gamma-ray burst light curve and the GRB luminosity (Feminore & Ramirez-Ruiz, [1]), and between the differential time lags for the arrival of burst pulses at different energies and the GRB luminosity (Norris, Marani & Bonnell, [2]), although these correlations are still tentative, they offer the possibility to derive independent estimates of the redshifts of GRBs.

Here we will discuss the properties of two GRB groups, one group includes the bursts with known redshifts, and another group consists of bursts whose redshifts are derived from the luminosity - variability relation. We will show that the properties of these two GRB groups are all correlated with redshift.

THE PROPERTIES OF TWO GRB GROUPS

Amati et al. ([3]) have analysed the spectral properties of the X-ray and gamma-ray emission from GRBs with known redshifts. In their sample total 12 gamma-ray bursts were included, however among them, there are three bursts (GRB980326, GRB980329 and GRB000214) whose redshifts are not determined accurately, so our one group contains the other 9 bursts with firm redshifts.

Several authors have suggested that there may be correlation between the properties of burst time structure and burst luminosity (e.g. Feminore & Ramirez-Ruiz, [1]; Reichart et al., [4]; Norris, Marani & Bonnell, [2]). Lloyd-Ronning & Ramirez-Ruiz ([5]) used 159 bursts to investigate the dependence of the burst spectra on variability. Our second

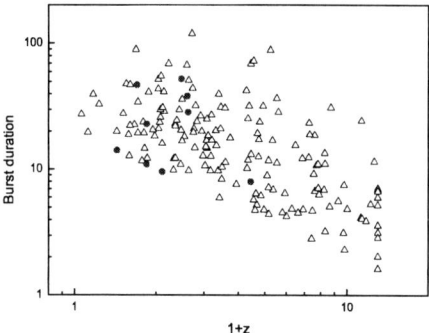

FIGURE 1. The duration of GRBs in their cosmological rest frame versus the burst redshift. The filled circles are bursts with secure redshifts estimates, while the empty triangles are bursts in which the redshifts are derived using the luminosity - variability distance indicator.

group consists of these 159 bursts with known peak energy, and their redshifts are derived from the luminosity - variability relation (Feminore & Ramirez-Ruiz, [1]).

Figure 1 gives the GRB duration in their cosmological rest frame, $T' = T/(1+z)$, versus the redshift. Figure 2 shows the luminosity of GRBs in their cosmological rest frame versus the redshift, Figure 3 shows the relation between the peak energy E_p of the νF_ν spectra in their cosmological rest frame and the redshift.

From Figure 1 we see that although the distribution of the burst duration is somewhat scattered, there is still a clear trend that the intrinsic duration decreases with the redshift, we find $T' \propto (1+z)^{-0.85 \pm 0.08}$. Figure 2 showes that the isotropic luminosity has a positive correlation with the redshift, excluding the 4 bursts with smallest redshifts, we have $L \propto (1+z)^{2.5 \pm 0.1}$. Figure 3 shows that the peak energy of νF_ν spectra also has a positive correlation with the redshift, the power law fit is $E_p \propto (1+z)^{0.76 \pm 0.07}$.

Now we use the Monte Carlo simulation method to test the flux truncation effect. We simulate the burst sample, assuming their luminosity satisfies $L = L_0(1+z)^\alpha$, and their redshifts are distributed uniformly between 0 and 10. We find that, when $\alpha = 1.7 \pm 0.5$, there is the relation $L \propto (1+z)^{2.50 \pm 0.08}$, which is consistent with the observed value.

Figure 4 shows the relation between the intrinsic peak energy of the νF_ν spectra and the luminosity for group one, the solid line represents the relation $E_p \propto L^{1/2}$. We also find that for group two, the relation $E_p \propto L^{1/2}$ can account for the observed data quite well.

DISCUSSION AND CONCLUSION

Figure 4 shows that there is a good correlation between the peak energy and luminosity for both groups of GRBs, and the relation $E_p \propto L^{1/2}$ can account for the observed

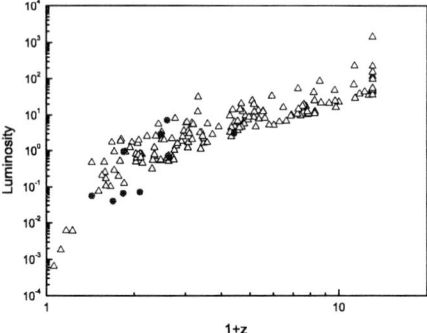

FIGURE 2. The luminosity of GRBs in their cosmological rest frame versus the burst redshift. The filled circles are bursts with secure redshifts estimates, while the empty triangles are bursts in which the redshifts are derived using the luminosity - variability relation.

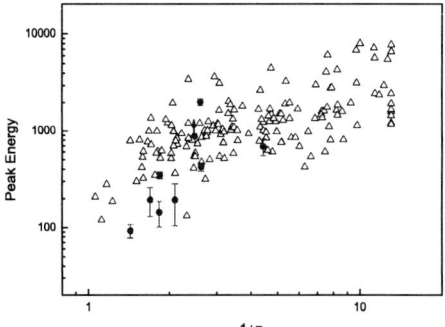

FIGURE 3. The peak energy of νF_ν spectra of GRBs in their cosmological rest frame versus the burst redshift. The filled circles are bursts with secure redshifts estimates, while the empty triangles are bursts in which the redshifts are derived using the luminosity - variability relation.

data quite well. Zhang & Meszaros ([6]) analysed various fireball models within a unified picture and investigated the E_p predictions of different models. It is known that for internal shock model, if the GRB bulk Lorentz factors are not correlated with the luminosities, then there is the relation $E_p \propto L^{1/2}$, while for external shock model $E_p \propto \Gamma^4$. So our results suggest that the gamma-ray burst emissions are more likely from the internal shock.

Frail et al. concluded that the GRB emission energy is nearly a constant, $E \sim 5 \times 10^{50}$ ergs (Frail et al., [7]). However, Figure 1 and Figure 2 show that the isotropic radiated

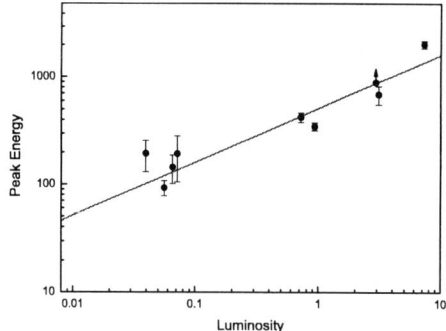

FIGURE 4. The peak energy of νF_ν spectra of group one GRBs in their cosmological rest frame versus the burst luminosity. The solid line represents the relation $E_p \propto L^{1/2}$.

energy increases with redshift, so if the conclusion of Frail et al. is true, then the jet opening angle should decrease with the redshift. We take the values of opening angle from the paper of Frail et al. ([7]) and find that, in fact, the opening angle does indeed decrease with the redshift. This point is very interesting, since it can put constraints on the central engines of GRBs. Of course, this phenomenon can also be explained within the framework of a structured universal jet model (Zhang & Meszaros, [8]; Rossi et al., [9]). In this model, an observer closer to the jet axis would detect a higher luminosity, thus at higher redshift, smaller viewing angle detections are preferred due to luminosity selection effect.

ACKNOWLEDGMENTS

This work is supported by the National Natural Science Foundation and the National 973 Project.

REFERENCES

1. Fenimore, E. E., & Ramirez-Ruiz, E., ApJ, submitted (astro-ph/0004176) (2001).
2. Norris, J. P., Marani, G. F., & Bonnel, J. T., ApJ, **534**, 248 (2000).
3. Amati, L., et al., A&A, **390**, 81 (2002).
4. Reichart, D.E., Lamb, D.Q., Fenimore, E.E., et al., ApJ, **552**, 57 (2001).
5. Lloyd-Ronning, N.M., Ramirez-Ruiz, E., ApJ, **576**, 101 (2002).
6. Zhang, B., & Meszaros, P., ApJ, **581**, 1236 (2002).
7. Frail, D.A., et al., ApJ, **562**, L55 (2001).
8. Zhang, B., & Meszaros, P., ApJ, **571**, 876 (2002).
9. Rossi, E., Lazzati, D., & Rees, M.J., MNRAS, **332**, 945 (2002).

Evidence for Different Spectral Behaviours for Long and Short GRBs

B. M. Belli

Istituto di Astrofisica Spaziale e Fisica Cosmica, area di ricerca del CNR,di Tor Vergata, via del Fosso del Cavaliere 100,00133 Roma, Italy

Abstract. An analysis of the spectral behaviour of the GRBs recorded by the BATSE experiment has been performed on the basis of the spectral hardness HR_{21} and HR_{32} evaluated in the energy ranges 25-50 50-100 and 50-100 100-300 keV respectively. The plot of the events in the HR_{21} and HR_{32} plane shows a behaviour different for the two Classes of short and long GRBs. The distribution of the long events follows on the average an increasing power law, on the contrary the short event distribution shows decrease in HR_{21} when HR_{32} increases. This result suggests important differences in the physical scenario at the basis of the two Classes of events.

INTRODUCTION

The esistence of at least two classes of GRBs in the BATSE catalog has been suggested by the histogram of the event duration [1] and by the clustering in two groups of the events in the plane duration hardness- ratio [2]. The averaged values of the durations and hardness-ratios are longer and softer for Class I and shorter and harder for Class II, respectively. Another observational difference is that measurable energy-dependent pulse lag have been found only in the time histories of long events, and not in short events [3]. No comparison for the behaviour of afterglows in the two Classes has been possible until now because no afterglow for Class II events has yet been observed.

In this paper we study the energy spectra of the BATSE catalog GRBs, analysing the event distribution in the HR_{21}-HR_{32} plane, i. e. the color-color diagram [4]. This method compared to a study based on the analysis of the parameters of the fits of all GRB energy spectra [5, 6] could appear not very powerful. On the contrary, differently then before, it provides the possibility of a very synthetic approch to the problem and it allows to compare the behaviour of the energy spectra at the low and high energies of the same event. Finally we compare our results with the behaviour of the energy spectra of three short GRBs of the KONUS catalog [7].

ENERGY SPECTRA OF GRBS OF THE BATSE CATALOG

In our analysis we used the fluences relative to the four consecutive channels in which the total BATSE energy range has been divided. These fluences are reported in the BATSE catalog for every GRB. These quantities have been calculated on the basis of the count rates recorded in each energy channel with a very simplifying hypothesis for the GRB

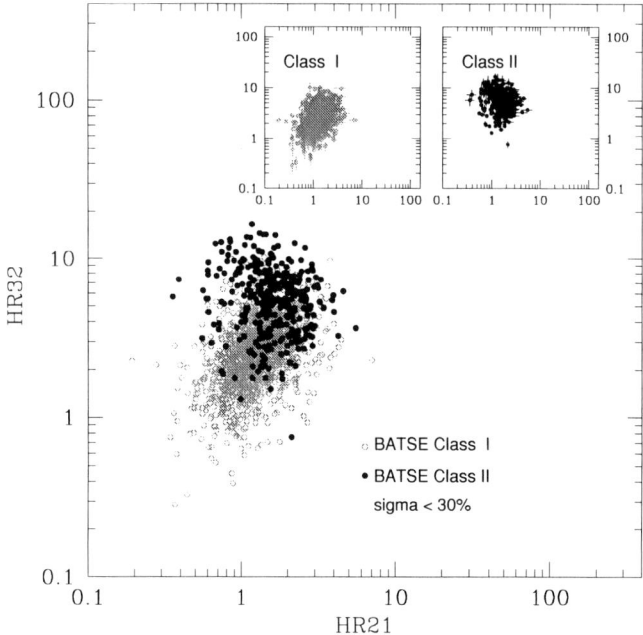

FIGURE 1. Plot in the HR_{21}-HR_{32} plane of the events of the two Classes reported in the BATSE catalog with errors for HR_{21} and HR_{32} slower than 30%. The events of the two Classes are represented with different labels and grey scales.

energy spectra; they represent only a rough estimate of the real values. Nevertheless we think that they are able to put in evidence the possible differences in the global characteristics of the two classes. We remind that HR marked with two index is the ratio of the two fluences relative to the energy channels with the same index, respectively. We have plotted in the plane HR_{21}-HR_{32} the BATSE catalog events with different marks and grey tones for the two Classes (Fig. 1). The event distribution appears different for Class I and for Class II. In Class I the hardness HR_{21} and HR_{32} grow on average together. Instead Class I events situated at the highest values of HR_{32} have lower values of HR_{21}.

ENERGY SPECTRA OF SOME SHORT GRBS OF THE KONUS CATALOG

The Konus catalog of short GRBs [7] contains data on 130 short events observed with the Konus-Wind experiment on the Wind spacecraft in the years 1994-2002. The catalog

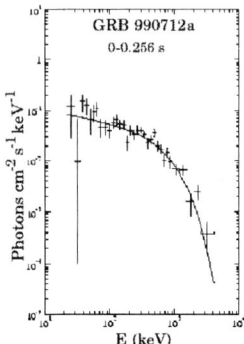 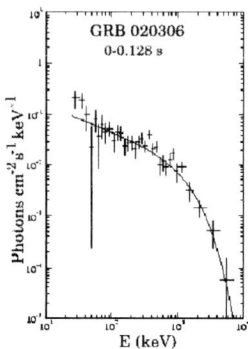 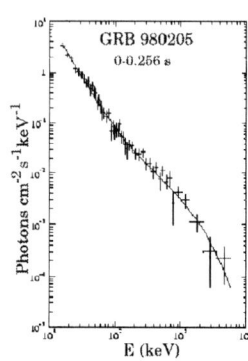

FIGURE 2. Energy Spectra of three GRBs with duration < 2s observed by the Konus-Wind experiment onboard the Wind satellite, taken from the Konus catalog (Mazets et al. 2002 [7]).

reports, together with many other data, the GRB averaged energy spectra. In this catalog the spectra have been fitted with the simpler laws commonly used to fit the energy spectra of GRBs: power law, Compton law, the four parameter law suggested by [8] or by a combination of these laws. In Fig. 2 we report three events from the Mazets catalog. When represented in the $HR_{21} - HR_{32}$ plane, they appear in the high left part of the BATSE event distribution. In the first two cases, the hardness-ratio values have been calculated by the fluences of each spectrum point directly. The shown best fits do not take into account the excess of the observed X-ray radiation.

CONCLUSIONS

As this rapid analysis has shown, the two Classes of GRBs show some meaningful differences, in the spectral behaviour. The presence of the Class II events in the hardeness plane zone with high H_{32} and low H_{21} implies an excess of X-ray also in a hard energy spectrum. This peculiar spectrum shape can provide important information both on the enviroment and on the physical properties [9]. A more detailed study performed for example on the BATSE event energy spectra directly, can provide more insight on the origin of short GRBs.

REFERENCES

1. Kouveliotou C. et al., *ApJ.* **413**, L101 (1993).
2. Belli B. M., *Ap&SS.* **231**, 43 (1995).
3. Norris J. P. et al., ESO Astrophysics Symposia, Costa E. Frontera F. and Hjiorth J. editors, 40 (2000).
4. Giblin T. W. et al., AIP Conference Proc.s **526**, Kippen R. M. Mallozzi R. S. and Fishman G. J. editors, 394 (2000).
5. Paciesas W. S. et al., AIP Conference Proc.s **662**, Ricker G. R. Vanderspek R. K. editors, 248 (2001).

6. Ghirlanda G. et al., these proceedigs, (2003)
7. Mazets E. P. et al., in press (2002).
8. Band D. et al., *ApJ.* **413**, 281 (1993).
9. Piro L., AIP Conference Proc.s **662**, Ricker G. R. Vanderspek R. K. editors, 372 (2001).

Gamma-Ray Bursts in Wavelet Space

Zsolt Bagoly[*], István Horváth[†], Attila Mészáros[**,‡] and Lajos G. Balázs[§]

[*]*Laboratory for Information Technology, Eötvös University, H-1117 Budapest, Pázmány P. s. 1./A, Hungary*
[†]*Department of Physics, Bolyai Military University, H-1456 Budapest, POB 12, Hungary*
[**]*Astronomical Institute of the Charles University, V Holešovičkách 2, CZ-180 00 Prague 8, Czech Republic*
[‡]*Stockholm Observatory, AlbaNova, SE-106 91 Stockholm, Sweden*
[§]*Konkoly Observatory, H-1525 Budapest, POB 67, Hungary*

Abstract. Gamma ray burst light curves have been analyzed using a special wavelet transformation. The applied wavelet base is based on a typical Fast Rise-Exponential Decay (FRED) pulse. The shape of the wavelet coefficients' total distribution is determined on the observational frequency grid. Our analysis indicates that the pulses in the long bursts' high energy channel lightcurves are more FRED-like than the lower ones, independently from the actual physical time-scale.

INTRODUCTION

The shapes of the gamma-ray burst's (GRB's) 64 ms resolution lightcurves in the BATSE Gamma-Ray Burst Catalog [1] carry immense amount of information. However, the chaging S/N ratio complicates the detailed comparative analysis of the lightcurves. During the morphological analysis of GRBs [2, 3] a subclass with Fast Rise-Exponential Decay (FRED) pulse shape was observed. This shape is quite attractive because of its phenomenological simplicity. Here we use a special wavelet transformation with a kernel function based on a FRED-like pulse. A similar approach has been used by [4], but their base functions were constructed differently.

THE FRED WAVELET TRANSFORM

We have used the Discrete Wavelet Transform (DWT) matrix formalism (see [5]): here for an input data vector v, the one step of the wavelet transform is a multiplication with a special matrix F:

$$F = \begin{bmatrix} c_0 & c_1 & c_2 & c_3 & & & & \\ c_3 & -c_2 & c_1 & -c_0 & & & & \\ & & c_0 & c_1 & c_2 & c_3 & & \\ & & c_3 & -c_2 & c_1 & -c_0 & & \\ \vdots & \vdots & & & & & \ddots & \\ c_2 & c_3 & & & & & c_0 & c_1 \\ c_1 & -c_0 & & & & & c_3 & -c_2 \end{bmatrix} \quad (1)$$

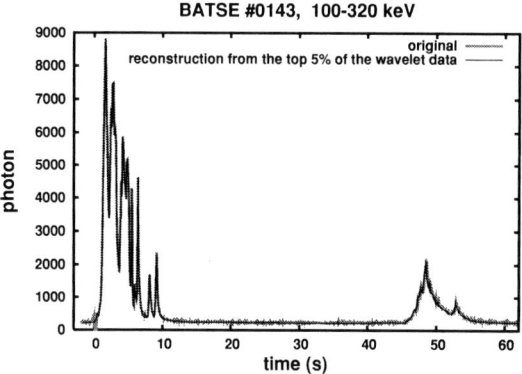

FIGURE 1. The original and the reconstructed ligthcurve of BATSE trigger 0143 - only 5% of the total wavelet data is used.

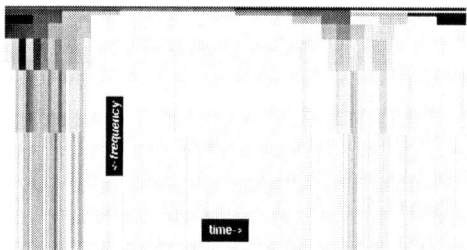

FIGURE 2. The wavelet phase-space density for BATSE trigger 0143.

where the $c_0, \ldots, c_3$ are the 4-stage FIR (Finite Impulse Response) filter parameters defining the wavelet. To obtain these values we require the matrix F to be orthogonal (i.e. no information loss), and the output of the even (derivating-like) rows should disappear for a constant and for a FRED-like $e^{-t/\tau}$ input signal. These requirements give two different solutions for $c_0, \ldots, c_3$: a rapidly oscillating one and a smooth one. In the following we will use the latter one.

Our filter process with the FRED wavelet transform consists of the usual *transform* → *filter/cut* → *inverse transform* digital filtering steps. During the filtering we will loose some information, however this could be quite small. To demonstrate the efficiency of the algorithm on Fig. 1. we reconstructed the 100-320 keV 64ms ligthcurve of BATSE trigger 0143 from the biggest 5% of the total wavelet coefficients. The excellent reconstruction of each individual pulse is obvious.

The wavelet transformation algorithm divides the phase-space into equal area regions. On Fig. 2. the wavelet phase-space density is shown. Here the dark segments are the really important coefficients - however they cover only a small portion of the total area which explains the high efficiency of the reconstruction.

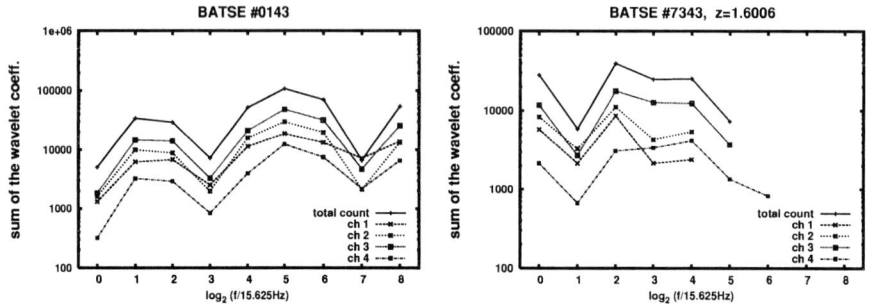

FIGURE 3. The wavelet signal's frequency distribution for BATSE triggers 0143 and 7343 respectively.

WAVELET SCALE ANALYSIS

For a frequency-like wavelet scale analysis we would like to create a power-spectrum like distribution along the frequency axis. However, one should be careful. In the classical signal processing one uses the power spectrum from the Fourier-transform, because the signals are electromagnetic-like usually, e.g. the power (or energy) is proportional to the square of the signal. Here the lightcurves measure photon counts — so the signal's energy is simply the sum of the counts. For this reason we approximate the signal's strength as a sum the magnitude of the coefficients along the given frequency rows.

This signal's strength indicates on Fig 3. (BATSE trigger 0143) the maximum power to be around $f \approx 500$Hz. In each energy channel the signals are similar (observe the logarithmic scale), because the signal is strong even at high energies (channel 4). For BATSE trigger 7343 (with optical redshift $z = 1.6006$) one can observe a strong high frequency cutoff: some of the signal's high frequency part is missing. However all the 4 channels are visible, while the maximum power is at $f \approx 62.5$Hz. It is interesting to remark that the signal's shape is quite similar to trigger 0143 if that is scaled down by a factor of $\approx 2.4 - 2.8$ in frequency.

WAVELET FILTERING AND SIMILARITY

The FRED wavelet transform measures the similarity between the different wavelet kernel functions (here all are FRED-based) and the actual signal. To quantify the similarity we define a magnitude cutoff in the wavelet space so, that the *reconstructed* T_{50} value from the filtered data should be similar to the original values. The T_{50} value and its σ_P error from the photon count statistics could be easily determined from the original ligthcurve. To keep only the important features we define the $T_{50\text{break}}$ breakpoint where

$$|T_{50\text{break}} - T_{50}| = 10.0\sigma_P$$

Using a cut-off point it is possible to define a Compressed Size (CS) for a burst: it is the number of bins (in the wavelet space) needed to restore the curve at the break.

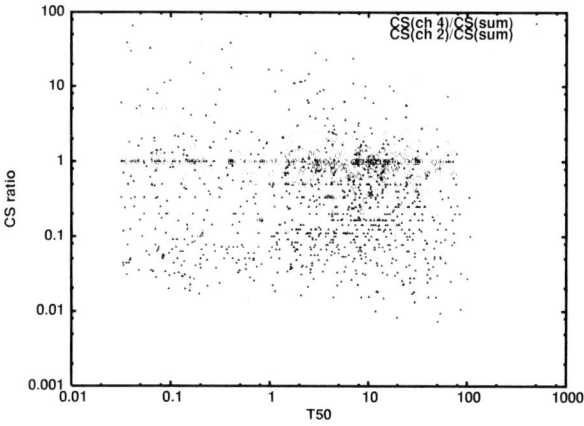

FIGURE 4. The relative value of the CS against the total lightcurves' CS for channels 2 and 4.

The CS value is a robust measure quantifying the similarity between the FRED kernel and the different channels' lightcurves. Our analysis suggest that all the low energy channels #1, #2 and #3 behave similarly, while the high energy (> 320 keV) channel is different (which is not very surprising, e.g. [6]). Fig. 4. shows the *ratio* of the CS's against the total count lightcurves' CS for channels 2 and 4. This distributions indicate that the pulse-shapes in the long bursts' high energy channel are more FRED-like than the lower ones - and this is *independent* from the actual FRED time-scale!

ACKNOWLEDGMENTS

This research was supported in part through OTKA grants T024027 (L.G.B.), and T034549, Czech Research Grant J13/98: 113200004 and by a grant from the Wenner-Gren Foundations (A.M.).

REFERENCES

1. Meegan, C., Malozzi, R.S., Six, F. & Connaughton, V. 2001, *Current BATSE Gamma-Ray Burst Catalog*, http://gammaray.msfc.nasa.gov/batse/grb/catalog
2. Norris, J. P., Nemiroff, R. J., Bonnell, J. T., Paciesas, W. S., Kouveliotou, C., Fishman, G. J., & Meegan, C. A. Bulletin of the American Astronomical Society, **26**, 1333 (1994).
3. Norris, J. P., Scargle, J. D., Bonnell, J. T., & Nemiroff, R. J., *Gamma-Ray Bursts, 4th Hunstville Symposium*, AIP **171** (1998).
4. Quilligan, F., McBreen, B., Hanlon, L., McBreen, S., Hurley, K. J. & Watson, D., Astronomy and Astrophysics, **385**, 377 (2002).
5. Press W.H., Teukolsky S.A., Vetterling W.T., Flannery B.P. *Numerical Recipes in Fortran, Second Edition*, Cambridge University Press, Cambridge (1992).
6. Bagoly, Z., Mészáros, A., Horváth, I., Balázs, L. G., & Mészáros, P., ApJ, **498**, 342 (1998).

X-RAY FLASHES

Origin of XRFs: low $\dot{E}$, low contrast of Γ or large viewing angle?

Céline Barraud*, Frédéric Daigne[†], Robert Mochkovitch[†] and Jean-Luc Atteia*

Observatoire de Toulouse, France
[†]*Institut d'Astrophysique de Paris, France*

Abstract. We have developed a toy model for internal shocks which has been used to produce a large number of synthetic GRB/XRF events in order to find the critical parameters for the production of X-ray flashes. The key factor appears to be a small contrast of the Lorentz factor in the relativistic outflow emitted by the central engine.

INTRODUCTION

The recently discovered X-ray flashes (XRFs) [1] have non-thermal spectra with $E_p < 50$ keV. In the definition used by the HETE 2 team their softness (defined as the ratio of the 2 - 30 keV to the 30 - 400 keV fluences) is larger than unity. Except for being softer, they appear to share many common properties with GRBs, and the confirmation that they are located at cosmological distances came recently with a redshift determination for XRF 020903 at $z = 0.251$ [2]. About one third of the events detected by HETE 2 are XRFs while "X-ray rich GRBs" (with a softness between 0.3 and 1) represent another third of the total. If GRBs and XRFs are related objects then the origin of the softness of XRFs remains controversial. XRFs can be intrinsically soft due to some specific values of the physical parameters affecting the energy of the emitted photons but they can be also standard GRBs observed under peculiar conditions, such as a large redshift or a large viewing angle [3]. If a large z seems to be excluded – the duration distribution of XRFs and GRBs are similar and the redshift of XRF 020903 is only 0.25 while its E_p is less than 5 keV – the intrinsic softness and the large viewing angle remain open possibilities.

We have checked if XRFs can be intrinsically soft in the context of the internal shock model. We have adopted a very simple treatment of the shocks and we have explored the parameter space of the relativistic outflow (injected energy, average value and contrast of the Lorentz factor, duration of the emission, etc.) to find out if and how XRFs can be produced.

HOW TO TRANSFORM A GRB INTO A XRF?

The internal shock model for GRBs predicts that the peak energy of the emitted spectrum is related to the flow parameters in the following way

$$E_p \sim C \frac{\dot{E}^x f_{xy}(\kappa)}{\tau^{2x} \bar{\Gamma}^{6x-1}} \qquad (1)$$

where $\dot{E}$, τ, $\bar{\Gamma}$ and κ are respectively the injected power, the duration of energy injection, the average Lorentz factor and the contrast of the Lorentz factor distribution. The constant C and the exponents x and y depend on the emission process for the energy dissipated in shocks; the function f_{xy} is steadily increasing with κ. For synchrotron radiation and equipartition parameters ε_B and ε_e for the post-shock magnetic field and electron Lorentz factor, the exponents are $x = 1/2$, $y = 5/2$ and $C \propto \varepsilon_e^2 \varepsilon_B^{1/2}$.

The intrinsic E_p given by Eq. (1) will be reduced by redshift and geometrical effects at large viewing angles leading to the observed E_p

$$E_p^{\text{obs}} = \frac{E_p}{(1+z)[1+\Gamma^2(\theta_0 - \Delta\theta)^2]} \qquad (2)$$

where $\Delta\theta$ is the half opening angle of the outflow and θ_0 the viewing angle. This geometrical term is present only if $(\theta_0 - \Delta\theta) > 0$.

We have first considered a simple single pulse GRB produced with $\dot{E} = 10^{53}$ erg.s^{-1}, $\tau = 10$ s, $\bar{\Gamma} = 300$, $\kappa = 4$ and $\Delta\Theta = 5°$. At $z = 1$, the E_p value for this burst when viewed face-on is 120 keV and its fluence in the 2 – 400 keV energy range is $S_{2-400} = 4.8\,10^{-6}$ erg.cm^{-2}. To transform this GRB into an XRF (of softness larger than unity) one can either:

– increase the difference $(\theta_0 - \Delta\theta)$ from 0 to 0.5° ;
– decrease the injected power from 10^{53} to $10^{51.5}$ erg.s^{-1} ;
– decrease the contrast κ from 4 to 1.3 ;
– decrease the constant C in Eq. (1) by a factor of 3.

In this last case the increase of softness takes place at nearly constant fluence while in the three other cases the fluence is reduced by nearly a factor of 30. Both geometrical and intrinsic causes may then result in the production of an XRF. We concentrate below on the different possible intrinsic causes in the context of the internal shock model.

A TOY MODEL

To simplify the study internal shocks in a relativistic outflow we developed a toy model where we only consider the collision of two shells of equal mass m. Shell 2 (of Lorentz factor Γ_2) is generated a time τ after shell 1 (Lorentz factor $\Gamma_1 < \Gamma_2$). The average power injected into the wind in this two shell approximation is given by

$$\dot{E} = \frac{mc^2}{\tau}(\Gamma_1 + \Gamma_2) = \dot{M}\bar{\Gamma}c^2 \qquad (3)$$

where $\dot{M} = 2m/\tau$ and $\bar{\Gamma} = \frac{1}{2}(\Gamma_1 + \Gamma_2)$ are the average mass loss rate and Lorentz factor. Shell 2 will catch up with shell 1 at the shock radius $r_s = 2c\tau\frac{\Gamma_1^2\Gamma_2^2}{\Gamma_2^2-\Gamma_1^2}$. The two shells merge and the energy dissipated in the collision E_{diss} is radiated with a characteristic broken power law spectrum, the break energy being given by Eq. (1).

From the values of E_{diss}, τ, E_{break} and assuming a Band spectrum with low and high energy indices $\alpha = -1$ and $\beta = -2.5$ it is possible to estimate (for a given redshift) the average flux or count rate in any spectral band. The simplicity of the two shell approximation then allows to construct a large number of synthetic bursts to check if XRFs can be formed for some specific choice of the wind parameters.

We generate synthetic events by drawing the burst parameters, following distributions which include our basic knowledge on the burst physics and origin. The best constrained parameters are the redshift and the duration. If long GRBs (and XRFs) are related to the explosive death of massive stars their rate is directly proportional to the cosmic star formation rate ψ_* and their distribution in redshift can be deduced from $\psi_*(z)$ for which we have adopted the analytical expression given by [4] with a maximum at $z \sim 1.5$ (their SFR1).

The distribution of the observed duration t_{90} for long BATSE bursts is approximately log-normal with a maximum at $t_{90} \sim 20$ s. We have then also adopted a log-normal distribution for τ with a maximum at $\tau_{\text{max}} = 10$ s and we have checked a posteriori that the resulting distribution of $\tau_{\text{obs}} = (1+z)\tau$ for the synthetic bursts which would have been detected by BATSE agrees with that of t_{90}.

The four parameters $\bar{\Gamma}$, κ and $\dot{E}$ and C_p are much less constrained and we simply take for them uniform distributions between 100 and 500 for $\bar{\Gamma}$, 0 and 1 for $\text{Log}\,\kappa$, 51 and $\dot{e}_{\text{max}} = 53.4$ for $\text{Log}\,\dot{E}$ and $\text{Log}\,C_p = \text{Log}\,C_p^{100} \pm 0.5$ where C_p^{100} is the value of C_p which produces a burst having $E_p = 100$ keV if $\dot{E} = 10^{52}$ erg.s^{-1}, $\bar{\Gamma} = 300$ and $\kappa = 4$. The upper limit $\dot{e}_{\text{max}}$ for $\text{Log}\,\dot{E}$ has been estimated from the requirement that the synthetic $\text{Log}\,N$-$\text{Log}\,P$ relation agrees with the BATSE data [5].

Results

Softness - fluence distribution

We have plotted in Fig.1 our synthetic events in the softness/total fluence (2 - 400 keV) diagram. It can be seen that GRBs, X-ray rich GRBs as well as XRFs can be produced. Unfortunately, the present model cannot be used to fix the relative proportion of XRFs and GRBs since we don't know the true distributions of the wind parameters. What we can do however is to compare these different parameters in the two populations to find out which combinations of $\dot{E}$, τ, $\bar{\Gamma}$, κ and C favor the production of XRFs.

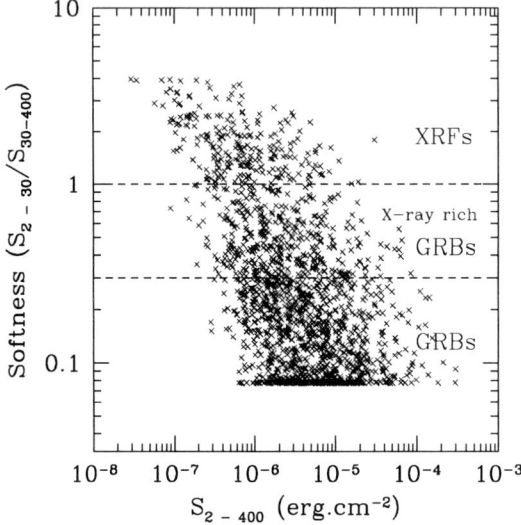

FIGURE 1. Softness-fluence diagram of synthetic events.

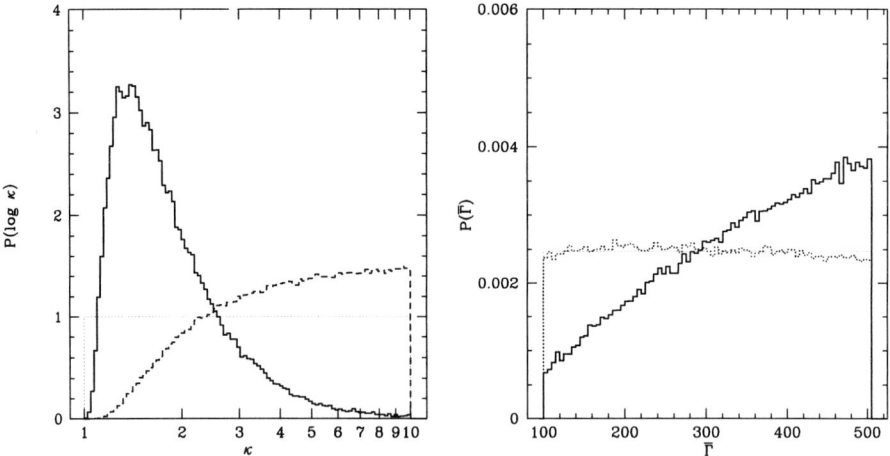

FIGURE 2. Distributions of the contrast κ of the Lorentz factor distribution and of the average Lorentz factor $\bar{\Gamma}$ in XRFs (full line) and GRBs (dashed line). In each diagram the thin line corresponds to the initial distribution used to generate the synthetic population.

Physical parameters in XRFs compared to GRBs

The distribution of duration and injected power in synthetic XRFs and GRBs are rather similar (we always consider in this discussion the sub-sample of events which would have been detected by HETE). We find that the average duration of XRFs is about 30% larger than that of XRFs but this small difference cannot be tested with the presently limited sample of XRFs.

The main differences between the two populations are illustrated in Fig.2 where the distributions of κ and $\bar{\Gamma}$ have been represented. In GRBs, $\bar{\Gamma}$ closely follow the initial uniform distribution used to generate the synthetic events. Large contrasts κ of the Lorentz factor are favored with few bursts having $\kappa < 2$. Conversely XRFs are obtained for small contrasts (the maximum of the distribution being located at $\kappa = 1.6$) and large $\bar{\Gamma}$ (with 6 times more XRFs at $\bar{\Gamma} = 500$ than at $\bar{\Gamma} = 100$). Within the internal shock model a clean (rather than a dirty) fireball leads to a reduction of E_p and an increase of softness.

CONCLUSION

Our analysis confirms that XRFs can be generated in the context of the internal shock model. Assuming uniform input distributions of C, $\bar{\Gamma}$ and $\log \kappa$ to produce our synthetic sample we get a fraction $f_{XRF} \sim 10\%$ of XRFs relative to GRBs. This may not be enough but f_{XRF} can be easily increased if one supposes for example that relativistic outflows with small contrasts of the Lorentz factor are more frequent than those with high contrasts. If however the flows are always highly variable, one should then rely on viewing angle effects only to make the XRFs.

REFERENCES

1. Heise, J., in't Zand, J., Kippen, R. M., and Woods, P. M., "X-Ray Flashes and X-Ray Rich Gamma Ray Bursts," in *Gamma-ray Bursts in the Afterglow Era*, 2001, p. 16.
2. Soderberg, A. M. et al, *astro-ph/0311050* (2003).
3. Yamazaki, R., Ioka, K., and Nakamura, T., *ApJL*, **571**, L31 (2002).
4. Porciani, C., and Madau, P., *ApJ*, **548**, 522 (2001).
5. Stern, B. E., Atteia, J.-L., and Hurley, K., *ApJ*, **578**, 304 (2002).

HETE-2 Observation of the Extremely Soft X-Ray Flashes, XRF010213 and XRF020903

T. Sakamoto*†, M. Suzuki*, N. Kawai*, Y. Nakagawa**, A. Yoshida**, Y. Shirasaki‡, T. Tamagawa†, K. Torii†, M. Matsuoka§, E. E. Fenimore¶, M. Galassi¶, D. Q. Lamb‖, C. Graziani‖, T. Q. Donaghy‖, J-L. Atteia††, C. Barraud††, M. Boer‡‡, J-P. Dezalay‡‡, J-F. Olive‡‡, G. Ricker§§, J. Doty§§, R. Vanderspek§§, G. B. Crew§§, J. Villasenor§§, N. Butler§§, J. G. Jernigan¶¶, K. Hurley¶¶, S. E. Woosley***, G. Pizzichini††† and HETE-2 science team‡‡‡

*Tokyo Institute of Technology
†The Institute of Physical and Chemical Research (RIKEN)
**Aoyama Gakuin University
‡National Astronomical Observatory
§Tsukuba Space Center, National Space Development Agency of Japan
¶Los Alamos National Laboratory
‖University of Chicago
††Observatoire Midi-Pyrenéés
‡‡CNRS
§§Massachusetts Institute of Technology
¶¶University of California at Berkeley
***Univesity of California at Santa Cruz
†††Consiglio Nazionale delle Ricerche (IASF)
‡‡‡

Abstract. We report HETE-2 WXM and FREGATE observations of two X-ray flashes (XRFs), XRF010213 and XRF020903. The signal is only seen in < 25 keV and < 10 keV for XRF010213 and XRF020903 respectively. Both events show a double-peak structure in their light curves. The durations of the bursts are > 10 seconds, and this feature is similar to that of the "long" GRBs.

According to the time-averaged spectral analysis using both WXM and FREGATE data, the fluence ratio of 2–30 keV to 30–400 keV energy band is 11.4 and 5.6 for XRF010213 and XRF020903 respectively. The E_{peak} energy in the Band function is < 10 keV. They are likely to belong to the same class as the X-ray flash events detected with GINGA and BeppoSAX.

In this paper, we will present the detail study of the prompt emission of XRF010213 and XRF020903, and compare with the characteristics of classic GRBs.

XRF010213

The bright X-ray Flash XRF010213 was detected on 13 February 2001 at 12:35:35 UTC [1]. Since HETE was in the performance verification phase at that period, the localization was performed by the ground analysis. The celestial coordinate of this source is (R.A., Dec.) = ($10^h31^m36^s$, $5°30'30''$) (J2000) with a 95% error radius of $30'$. This large uncertainty is due to the unstable aspect of the spacecraft. No afterglow candidate was found for XRF010213.

TABLE 1. The spectral parameters of XRF010213 in the power-law, the power-law times exponential cutoff, and the Band function.

Model	parameters	region 1	region 2	region 3	all
Power-law	α	$-1.23^{+0.54}_{-0.51}$	$-2.19^{+0.10}_{-0.10}$	-2.89	$-2.47^{+0.08}_{-0.09}$
	K_{15}*	$5.65^{+2.99}_{-3.19} \times 10^{-3}$	$2.64^{+0.35}_{-0.34} \times 10^{-2}$	1.85×10^{-2}	$1.75^{+0.23}_{-0.21} \times 10^{-2}$
	χ^2_ν/DOF	0.964/19	1.293/41	2.363/36	1.561/45
Cutoff power-law	α		-1.00 (fixed)	-1.00 (fixed)	-1.00 (fixed)
	E_{peak} [keV]		$5.45^{+0.82}_{-0.69}$	$2.57^{+0.22}_{-0.21}$	$3.60^{+0.34}_{-0.31}$
	K_{15}		$0.29^{+0.04}_{-0.04}$	$1.26^{+0.18}_{-0.17}$	$0.42^{+0.06}_{-0.05}$
	χ^2_ν/DOF		1.057/41	1.114/36	1.144/45
Band	α		-1.00 (fixed)	-1.00 (fixed)	-1.00 (fixed)
	β		$-2.55^{+0.22}_{-0.60}$	< -3.61	$-2.96^{+0.23}_{-0.53}$
	E_{peak} [keV]		$4.81^{+1.06}_{-0.78}$	$2.56^{+0.22}_{-0.23}$	$3.41^{+0.35}_{-0.40}$
	K_{15}		$0.32^{+0.06}_{-0.06}$	$1.26^{+0.21}_{-0.15}$	$0.45^{+0.07}_{-0.06}$
	χ^2_ν/DOF		0.949/40	1.120/35	0.940/44

* normalization at 15 keV

Light curve and spectrum of XRF010213

As we can see in figure 1 (left), two peaks are see in the WXM energy range. The first peak is visible for all four WXM energy bands. On the other hand, the second peak is only seen in below 10 keV. The double peak structure is also seen in the FREGATE 6–40 keV light curve. There is no significant emission in the FREGATE 32–400 keV band. The duration in the X-ray range is ~ 40 seconds.

The spectral analysis is performed in the four time intervals including the whole burst region. The results of the spectral fitting in three spectral models (the power-law model, the cutoff power-law model, and the Band function) are summarized in table 1. The energy fluence ratio of 2–30 keV and 30–400 keV is 11.4. Since our definition of XRF is $S_{2-30} / S_{30-400} > 1$, where S_{2-30} and S_{30-400} are the energy fluence in 2–30 keV and 30–400 keV respectively, this event is classified as XRF. The spectrum at the second, the third, and the whole time regions apparently requires the break in the observed energy range. The E_{peak} energy is a few keV and its softening is seen from the second to the third time interval.

XRF020903

The event XRF020903 was detected by WXM and SXC at 10:05:37.96 UT on 2002 September 3 [2]. The WXM location can be express as a 90% confidence circle that is $16.6'$ in radius and is centered at (R.A., Dec.) = $(22^h49^m25^s, -20°53'59'')$ (J2000). The one-dimensional localization was possible using the SXC data. The combined the SXC and the WXM localization can be described as a 90% confidence quadrilateral (see figure 2 left).

Soderberg et al. [3] discovered an optical transient within the HETE-2 SXC + WXM

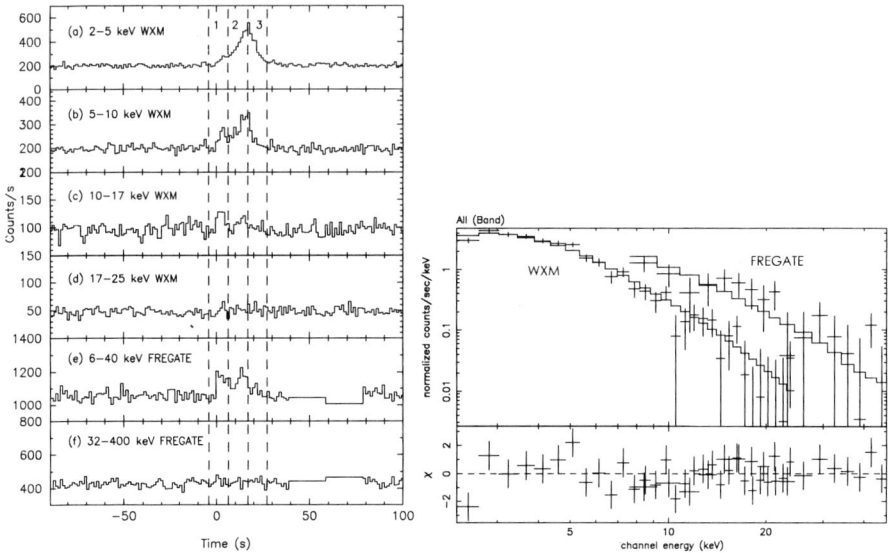

FIGURE 1. Left: The energy resolved light curves of XRF010213. The dashed lines represent the time intervals used in time-resolved spectral analysis. Right: The WXM+FREGATE joint fit spectrum of whole burst region. The spectral model is the Band function.

localization region at R.A. = $22^h48^m42.34^s$, Dec = $-20°46'09.3''$, using the Palomar 200-inch telescope. Spectroscopic observations of the optical transient, using the Magellan 6.5 m Baade and Clay telescopes, detected narrow emission lines from an underlying galaxy at the redshift $z = 0.25 \pm 0.01$, suggesting that the host galaxy of the optical transient is a star-forming galaxy. A fading bright radio source at the position of the optical transient was detected using the Very Large Array. Hubble Space telescope observations of the XRF 020903 field reveal the optical transient and show that its host galaxy is an irregular galaxy, possibly with four interacting components. These detections likely represent the first discoveries of the optical and radio afterglows of an XRF, together with its host galaxy.

Light curve and spectrum of XRF020903

The energy resolved light curves are shown in figure 2. There is no significant emission above 10 keV. The double peak structure is seen in the light curve.

The X-ray to the γ-ray fluence ratio is 5.6, therefore this burst is classified as an X-ray flash. Although the power-law model is acceptable for the prompt emission, the photon index larger than -2 is rejected with a high significance. If we assume that this XRF has the same spectral shape as ordinary GRBs, WXM and FREGATE observed the high energy portion of the Band function (photon index β). We applied the *constrained* Band function [2] to calculate the upper limit of the E_{peak} energy. The posterior probability

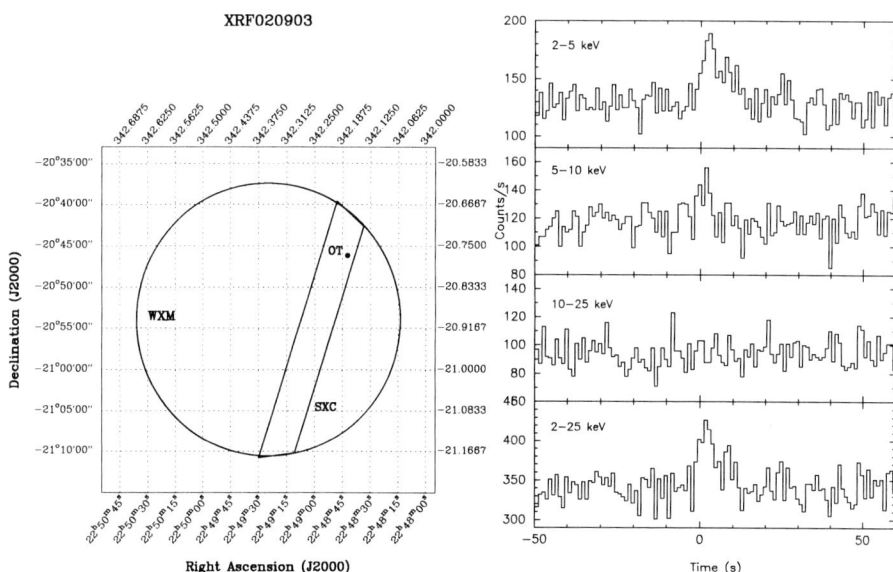

FIGURE 2. Left: The HETE-2 WXM/SXC localization for XRF020903. The point labeled "OT" is the location of the candidate optical and radio afterglow [3]. Right: The energy resolved WXM light curves of XRF020903.

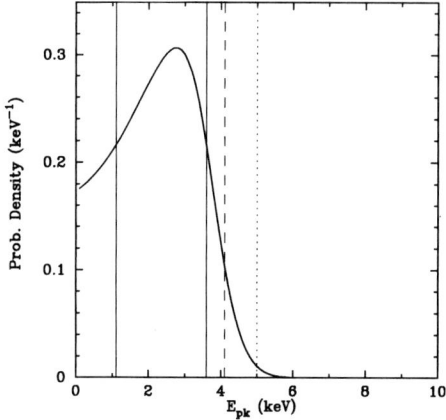

FIGURE 3. The posterior probability density distribution for E_{peak} in XRF020903. The solid lines define the 68% probability interval for E_{peak}, while the dashed and dotted lines show the 95% and 99.7% probability upper limits on E_{peak}.

density distribution is shown in the figure 3. We find a best-fit value $E_{peak} = 2.7$ keV, that 1.1 keV $< E_{peak} <$ 3.6 keV with 68% confidence limit, and that $E_{peak} <$ 4.1 and 5.0 keV with 95% and 99.7% confidence limit.

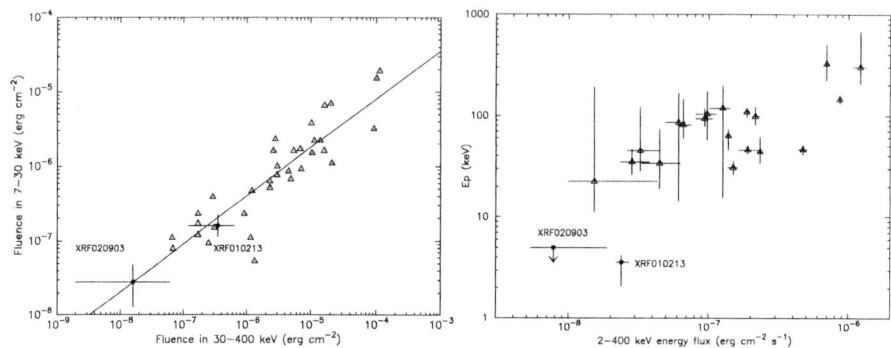

FIGURE 4. Left: The scatter plot of 30–400 keV (S_{30-400}) and 7–30 keV (S_{7-30}) fluences. The triangles are the 35 HETE/FREGATE GRBs studied by [4]. The solid line is the relation, $S_{7-30} = 3 \times 10^{-3} S_{30-400}^{0.643}$, found by [4]. Right: The 2-400 keV total energy flux vs. E_{peak}. The triangles are the HETE GRBs.

DISCUSSION AND SUMMARY

Figure 4 (left) shows the comparison of these two XRFs, XRF010213 and XRF020903, with ordinary GRBs in the (S_{30-400}, S_{7-30})-plane, where S_{30-400} and S_{7-30} are the energy fluence in the 30–400 and 7–30 keV bands. In figure 4 (right), the 2–400 keV energy fluxes and E_{peak} of two XRFs and GRBs are shown. XRF010213 and XRF020903 follow the same correlation with the GRBs.

XRF010213 and XRF020903 show similar characteristics with long GRBs. The HETE XRFs confirmed the spectral characteristics suggested by [5] that XRF has a lower E_{peak} energy. We also found the XRFs and GRBs follow the same correlation (figure 4). These observational properties may suggest that XRFs, X-ray-rich GRBs, and GRBs form a continuum and are a single phenomenon.

REFERENCES

1. G. Ricker, et al., GCN Circ. 934
2. T. Sakamoto, et al., ApJ, **602**, 875 (2004).
3. A. M. Soderberg, et al., submitted to ApJ, astro-ph/0311050 (2003).
4. C. Barraud, et al., A&A, **400**, 1021 (2003).
5. M. Kippen, et al., in proc. Gamma-Ray Bursts and Afterglow Astronomy, AIP Conf. Proc. **662**, 244 astro-ph/0111246 (2001).

Optical and X-ray Observations of the Afterglow to XRF 030723

N. Butler*, A. Dullighan*, P. Ford*, G. Ricker*, R. Vanderspek*, K. Hurley†, J. Jernigan†, D. Lamb** and C. Graziani**

*Center for Space Research, Massachusetts Institute of Technology, MA
†Space Sciences Laboratory, Berkeley, CA
**Department of Astronomy, University of Chicago, IL

Abstract. The X-ray-flash XRF 030723 was detected by the HETE satellite and rapidly disseminated, allowing for an optical transient to be detected ~ 1 day after the burst. We discuss observations in the optical with Magellan, which confirmed the fade of the optical transient. In a 2-epoch ToO observation with Chandra, we discovered a fading X-ray source spatially coincident with the optical transient. We present spectral fits to the X-ray data. We also discuss the possibility that the source underwent a rebrightening in the X-rays, as was observed in the optical. We find that the significance of a possible rebrightening is very low ($\sim 1\sigma$).

OBSERVATIONS

The X-ray-flash (i.e. for the fluence S, $\log[S_X(2-30 \text{ kev})/S_\gamma(30-400 \text{ kev})] > 0.0$) XRF 030723 was detected by the *HETE* satellite [1] with a $2'$ radius (90% confidence) SXC localization. An optical transient (OT) was reported approximately three days after the burst by Fox et al. [2]. These authors observed a fade from $R \sim 21.3$ by 1.1 mag between 1.23 and 2.23 days after the burst.

On 25 July 2003, the *Chandra Observatory* targeted the field of XRF 030723 for a 25 ksec (E1) observation spanning the interval 09:52-17:05 UT on 25 July, 51.4 - 59.0 hours after the burst. The SXC error circle from Prigozhin et al. [1] was completely contained within the field-of-view of the *Chandra* ACIS-I array. On 4 August 2003, *Chandra* re-targeted the field of XRF 030723 for an 85 ksec followup (E2) observation, spanning the interval 4 August 22:22 UT to 5 August 22:27 UT, 12.69 to 13.67 days after the burst. For this observation, the SXC error circle from Prigozhin et al. [1] was completely contained within the field-of-view of the *Chandra* ACIS-S3 chip.

From 24.8 hours to 25.2 hours after the burst (centered on July 24.31 UT), we observed the SXC error circle with the LDSS2 instrument on the 6.5m Magellan Clay telescope at Las Campanas Observatory in Chile. Four 6-minute Harris R-band exposures were taken in $\sim 0.6''$ seeing. Coaddition of the images gives a limiting magnitude of $R = 24.5$. On July 28.385 UT, 5.13 days after the burst, we again observed the the SXC error circle with Magellan. We obtained two 200-second exposures with the MagIC instrument in $\sim 0.8''$ seeing, reaching a limiting magnitude of $R = 24.3$.

TABLE 1. Source ("Cts") and background ("bg") counts and positions for the three *Chandra* sources detected within the SXC error region. We estimate a position uncertainty of 1.4″. Astrometry was performed using six stars from the USNO-A2 catalog.

#	Chandra Name	E1 Cts (bg)	E2 Cts (bg)
1	CXOU J214924.4-274248	78.5 (1.5)	75.6 (2.4)
3	CXOU J214926.9-274146	19.9 (3.1)	121.8 (4.2)
4	CXOU J214928.7-274211	16.2 (3.8)	98.1 (4.9)

CHANDRA DETECTION AND FITS

As reported in Butler et al. [3], 3 candidate sources were detected within the revised SXC error region in our E1 observation. Positions and other data for these sources are shown in Table 1. None of the sources were anomalously bright relative to objects in *Chandra* deep field observations [see, e.g., 4]. The brightest object within the SXC error circle (source #1), lies 62″ from the center of the SXC error circle, and is within 0.7″ of the optical transient reported by Fox et al. [2].

Table 1 shows the number of counts detected in E1 and in E2. The E2 observations were reported in Butler et al. [5]. Accounting for the difference in exposure times and sensitivity, the number of counts detected for a steady source in E2 should be ~ 6 times the number of counts detected in E1. Thus, sources 3 and 4 appear to have remained constant, while source 1 has faded. The number of counts detected in E2 corresponds to a $\sim 7\sigma$ significance decrease (i.e. factor of 6) in flux since the E1 observation.

To properly determine the fade factor we fit the E1 and E2 spectral data jointly. We reduce the spectral data using the standard CIAO[1] processing tools. We use "contamarf"[2] to correct for the quantum efficiency degradation due to contamination in the ACIS chips, important for energies below ~ 1 keV. We bin the data into 12 bins, each containing 12 or more counts, and we fit an absorbed power-law model by minimizing χ^2. The model has three parameters: two normalizations, and one photon index Γ. The absorbing column has been fixed at the Galactic value in the source direction, $N_H = 2.4 \times 10^{20}$ cm^{-2}. The model fits the data well ($\chi^2_\nu = 8.9/9$, Figure 1). The best fit photon number index is $\Gamma = 1.9 \pm 0.3$, which is a typical value for the X-ray afterglows of long duration GRBs [6]. Using this model, we find that the E1 flux is $(2.2 \pm 0.3) \times 10^{-14}$ erg cm^{-2} s^{-1} (0.5-8.0 keV band), while the E2 flux is $(3.5 \pm 0.5) \times 10^{-15}$ erg cm^{-2} s^{-1} (0.5-8.0 keV band). The decrease in flux between the two epochs can be described by a power-law with a decay index of $\alpha = -1.0 \pm 0.1$. This value of α is consistent with the power-law decline reported in the optical by Dullighan et al. [7] for $t \lesssim 1.5$ day after the GRB; however, the index is considerably flatter than the index at $t > 1.5$ days reported by Dullighan et al. [7]. This flatter X-ray decay may possibly be related to the rebrightening of the optical afterglow reported by Fynbo et al. [8].

---

[1] http://cxc.harvard.edu/ciao/
[2] http://space.mit.edu/CXC/analysis/ACIS_Contam/script.html

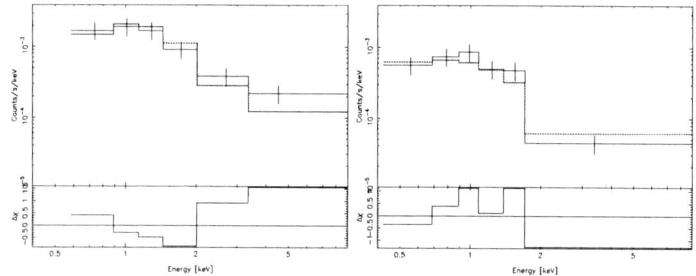

FIGURE 1. The counts in Epoch 1 (ACIS-I, left plot) and Epoch 2 (ACIS-S, right plot) are fitted simultaneously using an absorbed power-law model ($\chi_\nu^2 = 8.9/9$).

OPTICAL FADE, BREAK

We detected the OT of Fox et al. [2] in 2-epochs with Magellan, confirming those authors claims. Including the other detections reported over the GCN (Figure 2 (a)), we estimate a late time power law decay index of $\alpha \sim -2$, and an early power law decay of $\alpha \sim -0.9$. The break in the light curve occurs between 30-50 hours after the burst. Our measurements have been calibrated against the USNO photometry data reported Henden [9].

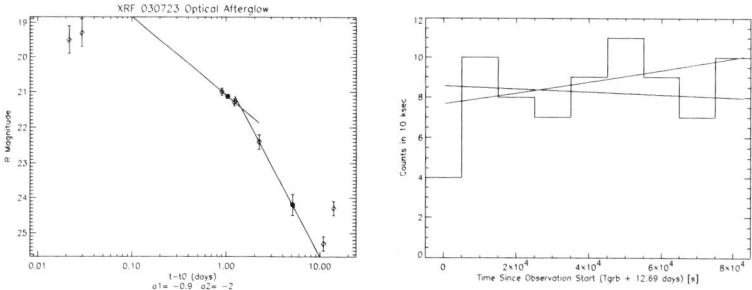

FIGURE 2. Left Plot: Optical light-curve in R-band taken from reports to the GCN [2, 10, 7, 11, 12, 8]. Our data are marked with stars. A temporal break may be present in the spectrum at $t \sim 1$ day [7]. The rightmost points have been argued to imply a rebrightening [8]. Right Plot: The Count rate during the *Chandra* E2 observation may be rising as was the optical flux during the same period. The significance of any rise is, however, $\lesssim 1\sigma$.

X-RAY REBRIGHTENING?

The afterglow emission in the optical was apparently rebrightening during our E2 *Chandra* observation [8]. We speculated above that the flat decay law we measure between E1 and E2 with *Chandra* may have been in part due to a rebrightening. To test whether this is or is not true, we test the E2 data against two hypotheses: (1) the count rate versus time is described by the power-law model which fits the overall

E1,E2 *decay* $r(t) = 8.56 \times 10^{-4} * (\frac{12.69\text{days}}{t})$ cts/s, (2) the count rate versus time is described by the power-law model which fits the optical *rise* during the E2 observation $r(t) = a * (\frac{t}{12.69\text{days}})^{3.7} cts/s$, where a is a free parameter. Using the arrival times t_i for 75 photons, we choose the model which maximizes the likelihood:

$$\mathscr{L}(t_1, t_2, ..., t_n) = r(t_1) \cdot r(t_2) \cdot ... \cdot r(t_n) \cdot \exp\left\{-\int_{t_0}^{t_n} r(\tau) d\tau\right\},$$

where the integral in the exponential is carried out for the good time intervals of *Chandra* data acquisition. We find a best fit value for a of 7.67×10^{-4} cts/s. (Figure 2 (b) shows the E2 counts in 10 ksec bins, with models (1) and (2) overplotted.) The corresponding difference in $\log(\mathscr{L})$ is 0.379. Simulating arrival times for 75 photons using model (1), a more extreme value of $\delta \log(\mathscr{L})$ found from fitting both models is observed to occur for approximately 1/3 of the trials. Thus, a rebrightening is preferred by the data, but at only 1σ significance.

CONCLUSIONS

We have derived power-law spectral parameters to X-ray data taken in two epochs with *Chandra* for XRF 030723. The photon index Γ we derive is a typical value for long-duration GRBs, possibly indicating a similarity between these objects an XRFs. The decrease in model normalization between the two epochs ($\Delta\chi^2 = 43.6$, for 1 additional degree of freedom; i.e. 6.6σ) confirms that source #1 is the X-ray counterpart to XRF 030723 and to the OT discovered by Fox et al. [2]. Our optical observations, along with the other observations reported over the GCN (Figure 2 (a)), imply a break in the R-band light curve at $t \sim 1$ day after the burst. We have tested for an X-ray rebrightening, but we find only very weak evidence for a rebrightening similar to that observed in the optical by Fynbo et al. [8].

REFERENCES

1. Prigozhin, G., et al., *GCN*, 2313 (2003).
2. Fox, D. B., et al., *GCN*, 2323 (2003).
3. Butler, N., et al., *GCN*, 2328 (2003).
4. Rosati, P., et al., *ApJ* 566, 667 (2002).
5. Butler, N., et al., *GCN*, 2347 (2003).
6. Costa, E., et al., *A&AS*, 138, 425 (1999).
7. Dullighan, A., et al., *GCN*, 2336 (2003).
8. Fynbo, J. P. U., et al., *GCN*, 2345 (2003).
9. Henden, A. 2003, *GCN*, 2343
10. Dullighan, A., et al., *GCN* 2326 (2003).
11. Bond, H. E. 2003, *GCN*, 2329
12. Smith, D. A., et al., *GCN*, 2338 (2003).

Cosmological X-Ray Flashes from Off-Axis Jets

Ryo Yamazaki[*], Kunihito Ioka[†] and Takashi Nakamura[*]

[*]*Department of Physics, Kyoto University, Kyoto 606-8502, Japan*
[†]*Department of Earth and Space Science, Osaka University, Toyonaka 560-0043, Japan*

Abstract. The $\langle V/V_{\max}\rangle$ of the cosmological X-ray flashes detected by WFC/*BeppoSAX* is calculated theoretically in a simple jet model. The total emission energy from the jet is assumed to be constant. We find that if the jet opening half-angle is smaller than 0.03 radians, off-axis emission from sources at $z \leq 4$ can be seen. The theoretical $\langle V/V_{\max}\rangle$ is less than 0.4, which is consistent with the observational result of 0.27 ± 0.16 at the 1 σ level. This suggests that the off-axis GRB jet with the small opening half-angle at the cosmological distance can be identified as the cosmological X-ray flash.

INTRODUCTION

The X-ray flash (XRF) is a class of X-ray transients, whose peak energy of νF_ν spectra is small but the other properties are roughly similar to those of GRBs [1, 2, 3]. The observational value of $\langle V/V_{\max}\rangle$ has been updated from 0.56 ± 0.12 [4] to 0.27 ± 0.16 [5]. The updated value of $\langle V/V_{\max}\rangle$ suggests that XRFs take place at a cosmological distance. Various models accounting for the nature of the XRFs have been proposed (see [6] and references therein; see also [7, 8]). In our *off-axis jet model*, if we observe the GRB jet with a large viewing angle, it looks like an XRF [9, 6]. In Yamazaki et al. [9], the value of the jet opening half-angle was adopted as $\Delta\theta = 0.1$. Then the distance to the farthest XRF ever detected is about 2 Gpc ($z \sim 0.4$) so that the cosmological effect is small and $\langle V/V_{\max}\rangle \sim 0.5$. Recent observations suggest that GRBs with relatively small opening angle exist, while the distribution of $\Delta\theta$ is not yet clear [10]. If we assume the total emission energy to be constant, the intrinsic luminosity is larger for the smaller $\Delta\theta$. Such GRBs at the cosmological distance observed from off-axis viewing angle may be seen as XRFs and $\langle V/V_{\max}\rangle$ is expected to be smaller than 0.5.

In this paper, we will show that our off-axis model has a possibility of accounting for the observational value of $\langle V/V_{\max}\rangle$ if we change some of the model parameters used in Yamazaki et al. [9].

CALCULATION OF $\langle V/V_{\max}\rangle$

We consider a simple jet model of XRFs [9, 6] taking into account the cosmological effect. The uniform jet with sharp edges is assumed. See Yamazaki et al. [6] for details. In order to study the dependence on the viewing angle θ_v and the jet opening half-angle $\Delta\theta$, we fix the other parameters as $\alpha_B = -1$, $\beta_B = -3$, $\gamma\nu'_0 = 200\,\text{keV}$, $r_0/c\beta\gamma^2 = 10\,\text{s}$,

TABLE 1. Results of calculation for fixed $\Delta\theta$

$\Delta\theta$	A_0*	$\theta_{v,p}$†	$z_{\max}(\theta_{v,p})$	$z_{\min}(\theta_{v,p})$	$\langle V/V_{\max}\rangle_{\Delta\theta}$**
0.10	0.84	0.103	2.8	1.5	0.46
0.09	1.0	0.095	2.9	1.4	0.45
0.08	1.3	0.086	3.0	1.4	0.44
0.07	1.7	0.077	3.1	1.3	0.44
0.06	2.3	0.068	3.3	1.2	0.44
0.05	3.4	0.060	3.5	1.2	0.44
0.04	5.2	0.052	3.6	1.1	0.43
0.03	9.3	0.045	3.8	0.99	0.40
0.02	22	0.038	4.0	0.89	0.38
0.01	109	0.034	4.1	0.77	0.35

* In units of erg cm^{-2} Hz^{-1}
† The viewing angle where $W(\theta_v)$ becomes maximum.
** For the XRFs detected by WFCs on BeppoSAX.

and $\gamma = 100$. We fix the amplitude A_0 so that the isotropic γ-ray energy $E_{\rm iso} = 4\pi d_L^2(1+z)^{-1}S_\gamma$ satisfies $(\Delta\theta)^2 E_{\rm iso}/2 = 0.5 \times 10^{51}$ ergs, when $\theta_v = 0$ and $z = 1$ [11]. The values of A_0 for different opening angles are summarized in Table 1. When the jet opening half-angle $\Delta\theta$ becomes smaller, A_0 becomes larger.

The $\langle V/V_{\max}\rangle$ for fixed opening half-angle $\Delta\theta$ is calculated as

$$\langle V/V_{\max}\rangle_{\Delta\theta} = \frac{\int \langle V/V_{\max}\rangle_{\Delta\theta,\theta_v} W(\theta_v)\,d\theta_v}{\int W(\theta_v)\,d\theta_v}, \quad (1)$$

where $\langle V/V_{\max}\rangle_{\Delta\theta,\theta_v}$ is for fixed $\Delta\theta$ and θ_v (See Yamazaki et al. [6] for details). The weight function $W(\theta_v)$ is the product of the solid angle factor and the volume factor:

$$W(\theta_v) = 2\pi \sin\theta_v \int_{z_{\min}(\theta_v)}^{z_{\max}(\theta_v)} dz \frac{n(z)}{1+z} 4\pi \left(\frac{d_L}{1+z}\right)^2 \frac{d}{dz}\left(\frac{d_L}{1+z}\right), \quad (2)$$

where $n(z)$ and d_L are the comoving GRB rate density and the luminosity distance, respectively. Here $z_{\max}$ ($z_{\min}$) is the maximum (minimum) redshift of the XRF for given $\Delta\theta$ and θ_v. In determining $z_{\min}$ and $z_{\max}$, we should note that the operational definition of the *BeppoSAX*-XRF is the fast X-ray transient with duration less than $\sim 10^3$ seconds which is detected by WFCs and not detected by the GRBM [1]. Therefore, if the sources are nearby such that $z < z_{\min}$, they are observed as GRBs because the observed fluence in the γ-ray band becomes larger than the limiting sensitivity of GRBM ($\sim 3 \times 10^{-6}$ ergs cm^{-2}). If the sources are too far such that $z > z_{\max}$, they cannot be observed by WFCs with a limiting sensitivity of about 4×10^{-7} ergs cm^{-2}. The behavior of $z_{\max}$, $z_{\min}$, $\langle V/V_{\max}\rangle_{\Delta\theta,\theta_v}$, and $W(\theta_v)$ for $\Delta\theta = 0.03$ are shown in Figure 1.

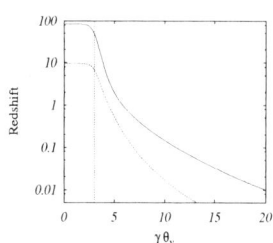 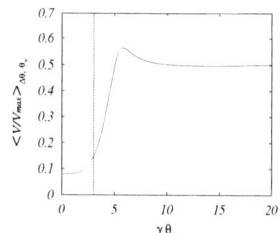 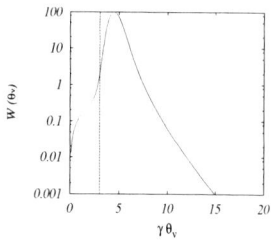

FIGURE 1. (Left panel): The maximum (minimum) redshift, $z_{\max}$ ($z_{\min}$), of the XRF as a function of the viewing angle $\gamma\theta_v$ is shown as the solid line (dashed line). The jet emission is observed as the XRF if the source has a redshift z in the range $z_{min} < z < z_{max}$. (Midle panel): $\langle V/V_{\max}\rangle_{\Delta\theta,\theta_v}$ for the XRF detected by the WFCs/*BeppoSAX* is shown as a function of $\gamma\theta_v$. (Right panel): The weight function $W(\theta_v)$ (arbitrary normalization), which is the relative observed event rate, is shown as a function of $\gamma\theta_v$. All figures are for the case of $\Delta\theta = 0.03$. The vertical dashed lines represent $\theta_v = \Delta\theta = 0.03$.

DISCUSSION

The results of the numerical integration are summarized in Table 1. For each $\Delta\theta$, $z_{\max}(\theta_{v,p})$ [$z_{\min}(\theta_{v,p})$] means the maximum (minimum) redshift where $W(\theta_v)$ takes the maximum value. If we take the jet opening half-angle as $\Delta\theta \leq 0.03$, $\langle V/V_{\max}\rangle_{\Delta\theta}$ is smaller than ~ 0.4, which is consistent with the observational result at the 1 σ level. The value of $\Delta\theta \sim 0.03$ is as low as the minimum of those having ever been inferred from afterglow light curve. The jet break time is given by $t_j \sim 13 \min (\Delta\theta/0.01)^{8/3}$ [11] and so it requires fast localization to observe the jet break for a narrow jet. Therefore, at present, the small number of GRBs with small $\Delta\theta$ may come from the observational selection effect. In the context of this scenario, we might be able to account for the fact that afterglows of XRFs have been rarely observed since the afterglow at a fixed time gets dimmer for an earlier break time. Furthermore, some "dark GRBs" might be such a small opening angle jet observed with an on-axis viewing angle for the same reason.

Table 1 shows that the sources with the viewing angle $\theta_{v,p} \sim \Delta\theta + 0.02$, where $W(\theta_v)$ takes maximum, are the most frequent class of the XRFs in the population for $\Delta\theta \leq 0.03$. The typical observed photon energy is estimated as [9, 6]

$$E_p \sim 2\gamma v_0'(1+z)^{-1}[1+(\gamma\theta_{v,p}-\gamma\Delta\theta)^2]^{-1} \sim 30 \text{ keV } [(1+z)/2.5]^{-1}, \quad (3)$$

which is the typical observed peak energy of the XRFs [3]. We can propose from our argument that the emissions from the jets with a small opening half-angle such as $\Delta\theta \leq 0.03$ are observed as XRFs when they are seen from off-axis viewing angle. If one can detect the afterglow of the XRF, which has the maximum flux at about several hours after the XRF, the fitting of light curve may give us the key information about the jet opening angle [12]. For example, the light curve of afterglow of XRF 030723 was unusual in early epoch, which can be well explained if the jet is seen from off-axis viewing angle [13].

We can estimate the observed event rate of the XRF for fixed $\Delta\theta$ as $R^{\text{XRF}}_{\Delta\theta} = (1/4\pi)\int W(\theta_v)d\theta_v$. For a reasonable proportionality constant, we derive $R^{\text{XRF}}_{\Delta\theta=0.03} \sim 10^2$ events yr^{-1}, which is consistent with the observation. The value of $R^{\text{XRF}}_{\Delta\theta}$ remains unchanged within a factor of 2 when we vary $\Delta\theta$ from 0.01 to 0.07.

When the jet opening half-angle $\Delta\theta$ has a distribution $f_{\Delta\theta}$, we integrate $\langle V/V_{\max}\rangle_{\Delta\theta}$ and $R_{\Delta\theta}^{\mathrm{XRF}}$ over the distribution of $\Delta\theta$ as

$$\langle V/V_{\max}\rangle \propto \int d(\Delta\theta)\, f_{\Delta\theta} R_{\Delta\theta}^{\mathrm{XRF}} \langle V/V_{\max}\rangle_{\Delta\theta}, \qquad (4)$$

$$R_{\mathrm{XRF}} \propto \int d(\Delta\theta)\, f_{\Delta\theta} R_{\Delta\theta}^{\mathrm{XRF}}, \qquad (5)$$

respectively [6]. When we adopt a power-low distribution as $f_{\Delta\theta} \propto (\Delta\theta)^{-q}$, with $q = 4.54$ [11] and integrate over $\Delta\theta$ from 0.01 to 0.2 rad, we find $\langle V/V_{\max}\rangle = 0.36$ and $R_{\mathrm{XRF}} \sim 10^2$ events yr^{-1}. These values mainly depend on the lower cut-off of $f_{\Delta\theta}$. For example, we obtain $\langle V/V_{\max}\rangle = 0.43$ and $R_{\mathrm{XRF}} \sim 3$ events yr^{-1} if the integration is done over $\Delta\theta$ from 0.03 to 0.2 rad. Hence, we might be able to determine the lower cut-off if the other uncertain factors are fixed by other arguments. Since the statistics of the observational data will increase in the near future owing to instruments such as *HETE-2* and *Swift*, we will be able to say more than in the above discussion, including giving a more accurate functional form of $f_{\Delta\theta}$ than that we have considered above, as well as the relation to the GRB event rate.

ACKNOWLEDGMENTS

This work was supported in part by Grant-in-Aid for Scientific Research of the Japanese Ministry of Education, Culture, Sports, Science and Technology, No.05008 (R.Y.), No.00660 (K.I.), No.14047212 (T.N.), and No.14204024 (T.N.).

REFERENCES

1. Heise, J. et al., in Proc. 2nd Rome Workshop Gamma-Ray Bursts in the Afterglow Era, asto-ph/0111246 (2001).
2. Kawai, N. 2004, in these proceedings (2004).
3. Kippen, R. M., et al., in Proc. Woods Hole Gamma-Ray Burst Workshop, astro-ph/0203114 (2002).
4. Heise, J., talk in the 2nd Workshop "Gamma-Ray Burst in the Afterglow Era", Rome (2000).
5. Heise, J., talk in the 3rd Workshop "Gamma-Ray Burst in the Afterglow Era", Rome (2002).
6. Yamazaki, R., Ioka, K., & Nakamura, T., ApJ, **593**, 941 (2003).
7. Lamb, D.Q. 2003, in these proceedings.
8. Mochkovitch, R., in these proceedings.
9. Yamazaki, R., Ioka, K., & Nakamura, T., ApJ, **571**, L31 (2002). o
10. Panaitescu, A. & Kumar, P., ApJ, **571**, 779 (2002).
11. Frail, D., A., et al., ApJ, **562**, L55 (2001).
12. Granot, J., Panaitescu, A., Kumar, P.,& Woosley, S. E., ApJ, **570**, L61 (2002).
13. Huang, Y. F., et al., astro-ph/0309360 (2003).

Comparing Prompt Emission from X-ray Flashes and Gamma-ray Bursts

R. M. Kippen[*], J. J. M. in 't Zand[**], P. M. Woods[§], J. Heise[**], R. D. Preece[§] and M. S. Briggs[§]

[*]*Los Alamos National Laboratory, ISR-2, MS B244, Los Alamos, NM 87545, USA*
[†]*SRON National Institute for Space Research, Sorbonnelaan 2, 3584 CA Utrecht, The Netherlands*
[**]*Astronomical Institute, Utrecht University, P.O. Box 80 000, 3508 TA Utrecht, The Netherlands*
[‡]*Universities Space Research Association, Huntsville, AL 35806, USA*
[§]*National Space Science & Technology Center, 320 Sparkman Dr., Huntsville, AL 35805, USA*
[¶]*University of Alabama in Huntsville, Huntsville, AL 35899, USA*

Abstract. The final year of the *Beppo*SAX mission provided a much needed clue as to the nature of X-ray flashes. The detection of afterglow counterparts and their underlying hosts provides strong evidence that X-ray flashes and gamma-ray bursts originate from similar sources in cosmologically distant galaxies. These observations support findings that the prompt emission characteristics of X-ray flashes are similar to those of traditional gamma-ray bursts. Using wide-band observations from *Beppo*SAX and BATSE, we present the latest results in our on-going effort to quantify the similarities and differences in prompt emission characteristics.

INTRODUCTION

The revolution in gamma-ray burst (GRB) astronomy prompted by the discovery of multiwavelength afterglow counterparts has brought tremendous progress in understanding the nature of burst progenitors, their surrounding environments, and host galaxies. Understanding of the mechanisms giving rise to the prompt burst emission itself, however, is comparatively confused despite a large amount of observational data. This paper summarizes our ongoing efforts [1, 2, 3, 4] to investigate one revealing characteristic that could help this situation — the lower energy form of the GRB emission process — by comparing prompt GRB properties to those of the so-called X-ray flashes (XRFs). The analysis is based on a sample of XRFs that were selected using *Beppo*SAX Wide Field Camera (WFC; 2–26 keV) X-ray observations, but also detected in 20–300 keV gamma rays through an off-line scan of BATSE data.

OBSERVATIONS

Our test sample is based on the events identified, selected, and classified using the *Beppo*SAX WFC and GRB Monitor (GRBM) instruments. In this scheme, XRFs are differentiated from GRBs based on the lack of GRBM (40–400 keV) detection. Using simultaneous BATSE observations we can reveal wide-band spectral properties and make a more quantitative comparison between XRFs and "traditional" GRBs.

TABLE 1. WFC/BATSE Observation Summary 21-Apr-91 to 26-May-00.

WFC Classification	WFC Detections	Observable with BATSE	BATSE Triggers	BATSE Off-line Detections
GRB	32	21	18	21
XRF	15	9	0	9
Questionable	7	4	0	4
Total	54	34	18	34

Apart from observational outages, the WFC and BATSE instruments operated simultaneously for 3.8 years, ending with the termination of BATSE science operations on 26-May-2000. For all of the GRB-like transient events detected by WFC, we performed an off-line search of the >20 keV BATSE data. Results of the search are listed in Table 1. Not all events were observable with BATSE due to data gaps and Earth occultation. The result is that all GRB-like WFC events observable with BATSE were detected with $\geq 5\sigma$ statistical significance in the off-line search. The list of detections includes four questionable events, three of which are likely long-duration (~ 1000 s) GRBs [5], and one is likely a Type I X-ray burst. These questionable events are excluded from further analysis. For the remaining 21 GRB and 9 XRF we have a complete set of WFC+BATSE data to use in comparing XRF and GRB properties.

SPECTRAL ANALYSIS

For each of the 30 XRF and GRB events we computed standard parameters (peak flux, fluence, and duration) using the same processes developed for the BATSE GRB catalogs. Furthermore, the WFC and BATSE data were used to jointly estimate the time-averaged, 2 keV to 2 MeV spectrum of each event. Four separate spectral models were used in this process: black body (BB), power law (PL), power law times exponential (COMP), and Band's GRB function.

Based on the chi-squared values for the various models, we make the following conclusions. First, none of the GRB or XRF events are consistent with the BB model, as is expected for non-thermal GRB-like spectra. Second, for most GRB events (19 of 21), and three of the XRF events, a single PL model can be rejected with good confidence. This is typical of GRBs, which usually have strongly curved (i.e., non power-law) broadband spectra. Finally, the change in chi-squared from a power law to a COMP or Band model is statistically significant for most of the GRB and XRF events. This is an important indication that curved spectra are favored for XRFs just as they are for traditional GRBs.

XRFS VS. BRIGHT GRBS

To compare the spectral properties of XRFs to those of well-measured (i.e., bright) GRBs we use the 21 WFC-selected GRBs and the Preece et al. [6] catalog of 156 bright BATSE GRBs (BBGs). Figure 1 compares the distributions of Band-model spectral parameters for the three event samples (similar results were obtained with the COMP model). We use the Kolmogorov-Smirnov (KS) test on unbinned data to compare the different distributions. The statistical significance of the observed deviations between distributions is evaluated through Monte Carlo simulations that account for the sample sizes as well as the measured statistical uncertainties in spectral parameters. The results are that the α and β parameters are reasonably consistent between XRFs and GRBs, but XRFs have significantly lower E_{peak} than GRBs.

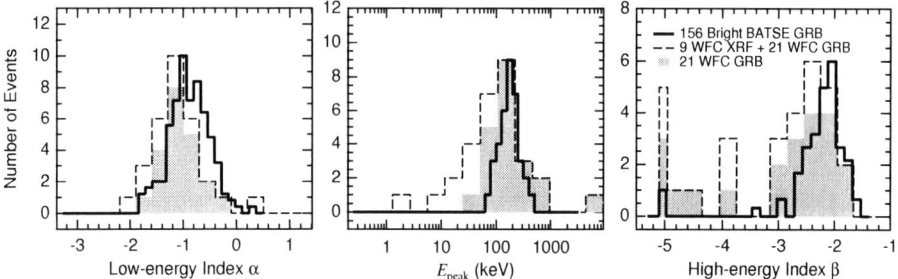

FIGURE 1. Distributions of time-averaged Band model spectral parameters for WFC-selected XRFs and GRBs compared to bright BATSE GRBs (normalized to peak of dashed curves).

XRFS VS. DIM GRBS

The above bright burst comparison ignores the known GRB hardness–intensity correlation. It is therefore important to compare XRFs to weak GRBs that have similar (gamma-ray) brightness. To do this we use the Mallozzi et al. [7, MAL] burst sample, which includes 523 long-duration ($T_{50} > 1$ s) BATSE bursts that were fit using the Band spectral model. Figure 2 compares these bursts with the WFC-selected events. The hardness–intensity correlation is evident in WFC GRBs and MAL GRBs.

To compare XRFs and dim GRBs we first modeled the hardness-intensity (E_{peak} vs. peak flux) correlation using a power-law fit to binned GRB data (including statistical uncertainties). The power-law was then extrapolated into the intensity regime of the XRFs, assuming different models for the GRB intensity (LogN–LogP) distribution. Finally, the KS test was used to compare the unbinned XRF data to the extrapolated GRB E_{peak} distribution. This analysis indicates that (within sizable uncertainties) XRFs and extrapolated dim WFC GRBs are statistically consistent, with chance KS probability $P_{\text{KS}} \approx 0.2$–$0.4$ (depending on the choice of LogN–LogP). The extrapolated BATSE GRBs, however, have significantly larger E_{peak} than XRFs.

The above comparison ignores differences in selection biases between the WFC and BATSE samples. Starting with the BATSE power-law fit, we simulated the effect of the WFC selection bias assuming (1) α and β are independent of burst intensity and described by the BBG distributions of Figure 1, (2) random burst directions over the WFC field of view, and (3) the approximate WFC trigger criteria. The effect of this simulated bias on the power-law hardness–intensity correlation is indicated on the right-most plot in Figure 2. Including this bias, the E_{peak} distribution of XRFs and extrapolated

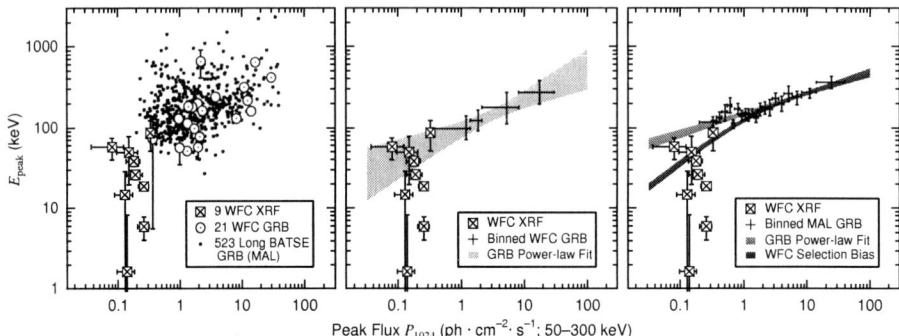

FIGURE 2. Long-duration GRB hardness–intensity data (left) binned and fit to a power law (center). The rightmost plot shows BATSE GRB data with a power-law fit. The lower most curve indicates the effect of a simulated WFC selection bias on the power-law correlation.

BATSE-selected dim GRBs are statistically consistent, with $P_{KS} \sim 0.1$ (depending on the choice of LogN–LogP).

CONCLUSION

While XRF-like events have been detected by *Ginga* in the past and HETE-II at present, the WFC+BATSE sample probably offers the greatest broad-band sensitivity. The nine jointly observed XRFs therefore represent a unique resource for comparing prompt XRF and GRB behavior.

Our results indicate that the prompt, broad-band emission from XRFs is quantitatively consistent with that expected from weak, long-duration, traditional GRBs. Combined with their similar temporal properties, this strongly suggests that XRFs and long GRBs are produced by a continuous variation of the same phenomenon. The detailed hardness–intensity correlation over the full range spanned by XRFs and GRBs presumably provides a valuable clue as to the nature of this phenomenon.

REFERENCES

1. Heise, J., in 't Zand, J., Kippen, R. M., & Woods, P. M., in *Gamma-Ray Bursts in the Afterglow Era*, ed. E. Costa, F. Frontera, & J. Hjorth, Springer ESO Astrophysics Symposia, Berlin, 2001, p. 16.
2. Kippen, R. M., Woods, P. M., Heise, J., et al., in *Gamma-Ray Bursts in the Afterglow Era*, ed. E. Costa, F. Frontera, & J. Hjorth, Springer ESO Astrophysics Symposia, Berlin, 2001, p. 22.
3. Kippen, R. M., Woods, P. M., Heise, J., et al., in *Gamma-Ray Burst and Afterglow Astronomy 2001*, ed. G. R. Ricker & R. K. Vanderspek, AIP Conf. Proc. 662, New York, 2003, p. 244.
4. Kippen, R. M., et al., *ApJ* in preparation.
5. in 't Zand, J. J. M., Guidorzi, C., & Kippen, R. M., in *Proc. Gamma-Ray Bursts in the Afterglow Era: 3rd Workshop*, in press (astro-ph/0305361).
6. Preece, R. D., Briggs, M. S., Mallozzi, R. S., et al., *ApJSS* **126**, 19 (2000).
7. Mallozzi, R. S., Pendleton, G. N., Paciesas, W. S., et al., in *Gamma-Ray Bursts: 4th Huntsville Symp.*, ed. C. Meegan et al., AIP Conf. Proc. 428, New York, 1998, p. 273.

ULTRA-HIGH ENERGY GAMMA-RAYS, NEUTRINOS, GRAVITY WAVES

Ultra-high Energy Gamma-rays, Neutrinos, and Gravitational Waves from GRBs

P. Mészáros*†, S. Kobayashi*, S. Razzaque* and B. Zhang*

Dept. Astronomy & Astrophysics Pennsylvania State University
†*Institute for Advanced Study, School of Natural Sciences, Princeton,*

Abstract. Most current knowledge about GRBs is based on sub-MeV electromagnetic signals. However, some observations and recent theoretical work indicate that important clues may be gleaned from their GeV-TeV photon spectra. In addition, one expects GRB to be prominent in two other potentially important non-electromagnetic channels, gravitational waves and TeV or higher energy neutrinos. The former is a natural consequence of compact merger or collapse scenarios, while the latter is associated with a hadronic component in the outflow. The information in these channels and prospects for their detection are discussed.

UHE PHOTONS FROM GRB

Ultra-high energy emission, in the range of GeV and harder, is expected from electron inverse Compton (IC) in external [1] and internal shocks [2] in the prompt phase. The combination of prompt MeV radiation from internal shocks and a more prolonged GeV IC component for external shocks [3] is a likely explanation for the delayed GeV emission seen in some GRB [4]. GeV photon emission should also arise from the long-term IC component in external afterglow shocks [5, 6, 7]. Another possible contributor at these energies may be π^0 decay from $p\gamma$ interactions between shock-accelerated protons and photons in the GRB shocks [8, 9, 10]. However, under the conservative assumption that the relativistic proton energy does not exceed the energy in relativistic electrons and that the proton spectral index is -2.2 instead of -2, both the proton synchrotron and the $p\gamma$ components are substantially less at GeV-TeV than the IC component [5]. Another GeV photon component is expected in baryonic GRB outflows when neutrons decouple from protons, before any shocks occur. The *pn* inelastic collisions lead to pions, including π^0, resulting in UHE photons which cascade down to the GeV range [11, 12]. The final GeV spectrum results from a complex cascade, but a rough estimate indicates that 1-10 GeV flux should be detectable with GLAST for bursts at $z \lesssim 0.1$ [12].

At the high photon densities implied by GRB models, $\gamma\gamma$ absorption inside the source is important [13, 14]. The observation of photons up to 10-20 GeV with EGRET puts a lower limit on the bulk Lorentz factor, since the compactness parameter (optical depth to $\gamma\gamma$) is proportional to the comoving photon density, which depends on the bulk Lorentz factor. This gives [14] lower limits on $\Gamma \sim 300 - 600$ for a number of bursts observed with EGRET.

A tentative $\gtrsim 0.1$ TeV detection at the 3σ level of GRB970417a has been reported with the water Cherenkov detector Milagrito [15]. Another possible TeV detection [16]

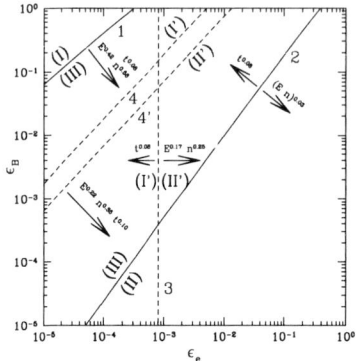
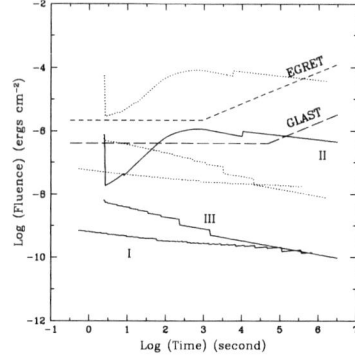

FIGURE 1. *Left panel*: Parameter regimes for GRB GeV emission dominated by proton synchrotron (I), electron synchrotron (III) and electron inverse Compton (II). *Right panel*: GeV lightcurves for the three types of bursts in the range 400 MeV - 200 GeV. The solid curves indicate bursts in regimes I (p-sy), II (e-IC) and III (e-sy), at a distance $z = 1$. The three dotted unmarked curves are the same but located at $z = 0.1$. Also shown are the sensitivity curves for *EGRET* and *GLAST*

of GRB971110 has been reported with the GRAND array, at the 2.7σ level. Better sensitivity is expected from the upgraded MILAGRO, as well as from atmospheric Cherenkov telescopes under construction such as VERITAS, HESS, MAGIC and CANGAROO-III [17]. However, GRB detections in the TeV range are expected only for rare nearby events, since at this energy the mean free path against $\gamma\gamma$ absorption on the diffuse IR photon background is $\sim$ few hundred Mpc [18, 19]. The mean free path is much larger at GeV energies, and based on the handful of GRB reported in this range with EGRET, several hundred should be detectable with large area space-based detectors such as GLAST [20, 5].

GRAVITATIONAL WAVES FROM GRB

The ultimate energy source of GRB is thought to be either stellar collapse or compact stellar mergers, and both of these are expected to be sources of gravitational waves (GWs). A time-integrated GW luminosity of the order of a solar rest mass ($\sim 10^{54}$ erg) is predicted from merging NS-NS and NS-BH models [21, 22, 23], while the luminosity from collapsar models is more model-dependent, but expected to be lower ([24, 25]; c.f. [26]). We have estimated the strains of gravitational waves from current GRB progenitor systems [27]. The expected detection rates of gravitational wave events with LIGO from compact binary mergers, in coincidence with GRBs, has been estimated by [28, 29]. If some fraction of GRBs are produced by double neutron star or neutron star - black hole mergers, the chirp signal of the in-spiral phase should be detectable by the advanced LIGO within one year, associated with the GRB electromagnetic signal. We have also estimated the signals from the black hole ring-down phase, as well as the possible contribution of a bar configuration from gravitational instability in the accretion disk following tidal disruption or infall in GRB scenarios. Other binary progenitor scenarios, such black hole – Helium star and black hole – white dwarf merger GRB progenitors are unlikely to be detectable, due to the low estimates obtained for the maximum non-axisymmetrical perturbations.

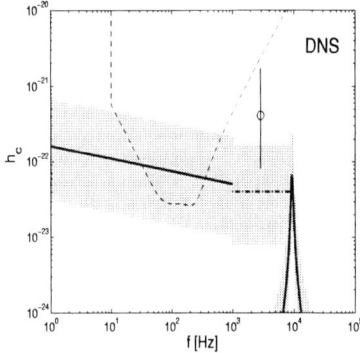
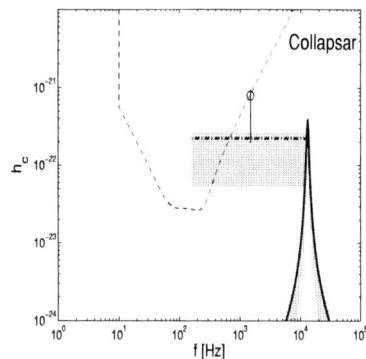

FIGURE 2. *Left panel:* Characteristic GW strains for merging double neutron stars: in-spiral (solid line), merger (dashed dotted line), bar (circle), ring-down (solid spike). Also shown is the advanced LIGO nose curve $\sqrt{fS_h(f)}$ (dashed curve). The shaded region and the vertical line reflect the uncertainty of the formation rate (see [27] for details). *Right panel*: Characteristic GW strains for collapsars: blob merger (dashed dotted line), bar (circle) and ring-down (solid spike).

For the collapsar scenario of GRB, the non-axisymmetrical perturbations may be stronger [30, 31, 32], and the estimated formation rates are much higher than for other progenitors [31, 33], with typical distances correspondingly much nearer to Earth. Collapsars are of special interest, since they have received observational support from GRB afterglow observations. In the absence of detailed numerical 3D calculations of collapsars, we have estimated [27] the strongest signals that might be expected for bar instabilities in the accretion disk around the resulting black hole, and in the maximal version of the fragmentation scenario of the infalling core. Although the waveforms of the gravitational waves produced in the break-up, merger and/or bar instability phase of collapsars are not known, a cross-correlation technique can be used making use of two co-aligned detectors. Under these assumptions, collapsar GRB models would be expected to be marginally detectable as gravitational wave sources by the advanced LIGO within one year of observations.

For binaries a matched filtering technique can be used, while for collapsars, where the wave forms are uncertain, the simultaneous detection by two elements of a gravitational wave interferometer, coupled with electromagnetic coincident detections, provides a possible detection technique. We have made [27] specific detection estimates for both the compact binary scenarios and the collapsar scenarios,

UHE NEUTRINOS FROM GRB

Stellar merger or collapse scenarios for GRB naturally result in large thermal ($\sim 10-30$ MeV) neutrino luminosities comparable to those in supernovae. At typical redshifts

$z \sim 1$ these are very hard to detect due to the low cross neutrino sections at these energies. However, the neutrino detection cross section increases with energy, and near TeV energies there are realistic chances for detection [34] with cubic kilometer ice or water Cherenkov detectors, such as the planned ICECUBE or ANTARES.

Neutrinos of energy $\gtrsim 100$ TeV are expected from p, γ interactions between relativistic protons accelerated in internal or external shocks, a high collision rate being ensured by the large photons density. The interaction between MeV photons in internal shocks and relativistic protons accelerated by the same shocks [35] leads to charged pions, muons and neutrinos. This p, γ reaction peaks at the Δ resonance in the fluid frame moving with Γ, satisfying the condition $\varepsilon_p \varepsilon_\gamma \gtrsim 0.2 \text{GeV}^2 \Gamma^2$. For observed 1 MeV photons this implies $\gtrsim 10^{16}$ eV protons, and neutrinos with $\sim 5\%$ of that energy, $\varepsilon_\nu \gtrsim 10^{14}$ eV in the observer frame. Above this threshold, the typical spectrum per decade is flat, $\varepsilon_\nu^2 \Phi_\nu \sim$ constant. Synchrotron and adiabatic losses limit the muon lifetimes [36], leading to a suppression of the flux above $\varepsilon_\nu \sim 10^{16}$ eV. In external shocks, the most targets are O/UV photons from the afterglow reverse shock (e.g. as in the GRB 990123 prompt flash [37]). Here the resonance condition implies higher energy protons, and neutrinos of $10^{17} - 10^{19}$ eV [38, 39]. These neutrino fluxes are detectable above the atmospheric background with the planned ICECUBE detector [34]. Limits to their total contribution to the diffuse ultra-high energy neutrino flux can be derived from observed cosmic ray and diffuse gamma-ray fluxes [40, 41, 42].

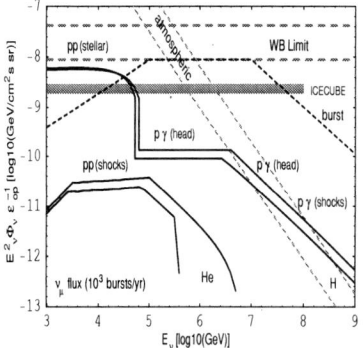 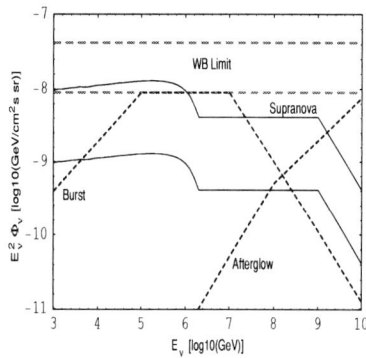

FIGURE 3. *Left*: Diffuse muon-neutrino flux $E_\nu^2 \Phi_\nu \varepsilon_{op}^{-1}$ shown as solid lines from sub-stellar jet shocks in two progenitor models, $r_{12.5}$ (H) and r_{11} (He). These neutrinos are precursors (10-100 s before) of the γ-ray bright bursts. Also shown are the diffuse neutrinos arriving simultaneously with the γ-rays from shocks outside the stellar surface in observed GRB (dark short-dashed curve); the Waxman-Bahcall (WB) diffuse cosmic ray bound (light long-dashed curves); and the atmospheric neutrino flux (light short-dashed curves). For a hypothetical 100:1 ratio of γ-ray dark (in which the jets do not emerge) to γ-ray bright collapses, the solid neutrino spectral curves would be 100 times higher. *Right*: Diffuse neutrino flux $(E_\nu^2 \Phi_\nu)$ from supra-nova models of GRBs (solid curves), assuming that (top curve) all GRBs have an SNR shell, or (bottom) 10% of all GRBs have an SNR shell, and 10^{-1} of the fireball protons reach the shells. Long dashed lines correspond to the WB cosmic-ray limit, short dashed curves are the diffuse ν flux from GRB internal shocks and afterglows.

One also expects p, γ interactions occuring *inside* collapsars as the jet is burrowing its way out of the star [43], before it breaks through the stellar envelope to produce a GRB outside (or also if the jet stalls). The buried jet produces $\sim$ keV thermal X-rays which interact with $\gtrsim 10^{14}$ eV protons accelerated in internal shocks occurring in the jet inside the star, producing $\gtrsim$ few TeV neutrinos for tens of seconds, which penetrate the envelope. Another contribution of TeV ν_μ is due to pp, pn collisions between buried jet relativistic protons and thermal nucleons in the envelope [44]. At TeV energies the number of neutrinos is larger for the same total energy output, which improves the detection statistics. The rare bright, nearby or high γ collapsars could occur at the rate of $\sim$ 10/year, including both γ-ray bright GRBs (where the jet broke through the envelope) and γ-ray dark events where the jet is choked (failed to break through), and both such γ-bright and dark events could have a TeV neutrino fluence of $\sim$ 10/neutrinos/burst, detectable by ICECUBE in individual bursts.

Supernovae are expected to be associated with massive stellar collapse GRBs. In one extreme version, a SN explosion is postulated to occur weeks before the GRB ("supranova" model, [45]). In this scenario the pre-ejected SNR shell provides nucleon targets for pp interactions with protons accelerated in the MHD wind of a pre-GRB pulsar [46], leading to a 10 TeV neutrino precursor to the GRB. The SNR also provides additional target photons for $p\gamma$ interactions [46] with internal shock-accelerated protons, resulting in $\sim 10^{16}$ eV neutrinos. We have calculated the neutrino fluxes and muon event rates from individual bursts as well as the diffuse contribution [47]. The neutrinos from pp and $p\gamma$ between GRB relativistic protons and SNR shell target protons and photons will be contemporary with the GRB electromagnetic event. Depending on the fraction of GRB with SNR shells, their contribution to the GRB diffuse neutrino flux has a pp component which is relatively stronger at TeV-PeV energies than the internal shock $p\gamma$ component of [35], and a shell $p\gamma$ component which is a factor 1 (0.1) of the internal shock $p\gamma$ component (Fig. 3) for a fraction 1 (0.1) of GRB with SNR shells. Our pp component is caused by internal shock-accelerated power-law protons contemporaneous with the GRB event, differing from [46] who considered quasi-monoenergetic $\gamma_p \sim 10^{4.5}$ protons from an MHD wind over 4π leading to a $\sim$ 10 TeV neutrino months-long precursor of the GRB. Our $p\gamma$ component arises from the same GRB-contemporaneous internal shock protons interacting with thermal 0.1 keV photons within the shell wall, whereas [46] consider such protons interacting with photons from the MHD wind inside the shell cavity. These neutrino fluxes provide a test for the supranova hypothesis, the predicted event rates being detectable with kilometer scale planned Cherenkov detectors.

ACKNOWLEDGMENTS

This research is supported partly through NASA NAG5-13286, NSF AST-0098416, NSF PHY 01-14375, and the Monell Foundation.

REFERENCES

1. Mészáros, P., Rees, M. J. & Papathanassiou, H. ApJ, **432**, 181 (1994)
2. Papathanassiou, H. & Mészáros, P. ApJ, **471**. L91 (1996)
3. Mészáros, P. & Rees, M. J. MNRAS, **269**, L41 (1999)
4. Hurley, K. et al. Nature, **372**, 652 (1994)

5. Zhang, B. & Mészáros, P. ApJ, **559**, 110 (2001)
6. Sari, R. & Esin, A. A. ApJ, **548**, 787 (2001)
7. Derishev, E. V., Kocharovsky, V. V. & Kocharovsky, Vl. V. A&A, **372**, 1071 (2001)
8. Böttcher, M. & Dermer, C. D. ApJ, **499**, L131 (1998)
9. Totani, T. ApJ, **502**, L13 (1998)
10. Fragile, P. C. et al., AstroParticle Phys. in press (astro-ph/0206383) (2003)
11. Derishev, E. V., Kocharovsky, V. V. & Kocharovsky, Vl. V. ApJ, **521**, 640 (1999)
12. Bahcall, J. N. & Mészáros, P. Phys. Rev. Lett., **85**, 1362 (2000)
13. Baring, M. G. in GeV-TeV Gamma-Ray Astrophysics Workshop (eds. Dingus, B. L. et al.), AIP Conf. Proc., 515, 238 (2000)
14. Lithwick, Y. & Sari, R. ApJ, **555**, 540 (2001)
15. Atkins, R. et al. ApJ, **533**, L119 (2000)
16. Poirier, et al. Phys. Rev. D., in press (astro-ph/0004379) (2003)
17. Weekes, T. in *Heidelberg 2000, High energy gamma-ray astronomy*, AIP Conf. Proc. 558, eds. Aharonian, F. & Völk, H. (AIP:NY), 15 (2000)
18. Coppi, P. & Aharonian, F. ApJ, **487**, L7 (1997)
19. de Jager, O. C. & Stecker, F. W. ApJ, **566**, 738 (2002)
20. Gehrels, N. & Michelson, P. AstroParticle Phys. **11**, 277 (1999)
21. Eichler, D., Livio,M., Piran,T. & Schramm,D.N., Nature, **340**, 126 (1989)
22. Paczynski,B., Acta Astronomica, **41**, 257 (1991)
23. Ruffert, M. & Janka, H. T. A&A, **338**, 53 (1998)
24. Fryer, C., Woosley, S. & Heger, A. ApJ, **550**, 327 (2001)
25. Dimmelmeier, H., Font, J. & Mueller, E. ApJ, **560**, L163 (2001)
26. van Putten, M. H. P. M., ApJ, **562**, L51 (2001)
27. Kobayashi, S., Mészáros, P., ApJ, **589**, 861 (2003)
28. Finn, L. S., Mohanty, S. D., Romano, J. D., Phys. Rev. D, **60**, 121101 (1999)
29. Finn, L. S., Krishnan, B. & Sutton, P. J. Class. Quant. Grav., **20**, S815 (2003)
30. Davies, M. B., King, A., Rosswong, S., & Wynn, G. ApJ, **579**, L63 (2002)
31. Fryer, C. L., Holz, D. E. & Hughes, S. A., ApJ, **565**, 430 (2002)
32. van Putten, M. H. P. M., ApJ, **575**, L71 (2002)
33. Belczynski, K., Bulik, T., Rudak, B., ApJ, **571**, 394 (2002)
34. Halzen, F., in *Weak Interactions & Neutrinos*, Procs 17th Int. Wkshp (World Sci:Singapore), p.123 (2000)
35. Waxman, E., & Bahcall, J.N. Phys. Rev. Lett., **78**, 2292 (1997)
36. Rachen, J. & Mészáros, P. Phys. Rev. D, **58**, 123005 (1998)
37. Akerlof, C. W. Nature, **398**, 400 (1999)
38. Waxman, E., & Bahcall, J.N. ApJ, **541**, 707 (1999)
39. Vietri, M. Phys. Rev. Lett., **80**, 3690 (1998)
40. Waxman, E., & Bahcall, J.N. Phys. Rev. D, **59**, 023002 (1999)
41. Bahcall, J.N. & Waxman, E. Phys. Rev. D. **64**, 023002 (2001)
42. Mannheim K. in *Heidelberg 2000, High energy gamma-ray astronomy*, AIP Conf.Proc. 558, eds. Aharonian F & Völk H (AIP:NY) p.417 (2001)
43. Mészáros, P. & Waxman, E. Phys. Rev. Lett. **87**, 171102 (2001)
44. Razzaque, S., Mészáros, P. & Waxman, E. Phys. Rev. D68, 083001 (2003)
45. Vietri, M. & Stella, L. ApJ, **507**, L45 (1998)
46. Granot, J. & Guetta, D. Phys. Rev. Lett. **90**, 191102 (2003); Guetta, D. & Granot, J. Phys. Rev. Lett. **90**, 201103 (2003)
47. Razzaque, S., Mészáros, P. & Waxman, E. Phys. Rev. Lett. **90**, 241103 (2003)

Milagro–A TeV Observatory for Gamma Ray Bursts

B.L. Dingus* and the Milagro Collaboration[†]

Los Alamos National Laboratory
[†]*University of Maryland, University of California Santa Cruz, University of California Irvine, New York University, University of Wisconsin, University of New Hampshire, George Mason University*

Abstract. Milagro is a large field of view ($\sim$ 2 sr), high duty cycle ($\sim$90%), ground-based observatory sensitive to gamma-rays above $\sim$100 GeV. This unique detector is ideal for observing the highest energy gamma-rays from gamma-ray bursts. The highest energy gamma rays supply very strong constraints on the nature of gamma-ray burst sources as well as fundamental physics. Because the highest energy gamma-rays are attenuated by pair production with the extragalactic infrared background light, Milagro's sensitivity decreases rapidly for bursts with redshift > 0.5. While only 10 % of bursts have been measured to be within z=0.5, these bursts are very well studied at all wavelengths resulting in the most complete understanding of GRB phenomena. Milagro has sufficient sensitivity in units of E^2 dN/dE to detect VHE luminosities lower than the observed luminosities at $\sim$ 100 keV for these nearby bursts. Therefore, the launch of SWIFT and its ability to localize and measure redshifts of many bursts points to great future possibilities.

IMPORTANCE OF HIGH ENERGY OBSERVATIONS

The observations of very high energy (VHE), >$\sim$ 100 GeV, flux from GRBs are few. At slightly lower energies, EGRET observed GRBs emission up to $\sim$ 20 GeV [1] and no cutoff up to 10 GeV in the average prompt spectrum of differential spectral index -1.95 $\pm$ 0.25 [2]. Recently a GRB has been reported with an even brighter (at least 3 times the fluence) higher energy component that begins at a few MeV and extends to at least 200 MeV with a differential spectral index of -1.0 $\pm$ 0.3 [3]. Milagrito, a small prototype of Milagro, observed an excess from the GRB on 17 April 1997 which implied more VHE fluence than the MeV fluence detected by BATSE [4], but the significance of the detection was marginal ($\sim$ 3 σ). Also, air Cherenkov telescopes have attempted to look for VHE afterglow emission, but have not detected any (e.g. [5]).

However, VHE gamma rays are a natural byproduct of most GRB production models and are often predicted to have comparable fluence at TeV and MeV scales (e.g. [6], [7], [8]). This is due to the fact that the MeV emission from GRBs is likely due to synchrotron radiation produced by highly accelerated electrons within the strong magnetic field of a jet with bulk Lorentz factors exceeding 100. In such an environment, the inverse Compton mechanism for transferring energy from electrons to gamma rays is likely to complement synchrotron radiation and produce a second VHE component of GRB emission with fluence possibly peaked at 1 TeV or beyond. Whether or not the inverse Compton mechanism contributes minimally or even dominates the energy production depends on the environment of the particle acceleration and the gamma-ray production. VHE measurements may be critical to the understanding of gamma-ray production in GRBs similar to the manner in the TeV measurements have resolved the degeneracy between magnetic field and electron energy in blazars. In addition, GRBs will likely accelerate hadrons-maybe even producing the ultra high energy cosmic rays [9], [10]. Hadrons will create TeV gamma rays via cascades made by photo-pion production or possibly through synchrotron emission of protons [11].

VHE emission from GRBs is attenuated by interactions with the extragalactic infrared background light producing electron-positron pairs. The amount of absorption is uncertain because the infrared photon density cannot be directly measured due to the foreground of our own galaxy. Several models exist for the infrared light which is due to reprocessed starlight and therefore depends on stellar and galactic evolution at earlier epochs. Unfortunately, this reduces the sensitivity of VHE observatories, especially at higher energies. Fortunately, VHE detections and spectra also constrain the infrared photon density. Thus, VHE observations at different redshifts contribute to our fundamental understanding of the evolution of the Universe.

Fundamental physics can also be probed by the detection of VHE emission from a GRB, by providing the most sensitive measurement to date for the constancy of the speed of light as a function of energy. Some quantum gravity theories predict a breakdown of Lorentz invariance observable as an energy dependency of the speed of light, which can be written as $v = c(1 - \xi E_\gamma / E_{QG})$ [12], where ξ is of order 1 and E_{QG} is the energy scale at which quantum gravity becomes important. Based on this formula the following figure of merit may be derived – $Q = 4 \times 10^{17} z E_{GeV} / \Delta t_{sec}$, where z is the redshift to the source, E_{GeV} is the energy of the photons detected in GeV, and Δt_{sec} is the duration of the event in seconds. The best limit on E_{QG} ($> 6 \times 10^{16}$ GeV) was derived from a 30 minute flare from the active galaxy Mrk 421 detected in the TeV energy band by the Whipple air Cherenkov telescope [13]. A 1-second gamma-ray burst at a redshift of 0.3, detected above 300 GeV would be sensitive to effects where E_{QG} is above 10^{19} GeV, the Planck scale. Thus if a satellite-based gamma-ray observatory can measure both the redshift and lightcurve of the burst, a VHE detection would probe the nature of spacetime at energies near or beyond the Planck scale.

MILAGRO'S CAPABILITIES

Milagro is a new type of ground-based gamma-ray observatory which detects the particles in the extensive air showers created when a VHE gamma ray interacts with the Earth's atmosphere. This technique has been proven successful with Milagro's detection and VHE flux measurement of the Crab nebula [14] and Mrk 421 [15]. Milagro is located in the Jemez mountains near Los Alamos, New Mexico and has been operational since January 2000. Recently the detector has been upgraded to improve the angular resolution, the energy resolution, the cosmic-ray background rejection efficiency, and to lower the energy threshold. Individual showers are reconstructed with an average angular resolution of 0.5^o, which depends on the number of the particles in the shower and the primary energy of the gamma ray.

Milagro is ideally suited to observe VHE emission from GRBs. The alternative technique of detecting VHE gamma rays from the Cherenkov light created in the atmosphere by the particles in the extensive air show is more sensitive, but less well suited to observations of GRBs due to the small field of view of a few square degrees and the low duty cycle of 5-10%. In contrast, Milagro's average uptime was > 90% for the last year, and the field of view is $\sim$ 2 sr. Milagro's effective area is $\sim$ 10 m^2 at 100 GeV increasing to greater than 10^4 m^2 at a few TeV. The background rate of cosmic-rays is $\sim$ 1700/sec within this 2 sr field of view, but only $\sim$ 3/sec are consistent with any individual point source.

Using the known background rate and the Monte Carlo simulated effective area (which was confirmed by the determination of the known Crab flux), Milagro's sensitivity to a 100 second duration gamma-ray burst is quantified in Figure 1 as a function of the GRB redshift. The extragalactic, infrared background light model of [16] is used to attenuate the observed spectrum at different redshifts. The plotted lines are for different zenith angle ranges and give the minimum required luminosity in units of E^2 dN/dE emitted at the source at 100 GeV assuming a differential photon spectrum of E^{-2}. The field of view decreases slightly at higher redshifts because the energy threshold of Milagro increases as the effective depth of the atmosphere increases at larger zenith angles.

The VHE luminosity of GRBs in unknown, but if the VHE luminosity is comparable to the luminosity emitted at $\sim$ 100 keV (shown as triangles in Figure 1) and if GRB luminosity is not strongly correlated with redshift, then Milagro has excellent sensitivity to GRBs with $z < 0.5$ and some sensitivity to $z \sim 1$. Over 10% of GRBs with measured redshifts are nearer than $z = 0.5$, and given Milagro's field of view is one sixth of the sky, Milagro's sensitivity in units of E^2 dN/dE is lower the detected isotropic luminosities for a few percent of the GRB detected.

Milagro has searched for VHE emission from GRBs detected by satellites; however, fewer than 40 bursts have been localized to be within Milagro's field of view during its operation from January 2000 until present. This number 40 can be compared with the 54 bursts searched for VHE emission with the Milagrito VHE gamma-ray detector, of which one candidate was found, GRB970417a [4]. None of these bursts have been detected by Milagro, but one of the bursts was especially interesting due to its nearby redshift. An upper limit was derived from the Milagro data for this GRB at a redshift of 0.45 on 21 September 2001 assuming the infrared extragalactic background absorbs all gamma-rays above 150 GeV [17]. Milagro's upper limit for this burst is below the extrapolation of the spectrum measured by HETE, and this upper limit was announced promptly via a GCN Circular.

The Milagro data itself is also being searched online for transients of duration from 250 microseconds up to 2 hours. The results of the search are emailed within a few seconds to Milagro collaboration members for verification. No significant transient sources have been detected, and the nondetection has been used to place an upper limit on the number of TeV emitting GRBs which depends on the redshift distribution of GRBs [18] [19].

Milagro will continue operation during the SWIFT mission. Milagro's continuous operation and large field of view guarantees approximately one sixth of GRBs will be observed before, during, and after the SWIFT detection. Milagro will use the GCN to disseminate VHE detection information such as photon flux and duration. We will also

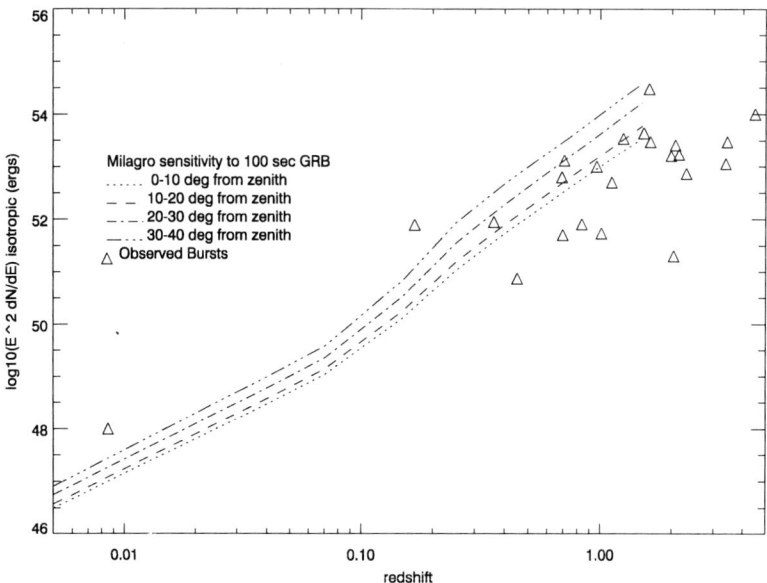

FIGURE 1. Isotropic luminosity required for a 5 σ detection by Milagro for a 100 sec duration gamma-ray burst at different redshifts. The differential photon GRB spectrum is assumed to be E^{-2} and is attenuated by extragalactic, infrared, background light as predicted in [16]. Triangles indicate the isotropic luminosity of GRBs observed at $\sim$100 keV.

directly notify other interested observers–for example, TeV air Cherenkov telescopes, which have sensitivities (in E^2 dN/dE units) comparable to past X-ray afterglow detections. For nondetections of VHE emission, we will update a web page listing all SWIFT GRBs observed giving flux upper limits for different durations. Also, we will continue searching our own data for any evidence of a VHE-detected GRB, and will promptly notifying the community of any positive detections.

CONCLUSION

Milagro is a new type of large field of view, VHE gamma-ray detector that has been proven to work with the detection of astrophysical sources. This new technique is essential to detect VHE gamma rays from gamma-ray bursts, and Milagro has sufficient sensitivity to test theoretical predictions of GRB emission, as well as the possibility to make fundamental observations. We eagerly await the launch of SWIFT, and the many localizations of GRBs to follow.

ACKNOWLEDGMENTS

We acknowledge Scott Delay and Michael Schneider for their dedicated efforts in the construction and maintenance of the Milagro experiment. This work has been supported by the National Science Foundation (under grants PHY-0070927, -0070933, -0075326, -0096256, -0097315, -0206656, -0302000, and ATM-0002744) the US Department of Energy (Office of High-Energy Physics and Office of Nuclear Physics), Los Alamos National Laboratory, the University of California, and the Institute of Geophysics and Planetary Physics.

REFERENCES

1. Hurley, K., et al., Nature, **372**, 652. (1994).
2. Dingus,B.L., *High Energy Gamma Ray Astronomy*, ed. Aharonian, F.A., Volk, H.J., AIP Vol. **558**, 383 (2001).
3. Gonzalez, M.M., Dingus, B.L., Kaneko, Y., Preece, R.D., Dermer, C.D., & Briggs,M.S., Nature, **424**, 749 (2003).
4. Atkins, R., et al. (The Milagro Collaboration), 2003, ApJ, **583**, 824.
5. Connaughton, V. et al., ApJ **479**, 859 (1997).
6. Dermer, C.D., Chiang, J., & Mitman, K.E., ApJ, **537**, 785 (1999).
7. Pilla, R.P., & Loeb, A. ApJ Lett. **494**, L167. (1998).
8. Zhang, B. & Meszaros, P., ApJ, **559**, 110. (2001).
9. Waxman, E., Phys.Rev.Lett., **75**, 386 (1995).
10. Wick, S.D. & Dermer, C.D., these proceedings (2003).
11. Dermer, C.D., these proceedings (2003).
12. Amelino-Camelia, G., et al., Nature, **393**, 319 (1998).
13. Biller, S.D., et al., Phys Rev Lett, **83** (11), 2108 (1999).
14. Atkins, R., et al. (The Milagro Collaboration), ApJ, **595**, 803 (2003).
15. Sinnis, C. et al. (The Milagro Collaboration), *28th International Cosmic Ray Conference*, Tsukuba, Japan OG2.3, 2583 (2003).
16. Primack, J., et al., *High Energy Gamma Ray Astronomy*, ed. Aharonian, F.A., Volk, H.J., AIP Vol. **558**, 463 (2001).
17. McEnery, J.E., et al. (The Milagro Collaboration), *Gamma-Ray Burst and Afterglow Astronomy*, eds Ricker, G. & Vanderspeck, R., AIP Vol 662, 529 (2003).
18. Atkins, R. et al., (the Milagro Collaboration), submitted to ApJ Lett. (astro-ph/0311389) (2003).
19. Morales, M.M. et al., (the Milagro Collaboration), these proceedings (2003).

Gravitational Radiation from Gamma-Ray Burst Progenitors

Shiho Kobayashi* and Peter Mészáros*

*Center for Gravitational Wave Physics, Pennsylvania State University, University Park, PA 16802
Department of Astronomy & Astrophysics Pennsylvania State University, University Park, PA 16802*

Abstract. We study gravitational radiation from various proposed gamma-ray burst progenitor models, in particular compact mergers and massive stellar collapses. These models have in common a high angular rotation rate, and the final stage involves a rotating black hole and accretion disk system. We consider the in-spiral, merger and ringing phases, and for massive collapses we consider the possible effects of asymetric collapse and break-up, as well bar-mode instabilities in the disks. We evaluate the order-of-magnitudes of the strain and frequency of the gravitational waves expected from various progenitors, at distances based on occurrence rate estimates. Based on simplifying assumptions, we give estimates of the probability of detection of gravitational waves by the advanced LIGO system from the different gamma-ray burst scenarios. We discuss possible correlations between the burst photon luminosity, or the delay between gravitational wave bursts and X-ray flashes, and the polarization degree of the gravitational waves.

INTRODUCTION

It is widely accepted that Gamma-Ray Bursts (GRBs) are the results of catastrophic events involving either compact stellar mergers or massive stellar collapses . Some arguments suggest that GRBs are powered by accretion disks and that the accretion timescales determine the durations (for reviews, see [1, 2]). The observed energy in GRBs requires a massive ($\geq 0.1 - 1 M_\odot$) disk. Such a massive disk can form from the fall-back of debris during the formation of the compact object itself, which ultimately is likely to be a newborn black hole. Several scenarios could lead to a black hole-massive accretion disk system. This includes the merger of double neutron star binaries, neutron star - black hole binaries, black hole - white dwarf binaries, black hole - helium star binaries.

A coincidence between a gravitational wave signal and a gamma-ray signal would greatly enhance the statistical significance of the detection of the gravitational wave signal [3, 4]. Therefore, it is of interest to study the gravitational wave emission from GRB associated with specific progenitors. Another reason for doing this is that, since the γ-rays and the afterglow are thought to be produced at very large distances ($\geq 10^{13}$cm) from the central engine, we have only very indirect information about the nature of the latter. However, gravitational waves should be emitted from the immediate neighborhood of the GRB central engine itself, and their observation should give valuable information about its identity.

TABLE 1. Formation Rates and Distances

	Formation Rate [Myr^{-1}galaxy^{-1}]		Distance [Mpc]	
	Standard	Range	Standard	Range
DNS	1.2	0.01-80	220	53-1100
BH-NS(a)	2.6	0.001-50	170	62-2300
(b)	0.55	0.001-50	280	62-2300
BH-WD	0.15	0.0001-1	430	230-4900
BH-He	14	0.1-50	95	62-490
Collapsar	630	10-1000	27	23-110

EMISSION MECHANISMS

The process of binary coalescence as a gravitational wave source is in principle simpler to analyze than that of a massive stellar collapse, although they share some common features, especially in the later phases. The binary coalescence process can be divided into three phases: in-spiral, merger and ring-down (e.g., [5, 6]).

Collapsars, i.e. massive stellar collapses leading to a GRB, require a high core rotation rate, which may be easier to achieve if the star is in a binary system, although this is not necessary (e.g. [7]). The high rotation rate is required to form a centrifugally supported disk around a central, possibly spinning black hole, to power a GRB jet. A high rotation rate, however, may be conducive to the development of bar or fragmentation instabilities in the collapsing core or/and in the massive disk around the central object [8, 9, 10, 11, 12, 13] (however, see [14, 15] also). The asymmetrically infalling matter also perturbs the black hole's geometry, which leads to ring-down gravitational radiation. The gravitational wave emission from collapsars can thus in principle be estimated in a similar way to what is done in binaries during the in-spiral, merger and ring-down phases, although with considerably larger uncertainties. (see [6] for details).

DETECTION

Recently the formation rate of GRB progenitors has been estimated by using population synthesis methods [16, 17]. The results of [16] are summarized in Table 1, where the standard values of the formation rates and the uncertainty ranges are listed. Assuming the galaxy density $n_{glx} = 0.02$ Mpc^{-3}, we can estimate the distance inside which an event is expected to happen within in a year from the formation rates R. The estimates on the formation rates of [17] are consistent with the results of [16] and within the uncertainty range in table 1 in most of their models. Though some of their models predict higher formation rates by a factor of a few than the upper limits in table 1, the uncertainty range of the distances are similar because the distances are rather insensitive to the rate $d \propto R^{-1/3}$. The leading models for the ultimate energy source of GRB are stellar collapse or compact stellar mergers, and these are expected to be sources of gravitational waves (GWs). A time-integrated GW luminosity of the order of a solar rest mass ($\sim 10^{54}$ erg) is predicted from merging NS-NS and NS-BH models, while the luminosity from

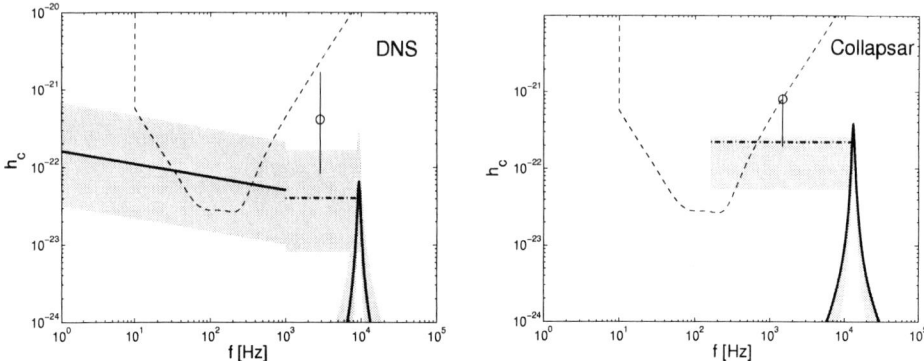

FIGURE 1. *Left panel:* Characteristic GW strains for merging double neutron stars : in-spiral (solid line), merger (dashed dotted line), bar (circle), ring-down(solid spike). Also shown is the advanced LIGO nose curve $\sqrt{fS_h(f)}$ (dashed curve). The shaded region and the vertical line reflect the uncertainty of the formation rate R in Table 1. *Right panel:* Characteristic GW strains for collapsars: blob merger (dashed dotted line), bar (circle) and ring-down(solid spike).

collapsar models is more model-dependent, but expected to be lower. We have estimated the strains of gravitational waves from some of the most widely discussed current GRB progenitor stellar systems [6]. The expected detection rates of gravitational wave events with the Laser Interferometric Gravitational Wave Observatory (LIGO) from compact binary mergers, in coincidence with GRBs, has been estimated by. If some fraction of GRBs are produced by double neutron star or neutron star – black hole mergers, the gravitational wave chirp signal of the in-spiral phase should be detectable by the advanced LIGO within one year, associated with the GRB electromagnetic signal. We have also estimated the signals from the black hole ring-down phase, as well as the possible contribution of a bar configuration from gravitational instability in the accretion disk following tidal disruption or infall in GRB scenarios. Other binary progenitor scenarios, such black hole – Helium star and black hole – white dwarf merger GRB progenitors, are unlikely to be detectable, due to the low estimates obtained for the maximum non-axisymmetrical perturbations.

For the massive rotating stellar collapse (collapsar) scenario of GRB, the non-axisymmetrical perturbations may be stronger, and the estimated formation rates are much higher than for other progenitors, with typical distances correspondingly much nearer to Earth. This type of progenitor is of special interest, since it has so far received the most observational support from GRB afterglow observations. For collapsars, in the absence of detailed numerical 3D calculations specifically aimed at GRB progenitors, we have estimated the strongest signals that might be expected in the case of bar instabilities occurring in the accretion disk around the resulting black hole, and in the maximal version of the recently proposed fragmentation scenario of the infalling core. Although the waveforms of the gravitational waves produced in the break-up, merger and/or bar instability phase of collapsars are not known, a cross-correlation technique can be used making use of two co-aligned detectors. Under these assumptions, collapsar GRB models would be expected to be marginally detectable as gravitational wave sources by the advanced LIGO within one year of observations.

In the case of binaries the matched filtering technique can be used, while for sources such as collapsars, where the wave forms are uncertain, the simultaneous detection by two elements of a gravitational wave interferometer, coupled with electromagnetic simultaneous detection, provides a possible detection technique. We have made more specific detection estimates for both the compact binary scenarios and the collapsar scenarios.

CORRELATIONS

Both the compact merger and the collapsar models have in common a high angular rotation rate, and observations provide evidence for jet collimation of the photon emission, with properties depending on the polar angle, which may also be of relevance for X-ray flashes. We have considered the gravitational wave emission and its polarization as a function of angle which is expected from such sources [18]. The GRB progenitors emit gravitational waves, which are circularly polarized on the polar axis, while the + polarization dominates on the equatorial plane [4]. Recent GRB studies suggest that the wide variation in the apparent luminosity of GRBs are caused by differences in the viewing angle, or possibly also in the jet opening angle. Since GRB jets are launched along the polar axis of GRB progenitors, correlations among the apparent luminosity of GRBs ($L_\gamma(\theta) \propto \theta^{-2}$) and the amplitude as well as the degree of linear polarization P degree of the gravitational waves, $P \propto \theta^4 \propto L_\gamma^{-2}$, are expected. At a viewing angle larger than the jet opening angle θ_j the GRB γ-ray emission may not be detected. However, in such cases an "orphan" long-wavelength afterglow could be observed, which would be preceded by a pulse of gravitational waves with a significant linearly polarized component. As the jet slows down and reaches $\gamma \sim \theta_j^{-1}$, the jet begins to expand laterally, and its electromagnetic radiation begins to be observable over increasingly wider viewing angles. Since the opening angle increases as $\sim \gamma^{-1} \propto t^{1/2}$, at a viewing angle $\theta > \theta_j$, the orphan afterglow begins to be observed (or peaks) at a time $t_p \propto \theta^2$ after the detection of the gravitational wave burst. The polarization degree and the peak time should be correlated as $P \propto t_p^2$.

A new type of fast transient source, called "X-ray flashes", have recently been observed with the Beppo-SAX satellite. The nature of these sources is not yet fully understood. Although they could be a totally different astrophysical phenomenon, it has been suggested that these events may be GRBs with large viewing angles. If this is the case, linearly polarized gravitational waves should be observed prior to the X-ray flashes. The degree of polarization should be positively correlated with longer delays and with the softness of the X-ray flashes, which increase with angle. Since the degrees of linear and circular polarization depend on the viewing angle, a determination of the polarization degree would be a measure of the viewing angle. Such measurements, which are likely to require the advent of a future generation of detectors, could provide a new a tool for estimating the absolute luminosity of GRBs, including its photon component. By comparing the estimated absolute photon luminosity with the apparent luminosity, the distance to the source may be estimated independently of any redshift measurement. No optical afterglows have been found for about half of all the GRBs detected by Beppo-SAX (the so called "dark GRBs"), and the present method would have the potential to help determine or constrain the distances to such dark GRBs.

ACKNOWLEDGMENTS

We acknowledge support through the Center for Gravitational Wave Physics, which is funded by NSF under cooperative agreement PHY 01-14375, and through NSF AST0098416 and NASA NAG5-9192.

REFERENCES

1. Piran,T., Phys. Rep., **333**, 529 (2000)
2. Mészáros ,P., ARA&A, **40**, 137 (2002)
3. Finn, L. S., Mohanty, S. D., Romano, J. D., Phys. Rev. D, **60**, 121101 (1999)
4. Kochanek, C., Piran, T., ApJ, **417**, L17 (1993)
5. Flanagan,E.E., Hughes, S.A. Phys. Rev. D, **57**, 4535 (1998)
6. Kobayashi, S, Mészáros , P., ApJ, **589**, 861 (2003)
7. Woosley, S., in Gamma-Ray Bursts in the Afterglow Era, ed. E. Costa, F.Frontera & J. Hjourth, Springer, 257 (2001)
8. Nakamura,T., Fukugita,M., ApJ, **337**, 466 (1989)
9. Bonnell, I.A., Pringle, J.E., MNRAS, **273**, L12 (1995)
10. van Putten, M.H.P.M., ApJ, **562**, L51 (2001)
11. van Putten, M.H.P.M., ApJ, **575**, L71 (2002)
12. Davies, M.B., King,A., Rosswong,S., Wynn,G., ApJ, **579**, L63 (2002)
13. Fryer, C. L., Holz, D. E. & Hughes, S. A., ApJ, **565**, 430 (2002)
14. Dimmelmeier,H., Font,J.A., Muller,E., A&A, **393**, 523 (2002)
15. Rampp,M., Muller,E., Ruffert,M., A&A, 332, 969 (1998)
16. Fryer, C.L., Woosley,S.E., Herant,M., Davies,M.B., ApJ, **520**, 650 (1999)
17. Belczynski, K., Bulik, T., Rudak, B., ApJ, **571**, 394 (2002)
18. Kobayashi, S, Mészáros , P., ApJ, **585**, L89 (2003)

High-Energy Cosmic Rays from Galactic and Extragalactic Gamma-Ray Bursts

S. D. Wick*, C. D. Dermer[†] and A. Atoyan**

NRL/NRC Research Associate, Code 7653, NRL, Washington, D.C. 20375-5352
[†]*Code 7653, Naval Research Laboratory, Washington, D.C. 20375-5352*
**CRM, Université de Montréal, Montréal, Canada H3C 3J7*

Abstract. A model for high energy ($\gtrsim 10^{14}$ eV) cosmic rays (HECRs) from galactic and extragalactic gamma-ray bursts is summarized. Relativistic outflows in gamma-ray bursts (GRBs) are assumed to inject power-law distributions of CR protons and ions to the highest ($\gtrsim 10^{20}$ eV) energies. A diffusive propagation model for HECRs from a single recent GRB within ≈ 1 kpc from Earth explains the CR spectrum near and above the knee. The CR spectrum at energies above $\sim 10^{18}$ eV is fit with a component from extragalactic GRBs. By normalizing the energy injection rate to that required to produce the CR flux from extragalctic sources observed locally, we determine the amount of energy a typical GRB must release in the form of nonthermal hadrons. Our interpretation of the HECR spectrum requires that GRBs are hadronically dominated, which would be confirmed by the detection of HE neutrinos from GRBs.

INTRODUCTION

We describe our model for HECRs, which assumes that HECRs are accelerated in the relativistic shocks in GRBs [1]. GRBs inject HECRs with a single power-law from a low-energy cutoff $E_{min} \approx 10^{14}$ eV to a high-energy cutoff $E_{max} \gtrsim 10^{20}$ eV. We extend previous hypotheses [2, 3] that UHECRs originate from GRBs to include HECR origin from GRBs within our Galaxy [4].

GRBs are located in the disks of active star-forming galaxies such as the Milky Way. HECRs with energies $\lesssim 10^{18}$ eV diffuse through their host galaxy. UHECRs with energies $\gtrsim 10^{18}$ eV escape from their host galaxy and propagate almost rectilinearly in extragalactic space. UHECRs travelling over cosmological distances have their spectrum modified by energy losses. An observer in the Milky Way will measure a superposition of UHECRs from extragalactic GRBs and HECRs produced in our Galaxy.

By fitting to the measured KASCADE [5] spectra of HECRs in the knee region, we determine the properties of a Galactic GRB that produces CRs. Our fits to the UHECR spectrum [6] measured with the High-Res experiment imply that GRBs must be strongly baryon-loaded, implying a detectable number of high energy neutrinos with a km-scale neutrino detector such as IceCube.

THE MODEL

Our model of diffusive propagation of HECRs from a single nearby GRB in the Galaxy assumes pitch-angle scattering of CR trajectories in Galactic magnetic fields on which is superposed a spectrum of magneto-hydrodynamic (MHD) turbulence described by a distribution in wave-number k [1]. The Larmor radius of a CR propagating in a magnetic field of strength $B = B_{\mu G}$ μG is $r_L \cong (A\gamma_6)/(ZB_{\mu G})$ pc, where $\gamma = 10^6 \gamma_6$ is the Lorentz factor of CR with atomic (mass,charge)= (A,Z). The model assumes that the mean-free-path λ between pitch-angle scatterings of a CR with Larmor radius r_L is inversely proportional to the energy density in the MHD spectrum at wave-number $k \sim r_L^{-1}$. A two component turbulence spectrum is assumed with wave-number index $q = 5/3$ for a Kolmogorov-type (for large wave-number) and $q = 3/2$ for a Kraichnan-type (for small wave-number) turbulence. The two components give an energy-dependent break λ at energy $E_Z(\text{PeV}) \cong ZB_{\mu G}b_{pc}$ where $B_{\mu G} = 3$ and $b_{pc} = 1.6$ is the wavelength in parsecs of the MHD waves where the spectrum changes from Kraichnan to Kolmogorov turbulence.

The diffusion radius $r_{dif} \cong 2\sqrt{\lambda ct/3}$. When $r \ll r_{dif}$, the number density of HECRs $n(\gamma; r,t) \propto t^{-3/2} \times \gamma^{-p-1/2(3/4)}$ for $q = 5/3$ (3/2). The measured spectrum is steepened by $\frac{3}{2}(2-q)$ units as the diffusion coefficient $D \propto \lambda \propto \gamma^{2-q}$ for an impulsive source [7] and depending on the CR energy and charge Z. An injection spectrum with $p = 2.2$ gives a measured spectrum $n_{Z,A}(\gamma; r,t) \propto \gamma^{-s}$, with $s = 2.7$ at $E \ll E_Z$ and $s = 2.95$ at $E \gg E_Z$. Because these indices are similar to the measured CR indices above and below the knee energy, we adopt this model for CR transport and investigate the implications of an injection spectrum $p = 2.2$.

Typical GRBs are thought to be beamed by a factor 500, meaning that they are $1/500\times$ as energetic as observations imply and 500× more frequent. Accounting for the beaming factor, star formation rate (SFR) evolution, and that dirty and clean fireball transients [8] may not be detected as GRBs, the GRB rate per L^* galaxy is $(1-3) \times 10^{-4}\ L^{*-1}$ yr^{-1}. In the Milky Way we expect one GRB every 3,000 - 10,000 years with a high probability that it will be beamed away from Earth.

UHECRs produced by extragalactic GRBs are assumed to propagate and lose energy from momentum red-shifting and photo-pair and photo-pion production on the CMBR. Attenuation produces features in the UHECR flux at characteristic energies $\sim 4 \times 10^{18}$ eV and $\sim 5 \times 10^{19}$ eV for photo-pair and photo-pion, respectively, from sources at redshift $z \sim 1$. We take the local GRB CR luminosity density to be $\dot{\varepsilon}_{CR} = f_{CR}\dot{\varepsilon}_{GRB,X/\gamma}$ where $\dot{\varepsilon}_{GRB,X/\gamma} = 10^{44}$erg Mpc^{-3}yr^{-1} [4, 9] and f_{CR} is the nonthermal baryon-loading fraction required of the model.

The GRB cosmic rate-density evolution is assumed to follow the SFR history derived from the blue and UV luminosity density of distant galaxies (see [1]). To accommodate uncertainty in the SFR evolution we take two models, one based on optical/UV measurements without extinction corrections (lower SFR) and with extinction corrections [10] (upper SFR). The upper SFR is roughly a factor of 3(10) greater than the lower SFR at red-shift $z = 1(2)$. For $> 10^{20}$ eV CRs, both evolution models give the same flux, but the upper SFR contributes a factor ~ 3 more CR flux over the lower SFR at energies $\lesssim 10^{18}$ eV.

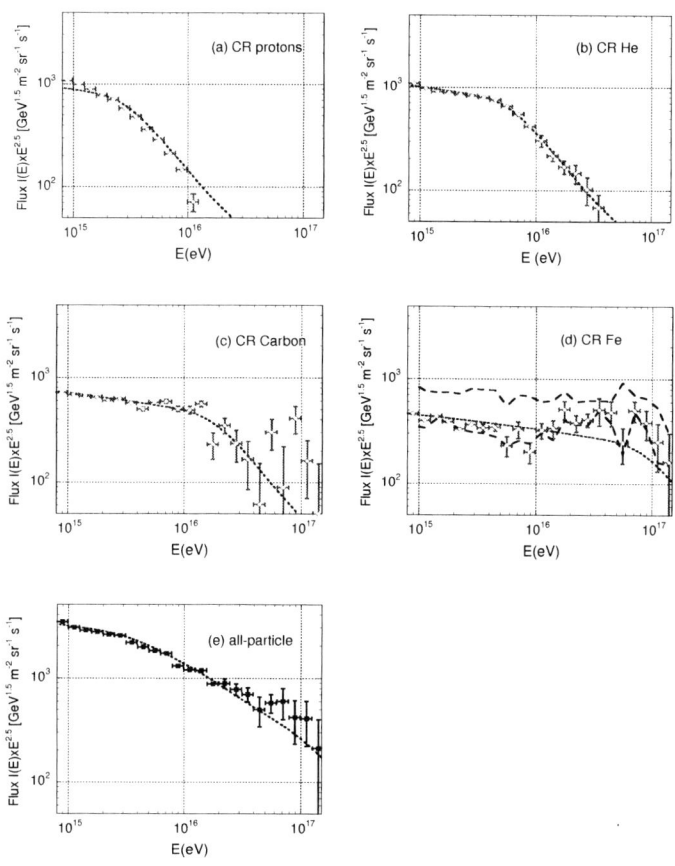

FIGURE 1. Data points show preliminary KASCADE measurements of the CR proton (panel a), He (panel b), Carbon (panel c), Fe (panel d), and the all-particle spectrum (panel e), along with model fits (dotted curves) to the CR ionic fluxes. In the model, a GRB that occurred 2.1×10^5 years ago and at a distance of 500 pc injects 10^{52} ergs in CRs. The CRs isotropically diffuse via pitch-angle scattering with an energy-dependent mean-free-path λ in an MHD turbulence field.

RESULTS

The preliminary (2001) KASCADE data [5] are fit in Fig. 1 using the diffusion model with a GRB source a distance $r \approx 500$ pc away that exploded $\approx 2 \times 10^5$ yrs ago. The break in the CR particle spectrum $\propto$ ionic charge Z is apparent in the plots. We

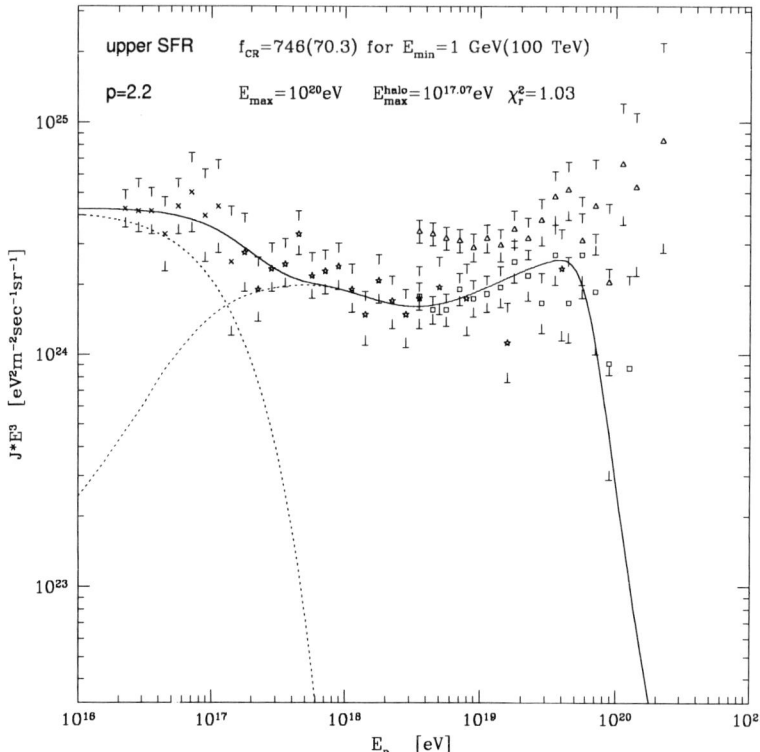

FIGURE 2. Best fit to the Kascade (crosses), HiRes-I Monocular (squares), and HiRes-II Monocular (stars) data assuming a spectral cutoff at the source of $E_{max} = 10^{20}$ eV and using the upper limit to the SFR evolution. We also show the AGASA data [11] (triangles) but do not include these in our fits. The cutoff energy for the halo component is $E_{max}^{halo} = 10^{17.07}$ eV and $\chi_r^2 = 1.03$. The requisite baryon loading factor for this fit is $f_{CR} = 746(70.3)$ for a low energy cutoff at the source of $E_{min} = 10^9(10^{14})$ eV. This fit implies that the transition from Galactic to extragalactic CRs occurs near the second knee and that the ankle is associated with photo-pair production.

did not perform a rigorous fit to the data, but instead adjusted the wavenumber k_1 (where the turbulence spectrum breaks) and the compositions of the ionic species until a reasonable fit was obtained. The best fits were obtained with $k_1 \cong 1/1.6$ pc and composition enhancements by a factor of 50 and 20 for C and Fe, respectively, over Solar photospheric abundances.

The combined KASCADE, HiRes-I and HiRes-II Monocular data between $\approx 2 \times 10^{16}$ eV to 3×10^{20} eV are fit in Fig. 2. We investigated 8 separate cases [1] with cutoff energy $E_{max} = 10^{20}$ eV and 10^{21} eV, spectral indices of $p = 2.0$ and $p = 2.2$ (for optimizing KASCADE fits), and upper vs. lower SFR evolution. The free parameters we vary are the galactic-halo-cutoff E_{max}^{halo}, the baryon loading f_{CR}, and the intensity of the galactic halo CR component. We give values of f_{CR} corresponding to $E_{min} = 10^9$ eV and 10^{14} eV. Fig. 2 shows our best case, with $p = 2.2$, $E_{max} = 10^{20}$ eV, and the upper SFR. The $p = 2.0$ spectrum provides a worse fit than the $p = 2.2$ case, although the CR energy demand is less in this case because CRs are injected equally per unit decade in paricle energy. The transition between galactic and extragalactic CRs is found in the vicinity of the second knee ($10^{17.6}$ eV), consistent with a heavy-to-light composition change [12]. The ankle ($10^{18.5}$ eV) is interpreted as a suppression from photo-pair losses, analogous to the GZK suppression.

Our results imply that GRB blast waves are baryon-loaded by a factor $f_{CR} \gtrsim 60$ compared to the energy injected and emitted by the primary electrons that is inferred from hard X-ray and soft γ-ray measurements of GRBs. For the large baryon load required for this model, calculations show that 100 TeV – 100 PeV neutrinos could be detected several times per year from all GRBs with kilometer-scale neutrino detectors such as IceCube [13, 1]. Detection of even 1 or 2 neutrinos from GRBs with IceCube or a northern hemisphere neutrino detector would unambiguously demonstrate the high nonthermal baryon load in GRBs, and would provide compelling support for this scenario for the origin of cosmic rays.

The work of S.D.W. was performed while he held a National Research Council Research Associateship Award at the Naval Research Laboratory (Washington, D.C). The work of C.D.D. is supported by the Office of Naval Research and NASA *GLAST* science investigation grant DPR # S-15634-Y. A.A. acknowledges support and hospitality during visits to the High Energy Space Environment Branch.

REFERENCES

1. S. D. Wick, C. D. Dermer, and A. Atoyan, Astropart. Phys., submitted (astro-ph/0310667) (2003).
2. M. Vietri, Astrophys. J. **453**, 883 (1995).
3. E. Waxman, Phys. Rev. Lett. **75**, 386 (1995).
4. C. D. Dermer, Astrophys. J. bf 574, 65 (2002).
5. K.-H. Kampert et al., in: 27th International Cosmic Ray Conference (Copernicus Gesellschaft, Hamburg, Germany) Invited, Rapporteur, and Highlight Papers, 240 (2001).
6. T. Abu-Zayyad et al., astro-ph/0208243 (2003).
7. A. M. Atoyan, F. A. Aharonian, H. J. Völk, Phys. Rev. D **52**, 3265 (1995).
8. M. Böttcher, C. D. Dermer, Astrophys. J. **529**, 635 (2000).
9. M. Vietri, D. De Marco, D. Guetta, Astrophys. J. **592**, 378 (2003).
10. A. W. Blain et al., Mon. Not. R. Astron. Soc. **309**, 715 (1999).
11. M. Takeda et al., Phys. Rev. Lett. **81**, 1163 (1998).
12. D. J. Bird et al., Phys. Rev. Lett. **71**, 3401 (1993).
13. C. D. Dermer and A. Atoyan, Phys. Rev. Lett. **91**, 071102 (2003).

Method for Detecting Neutrinos from Internal Shocks in GRB Fireballs with AMANDA

Michael Stamatikos* and the AMANDA Collaboration[†]

*Department of Physics, University of Wisconsin, Madison, WI, USA[1]
[†]For an author list, see Phys. Rev. Lett. 92: 071102, 2004 (astro-ph/0309585)

Abstract. Neutrino-based astronomy provides a new window on the most energetic processes in the universe. The discovery of high-energy ($E \geq 10^{14}$ eV) muonic neutrinos (ν_μ) from gamma-ray bursts (GRBs) would confirm hadronic acceleration in the relativistic GRB-wind, validate the phenomenology of the canonical fireball model and possibly reveal an acceleration mechanism for the highest energy cosmic rays (CRs). The *Antarctic Muon and Neutrino Detector Array* (AMANDA) is the world's largest operational neutrino telescope with a PeV muon effective area (averaged over zenith angle) $\sim 50,000$ m^2. AMANDA uses the natural ice at the geographic South Pole as a Cherenkov medium and has been successfully calibrated on the signal of atmospheric neutrinos (ν_{atm}). Contrary to previous diffuse searches, we describe an analysis based upon confronting AMANDA observations of individual GRBs, adequately modeled by fireball phenomenology, with the predictions of the canonical fireball model. The expected neutrino flux is directly derived from the fireball model description of the photon spectrum. The expected neutrino event rate is a function of the distribution of each individual burst in measured (or best-estimated) red shift. Strict spatio-temporal constraints (based upon satellite detection) and selection criteria (optimized for sensitivity) will be leveraged to realize a nearly background-free search.

INTRODUCTION: THE CASE FOR NEUTRINO ASTRONOMY

As we explore the universe with new precision in the regimes of the electromagnetic spectrum, we are fundamentally limited by the interaction and absorption of photons. Neutrino astronomy may hold the promise of a new frontier of discovery. Neutrinos are unique cosmic messengers, since they rarely interact and are unaffected by magnetic fields. Hence, the trajectories of astrophysical neutrinos (ν_{astro}) provide a direct path to their source. AMANDA's function as a high-energy neutrino observatory may provide us with a glimpse of the internal processes of otherwise enigmatic astrophysical phenomena such as GRBs. The operation of AMANDA [1] and the results of previous GRB analyses [2, 3] have been described elsewhere. Here we present a preliminary framework, motivating techniques that will bridge the gap between coincident neutrino observations and satellite GRB detection. A positive GRB-ν coincidence may provide an acceleration mechanism for CRs while testing the underlying physics of the standard fireball model [4]. An absence of any GRB-ν signal would result in neutrino event upper limits, constraining the fireball paradigm and associated predictions of neutrino flux.

---

[1] Correspondence to: ms25@amanda.wisc.edu

TABLE 1. Neutrino energy (ε_v) regimes and their production mechanisms in GRBs.

Regime	ε_v (eV)	Mechanism/Comments
1	$\sim 10^7$	Collapse/merger of progenitor event (quasi-thermal)
2	$\sim 10^9 - 10^{10}$	Longitudinal decoupling of the baryonic (i.e. n, p) flow in fireball
3	$\sim 10^{12} - \leq 10^{14}$	Precursors $\sim 10 - 100$ sec before γ_{Prompt} ($pp, p\gamma$ between jet/star)
4	$\sim 10^{14} - 10^{15}$	Photomeson interactions/internal shock, simultaneous* with γ_{Prompt}
5	$\sim 10^{17} - 10^{18}$	Afterglow, $p\gamma$/external (reverse) shock, ~ 10 seconds after γ_{Prompt}

* *Flight time delay due to neutrino mass is negligible for $\varepsilon_v \sim PeV$.*

FIREBALL PHENOMENOLOGY & THE GRB-ν CONNECTION

The generic mechanism responsible for generating the large energy ($\sim 10^{52}$ ergs) released on the order of seconds in GRBs is the dissipation, via shocks, of highly relativistic kinetic energy, acquired by protons and electrons Fermi-accelerated in an optically thick, relativistically expanding plasma of electrons and positrons (commonly referred to as a *fireball* [4]). The acceleration of electrons in the intense magnetic field of the fireball leads to the emission of prompt non-thermal γ-rays ($\bar{E}_\gamma \sim 250$ keV) via synchrotron radiation and possibly inverse Compton scattering. Photomeson interactions involving relativistically shock-accelerated protons ($E_p \geq 10^{16}$ eV) and synchrotron γ-ray photons in the fireball wind generate pions. High energy neutrinos ($E_v \sim 10^{14} - 10^{15}$ eV) are among the leptonic decay products of the photo-produced charged pions, that is,

$$p + \gamma \to \Delta^+ \to \pi^+ + [n] \to \mu^+ + \nu_\mu \to e^+ + \nu_e + \bar{\nu}_\mu + \nu_\mu \tag{1}$$

The prompt γ-rays are detected by satellite experiments such as BATSE [5] and TeV-PeV neutrinos could be detected by AMANDA. The hardness of the GRB spectra and the necessity for the expanding fireball shell to be optically thin, require $\Gamma_{Bulk} \sim 300$. The variability of the internal shocks is manifested in the complexity of the time profile of GRB light curves, while the interaction of external shocks with the surrounding matter produces multi-wavelength afterglow emissions. The cosmological distance scales of GRBs ($z \sim 1$) coupled with their measured flux and rapid variability time (~ 10 ms) implies the release of energies ~ 1 $M_\odot$ in compact regions with linear scale $\leq 10^8$ cm. The two main progenitor models include the merger of binary compact objects (e.g. NS/NS, NS/BH, etc.) and the death of massive stars (e.g. collapsars). Ultimately, both progenitor models culminate in the birth of a black hole.

The average rate of GRB energy emission in gamma-rays is comparable to the energy generation rate of the highest energy CRs (i.e. above the ankle) in models where high energy CRs are produced via a cosmological distribution of sources. Observations of the CR flux imply that the energy density of their sources is similar to the energy density requirement of protons ($E_p \sim 10^{20}$ eV) within the wind of the cosmologically distributed GRBs. This suggests that the highest energy CRs and GRBs may have a common origin. The assumption that GRBs are the source of the highest energy CRs [6] requires equal dissipation of energy among electrons and protons within the fireball.

EXPERIMENTAL METHODS: DATA SETS AND ANALYSIS

The primary objective of AMANDA is to detect high energy v_{astro}. AMANDA's function as a neutrino telescope has been demonstrated by the observation of v_{atm} [1]. From 1997-1999, the detector, known as AMANDA-B10, was comprised of 10 strings and 302 optical modules (OMs). By 2000, it had grown to 19 strings and 677 OMs and is referred to as AMANDA-II. The background, a consequence of atmospheric spallations of CRs, primarily consists of down-going atmospheric muons (μ_{down}^{atm}), detected at a rate ~ 100 Hz, and v_{atm}, detected at a rate $\sim 10^{-4}$ Hz. Neutrino induced muons, via charged current interactions such as: $v_\mu + N \rightarrow \mu^\pm + X$, represent signal events, and are isolated via the exclusive use of up-going muon (μ_{up}) reconstructed events in concert with optimized selection criteria (quality cuts). Hence, the entire bulk of the Earth is used as an atmospheric muon filter. The topology and timing of the OM data is used to reconstruct the μ_{up} tracks via a maximum likelihood method. At TeV energies, the offset between the direction of the incident v_μ and the secondary μ is $\sim 1°$. For point sources, the mean angular resolutions of AMANDA-B10 and AMANDA-II are $\sim 3.9°$ and $\sim 2.1°$ respectively. For previous GRB analyses, we found a PeV muon effective area (averaged over zenith angle) $\sim 50,000$ m^2 [2].

A detailed study of the application of the fireball model to BATSE GRBs in conjunction with AMANDA data has been published elsewhere [7] and forms the basis of this analysis. A survey of the current BATSE catalog [5] has resulted in the selection of ~ 100 northern hemisphere GRB candidates which are adequately modeled by fireball phenomenology. The data set spans from April 1997 to May 2000 and has been localized to a mean total positional error box radius of $\sim 2.5°$.

GRBs are expected to produce a wide energy range of neutrinos, as seen in Table 1. A nearly background free search is realized by using spatial and temporal data. The search is optimized for high energy neutrinos arriving simultaneously with the prompt γ-ray emission, since the spatial position, emission time, and duration are known. Hence, the proposed analysis will initially search for neutrinos from the fourth regime. Searches for neutrinos from other energy regimes would constitute different analyses, due to the difference in the energy spectra and arrival times of the neutrinos. On-source/off-time data will be used to estimate the background of each GRB and achieve blindness, facilitating an unbiased analysis. Another measure of the effectiveness of this analysis may be found in the stability of the detector. It has been demonstrated that the event counting rate per 10 second bins in a time window of ± 1 hour centered around a GRB is Gaussian [2]. The stability of the detector has been further corroborated via the observation of a $\sim 10\%$ seasonal variation in the background atmospheric muon flux over multiple annual data sets. Negligible expected background rates make the detection of only a few events on-time and on-source a statistically significant observation.

The neutrino energy spectrum is expected to trace the measured γ-ray photon emission spectrum. The fit parameters of the measured photon spectrum consist of a normalization constant (A_γ), low spectral index (α), high spectral index (β) and a photon break energy (ε_γ^b) as defined by the Band function [8]. Although the Band function provides a good fit to GRBs, the values of the fit parameters are not universal. In addition to the spectral fit parameters, the red shift (z) and Lorentz boost factor (Γ) of each GRB also determine its corresponding neutrino energy spectrum. The main parameters that define the energy spectrum of neutrinos from the fourth regime are a normalization constant (A_v), first spectral index (α_v), a neutrino break energy (ε_v^b) a synchrotron break energy (ε_s^b) and a second spectral index (β_v). Details involving the relationships between the γ-ray photon parameters and the parameters which determine the corresponding neutrino energy spectrum may be found in [7]. Individualized neutrino energy spectra, resulting in individual neutrino event rates, are a natural consequence of using neutrino input parameters that are determined by the measured photon parameters of each GRB.

FUTURE WORK

Current efforts are focused on generating spectral fits for the data sample. Since only a small fraction of GRBs in the data set have measured red shifts, empirical determinations of red shift indicators [9, 10] are being investigated. Simulations of μ_{up} will be generated by propagating $\sim$ 200,000 neutrinos and anti-neutrinos from regions in the sky defined by the total positional error box of each GRB. The expected neutrino event rate for each GRB will involve a convolution with the effects of neutrino oscillations, the propagation of neutrinos through the Earth, and the AMANDA response. The absolute normalization of the detector simulation is based upon the measurement of v_{atm}. Selection criteria will be optimized for sensitivity. In the event of no GRB-v signal, event upper limits will be calculated for each individual GRB. These techniques will be extended to analyses involving more sensitive instruments such as Swift [11] and IceCube [12].

ACKNOWLEDGMENTS

The author would like to thank Jean-Luc Atteia, David Band, Joshua Bloom, Charles Dermer, Neil Gehrels, Daniel Hooper, Kevin Hurley, Ray Klebasadel, Charles Meegan, Peter Mészáros, Tsvi Piran, Robert Preece, Virginia Trimble, Ralph Wijers and Stan Woosley for their support and valuable comments.

REFERENCES

1. Ahrens, J., et al., *Phys. Rev. D*, **66**, 012005 (2002).
2. Hardtke, R., Kuehn, K., and Stamatikos, M. et al., "28th ICRC Proceedings," 2003, pp. 2717–2720.
3. Hardtke, R., et al., "Proceedings of the 2003 GRB Symposium," AIP, 2003.
4. Piran, T., *Phys. Rept.*, **314**, 575–667 (1999).
5. (2003), URL http://www.batse.msfc.nasa.gov/batse/grb/catalog/current/.
6. Waxman, E., *Phys. Rev. Letters*, **75**, 386–389 (1995).
7. Guetta, D., et al., *Astropart. Phys.*, **20**, 429–455 (2004).
8. Band, D., et al., *ApJ*, **413**, 281–292 (1993).
9. Fenimore, E., and Ramirez-Ruiz, E., *astro-ph/0004176* (2000).
10. Band, D., et al., *astro-ph/0403220* (2004).
11. (2003), URL http://swift.gsfc.nasa.gov/.
12. Ahrens, J., et al., *Astropart. Phys.*, **20**, 507–532 (2004).

The ARGO−YBJ Sensitivity to GRBs

T. Di Girolamo*, G. Di Sciascio* and S. Vernetto[†]

*INFN, Napoli, Italy
[†]IFSI-CNR and INFN, Torino, Italy

Abstract. ARGO−YBJ is a "full coverage" air shower detector under construction at the YangBa-Jing Laboratory (4300 m a.s.l., Tibet, P.R. of China). Its main goals are γ-ray astronomy and cosmic ray studies. In this paper we present the capabilities of ARGO−YBJ in detecting the emission from Gamma Ray Bursts (GRBs) at energies $E > 10\,GeV$.

THE EXPERIMENT

The ARGO−YBJ detector is currently under construction at the YangBaJing High Altitude Cosmic Ray Laboratory in Tibet (P.R. of China), 4300 m above the sea level. It is a full coverage array of dimensions $\sim 74 \times 78\,m^2$, built with a single layer of Resistive Plate Counters (RPCs). The area surrounding this central detector ("carpet"), up to $\sim 100 \times 110\,m^2$, is partially ($\sim 50\%$) instrumented with RPCs ("guard ring"), for a total active area of $\sim 6400\,m^2$ (see left side of Figure 1). The detector basic element is the "pad", of dimensions $56 \times 62\,cm^2$, which defines its space-time granularity in observing shower fronts. Moreover, the detector is divided in 6×2 RPCs units ("clusters") and covered by a 0.5 cm thick layer of lead, in order to convert a fraction of the secondary γ-rays in charged particles, and to reduce the time spread of the shower particles [1].

OBSERVATIONAL TECHNIQUES

ARGO−YBJ will perform two different types of measurements:
a) Shower technique
Detection of showers with a trigger threshold $N_{pad} \geq 20$, where N_{pad} is the number of fired pads on the detector ("multiplicity"). For these events the position and time of any fired pad will be recorded. An example of shower with $N_{pad} \sim 500$, recorded with $\sim 10\%$ of the whole carpet during a test run, is shown in the right side of Figure 1: note the very detailed view of the shower front pattern. From the pads data the shower parameters (core position, arrival direction and size) can be reconstructed.

The trigger condition $N_{pad} \geq 20$ corresponds to a primary γ-ray energy threshold of a few hundreds GeV, the exact value depending on the source spectrum and on the zenith angle of the observation. A vertical γ-ray of energy $E \sim 120\,GeV$ gives a mean number of particles $N_e = 20$ at the ARGO−YBJ altitude.

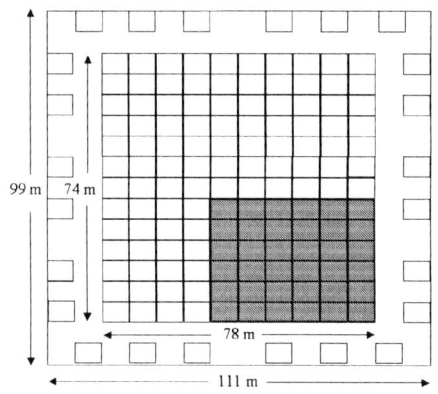
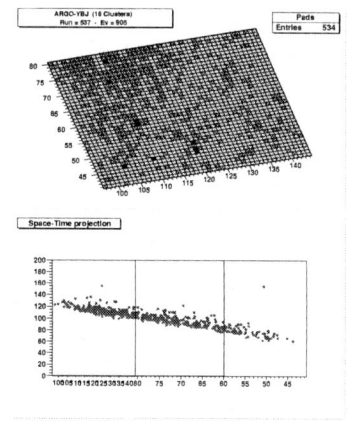

FIGURE 1. Left: Layout of ARGO−YBJ, showing the central RPCs carpet and the outer ring. The rectangles are subsets of the detector ("clusters") of area $\sim 42\,m^2$. The shaded area (36 clusters) is already installed. Right: Example of a shower recorded with 16 clusters.

b) Single particle technique
Every $0.5\,s$ the rate of the single particles hitting the detector is recorded. This measurement allows the detection of the secondary particles from very low energy showers ($E > 10\,GeV$) that reach the ground in a number insufficient to trigger the detector operating with the shower technique.

HIGH ENERGY EMISSION FROM GRBS

So far the only existing data reporting high energy γ-rays from GRBs come from the observations of EGRET (however, there are also possible TeV emissions claimed in the past by various ground-based experiments, in particular that from GRB970417a recorded by the Milagrito detector [2]). During its lifetime it detected 16 intense events, with a maximum photon energy of $18\,GeV$ [3]. All their spectra show a power law behaviour without any cutoff, suggesting that a large fraction of GRBs could emit GeV or even TeV γ-rays.

However, high energy γ-rays undergo pair production with infrared and optical stellar photons in the intergalactic space, and are strongly absorbed during their travel towards the Earth. The optical depth of this process $\tau(E,z)$ increases with the source redshift z and the γ-ray energy E. The majority (57%) of the GRB redshifts measured so far (November 2003) is located at $z > 1$ and only 2 GRBs have a redshift $z < 0.2$. According to [4], at a distance of $z = 1(0.1)$ the optical depth becomes larger than 1 for γ-ray energies $E > 50(800)\,GeV$. Therefore it seems unlikely to detect TeV emission from GRBs and most of the efforts must be concentrated in the $10 - 1000\,GeV$ energy range.

In order to evaluate the ARGO−YBJ sensitivity to GRBs we consider a simple model in which the GRB high energy flux is described by a power law spectrum with photon

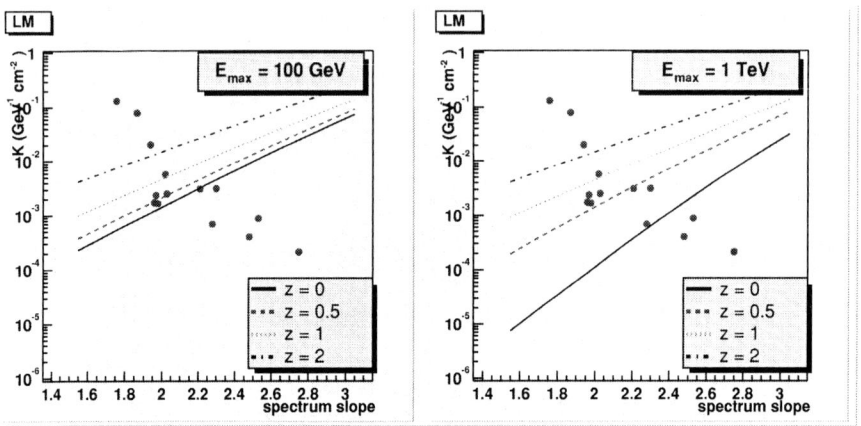

FIGURE 2. The value of the GRB spectral normalization factor K necessary to give a 4σ signal with the LM technique, as a function of the spectrum slope for different values of E_{max} and z. The points represent 14 GRBs observed by EGRET.

index Γ extending up to a maximum energy E_{max} (intrinsic cutoff), and affected by an exponential cutoff due to the intergalactic absorption: $dN/dE = KE^{-\Gamma}e^{-\tau(E,z)}$.

The GRB is assumed at a zenith angle $\theta = 20^o$ with an intrinsic energy cutoff in the range $100\,GeV < E_{max} < 1\,TeV$ and a distance in the range $0 < z < 2$. The absorption factor is calculated exploiting the values of $\tau(E,z)$ given in [4].

The sensitivity has been obtained by comparing the number of events expected from the GRB with the number of background events, according to both detection techniques, varying the GRB parameters: spectral normalization factor K, spectrum slope Γ, cutoff energy E_{max}, redshift z.

RESULTS

In the case of observation technique a), a GRB candidate will appear as a statistically significant excess of Low Multiplicity (LM) events clustered in time and arrival direction. The angular resolution of the detector (*i.e.*, the opening angle containing 71.5% of the γ-ray events) for showers with $N_{pad} \geq 20$ is $\sim 2.7^o$.

In the case of observation technique b), no reconstruction of the shower parameters is possible and a GRB is observed as an excess in the Single Particle (SP) background rate, possibly in time coincidence with a GRB satellite detection [5].

Our results are summarized in Figure 2 (for the LM technique) and in Figure 3 (for the SP technique), where the value of K necessary to give a signal with a statistical significance of 4σ is shown as a function of the spectrum slope Γ for different values of E_{max} and z. In these calculations a GRB duration $\Delta t = 1\,s$ is assumed. The sensitivity for different durations can be easily obtained by multiplying K by $\sqrt{\Delta t}$. To compare the ARGO−YBJ expected sensitivity with real GRBs, the K vs. Γ values of 14 EGRET

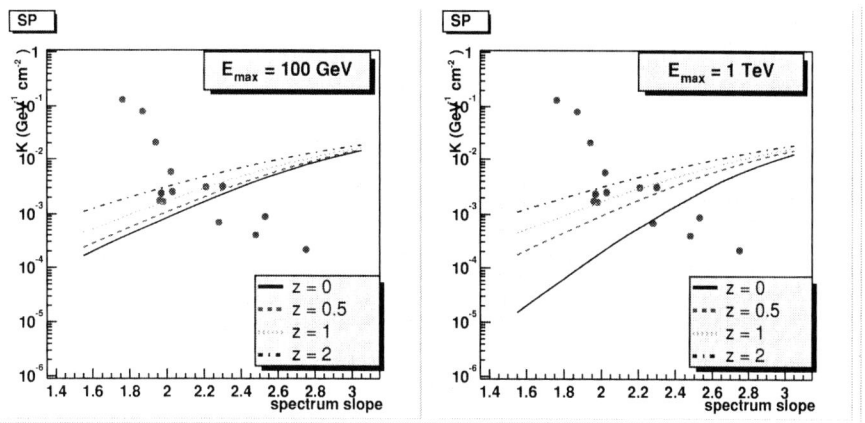

FIGURE 3. Same as Figure 2 but in the case of the SP technique.

GRBs [3] are plotted in the same figures.

These results show that the SP technique is in general more sensitive to GRBs, in particular for high z where the intergalactic absorption strongly affects the high energy tail of the spectrum. Only in the case $z = 0$ and $E_{max} = 1\,TeV$ the LM shower technique is slightly better.

CONCLUSIONS

ARGO–YBJ could observe the high energy emission of the most intense GRBs. Since γ-rays of energy $E > 1\,TeV$ emitted by cosmological sources are strongly absorbed during their travel towards the Earth, the SP technique provides the best approach to detect GRBs, being particularly sensitive in the $10 - 1000\,GeV$ energy range.

The sensitivity and the event rate depend critically on the shape of the GRBs spectra above 10 GeV. This shape is determined by the possible existence of an intrinsic cutoff and by the absorption of γ-rays in the intergalactic space.

The analysis of 14 EGRET GRBs indicates that if the intrinsic cutoff is not too low ($E_{max} > 100\,GeV$) and the sources redshift is $z < 2$, a fraction of the events ranging from $\sim 20\%$ to $\sim 80\%$ would be detectable.

REFERENCES

1. Surdo, A. et al., *Proc. of the 28th ICRC*, HE 1.1, p. 5 (2003) (see also http://argo.na.infn.it).
2. Atkins, R. et al., ApJ, **533**, L119 (2000).
3. Catelli, J.R., et al, *AIP Conference Proceedings*, **428**, p. 309 (1998).
4. Salamon, M.H., and Stecker, F.W., ApJ, **493**, 547 (1998).
5. Vernetto, S., Astroparticle Physics, **13**, 75 (2000).

Neutrino Oscillation in Gamma Ray Burst Fireball

Juan Carlos D'Olivo*, José F. Nieves† and Sarira Sahu*

*Instituto de Ciencias Nucleares Universidad Nacional Autónoma de México Circuito Exterior,C.U., A. Postal 70-543 04510 Mexico DF, Mexico
†Laboratory of Theoretical Physics, Department of Physics, P.O. Box 23343 University of Puerto Rico, Río Piedras, Puerto Rico 00931-3343

Abstract. The neutrino oscillation in GRB fireballs is studied by assuming the fireball as a plasma of photons and electrons-positrons. We show that, in this medium, only the $v_e \leftrightarrow v_{\mu,\tau}$ oscillation can take place resonantly depending on the fireball parameters. On the other hand the $\bar{v}_e \leftrightarrow \bar{v}_{\mu,\tau}$ oscillation is suppressed. In most of the models for producing the GRB, the central engine will produce more v_e and $\bar{v}_e$ then v_μ, v_τ and the corresponding anti-neutrinos. When these propagate through the fireball, they will undergo resonant conversion from one type to another and because of the higher flux of v_e, the probability of obtaining v_μ and v_τ will be higher then in the reverse process, thus enhancing the fluxes of v_μs and v_τs.

INTRODUCTION

GRBs are short, non-thermal bursts of low energy ($\sim$ 100 KeV-1 MeV) photons and release about 10^{51}-10^{53} ergs in a few seconds, which makes them the most luminous objects in the universe. A class of models called *the fireball model* seem to explain the temporal structure of the bursts and the non-thermal nature of their spectra[1, 2]. Sudden release of copious amount of γ rays into a compact region with a size $R_0 \sim$ 100 Km creates an opaque photon-lepton fireball due to the process $\gamma + \gamma \to e^+ + e^-$. The optical depth is so large that photons cannot escape. Even if there are no pairs to begin with, they will form very rapidly and will Compton scatter lower energy photons. In fireballs the γ and $e^\pm$ pairs will thermalize with a temperature 3-10 MeV.

In addition to γ and $e^\pm$ pairs, fireballs may also contain some baryons, both from the progenetor and the surrounding medium and these will affect the expansion of the fireball. Irrespective of it, the baryonic load has to be very small, otherwise, the expansion of the fireball will be Newtonian, which is inconsistent with the present observations. The neutrino oscillation may overcome the baryon loading problem! [3] The evolution of the pure fireball (with no baryons) has been studied in ref.[4, 5].

Observation suggests that the hidden central engine which powers the fireball must be compact. The prime candidates are the merger of neutron star with neutron star (NS-NS), black hole-neutron star binaries (BH-NS), and hypernova/collapsar models involving a massive stellar progenitor[1, 2]. In all these models, the gravitational energy is released mostly in the form of $v\bar{v}$ and gravitational radiation, while a small fraction ($\sim 10^{-3}$) is responsible for powering the GRB. Neutrinos of 10-30 MeV, which are generated due to the stellar collapse or merger event that triggers the burst, will pass through the fireball.

Maybe due to nucleonic bremsstrahlung $NN \to NN\nu\bar{\nu}$ and $e^+e^- \to \nu\bar{\nu}$ processes, muon and tau type neutrinos can also be produced during the merger process, but their flux will be very small. Also, because of the weak interaction process $p + e^- \to n + \nu_e$, neutrinos can be generated within the fireball.

NEUTRINO OSCILLATION

The early universe hot plasma is supposed to have almost equal populations of particles and anti-particles, thus the leading contribution to the neutrino effective potential will be proportional to $1/M^4$, where M is the W-boson mass. A similar situation can arise in GRB fireballs if one considers mostly photon-lepton fireballs[1, 2]. In the present work we assume a photon-lepton spherical fireball, which mimics the radiation dominated era in the early universe hot plasma, and study the neutrino propagation within it.

We consider the neutrino oscillations $\nu_e \leftrightarrow \nu_{\mu,\tau}$ and $\bar{\nu}_e \leftrightarrow \bar{\nu}_{\mu,\tau}$ in the relativistc and non degenerate plasma of the GRB fireball. The conversion probability for the above process at a distance R is given by

$$\mathscr{P}(R) = \frac{\Delta^2 \sin^2 2\theta}{\omega_f^2} \sin^2\left(\frac{\omega_f R}{2}\right), \tag{1}$$

with

$$\omega_f^2 = (V - \Delta\cos 2\theta)^2 + \Delta^2 \sin^2 2\theta, \tag{2}$$

where $\Delta = \delta m^2/2E_\nu$ and θ is the mixing angle. For a γ, $e^\pm$ plasma the effective potentials are given by

$$V_{\nu_e(\bar{\nu}_e)} = \sqrt{2} G_F N_\gamma \left[\pm L_e - \left(\frac{7\xi(4)}{\xi(3)}\right)^2 \frac{T^2}{M_W^2} \right], \tag{3}$$

where N_γ is the photon number density. For the anti-neutrinos the potential is always negative, and the first term in Eq.(2) will never vanish. On the other hand, for neutrinos, depending on the fireball parameters, the first term may vanish, giving rise to a resonant conversion. In the later case the resonance condition is

$$V = \Delta\cos 2\theta. \tag{4}$$

The oscillation length is given by

$$L_{osc} = L_\nu \left(\cos^2 2\theta \left(1 - \frac{V}{\Delta\cos 2\theta}\right)^2 + \sin^2 2\theta \right)^{-1/2}, \tag{5}$$

where $L_\nu = 2\pi/\Delta$ is the vacuum oscillation length. Recent analysis of the solar neutrino data from SNO[6], strongly favor the oscillation of ν_e to other active flavors ν_μ and/or ν_τ with a maximal mixing of $\sin 2\theta \sim 0.8$[7]. In our calculation we take this value of the

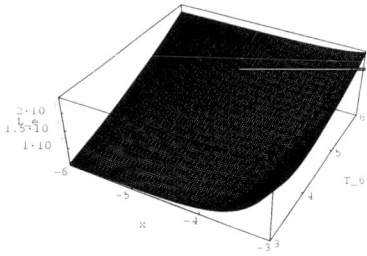

FIGURE 1. The lepton asymmetry L_e is plotted as functions of T_6 and x.

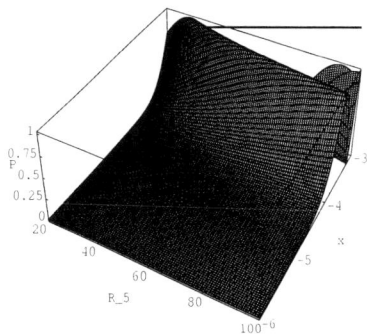

FIGURE 2. The resonant conversion probability $\mathscr{P}(R)$ is plotted as funcions of R_5 in Km and x.

oscillation parameters into account. From Eq.(4), at the resonance

$$L_e = 7.47 \times 10^{-2} \frac{\tilde{\Delta}m^2}{E_6 T_6^3} + 6.14 \times 10^{-9} T_6^2, \tag{6}$$

where T_6 and E_6 are in units of 10^6 eV and $\tilde{\delta}m^2$ is in units of eV^2. As we are considering the propagation of neutrinos in the fireball medium, the only neutrinos avilable are the low energy ones (about $10-30 MeV$) produced during the stellar collapse or merger events. Thus for the neutrino energy we take $E_6 \sim 20$. Now the only parameters left are the $\tilde{\delta}m^2$ and the fireball temperature. It has been discussed that, the $e^{\pm}$ plasma will thermalize with a temperature of about $3-10$ MeV. In Fig. 1 we show the variation of L_e as a function of $\tilde{\delta}m^2$ and the fireball temperature. We consider the range of $\tilde{\delta}m^2$ from 10^{-6} to 10^{-3} which basically covers the both atmospheric (2.5×10^{-3}) and solar (7×10^{-5}) mass sqare differences. We find that, for the fireball temperature in the range 3-6 MeV, to satisfy the resonance condition the lepton asymmetry has to be of order 10^{-7}. In Fig. 2 we show the resonance conversion probability as a function of R_5 and x ($\tilde{\delta}m^2 = 10^x$). From this plot it can be see that there is a significant conversion for large values of x and for all values of R_5, but for small x, $\mathscr{P}(R)$ is very small.

CONCLUSIONS

The neutrinos, passing through the fireball and/or produced within the fireball may resonantly convert from one species to another depending on the lepton asymmetry of the fireball and its temperature. For anti-neutrino it is very much supressed. From the merger model of SN with SN and/or from the collison of SN with a black-hole, normally the fluxes of v_e and $\bar{v}_e$ are much larger than the fluxes of v_μ, v_τ and the corresponding anti-neutrinos. The v_e flux being much higher than the v_μ and v_τ, the probability of conversion of v_e to v_μ and v_τ is high. So if the fireball parameters described above are correct, due to the resonant conversion we will get more v_μ and v_τ, then originally produced.

ACKNOWLEDGMENTS

JCD and SS are partially supported by the projects DGAPA-UNAM (PAPIIT) No. IN109001 and CONACyT (Mexico) grant No. 32279E.

REFERENCES

1. T. Piran, Phys. Rep. **314**, 575 (1999); T. Piran, Phys. Rep. **333-334**, 529 (2000).
2. P. Mészáros, Nucl. Phys. B (proc. Suppl.) **80**, 63 (2000); P. Mészáros, Annu. Rev. Astron. Astrophys. **40**, 137 (2002).
3. W. Kluźniak, Astrophys. J. Lett. **508**, L29 (1998); Raymond R. Volkas and Yvonner Y. Y. Wong, Astropart. Phys. **13**, 21 (2000).
4. B. Paczyński, Astrophys. J. Lett. **308**, L43 (1986).
5. J. Goodman, Astrophys. J. **308**, L47 (1986).
6. The SNO Collaboration, Phys. Rev. Lett. **89**, 011301 (2002).
7. John N. Bahcall, Concepción M. Gonzalez-Garcia, and Carlos Peña-Garay, JHEP **07** (2002) 054.

The AMANDA Search for High Energy Neutrinos From Gamma Ray Bursts

Rellen Hardtke* and the AMANDA Collaboration[†]

Department of Physics, University of Wisconsin, Madison, WI 53706, USA
[†]*For the full author list, please see astro-ph/0309585*

Abstract. We have searched three and a half years of AMANDA data for high energy muon neutrinos from gamma-ray bursts (GRBs). The data was recorded from 1997 through 1999 by the AMANDA-B10 detector and in 2000 by the AMANDA-II detector. AMANDA is a Čerenkov detector embedded 1.5 to 2 km deep in the transparent ice of the South Polar plateau. We searched for neutrino candidates from the direction of, and coincident with, GRBs detected by the Burst and Transient Source Experiment (BATSE). The current result is consistent with no signal. A preliminary event upper limit for GRB neutrino emission is presented as well as a description of AMANDA's cubic-kilometer successor, IceCube.

AMANDA

The Antarctic Muon and Neutrino Detector Array (AMANDA) is a three-dimensional array of photo-multiplier tubes (PMTs) deployed in the deep, transparent ice of the South Polar plateau. See Table 2 and Figure 1. The array of PMTs detects the light produced by the Čerenkov emission of muons that are traveling through the ice at speeds greater than c/n, where n is the index of refraction of the ice, about 1.33. AMANDA detects muons induced by cosmic rays bombarding the southern hemisphere, but it also detects muons that have been produced by neutrinos from the northern skies that have traveled through the earth and interacted in the ice or rock near the array. Since only neutrinos can pass through the entire earth, the earth can be used as a filter to tag upgoing neutrino-induced muons. At TeV energies, the path of the muon is aligned with the direction of its parent neutrino to about 1° and neutrino astronomy is possible.

If GRBs produce muon neutrinos or electron neutrinos that oscillate to muon neutrinos, then AMANDA can search the northern hemisphere for neutrinos that are coincident in time and location with known GRBs[1]. For a detailed discussion of the mechanism that would produce GRB neutrinos, please see [1] in these proceedings.

When the time and location of GRBs are known, the AMANDA search for high energy muon neutrinos from GRBs becomes an almost background-free search. The work described here utilizes GRB detections made by BATSE from 1997 until it was turned off in May of 2000.

---

[1] Electron and tau neutrinos leave different, unique signatures in the detector and will be studied in other analyses.

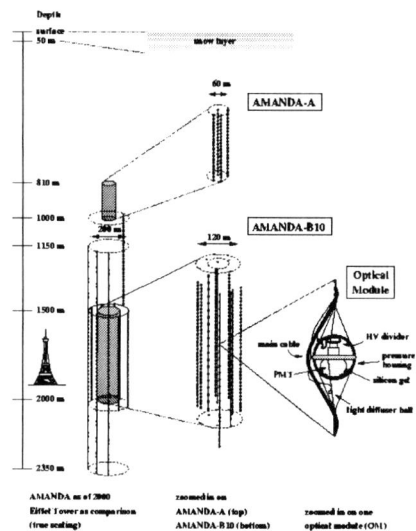

FIGURE 1. The Antarctic Muon and Neutrino Detector Array.

ANALYSIS

AMANDA data recorded during the hour before and hour after a BATSE GRB were extracted. These two hours of data are used to characterize the background event rate at the time of the GRB. Ten minutes of data surrounding the BATSE trigger time are strictly excluded from the analysis until final quality cuts have been determined. This statistically blind approach to data analysis ensures that quality cuts are not biased by the presence or absence of signal events. The data recorded around the GRB time was required to be complete and stable.

The sensitivity of AMANDA-B10 varied with zenith angle and quality cuts were optimized for ten different zenith bins, from directly upgoing to the horizon. Most background events are due to misreconstructed downgoing muons created by cosmic rays striking the southern hemisphere. The temporal and directional information provided by BATSE reduces this background rate.

Quality cuts remove almost all of the remaining events. A "direct hit" in AMANDA is one that is minimally scattered in the ice (<75 nanoseconds) before being detected by a PMT. In AMANDA-II, each event was required to have at least 15 direct hits. The direct hits in each event were also required to be smoothly distributed along the muon's reconstructed path. This requirement further increases confidence in the track's direction.

Additionally, GRB neutrinos will have energies greater than atmospheric muons. See Figure 2. The energy deposited in the detector is related to the number of PMTs that see light in each event. In this analysis, at least 26 PMTs were required to participate. Muon and neutrino effective areas are shown in Figure 3. The particle's energy is reported at its closest approach to the center of the detector.

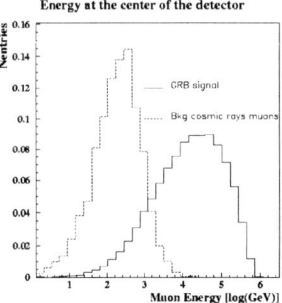 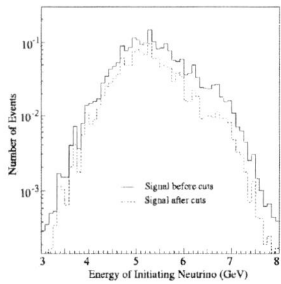

FIGURE 2. On the left, the energy of muons at the center of the AMANDA-B10 detector, from background cosmic rays and from the predicted Waxman-Bahcall [2] GRB neutrino flux. On the right, GRB signal generated by Monte Carlo simulations before (solid) and after (dashed) quality cuts.

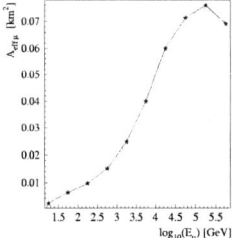 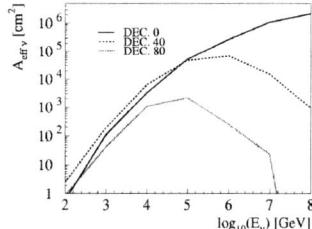

FIGURE 3. On the left, the muon effective area as a function of energy averaged over zenith angle. On the right, the effective neutrino area as a function of energy for a few declinations. Both plots are for AMANDA-II.

RESULTS

Results are shown in Table 1. Quality cuts were optimized via the Model Rejection Potential technique [3]. In the absence of a positive detection, this technique predicts an event upper limit sensitivity of 3.4, based only on the expected background and averaged over all possible no-signal outcomes. The total event upper limit is based on the expected background and the actual number of events observed. It is calculated according to [4]. The current result is consistent with no GRB muon neutrino signal. Two different studies of the AMANDA-II data from 2000 were conducted. The results of these analyses, labeled 2000A and 2000B, are consistent and both are shown in Table 1.

Several models for coincident neutrino emission lie below our current upper limit. Additional searches by AMANDA-II and the increased effective area of IceCube (see below) will confirm or rule out these predictions. In addition, some GRB models predict neutrinos at times distinctly before the onset of the gamma-ray emission, after it has stopped, or at less than TeV energies [5, 6]. These predictions will require differently optimized analyses.

TABLE 1. Preliminary results of the 1997-2000 GRB neutrino analysis.

Year	GRBs Examined	Total Background	Events Observed	Event Upper Limit
1997	78	0.06	0	2.41
1998	94	0.20	0	2.24
1999	96	0.20	0	2.24
2000A	44	0.83	0	1.72
2000B	44/68	0.40/0.64	0/0	2.05/1.90
Total	312	1.29	0	1.45
Sensitivity	312	1.29		3.4

TABLE 2. South Pole neutrino telescope.

Detector	Deployed	Number of PMTs	Strings	Volume $[m]^3$	Approx. Effective Area $[m]^3$	Angular Resolution
AMANDA-B10	1995-1997	302	10	2×10^7	1×10^4	$\sim 3.5°$
AMANDA-II	1997-2000	677	19	6×10^7	3×10^4	$\sim 2°$
IceCube	2004-2010	4,800	80	1×10^9	1×10^6	$\sim 0.7°$

ICECUBE AND FUTURE WORK

IceCube [7] is the cubic-kilometer successor of AMANDA. See Table 2 for details. Ice-Cube will be an ideal instrument to search for neutrinos from GRBs detected by SWIFT and GLAST [8]. In the mean time, further studies are underway with AMANDA-II. One analysis [9] utilizes post-BATSE GRB detections made by the Third Interplanetary Network (IPN3). Another effort [1] is being conducted to more accurately predict the neutrino emission of GRBs by accounting for the burst's distance, internal Lorentz boost factor, spectrum, and inherent energy.

REFERENCES

1. Stamatikos, M., et al., these proceedings (2003).
2. Waxman, E., and Bahcall, J., *Phys. Rev. Letters* **78** 2292 (1999).
3. Hill, G. C., and Rawlins, K., *Astropart. Phys.* **19** 393 (2003).
4. Feldman, G. J., and Cousins, R. D., *Phys. Rev. D* **57** 3873 (1998).
5. Dermer, C. D., and Atoyan, A., *Phys. Rev. Letters* **91** 071102 (2003).
6. Guetta, D., and Granot, J., *Phys. Rev. Letters* **90** 191102 (2003).
7. Ahrens, J., et al., accepted by *Astropart. Phys.* and available at astro-ph/0305196 (2003); http://icecube.wisc.edu/
8. http://swift.gsfc.nasa.gov/ and http://glast.gsfc.nasa.gov/
9. Kuehn, K., et al., to be published in TAUP proceedings, Seattle, WA (2003).

Limits on Very High Energy Emission from Gamma-Ray Bursts with the Milagro Observatory

Miguel F. Morales for the Milagro Collaboration

Massachusetts Institute of Technology, Cambridge, MA 02139, mmorales@space.mit.edu

Abstract. The Milagro telescope monitors the northern sky for 100 GeV to 100 TeV transient emission through continuous very high energy wide-field observations. The large effective area and ~100 GeV energy threshold of Milagro allow it to detect very high energy (VHE) gamma-ray burst emission with much higher sensitivity than previous instruments and a fluence sensitivity at VHE energies comparable to that of dedicated gamma-ray burst satellites at keV to MeV energies. Even in the absence of a positive detection, VHE observations can place important constraints on gamma-ray burst (GRB) progenitor and emission models. We present limits on the VHE flux of 40 s – 3 h duration transients nearby to earth, as well as sensitivity distributions which have been corrected for gamma-ray absorption by extragalactic background light and cosmological effects. The sensitivity distributions suggest that the typical intrinsic VHE fluence of GRBs is similar or weaker than the keV – MeV emission, and we demonstrate how these sensitivity distributions may be used to place observational constraints on the absolute VHE luminosity of gamma-ray bursts for any GRB emission and progenitor model.

INTRODUCTION

The heart of the Milagro observatory is a large water reservoir located at 2600 m altitude in the Jemez mountains of New Mexico. Very high energy gamma rays incident at the earth pair produce in the upper atmosphere and produce extensive air showers (EAS) which propagate to lower altitudes. Milagro uses the water Cherenkov technique to detect EASs by converting the front of relativistic particles in the EAS into a front of Cherenkov light which is detected by the 723 photomultiplier tubes (PMTs) instrumenting the water. Milagro is a very sensitive detector with a low ~ 100 GeV energy threshold and wide >1.8 sr field-of-view and is fully capable of autonomous identification of VHE GRB emission. The low energy threshold is particularly important because of the reduced attenuation by extragalactic background light near 100 GeV, which dramatically increases the volume of space observed.

THE 40 S TO 3 HOUR GRB SEARCH IN MILAGRO

The 40 s to 3 hour transient search implements an analysis developed by Morales et al. [1] which enhances the sensitivity of wide-angle gamma-ray observatories. The 40 s – 3 hour duration window was analyzed using nine separate logarithmically spaced search

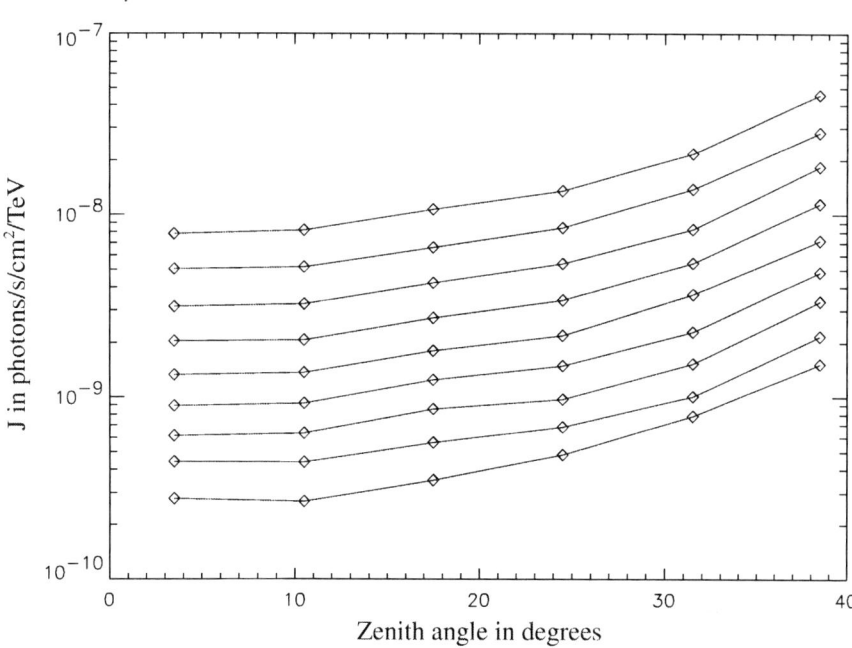

FIGURE 1. The 90% confidence upper limits for an $E^{-2.0}$ spectrum for all nine time scales as a function of zenith angle. The diamonds indicate the calculated limits on the normalization factor J in photons s^{-1} cm^{-2} TeV^{-1} for the spectrum $\frac{dN}{dE} = J(\frac{E}{1\text{TeV}})^{-2.0}$. The time scales are from top to bottom: 40 s, 80 s, 160 s, 320 s, 640 s, 1280 s, 2560 s, 5120 s, and 10240 s. Monte Carlo statistics lead to an error in the upper limits of 19%. Systematic errors are due principally to uncertainties in the Monte Carlo simulation and are estimated to be +40%/-20%, for a total estimated error of +44%/-27%.

windows, each covering a 1.84 sr field of view for data taken between 2001 May 2nd and 2002 May 22nd. The search over-samples in both space and time and has ∼ 290 days of total observation (see Atkins et al. [2] for details). The observations are entirely consistent with the expected trials factors, and no evidence was observed for transient VHE emission of 40 s to 3 hours duration.

FLUX LIMITS

The flux limits in Figure 1 represent the VHE transient flux at the earth for a concident event which can be excluded by this study at the 90% confidence level. For this paper an $E^{-2.0}$ power-law spectra (as observed local to the earth) from 100 GeV to 21 TeV was chosen. This spectrum serves as a reasonable model for an inverse Compton bump at TeV energies or a hard VHE extension of the observed GRB spectrum past the multi-GeV observations by EGRET. These limits represent the strongest flux limits on VHE GRB emission obtained to date.

CONSTRAINTS ON THE INTRINSIC LUMINOSITY OF GRBS

In an effort to place the current observations in context, a set of assumptions about the emitted spectrum, extragalactic background light (EBL) absorption, and cosmology have been chosen and sensitivity distributions calculated within this theoretical framework. For a set of source luminosities and durations, simulated GRBs were created with an $E^{-2.0}$ emission spectrum from 100 GeV – 21 TeV, an isotropic sky position, and following the star formation rate in z. The EBL absorption was determined by Bullock et al. [3] (similar to Primack et al. [4]) using a ΛCDM cosmology ($\Omega_M = 0.3$, $\Omega_\Lambda = 0.7$, h= 0.65). The results are rather insensitive to the emitted spectrum due to the strong EBL absorption which eliminates nearly all emission above 300 GeV at redshift ~ 0.3, and limits the distance to which Milagro can observe VHE emission to redshift < 0.7.

The probability that Milagro would have observed a VHE GRB as a function of the isotropic luminosity, distance, and duration of the source is given in Table 1 of Atkins et al. [2]. The results for a GRB with VHE emission of 80 s duration is plotted as an example in Figure 2.

As an example of the limits on VHE emission which can be made using these probabilities, consider a model which predicts that GRBs follow the star formation rate with all GRBs emitting a characteristic 80 s pulse of VHE emission. The resulting 90% confidence upper limits for this model are 6.2×10^{-8} GRBs/$M_\odot$ of star formation (an average of 4.8 GRBs/Gpc^3/year over 0<z<0.5) for an isotropic luminosity of 10^{51} ergs/s, or 1.1×10^{-8} GRBs/$M_\odot$ of star formation (an average of 0.8 GRBs/Gpc^3/year over 0<z<0.5) for a luminosity of 10^{52} ergs/s, with significantly tighter constraints if the GRB distribution trails the star formation rate (i.e. there are more low redshift GRBs). Of the thirty-six GRBs with known distances, five have a redshift below 0.5. If the GRBs detected by BATSE follow the same distance distribution, a rough estimate of the observed GRB rate yields $\sim$2.6 GRBs/Gpc^3/year, or $\sim$3.4 $\times 10^{-8}$ GRBs/$M_\odot$ of star formation if they follow the SFR. While detailed model calculations are needed to convert the probabilities in Atkins et al. [2] into meaningful upper limits, comparison with the limits from this simple model suggests that if GRBs follow the star formation rate, the typical luminosity of 40 s – 3 h VHE GRB counterparts is constrained to be similar to or less than the prompt keV – MeV emission.

ACKNOWLEDGMENTS

We acknowledge Scott Delay and Michael Schneider for their dedicated efforts in the construction and maintenance of the Milagro experiment. This work has been supported by the National Science Foundation (under grants PHY-0070927, -0070933, -0075326, -0096256, -0097315, -0206656, -0302000, and ATM-0002744) the US Department of Energy (Office of High-Energy Physics and Office of Nuclear Physics), Los Alamos National Laboratory, the University of California, and the Institute of Geophysics and Planetary Physics. MFM was a NASA Graduate Student Researcher.

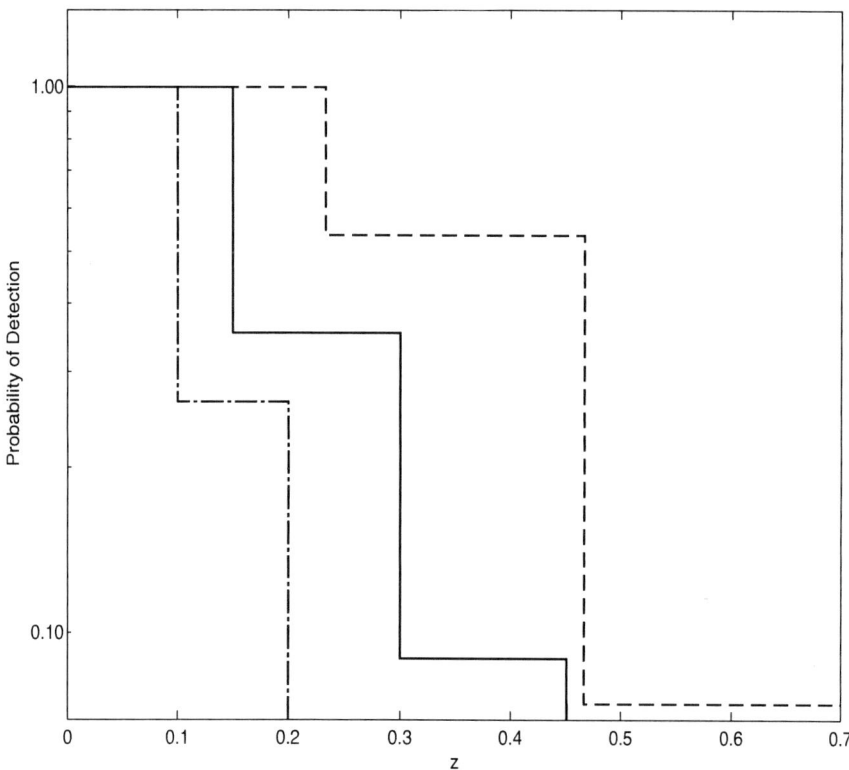

FIGURE 2. The probability of detection for GRBs of 80 seconds duration within the 1.84 sr field of view of Milagro with an isotropic luminosity 10^{50} ergs/s (dash-dot line), 10^{51} ergs/s (solid line), and 10^{52} ergs/s (dashed line) as a function of z. Table 1 in Atkins et al. [2] lists the probabilities for many different luminosity and burst durations

REFERENCES

1. Morales, M. F., Williams, D. A., & DeYoung, T., *Astropart. Phys.*, in press, astro-ph/0303178 (2003).
2. Atkins, R., et al., *submitted to ApJL*, astro-ph/0311389 (2004).
3. Bullock, J. R., Somerville, R. S., Primack, J. R., *in prep.* (2003).
4. Primack, J. R., Bullock, J. S., Somerville, R. S., & MacMinn, D., *Astropart. Phys.*, **11**, 93 (1999).

GRB Observations around 100 GeV with STACEE

D. A. Williams*, L. M. Boone*[†], D. Bramel**, J. Carson[‡], C. E. Covault[§],
P. Fortin[¶], D. M. Gingrich[||], D. Hanna[¶], A. Jarvis[‡], J. Kildea[¶], T. Lindner[¶],
C. Mueller[¶], R. Mukherjee**, R. A. Ong[‡], K. Ragan[¶], R. A. Scalzo[††] and J. Zweerink[‡]

*Santa Cruz Institute for Particle Physics, University of California, Santa Cruz, California USA
[†]Physics Department, College of Wooster, Wooster, Ohio USA
**Barnard College and Columbia University, New York, New York USA
[‡]Division of Astronomy and Astrophysics, University of California, Los Angeles, California USA
[§]Department of Physics, Case Western Reserve University, Cleveland, Ohio USA
[¶]Department of Physics, McGill University, Montreal, Quebec Canada
[||]Centre for Subatomic Research, University of Alberta, Edmonton, Alberta Canada
[††]Department of Physics, University of Chicago, Chicago, Illinois USA

Abstract. STACEE is an atmospheric Cherenkov detector using the large mirror area of a solar research facility to obtain a low energy threshold. The peak of a detected power law spectrum is around 100 GeV. An exciting possibility for STACEE is to follow up gamma-ray burst alerts from satellites. The low energy threshold of STACEE allows detection of gamma rays from higher redshifts than most other ground-based experiments. The STACEE instrument can be re-targeted to the position of a GRB within about four minutes of an alert to search for emission above 50 GeV. We discuss the STACEE sensitivity to high energy gamma-ray emission from GRBs.

THE STACEE TELESCOPE

STACEE uses the National Solar Thermal Test Facility (NSTTF) at Sandia National Laboratories outside Albuquerque, New Mexico, USA. The NSTTF is located at 34.96° N, 106.51° W and is 1700 m above sea level. The facility has 220 heliostat mirrors designed to track the sun across the sky, each with 37 m^2 area. STACEE uses 64 of these heliostats to collect Cherenkov light produced by cascades in the atmosphere.

STACEE employs five secondary mirrors on the solar tower to focus the Cherenkov light onto photomultiplier tube (PMT) cameras, as shown in Figure 1. The light from each heliostat is detected by a separate PMT and the waveform of the PMT signal is recorded by a flash ADC. A programmable digital delay and trigger system[1] selects showers for acquisition while eliminating most random coincidences of night sky background photons. Details about an early version of the instrument can be found in D. S. Hanna et al. [2].

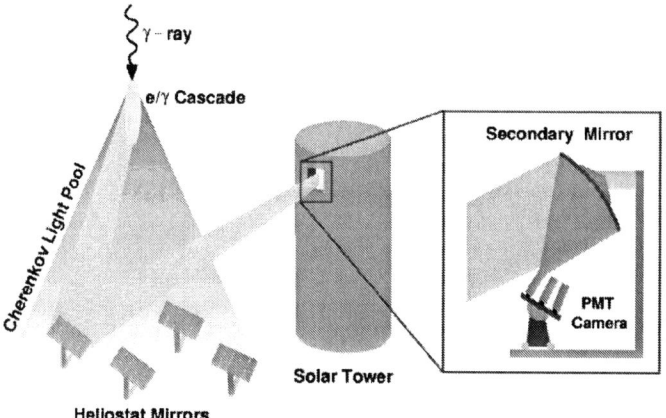

FIGURE 1. The STACEE technique. Cherenkov light produced in the atmosphere is reflected by the heliostat mirrors, which track the candidate source, to stationary secondary mirrors on the solar tower. The secondary mirrors focus the light from each heliostat onto a distinct PMT in the PMT camera.

STACEE PERFORMANCE

The large mirror area used by STACEE leads to an energy threshold – defined as the peak of a detected power law spectrum – around 100 GeV, with significant effective area as low as 50 GeV. This threshold is lower than most other current ground-based detectors. The low energy threshold opens up the possibility of detecting more distant sources, as shown in Figure 2. Collisions of high energy gamma rays with starlight photons to create electron-positron pairs attenuate the gamma-ray flux from more distant sources. The extinction becomes more severe with increasing energy, producing an energy-dependent horizon for gamma-ray observations.

The construction of the STACEE experiment is complete. However, analysis methods, including background rejection techniques, are still under development. STACEE operates with a trigger rate of about 8 Hz and a trigger threshold around 4 photoelectrons per heliostat, corresponding to a trigger threshold of about 50 GeV. Including the anticipated contributions from analysis methods under development, approximately 25 hours would be required for a 10σ detection of the Crab – analysis techniques to reject background cosmic ray events should improve this to 4 hours.

The STACEE sensitivity to a GRB will be about 2×10^{-9} cm^{-2} s^{-1} above 70 GeV (5σ in a 30 minute observation). STACEE would easily detect the flux estimated by power-law extrapolations of the EGRET data. For example, the flux from GRB940217 [3] extrapolated to STACEE energies is $\sim$50 times higher than this sensitivity.

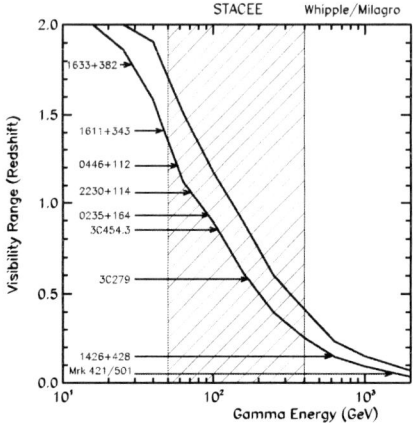

FIGURE 2. Expected gamma-ray horizon *vs.* energy. As source distance increases, lower gamma-ray energies are required to evade absorption by the extragalactic background light (EBL). The curves represent a range of plausible EBL models. The hatched band shows the region covered by STACEE that is largely inaccessible to other present generation experiments, *e.g.* Whipple and Milagro.

TABLE 1. Summary of prompt burst alerts received by STACEE since September 1, 2002

GRB	UTC Time	Spacecraft Providing Alert	Notice Delay (minutes)	Delay until Observable (hours)	STACEE Observations
021004	12:06:14	HETE	0.8	14.1	None
021112	03:28:16	HETE	81	3.2	Starting 219 min after burst; 112 min on burst position
021211	11:18:34	HETE	0.4	0.0	None; bad weather
030115	03:22:34	HETE	71	8.4	None; full moon*
030227	08:42:16	INTEGRAL	48	17.7	None
030324	03:12:43	HETE	0.4	2.0	Starting 123 min after burst; 56 min on burst position
030328	11:20:58	HETE	53	16.7	None
030329	11:37:15	HETE	73	15.2	None
030418	09:59:19	HETE	3.6	17.2	None
030501A	03:10:19	INTEGRAL	0.3	4.6	Starting 369 min after burst; 28 min on burst position
030519	14:04:54	HETE	0.6	Infinite	None
030528	13:03:03	HETE	0.6	17.6	None
030723	06:28:18	HETE	0.8	Infinite	None
030824	16:47:35	HETE	60	11.7	None

* No observations scheduled that night because there were less than 3 hours of darkness

GRB OBSERVATIONS

Observing gamma-ray bursts is a high priority for STACEE. The GCN burst alerts are monitored with a computer program which alerts the STACEE operators if one is visible from the STACEE site. The computer network link to the STACEE site has occasional outages which do not otherwise interfere with STACEE operations. To insure that burst alerts are not missed as a result of such an outage, we have recently equipped the STACEE operators with a pager which receives alerts directly from GCN. The STACEE instrument can be re-targeted to the position of a GRB within about four minutes to search for emission above 50 GeV. We also search for afterglow emission from bursts that have occurred within the previous 12 hours.

The ability of STACEE to observe the GRB source position within minutes of the first emission is very significant. EGRET detected GeV emission, including an 18 GeV photon, from GRB940217 up to ninety minutes after the start of the burst [3].

Since September 1, 2002, we have received notices containing a burst localization within 90 minutes of the burst onset for 14 GRBs, listed in Table 1. STACEE can only observe bursts within 60° of zenith and when both the sun and the moon are below the horizon. The time until these three conditions were met is given in the table. We have taken data for three of the bursts, as indicated. Preliminary analysis of the data from GRB 021112 and GRB 030501 shows no evidence for a detection. The data from the third burst were compromised by difficulties with the data acquisition system and will require further effort before a result can be obtained.

STACEE operates with a duty cycle of approximately 8%. The primary constraints are daylight and moonlight, with most of the remaining time lost because of bad weather. Once Swift is launched, we expect to have rapid observations for about 2% of the bursts, or about 3 per year, with afterglow observations within the first 24 hours for an additional 10% of the bursts, about 15 per year. These numbers assume that the bursts are found isotropically on the sky. To the extent that the Swift field of view is aligned in the anti-solar direction, the number of Swift bursts visible to STACEE could as much as double.

ACKNOWLEDGMENTS

We are grateful to the staff at the National Solar Thermal Test Facility, who continue to support our science with enthusiasm and professionalism. This work is supported in part by the National Science Foundation, the Natural Sciences and Engineering Research Council, FQRNT (Fonds Quebecois de la Recherche sur la Nature et les Technologies), the Research Corporation, and the California Space Institute.

REFERENCES

1. Martin, J.-P., and Ragan, K., "A Programmable Nanosecond Digital Delay and Trigger System," in *Proc. IEEE Nuclear Science Symposium*, 2000, vol. 8, pp. 12–141–12–144.
2. D. S. Hanna *et al.*, *Nucl. Instrum. Methods Phys. Res. A*, **491**, 126–151 (2002).
3. K. Hurley *et al.*, *Nature*, **372**, 652–654 (1994).

Neutrinos and Gamma Rays from Photomeson Processes in Gamma Ray Bursts

Armen Atoyan* and Charles D. Dermer[†]

*Centre de Recherches Mathématiques, Université de Montréal, Montréal, Canada H3C 3J7
[†]Code 7653, Naval Research Laboratory, Washington, DC 20375-5352 USA

Abstract. Acceleration of high-energy hadrons in GRB blast waves will be established if high-energy neutrinos are detected from GRBs. Recent calculations of photomeson neutrino production are reviewed, and new calculations of high-energy neutrinos and the accompanying hadronic cascade radiation are presented. If hadrons are injected in GRB blast waves with an energy corresponding to the measured hard X-ray/soft γ-ray emission, then only the most powerful bursts at fluence levels $\gtrsim 3 \times 10^{-4}\,\mathrm{erg\,cm^{-2}}$ offer a realistic prospect for detection of ν_μ. Detection of high-energy neutrinos are likely if GRB blast waves have large baryon loads and Doppler factors $\lesssim 200$. Significant limitations on the hadronic baryon loading and the number of expected neutrinos are imposed by the fluxes from pair-photon cascades initiated in the same processes that produce neutrinos.

INTRODUCTION

In recent work [1], we considered neutrino production for two leading scenarios for the sources that power long-duration GRBs, namely the collapsar [2] and supranova (SA) [3] models. In the collapsar model, the core of a massive star collapses directly to a black hole, and the most important radiation field for photomeson neutrino production is the internal synchrotron radiation field [4]. In the supranova (SA) model, a pulsar nebula synchrotron radiation within an expanding supernova remnant (SNR) shell [5] provides additional external photon target for photomeson interactions (see also Ref. [6]).

We found that the presence of the external field in the SA model can increase the number of detectable neutrinos by an order of magnitude or more over the collapsar model when the Doppler factor $\delta \gtrsim 200$. When $\delta \lesssim 200$, the internal synchrotron field can become effective for photomeson interactions. In our calculations, we assumed that the energy injected in protons is equal to the energy of electrons producing the photon fluence measured at X-ray and γ-ray energies. In both models, the likelihood of detecting a neutrino from a GRB with a km-scale detector such as IceCube is small except for rare GRBs with fluence $\gtrsim 3 \times 10^{-4}\,\mathrm{erg\,cm^{-2}}$. If GRB blast waves are strongly baryon-loaded, however, as required in a GRB model for high-energy cosmic rays [7], then we predict that 100 TeV – 100 PeV neutrinos will be detected several times per year with IceCube when $\delta \lesssim 200$.

Here we summarize our calculations of photomeson neutrino production for the collapsar and SA models, and present new calculations for the hadronic cascade radiation from a GRB.

MODEL

The GRB model is adapted from our photo-hadronic model for blazar jets [8], and takes into account the injection of nonthermal protons which lose energy through photomeson interactions. Protons are injected with a number spectrum $\propto E_p^{-2}$ at comoving proton energies $E_p > \Gamma$ GeV up to a maximum proton energy determined by the condition that the particle Larmor radius is smaller than both the size scale of the emitting region and the photomeson energy-loss length. Here Γ is the Lorentz factor of the blast wave. The observed synchrotron spectral flux in the prompt phase of the burst is parameterized by the expression $F(v) \propto v^{-1}(v/v_{br})^\alpha$, where $hv_{br} = 300$ keV, $\alpha = -0.5$ above v_{br} and an exponential cutoff at 10 MeV, and $\alpha = 0.5$ when $30\,\text{keV} \leq hv \leq hv_{br}$. At lower energies, $\alpha = 4/3$. The observed total hard X-ray/soft γ-ray photon fluence $\Phi_{tot} \cong t_{dur} \int_0^\infty dv F(v)$, where t_{dur} is the characteristic duration of the GRB. We consider a source at redshift $z = 1$ and take the hard X-ray/soft γ-ray fluence $\Phi_{tot} \gtrsim 3 \times 10^{-5}$ erg cm^{-2}. Two or three GRBs should occur each month above this fluence level.

A total amount of energy $E' = 4\pi d_L^2 \Phi_{tot} \delta^{-3}(1+z)^{-1}$ is injected in the form of accelerated proton energy into the comoving frame of the GRB blast wave. Here z is the redshift and d_L is the luminosity distance. The energy deposited into each of N_{sp} light-curve pulses (or spikes) is therefore $E'_{sp} = E'/N_{sp}$ ergs. We assume that all the energy E'_{sp} is injected in the first half of the time interval of the pulse (to ensure variability in the GRB light curve), which effectively corresponds to a characteristic variability time scale $t_{var} = t_{dur}/2N_{sp}$. The proper width of the radiating region forming the pulse is $\Delta R' \cong t_{var} c\delta/(1+z)$, from which the energy density of the synchrotron radiation can be determined [8]. We set the GRB prompt duration $t_{dur} = 100$ s, and let $N_{sp} = 50$, corresponding to $t_{var} = 1$ s. The magnetic field is determined by assuming equipartition between the energy densities of the magnetic field and the electron energy.

For the SA model, we assume the existence of an external radiation field given by the expression $vL_v \propto v^{1/2} \exp(-v/v_{ext})$, with $hv_{ext} \approx 1$ keV [5]. The intensity of this field is determined by the assumption that the integral power $L_{ext} = \int_0^\infty L_v dv$ is equal to the power of the pulsar wind $L_{pw} \approx (10^{53}\,\text{erg})/t_{delay}$, assuming that a total of $\approx 10^{53}$ erg of pulsar rotation energy is radiated during the time t_{delay} (which is here set equal to 0.1 yr) from the rotating supramassive neutron star before it collapses to a black hole. The energy $hv_{ext} \simeq 0.1$ keV is the characteristic energy of synchrotron radiation emitted by electrons (of the pulsar wind) with Lorentz factors $\gamma_{pw} \sim 3 \times 10^4$ in a randomly ordered magnetic field of strength ≈ 10 G. The radius $R = 0.05ct$ is determined by assuming that $0.05c$ is the mean speed of the SNR shell, and that the external photon energy density $\propto L/2\pi R^2$.

Fig. 1 shows the total v_μ fluences expected from a model GRB with $N_{sp} = 50$ pulses. The thin curves show collapsar model results at $\delta = 100, 200$, and 300, with $\Gamma = \delta$. The expected numbers of v_μ that a km-scale detector such as IceCube would detect are $N_v = 3.2 \times 10^{-3}$, 1.5×10^{-4}, and 1.9×10^{-5}, respectively. There is no prospect to detect v_μ from GRBs at these levels. The heavy solid and dashed curves in Fig. 1 give the SA model predictions of $N_v = 0.009$ for both $\delta = 100$ and $\delta = 300$. The equipartition magnetic fields are 1.9 kG and 0.25 kG, respectively. The external radiation field in the SA model makes the neutrino detection rate insensitive to the value of δ (as well

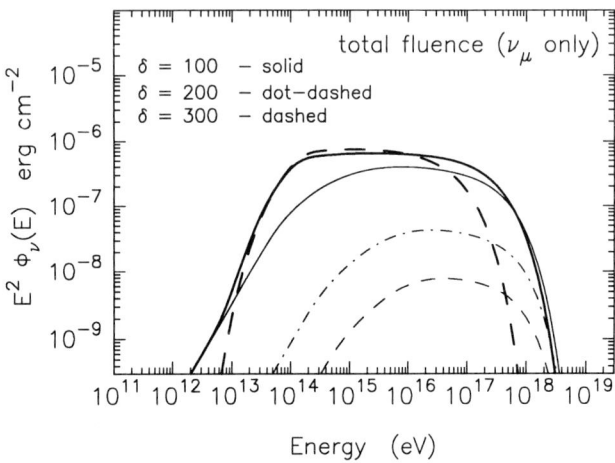

FIGURE 1. Energy fluence of photomeson muon neutrinos (ν_μ) for a model GRB with hard X-ray fluence $\Phi_{tot} = 3 \times 10^{-5}\,\mathrm{erg\,cm^{-2}}$ for different Doppler factors δ. The thin curves show collapsar model results where only the internal synchrotron radiation field provides a source of target photons.

as to $t_{var} \gtrsim 0.1\,\mathrm{s}$, as verified by calculations), but there is still little hope that a km^3 detector could detect such GRBs. Neutrino production efficiency would improve by at most a factor of 3 in the collapsar model if $t_{var} \sim 1\,\mathrm{ms}$ (and $N_{sp} = 5 \times 10^4$ to provide the same total fluence). Such narrow spikes are, however, then nearly opaque to γ rays with energies $\gtrsim 100\,\mathrm{MeV}$ [1]. Even in this case, there is little hope to detect neutrinos except from rare GRBs with fluence $\Phi_{tot} \gtrsim 3 \times 10^{-4}\,\mathrm{erg\,cm^{-2}}$.

Fig. 2 shows new calculations of photomeson neutrino production for a GRB with $\Phi_{tot} = 3 \times 10^{-4}\,\mathrm{erg\,cm^{-2}}$ and $\delta = 100$, as well as the accompanying electromagnetic radiation induced by pair-photon cascades from the secondary electrons and γ rays from the same photomeson interactions. The total number of ν_μ expected with IceCube is $\cong 0.1$. The total fluence of cascade photons shown here is contributed by lepton synchrotron (dot-dashed) and Compton (dashed) emissions. For comparison, the dotted curve shows the primary lepton synchrotron radiation spectrum assumed for the calculations. The level of the fluence of the cascade photons is $\approx 10\%$ of the primary synchrotron radiation. This means that the maximum allowed baryon loading for these parameters cannot exceed a factor of ≈ 30 in order not to overproduce the primary synchrotron radiation fluence. This limits the maximum number of ν_μ to ≈ 3 even in the case of large baryon loading for rare, powerful GRBs. These numbers cannot be further increased in the SA model because of the efficient extraction that is already provided by internal photons alone for these parameters.

In conclusion, we predict that at most a few high-energy ν_μ can be detected with IceCube even from a very bright GRB at the fluence level $\Phi_{tot} \gtrsim 3 \times 10^{-4}\,\mathrm{erg\,cm^{-2}}$, and only when the baryon loading is high [1, 7]. This is because the detection of a single ν_μ requires a ν_μ fluence $\gtrsim 10^{-4}\,\mathrm{erg\,cm^{-2}}$ above 1 TeV. Since the energy release in high-energy neutrinos and electromagnetic secondaries is about equal, this energy will

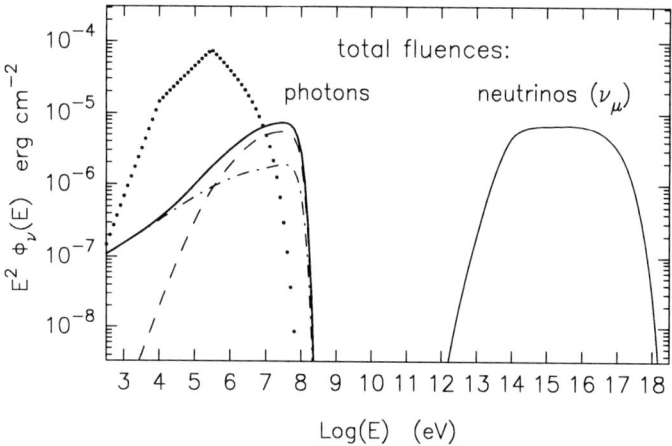

FIGURE 2. Energy fluence of photon and photomeson muon neutrinos for a collapsar-model GRB with hard X-ray fluence $\Phi_{tot} = 3 \times 10^{-4}$ erg cm^{-2} and $\delta = 100$. The dotted curve shows the fluence of a GRB used for calculations, and the dashed and dot-dashed curves show the Compton and synchrotron contributions to the photon fluence from the electromagnetic cascade initiated by secondaries from photomeson processes, respectively.

be reprocessed in the pair-photon cascade and emerge in the form of observable radiation at hard X-ray/soft γ-ray energies, and this radiation cannot exceed the measured fluence in this regime. This imposes a robust limit on the maximum number of ν_μ even from a GRB with very high baryon loading.

ACKNOWLEDGMENTS

AA thanks the NRL High Energy Space Environment Branch for support and hospitality during visits. The work of CD is supported by the Office of Naval Research and NASA GLAST science investigation grant DPR # S-15634-Y.

REFERENCES

1. C. D. Dermer and A. M. Atoyan, *Phys. Rev. Lett.* **91**, 071102, (2003).
2. C. L. Fryer, S. E. Woosley, and D. H. Hartmann *Astrophys. J.* **526**, 152 (1999).
3. M. Vietri and L. Stella, *Astrophys. J.* **507**, L45 (1998).
4. E. Waxman and J. N. Bahcall, *Phys. Rev. Lett.* **78**, 2292 (1997).
5. A. Königl and J. Granot, *Astrophys. J.* **574**, 134 (2002).
6. Razzaque, S., Meszaros, P., and Waxman, E. *Phys. Rev. Lett.* **90**, 241103 (2003).
7. S. D. Wick, C. D. Dermer, and A. Atoyan, *Astroparticle Physics*, submitted (astro-ph/0310667).
8. A. M. Atoyan and C. D. Dermer, *Astrophys. J.* **586**, 79, (2003).

Gravitational Waves and GRBs from Tidal Disruption of Stars in the Center of Galaxies

P. Fortini* and A. Ortolan[†]

Department of Physics, University of Ferrara and INFN Sezione di Ferrara, via Paradiso 12, 44100 Ferrara, Italy
[†]*INFN - National Laboratories of Legnaro, viale dell'Università 2, 35020 Legnaro (PD), Italy*

Abstract. Recent measurements by the Chandra satellite have shown that a supermassive black hole of $M = 2.6 \times 10^6 M_\odot$ is located in the Galactic Center; it seems probable from other observations that this fact is common in the majority of galaxies. On the other hand, GRB explosions are typical phenomena linked to galactic dynamics. In the present paper we discuss the possibility that GRBs are tidal disruption of stars by supermassive black holes located in the center of galaxies. This conjecture can be tested by a gravitational wave detector of the class of AURIGA.

INTRODUCTION

GRB engines are characterized by the following facts: 1) they are point-like sources of electromagnetic energy ($10^{51 \div 54}$ *erg*) comparable to that of a single galaxy; 2) they are always associated with a host galaxy (see ref. [1]). A plausible explanation can be found if there is a Supermassive Black Holes (SBH) in the Galactic Center with a mass of about $2 \times 10^6\ M_\odot$ as it was recently discovered by satellite Chandra (see e http://science.nasa.gov/headlines/y2000/ast29feb_1m.htm of Feb. 29, 2000). Moreover, it is very likely that the globular clusters have in their centers black holes with smaller masses.

The detailed study of the dynamics of the Galactic Center (see ref. [2]) points out that near the horizon of the SBH all the galactic objects are crushed by tidal disruption so that they can emit both electromagnetic and gravitational waves. For instance according to this view a GRB can be considered as the end of a star while it is swallowed by the SBH.

If this conjecture is correct, it should be possible to detect gravitational radiation by means of gravitational wave (gw) detectors using their directional sensitivity (antenna pattern) towards the candidate sources – either the Galactic Center or the center of nearby galaxies (M31, M84, M87, etc.) or the globular clusters. The basic idea is to separate the collected events recorded by a gravitational wave detector into the *"on source"* and *"off source"* sets depending on the sidereal hours when the detector is pointing toward a given source. The "off source" events will give an estimate of the background events (from unmodeled noise sources) while gw bursts can be detected as a mean excess of event energy in the "on source" set.

The paper is divided into two parts: in Section 2 we discuss some theoretical problems connected with the SBH model and in Section 3 we describe a method for the measurement of gw emission from the Galactic Center by means of one or more gw detectors.

THEORETICAL PROBLEMS OF SBH

The paper by Ayal et al. [3] is very important because, in our opinion, it was the first attempt to calculate, using the 1PN approximation of General Relativity, what happens to a star falling into the SBH existing in the center of the Galaxy. However, these calculations are in some ways too rough, in particular:

1. An object like the sun is modeled as a polytropic fluid in a spherical configuration: only in this way it can be treated mathematically, but its physical structure is completely ignored.
2. Near a few Schwarzschild radii the classical mechanics break down and the system of a star and a black hole, which has well known obvious closed form solutions in the classical limit, has no relativistic closed form solution
3. The various approximations in general don't converge and the usual result is the appearance of divergences which make the two body problem mathematically intractable. These issues are covered in Chapter 9 of ref. [2]. Here we quote the main results. The singularities are of three kinds: i) a singularity in the density: stars cannot exist nearer than the horizon, and effectively they are destroyed before that point, either by collisions or by the SMB tidal field. In ref. [3] it is shown with explicit calculation that a polytropic sphere similar to the sun is destroyed at a distance of approximately $15R_{Sch}$ where R_{Sch} is the SMB horizon; ii) a singularity in the velocity which makes the dynamics mathematically intractable because of the Keplerian velocity divergence; iii) an optical singularity due to the fact that any mass bends light and, close to massive black hole and in small regions, some caustics appear, which cause the kinematics to be equally divergent.
4. Due to the severe limitation of the singularity of the calculations it is very difficult to believe the numbers put forth by Ayal et al. [3]. For instance, the energy emitted in gw is $\sim 1.6 \cdot 10^{46} erg$ and it could be underestimated by orders of magnitude. This value ensures that, on one hand the 1PN approximation holds, and on the other the gravitational interaction is sufficiently strong to destroy the star. It should be noted that the energy converted in gw is calculated by the usual quadrupole formula and therefore, if it were possible to push the approximation closer to the black hole horizon, one would expect a greater gw emission.

Having made the above points, it would be highly desirable to pass directly to the measurement of gws, in order to test the various approximations and the effects of non-linearities in the mechanism of gw emission.

DESCRIPTION OF THE MEASUREMENT

The gravitational wave detector AURIGA, which is operating at the National Laboratories of Legnaro (Italy), is essentially an aluminum cylinder equipped with a resonant capacitive transducer which is coupled to a high sensitivity dc-squid amplifier [4]. Due

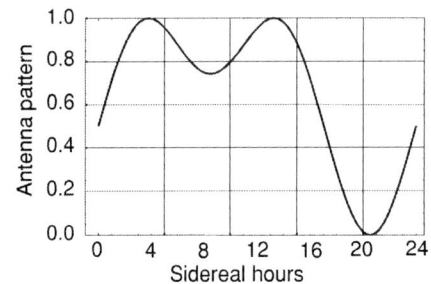

FIGURE 1. Relative sensitivity of the AURIGA detector towards the Galactic Center.

to the combination of the Earth's rotation and the antenna pattern, two times a day AURIGA is much more sensitive to the gw flux from the Galactic Center (see Figure 1). This particular combination happens at the same sidereal hours, when the Galactic Center direction is orthogonal to the detector symmetry axis. The modulation of the gw flux must have a period of one sidereal day with the two characteristic maxima of Figure 1 occurring at sidereal hours 4 and 13. This peculiar signature of the gw signals can be used to test if the conjecture we presented in the preceding Sections is true. In fact, the Galactic Center is the place where a high flux of gravitational waves should be emitted by stars and gas swallowed by the black hole. The cumulative counts of detected events should exhibit two peaks separated by 11 hours. If we collect data for a sufficiently long time (~ 100 sidereal days), we can form two distinct set of events: the "on source" set which collects all the events occurring around sidereal hours 4 and 13, and the "off source" sets corresponding to events around sidereal hour 20. The "off source" events will give an estimate of the background events (from unmodeled noise sources) while gravitational wave bursts (GWBs) can be detected as an excess of event energy (in average) for events belonging to the "on source" set.

The Mann–Whitney test, also known as the rank sum test or U–test [5], can be used to falsify the "null hypothesis" that the two populations "on" and "off" are identical, i.e. that the gw candidate events (belonging to the "on source" set) have the same mean energy $\langle E \rangle$ of background events (belonging to the "off source set"). The mean energy captured by the detector for the two sets can be expressed as

$$\langle E_{on} \rangle = \langle E_{off} \rangle + h_{RMS}^2 , \qquad (1)$$

where h_{RMS}^2 is the averaged GWB amplitude associated with the activity of the Galactic Center. The sensitivity of the search depends on three main factors [6], namely: i) the minimal GWB amplitude detectable at unit signal-to-noise ratio $h_{min} \equiv (\tau_s (\int_{-\infty}^{+\infty} 1/S_h(\nu) d\nu)^{-1/2}$, where τ_s is the duration of the GWB (~ 1 $msec$) and $S_h(\nu)$ is the power spectral density of the noise expressed in terms of gravitational wave amplitude at the detector input; ii) the event search threshold, usually set to $5 \times h_{min}$ and iii) the number of the events in the "on" and "off" sets, N_{on} and N_{off} respectively.

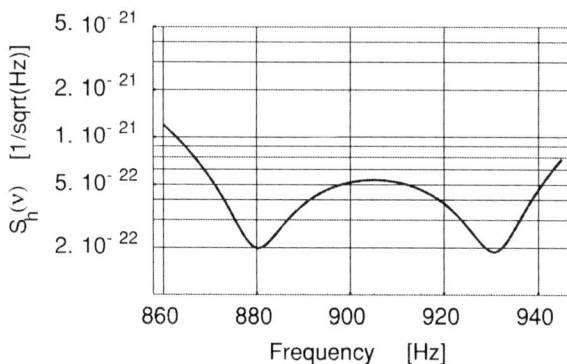

FIGURE 2. Power spectral density of the AURIGA noise expressed in terms of gravitational wave amplitude at the detector input.

A similar technique has been applied by our group (see ref. [6] and [7]) to test if there exists a concomitant emission of GRBs and gw bursts. Unfortunately, as most GRBs are at cosmological distances, the AURIGA sensitivity ($h_{min} \sim 10^{-19}$) and duty cycle ($N_{on} \approx 100$) were not sufficient to detect any positive effect. However, in the present paper we limit our analysis at the distance of the Galactic Center (i.e. $\sim 8\ kpc$), and so the probability to detect GWBs is higher.

In addition, the new experimental setup of AURIGA [4] promises an increase of its sensitivity and bandwidth as reported in Figure 2, which translates into a burst sensitivity of $h_{min} \sim 10^{-20}$. The same procedure can be applied to the events in coincidence among two or more parallel detectors. To conclude we must recall that the group of Rome have already tried to detect GWBs coming from the galactic plane (see ref. [8] and the subsequent criticism in ref. [9]).

REFERENCES

1. van Paradijs, J., Kouveliotou, C., and Wijers, R. A. M. J., *Annal Rev. of Astron. and Astroph.*, **38**, 379–425 (2000).
2. Falcke, H. F., and Hehl, F. W., *The galactic black hole*, I.O.P., 2003, pp. 212–213.
3. Ayal, S., Livio, M., and Piran, T., *Astrophys.J.*, **545**, 772–780 (2000).
4. Zendri, J. P. et Al.., *Class. Quantum Grav.*, **19**, 1925–1933 (2002).
5. Mann, H. B., and Whitney, D. R., *Ann. Math. Stat.*, **18**, 52–54 (1947).
6. Tricarico, P. et Al.., *Phys. Rev. D*, **63**, 082002–7 (2001).
7. Tricarico, P., Ortolan, A., and Fortini, P., *Class. Quantum Grav.*, **20**, 3523–3531 (2003).
8. Astone, P. et Al.., *Class. Quantum Grav.*, **19**, 5449–5463 (2002).
9. Finn, L. S., *Class. Quantum Grav.*, **19**, L37–L44 (2002).

PROMPT EMISSION AND EARLY AFTERGLOWS

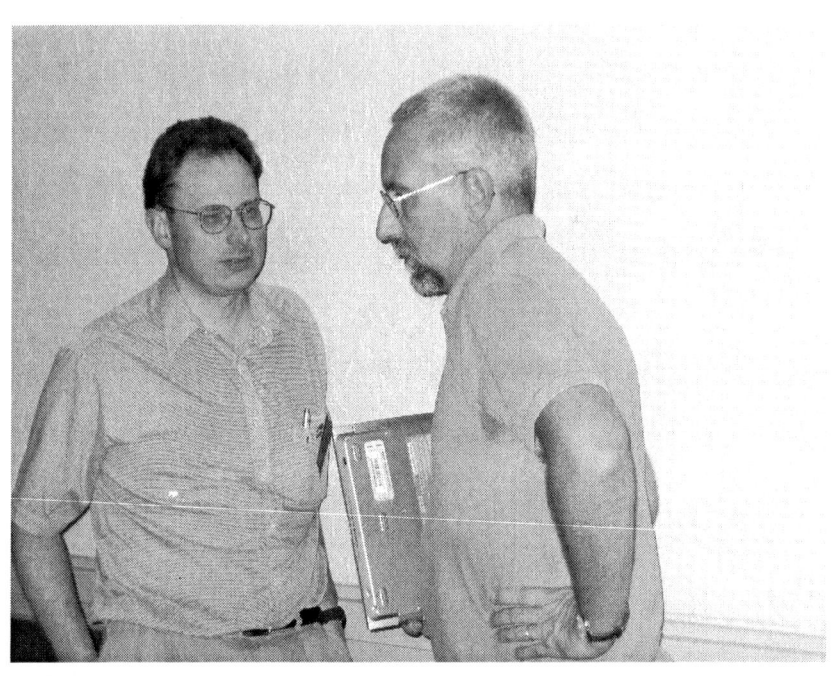

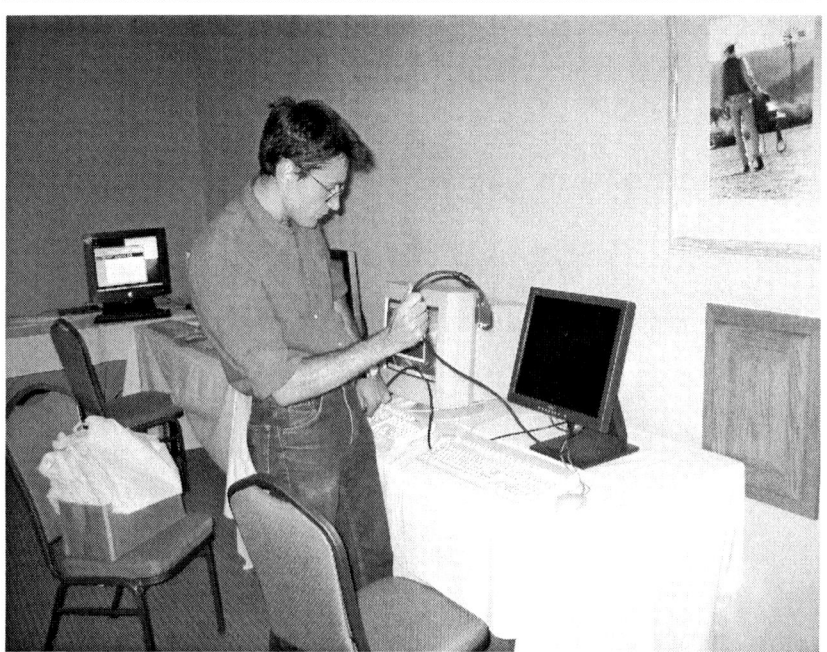

Some Recent Peculiarities of the Early Afterglow

Tsvi Piran*, Ehud Nakar* and Jonathan Granot[†]

Racah Institute for Physics, The Hebrew University, Jerusalem, 91904, Israel
[†]*Institute for Advanced Study, Einstein Drive, Princeton, NJ 08540, USA*

Abstract. We consider some recent developments in GRB/afterglow observations: (i) the appearance of a very hard prompt component in GRB 941017, and (ii) variability in the early afterglow light curves of GRB 021004 and GRB 030329. We show that these observations fit nicely within the internal-external shocks model. The observed variability indicates that the activity of the inner engine is more complicated than was thought earlier and that it involves patchy shells and refreshed shocks. We refute the claims of Berger et al. and of Sheth et al. that the radio and mm observations of GRB 030329 are inconsistent with refreshed shocks.

INTRODUCTION

The early afterglow and the GRB/afterglow transition are among the unexplored regimes of GRBs. Great progress on this front was achieved during the last year when, following fast HETE-2 identifications, two afterglows (GRBs 021004, 030329) were followed from very early on showing remarkable variability and rich structure. Also, somewhat unexpectedly, a search within the BATSE/EGRET archives revealed a new very hard long lasting (~ 200 s) component of GRB 941017. This is most likely a manifestation of the early afterglow and of the GRB/afterglow transition. We discuss these developments and their implication within the internal-external shocks model.

THE PROMPT HIGH ENERGY EMISSION FROM GRB 941017 AS A GRB/AFTERGLOW TRANSITION

Recently, González et al. [1] discovered a high energy tail that extended up to 200 MeV in the combined BATSE and EGRET data of GRB 941017. The tail had a hard spectral slope ($F_\nu \propto \nu^0$) up to 200 MeV. It appeared $\sim 10\text{-}20$ s after the beginning of the burst and displayed a roughly constant flux while the lower energy component decayed. At late times (~ 150 s after the trigger) the very high energy ($\sim 10\text{-}200$ MeV) tail had a luminosity ~ 50 times higher than the "main" γ-ray energy band (~ 30 keV-2 MeV).

Granot & Guetta [2] where the first to suggest that we see here another manifestation of the very early afterglow. Sari [3] has shown that for long bursts the external shocks (and hence the afterglow) begin $\sim R/c\Gamma^2$ after the beginning of the burst while the internal shocks are still going on and hence the burst is still active. This behavior was seen in the transition from the harder initial burst to the softer early afterglow [4, 5, 6]. It was also seen in the prompt optical flash of GRB 990123 where the lower (optical) energy component did not trace the ∼MeV γ-rays and a pronounced hard to soft evolution was seen in the γ-ray signal.

The very high energy of this emission suggests that it is inverse Compton. There are two relevant emitting regions in the early afterglow: the forward shock and the reverse shock. With typical parameters [7] the energy of the synchrotron photons and the electrons' Lorentz factor in the forward and reverse shocks are:

$$\nu_{\text{synch,F}} \approx 0.1 \text{ MeV } (\Gamma/300)^4 \quad ; \quad \gamma_{e,F} \approx 10^4 (\Gamma/300)$$
$$\nu_{\text{synch,R}} \approx 1 \text{ eV } (\Gamma/300)^2 \quad ; \quad \gamma_{e,R} \approx 300$$

There are four possible combinations of seed photons and scattering electrons:

	Rev. shock electrons	For. shock electrons
Rev. shock photons	$\sim 0.1 \text{ MeV}(\Gamma/300)^2$	$\sim 100 \text{ MeV}(\Gamma/300)^3$
For. shock photons	$\sim 10 \text{ GeV}(\Gamma/300)^4$	$\sim 10 \text{ TeV}(\Gamma/300)^5$ (within Klein Nishina)

While these approximate results are very sensitive to Γ, they indicate that inverse Compton scattering of the reverse shock photons on the forward shock electrons yeilds the right energy range. The detailed calculations of Pe'er & Waxman [8] confirm these naive estimates. The main problem, however, is not to explain the location of the spectral peak but to explain the spectral slope ($F_\nu \propto \nu^0$) and the temporal slope ($F_\nu \propto t^0$). Pe'er & Waxman [8] reproduce the spectral slope by requiring that the synchrotron self absorption frequency of the reverse shock emission would be high enough to effect the observed spectrum. Granot & Guetta [2] reproduce both the temporal and spectral behaviors by considering a slightly different scenario where the high energy component is produced by Synchrtron self Compton within the reverse shock. They require a slightly higher external density with a somewhat unusual profile, $\propto R^{-1}$, for a uniform ejecta shell (which may explain the rareness of the event). In both models the high energy component is a clear manifestation of the onset of the afterglow and the GRB/afterglow transition.

AFTERGLOW LIGHT CURVE VARIABILITY

Theory: The different scenarios that lead to afterglow temporal variability can be distinguished according to their characteristic features. Density variations produce only weak fluctuations above the cooling frequency, ν_c, and cannot produce sharp changes in the light curve. Energy variations produce variability both above and below ν_c, and can arise either due to refreshed shocks or due to a patchy shell structure. These two mechanisms produce very different light curves. While the former produce a step-wise increase in the light curve, the later produces random fluctuations with a decreasing amplitude.

Rees & Mészáros [9], Kumar & Piran [10] and Sari & Mészáros [11] suggested that slow shells take over the slowing down matter behind the afterglow shock and produce *refreshed shocks*. Slow shells with Γ_s emitted from the source right after the fast ejecta, catch up and collide with the slowing down ejecta at an observer time $t \sim 0.25(\Gamma_s/10)^{-8/3}(E_{\text{iso},52}/n_0)^{1/3}$ days (where $E_{\text{iso},52}$ is the isotropic kinetic energy in units of 10^{52} ergs and n_0 is the external density in cm^{-3}) when the ejecta's Lorentz factor drops slightly below Γ_s. The clearest feature of refreshed shocks is a monotonous increase in the overall energy. Therefore the observed flux can only increase (relative to the expected decay). The light curve has a step wise form with each step produced by the arrival of a single shell. This step wise structure is seen both above and below the cooling frequency with a similar amplitude. Each step (in the optical light curve) should be accompanied by a flare in low frequencies that is produced by the reverse shock which propagates back into the slow shell [10, 11]. The time scale, Δt of the steps and the corresponding flares depends on their timing relative to the jet break. Before the jet break the refreshed shocks are "locally" spherically symmetric and therefore the angular time imposes $\Delta t \sim t$ [10]. The intensity of the reverse shock flare in this regime is calculated in [10], and the decay after the peak is $\propto t^{-2}$ [7]. In the post-break case the cold slow shells may not expand sideways (if cold enough). Then they keep their original angular size, θ_j, which is smaller than Γ_s^{-1}. In this case $\Delta t \approx t_j < t$, where t_j is the jet break time [12] and the transition is fast. A reverse shock flare is expected in this case as well. However, its frequency, intensity and temporal decay (which is expected to be steeper than in the spherical case) are much harder to calculate.

Kumar & Piran [13] suggested, in the *Patchy shell model*, that the shells have an intrinsic angular structure. As the blast wave decelerates the angular size of the observed region ($\sim 1/\Gamma$) increases. The effective (average) energy of the observed region and hence the observed flux, relative to the expected decay, varies with time depending on the angular structure. The variability time scale is $\Delta t \sim t$ [14]. The averaging over a larger and larger random structure leads to a decay of the envelope as $t^{-3/8}$ [14, 15]. An important feature of this scenario is the break of the axial-symmetry and therefore the production of a linear polarization. The variation of the polarization, both in degree and in angle, are correlated with the light curve variations [14, 16]. The variability will be observed both above and below the cooling frequency v_c with a similar amplitude.

Wang & Loeb [17], Lazzati et al. [18] and Nakar et al. [15] considered *External density variations*. Such variations may result from ISM turbulence or from a variable pre-burst stellar wind. Wang & Loeb [17] analyzed the light curve resulting from mild density fluctuations due to ISM turbulence. They show that these density fluctuations can produce short time scale ($\Delta t < 0.1t$) and low amplitude ($\sim 10\%$) fluctuations in the light curve. Lazzati et al.[18], Nakar et al. [15] and Nakar & Piran [19] considered large amplitude spherical density fluctuations. A basic feature of the resulting light curve that distinguishes it from energy variations is that in the former the light curve is different above and below the cooling frequency, v_c. Density variations produce only weak fluctuations above v_c. The amplitude of the fluctuations above v_c is at most tens of percents and it is much smaller than the amplitude of the fluctuations below v_c [19].

A second feature of density fluctuations is their inability to produce a sharp variation (either increase or decrease) in the light curve [20]. First, we note that because of angular spreading, spherical density drops cannot produce decays sharper than $t^{-2.6}$ [19, 21] and even this decay is reached very slowly. More interesting is the fact that even a sharp density enhancement cannot produce a steep increase in the light curve. The earlier calculations [15, 18, 19] assumes that the ejecta can be described by a Blandford-McKee solution whose density profile varies instantaneously according to the external density. These calculations do not account, however, for the reverse shock resulting from density enhancement and its effect on the blast-wave. Thus the above models are limited to slowly varying and low contrast density profiles. Now, the observed flux depends on the external density, n, roughly as $n^{1/2}$. Thus, a large contrast is needed to produce a significant re-brightening. Such a large contrast will, however, produce a strong reverse shock which will sharply decrease the Lorentz factor of the emitting matter behind the shock, Γ_{sh}, causing a sharp drop in the emission below ν_c and a long delay in the arrival time of the emitted photons (the observer time is $\propto \Gamma_{sh}^{-2}$). Both factors combine to suppresses the flux and to set a strong limit on the steepness of the re-brightening events caused by density variations. Note that, while nonspherical density fluctuations may lead to a steeper decline, they usually do not lead to a steeper increase in the flux.

Implications: The early afterglow of **GRB 021004** showed clear deviations from a smooth power law decay, lasting from 0.04 days to 3 days. The fluctuations in the light curve where accompanied by fluctuations both in the degree and in the angle of the polarization ([14] and references therein).

The steep decays after each bump imply that the variations do not result from refreshed shocks. Thus, variable external density variations [15, 18, 22] and the patchy shell model [14, 15] were considered as possible explanations. Unfortunately, the X-ray observations are not detailed enough to clearly distinguish between density and energy variations (although the former are favored by [22]). However, the sharp decays cannot be produced by "locally" spherical density variations [19]. Furthermore, the first bump requires, using the instantaneous Blandford-McKee approximation, an increase in the external density by a factor of ~ 10 over $\Delta R/R \approx 0.05$ [18, 19]. Such a density contrast produces a mildly relativistic reverse shock which reduce Γ_{sh} by a factor of ≈ 2, making the approximation inconsistent. Preliminary results [20] of detailed numerical simulations (including both hydrodynamics and synchrotron radiation) suggest that this bump cannot be produced by density enhancement (due to the suppressed forward shock emission, caused by the reverse shock). These results leave the patchy shell as the only viable explanation. Indeed, Nakar & Oren [14] show that patchy shell can reproduce the light curve (including the sharp rise and steep decay of the first bump). They show that angular energy profiles which produce the observed light curve produce also a polarization curve that fits the observed polarization. We conclude that angular energy fluctuations are the dominant process that produce the observed fluctuations in GRB 021004.

In addition to the remarkable supernova signature, the optical afterglow **GRB 030329** has shown also a unique variability ([12] and references therein). Several step-wise bumps, at $t = 1.5, 2.6, 3.3$ and 5.3 days, are seen after the jet break at $t_j \approx 0.5$ days. The first bump was the largest and best monitored, but even with the less dense monitoring of the later bumps the step-wise profile (where after each bump the original decay slope is resumed) is clear. All the bumps had a short rise time $\Delta t \approx 0.4$-0.8 d $< t$. The step-wise profile seems like a clear signature of post-break refreshed shocks [12], where $\Delta t \approx t_j < t$ because the later slower shells did not expand sideways before colliding with the faster earlier ejecta. Moreover, the energy injected in these shocks is 10 times the energy in the original blast-wave. This late (or rather slow) energy injection explains an additional peculiarity of this GRB: the low energy output in γ-rays and in the early X-ray afterglow.

Berger et al. [23] suggested that the first bump and the energy deficiency can be explained by a two component jet: A slow and energetic component with a wide half-opening angle (17°) dominates the afterglow after 1.5 days, and a fast component with a narrow half-opening angle (5°) dominates the afterglow before 1.5 days. The slow component is observed only after 1.5 days$\approx t_{\rm dec} \approx 0.5(\Gamma_s/10)^{-8/3}(E_{\rm iso,52}/n_0)^{1/3}$ days, since only at this time its reverse shock consumes the slow shell. However, this model, which received great publicity, predicts that the rise time, Δt, of the first bump should be of the same order as the observed time, $t_{\rm dec}$. Furthermore, it predicts a smooth light curve after the first bump. Both predictions are contradicted by the observations.

Millimeter observations of GRB 030329 [24] show that at 100 & 250 GHz the flux is rather constant during the first week. Most surprising are two measurements at 100 GHz, one before the first bump (0.6-1 d) and one after (1.7-1.9 d), which show a constant flux. Sheth et al. [24] and Berger et al. [23] claim that these results support the two component jet model and reject the refreshed shocks model due to the lack of radio flares. However this analysis overlooks the fact that the encounter of the slow component in the two-components jet model with the external matter produces a reverse shock which should produce a radio flash. This flash is analogous to the optical flash produced by the deceleration of the fast component [7], and to the radio flare expected in the refreshed shocks model. The timing of this flash is $t_{\rm dec} = 1.5$ d and its magnitude is easily calculated. The contribution from this reverse shock at $t_{\rm dec} = 1.5$ d is larger by a factor of up to $\Gamma_s^{5/3} \sim 15$ than the flux from the forward shock. Thus a very bright and fast fading millimeter flash is expected in the two component jet model. A more detailed calculation (using [11] and the parameters of the wide jet presented in [23]) shows that both v_a and v_m of the reverse shock at $t_{\rm dec}$ are around 100-200 GHz and that the flux at v_m is ~ 500 mJy which is an order of magnitude larger than the expected flux from the forward shock and than the observed fluxes at 100 & 250 GHz (~ 50 mJy) at this time.

The main argument of [23] in favor of a two-components jet is the existence of a second jet break in the radio after $t_{j,2} \sim 10$ d. However, even this argument is not strongly supported by the data. According to this model the flux below v_m should rise as $t^{1/2}$ before $t_{j,2}$ and decay as $t^{-1/3}$ after $t_{j,2}$. However, the data of [24] contradict this prediction. At 100 & 250 GHz the flux is constant before the passage of v_m and the decay after this passage (at $t = 6$ & 8 d $< t_{j,2}$, respectively) is steeper than $t^{-1.7}$ in both bands. This looks like a clear signature of a post break behavior in the radio at $t > 1$ d. Now the radio observations at lower frequencies (≤ 22 GHz) do not conform with the simple post-break model ($F_v \propto t^{-1/3}$). The flux at these frequencies rises with time before the passage of v_m. Interestingly enough, this radio behavior is exactly the one predicted as the post break radio behavior by [25], using a 2D relativistic hydrodynamical simulations. We find the striking similarity between the radio observations and Fig. 2 of [25] as a very strong support that the broad band data are totally consistent with a single jet.

Sheth et al., [24] emphasize the fact that radio flare [10] was not detected during the first bump at $t \sim 1.5$ days and argue that this rules out the refreshed shocks model. However, in the post-break refreshed shocks model, the radio flash is expected to be fainter and to decay faster than the radio flash in the two-components jet model (due to the lower energy and the lateral spreading of the slow shell as opposed to the more energetic and "locally" spherical wide jet). Thus, by assuming spherical symmetry, Sheth et al., [24] over estimate the expected flash in the post-break refreshed shocks scenario. It may be that the flash was missed by the sparse measurements due to its lower intensity and faster decay. A detailed (and highly nontrivial) calculations should be done in order find out.

SUMMARY

With an increasing flow of new observations we discover that GRBs and afterglows are richer than what was previously thought. The simple spherical theory had to be modified, first with jets and now with additional angular structure (patchy shells) and more extended velocity structure (refreshed shocks). The simple synchrotron theory has to be modified with inverse Compton scattering. One can worry, are we adding epicycles trying to revive a wrong theory? We don't believe so. Complications and variation are common in astrophysics and are found everywhere in nature. Moreover, the three main themes that have been introduced here: patchy shells, refreshed shocks and inverse Compton, were not invoked aposteriori to explain the new observations. On the contrary, all three have been suggested long ago. It is just natural and even reassuring to discover them when better data become available.

The research was supported by US-Israel BSF.

REFERENCES

1. González M. M., et al., Nature **424**, 749 (2003).
2. Granot, J. & Guetta, D., ApJ **598**, L11 (2003).
3. Sari, R., ApJL **489**, L37 (1997).
4. Burenin, R. A., et al., A&A **344**, L53 (1999).
5. Giblin, T. W., et al., ApJL **524**, L47 (1999).
6. Piro., L. in GRB in the afterglow era, Rome 17-20 2000, E. Costa, F. Frontera, J. Hjorth eds, Springer, p.97 (2000).
7. Sari, R. & Piran, T., ApJ **520**, 641 (1999).
8. Pe'er, A. & Waxman, E., astro-ph/0310836 (2003).
9. Rees, M. J.& Meszaros, P., ApJL **496**, L1 (1998).
10. Kumar, P., & Piran, T., ApJ **532**, 286 (2000).
11. Sari, R. & Meszaros, P., ApJL **535**, L33 (2000).
12. Granot, J., Nakar, E., & Piran, T., Nature **426**, 138 (2003).
13. Kumar, P.,& Piran, T., ApJ **535**, 152 (2000).
14. Nakar, E. & Oren, Y., astro-ph/0310236 (2003).
15. Nakar, E. , Piran, T, & Granot, J., New Astronomy **8**, 495 (2003).
16. Granot, J., & Königel, A., ApJL **594**, L83 (2003).
17. Wang, X. & Loeb, A., ApJ **535**, 788 (2000).
18. Lazzati, D., Rossi, E., Covino, S., Ghisellini, G., & Malesani D., AA **396**, L5 (2002).
19. Nakar E., & Piran T., ApJ **598**, 400 (2003).
20. Nakar E., Piran T. & Granot J, in preparation (2003).
21. Kumar, P. & Panaitescu, A., ApJ **541**, L51 (2000).
22. Heyl, J. & Perna, R., ApJL **586**, L13 (2003).
23. Berger, E. et al., ApJ **426**, 154 (2003).
24. Sheth, K. et al., ApJ **595**, L33 (2003).
25. Granot, J., Miller, M., Piran, T., Suen, W.M. & Hughes, P.A. in Gamma-Ray Bursts in the Afterglow Era, ed. E. Costa, F. Frontera, & J. Hjorth (Berlin; Springer) p. 312 (2001).

Early Stages of the GRB Explosion

A. M. Beloborodov

Physics Department, Columbia University, 538 West 120th Street New York, NY 10027

Abstract. Physics of GRB blast waves is discussed with a focus on two effects: (1) pair creation in the external medium by the gamma-ray front and (2) decay of neutrons ahead of the decelerating blast wave. Both effects impact the afterglow mechanism at radii up to 10^{17} cm.

INTRODUCTION

GRB afterglow is well explained as emission from a decelerating relativistic blast wave. Most of the afterglow data collected to date are obtained relatively late, hours or days after the prompt GRB, when the blast wave is already at final stages of deceleration. *Swift* satellite will provide the missing data on the early stage when the afterglow is most luminous.

The blast wave deceleration begins at a radius $R_{\text{dec}} \sim 10^{15} - 10^{17}$ cm, which depends on the ambient density n_0 and the initial Lorentz factor Γ_0 of the blast wave.[1] R_{dec} does not exceed 10^{17} cm, the estimated fireball size during the late afterglow, and can be one or two orders smaller than 10^{17} cm, especially if the circumburst medium has R^{-2} density profile. Three effects are predicted to occur at $R \sim 10^{16}$ cm:

- The reverse shock crosses the ejecta and can produce a detectable flash of soft emission (if the ejecta are not dominated by the Poynting flux) [1, 2].
- The external medium ahead of the forward shock is loaded with a large number of $e^{\pm}$ pairs by the prompt γ-ray front [3, 4, 5]. The leptonic component of the preshock medium is enriched by $e^{\pm}$, which leads to a dramatic softening of the early afterglow.
- The neutron component of the GRB ejecta overtakes the decelerating blast wave and deposits energy and momentum into the external medium by β-decay [6]. The leading neutron front leaves behind a relativistic trail — a hot mixture of the decay products and ambient particles. The forward shock of the blast wave propagates in this trail instead of the customary ambient medium. The impact of neutrons lasts about 10 e-foldings of the β-decay, and at $R_{\text{trail}} \approx 10^{17}$ cm the trail becomes static and cold, i.e. indistinguishable from a normal ambient medium.

The paradox of relativistic explosions is that even a small fraction of their energy deposited by a precursor into the external medium impacts the ensuing blast wave and

---

[1] Γ_0 can be smaller than the ejecta Lorentz factor Γ_{ej} if the reverse shock in the ejecta is relativistic, which is the case for a dense external medium.

the afterglow radiation. The importance of both γ-ray and neutron precursors is entirely due to the high Lorentz factor of the explosion. The study of both precursors together is a complicated physical problem and we discuss them separately here.

ELECTRON-POSITRON LOADING

A medium overtaken by a front of collimated γ-rays is inevitably $e^{\pm}$-loaded [3, 4, 5]. This happens because some γ-rays Compton scatter off the medium and get absorbed by the primary collimated radiation via reaction $\gamma + \gamma \to e^+ + e^-$. The medium is optically thin, so only a tiny fraction of the prompt GRB scatters, however, the number of scattered γ-rays and created $e^{\pm}$ *per ambient electron* is huge, $n_{\pm}/n_0 \gg 1$.

The pair loading factor $Z = 1 + n_{\pm}/n_0$ does not depend on the ambient density and can be calculated starting with just one ambient electron. The number of γ-rays scattered by the electron at a radius R is proportional to the column density of the γ-ray front at this radius, $E_\gamma/4\pi R^2 \propto R^{-2}$, and therefore Z is high at small R. $Z(R)$ is determined by the isotropic equivalent of the prompt GRB energy E_γ (and slightly depends on the exact spectral shape of the prompt GRB). For the brightest bursts $E_\gamma > 10^{54}$ erg, and a typical $E_\gamma \sim 10^{53}$ erg.

The created $e^{\pm}$ also scatter radiation, which can lead to an exponential runaway of pair creation. This runway takes place at radii

$$R < R_{\text{load}} \approx 2 \times 10^{16} (E_\gamma/10^{53})^{1/2} \text{ cm}, \tag{1}$$

leading to $Z \gg 1$ [5]. The calculation shows also that at $R < R_{\text{acc}} = R_{\text{load}}/2.3$ the $e^{\pm}$-loaded medium is pushed to relativistic velocities $\beta\gamma > 1$ by the γ-ray front. The scattered fraction δE_γ of the GRB radiation is proportional to the optical depth of the swept-up ambient mass m, which is proportional to n_0. However, the medium acceleration $\gamma = \delta E_\gamma/mc^2$ does not depend on n_0 (m cancels out), so only E_γ determines R_{acc}.

The pair-loading factor $Z(R)$ and the Lorentz factor $\gamma(R)$ of a medium overtaken by the GRB radiation front are shown in Fig. 1. These parameters were calculated numerically and approximated by analytical formulae in [5]. Z varies exponentially between R_{acc} and $R_{\text{load}} = 2.3 R_{\text{acc}}$ and $Z(R_{\text{acc}}) \approx 74$. At $R < R_{\text{acc}}$ both γ and Z vary as power-laws with radius. An interesting phenomenon takes place at small radii $R < R_{\text{gap}} \approx R_{\text{acc}}/3$: here the external shock may not exist at all because the medium gains so high γ that it runs away from the ejecta and a gap is opened.

The main observational effect of $e^{\pm}$-loading is a strong softening of synchrotron emission from the forward shock. Indeed, the shock energy per proton, $\Gamma m_p c^2$, can now be shared by Z leptons, and the energy per lepton is reduced by the factor $Z^{-1} \ll 1$. Then the frequency of synchrotron radiation v_s is reduced as Z^{-2}. Accurate calculation shows that the preacceleration γ also softens the emission because it reduces the pressure in the blast wave; as a result $v_s \propto Z^{-2}\gamma^{-5/2}$. This has a strong effect on the early afterglow: it starts as a very soft signal and later evolves to the normal X-ray emission.

The energy dissipated in the forward shock at $R < R_{\text{load}}$ depends on $R_{\text{dec}}/R_{\text{load}}$ and can vary. For example, in a medium with $n_0 \sim 1 - 10$ cm^{-3} (standard ISM) it can be estimated as $E_{\text{soft}} \approx E(R_{\text{load}}/R_{\text{dec}})^3 = (10^{-3} - 10^{-1})E$ where E is the ejecta energy

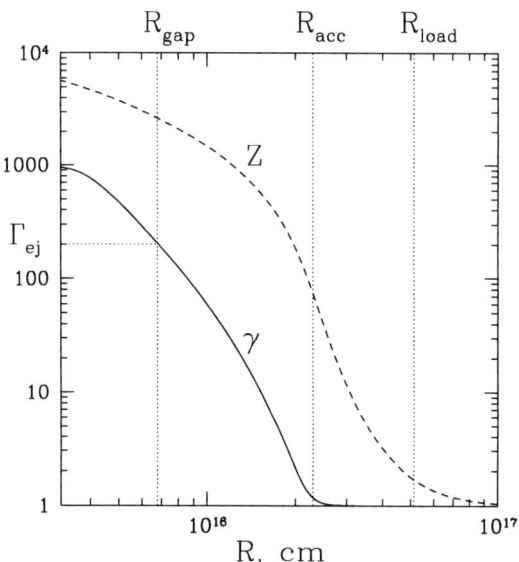

FIGURE 1. Pair-loading factor $Z = 1 + n_{\pm}/n_0$ and acceleration γ of the external medium by the γ-ray front. As the front propagates to larger radii R, the $e^{\pm}$ loading and acceleration effects are reduced and become negligible at $R \approx 10^{17}$ cm. The figure shows the results for GRBs with isotropic energy $E_{\gamma} = 10^{54}$ erg. The corresponding curves for different E_{γ} are obtained by re-scaling radius $R \to R(E_{\gamma}/10^{54})^{1/2}$. Example ejecta Lorentz factor $\Gamma_{\rm ej} = 200$ is indicated; $\gamma = \Gamma_{\rm ej}$ defines the gap radius $R_{\rm gap}$.

[5, 7]. The initial soft flash rises sharply at the observer time $t_{\rm obs} \sim R_{\rm acc}/\Gamma_{\rm ej}^2 c$, which is before 10 s in most cases, so the rise may be difficult to catch with the current instruments. However, the emission can last much longer after the rise, more than one minute, and then would be easily observed by *Swift*. Estimates of the expected emission have been done in [5, 8, 9], however, the problem requires a more accurate calculation that keeps track of the evolution of each shell in the blast material.

Blast waves in wind-type media ($n_0 \propto R^{-2}$) have small $R_{\rm dec}$ [10, 11, 12], and the neglect of $e^{\pm}$ loading and preacceleration by γ-rays is inconsistent in this situation. The inclusion of these effects leads to a very powerful prompt soft flash because most of the blast wave energy dissipates at $R < R_{\rm load}$ [5]. Its detection would be a clear signature of a massive progenitor. On the other hand, the existing upper limits on the early optical emission in several GRBs exclude high-density winds in these bursts.

Optical flashes are also expected from the reverse shocks in the GRB ejecta [1, 2]. The reverse shock with magnetic field B comparable to that in the forward shock produces an optical flash like the one observed in GRB 990123 [13]. This can be used to probe the composition of the GRB ejecta as the reverse shock emission is sensitive to B. A magnetic field near or above equipartition with the fluid pressure is likely to suppress the reverse shock component of the afterglow, and then only the $e^{\pm}$ emission from the forward shock contributes to the optical flash. The forward-shock component does not depend on the nature of the ejecta and can help to determine the ambient density, magnetic field, and Lorentz factor of the blast wave.

NEUTRON DECAY

A significant fraction of baryons in the GRB ejecta are neutrons [14, 15, 16]. They are collisionally coupled to the ions when the fireball is accelerated by radiation pressure and develop a high $\Gamma_n = 10^2 - 10^3$. Then the neutrons decouple and coast with $\Gamma_n = const$.

The β-decay depletes exponentially the neutron component outside the mean-decay radius $R_\beta \approx 10^{16}(\Gamma_n/300)$ cm, which is comparable to the expected radius of the early afterglow. However, the neutrons impact the blast wave at radii significantly larger than R_β, even though their number is exponentially reduced at large radii.

The front of survived neutrons overtakes the decelerating blast wave at some radius R_* where the blast-wave Lorentz factor Γ decreases below Γ_n. Using the Blandford-McKee solution $\Gamma^2(R) = (17 - 4k)E/8\pi\rho_0 c^2 R^3$ for adiabatic blast waves in a medium with density $\rho_0 \propto R^{-k}$ we have

$$R_*^3 = \frac{(17-4k)E}{8\pi\rho_0 c^2 \Gamma_n^2}. \qquad (2)$$

At $R > R_*$ the β-decay takes place in the external medium *ahead* of the forward shock. The impact of this decay can be understood by comparing the energy of neutrons, $E_n = X_n E \exp(-R/R_\beta)$ (X_n is the initial neutron fraction of the explosion) with the ambient mass $mc^2 = \frac{4\pi}{3-k} R^3 \rho_0 c^2 = \frac{17-4k}{2(3-k)} (E/\Gamma^2)$ they interact with,

$$\frac{E_n}{mc^2} = \frac{2(3-k)}{17-4k} X_n \Gamma^2 \exp\left(-\frac{R}{R_\beta}\right). \qquad (3)$$

Just after R_* this ratio can be as large as Γ_n^2 depending on R_*/R_β. The decaying neutron front with $E_n > mc^2$ deposits huge momentum and energy into the ambient medium, leaving behind a relativistic trail. The exact parameters of this trail are found from energy and momentum conservation applied to the collision of β-decay products with the ambient medium [6].

The ratio E_n/mc^2 becomes smaller than unity only after ≈ 10 e-foldings of β-decay. Therefore, the impact of neutrons lasts until $R_{\text{trail}} \approx 10 R_\beta \approx 10^{17}$ cm, and one expects an observational effect if $R_* < R_{\text{trail}}$. For a homogeneous medium ($k = 0$) this requires $n_0 > 0.1 E_{52}(\Gamma_n/300)^{-5}$ cm^{-3}. For a wind medium ($k = 2$) $R_* < R_{\text{trail}}$ for all plausible parameters of the wind if $\Gamma_n \sim 10^2$ or higher. Besides, the forward shock in a dense wind is likely to be slow from the very beginning (the reverse shock in the ejecta is relativistic). Then the neutrons can overtake the forward shock immediately, before the self-similar deceleration sets in.

The β-decay ahead of the shock transforms the cold static external medium into a hot, dense, relativistically moving, and possibly magnetized, material. The ion ejecta follow the neutron front and drive a shock wave in the trail material. Dynamics and dissipation in the shock are discussed in [6]. Like the neutron-free shocks, it is difficult to calculate the expected synchrotron emission from first principles because the electron acceleration and magnetic field evolution are poorly understood. One can apply a phenomenological shock model with the customary parameters ε_e and ε_B and fit the data with the model. This may enable an observational test for the β-decay.

Any neutron signature revealed in a GRB afterglow emission would confirm that the ejected baryonic material has gone through a hot high-density phase in the central engine. Neutrons thus provide a unique link between the physics of the central engine and the observed afterglow. Numerical simulations of neutron-fed blast waves may help to identify such signatures. One possibility, for instance, is an exponentially decaying emission component. Another possible signature is a spectral transition or a bump in the afterglow light curve at $R \approx R_{\text{trail}}$ [6].

Absence of neutron signatures would indicate that the GRB ejecta are dominated by magnetic fields. In such a low-density fireball, the neutrons would decouple early with a modest Lorentz factor and decay quickly. Two-component ejecta with less collimated and less energetic neutrons is possible in the MHD acceleration scenario [17].

ACKNOWLEDGMENTS

This research was supported by NASA grant NAG5-13382.

REFERENCES

1. Mészáros, P., & Rees, M. J., ApJ **418**, L59 (1993).
2. Sari, R., & Piran, T. 1999, ApJ, 517, L109 (1999).
3. Thompson, C., & Madau, P., ApJ **538**, 105 (2000).
4. Mészáros, P., Ramirez-Ruiz, E., & Rees, M. J., ApJ **554**, 660 (2001).
5. Beloborodov, A. M., ApJ **565**, 808 (2002).
6. Beloborodov, A. M., ApJL **58**, L19 (2003).
7. Beloborodov, A. M., in "Beaming and Jets in Gamma-Ray Bursts", "The Gamma-Ray Universe", astro-ph/0305118 (2003).
8. Li, Z., Dai, Z. G., & Lu, T., MNRAS, in press, astro-ph/0307388 (2003).
9. Kumar, P., & Panaitescu, A., MNRAS, submitted, astro-ph/0309161 (2003).
10. Li, Z.-Y., & Chevalier, R. A., ApJ **589**, L69 (2003).
11. Kobayashi, S., & Zhang, B., ApJ **597**, 459 (2003).
12. Wu, X. F., Dai, Z. G., Huang, Y. F., & Lu, T., MNRAS **342**, 1131 (2003).
13. Akerlof, C. et al., Nature **398**, 400 (1999).
14. Derishev, E. V., Kocharovsky, V. V., & Kocharovsky, Vl. V., ApJ **521**, 640 (1999).
15. Pruet, J., Woosley, S. E., & Hoffman, R. D., ApJ **586**, 1254 (2003).
16. Beloborodov, A. M., ApJ **588**, 931 (2003).
17. Vlahakis, N., Peng, F., & Königl, A., ApJ **594**, L23 (2003).

Broad-band (2-400 keV) Spectra of Gamma-Ray Bursts and X-Ray Flashes based on HETE-2 Observations

N. Kawai[*,†], T. Sakamoto[*,†], M. Suzuki[*], M. Matsuoka[**], A. Yoshida[‡], Y. Shirasaki[§], T. Tamagawa[¶], Y. Nakagawa[‡], Y. Yamazaki[‡], R. Sato[*], K. Torii[¶], E. E. Fenimore[||], M. Galassi[||], D. Q. Lamb[††], C. Graziani[††], T. Q. Donaghy[††], G. Ricker[‡‡], J. Doty[‡‡], R. Vanderspek[‡‡], G. B. Crew[‡‡], J. Villasenor[‡‡], N. Butler[‡‡], J-L. Atteia[§§], C. Barraud[§§], M. Boer[¶¶], J-P. Dezalay[***], J-F. Olive[***], J. G. Jernigan[†††], K. Hurley[†††], S. E. Woosley[‡‡‡], G. Pizzichini[§§§] and HETE-2 Science Team[¶¶¶]

[*]*Department of Physics, Tokyo Institute of Technology, Japan*
[†]*RIKEN (The Institute of Physical and Chemical Research)*
[**]*Tsukuba Space Center, Japan Aerospace Exploration Agency*
[‡]*Aoyama Gakuin University, Japan*
[§]*National Astronomical Observatory of Japan*
[¶]*The Institute of Physical and Chemical Research (RIKEN), Japan*
[||]*Los Alamos National Laboratory, USA*
[††]*Department of Astronomy and Astrophysics, University of Chicago, USA*
[‡‡]*Center for Space Research, Massachusetts Institute of Technology, USA*
[§§]*Observatoire Midi-Pyrenées, France*
[¶¶]*Observatoire de Haute Provence, France*
[***]*Centre d'Etude Spatiale des Rayonnements, France*
[†††]*University of California at Berkeley, USA*
[‡‡‡]*Univesity of California at Santa Cruz, USA*
[§§§]*Consiglio Nazionale delle Ricerche (IASF)/TESRE, Italy*
[¶¶¶]

Abstract. One of the primary goals of HETE-2 is the spectroscopy of the prompt emission of gamma-ray bursts (GRBs) in a wide energy range. A large overlapping energy range of WXM and FREGATE ensures reliable spectral measurements. Thanks to the enhanced sensitivity in the X-ray range, HETE-2 has detected large numbers of X-ray rich events in addition to the classical GRBs.

We have examined the spectral properties of these cosmological bursts localized with HETE-2. They have much variety in their properties, with the spectral peak energy ranging from a few keV to ~400 keV, and the gamma-ray (30–400 keV) fluence from $\sim 10^{-8}$ to $\sim 10^{-4}$ erg cm^{-2}. According to the ratio of fluence in the X-ray range (2–30 keV) to that in the gamma-ray range (30–400 keV), we have classified the localized cosmological bursts into three classes, hard GRBs, X-ray rich GRBs and X-ray flashes. When the nature of the X-ray rich GRBs and X-ray flashes are studied systematically, we find that they have many properties in common with the classical GRBs. In fact, there is no clear separation in properties of these three classes, suggesting that they are a single phenomenon.

The relation between the spectral peak energy and the isotropic-equivalent radiated energy (both in the source frame) has been confirmed for the HETE-localized GRBs and extended for XRFs.

INTRODUCTION

One of the primary goals of HETE-2 is the spectroscopy of the prompt emission of gamma-ray bursts (GRBs) in a wide energy range. It is the First experiment since Ginga GBD that achieves continuous coverage from the soft X-ray band up to 400 keV within a single satellite. The Wide-Field X-ray Monitor (WXM) consists of Xe-filled proportional counters which cover the energy range of 2–25 keV [1], while the French Gamma Telescope (FREGATE) employing NaI (Tl) scintillators is sensitive in the 6–400 keV range [3]. A large overlapping energy range of WXM and FREGATE ensures reliable spectral measurements. We have extensively cross-calibrated their energy response in flight using the observations of the Crab Nebula [2]. The localization capability of HETE-2 eliminates uncertainties in the energy response due to unknown incidence angles, which often plagued earlier missions. Thanks to the enhanced sensitivity in the X-ray range, HETE-2 has detected significant numbers of X-ray rich events in addition to the classical GRBs. The spectral properties and the evolution of these bursts are presented.

X-ray Flashes

HETE-2 is detecting X-ray flashes (XRFs), which are similar to regular "classical" GRBs in many ways except that XRFs have larger fluence in the X-ray band (2–30 keV) than in the gamma-ray band (30–400 keV). XRFs have received increasing attention in the past several years [4, 5].

XRFs have t_{90} durations between 10 and 200 sec [6] and their sky distribution is consistent with isotropy. In these respects, XRFs are similar to "classical" GRBs. A joint analysis of WFC/BATSE spectral data showed that the low-energy and high-energy photon indices of XRFs are -1 and ~ -2.5, respectively, which are similar to those of GRBs, but that the XRFs had spectral peak energies E^{obs}_{peak} that were much lower than those of GRBs [5]. The only difference between XRFs and GRBs therefore appears to be that XRFs have lower E^{obs}_{peak} values. It has therefore been suggested that XRFs might represent an extension of the GRB population to bursts with low peak energies.

Some of the possibilities popularly discussed are: i) XRFs represent a very high-z population of GRBs with its photon energies redshifted down into the X-ray range; ii) XRFs have jets with intrinsically lower Lorentz factor; iii) lower Lorentz factor components of strucutued jets are seen as XRFs; and iv) XRFs have lower relativistic beaming due to large off-axis viewing angles. Clarifying the nature of XRFs and X-ray-rich GRBs, and their connection to GRBs, could provide a breakthrough in our understanding of the prompt emission of GRBs.

The low energy threshold of WXM, SXC (2 keV) and FREGATE (6 keV), and the effective areas at X-ray energies of these instruments, make HETE-2 ideal for detecting and studying XRFs. Unlike previous missions (*Ginga*, BATSE, and *Beppo*SAX), HETE-2 has the ability to trigger on and localize XRFs, and can carry out detailed studies of their spectral properties using FREGATE and the WXM.

One of the softest event HETE-2 localized is XRF020903: The upper limit $E^{obs}_{peak} <$

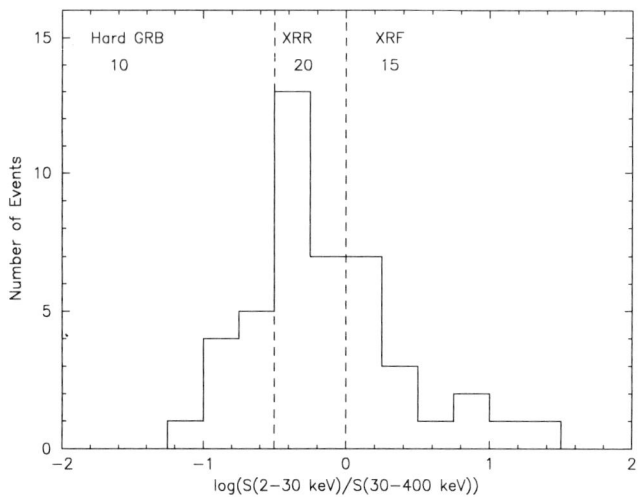

FIGURE 1. Distribution of the fluence ratio of 2–30 keV to 30–400 keV in the logarithmic scale. The dashed lines are the borders of hard GRB/X-ray rich GRB, and X-ray rich GRB/XRF.

5 keV (99.7% confidence level) makes this event one of the softest bursts seen so far by HETE-2, and no photons were significantly detected above $\sim$ 10 keV [7, 8]. Its light curve exhibits a double-peak structure, and the burst duration is shorter at higher energies [7]. These are commonly seen in "classical" long GRBs. Follow-up observations made possible by the HETE-2 localization identified the likely optical afterglow of the XRF [9]. Later observations determined that the optical transient occurred in a star-forming galaxy at a distance $z = 0.25$ [10, 11]; both of these properties are typical of GRB host galaxies.

General properties

The fluence ratio distribution between 2–30 keV (S_X) and 30-400 keV (S_γ) is shown in Figure 1 for a sample of HETE-localized bursts which have sufficient photon statistics in WXM and/or FREGATE for spectral analysis. Here we define XRFs, X-ray rich GRBs, and hard GRBs as those events for which $\log(S_X/S_\gamma) > 0$, $-0.5 < \log(S_X/S_\gamma) \leq 0$, and $\log(S_X/S_\gamma) \leq -0.5$ respectively. According to this working definition, the HETE-2 localized bursts are classified to approximately equal numbers of hard GRBs, X-ray rich GRBs and XRFs.

When the 30–400 keV fluence is plotted against the 2–30 keV fluence (Figure 2), we find that not only do the XRFs have lower gamma-ray fluence, but also they tend to have lower X-ray fluence among the entire sample. Accordingly, the spectral peak energies (E_{peak}) and the peak flux in the gamma-ray band (50–300 keV) shows a strong correlation (Figure 3). We find that the spectral properties of XRFs and "X-ray rich" GRBs form a continuum with those of ordinary GRBs and suggest that XRFs may

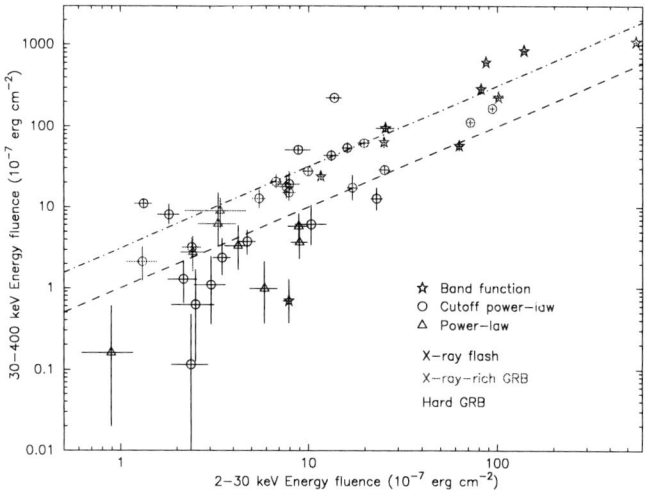

FIGURE 2. Distribution of HETE-2 bursts in the (S_X, S_γ)-plane. S_X and S_γ are the energy fluence in the 2-30 keV and the 30-400 keV energy ranges respectively. The marks below the dashed lines are XRFs, the marks above the dot-dashed lines are hard (classical) GRBs, and the marks between are X-ray rich GRBs.

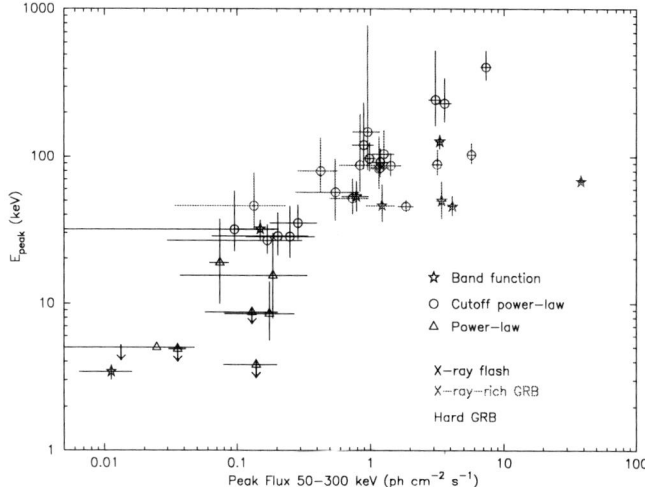

FIGURE 3. Distribution of HETE-2 bursts in the peak photon flux (50-300 keV) vs. E_{peak} plane. The flux is measured as the average over the 1 s at the peaks. A strong correlation is seen despite the fact that these bursts are sampled from various redshifts.

represent a further extension of this continuum. XRF 020903 lies on the hard/soft fluence correlation for the other GRBs and X-ray-rich GRBs, and appears to extend by a decade the hardness-intensity correlation [12].

This correlation is not easy to understand, since the GRBs and XRFs plotted here are

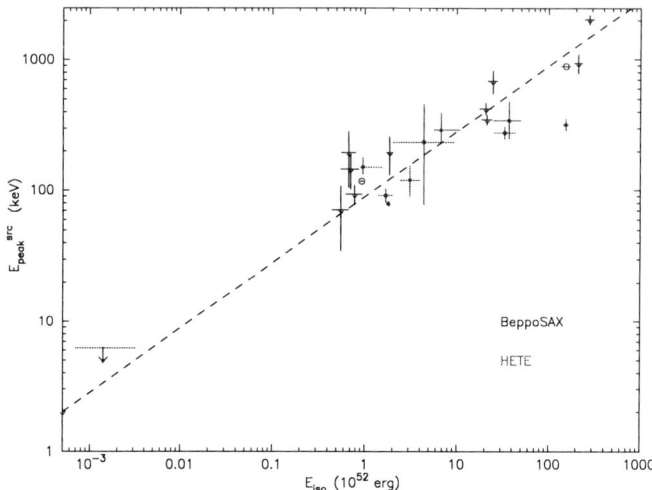

FIGURE 4. Distribution of HETE-2 (filled dots) and *Beppo*SAX (open circles) bursts in the ($E_{\rm iso}, E_{\rm peak}$)-plane, where $E_{\rm iso}$ and $E_{\rm peak}$ are the isotropic-equivalent GRB energy and the peak of the GRB spectrum in the source frame. The HETE bursts confirm the relation between $E_{\rm iso}$ and $E_{\rm peak}$ found by Amati et al. [13], and extend it by a factor ~ 300 in $E_{\rm iso}$. The bursts with the lowest value of $E_{\rm iso}$ is XRF020903. The dashed line is $E_{\rm peak}^{\rm src} = 89\,(E_{\rm iso}\,/\,10^{52}\,{\rm erg})^{0.5}$.

supposedly sampled from a wide range of redshifts from $z < 0.5$ to $z > 3$; the effect of distance (redshift) on the observed flux, which should be much larger than that on $E_{\rm peak}$, would weaken the intrinsic correlation.

In fact, the properties in the source frame of the GRBs have a surprisingly tight correlation. Using twelve *Beppo*SAX GRBs with measured redshifts, Amati et al. [13] showed that the spectral peak energy at the source frame $E_{\rm peak}^{\rm src}$ and the isotropic-equivalent radiated energy $E_{\rm iso}$ are tightly correlated, and follows a relation $E_{\rm peak}^{\rm src} \propto E_{\rm iso}^{1/2}$. With the 10 HETE GRBs/XRFs with measured redshifts (Figure 4) we have confirmed this relation. Furthermore, we extended this relation by three orders of magnitude in $E_{\rm iso}$.

These results provide strong evidence that XRFs and X-ray-rich GRBs form a continuum, and are a single phenomenon. The extended Amati et al relation ($E_{\rm peak}^{\rm src} \propto E_{\rm iso}^{1/2}$) suggest that the $E_{\rm peak}^{\rm src}$ and $E_{\rm iso}$ are controlled by some single parameter, which differentiate XRFs and GRBs. Understanding this key parameter should certainly lead to the understanding of the energetics and radiation mechanism of GRBs. There are several theoretical proposals for the the nature of XRFs: "off-axis jet" model ([14]; [15]), "structure jet" model ([16], "unified jet" model ([17]), and so on. In order to understand the nature of GRBs and XRFs, it is essential to obtain a larger sample of XRFs and GRBs with measured redshifts. In particular, additional redshift determinations are clearly needed for XRFs with $1\,{\rm keV} < E_{\rm peak} < 30\,{\rm keV}$ in order to confirm these results and to test the theories.

CONCLUSION

We have examined the spectral properties of the GRBs and "X-ray flashes" (XRFs). HETE-2 has detected comparable number of "classical" (hard) GRBs, X-ray rich GRBs, and XRFs, as defined by the ratio of the fluence in X-rays to that in gamma rays. We have confirmed the general correlation between the spectral hardness and the gamma-ray fluence, and that the same relation also holds for X-ray flashes.

When we compare the properties of GRBs and XRFs, such as peak flux, spectral peak energies, duration, we find that there are no distinctly separated populations. Instead, we find that hard GRBs, X-ray rich GRBs and XRFs form a continuum, and that they are a single phenomenon.

The apparent correlation between the gamma-ray flux and the spectral peak energy has been confirmed and extended toward lower peak energies and lower fluxes. The relation between the spectral peak energy and the isotropic-equivalent radiated energy (both in the source frame) has been confirmed for GRBs and extended for XRFs. These new results from HETE-2 have shown that XRFs provide unique insights into the structure of GRB jets. The wide energy coverage (2–400 keV) of HETE-2 makes it ideal for detecting and studying most of XRFs and hard GRBs at the same time. HETE-2 will remain as a useful probe for studying the spectral peak energy E_{peak} for very soft XRFs and very hard GRBs even after Swift is in operation. More samlple of XRFs with measured redshift are clearly needed for the study the GRB/XRF phenomenon.

REFERENCES

1. Kawai, N., et al., in *Gamma-Ray Bursts and Afterglow Astronomy*, eds. G. R. Ricker and R. Vanderspek (New York: AIP), p. 25 (2002).
2. Shirasaki, Y. et al., PASJ **55**, 1033 (2003).
3. Atteia, J-L, et al., in *Gamma-Ray Bursts and Afterglow Astronomy*, eds. G. R. Ricker and R. Vanderspek (New York: AIP), p. 17 (2002).
4. Heise, J., et al., in Proc. 2nd Rome Workshop: *Gamma-Ray Bursts in the Afterglow Era*, eds. E. Costa, F. Frontera, J. Hjorth (Springer-Verlag), p. 16 (2000).
5. Kippen, R. M., et al., Astro-ph/0203114 (2002).
6. Suzuki, M., et al., this proceedings (2004).
7. Sakamoto, T. et al.,, this proceedings (2004).
8. Sakamoto, T. et al., ApJ **602**, 875 (2004).
9. Soderberg, A. M., et al., GCN Circular **1554** (2002).
10. Soderberg, A.M., et al., ApJ **606**, 994 (2004).
11. Chornock, R. & Filippenko, A. V., GCN Circular **1609** (2002).
12. Mallozzi, R. S., et al., ApJ, **454**, 597 (1995).
13. Amati, L. et al., Known Redshifts, A&A, **390**, 81 (2002).
14. Yamazaki, R., Ioka, K., & Nakamura, T., ApJL, **571**, L31 (2002).
15. Yamazaki, R., Ioka, K., & and Nakamura, T., ApJ, **593**, 941 (2003).
16. Rossi, E., Lazzati, D., & Rees, M.J., MNRAS, **332**, 945 (2002).
17. Lamb, D.Q., Donaghy, T.Q., & Graziani, C., in proc. *2nd VERITAS Symposium on TeV Astrophysics*, Chicago, Illinois (astro-ph/0309456) (2003).

Heating and Deceleration of GRB Fireballs by Neutron Decay

Elena M. Rossi[*], Andrei M. Beloborodov [†] and Martin J. Rees[*]

[*]*Institute of Astronomy, University of Cambridge, Madingley Road, Cambridge CB3 0HA, England*
[†]*Physics Department, Columbia University, 538 W 120th Street, New York, NY 10027, USA*

Abstract. Fireballs with high energy per baryon rest mass ($\eta > \eta_* \sim 400$) contain a relatively slow neutron component. We show here that in this situation the thermal history of fireballs is very different from the standard adiabatic cooling.

INTRODUCTION

Baryonic matter ejected in a GRB explosion is partially composed of free neutrons (see [1] and [2] for detailed calculations and references). This has a strong impact on the external blast wave and changes the afterglow mechanism at radii $r \lesssim 10^{17}$ cm [3]. We here investigate the dynamical effect of neutrons on the early evolution of a relativistic fireball, prior to the afterglow phase.

FIREBALL ACCELERATION AND NEUTRON-PROTON DECOUPLING

At the initial stage of the fireball acceleration, the neutrons are collisionally coupled to the protons and accelerated by radiation pressure at the same rate. If the energy-per-baryon, η, is high enough, neutrons decouple before the acceleration stage is completed. This happens if

$$\eta \geq \eta_* \simeq \begin{cases} 400 \left[\dfrac{L_{52}}{R_{07}^{\frac{1+\xi}{2}}}\right]^{1/4} & \xi \leq 1 \\ 400 \left[\dfrac{L_{52}\xi}{R_{07}^{\frac{1+\xi}{2}}}\right]^{1/4} & \xi > 1 \end{cases} \quad (1)$$

where ξ is the ratio of neutron and proton densities, $R_0 = 10^7 \,\text{cm}\, R_{07}$ is the initial size of the fireball, and $L = 10^{52}\,\text{erg}\, L_{52}$ is the equivalent isotropic luminosity of the explosion. Beyond the decoupling radius,

$$R_{np} \simeq R_0 \eta_* \left(\dfrac{\eta_*}{\eta}\right)^{1/3}, \quad (2)$$

neutrons coast with a constant Lorentz factor

$$\Gamma_n \simeq \frac{R_{np}}{R_0} \simeq \eta_* \left(\frac{\eta_*}{\eta}\right)^{1/3}. \tag{3}$$

The optically thick proton fireball continues to accelerate as $\Gamma_p \simeq r/R_0$ until the internal energy equals the proton rest-mass energy and acceleration is no longer efficient. If the fireball remains optically thick during the whole acceleration stage, its Lorentz factor saturates at

$$\Gamma_p \simeq \begin{cases} \eta & \eta < \eta_*, \\ \eta\left(1+\xi-\xi\left(\frac{\eta_*}{\eta}\right)^{4/3}\right) & \eta > \eta_*, \end{cases} \tag{4}$$

and the acceleration stage ends at radius

$$R_s \simeq \Gamma_p R_0. \tag{5}$$

ADIABATIC COOLING

We first consider the thermal history of the fireball without neutrons. Electrons, protons, and radiation maintain a common temperature T in the early dense fireball via Coulomb collisions and Compton scattering, and T decreases adiabatically to very low values. During such adiabatic expansion, the photon-to-baryon ratio $\phi = n_\gamma/n_b = const \sim 10^5$, and radiation completely dominates the internal energy and pressure. Therefore, T decreases according to the radiation adiabatic law with index $\hat{\gamma} = \frac{4}{3}$. This continues until the fireball becomes transparent to radiation at the photosphere radius R_τ. Assuming $R_\tau > R_s$, we have

$$R_\tau = \frac{L\sigma_T}{4\pi m_p c^3 \Gamma_p^3}, \tag{6}$$

here σ_T is Thomson cross section. After becoming transparent the plasma is still tracking the temperature of the (freely streaming) photons, which is constant in the plasma frame if the fireball coasts with a constant Lorentz factor. The electrons decouple thermally from radiation only when the Compton timescale ($t_C = \Gamma_p(3m_e c/4U_{rad}\sigma_T)$ where U_{rad} is the radiation energy density) exceeds the expansion timescale R/c. This happens at radius

$$R_{e\gamma} = \frac{\sigma_T L}{3\pi m_e c^3 \Gamma_p^{7/3}} \left(\frac{R_0}{R_\tau}\right)^{2/3}. \tag{7}$$

The protons decouple from the electrons when the Coulomb timescale $t_{ep} \approx 17 T^{3/2} n_e^{-1} \Gamma_p$ s exceeds R/c. The corresponding radius of e-p decoupling is

$$R_{ep} \simeq \left(68\pi m_p c^4\right)^{-1} \frac{L}{\Gamma_p^3 T_s^{3/2}} \left(\frac{R_\tau}{R_s}\right), \tag{8}$$

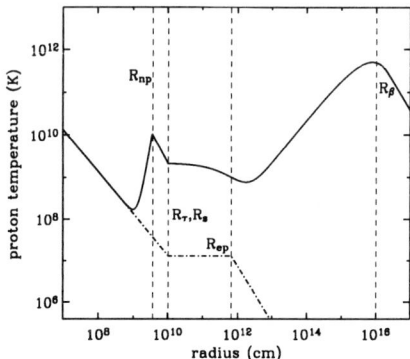

FIGURE 1. Proton temperature as a function of radius in a fireball with neutron-to-proton ratio $\xi = 1$ (solid curve). The relevant radii R_{ep}, R_{np}, R_τ, R_s, and R_β are shown by vertical dashed lines (note that $R_s = R_\tau$ in this case). For comparison, the dot–dashed curve shows the evolution of a pure proton fireball ($\xi = 0$) with the same luminosity $L = 10^{52}$ erg s^{-1}, initial radius $R_0 = 10^7$ cm and final Lorentz factor $\Gamma_p \simeq 1040$.

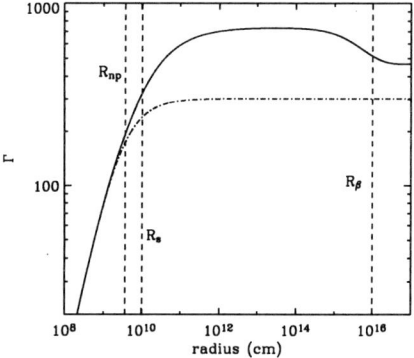

FIGURE 2. The bulk Lorentz factor of protons (solid line) and neutrons (dot-dashed line) as a function of radius for the same parameters as in Fig. 1.

where $T_s = T_0/\Gamma_p$ is the temperature at the saturation radius, and T_0 is the initial temperature of the fireball. The protons decouple thermally from the electrons before the electrons decouple from radiation if $R_{ep} < R_{e\gamma}$. One can show that $R_{ep}/R_{e\gamma} \approx 4 \times 10^{-2} T_{s,7}^{-3/2} (R_\tau/R_s)^{5/3}$ (with $T_s = T_{s7} \times 10^7$ K). At radii $R > \min(R_{e\gamma}, R_{ep})$, the protons are decoupled from radiation and cool adiabatically with index $\hat{\gamma} = \frac{5}{3}$.

Thus, in the absence of neutrons, the fireball cools down adiabatically as $T \propto r^{-1}$ during the acceleration stage and as $T \propto r^{-2/3}$ during the subsequent coasting stage ($\Gamma_p \sim \eta$) up to the transparency radius. After thermal decoupling from radiation the protons cool as $T_p \propto r^{-4/3}$. The thermal history of a pure proton fireball is shown by dot-dashed curve in Fig. 1. For most of the early evolution, the fireball is a cold coasting outflow. The presence of neutrons and their β-decay change this picture.

HEATING BY PROTON–NEUTRON COLLISIONS

Protons in the fireball are continuously heated via proton-neutron collisions [4]. The collisional heating reaches its peak at $R \approx R_{np}$: at $R < R_{np}$ the n-p collisions are frequent but the relative velocity of the neutron and proton components is small ($\beta_{np} \approx \frac{t_{np}}{R/c} < 1$) and at $R > R_{np}$ the relative velocity is relativistic $\beta_{np} \sim 1$ however the collisions are rare. The collisional heating, thus, peaks at R_{np} where the relative velocity becomes comparable to the speed of light and the rate of collisions still allows an efficient transfer of energy on a dynamical timescale. There are two sinks of heat gained by protons via n-p collisions: adiabatic cooling and Coulomb scattering off electrons. The competition between heating and cooling shapes the first peak in the proton temperature profile (Fig. 1). At $R = R_{ep}$ the energy transfer from the protons to the electrons becomes inefficient on the expansion timescale R/c. Then most of the heat gain by protons remains stored in the proton component, not given to the electrons and radiation.

FIREBALL HEATING AND DECELERATION BY DECAYED NEUTRONS

In a fireball with $\eta > \eta_*$, the neutrons have a smaller momentum than the protons after R_{np}. Their decay products e^- and p exchange momentum with the ion fireball and heat it up. The heating of an ion medium by β-decay of neutrons moving with respect to the medium was discussed in [3]. The decay particles exchange momentum with the medium through the two-stream instability or because of gyration in a transverse magnetic field frozen into the medium. Since the neutrons decay gradually at all radii, this momentum exchange and heating take place continuously from the moment of decoupling.

The heating rate of the fireball due to β-decay is

$$\frac{dE_h}{dr} = (\Gamma_{rel} - 1) \frac{dM_{np}c^2}{dr}, \qquad (9)$$

where Γ_{rel} is the neutron Lorentz factor in the fireball rest frame,

$$\frac{dM_{np}}{dr} = \frac{M_n}{R_\beta}, \qquad (10)$$

is the decayed neutron mass per unit radius, and

$$R_\beta \simeq 0.8 \times 10^{16} \text{ cm} \left(\frac{\Gamma_n}{300}\right) \qquad (11)$$

is the mean radius of decay. The heating by β-decay begins at R_{np} and significantly changes the thermal history of the fireball. It shapes the second peak in proton temperature (see Fig. 1).

As long as proton thermal energy is much below $m_p c^2$, adiabatic cooling is small compared to heating and the proton temperature increases linearly with radius,

$$T_p \propto r.$$

At a radius R_{dis} the dissipated bulk kinetic energy equals the total energy of the fireball and Γ_p begins to decrease. Between R_{dis} and R_β the β-decay heating balances the adiabatic cooling and

$$\Gamma_p \propto \left(\frac{R_\beta}{r}\right)^{1/2}.$$

When the fireball reaches R_β the neutron component is exponentially depleted and the dissipation process switches off. The adiabatic cooling then leads to a power-law decrease in temperature,

$$T_p \propto r^{-4/3}.$$

DISCUSSION

We have shown here that the presence of a neutron component greatly affects the early dynamics of GRB fireballs with $\eta > \eta_* \sim 400$. Contrary to the pure proton model, the fireball with a neutron component heats up significantly at $R \gtrsim R_{np}$. In a popular GRB scenario, the bulk energy is partially converted into the observed γ-rays through shock waves inside the fireball at $r \gtrsim 10^{12}$ cm [5]. The β-decay process described here should affect the internal shocks. It decelerates the fastest portions of the inhomogeneous fireball and reduces the contrast of Lorentz factors, which should significantly reduce the dissipation efficiency of internal shocks. A detailed study of these effects is in preparation.

REFERENCES

1. Derishev, E. V., Kocharovsky, V. V., and Kocharovsky, V. V., *ApJ*, **521**, 640–649 (1999).
2. Beloborodov, A. M., *ApJ*, **588**, 931–944 (2003).
3. Beloborodov, A. M., *ApJ*, **585**, L19–L22 (2003).
4. Pruet, J., Abazajian, K., and Fuller, G. M., *Phys. Rev.*, pp. 63002–+ (2001).
5. Rees, M. J., and Mészáros, P., *ApJ*, **430**, L93–L96 (1994).

Discovery of a Distinct Higher Energy Spectral Component in GRB941017

M.M. González*[†], B.L. Dingus[†], Y. Kaneko**, R.D. Preece**, C.D. Dermer[‡] and M.S. Briggs**

*Physics Department, University of Wisconsin, Madison
[†]Los Alamos National Laboratory
**Physics Department, University of Alabama in Huntsville, National Space Science and Technology Center
[‡]Naval Research Laboratory

Abstract. We report an observation of a multi-MeV spectral component in the burst of 17 October 1994 that is distinct from the previously observed lower energy gamma-ray component. This higher-energy component is described by a power law of differential photon number index ~ -1 up to 200 MeV. Its flux decays more slowly and its fluence is more than 3 times the fluence of the lower-energy component. Despite the unique behavior of this higher-energy component, the lower-energy component behaves similarly to most GRBs, presenting a hard-to-soft temporal evolution.

INTRODUCTION

Combining data from one of the BATSE's Large Area Detectors (LAD) and EGRET's calorimeter (TASC) yields prompt GRB spectra in the energy range of 0.03 − 200 MeV. We combined and analyzed LAD-TASC data for 26 bursts selected from the BATSE catalog because they were bright above 300 keV and their photon flux was above 10 ph/cm^2/s. All bursts, except GRB941017, presented high-energy spectra consistent with the single spectral component observed by BATSE. The spectroscopy of 8 of these bursts is presented in [1]. The spectral analysis of GRB941017 is presented in this paper as well as in [2].

GRB941017 was observed by BATSE, COMPTEL and TASC at 10:19:33.938 UT from the galactic coordinates $l = 50.5°$ and $b = -11.7°$. BATSE observed GRB941017 in the energy interval of 0.03 − 2 MeV with ninety percent of the emission occurring in 77s. The BATSE spectra were analyzed [3] using the Band function and a total energy fluence of 1×10^{-4} erg/cm^2 was estimated, placing this burst as the eleventh highest fluence one. A hard-to-soft temporal evolution of the Band parameter β was also observed. This burst was also a possible candidate [4] for low-energy spectral lines.

The burst direction was 66° with respect to the pointing-axis direction of CGRO, reducing COMPTEL's effective area and placing the burst outside the $\sim$1sr field of view of the EGRET spark chamber. The COMPTEL telescope-mode lightcurve is reported in [5] for the energy range of 0.7 − 30 MeV. The two COMPTEL burst modules also observed this burst in the energy range of 0.3 − 10 MeV. The analysis [6] of COMPTEL's burst module data have shown to be consistent with the results presented in this paper.

TABLE 1. Spectral fitting parameters of the differential photon flux. The first six rows contain the best spectral fit parameters for the five time intervals shown in Figure 2. The seventh row shows the probability that the improvement in χ^2 from the addition of the high-energy power law in the fit is due to chance, as determined by the χ^2 test.

Time from BATSE trigger (s)	-18 to 14	14 to 47	47 to 80	80 to 113	113 to 211
Band GRB function parameters					
$A(\frac{\times 10^{-2} \text{ph}}{\text{s cm}^{-2}\text{keV}})$	1 ± 0.05	6 ± 0.1	2 ± 0.1	0.7 ± 0.3	5^{+4}_{-2}
E_{peak} (keV)	505 ± 69	350 ± 8	240 ± 14	91 ± 15	10 fixed
α	-0.84 ± 0.09	-0.79 ± 0.02	-1.08 ± 0.06	-1.45 ± 0.27	-1 fixed
β	-2.22 ± 0.13	-2.46 ± 0.05	$-2.65^{+0.22}_{-0.85}$	$-3.13^{+0.63}_{-\infty}$	-2.73 ± 0.55
High-energy power law parameters					
$A_{PL}(\frac{\times 10^{-7} \text{ph}}{\text{s cm}^{-2}\text{keV}})$	NA	2.4 ± 0.6	2.4 ± 1.0	3.2 ± 0.7	1.8 ± 0.5
γ	NA	-1 fixed	$-1.06^{+0.70}_{-0.44}$	$-1.10^{+0.32}_{-0.17}$	$-0.96^{+0.36}_{-0.17}$
Probability	NA	1.4×10^{-4}	3.7×10^{-4}	6.5×10^{-8}	1.4×10^{-7}

TASC observed GRB941017 in the energy range of 1 – 200 MeV. The time integrated TASC spectrum in T90 was analyzed previously by [7].

ANALYSIS

The lightcurves of the emission observed by LAD and TASC are shown in Fig. 1, left panel. The LAD data show a weak precursor ~90s before the trigger. TASC detected seven 32.768 s spectra starting 18 s before the trigger. Each of the TASC spectra has more than 3σ excess over the background in the energy range of 1 – 200 MeV. Both lightcurves are correlated, for instance, the highest time interval in TASC data corresponds to the BATSE lightcurve's peak. The BATSE flux in the time interval from 113 – 211 s is still significantly detected with a 17σ excess over the background, while TASC flux is only 6σ excess above the background in both of the energy ranges shown in Fig. 1.

The TASC background was obtained by fitting an energy dependent 4th order polynomial to the spectra before and after the burst, and was checked against the background 15 orbits earlier or later when the spacecraft was located at the same geomagnetic rigidity. In Fig. 1 the TASC count rates around GRB941017 trigger and 15 orbits earlier are shown. The background is fairly smooth in time and well described by a polynomial, therefore, the count excess in all seven TASC spectra are related to GRB941017.

The LAD spectra were binned to match the 32.768s TASC time intervals, and the last 3 time intervals for the TASC and LAD were combined in a single bin. The TASC response matrix was obtained with the complete CGRO mass model and the EGS-4 Monte Carlo code. Background subtracted spectra of LAD and TASC data were fitted jointly using the BATSE spectral analysis software called RMFIT. The data was analyzed using the Band function plus a power law function described by

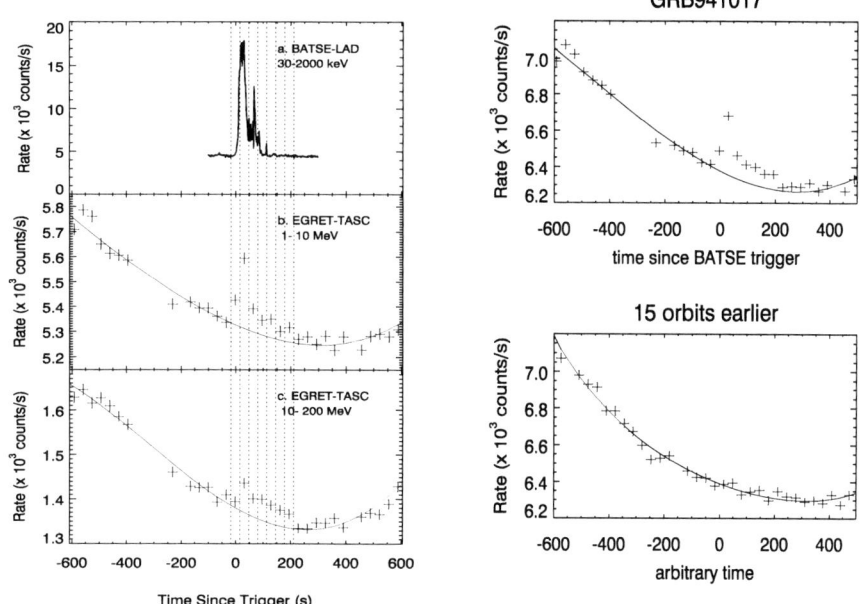

FIGURE 1. Counts rates for GRB941017. The BATSE lightcurve is shown in the top-left window labeled as a. The TASC lighcurves are shown in the middle-left and bottom-left windows labeled as b and c for the energy ranges of 0.03 – 2, 2 – 10 and 10 – 200 MeV respectively; and for the whole TASC energy range (1 – 200 MeV) in the top-right window. For comparison, the TASC count rates for 15 orbits earlier to GRB941017 are shown in the bottom-right window. When TASC data is plotted, the solid line corresponds to the 4th order polynomial that best fits the background.

$A_{PL}[E(\text{keV})/30\text{MeV}]^\gamma$ that was required to fit the higher energy γ-ray excess at later times. A normalization factor of 0.45 was applied to TASC data to account for the errors in the calculated effective area that was shown to be needed even in the pre-flight calibration [8]. The data and spectral fits for the five time intervals are shown in Fig. 2 with the best fit parameters given in Table 1. For more details in the analysis method, refer to [9].

In the first time interval, LAD-TASC data could be mainly described by the Band function, while in the last time interval the higher-energy power law describes most of the data. Thus, both spectral components, the Band function at lower energies and the higher-energy power law, are observed by both detectors. In the second time interval, TASC data below 10 MeV smoothly continues the lower-energy component described by the Band function, showing the correct energy dependence and normalization factor.

As observed in Fig. 2, the lower-energy component evolves with time while the higher-energy component remains fairly constant. Fig. 3 shows the hard-to-soft evolution of the Band parameters α and β. E_{peak} moves to lower energies while the amplitude of the lower-energy component decreases. This is the generalized evolution observed in other bursts [3, 1]. The higher-energy component stays bright and the spectral index does

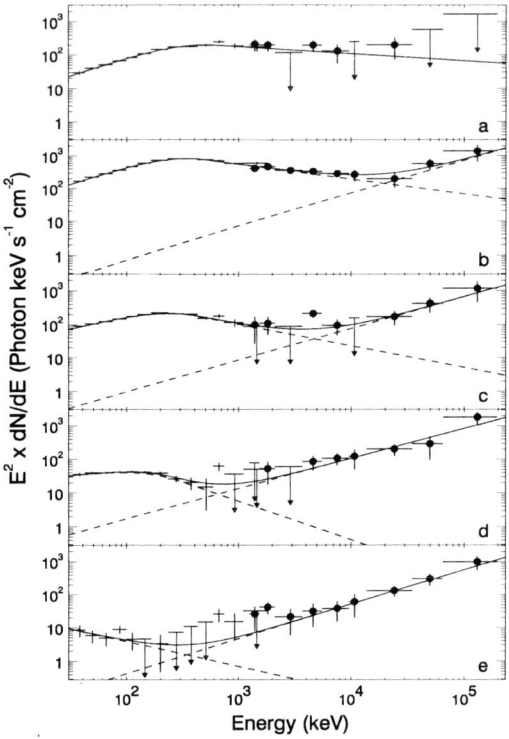

FIGURE 2. Energy fluxes for GRB941017, during the time intervals shown in Table 1. Crosses and circles correspond to BATSE-LAD and EGRET-TASC data respectively. For the purpose of the plot, but not for the spectral fit, the TASC data are binned in energy to give at least 2σ significance over the background. The upper limits correspond to 2σ deviation from the background. The solid lines represents the best fit to the joint LAD-TASC data using a Band function. The best fit parameters are shown in Table 1. The normalization factor between the data sets was fixed to 0.45 through the whole burst.

not change within the statistical uncertainties. The energy flux for three different energy ranges are also shown in Fig. 2 for each time interval. The temporal fit of the energy fluxes (F) in the last four time intervals when described by $F = At^{-\phi}$, yields $\phi = 2.8$, 1.45 and 0.2 for the energy ranges of $0.03 - 2$, $2 - 10$ and $10 - 200$ MeV, respectively, showing that the higher-energy component decays much slower than the lower-energy component. The additional higher-energy component results in a total >30 keV fluence greater than 6.5×10^{-4}erg/cm^2, since it peaks above 200 MeV. This total >30 keV fluence is more than three times that estimated from the BATSE energy alone and it places GRB941017 within the first 3 brightest bursts observed by BATSE.

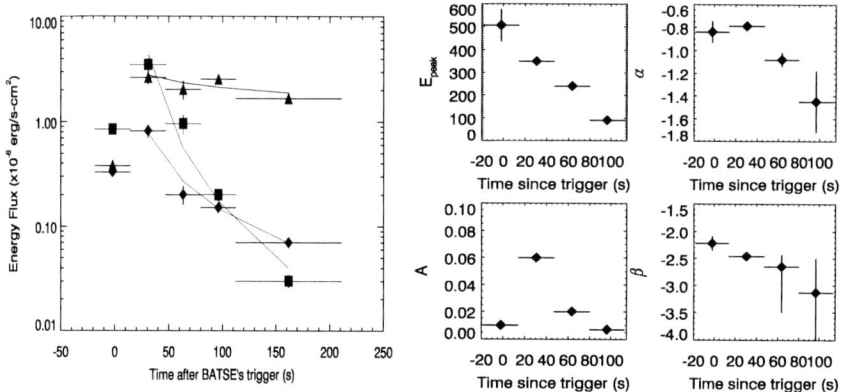

FIGURE 3. Time evolution of the energy fluxes observed in GRB941017 is shown in the big window on the left for the energy ranges of 0.03 – 2 MeV(square), 2 – 10 MeV(diamond) and 10 – 200 MeV(triangle). The Band parameters that best fit the spectra as function of time are shown in the four windows on the right.

CONCLUSIONS

We have observed a higher-energy component separated from the lower-energy component usually described as a Band function. This higher-energy component is described by a power law with differential photon index ~ -1 up to 200 MeV. This higher-energy component decays more slowly than the lower-energy component and its fluence is at least twice the fluence of the lower-energy component. For physical consequences and interpretations of this result refer to [10, 11].

REFERENCES

1. M.M. González, e. a., "Spectral Time Evolution for GRBs observed by BATSE and EGRET-TASC," in *this proceedings*, 2004.
2. M.M. González, et al., *Nature*, **424**, 749–751 (2003).
3. R.D. Preece, et al., *Astrophys. J. Supp.*, **126**, 19–36 (2000).
4. M.S. Briggs, et al., "BATSE evidence for GRB spectral features," in *Gamma-ray Bursts*, AIP Conference Proceedings, 1998, pp. 299–303.
5. http://wwwgro.unh.edu/bursts/cgrbdata.html (08/13/1997).
6. Y. Kaneko, et al., "COMPTEL Observation of GRB941017 with Distinct High-Energy Component," in *this proceedings*, 2003.
7. J. R. Catelli and B. L. Dingus, "EGRET Observations of Bursts at MeV Energies," in *Gamma-ray Bursts*, AIP Conference Proceedings, 1998, pp. 309–313.
8. D.J.Thompson, et al, *Astrophys. J. Supp.*, **86**, 629–656 (1993).
9. M.M. González, et al., "BATSE-EGRET combined spectral fits," in *Gamma-ray Bursts and Afterglow Astronomy 2001*, AIP Conference Proceedings, 2003, pp. 267–269.
10. C.D. Dermer and A. Atoyan, "On Hadronic Models for the Anomalous Gamma-Ray Emission Component in GRB 941017," in *this proceedings*, 2003.
11. J. Granot and D. Guetta, *ApJ*, **598**, L11–L14 (2003).

Early Afterglow, Magnetized Central Engine, and a Quasi-Universal Jet Configuration for Long GRBs

Bing Zhang*, Shiho Kobayashi*, Peter Mészáros*, Nicole M. Lloyd-Ronning[†] and Xinyu Dai*

Department of Astronomy & Astrophysics, Penn State University, University Park, PA 16802
[†] *Los Alamos National Laboratory, MS B244, Los Alamos, NM 87544, USA*

Abstract. Two separate topics are discussed. (1) We describe the classifications of the long GRB early afterglow lightcurves within the framework of the fireball shock model, focusing on the interplay between the reverse and forward shock emission components. We also provide evidence that the central engine of at least two bursts are entrained with strong magnetic fields, and discuss the implications of this result for our understanding of the GRB phenomenon; (2) We argue that the current gamma-ray burst (GRB) and X-ray flash (XRF) data are consistent with a picture that all GRB-XRF jets are structured and quasi-universal, with a typical Gaussian-like jet structure.

EARLY AFTERGLOWS

Classifications

A GRB fireball is eventually decelerated by an ambient medium. During the deceleration, a long-lived forward shock propagates into the medium, and a short-lived reverse shock propagates into the fireball shell[1]. The former is responsible for the long-term afterglow emission, while the latter contributes a noticeable emission component at the very early afterglow epoch[1, 2, 3]. So a GRB early afterglow is the interplay between the reverse and the forward shock emission components, and its diagnosis could reveal rich information about the GRB fireball and the ambient medium. Very early optical afterglows have now been detected for a handful of GRBs[4, 5, 6], and the *Swift* GRB mission, scheduled to be launched in June 2004, will greatly increase the body of GRB early afterglow data. Here we discuss the classifications of the early afterglow lightcurves within the framework of the fireball shock model. The predictions will be fully confronted by the abundant early afterglow data that will come in the near future.

In general, early afterglow lightcurves can be categorized according to the type of ambient medium. Two well-discussed types of medium include a constant density medium ($n =$const) which is applicable for interstellar medium (ISM), and a wind-type medium ($n \propto r^{-2}$), which is typical for the environment of a pre-burst massive star progenitor. The left panel of Figure 1 outlines the typical optical early afterglow lightcurves for ISM (bottom) and wind (top) cases, respectively[7]. In both cases, the reverse shock emission component peaks at the time when the reverse shock crosses the shell, and drops rapidly after the peak. For the forward shock component, there is a peak for the ISM case corresponding to the crossing of the typical synchrotron frequency across the band[8, 9], while for the wind case, the flux fades exclusively[10].

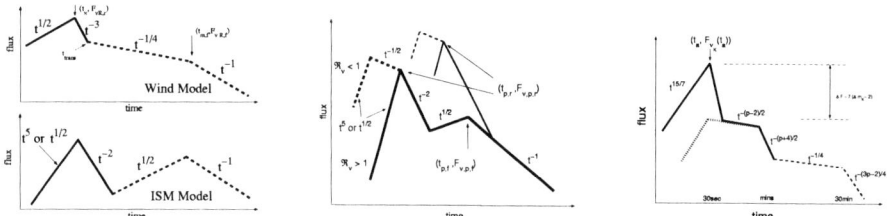

FIGURE 1. *Left panel*: Typical early optical afterglow lightcurves for the ISM (bottom) and the wind (top) cases[7]. Solid lines are for the reverse shock component, and the dashed lines are for the forward shock emssion component; *Middle Panel*: Two types of early optical afterglow lightcurves for the ISM case[11]. Type I (thick line) is the rebrightening type, with a distinct separation of the reverse and forward shock components. Type II (thin line) is the flattening type, with the forward shock peak buried beneath the reverse shock component. A Type II lightcurve is usually associated with a strongly magnetized fireball; *Right panel*: Another typical optical/IR lightcurve for the wind case[12]. This is applicable when the inverse Compton scattering is not important in the reverse shock region.

For the ISM case, the early afterglow lightcurves could be further categorized into two types[11] (see the middle panel of Figure 1). For typical parameters and assuming that the shock parameters (equipartition parameters ε_e and ε_B, as well as the power-law index of the particle distribution p) are the same in both shocks, the lightcurve (Type I) is characterized by a "re-brightening" feature, i.e., there are two distinct lightcurve peaks for both shocks. Conversely, under certain conditions, the forward shock peak is buried beneath the reverse shock emission component, and the lightcurve (Type II) is characterized by a "flattening" feature. A Type-II lightcurve usually requires a stronger magnetic field in the reverse shock region than in the forward shock region, i.e., it implies a strongly magnetized central engine.

For the wind case, the early afterglow lightcurves could be also further categorized into two types based on whether synchrotron self-inverse Compton (IC) emission is important. A wind-type medium implies a low cooling frequency and a high self-absorption frequency. The synchrotron self-absorption effect prevents the electrons from cooling, so that electrons are potentially piled up near the self-absorption energy[12]. If the IC effect is important, this additional cooling mechanism tends to destroy the pile-up bump, so that the treatment that neglects it gives the (approximately) correct description of the lightcurves[7] (top lightcurve, left panel, Figure 1). If the IC cooling is less important compared with the synchrotron cooling, as is expected for a strongly magnetized central engine[11], the electron pile-up effect is prominent, which implies a bump in the synchrotron spectrum and hence, another bump in the early afterglow lightcurve[12] (see right panel, Figure 1). Detections of such a bump would provide valuable information to estimate the wind mass loss and other fireball parameters[12].

A Strongly Magnetized GRB Central Engine

Early afterglow lightcurves could be used to constrain important fireball parameters, such as the initial Lorentz factor, wind mass loss, etc[11, 7, 12]. Another important piece of information is the magnetic content of the fireball. A strongly magnetized central engine is widely speculated to power GRBs – this is argued on many grounds (e.g. [13] for a review). If this is the case, the magnetic field in the reverse shock region is expected to be stronger than that in the forward shock region, since the fireball itself would carry some fields from the central engine. Defining $R_B = B_r/B_f$ as a free parameter (where B_r and B_f are the magnetic field strengths in the reverse shock and forward shock, respectively), one can use the early afterglow lightcurves to constrain R_B[11]. Using a straightforward analysis by combining both the reverse and the forward shock emission data, we have performed detailed case studies for the GRBs that have early afterglow detections. The results suggest that R_B is larger than unity for both GRB 990123 and GRB 021211[11]. The latter result is confirmed by a more detailed, independent study[14]. These results suggest that, at least for some bursts, the central engine is likely entrained with strong magnetic fields. The discovery of strong linear polarization[15] of gamma-ray emission in GRB 021206 is also consistent with such a picture.

An important question is how strong the magnetic energy density is, compared with the kinetic energy density. Conventionally one can define $\sigma = L_P/L_K$ to categorize the fireball (where L_P and L_K are the Poynting flux luminosity and kinetic energy luminosity, respectively). The canonical fireball is in the $\sigma \ll 1$ regime. For GRB 990123, broadband modeling suggests that $\varepsilon_{B,f} \sim 7.4 \times 10^{-4}$[16]. Our analysis indicates that $R_B = (\varepsilon_{B,r}/\varepsilon_{B,f}) \sim 15$ for this burst, so that $\varepsilon_{B,r} \sim 0.17$. This is still in the $\sigma \ll 1$ regime, which ensures self-consistency of our hydrodynamical treatment. In the meantime, it suggests that the magnetitized fireball is not Poynting-flux dominated at the deceleration radius.

A QUASI-UNIVERSAL STRUCTURED JET MODEL

Uniform vs. Universal Jets

One intriguing finding in the GRB afterglow observations is that the geometry-corrected total energy for different bursts is standard[17, 18]. There are two equivalent interpretations. One is that different GRBs collimate the same amount of energy in different solid angles, but with a uniform energy distribution within the jets. The other is that all GRBs have a same jet configuration, but the energy per solid angle decreases with angle from the jet axis in the form of $\varepsilon(\theta) \propto \theta^{-2}$, so that the inferred jet angles from the afterglow data correspond to the observers' viewing angles[19, 20]. The former is called "uniform jets", and the latter is called "universal jets".

These two models are two extremal presentations of what might happen in reality. In realistic simulations such as those in the collapsar model, the emerging GRB jets natually have a non-uniform angular structure[21]. On the other hand, it is unrealistic to expect that all GRB jets are exactly universal. Such an exactly universal picture is already disfavored by the $\log(E_{iso}) - \log(\theta_j)$ plot of the observed data, which indicate a large scatter around the $E_{iso} \propto \theta_j^{-2}$ line (see solid squares in Figure 2).

Quasi-Universal Jets: Power Law vs. Gaussian

A reasonable picture is that GRB jets preserve certain angular structure individually, and may have a "quasi-universal" pattern of the jet structure[22, 23]. The so-called quasi-universal jet model suggests that all GRBs have a more-or-less similar angular jet structure, with the model parameters (e.g. the power-law index for the power-law jets, the typical angle for the Gaussian jets, and the normalization parameters for both types of jets) being distributed around some standard values with a small scatter.

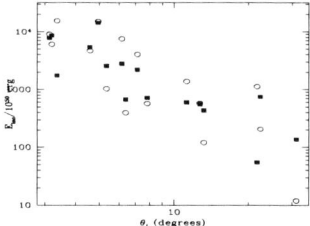
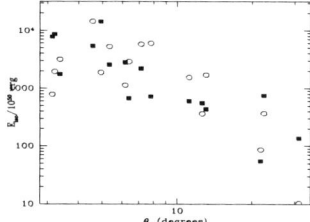

FIGURE 2. Simulated $E_{iso} - \theta_j$ data from the quasi-universal structured jet models[22] (open circles) as compared with the data[18] (solid squares). *Left panel:* Quasi-universal power-law model; *Right panel*: Quasi-universal Gaussian model.

When parameters are allowed to have some scatter, the $k = -2$ power law structure is no longer a pre-requisite for individual bursts. Other types of jet structure (such as Gaussian)[20] are also allowed, especially when the total energy within the jet is a quasi-constant. Figure 2 shows that both a quasi-universal power-law model and a quasi-universal Gaussian model can reproduce the $E_{iso} - \theta_j$ data[22].

Quasi-Universal Gaussian Jets: A Unified Model for GRBs and XRFs

X-ray flashes (XRFs) are the natural extension of GRBs towards the softer and fainter regime. Recent HETE-2 data reveal that an intriguing empirical relation $E_p \propto (E_{iso})^{1/2}$ (where E_p is the peak energy of the GRB-XRF spectrum)[24] is extended from GRBs to XRFs[25], and that the number ratio among GRBs, X-ray rich GRBs (XRGRBs) and XRFs is roughly 1:1:1. These facts pose severe constraints on both the universal[25] and the uniform[23] jet models. The current GRB-XRF prompt emission and afterglow data are, however, consistent with a quasi-universal Gaussian jet model[23].

Figure 3 presents some predictions of the quasi-universal Gaussian jet model. The GRB luminosity function (left panel) is predicted to be a broken power-law with indices changing from -1 to ~ -2[22]. This is consistent with some luminosity function studies. The GRB:XRGRB:XRF number ratio is roughly 1:1:1 (middle panel), and the afterglow

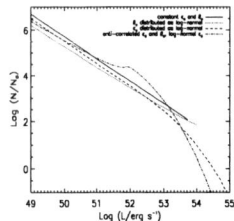 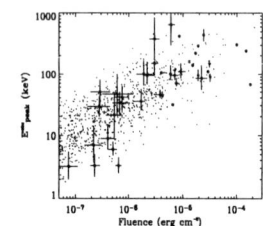 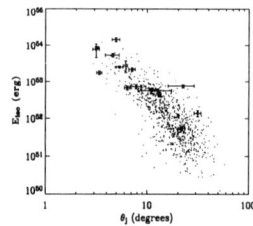

FIGURE 3. A quasi-universal Gaussian jet model confronted with the current data. *Left:* the predicted GRB luminosity functions[22]; *Middle:* the E_p-fluence diagram[23]; *Right:* the $E_{iso} - \theta_j$ diagram[23].

$E_{iso} - \theta_j$ correlation is consistent with the "standard energy reservoir" relation[23]. More rigorous tests of this model with a wider spectrum of data are being performed[26].

ACKNOWLEDGMENTS

This research is supported partly through NASA NAG5-13286, NSF PHY 01-14375, and the Monell Foundation.

REFERENCES

1. Mészáros, P. & Rees, M. J. ApJ, **476**, 232 (1997)
2. Sari, R. & Piran, T. ApJ, **517**, L109 (1999)
3. Kobayashi, S. ApJ, **545**, 807 (2000)
4. Akerlof, C. W. Nature, **398**, 400 (1999)
5. Fox, D. et al. Nature, **422**, 284 (2003)
6. Li, W. et al. ApJ, **586**, L9 (2003)
7. Kobayashi, S. & Zhang, B. ApJ, **597**, 455 (2003)
8. Sari, R., Piran, T. & Narayan, R. ApJ, **497**, L17 (1998)
9. Kobayashi, S. & Zhang, B. ApJ, **582**, L75 (2003)
10. Chevalier, R. A. & Li, Z.-Y. ApJ, **520**, L29 (1999)
11. Zhang, B., Kobayashi, S. & Mészáros, P. ApJ, **595**, 950 (2003)
12. Kobayashi, S. Mészáros, P. & Zhang, B. ApJ, **601**, L13 (2004)
13. Zhang, B. & Mészáros, P. Int. J. Mod. Phys. A, in press (astro-ph/0311321) (2003)
14. Kumar, P. & Panaitescu, A. MNRAS, **346**, 905 (2003)
15. Coburn, W. & Boggs, S. E. Nature, **423**, 415 (2003)
16. Panaitescu, A. & Kumar, P. ApJ, **571**, 779 (2003)
17. Frail, D. et al. ApJ, **562**, L155 (2001)
18. Bloom, J., Frail, D. A. & Kulkarni, S. ApJ, **594**, 674 (2003)
19. Rossi, E., Lazzati, D. & Rees, M. J. MNRAS, **332**, 945 (2002)
20. Zhang, B. & Mészáros, P. ApJ, **571**, 876 (2002)
21. Zhang, W., Woosley, S. E. & MacFadyen, A. I. ApJ, **586**, 356 (2003)
22. Lloyd-Ronning, N. M., Dai, X. & Zhang, B. ApJ, **601**, 371 (2004)
23. Zhang, B., Dai, X., Lloyd-Ronning, N. M. & Mészáros, P. ApJ, **601**, L119 (2004)
24. Amati, L. et al. A&A, **390**, 81 (2002)
25. Lamb, D. Q., Donaghy, T. Q. & Graziani, C. astro-ph/0309456 (2003)
26. X. Dai and B. Zhang, in preparation (2003)

Further Analysis of GRB 030501

Martin Topinka[†]

Charles University Prague, Faculty of Mathematics and Physics, Czech Republic
Astronomical Institute of the Academy of Sciences of Czech Republic, Ondřejov, Czech republic

Abstract. We analyze GRB 030501 using combined available data from INTEGRAL and RHESSI. We try to decompose the lightcurve into elementary peaks and show the correlation between lightcurve and spectral behavior of the burst within the framework of the fireball model. First we try to generate synthetic lightcurves and spectra and compare it with real bursts. We suggest non-constant secondary induced magnetic field.

INTRODUCTION

The gamma-ray burst (GRB) fireball model [1] predicts a synchrotron power-law spectrum and a fast rise and exponential decay (FRED) shape pulses in a gamma-ray lightcurve to be produced by internal shock in an interaction between two colliding shells (does not matter if it is $e^- p^+$ or Poynting flux dominated). We intent to check the theory on a complex structure of GRB 030501 detected by INTEGRAL and RHESSI satellites through a correlation between spectral and photometry evolution of the burst. While doing this, a need developed for data denoising and peak finding algorithms.

Thanks to the excellent work of INTEGRAL and RHESSI teams we have quite nice data sets, a lightcurve and spectral evolution (IBIS, ISGRI, SPI, RHESSI instruments), see Fig. 1 and Fig. 2.

INTERESTING FEATURES

There is many interesting points worth mentioning when analyzing GRB 030501:

a) No energy peak in the available spectral window, therefore the peak energy is beyond 300 keV

b) Burst was triggered while performing a galactic scan, therefore a huge extinction disabled an optical afterglow detection, any energy or local density estimation, host galaxy detection and we have no redshift information.

c) GRB 030501 has an almost "symmetric" peak, not a FRED, therefore a complex structure of overlapping pulses might play a role.

d) The peak is broad and the burst is relatively faint, Do we see a high redshift event (statistically fainter bursts originate at further distance)? Will we also see stretched periods in Fourier or wavelet image caused by cosmological time dilation?

e) There is bizarre "upside-down" spectral (photon) index behavior during a single peak. Is this a sign of a changing magnetic field (assuming synchrotron emission)?

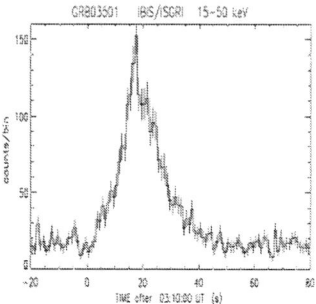

FIGURE 1. ISGRI GRB 030501 lightcurve [2]

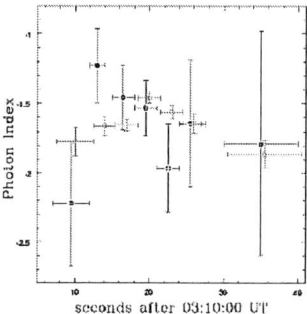

FIGURE 2. Evolution of photon index power-law slope [2]

f) INTEGRAL has provided a nicely time resolved spectral evolution during the peak. Is there a correlation between a spectral change and a lightcurve?

DENOISING AND DECONVOLUTION USING DISCRETE AND CONTINUOUS WAVELET TRANSFORMATION (DWT AND CWT)

Wavelet transformation [3] is both a time and a period resolved transformation of the signal x_n defined as a convolution with s scaled mother Ψ_0 wavelets that can be either orthogonal (DWT) or not (CWT)

$$W_n(s) = \sum_{n'=0}^{N-1} x_n \psi^* \left[\frac{(n-n')\delta t}{s} \right]$$

We tried to denoise the data using soft and hard WT coefficients suppressing. We used modified Lucy-Richardson, maximum entropy and classical deconvolution methods to decompose a lightcurve into pulses of a desired point-spread-function (PSF) shape. We tested the algorithm on a synthetic pregenerated lightcurve, "peak-rich" BATSE [4]

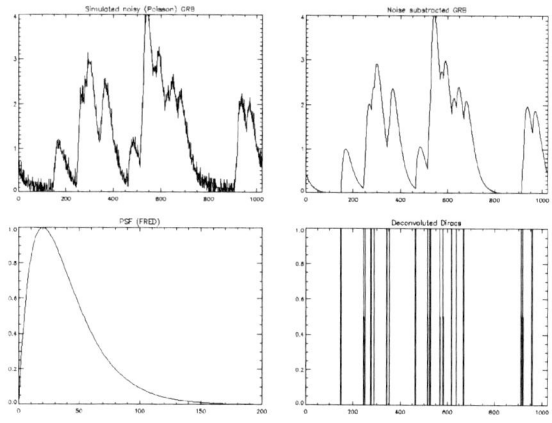

FIGURE 3. Synthetic GRB denoising and deconvolution

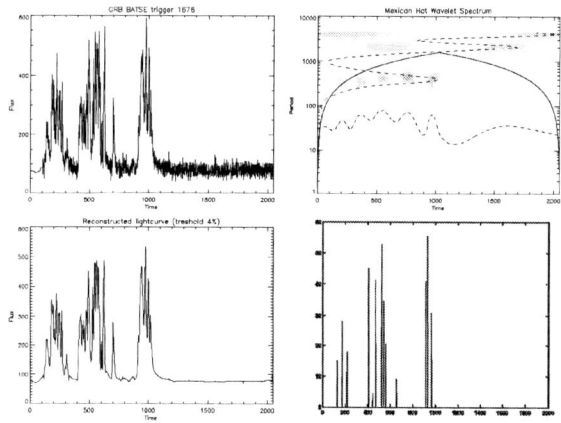

FIGURE 4. BATSE trigger 1676 denoising, CWT and deconvolution

triggers 1676 and 7906 before applying on GRB 030501. Preliminary results are shown in Fig. 3, Fig. 4 and Fig. 5.

FUTURE PLANS

Denoising process should be applied very carefully not to lose too much of real data, different types of noise will be tested.

Deconvolution using parametric PSF of more than one shape is mathematically impossible to do. A direct curve-fitting is applicable only if there is a small number of peaks. We will try to find a way out from this dead end. Any experience and help would

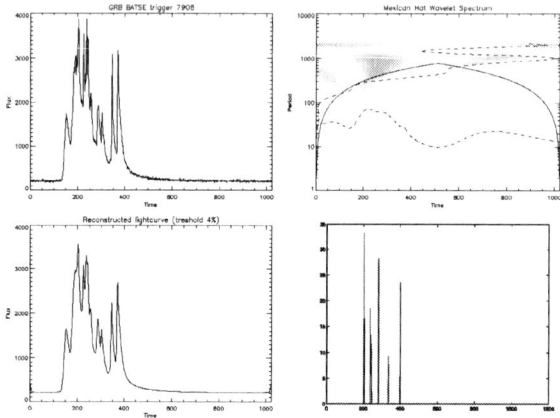

FIGURE 5. BATSE trigger 7906 denoising, CWT and deconvolution

be appreciated with pleasure. We will try to decompose the light curve using different PSF step by step and to weight the amount of each pulse by fitting. Another method is to find a FRED-like wavelet base already in a space of free parameters.

According to the fireball model we plan to derive energy, velocity, induced magnetic field and cooling regime of each merging shells, as well as the distribution of time-lag between each collision imprinted in each pulse profile and spectral behavior. Thus a correlation of the peak distribution and the spectrum will be computed.

ACKNOWLEDGMENTS

We acknowledge the support provided by the grant A3003206 provided by the Grant Agency of the Czech Republic and by the ESA PRODEX, Project 14527.

REFERENCES

1. T. Piran, *Gamma-ray Bursts and the Fireball Model*, astro-ph/9810256 (1998).
2. V. Beckmann, J. Borkowski, T.J.-L. Courvoisier, D. Goetz, R. Hudec, F. Hroch, N. Lund, S. Mereghetti, S.E. Shaw, C. Wigger, *Time resolved spectroscopy of GRB030501 using INTEGRAL*, A&A **411L**, 327B (2003).
3. C. Torrence and G. P. Compo, *A Practical Guide to Wavelet Analysis*, (1998).
4. http://cossc.gsfc.nasa.gov/batse/index.html

Durations of Gamma-ray Bursts and X-Ray Flashes in X-ray and Gamma-Ray Bands Observed with HETE-2

M. Suzuki[*], N. Kawai[*†], A. Yoshida[**†], Y. Shirasaki[‡], M. Matsuoka[§], T. Tamagawa[†], K. Torii[†], T. Sakamoto[*†], C. Graziani[¶], D.Q. Lamb[¶], J.L. Atteia[‖], E. E. Fenimore[††], M. Galassi[††], T. Donaghy[¶], G. Ricker[‡‡], J. Doty[‡‡], R. Vanderspek[‡‡], G. B. Crew[‡‡], J. Villasenor[‡‡], N. Butler[‡‡], J. G. Jernigan[§§], C. Barraud[‖], M. Boer[¶¶], J-P. Dezalay[¶¶], J-F. Olive[¶¶], K. Hurley[§§], S. E. Woosley[***] and HETE-2 Science Team[†††]

[*]*Department of Physics, Tokyo Institute of Technology*
[†]*RIKEN (Institute of Physical and Chemical Research)*
[**]*Department of Physics, Aoyama Gakuin University*
[‡]*National Astronomical Observatory Japan*
[§]*Japan Aerospace Exploration Agency*
[¶]*Department of Astronomy and Astrophysics, University of Chicago*
[‖]*Laboratoire d'Astrophysique, Observatoire Midi-Pyrénées*
[††]*Los Alamos National Laboratory*
[‡‡]*Center for Space Research, Massachusetts Institute of Technology*
[§§]*University of California at Berkeley*
[¶¶]*Centre d'Etude Spatiale des Rayonnements, Observatoire Midi-Pyrénées*
[***]*Department of Astronomy and Astrophysics, University of California at Santa Cruz*
[†††]

Abstract. We report timing properties of Gamma-ray bursts (GRBs) and X-ray flashes (XRFs) observed by HETE-2. We studied 1) duration (or emission time), 2) Variability and 3) spectral lag of the bursts and empirical relations of them. In particular, we examine whether the empirical relations hold or not in lower energy band below 25 keV.

INTRODUCTION

HETE-2 (High Energy Transient Explorer 2) satellite has localized more than 30 gamma-ray bursts (GRBs) and X-ray flashes (XRFs) since its launch. We studied durations of these bursts. It is empirically known that the durations of GRBs are shorter in higher energy bands [1]. It is interesting to check if this relation still holds in the X-ray bands of GRBs, and even with XRFs.

T90 or T50 has been used for these kind of studies. However T90 and T50 are not always determined reliably, in particular for the bursts with poor statistics.

We try avoiding this problem by using different methods for calculating durations and comparing results each other. Spectral lag is another interesting value obtained through timing analyses, because the spectral lags are known to be related with the peak luminosities of the bursts. The common way to calculate spectral lags is cross-

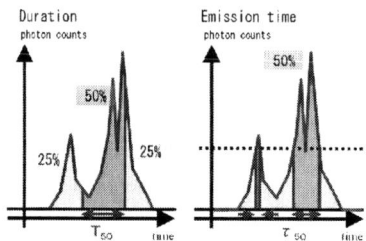

FIGURE 1. Duration T_{50} and emission time τ_{50}

correlation method [2]. This method is not always applicable, because cross-correlation method needs relatively good statistics. To calculate spectral lags for the data of bursts with poor statistics, we adopt the method which use the difference of the time when the half of the burst photons came. We examined the lag-luminosity relation using this definition.

DATA SELECTION

To study timing properties, we chose 17 bursts for which 1) both X-ray and gamma-ray data are available, 2) there is no variable X-ray source in the field of view, 3) there is no contamination from particle background. Among 17 samples, 7 have known redshift. We analyzed two type of data for X-rays: "TH" (1.2 sec resolution) and "TAG" (with 0.1 sec binning), both TH and TAG in 4 energy bands: (0) 2-25 keV, (1) 2-5 keV, (2) 5-10 keV and (3) 10-25 keV. There are also two type of data for gamma-rays: TH (1.3 sec resolution) and PH (with 0.1 sec binning). We use 3 energy band for TH: (A) 6-40 keV, (B) 6-80 keV and (C) 32-400 keV. For PH data, we set the energy boundary the same as BATSE: (4) 25-50 keV, (5) 50-100 keV and (6) 100-300 keV.

We use the data with higher time resolution as well as possible.

ANALYSIS

Durations and emission times. We calculate durations (T_{50} or T_{90}) and emission times (τ_{50} or τ_{90}) [3]. The definition of duration and emission time are shown in Figure 1.

Variability. We also examined variability-luminosity relation using the ratio of T_{90} and emission time τ_{90}. We define our variability V as $V = T_{90}/\tau_{90}$. This is quite different than variability has been defined in the past [4].

Spectral lags. To evaluate spectral lags, we calculated t_m which is median of photon arrival time in each energy band. Then we defined our spectral lags as the difference

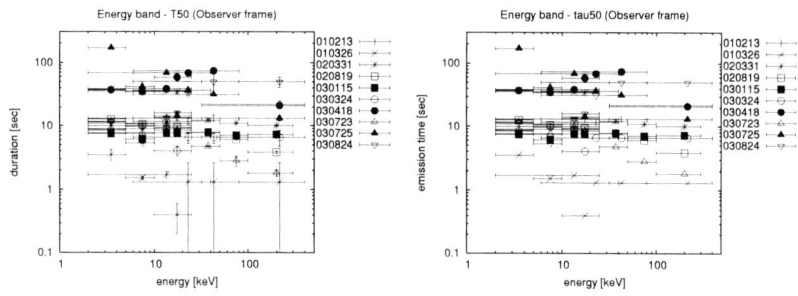

FIGURE 2. Duration T_{50} and emission time τ_{50}

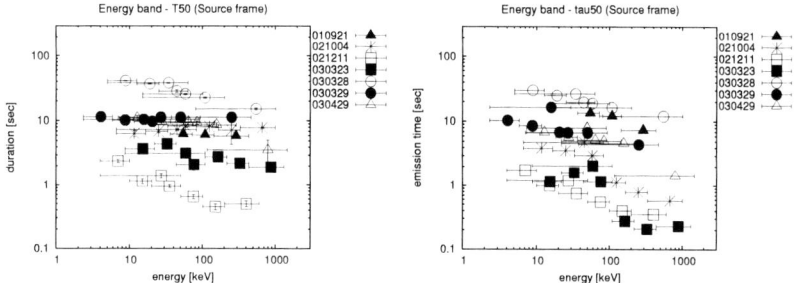

FIGURE 3. Duration T_{50} and emission time τ_{50} in source frame

between t_m of different energy band. This definition is quite different than has been used in the past.

RESULTS

Durations and emission times. We examined energy-duration relation first. Figure 2 shows that the relation between the duration and energy extended to the lowest energy band of our observation ($\sim$ 2keV).

We also calculated the durations and energy ranges in source frame (Figure 3). We can conclude that there is intrinsic variation in durations in the same energy range. However the relation between duration and energy band is always present.

Variability. We note that the relation can be clearly observed when we take the energy band the same as BATSE [4], however the relation cannot be seen clearly for the other bands. (Figure 4 left)

Spectral lags. In Figure 4 right, the peak luminosities are plotted as a function of the spectral lags. The thick solid line is the relation reported by Norris et al. [5] and thin

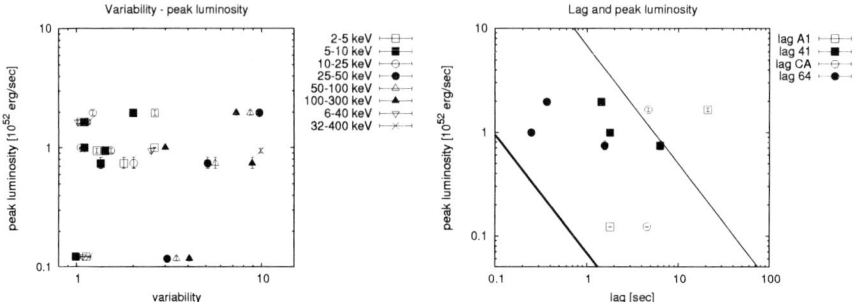

FIGURE 4. Variability-luminosity relation (left) and lag-luminosity relation (right). "lag A1" means lag between band A and band 1, which is defined as t_m(band 1) - t_m(band A).

solid line has the same slope as the reported value but its normalization is hundred times as large as the reported value.

We found that all the lags we calculated are larger than the reported relation [5]. Of course, our lag is defined differently. There are the cases that the bursts have several peaks and the last few peaks are only seen in lower energy bands below 10 keV. In these cases, the lags between lower bands and other bands become extremely large. It may be the reason why the relation between lag and luminosity is not observed.

CONCLUSION

We studied the 1) energy-duration, 2) variability-luminosity and 3) lag-luminosity relation of GRBs observed by HETE-2. We found 1) energy-duration relation extends to ~ 2 keV and 2) variability-luminosity relation is observed in some energy bands but not observed in the other bands. We adopt a new definition of spectral lags and studied lag-luminosity relation. The lags calculated with the new definition are relatively larger than the lags calculated with cross-correlation method and the lag-luminosity relation is not observed.

REFERENCES

1. Fenimore, E. E., in 't Zand, J. J. M., Norris, J. P., Bonnell, J. T., and Nemiroff, R. J., *ApJL*, **448**, L101+ (1995).
2. Band, D. L., *ApJ*, **486**, 928–+ (1997).
3. Mitrofanov, I. G., Anfimov, D. S., Litvak, M. L., Sanin, A. B., Saevich, Y. Y., Briggs, M. S., Paciesas, W. S., Pendleton, G. N., Preece, R. D., Koshut, T. M., Fishman, G. J., Meegan, C. A., and Lestrade, J. P., *ApJ*, **522**, 1069–1078 (1999).
4. Reichart, D. E., Lamb, D. Q., Fenimore, E. E., Ramirez-Ruiz, E., Cline, T. L., and Hurley, K., *ApJ*, **552**, 57–71 (2001).
5. Norris, J. P., Marani, G. F., and Bonnell, J. T., *ApJ*, **534**, 248–257 (2000).

Crude Limits on Prior and Prompt Optical Emission from GRBs from CONCAMs of the Night Sky Live Global Network

Robert J. Nemiroff*, Dolores Perez-Ramirez* and Daniel Cordell*

Michigan Technological University, Department of Physics, 1400 Townsend Drive, Houghton, MI 49931

Abstract. The developing global Night Sky Live network of nighttime fisheye web-cameras (CONtinuous CAMeras or CONCAMs) monitors much of the night sky down to a limiting visual magnitude as faint as 7. Several times now, a CONCAM has imaged the position of a gamma-ray burst (GRB) at trigger time. So far, however, no optical transient has been bright enough to be recorded by a CONCAM. Several example cases are discussed.

THE NIGHT SKY LIVE GLOBAL NETWORK

CONtinuous CAMeras (CONCAMs) are fisheye nighttime web cameras populating a developing Night Sky Live global network. CONCAMs are now welcomed by the world's major astronomical observatories because they help observers make important observing decisions more quickly and more reliably than ever before. Current CONCAMs record the night sky as deep as visual magnitude 7 in the image center over a (typically) three-minute exposure. CONCAMs were developed at Michigan Tech and deployed over the past three years to Mauna Kea Observatory in Hawaii, Kitt Peak Observatory in Arizona, Mt. Wilson Observatory in California, Rosemary Hill Observatory in Florida, The European Northern Observatory in the Carnary Islands in Spain, Wise Observatory in Israel, Siding Spring Observatory in Australia, and near the South African Large Telescope in South Africa. Ten (10) more CONCAMs are now waiting to be deployed, likely over the next year.

All images recorded by CONCAMs appear in real time on the Night Sky Live (NSL) website at http://nightskylive.net or http://concam.net . All NSL data is placed immediately in the public domain. This means that any observer can check for bright optical transients accompanying GRBs, and even write a paper about it, without consent or knowledge of the Night Sky Live collaboration.

The current Night Sky Live network continually monitors just over 50 % of the night sky. What does this mean for the likelihood that an NSL CONCAM has imaged the position of a GRB at trigger time? Assuming GRBs occur randomly over all angles, roughly 50 % of GRBs will occur at night over the location of a CONCAM and thus be potentially visible to a CONCAM. Of these roughly 50 % will occur when the Moon disrupts either the whole frame or the candidate GRB position. Of these remaining GRBs, about half will be lost to clouds and operational inefficiencies. Of these remaining

GRBs, only about half will be high enough in the sky to get a good look. This indicates, to a first approximation, a good prompt CONCAM observation will occur for about 1 in 16 GRBs at the time of trigger by the current Night Sky Live Network of 8 CONCAM nodes. As we expect further expansion of the NSL global network, this efficiency is likely to increase.

PROMPT GRB EMISSION MIGHT BE BRIGHT

There are theoretical reasons to expect strong optical flashes to precede the main part of some gamma-ray bursts. Meszaros et al. [1] and Belobodorov [2] suggested a standard GRB model that includes the possibility of generation of a low energy precursor to the proper GRB. Paczynski [3] and Kumar & Panaitescu [4] predicted the possibility of generation of a strong optical flash preceding the main GRB.

A bright optical transient surely accompanied GRB 990123, which appeared at visual magnitude 9 in a 5 second exposure that started 25 seconds after the gamma-ray trigger (Akerlof & McKay, IAUC 7100). Other prompt bright afterglows include GRB 021211 and GRB 030329, the later of which may have become as bright as visual magnitude 3 (Ofek et al., GCN 2031).

Evidence indicates that that at least some GRBs are extremely red with $R - I$ 2.8 [5, 6, 7]. Our approach for studying the upper limits started by studying the changes in magnitude of a very red star, iota Pisces (HD 223075, V = 5.01, R-I = 2.59) used as a prototype star visible in our field of view, measuring its changes in brightness as it moves from the center to the edge.

CANONICAL RESULTS FOR SPECIFIC GRBS

What follows below are brief highlights of three events that were imaged by a NSL CONCAM at or near GRB trigger time. These particular results were all submitted as GCN circulars. Other GRBs have occurred in CONCAM fields but they are no longer routinely reported to the GCN as they do not enable follow-up observations. For example, two relatively recent candidates of note were GRB 030328 which showed no concurrent optical flash to visual magnitude of about $V \sim 5$, and GRB 030528 which showed no concurrent optical flash to a visual magnitude of about $V \sim 3.5$.

CONCAM crude optical limits on GRB 001005: GCN Circular # 842

The Kitt Peak CONCAM imaged the field of this GRB just before and after the reported trigger time. The correspondent frames found in our archive that bracket the time reported for this GRB, 03:25:09 UT, (Hurley et al GCN 838) were recorded at 03:21:35 UT (before the event), 03:25:14 UT (closer to the end) and 03:28:42 UT (after the event). About 100s of dead time follows each 120s integration. We note that although the Moon was in the field, the position of GRB001005 was relatively unaffected. This position is placed on the east edge of the CONCAM field. No obvious optical counterpart down to about visual magnitude 4 over a two-minute exposure was visible on the frames.

CONCAM's null results during GRB 010126: GCN Circular # 927

The Kitt Peak CONCAM imaged the field of this GRB coincident with the reported trigger time. We found in our archive a relevant CONCAM frame which contained an integration from 33022s (09:10:22 UT) to 33202s (09:13:22 UT) that brackets 33048s (09:10:48 UT), the time of the GRB reported by Hurley et al. (GCN 922). Thus, we examined three key frames corresponding to times right before, during and after the GRB. The time recorded on the frames was behind the actual UT time by about 26s. We corrected for this resulting an uncertainty of about 2s. For these frames, the integration time was three-minutes exposure. No obvious counterpart down to about visual magnitude 6.3 was visible on the frames. This was, to the best of our knowledge, the first time an optical electronic instrument ever observed a GRB location during a GRB trigger.

CONCAM null result during GRB 030329: GCN Circular # 2031

This GRB occurred near the center of the Mauna Kea CONCAM field and near the edge of the Mt. Wilson CONCAM field and the Kitt Peak CONCAM field. Unfortunately, the Mauna Kea CONCAM was turned off during the day to fix a imaging problem on the MK CCD, and the MK field operators forgot to turn it back on until the following day. The MW CONCAM did not record useful information at that field edge due to the bright lights of the city of Los Angeles. The best images came (again) from the Kitt Peak CONCAM. The Michigan Tech group searched the fields but did not detect anything and decided not to issue a GCN circular. Astronomers at Tel Aviv University, however, did their own search on http://concam.net data, achieved similar results, and due to the free and public domain nature of CONCAM data issued their own GCN Circular (Ofek et al., GCN 2031). Three 180 s CONCAM images were taken at Mar 29, 11:29:57, 11:33:52, and 11:41:46 UT bracketing the burst time of March 29, 11:37:14 (Ricker et al., HETE II) and, although not directly incorporating the trigger time, the second cited integration ended only 17 seconds before the HETE trigger. The limiting visual magnitude is about 3.5.

ACKNOWLEDGMENTS

RJN acknowledges support from NASA and the NSF.

REFERENCES

1. Meszaros, P., Ramirez-Ruiz, E. & Rees, M.J., ApJ **554**, 668 (2001).
2. Beloborodov, A.M., astro-ph/0103321 (2001).
3. Paczynski, B., 2001, astro-ph/0108522 (2001).
4. Kumar, P. & Panaitescu, A., in preparation (2001).
5. Fruchter, A., ApJ **512**, L1 (1999).
6. Reichart, D.E., ApJ **495**, L99 (1998).
7. Reichart, et al.,, astro-ph/9806082 (1998).

INTEGRAL Spectrometer Analysis of GRB030227 & GRB030131

L. Moran*, L. Hanlon*, B. McBreen*, R. Preece[†], Y. Kaneko[†], O. R. Williams**, K. Bennett**, R. Marc Kippen[‡], A. Von Kienlin[§], V. Beckmann[¶], S. McBreen* and J. French*

*Department of Experimental Physics, University College Dublin, Ireland
[†]Department of Physics, University of Alabama at Huntsville, USA
**Science Operations and Data Systems Division of ESA/ESTEC, SCI-SDG, NL-2200 AG Noordwijk, The Netherlands
[‡]Space and Remote Sensing Sciences, Los Alamos National Laboratory, USA
[§]Max-Planck-Institut für Extraterrestrische Physik, 85748 Garching, Germany
[¶]NASA Goddard Space Flight Center, University of Maryland Baltimore County, USA

Abstract. The spectrometer SPI on board INTEGRAL is capable of high-resolution spectroscopic studies in the energy range 20 keV to 8 MeV for GRBs which occur within the fully coded field of view (16° corner to corner). Six GRBs occurred within the SPI field of view between October 2002 and November 2003. We present results of the analysis of the first two GRBs detected by SPI after the payload performance and verification phase of INTEGRAL.

INTRODUCTION

On October 17th 2002, ESA's gamma-ray observatory INTEGRAL was successfully launched from the Baïkonur Cosmodrome in Kazakhstan. INTEGRAL has a burst alert system (IBAS) which carries out rapid localisations for gamma-ray bursts (GRBs) incident on the IBIS detector. These co-ordinates are then distributed, allowing for fast follow up observations at other wavelengths [1]. The main instruments IBIS and SPI also contribute greatly to INTEGRAL's GRB capabilities. IBIS is a high resolution imager [2] with angular resolution of 12 arcminutes for sources within its 9° × 9° fully coded field of view and broadband spectral capabilities. SPI is optimised for spectroscopic study of gamma-ray sources, with some imaging capabilities. IBIS and SPI are complemented by two smaller instruments, JEM-X and OMC, which monitor gamma-ray sources at x-ray and optical wavelengths.

SPI consists of 19 high purity Germanium detectors actively cooled to a temperature of ~ 85 K to provide an energy resolution FWHM of 2.5 keV at 1 MeV [3] in the range 20 keV-8 MeV. SPI's imaging capabilities are due to a coded mask comprising of 127 tungsten elements, with a thickness of 30 mm, placed at a distance of ~ 1.7 m from the detection plane, providing an angular resolution of 2.8°. The good angular resolution combined with excellent spectral resolution make SPI an ideal instrument for spectral studies of the prompt emission of GRBs.

GRBs, first detected over 35 years ago, are an intriguing phenomenon and remain at the forefront of research in astrophysics. The discovery by BeppoSAX of afterglows in the x-ray [4] and subsequent discoveries at optical [5] and radio [6] wavelengths have led to redshift measurements [7] for ~ 40 bursts ranging from $z = 0.168 - 4.5$. A theory of gamma-ray bursts must provide a mechanism capable of a non-thermal energy output of the order of 10^{52}-10^{54} ergs by compact sources at cosmological distances.

With a large detector area of approximately $500\,cm^2$ and its high spectral resolution, SPI can address the long-standing controversy over the existence of short-lived spectral features in GRB spectra, previously searched for with varying degrees of success [8, 9]. A confirmation of line features and details of specific spectral features could contribute to the debate on the connection between GRBs and core-collapse supernovae [10]. In addition, the broad energy coverage of SPI (20 keV-8 MeV) is well suited to constrain the spectral shape, both below and above the energy at which the GRB power output is typically peaked ($\sim 250\,keV$). Study of the spectral shape of the prompt emission of a GRB at the onset of the afterglow from SPI data could reveal the activity of the central engine that leads to the production of an afterglow. At the high energy end, there may exist a hard spectral upturn as recently found by González et al. (2003) in archival CGRO data of GRB941017.

SPI DATA ANALYSIS OF GRBS

A standard SPI pointing has a duration of ~ 35 minutes. In this Science Window (ScW) the numbers of single, double and higher multiplicity events striking each detector are recorded according to the energy (16384 channels). Photon by photon information, which contains details of the detector struck, the energy deposited and the exact time, is available for multiple events and all events analysed by the Pulse Shape Discriminator (PSD), about 10% of the total. The standard SPI pipeline is designed to process sets of ScWs [12], whereas a GRB lasts only a fraction of this duration. Therefore, a modified analysis procedure (see Fig. 1) is used in which the start and end times and the best known position of the GRB are manually inserted.

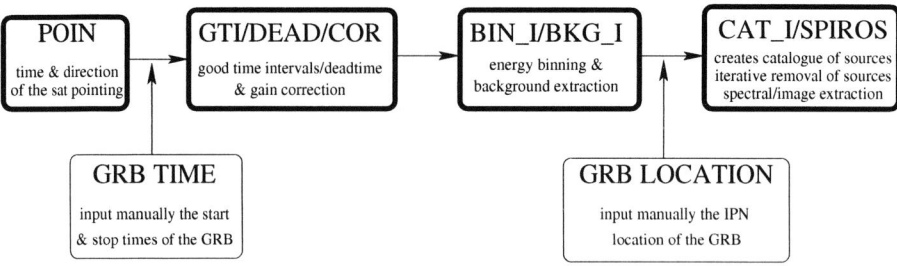

FIGURE 1. Flow chart of the SPI instrument specific software pipline, including the additional steps necessary for GRB analysis.

RESULTS

GRB030131 At 07:39:00 UT on January 31st 2003, SPI detected a gamma-ray burst of duration $\sim 60\,s$, the first since commencing full operational status. The corrected INTEGRAL position of the burst was given in *GCN 1847* as RA 202.13° and DEC

30.68° with a 5' radius error. The analysis of GRB030131 is complicated by the fact that 10 s after the GRB onset, the spacecraft started slewing and the remaining 50 s of the burst, including the brightest portion, occurred during this manoeuvre.

There are two main problems which arise when a SPI observation takes place during a satellite slew. The first is that not all types of events are recorded during a slew. To facilitate compression and transmission of the data from the preceding steady pointing, only multiple events and those single events analysed by the PSD are downlinked, together with the technical- and science-housekeeping data, with the loss of ∼90% of the single events.

The second complication is that to utilise the analysis procedure developed, the pointing direction of the instrument and details of the slew must be established and manually inserted. In particular, the RA and DEC of the satellite's orientation need to be chosen to reflect the motion of the spacecraft. As burst emission is evident for one third of the length of the slew, the co-ordinates of the SPI field of view were deemed to lie one sixth of the angular distance from the preceding steady pointing to the subsequent one.

Spectral anaysis was not possible with the limited telemetry received, but using the imaging capabilities of SPI the burst was located with a detection significance of 3.6 σ. The first row in *Table 1* gives the results obtained using the modified pipeline for the brightest 10 s of the burst.

TABLE 1. Results for the brightest 10 s of GRB030131.

RA	DEC	σ	$Flux(phs/cm^2/s)$	$Energy\,range\,(keV)$
200.650	30.210	3.6	0.86 ± 0.24	20-500
201.967	31.117	7.0	-	20-8000

Taking into account the SPI localisation precision of 2.8°, the SPI and IBIS locations are in agreement. The flux obtained is also consistent with that derived from IBIS data [13]. The second row in *Table 1* shows the results of analysis performed using the science-housekeeping data. This method does not require an input location and yet still locates a source consistent with the GRB location with a significance of 7 σ.

GRB030227 began at 08:42:04 UT on 27th Febuary 2003 and had a duration of 18 s. Though a weak burst [14], SPI images the burst with a detection significance of 7.5 σ and location in agreement with IBIS. A power law model fit to GRB030227 using XSPEC yields a photon index of 1.95 ± 0.17. Another fit using the same model in RMFIT [15] yields a photon index of 1.96 ± 0.18, where the data have been rebinned to increase the S/N (see Fig. 2). In both cases there is very good agreement between the results of SPI and the IBIS analysis in Mereghetti et al. (2003). In the energy range 20-200 keV the flux obtained for the burst is:

$$F_{20-200\,\text{keV}} = 5.5^{+5.1}_{-2.7} \times 10^{-8}\,\text{erg}/\text{cm}^2/\text{s}.$$

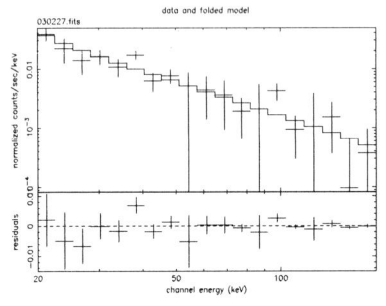 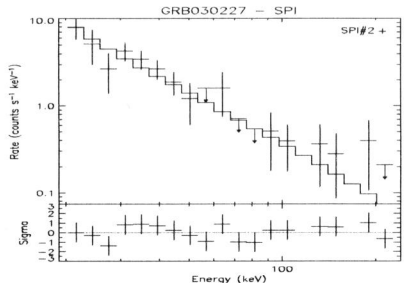

FIGURE 2. A power law model is fit to GRB030227 using *left*: XSPEC, yielding a photon index of 1.95 ± 0.17 and *right*: RMFIT, yielding a photon index of 1.96 ± 0.18.

CONCLUSIONS

GRB030131 was the first GRB in the FOV of the main instruments after the performance and verification phase of the satellite. The analysis of this burst shows the capabilities of the SPI instrument during a satellite slew, a frequent manoeuvre ($\sim 10\%$ of the time) due to the dither pattern employed by INTEGRAL. With limited telemetry it was still possible to image the burst and obtain a location and a flux. Although it was not feasible to carry out spectral analysis in this case, the imaging capability of SPI allows for cross calibration with IBIS. For GRB030227, though a very faint burst, a spectrum can be extracted in SPIROS and used to fit models in both XSPEC and RMFIT, obtaining a photon index and a flux in good agreement with previous work [14]. Therefore, SPI has demonstrated that when a burst of sufficient intensity is observed it will be possible to study the prompt emission of a GRB with the sensitivity necessary to determine the spectral evolution, identify hard spectral components and constrain models.

REFERENCES

1. Mereghetti, S., et al., "Real time localisation of Gamma Ray Bursts with INTEGRAL," in *Advanced Spectral Resolution*, Proceedings of the 34th COSPAR Scientific Assembly, Houston, 2002.
2. Ubertini, P., et al., *A&A*, **411**, L131–139 (2003).
3. Knödsleder, J., and Roques, J.-P., "SPI Science Prospects," in *The Gamma-Ray Universe*, Proceedings of the XXII Moriond Astrophysics Meeting, Les Arcs, 2002.
4. Costa, E., et al., *Nature*, **387**, 783–785 (1997).
5. van Paradijs, J., et al., *Nature*, **386**, 686–689 (1997).
6. Frail, D., et al., *Nature*, **389**, 261–263 (1997).
7. Metzger, M., *Nature*, **387**, 879–880 (1997).
8. Murakami, T., et al., *Nature*, **335** (1988).
9. Briggs, M., et al., "BATSE Evidence for GRB Spectral Features," in *Gamma-Ray Bursts*, edited by R. P. . T. K. C. Meegan, Proceedings of the 4th Huntsville Symposium 428, AIP, New York, 1997.
10. Hjorth, ., et al., *Nature*, **423**, 847–850 (2003).
11. González, M., et al., *Nature*, **424**, 749–751 (2003).
12. Courvoisier, T.-L., et al., *A&A*, **411**, L53–57 (2003).
13. Götz, D., et al., *A&A*, **409**, 831–834 (2003).
14. Mereghetti, S., et al., *ApJ Letters*, **590**, 73–78 (2003).
15. Preece, R., et al., *ApJSS*, **126**, 19–36 (2000).

XMM-Newton Observations of Gamma-Ray Burst Afterglows

N. Schartel

XMM-Newton Science Operations Centre, European Space Agency (ESA), Villafranca del Castillo, Apartado 50727, 28080 Madrid, Spain

Abstract. XMM-Newton observes X-ray afterglows of Gamma-Ray Bursts (GRB) with a reaction time of the order of 10 hours. Scientific highlights are the detection of an afterglow 67 days after the GRB occurred, Dark Bursts and the detection of soft X-ray emission lines. Based on the established reaction time and on the obtained results scientific objectives of future XMM-Newton observations of GRB afterglows are formulated.

OBSERVATIONS OF GRB AFTERGLOWS WITH XMM-NEWTON

XMM-Newton observes X-ray afterglows of GRBs by following-up GRB detections by other high energy satellites. XMM-Newton can observe GRBs detected by Integral, RXTE, IPN and BeppoSAX with a reaction time of the order of 10 hours. The observing strategy for GRBs is built on three pillars: immediate Target Of Opportunity (TOO[1]) observations, TOO observations based on the Director's decision (with a typical time delay of weeks) and Guest Observations of known GRB counterparts approved by the Observing Time Allocation Committee (with a typical time delay of years)

Data resulting from TOO observations, which do not violate data rights of another observation, become immediately public after standard processing. Up to now XMM-Newton has pointed in the direction of 8 GRBs performing 11 observations in total. In 9 observations an X-ray afterglow of the GRB was detected.

SCIENTIFIC RESULTS

The observations of GRB afterglows with XMM-Newton lead to several publications, including 9 GCN circulars, 1 IAU circular and 10 papers in refereed journals. In addition, several papers are in press. Three main scientific results may be outlined here in order to demonstrate the scientific potential of XMM-Newton observations for GRB afterglows:

- **An X-ray afterglow observed 67 days after the GRB:** The X-ray afterglow of the bright GRB 030329 was detected 37 and 61 days after the GRB in XMM-Newton observations [1], [2], [3]. This is the longest time delay between a GRB and an

---

[1] http://xmm.vilspa.esa.es/external/xmm_sched/too/index.shtml

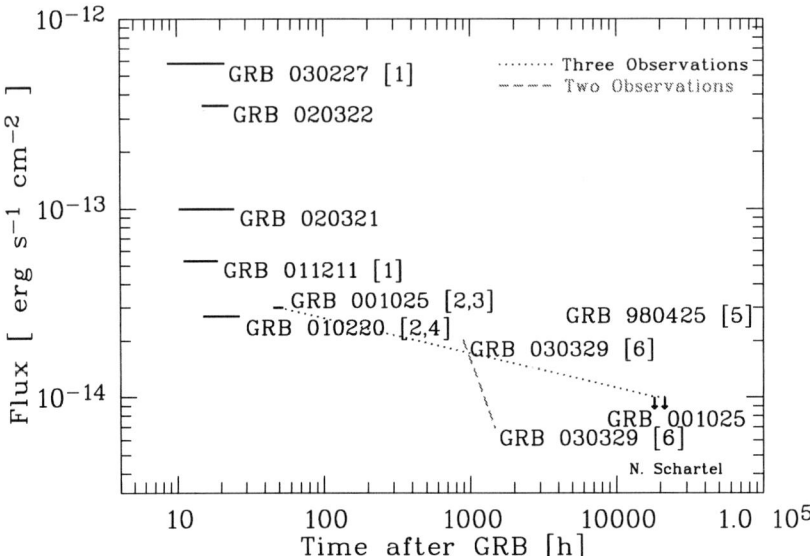

FIGURE 1. The observations of GRB afterglows with respect to the time of the GRB are shown versus the measured flux for the 2-10keV energy range. Features found are labeled: [1]: soft X-ray emission lines; [2]: possible soft X-ray emission lines; [3]: dark bursts; [4]: possible dark burst; [5]: complex light curve with contribution of supernova; [6]: break in light curve.

observation of its X-ray afterglow so far. The X-ray spectra show no thermal component, which excludes a contribution from SN2003dh. The combined Rossi-XTE and XMM-Newton measurements require a break at ≈ 0.5 days in the afterglow decay with a power law index increasing from 0.9 to 1.9.

- **Dark Bursts:** Despite several deep optical observations, no optical counterpart could be identified for GRB 001025A. GRBs for which no optical or radio counterpart can be identified are called "Dark Bursts". An XMM-Newton observation of GRB 001025A found an X-ray source in the error-box of the burst [4]. But the photon statistics were too poor to confirm a flux decrease with certainty. Therefore, it was not possible to exclude other types of variable X-ray sources, like AGN.

  The latter possibility, a variable X-ray source in the error box of the GRB, can be ruled out by the data taken in two follow-up observations (K. Hurley, in preparation). Consequently, we are sure that the X-ray source found in the error box is the X-ray afterglow of the (optically) Dark Burst GRB 001025A.

- **Soft X-ray Emission Lines:** The most surprising and scientifically interesting result of XMM-Newton observations of GRB afterglows so far is the detection of soft X-ray emission lines. Two observations of GRB afterglows show X-ray emission lines at a significant level: GRB 011211 ([5], [6], [7]) and GRB 030227 ([8], [9], [10], [11]). In addition, two further observations show evidence for emission lines. The description of the spectra requires a mekal spectrum in addition to a powerlaw

continuum. The spectra of both observations can be understood in the following terms:
- the lines are blueshifted with respect to the redshift of the host galaxy, implying an outflow velocity of one-tenth of the speed of light
- emission lines of lighter elements dominate (GRB 011211: Mg XI, Si XIV, S XV, Ar XVIII and CaXX; and GRB 030227: Mg XII, Si XIV, S XVI, Ar XVIII and Ca XX)
- emission lines of heavier elements, i.e. Ni and Fe, are not detected
- photoionization or reflection models fail to reproduce the spectra, as they favour heavier metals such as Fe or Ni

Although the observations of GRB 011211 and GRB 030227 differ in detail and require different interpretations with respect to the underlying geometry and time history of the late burst emission, the observations lead to the conclusions:
- both GRBs are a consequence of a recent supernova
- a "merging-of-compact-objects" scenario can be excluded for these two GRBs
- a thermal emission model is required for the description of the soft X-ray line emission
- the emission lines originate in a dense shell of material; i.e. the ejecta of a recent supernova is heated by the gamma-ray burst
- explanation for non-detection of Fe:
    * most of the illuminated ejecta is from the outer stellar layers implying domination of lower-atomic-core-number elements in recent supernovae
    * (stable) Fe is enriched later via beta-decay within months (Ni–>Co–>Fe, with reaction half-lives of 6 and 78 days, respectively)

SCIENTIFIC OBJECTIVES OF GRB OBSERVATIONS

As XMM-Newton can observe X-ray afterglows, with a short time delay with respect to the burst, only as TOOs, the scientific results cannot be predicted in advance. Nevertheless, one can estimate the possible impact of future XMM-Newton observations on currently open questions:

- **Soft X-ray emission lines:** The most obvious aim of XMM-Newton afterglow observations must be to extend the sample of spectra showing emission lines. Due to the high effective area of its instruments and the short time during which X-ray lines are emitted, XMM-Newton is the only facility which can perform such studies. The spectra, in combination with the details of the emission-line light curves, are fundamental for our knowledge, starting from identification of the progenitors, through constraining physical parameters of early supernova emission, up-to details of the early light curve and beam angle of the burst itself.
- **The Mystery of Dark Bursts:** The available observations suggest that all GRBs show an X-ray afterglow. But more than 50% of them do not show an afterglow at optical and radio wavelengths. Such bursts are called Dark Bursts. Due to the lack

of optical data our knowledge of Dark Bursts is very restricted: in particular, we do not know their distance. There are three main possibilities to explain the darkness:
1. the burst is emitted in a region which is highly obscured,
2. the burst occurs in surroundings where the medium is too tenuous for the forming of an optical afterglow,
3. the burst is located at very high redshift.

In the context of Dark Bursts, the observation of GRB 001025A is very promising [4]. On the one hand, scientists were able to show beyond doubt that XMM-Newton observed the X-ray afterglow. On the other hand the spectra of the X-ray afterglow shows evidence for soft X-ray emission lines. Future observation of GRB afterglows by XMM-Newton will make very important contributions by identifying soft X-ray emission lines and, with this, to determine the redshift, which will allow discrimination between the different discussed scenarios for Dark Bursts.

- **Short GRBs:** The analysis of Batse data of GRBs showed that there are two distinct populations of GRB. About 30% of the GRBs are characterised by an extremely short duration of the gamma-ray emission, in the time range of milliseconds. Whereas 70% of the GRBs have a duration >1 sec. Depending on the duration of the gamma-ray emission they are called either long bursts or short bursts [12]. All X-ray afterglows reported up to now are afterglows of long GRBs, - not a single afterglow of a short GRB has been detected so far. However, as Beppo-SAX, the workhorse for providing accurate positions of GRB, could not detect short GRBs, the lack of afterglows is most likely a selection effect. As there is no strong argument as to why short duration bursts should lack afterglows, their detection is a challenge for future XMM-Newton observations.

ACKNOWLEDGMENTS

I would like to acknowledge the XMM-Newton Science Operations Team for making rapid TOO observations possible at all, and the XMM-Newton Science Survey Centre for enabling a quick processing of data taken during TOO observations.

REFERENCES

1. A. Tiengo, e. a., *GCN*, **2241** (2003).
2. A. Tiengo, e. a., *GCN*, **2285** (2003).
3. A. Tiengo, e. a., *A&A*, **409**, 983 (2003).
4. D. Watson, e. a., *MNRAS*, **393**, 1 (2002).
5. J.N. Reeves, e. a., *Nature*, **416**, 512 (2002).
6. J.N. Reeves, e. a., *A&A*, **403**, 463 (2003).
7. M. Santos-Lleo, e. a., *GCN*, **1192** (2001).
8. D. Watson, e. a., *ApJ*, **595**, 29 (2003).
9. S. Mereghetti, e. a., *ApJ* (2003).
10. N. Loiseau, e. a., *GCN*, **1901** (2003).
11. R. Gonzalez-Riestra, e. a., *IAUC*, **8087** (2003).
12. C. Kouveliotou, e. a., *ApJ*, **413**, 110 (1993).

Particle Acceleration via Relativistic Magnetic-Dominated Expansion and GRBs

E. Liang

Rice University, Houston, TX 77005-1892

Abstract. We summarize recent results on the diamagnetic relativistic pulse accelerator (DRPA) and its potential applications to GRBs.

INTRODUCTION

Using 2-1/2-D Particle-in-Cell (PIC) [1] simulations to study the expansion of relativistic strongly magnetized plasmas, we recently demonstrated a new robust particle acceleration mechanism, which we call the diamagnetic relativistic pulse accelerator (DRPA) [2]. When a collisionless plasma with large ordered transverse magnetic energy/particle ($\Omega_e/\omega_{pe} >$ a few, where $\Omega_e =$ electron gyrofrequency and $\omega_{pe} =$ electron plasma frequency) is suddenly deconfined, the emergent electromagnetic (EM) pulse couples to the expanding plasma to create a strong cross-field transverse drift current [3], which both slows and reshapes the expanding electromagnetic (EM) pulse. This allows the EM pulse to trap and accelerate the surface particles via the pondermotive force [3], efficiently converting EM energy into directed particle energy. Remarkably this pulse structure persists as both the EM pulse and trapped surface particles become more and more relativistic with time. For e+e- plasmas, the DRPA eventually converts over 80% of the initial magnetic energy into the ultra-relativistic directed energy of the surface particles [2]. Hence DRPA represents a concrete and robust realization of the direct conversion of Poynting flux energy into ultra-relativistic directed particle energy. The Poynting flux scenario competes with the internal shock scenario as an alternative paradigm of prompt GRB gamma-ray emission [4, 5, 14].

It would be favored if GRB gamma-rays are indeed strongly polarized, suggesting the presence of large scale ordered magnetic fields. We have performed PIC simulations for both electron-positron ($m_i = m_e$) and electron-ion ($m_i/m_e = 100$) plasmas. In the e+e- case both species are energized equally. But in the e-ion case most of the energy gain goes to the ions due to charge separation. However, in the case of mixed e+e- and e-ion plasmas, the DRPA preferentially accelerates the e+e- component and leave the e-ion component behind. These results have important implications for both GRBs and ultra-high-energy cosmic ray production.

RELEVANCE OF DRPA TO GRBS

When the above PIC simulations are continued to >150 light-crossing times of the initial plasma, the plasma pulse exhibits many remarkable features which resemble the observed properties of cosmic gamma-ray bursts (GRBs) [6, 13]. These include the repeated bifurcation of the pulse profile into a complex chaotic structure at late times, the formation of a power-law momentum distribution with low-energy cut-off [7], and a hard-to-soft spatial profile of momentum distribution within the pulse. Due to time delays between photons emitted by different parts of the plasma pulse, and that emitted photon number should be proportional to the local electron density, we expect the detected time profile of photons emitted by the DRPA to roughly trace the spatial electron density profile at the time of radiation domination, which depends on the magnetic field strength. Hence the diverse morphology of the observed GRB light curves can be understood in terms of the radiaton age of the DRPA. If the DRPA has strong fields and radiate early, before any bifurcation, it produces a single smooth pulse (a FRED, [6]). But if the DRPA has weak fields and radiates late, after repeated bifurcation, it produces a complex multi-peak structure. We also find that the maximum attainable Lorentz factor (i.e. upper cutoff of the power-law) scales with the initial plasma size. The most remarkable and important property of the DRPA is that the group Lorentz factor of the EM pulse increases asymptotically as the square-root of the number of the gyroperiods: $\gamma_m = (f.\Omega_e(t).t)^{1/2}$ [7]. Parameters studies suggest that the coefficient $f = (\Omega_e/\omega_{pe})$ [8]. We also find that sustained particle acceleration occurs only when $\Omega_e/\omega_{pe} > 3$, i.e. when the plasma is magnetic-dominated. These and many other scaling properties of DRPA are being published in a series of papers [8, 9, 10]. We have also studied the radiative properties of the DRPA [11]. Since DRPA is primarily a linear accelerator, the intrinsic radiation mechanism resembles jitter radiation with the longitudinal jitter length scale given by the relativistic electron gyroradius. This gives a characteristic radiation frequency [12] $\omega_{cr} = \gamma_m \Omega_e$, similar to small-pitch-angle synchrotron radiation. Using this result together with the above square-root scaling, we find for typical long GRBs (with spectral break energy $E_{pk} = 500$ keV in source frame and radiation time $t = 300$ sec) $B = 10^6$ and $\gamma_m = few.10^7$. Remarkably, this translates into a total initial magnetic energy = 10^{51} ergs, consistent with GRB values. Using these numbers, we find that the complex pulse structure due to repeated bifuractions has a characteristic subpulse wavelength = fraction of a ms to a few ms. This is also consistent with observations.

The results obtained so far represent the tip of the iceberg of this extremely rich new plasma phenomenon. A key challenge is to find realistic physical settings under which the DRPA can be launched. Such realistic settings are mostly difficult to model with PIC simulations and likely require 3-D parallel PIC codes. At present we focus on more idealized initial conditions and geometries that can be simulated with 2-1/2-D codes. These include using initial configurations where the magnetic field deconfinement and current disruption occur on finite causal time scales, initial spatial profiles with more gradual surface gradients, and interactions of vacuum EM pulses with plasmas.

ACKNOWLEDGMENTS

This work is supported by NASA, LLNL and LANL.

REFERENCES

1. Birdsal, C. K. and Langdon, A. B., Plasma Physics via Computer Simulation, (IOP, Bristol, UK) (1991).
2. Liang, E., Nishimura, K. Li, H. and Gary, S.P.,, Phys. Rev. Lett. **90**, 085001 (2003).
3. Boyd, T. and Sanderson, J., Plasma Dynamics (Barnes and Noble, NY) (1969).
4. Lyutikov, M and Blandford, R., Proc. 1st N. Bohr Summer Inst., ed. Ouyed, R. et al. (Copenhagan), to appear (astroph020671) (2003).
5. Lyutikov, M. and Blackman, E., Mon. Not. Roy. Ast. Soc. **321**, 177 (2002).
6. Fishman G. and Meegan, C.A., Ann. Rev. Ast. Astrophys. **33**, 415 (1998).
7. Liang, E. and Nishimura, K., Phys. Rev. Lett., to appear (astroph0308301) (2003).
8. Liang, E., paper in preparation (2004).
9. Nishimura, K., and Liang, E., Phys. of Plasmas, to appear (astroph0308153) (2003).
10. Nishimura, K. Liang, E. and Gary, S.P.,, Phys. of Plasmas, to appear (astroph0307456) (2003).
11. Liang, E. and Nishimura, K., Phys. of Plasmas to be submitted (2004).
12. Landau, L. and Lifshitz, E.M., Classical Theory of Fields (Pergamon, London) (1965).
13. Preece, R.D. et al., Astrophys. J. Supp. **126**, 19 (1998).
14. Van Putten, M.H.P.M. and Levinson, A., Astrophys. J. **584**, 937 (2003).

Spectral Time Evolution for GRBs Observed by BATSE and EGRET-TASC

M.M. González*[†], B.L. Dingus[†], Y. Kaneko**, R.D. Preece** and M.S. Briggs**

Physics Department, University of Wisconsin, Madison
[†]*Los Alamos National Laboratory*
**Physics Department, University of Alabama in Huntsville, National Space Science and Technology Center*

Abstract. Analysis of the time evolution of GRB spectra yields important information about the radiation mechanisms taking place in Gamma-Ray Bursts. Spectroscopy of BATSE data using the Band function showed a generalized hard-to-soft temporal evolution of the GRB spectra. This analysis was limited to bursts whose spectrum peaks below 1 MeV. The EGRET's calorimeter TASC (Total Absorption Shower Counter) was sensitive to energies above 1 MeV and beyond 10 MeV for brighter bursts. In order to give better spectral fits over a broader energy range, we have combined BATSE and TASC data for 8 bursts that showed significant detection in more than one time interval in TASC data. We observed that the spectrum of a single peak evolves from hard-to-soft, except for one burst where the high-energy spectral index β stayed constant with time. For bursts with multiple peaks, no evident time evolution was observed.

INTRODUCTION

BATSE data has been analyzed [1, 2] to determine the temporal behavior of the GRB spectra. This analysis showed a generalized hard-to-soft spectral time evolution consistent with the synchrotron radiation as the mechanism responsible of the observed keV-emission. Because of the poor BATSE sensitivity near ~ 1 MeV, this analysis did not sufficiently constrain the fit parameters describing the spectra beyond its peak, thus a broader energy range is required.

The EGRET calorimeter, TASC, was triggered by BATSE and thus observed several bursts in the energy range of $1 - 200$ MeV. The analysis of combined LAD-TASC data has been developed and discussed previously by [3]. Spectroscopy of GRB941017 [4, 5] based on this analysis, uncovered a MeV-component that would be hidden in the brightness of the synchrotron component in a time integrated spectrum of the whole burst, while separation into shorter time intervals allowed a unique identification of a new component different from the synchrotron component. This is an excellent example of the importance of a spectral time analysis in a broad energy range.

We selected from the BATSE catalog a total of 43 bursts that were the brightest ones above 300 keV and had a photon flux higher than 10 ph/cm^2/s. 26 of the 43 bursts showed significant excess over the background in at least one time interval and only 16 in more than one time interval in the TASC data. We present in this paper the spectral

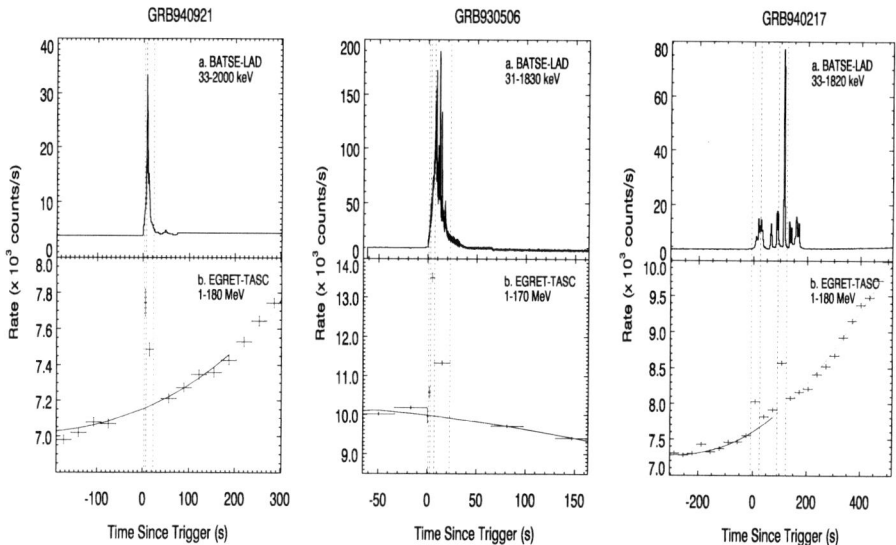

FIGURE 1. Count rates for GRB940921, GRB930506 and GRB940217 as observed by BATSE-LAD and EGRET-TASC. The dotted vertical lines indicate the time intervals used in the analysis and shown in Table 1. The 4th order polynomial that best fits the background observed by EGRET-TASC is shown as a solid line in all three bursts.

time evolution for 8 of the 16 bursts using joint BATSE-LAD and EGRET-TASC data.

ANALYSIS

EGRET-TASC spectra was generally accumulated in time intervals of 32.768 s (solar mode) and in sequential time intervals of 1,2,4 and 16 s (burst mode) when a BATSE trigger was received. Due to telemetry and dead time problems, accumulated spectra in burst mode were not always available, so then spectra in solar mode were used. The TASC background was obtained by fitting a 4th order polynomial to the spectra before and after each burst and was checked against the observed spectra 15 orbits earlier when the spacecraft was located at the same geomagnetic rigidity.

For each burst, we used the Band function to fit jointly the photon flux observed by both detectors. The time intervals were determined by EGRET-TASC data and the energy range was taken from 30 keV to the maximum energy shown in Table 1. The detector response matrices were obtained for each burst with the complete CGRO mass model and the EGS-4 Monte Carlo code. The best fit parameters and the normalization factor between BATSE-LAD and EGRET-TASC data sets for each burst are given in Table 1 and its caption respectively. For more details in the analysis refer to [3]

The bursts presented here last more than 20s. The observation by TASC was significant for most of the burst durations as determined by BATSE, except for GRB970202

TABLE 1. Joint BATSE-LAD and EGRET-TASC spectral fits to the Band function. The first column shows the burst date. The last four columns contain the parameters of the Band function that best-describes the differential photon flux observed by BATSE-LAD and EGRET-TASC in the time interval and up to the maximum energy shown in the third and second columns respectively. The normalization factors between BATSE-LAD and EGRET-TASC data set for the bursts shown in this table from top to bottom are 0.66, 0.54, 0.62, 0.35, 0.58, 0.48, 0.35 and 0.79 respectively.

Burst	Max. E (MeV)	Time from trigger (s)	A ($\times 10^{-3}$) ph/s-cm^2-keV	E_{peak} (keV)	α	β
910503	201	0 to 1	49.3 ± 0.9	1040 ± 74	-0.51 ± 0.04	-2.03 ± 0.04
		1 to 3	320.7 ± 2.7	727 ± 16	-0.6 ± 0.01	-2.22 ± 0.02
		3 to 7	132.8 ± 1.2	600 ± 15	-0.91 ± 0.01	-2.6 ± 0.05
		23 to 54	9.4 ± 0.1	1159 ± 132	-1.37 ± 0.02	$-2.6^{+0.24}_{-0.58}$
930506	167	1 to 3	41.3 ± 1.3	540 ± 54	-1.06 ± 0.04	-1.93 ± 0.06
		3 to 7	91.3 ± 0.6	1104 ± 41	-0.90 ± 0.01	-1.91 ± 0.02
		7 to 23	72.5 ± 0.4	871 ± 32	-1.24 ± 0.01	-1.92 ± 0.02
940217	177	-8 to 25	11.1 ± 0.1	1018 ± 32	-0.79 ± 0.02	$-3.17^{+0.22}_{-0.38}$
		25 to 90	12.9 ± 0.1	675 ± 26	-1.21 ± 0.01	$-3.70^{+0.53}_{-\infty}$
		90 to 123	40.2 ± 0.2	676 ± 14	-1.03 ± 0.01	-2.73 ± 0.07
940703	139	-1 to 32	11.1 ± 0.2	1121 ± 135	-1.09 ± 0.03	-1.92 ± 0.06
		32 to 64	110.6 ± 0.5	613 ± 9	-0.96 ± 0.01	-2.42 ± 0.03
940921	178	3 to 7	25.4 ± 0.5	810 ± 72	-1.01 ± 0.03	-2.21 ± 0.08
		7 to 23	27.0 ± 0.3	731 ± 31	-1.11 ± 0.02	-2.62 ± 0.11
970202	100	0 to 1	38.8 ± 1.9	500 ± 32	0.09 ± 0.08	-2.18 ± 0.09
		1 to 3	42.0 ± 0.9	644 ± 26	-0.21 ± 0.04	$-2.80^{+0.19}_{-0.31}$
990104	201	1 to 23	33.2 ± 0.8	476 ± 35	-1.14 ± 0.03	$-2.69^{+0.14}_{-0.21}$
		127 to 159	11.4 ± 0.4	2084 ± 289	-1.33 ± 0.06	$-3.13^{+0.40}_{-\infty}$
		159 to 192	131.0 ± 1.2	648 ± 27	-1.23 ± 0.01	-2.44 ± 0.03
990123	128	0 to 33	30.8 ± 0.2	734 ± 14	-0.62 ± 0.01	-2.49 ± 0.05
		33 to 66	44.6 ± 0.3	541 ± 10	-0.88 ± 0.01	-2.73 ± 0.08

where TASC data was significant for only for the first 3 s of T90.

GRB930506, GRB940921, GRB990123 and GRB910305 present a prompt single defined peak (with some variability) covered by more than one TASC time interval. All these peaks, except for the one in GRB930506, present a hard-to-soft evolution in α and β while E_{peak} moves to lower energies. GRB930506 presents a hard-to-soft evolution only in α while β stays fairly constant. GRB930506 has the shortest duration of these 4 bursts.

GRB940703 shows a hard-to-soft evolution only in β while E_{peak} moves to lower energies. The first time interval contains the first little peak and a small fraction of the second bright peak, therefore it is difficult to relate this hard-to-soft behavior to a peak-to-peak or single peak evolution.

Finally, for GRB990104, GRB940217 and GRB910503 there were TASC time intervals containing independent peaks. They do not show a monotonic temporal evolution through the burst meaning that there is no evident time evolution between peaks in the same burst.

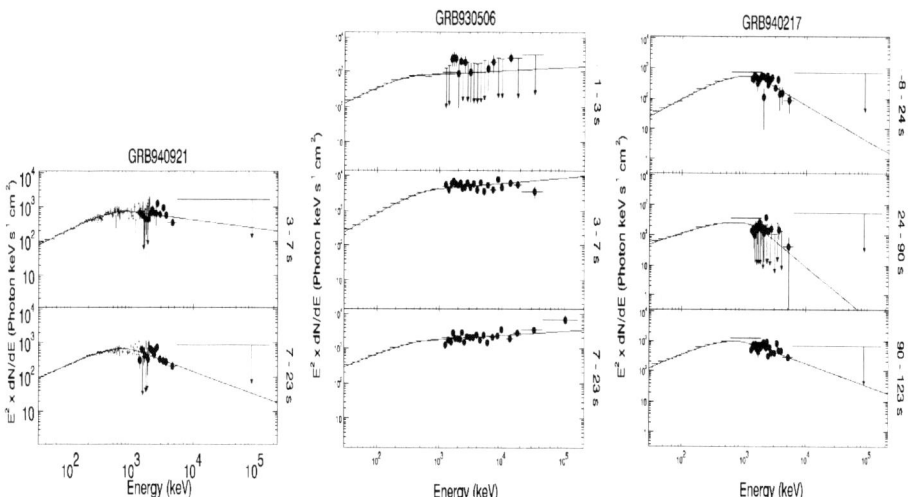

FIGURE 2. Energy fluxes for GRB940921, GRB930506 and GRB940217 during the time intervals indicated in Table 1. Crosses and circles represent BATSE-LAD and EGRET-TASC data respectively. The upper limits correspond to 2σ deviation from the background. The solid lines represent the best fit to the joint LAD-TASC data using a Band function. The best fit parameters and the normalization factor between the data sets are shown in Table 1 and its caption for each burst. These bursts represent the three different time evolution categories observed in our analysis, hard-to-soft, constant hard and non-monotonic evolution.

CONCLUSIONS

We have observed a generalized hard-to-soft time evolution behavior in single peaks. One burst with a single peak had a constant high-energy spectral index. When β presented a hard-to-soft time evolution, E_{peak} evolved toward lower energies. No generalized temporal evolution was observed for multi-peaked bursts.

A fit to the joint BATSE-TASC data was performed using only a Band function through the whole energy range, and every time interval, for all 8 bursts that are consistent with a single spectral component upto 150 − 200 MeV. No second spectral component in MeV energies similar to the one in GRB941017 was observed in any of these 8 bursts.

REFERENCES

1. R.D. Preece, et al., *Astrophys. J. Supp.*, **126**, 19–36 (2000).
2. R.D. Preece, et al., *Astrophys. J.*, **496**, 849–862 (1998).
3. M.M. González, et al., "BATSE-EGRET combined spectral fits," in *Gamma-ray Bursts and Afterglow Astronomy 2001*, AIP Conference Proceedings, 2003, pp. 267–269.
4. M.M. González, et al., *Nature*, **424**, 749–751 (2003).
5. M.M. González, et al., "Discovery of a Distinct Higher Energy Spectral Component in GRB941017," in *In this proceedings*, 2003.

GRB Optical Prompt Emission: The Role of Monitors

R. Hudec

Astronomical Institute, Academy of Sciences of the Czech Republic, CZ–251 65 Ondřejov, Czech Republic

Abstract. The prompt optical emission of Gamma Ray Bursts was detected only in one case so far. We discuss the possibility to detect this type of optical emission, which is still difficult to record by follow-up experiments, by wide-field optical monitors.

INTRODUCTION

The attempts to detect the direct (prompt optical emission, Optical Transients (OTs)) of Gamma Ray Bursts (GRBs) are still, despite of recent efforts, not very successful. So far, only in one case (GRB990123) the prompt optical emission has been recorded and investigated [1]. This is not necessarily due to the absence of optical emission, but rather due to instrumental limitations, and the still low rate of precisely positioned GRBs with immediately (within seconds) communicated positions. Even the recent sophisticated robotic alert systems specially designed to detect prompt optical emission of GRBs rely on rapid communication of GRB localizations.

Much less sensitive to the quality of satellite GRB data are the optical monitors. They usually monitor wide fields of sky independently on satellites. The preferences of sky monitors in optical GRB analyses (prompt emission) are as follows. The time 0 sec after the GRB onset can be never achieved by alert systems. Further, the alert systems will never achieve coverage for times before the GRB triggers. Both these time coverages can be, however, easily achieved by sky monitors. This is essential since there are theoretical expectations that optical flashes may precede GRBs (e.g. [2]).

THE OPERATIONAL ALL–SKY MONITORS

There are only few operational all–sky monitors. **The European Observational Network (EN)** operates 11 stations in the Czech Republic. Although designed primarily for meteor observations, the system is able to provide simultaneous optical data for various other projects, based on the complete sky monitoring (180 degrees diameter field of view). The optics of the system is based on the Fish-Eye Objective F-Distagon 3.5/30. Planfilm FOMAPAN 400 ASA or 100 ASA (panchromatic emulsion) serves as a detector, with area of 90 x 120 mm. The sky diameter on the film is 80 mm. The typical exposure time amounts to 3 hrs for guided cameras, and the whole night for fixed cam-

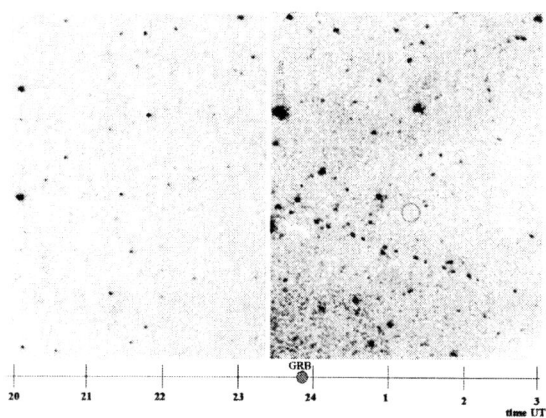

FIGURE 1. The example of GRB analyses for prompt OT emission: the case of the GRB000926 on the images by the photographic EN network covering the time before, during and after the trigger. Left: pre-burst image (end of exposure 29 minutes before the GRB trigger), limiting magnitude 10. Right: simultaneous image, limiting magnitude 8. The position of the GRB is indicated by a circle, time axis is below. Both images contain airplane trails/lights.

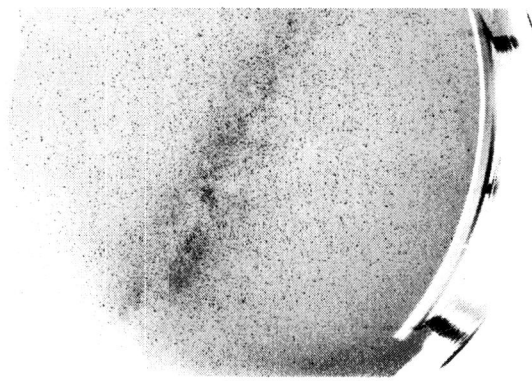

FIGURE 2. The all sky CCD image, Sonneberg Observatory.

eras. There are two stations equipped with guided and fixed cameras, while the others have only fixed cameras. The sensitivity for brief 1 sec triggers is 2-3 mag, for stars up to mag 12. The response is limited, due to the use of fish-eye lenses, to the red light above 400 nm. The system has large sky coverage (full visible hemisphere), as well as a large fraction of observation time: 2 400 to 6 000 sr.h for one station/year. The multiplicity of data eliminates background triggers easily, and allows the classification of detected triggers by parallax. The network will be soon operated as a fully remotely controlled network - without human assistance. The access to the plate data is facilitated by the new high quality flatbed Heidelberg CCD film scanner with optical resolution of 3 000 dpi connected to a powerful computer/graphic station.

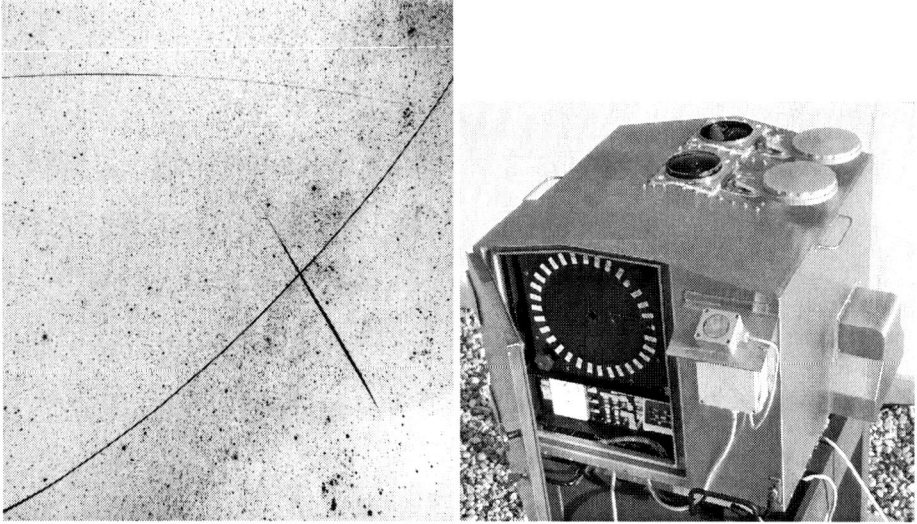

FIGURE 3. Left: All-sky digitized photographic EN image (central part of about 30 ×30 degrees). Right: The newly developed automated EN camera (photo courtesy P. Spurny).

The **Sonneberg all-sky CCD camera** is based on analogous optical system (Zodiak 3.5/30 mm) as the Czech EN network but with a 7k x 4k CCD camera OES MM7k4k (Philips chip 7168x4096 pixels 12 microns each, 16 bit, binning 4x4, readout time 60 sec) instead of film detector and is operated at the Sonneberg Observatory. The camera achieves a 9 mag sensitivity limit for a 1 min exposure.

SUMMARY OF GRB RELATED RESULTS

No optical emission above mag 5 (1 sec duration assumed) or mag 13 (full exposure time) or $L_g/L_o > 100$ to 300 has been detected from GRBs. The faintest limit (320) exists for GRB 830313 (Hudec, 1993). No optical emission above magnitudes 0...3 (1 sec duration assumed), mag 3–6 (1 min duration) or 4–11 (full exposure time) or $L_g/L_o > 0.1 ... 10$ has been detected for many (~140) GRBs.

CONCLUSION

The all-sky sky monitors operated at the Ondrejov and Sonneberg Observatories are able to provide valuable optical real-time and pre-burst data for GRBs. The recent analyses provide valuable limits for simultaneous optical emission of GRBs, for a 1 min emissions, there limits are between mag 3 and 6. Better limits are expected to be achieved for triggers occurring at clear nights high above the local horizon. The All Sky CCD Camera at the Sonneberg Observatory achieves limits of order of mag 9 for 1 min

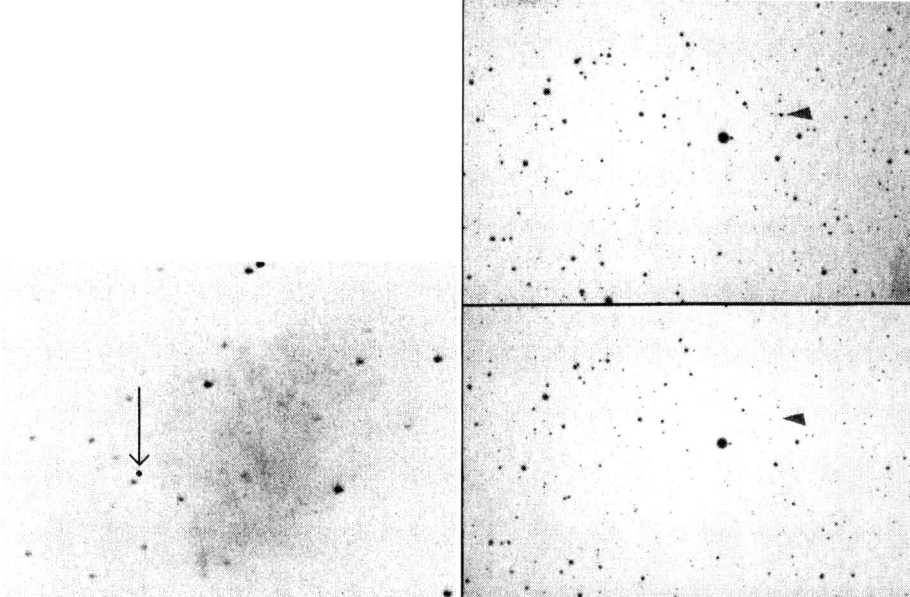

FIGURE 4. Two examples of OT images. Left: The OT image (left, the image is sharper than stars since the OT duration was shorter than the image exposure) of an astrophysical trigger of about 1 min duration, mag 10, recorded by the Brno Observatory 60 cm CCD telescope by F. Hroch. Right: The OT image of an object (OT Triangulum) 6 mag above the plate background recorded on the Sonneberg Observatory astrograph plate (top: the OT indicated by an arrow on the discovery plate, bottom: the comparison plate).

duration, hence is capable of detecting OTs analogous to the prompt optical emission of GRB990123. The importance of optical all sky monitors will increase in the Swift satellite era when numerous ($\sim$100/year) precisely localised GRBs are expected. The monitors will also allow better analyses and understanding of astrophysical OT triggers in general.

ACKNOWLEDGMENTS

The Czech GRB analyses on all-sky images are supported by the grant provided by the Grant Agency of the Academy of Sciences of the Czech Republic No. A3003206.

REFERENCES

1. Akerlof, C., and et al., *Nature*, **398**, 400 (1999).
2. Paczynski, B., *astro-ph/0108522* (2001).

COMPTEL Observation of GRB941017 with Distinct High–Energy Component

Y. Kaneko*, L. Hanlon[†], R.D. Preece*, M.M. González**, B.L. Dingus**, M.S. Briggs*, O.R. Williams[‡], K. Bennett[‡] and C. Winkler[‡]

*University of Alabama in Huntsville / NSSTC, Huntsville, AL
[†]University College Dublin, Dublin, Ireland
**University of Wisconsin, Madison, WI / LANL, Los Alamos, NM
[‡]ESA/ESTEC, Noordwijk, the Netherlands

Abstract. The joint spectral analysis of GRB941017 with BATSE and EGRET data revealed the existence of a distinct MeV spectral component that decayed slower than the lower energy component. The event was also observed with COMPTEL burst modules, which provides burst spectra in the energy range of 300 keV to 10 MeV. Due to the limited energy overlap between the BATSE Large Area Detector and the EGRET Total Absorption Shower Counter spectra, the relative normalization between the two instruments is poorly constrained. The COMPTEL spectra complement the energy ranges of the BATSE and EGRET data and are used herein to confirm and improve upon the previous analysis. Using the data from all three instruments, we present the result of joint spectral analysis for GRB941017. Including the COMPTEL data improved the statistics for the time interval in which the high energy component is more apparent.

INTRODUCTION

González et al. [1] (G03 hereafter) reported the existence of a distinct high–energy component in the spectra of GRB941017 in the MeV energy band in addition to a sub–MeV component observed with BATSE. There are no previous observation of a distinct component with similar characteristics; the high-energy component of GRB941017 cannot be explained by the standard synchrotron shock model of GRB prompt emission. The joint spectral analysis of G03 was performed using data from BATSE LAD (Large Area Detector) and EGRET TASC (Total Absorption Shower Counter), which together provided broadband spectra spanning four decades in energy. Although T_{90} determined by BATSE was ~ 80 seconds and the BATSE time profile lacks a tail, the high–energy component was visible in the TASC lightcurve as an excess that persisted for more than 200 seconds. The extra component was well described by an additional power law ($A_{PWL}[E(keV)/30MeV]^\gamma$), of photon index $\gamma \sim -1$ on top of the GRB function [2] with $E_{peak} \leq 500$ keV. The photon index seemed to remain constant while the sub–MeV component flux decayed, displaying hard–to–soft evolution.

Apart from the BATSE LAD and the EGRET TASC data, data from COMPTEL's burst module are also available for this event. We present spectral analysis of GRB941017 using data from three CGRO instruments, adding COMPTEL data to the LAD and TASC data used in G03.

TABLE 1. Detector Parameters

	BATSE	EGRET	COMPTEL *	
	LAD (1 of 8)	TASC	Low	High
Energy Range (MeV)	0.03 – 2	1 – 200	0.3 – 1.3	0.8 – 10.6
Time Resolution (s)	2.048[†]	32.768	1.0 (Burst Mode) 6.0 (Tail Mode)	
No. of Energy Channels	16[†]	256	122	128

* Effective areas: Low – 220 cm^2 at 1 MeV and High – 125 cm^2 at 8 MeV
[†] Continuous (CONT) Data

COMPTEL OBSERVATION

COMPTEL's burst module consisted of two NaI detectors that provided spectra in the energy range of 300 keV – 1.3 MeV ("Low") and 800 keV – 10.6 MeV ("High"). The Burst Spectrum Analyzer (BSA) accumulated data when activated by a BATSE trigger, providing 6 high–temporal resolution "burst mode" spectra and subsequent "tail mode" spectra with longer integration times (see table 1). A more detailed COMPTEL instrument description can be found elsewhere [3]. For this event, the early "burst mode" data from the BSA do not show significant signal above background. The COMPTEL signal faded after about 80 seconds, leaving 12 tail–mode spectra from 6 to 78 s with useful signal. Due to the large incident angle of the event (zenith angle = 66°), the normalization factor between the Low data and High data was found to be large: it is hoped to generate a response matrix which corrects for this, but this is not currently available.

SPECTRAL ANALYSIS

The analysis was performed using four datasets: BATSE LAD, EGRET TASC, COMPTEL Low, and COMPTEL High. The LAD data and the TASC data, as well as their response matrices and the spectral models used here, are identical to those used by G03. The COMPTEL data were available only in two out of five 33-second time intervals; 14 to 47 sec and 47 to 80 sec (2nd and 3rd intervals in G03). The COMPTEL accumulation intervals which best match these 2 intervals were 12 to 48 sec and 48 to 78 sec; the accuracy of matching the intervals is limited by the 6 s resolution of the BSA tail mode. For each of these two time intervals, the data from all detectors were jointly fitted to a single photon model. Fits were made to determine the best normalization factors between the instruments. The final fits were made using the same normalization factors for both time intervals. The photon index of the highest energy power law (γ) was fixed to -1 for the 14 to 47 sec time interval, as was done by G03, due to the weak constraint on this parameter by the data.

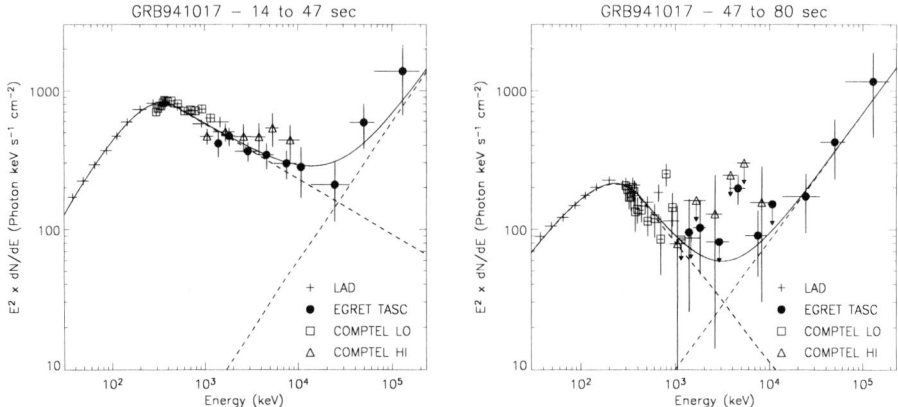

FIGURE 1. νF_ν spectra corresponding to the 2nd and 3rd time intervals of G03. GRB function and the additional high–energy power–law are shown separately with dashed lines. COMPTEL and TASC spectra have been binned for display purpose.

RESULTS

The deconvolved νF_ν spectra of the two time intervals for which the analysis was performed are shown in figure 1, along with the best fit parameters in table 2. For comparison, the spectral parameter determined only with LAD and TASC data (as in G03) are also listed. The best fit parameters to all four datasets are consistent with those found with only LAD and TASC data. The data from all three instruments show an especially significant improvement in the change in χ^2, and thus reduced chance probability, for the interval of 47 to 80 sec in which the high–energy component is more evident. Therefore, having the COMPTEL spectra along with the TASC spectra strengthen the evidence for the high–energy component described by the power law.

In addition to the fits presented here, the joint spectra of COMPTEL data only (High and Low) as well as COMPTEL and LAD were also fitted with the same photon models. In the case of COMPTEL–only fits, the values of E_{peak} and of the the low–energy photon index (α) were not well constrained due to the fact that the peak energy lies very close to the lower limit of the COMPTEL energy range. Nonetheless, the high–energy photon index (β) values agreed with those found with LAD–only spectral fits. Consequently, fitting the COMPTEL spectra along with the LAD spectra provided better constraints to all of the spectral parameters; however, the additional high–energy power law did not improve the fits although there was a slight indication of possible high–energy excess at least for the later time interval (47 to 80 sec) in the COMPTEL High data.

In the energy range of 1 to 10 MeV, the COMPTEL High data and the TASC data were fitted separately with a single power law, to test the consistency of the two datasets. This was only possible for the first time interval, in which the COMPTEL High data contained enough signals to produce an acceptable fit. The photon indices agreed within 1 σ, and thus the two datasets were confirmed to be consistent within the overlapping energy range.

TABLE 2. Best fit parameters for the 2 time intervals, with only LAD and TASC data ("LAD+TASC"), and with data from all three instruments ("+COMPTEL").

	14 – 47 s		47 – 80 s	
	LAD+TASC	+COMPTEL	LAD+TASC	+COMPTEL
Band's GRB Function				
A_{GRB}*	0.061 ± 0.001	0.060 ± 0.001	0.024 ± 0.001	0.025 ± 0.001
E_{peak} (keV)	349.7 ± 7.93	350.1 ± 6.96	240.4 ± 13.6	229.1 ± 9.40
α	-0.79 ± 0.02	-0.79 ± 0.02	-1.08 ± 0.06	-1.03 ± 0.05
β	-2.46 ± 0.05	-2.40 ± 0.03	$-2.65^{+0.22}_{-0.85}$	$-2.85^{+0.20}_{-0.40}$
Power Law (E_{piv} = 30 MeV)				
A_{PWL}*/10^{-7}	2.40 ± 0.63	1.97 ± 0.60	2.36 ± 0.93	2.51 ± 0.76
γ	-1.00 fixed	-1.00 fixed	$-1.06^{+0.70}_{-0.44}$	-1.09 ± 0.37
Normalization (Fixed)				
TASC/LAD	0.45	0.45	0.45	0.45
Low/LAD	–	1.15	–	1.15
High/LAD	–	2.30	–	2.30
χ^2/dof	259.3/214	485.4/400	235.2/213	388.1/399
$\Delta\chi^{2\dagger}$	14.5	10.4	15.8	20.1
Probability**	1.4E-4	1.3E-3	3.7E-4	4.3E-5

* In photons s^{-1} cm^{-2} keV^{-1}
† Change in χ^2 with and without the high–energy power law (Δdof = 1 for the 1st time interval and Δdof = 2 for the 2nd time interval)
** Chance probability for improvement of χ^2 by adding the high–energy power-law, determined by χ^2 probability function

CONCLUSION

The COMPTEL observation of GRB941017 was found to be consistent with the BATSE LAD and EGRET TASC observations previously reported by G03. The joint spectral fit using all three CGRO instruments resulted in better evidence (larger improvement in χ^2) for the additional power law in the later time interval, where the component was more distinct in the COMPTEL energy passband. These analysis results, obtained using three independent instruments, provides strong evidence for a prompt high–energy spectral component, which cannot be explained with the standard synchrotron shock model.

REFERENCES

1. González, M.M., et al, *Nature*, **424**, 749 – 751 (2003).
2. Band, D., et al, *ApJ*, **413**, 281 – 292 (1993).
3. Schönfelder, V., et al, *ApJS*, **86**, 657 – 692 (1993).

RELATIVISTIC JETS AND POLARIZATION

Linear Polarization on Gamma-Ray Bursts: from the Prompt to the Late Afterglow

Davide Lazzati

Institute of Astronomy, University of Cambridge, Madingley Road, CB3 0HA, Cambridge, UK

Abstract. The past year has witnessed a large increase in our knowledge of the polarization properties of Gamma-Ray Burst (GRB) radiation. In the prompt phase, the measurement (albeit highly debated) of a large degree of linear polarization in GRB 021206 has stimulated a deep theoretical study of polarization from GRB jets. The optical afterglow of GRB 030329, on the other hand, has been followed thoroughly in polarimetric mode, allowing for an unprecedented sampling of its polarization curve. I will review the present status of theories and observations of polarization in GRBs, focusing on how polarimetric observations and their modelling can give us informations on the structure and magnetisation of GRB jets which is not possible to obtain from their light curve.

INTRODUCTION

Linear polarization has revealed to be a characteristics of GRBs throughout their entire evolution. The recent claim by Coburn & Boggs[1] that the prompt emission of GRB 021206 was polarized at a very high level has stimulated a thorough analysis of the polarizing properties of the jet geometries in GRB outflows, drawing attention to the possibility that magnetic fields may be advected from the central source rather than generated by the internal shocks. On the other hand, observational and theoretical studies of afterglow polarization has revealed a much more complicated picture than previously thought, emphasising the importance of the jet structure, its dynamics and the properties of the ISM in shaping the polarization curves. Even though direct observations lack, the optical flash may also be highly polarized, especially if due to a reverse shock in the burst ejecta rather than to the pair enrichment of the nearby ISM.

In this paper I review the status of theories and observations of linear polarization in GRBs. The three phases are analysed initially separately and then their relation discussed. The importance and insight of polarization is emphasised in all phases.

THE PROMPT PHASE

Analysing the scattering geometry of photons in the RHESSI detector, Coburn & Boggs [1] were able to measure the average linear polarization of the prompt emission of GRB 021206 in the [150 keV–2 MeV] energy range. They find that the prompt emission of the burst is highly polarized, with $\Pi = 0.8 \pm 0.2$. This measurement was subsequently heavily criticised by Rutledge & Fox [2], who performed an independent analysis of the same dataset, obtaining a much smaller number of double-scattered photons and, as a consequence, merely an upper limit on the linear polarization of the event. Despite that, the result [1] has stimulated a vast theoretical effort in order to understand under what conditions such a large polarization could be obtained. Two classes of models have emerged. In the first class, the origin of polarization is ascribed to the presence of a large scale ordered magnetic field, which ought to be advected from the central engine and may play a role in the launching of the jet itself [3]. In the second class of models, the magnetic field is supposed to be shock generated and tangled on small timescales, and the asymmetry required to produce polarization is due to a particular location of the observer with respect to the jet axis [4]. This second class of models can be extended to different emission mechanisms, such as inverse bulk Compton scattering [5].

Magnetic models

We define here "magnetic models" to be those in which polarization is due to the large scale geometry of the magnetic field. The magnetic field is likely to be dominated by a toroidal component, since the radial field decays faster than the tangential one ($B_r \propto r^{-2}$ while $B_\perp \propto r^{-1}$). One important ingredient of these models is that the observer, due to the relativistic aberration of photons, cannot see the whole jet. In fact it is only a small $1/\Gamma$ region of the jet that is observable and therefore the observer does not detect the overall toroidal structure of the field (which would wash out the polarization signal) but a highly ordered patch. This is most important if the magnetic field is not the dominant component of the outflow. Since regions of the jet separated by more than $1/\Gamma$ are causally disconnected, it is difficult to envisage a coherent magnetic field on scales larger than $1/\Gamma$, unless the structure has been created before the acceleration of the jet (when it was still connected) and frozen into it. Such a transport seem easier to attain in a magnetic dominated outflow [6] and even natural in a force-free subsonic bubble [3].

Even if the observer has access only to a fully ordered region of field, the polarization cannot be as large as that expected from a non relativistic flow. In the classic case: $\Pi = (p+1)/(p+7/3)$ where p is the power-law index of the electron energy distribution. The reduction of the observed polarization in the relativistic case is due to the aberration of photon trajectories. In order to keep the electric and magnetic field of the wave orthogonal to each other and to the wave propagation direcion, the position angle of polarization is rotated in different ways as a function of the distance from the line of sight. After integration, the polarization is reduced by a factor that depends on the spectral slope of the radiation, and spans between 10% and 20%. The nonrelativistic solution is shown with the relativistic maximum polarization in the left panel of Figure: 1 (see also [7]).

A characteristic feature of these models is that any observer located within the opening angle of the fireball detects a highly polarized signal, with the exception of those observing the fireball within $\theta = 1/\Gamma$ from the symmetry axis. A level of polarization comparable to that detected by Coburn & Boggs cannot be achieved; a definitive conclusion cannot be drawn even given the large uncertainties in the observations.

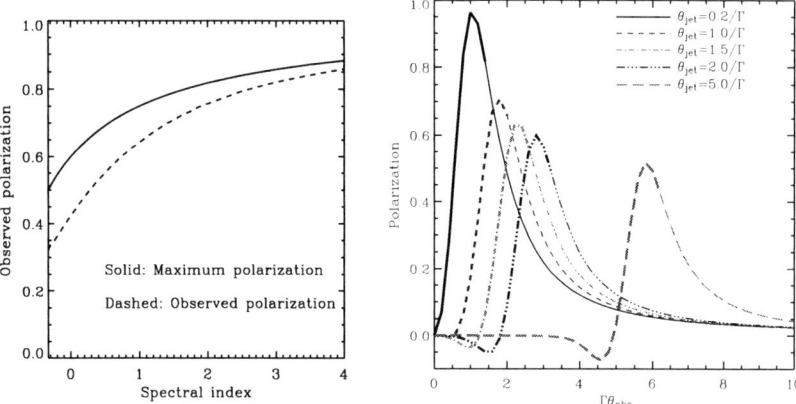

FIGURE 1. Left panel: Maximum synchrotron polarization for a non relativistic uniform field as a function of the spectral index α (solid line) compared to the maximum observable polarization from a relativistically moving uniform field (dashed line). The electron pitch angle distribution is uniform in both cases.. **Right Panel:** Inverse Compton polarization as a function of the observing angle θ_o in units of $1/\Gamma$ for a uniform jet with sharp edges. Different line styles show the polarization for jets with different opening angles. The lines are thicker in the region where the efficiency is larger than 2.5%.

Geometric models

It has been traditionally assumed that the magnetic field responsible for the synchrotron emission in GRBs is generated at the shock front. This is a robust conclusion in the afterglow phase, where the compression of the interstellar field is far too low to produce the observed radiation. It may hold true also for the prompt phase, in which case a tangled field would be responsible for the observed radiation. If this field is tangled in a plane, but compressed in the direction perpendicular to the plane itself [8, 9] it is possible to observe polarized synchrotron radiation since radiation emitted in the plane, which is maximally polarized in the comoving frame, is then aberrated toward the observer with an angle $\theta = 1/\Gamma$ [10].

In the afterglow phase, this configuration leads to polarization of up to several tens of per cent [10, 11], but in the prompt phase, it is usually negligible, unless a narrow jet is observed along the required direction [4]. Polarization has been analyzed in several papers with synchrotron as the radiation mechanism [7, 12] as well as if the photons are produced by bulk inverse Compton scattering [5]. The difference between the two cases is not of fundamental nature, since the dependence of polarization on angle is the same for the two mechanisms. Inverse Compton, on the other hand, can produce larger polarization since it can be maximally (100%) polarized in the comoving frame.

The polarization produced by a narrow jet is shown (for the case of inverse Compton) in the right panel of Fig. 1. In these models a narrow jet is fundamental since the number of observers that see a polarized event is limited to those lying in the region between the edge of the jet and $1/\Gamma$ from it. This region becomes vanishingly small for $\theta_{\rm jet} > 10/\Gamma$. GRB 021206 was exceptionally bright and, assuming it was at cosmological redshift, would have had a narrow jet with opening angle of few degrees at most.

To distinguish between magnetic and geometric models is quite easy once a reasonable number of measurement is available. In the geometric case only a small fraction of the brightest bursts should be polarized, while in the magnetic case most of them should be.

THE FLASH

It has been suggested by Granot & Königl [13] that the reverse shock emission could be as highly polarized as the prompt GRB, since the plasma responsible for this emission is the same one that produced the gamma-ray photons. This consideration can be included in a more general discussion on optical flashes, that is, bright optical components that appears at the beginning of the afterglow phase with a fast decay.

Optical flashes can be produced in two ways: by a reverse shock in a baryonic jet or by pair enrichment of the external medium in the vicinity of the GRB [14]. In most cases the optical flash is expected to be polarized at a level comparable to the prompt emission. There is actually only one case in which the optical flash following a polarized GRB can be unpolarized. In a magnetic model, if the flash is due to pair enrichment of the ISM, the flash comes from a shock generated field without the geometric constraints that produce polarization. In all geometric models the geometry of the prompt phase is preserved during the flash emission. Finally, if the flash is due to reverse shock, it should be polarized as discussed by Granot & Königl [13].

AFTERGLOW

The discovery of linear polarization in GRB afterglows dates back to 1999, when a small but highly significant level of polarization was detected in GRB 990510 [15]. The detection took place amid a theoretical effort to predict and/or explain it.

Guzinov & Waxman [16] discussed the possibility that the shock generated field organises in coherent patches that expand at a sizable fraction of the speed of light. They calculated that an observer should see approximatively $N \gtrsim 50$ patches. If the polarization inside a patch is $\Pi_0 \sim 70\%$, the observed one is $\Pi = \Pi_0/\sqrt{N} \lesssim 10\%$. The degree of linear polarization observed in GRB afterglows is in the per cent range[17]. However, given the random nature of the model, the degree and position angle of polarization fluctuates in time. This is not, at least in some cases, detected in polarization curves [18, 19] and for this reason this model is now not considered particularly promising.

Shortly after the discovery of polarization Ghisellini & Lazzati [10] and, independently, Sari [11] proposed a model based on the assumption that the fireball is beamed in a cone and that the shock generated field is either compressed in the shock plane or elongated in the axial direction. Polarization is observed if the observer is not coincident with the cone axis and has a definite and testable pattern. Polarization at early times is null, increases slowly with time until it reaches a maximum and then starts to decrease again until it vanishes. At this moment, which is roughly coincident with the jet

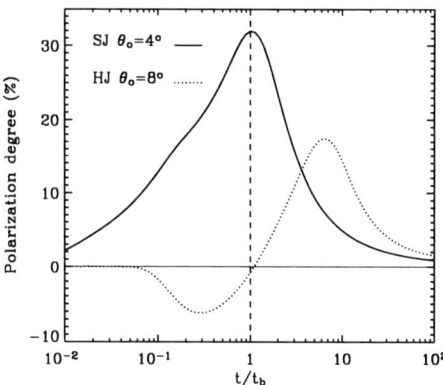

FIGURE 2. Comparison between the polarization curves of a homogeneous jet (dotted line) and a structured jet (solid line). The two light curves are virtually indistinguishable, while polarization behaves in a markedly different way.

break in the unpolarized lightcurve, the position axis of polarization rotates by 90°. Then the polarization curve is characterised by a second peak, of higher intensity, eventually vanishing to an unpolarized flux at long times. The intensity of the polarization signal depends on the off-axis angle: the larger the off-axis angle the larger the polarization.

These models have been further analyzed and extended by Rossi et al. [20]. They studied the effect of different assumptions on the jet sideward expansion showing that the ratio between the peaks of polarization is smaller for faster expansion speed. They also generalised the model to nonuniform jets, in which the energy per unit solid angle decreases as $E_\Omega \propto \theta^{-2}$ where θ is the angle with respect to the flow symmetry axis. In this case the polarization curve is largely different. While it is null for early and late times, as in the uniform case, it has a single peak, correspondent in time with the jet break time, and constant position angle. In Fig 2 we show the comparison between the polarization curves for the homogeneous and the structured models. It is clear that the two curves are different in an easily testable way. This is particularly important if we consider that the light curves of the two models are almost indistinguishable. Further complications to the models have been added by [13] who consider the presence of a coherent component of the magnetic field in the ISM. The propagation of the polarized light of the OT in the host and Galactic ISM have been instead discussed by [18].

Comparison of the models with the data has proven difficult. The main limitation of these models is that they assume that the emissivity of the fireball is uniform (or strictly dictated by the θ^{-2} law). Any deviation from this assumption, or inhomogeneity of the external medium, causes a noise on top of the models in both polarization and position angle. Usually this situation is recognisable in the light curve through the presence of bumps and wiggles on top of the regular power-law decay. Indeed, every time the light curve is complex, the polarization curve has a complex structure, such as in GRB 021004 [21, 18] and in GRB 030329 [19]. On the other hand, simple polarization curves seem to be associated to power-law afterglows (GRB020813 [22]).

COMPARISON OF THE THREE PHASES

The position angle of polarization should be related in the three phases. In the intrinsic models (i.e. neglecting polarization induced by the ISM and by an external magnetic field) the position angle of the polarization can be either contained in or orthogonal to the plane containing the jet axis and the line of sight. It is therefore expected that, should polarization be measured in the future in the three phases of a single GRB event, the position angle should either remain constant throughout the whole evolution or rotate by 90° between the optical flash and afterglow. It may eventually rotate back to the original position. Any difference from this simple behaviour should be considered a sign of an external component in the generation of polarization.

SUMMARY AND CONCLUSIONS

The study of polarization evolution in GRBs is highly informative, albeit difficult. It carries important informations about the jet structure and dynamics that are hidden in degeneracies of the light curve, but are emphasised in the polarization curve. Observationally, the afterglow phase is the most simple to investigate, even though is may be affected by small-scale inhomogeneities, and constraining the smoothness of the light curve is of fundamental importance in order to model a polarization curve. Polarization in the prompt and optical flash emission is highly informative of the structure of the ejecta, even though further observations are required in order to establish the mere existence of polarization in this phases.

REFERENCES

1. Coburn, W. & Boggs, S. E., Nature **423**, 415 (2003).
2. Rutledge, R. E. & Fox, D. B., MNRAS subm., astro-ph/0310385 (2003).
3. Lyutikov, M., Pariev, V. I. & Blandford, R. D., ApJ **597**, 998 (2003).
4. Waxman, E., Nature **423**, 388 (2003).
5. Lazzati, D., Rossi, E. M., Ghisellini, G. & Rees, M. J., MNRAS, **347**, L1 (2004).
6. Proga, D., MacFadyen, A. I., Armitage, P. J. & Begelman, M. C., ApJ **599**, L5 (2003).
7. Granot, J., ApJ **596**, L17 (2003).
8. Laing, R. A., MNRAS **193**, 439 (1980).
9. Medvedev, M. V. & Loeb, A., ApJ **526**, 697 (1999).
10. Ghisellini, G. & Lazzati, D., MNRAS **309**, L7 (1999).
11. Sari, R., ApJ **524**, L43 (1999).
12. Nakar, E., Piran, T. & Granot, J., JCAP **10**, 5 (2003).
13. Granot, J. & Königl, A., ApJ **594**, L83 (2003).
14. Beloborodov, A. M., ApJ **565**, 808 (2002).
15. Covino, S. et al., A&A **348**, L1 (1999).
16. Gruzinov, A. & Waxman, E., ApJ **511**, 852 (1999).
17. Covino, S., Ghisellini, G., Lazzati, D. & Malesani, D., astro-ph/0301608 (2003).
18. Lazzati, D. et al., A&A **410**, 823 (2003).
19. Greiner, J. et al., Nature **426**, 157 (2003).
20. Rossi, E., Lazzati, D., Ghisellini, G. & Salomonson, J. D., MNRAS subm., (2004)
21. Rol, E. et al., A&A **405**, L27 (2003).
22. Gorosabel, J. et al., A&A in press., astro-ph/0309748 (2003).

Magnetic Acceleration and Collimation of Gamma-Ray Burst Jets

Arieh Königl

Department of Astronomy & Astrophysics, University of Chicago, 5640 S. Ellis Ave., Chicago, IL 60637, U.S.A.

Abstract. Exact semianalytic solutions for GRB outflows were recently derived using the equations of special-relativistic ideal MHD (see the contribution by Vlahakis & Königl for a summary). This contribution focuses on the implications of these results to various modeling and observational issues in GRB sources, including the baryon loading problem, polarization measurements of the prompt and reverse-shock emission, and the possible existence of a two-component outflow.

MAGNETIC DRIVING OF GRB OUTFLOWS

Gamma-ray burst (GRB) outflows are likely powered by the extraction of rotational energy from a newly formed stellar-mass black hole or neutron star, or from a surrounding debris disk established in the course of the central object's formation [e.g., 1]. Magnetic fields threading the central object or disk provide the most plausible means of extracting the inferred amount of energy on the timescale of the burst; they can also guide, collimate, and accelerate the flow [see 2, and references therein]. This picture is supported by a recent measurement of a high ($80 \pm 20\%$) linear polarization in the prompt γ-ray emission from GRB 021206 [3], which can be plausibly interpreted in terms of a large-scale magnetic field advected from the origin [e.g., 4] and is consistent with magnetic driving by an ordered field that threads the source. Although purely hydrodynamic driving powered by neutrino emission or magnetic field dissipation at the source can probably be ruled out [e.g., 5, 6], thermal effects may nonetheless dominate the initial acceleration of magnetic jets [e.g., 7, 2].[1]

Motivated by the above considerations, Vlahakis & Königl [9, 10, hereafter VK03a and VK03b, respectively] constructed a general formalism for special-relativistic ideal MHD, allowing for the presence of a baryonic component as well as of a "hot" electron-positron/radiation component that can dominate the pressure. They showed how one can derive exact semianalytic solutions for axisymmetric outflows under the assumption of radial self-similarity and presented illustrative results for representative GRB parameters. Vlahakis, Peng, & Königl [11, hereafter VPK03] further generalized this scheme by obtaining solutions for initially neutron-rich outflows, which they used to address the baryon loading problem in GRB source models. A general description of the formal-

---

[1] It has also been argued [e.g., 8] that electromagnetic energy dissipation could contribute to the conversion of Poynting energy into kinetic energy throughout the acceleration region of such flows.

ism and of the derived solutions is given in Vlahakis & Königl's contribution in these Proceedings. The present contribution provides a brief overview of the main results and focuses on their observational implications.

RELATIVISTIC MHD SOLUTIONS

The initial (subscript i) magnetic field amplitude can be inferred from an estimate of the injected energy, $\mathscr{E}_i$ = (Poynting flux) × (surface area) × (burst duration). In a disk geometry [with initial cylindrical radius ϖ_i and radial width $(\Delta\varpi)_i$], $\mathscr{E}_i \approx cE_iB_{\phi,i}\varpi_i(\Delta\varpi)_i\Delta t$, where the electric field is given by $E = B_p V_\phi/c - B_\phi V_p/c$ (with the subscripts p and ϕ denoting the poloidal and azimuthal components, respectively). For characteristic parameter values [$\mathscr{E}_i \approx 10^{52}$ ergs, $\varpi_i \sim (\Delta\varpi)_i \approx 10^6$ cm, $\Delta t \approx 10$ s], one obtains $B_i \sim 10^{14} - 10^{15}$ G. This field is most plausibly generated by differential-rotation amplification of a much weaker poloidal field component that originally threads the source.

If $|B_{p,i}/B_{\phi,i}| > 1$, a *trans-Alfvénic* outflow is produced, whereas if $|B_{\phi,i}/B_{p,i}| > 1$, the outflow is *super-Alfvénic* from the start. The latter situation may correspond to amplified toroidal flux loops that have been disconnected by magnetic reconnection and escape from the disk surface in a nonsteady fashion. Exact solutions for these two situations were derived in VK03a and VK03b, respectively. It was demonstrated that, in either case, Poynting flux-dominated jets can transform $\gtrsim 50\%$ of their magnetic energy into baryon kinetic energy (with $E_K \sim 10^{51}$ ergs and terminal Lorentz factors $\gamma_\infty \sim 10^2 - 10^3$). If relativistic e^+e^- pairs and radiation dominate the initial enthalpy, then a thermal acceleration zone develops at the base of the flow and remains dominant until the specific enthalpy drops below $\sim c^2$, at which point magnetic acceleration takes over. In contrast to the trans-Alfvénic solutions, part of the enthalpy flux in the super-Alfvénic flows is transformed into Poynting flux during the thermal acceleration phase. Furthermore, the subsequent, magnetically dominated acceleration in these flows can be significantly less rapid than in the trans-Alfvénic case.

The derived solutions have a free parameter, F, which controls the distribution of the poloidal current $I = c\varpi B_\phi/2$. For $F > 1$ the flow is in the current-carrying regime, with the poloidal current density being antiparallel to the magnetic field. In this case the current tends to zero as the symmetry axis is approached, so such solutions should provide a good representation of the conditions near the axis of a highly collimated flow. Conversely, solutions with $F < 1$ correspond to the return-current regime (in which the poloidal current density is parallel to the field) and are most suitable at larger cylindrical distances. Although the detailed global current distribution cannot be modeled using the self-similarity approach, one can nevertheless generate "hybrid" flow configurations that combine a current-carrying solution for low values of ϖ and a return-current solution for high values of ϖ (see Fig. 1 below for an example). Initially Poynting-dominated flows that attain a rough equipartition between the kinetic and Poynting energy fluxes at large distances from the origin have F close to 1. When $F > 1$ the Lorentz force can collimate the flow to cylindrical asymptotics. For $F < 1$ the collimation is weaker and the flow only reaches conical asymptotics; however, the acceleration is more efficient in this case in that a larger fraction of the Poynting flux is converted into kinetic energy.

IMPLICATIONS TO THE BARYON LOADING PROBLEM

As an illustration of the unique properties of the relativistic MHD solutions, consider the ramifications of a hydromagnetic jet model to the baryon loading problem in GRB outflows. The apparent difficulty stems from a comparison between the estimated mass of protons in the jet, $M_{\text{proton}} = 3 \times 10^{-6}(E_K/10^{51} \text{ ergs})(\gamma_\infty/200)^{-1} M_\odot$, and the minimum mass of the debris disk from which the jet is thought to originate, obtained under the assumption that at most $\sim 10\%$ of the disk gravitational potential energy could be converted into outflow kinetic energy. This comparison implies that the outflow can comprise at most $\sim 10^{-4}$ of the disk mass, whereas disk outflow models that utilize a large fraction of the disk potential energy typically also entail substantial mass loading. One approach to this issue has been to postulate that the outflow emerges along magnetic field lines that thread the black-hole event horizon and not the disk, but then the converse problem — how to avoid having too few baryons — must be addressed [e.g., 12]. A possible resolution of the problem in the context of disk-fed jet models was proposed in [13], where it was noted that such outflows are expected to be neutron-rich [neutron/proton ratios as high as $n/p \sim 20-30$; e.g., 14, 15, 11]. Since only the charged outflow component couples to the electromagnetic field, the neutrons could potentially decouple from the protons before the latter attain their terminal Lorentz factor. In this picture, the inferred value of M_{proton} may represent only a small fraction of the total baryonic mass ejected from the disk, which would alleviate the loading problem. However, it can be shown that, for purely hydrodynamic outflows, the Lorentz factor γ_d at decoupling is at least a few times 10^2 [e.g., 16, 15, 11]. This implies that $\gamma_d/\gamma_\infty \sim 1$ and hence that the protons end up with only a small fraction of the injected energy, which is *not* a satisfactory resolution of the problem.

As demonstrated by VPK03, the incorporation of magnetic fields makes it possible to attain $\gamma_d \ll \gamma_\infty$ and thereby reclaim the promise of the Fuller et al. proposal. They wrote down the equations of motion for the neutron component (which couples to the protons through a collisional drag) and for the charged component (incorporating protons and their neutralizing electrons as well as initially "hot" pairs and radiation), and simplified them by considering a well-coupled neutral/charged fluid for $\gamma \leq \gamma_d$ and only the charged fluid component for $\gamma > \gamma_d$. The pre-decoupling region was described by a super-Alfvénic outflow solution. As noted in § 2, in this case part of the enthalpy flux is converted into Poynting flux during the initial thermal acceleration phase. This reduces the acceleration rate, so at the point of decoupling (when $V_{\text{proton}} - V_{\text{neutron}} \sim c$) the Lorentz factor is still comparatively low. The energy deposited into the Poynting flux is returned to the matter beyond the decoupling point as kinetic energy, thereby enhancing the acceleration efficiency of the proton component. The end result is a large γ_∞/γ_d ratio *and* comparable terminal kinetic energies in the proton and neutron components, in clear contradistinction to the purely hydrodynamic solutions.

An illustrative solution is shown in Fig. 1.[2] The top panel shows the behavior of the

---

[2] In this example $n/p = 30$, the pre-decoupling and post-decoupling regions correspond to the current-carrying ($F = 1.05$) and return-current ($F = 0.1$) regimes, respectively, and the flow collimates from an initial opening half-angle of $55°$ to $\theta_j \approx 20°$.

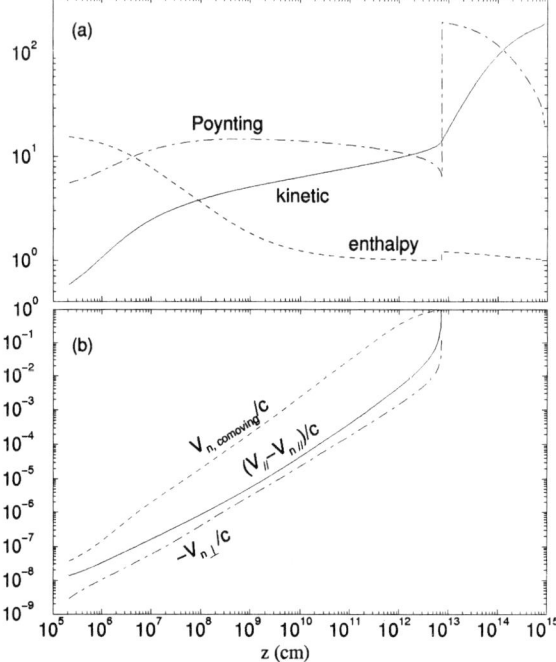

FIGURE 1. Illustrative relativistic-MHD solution of a neutron-rich outflow. (*a*) Components of the total energy flux, normalized by the mass flux $\times\, c^2$, as functions of height along a fiducial magnetic field line. The Poynting and enthalpy curves are discontinuous at the decoupling point, reflecting the decrease in the mass flux of field-coupled gas above that point. (*b*) Components of the proton–neutron drift velocity.

various components of the energy flux, corroborating the qualitative description given above. The thermal acceleration effectively terminates at a height $z \approx 10^9$ cm above the disk, and the neutrons decouple from the protons at $z_d \approx 10^{13}$ cm, corresponding to $\gamma_d \approx 15$. By the time of decoupling the neutrons have acquired $\sim 2/3$ of the injected energy, with the remainder residing predominantly in the electromagnetic field. The latter portion is then transferred with almost 100% efficiency into proton kinetic energy, so that, ultimately, the protons have $\gamma_\infty = 200$ and $E_{\mathrm{K,proton}} \approx 10^{51}$ ergs $\approx 0.5\, E_{\mathrm{K,neutron}}$. The proton jet thus carries $\sim 1/3$ of the injected energy but only $\sim 3\%$ of the injected mass. The lower panel of Fig. 1 shows that, even though the decoupling in this case is initiated by the growth of the n–p drift velocity along the poloidal magnetic field, there is also a transverse drift component (induced by the ongoing magnetic collimation), which at the time of decoupling is $V_{\mathrm{neutron},\perp} \sim 0.1\, c$.[3]

---

[3] The exact value of the angle between $\mathbf{V}_n$ and $\mathbf{V}_p$ at decoupling can only be obtained by solving the equations of motion without the "strong coupling" approximation adopted in the solution shown in Fig. 1.

ADDITIONAL IMPLICATIONS

Polarization — Magnetic driving of GRB outflows by large-scale, ordered magnetic fields would naturally lead to a large linear polarization P in the prompt γ-ray emission, and a high value of P is also predicted for the emission from the *reverse shock* (the "optical flash" and "radio flare") [17]. As shown in [4], in this picture the prompt emission may be expected to exhibit $P \sim 43\% - 61\%$ for typical values of the synchrotron-radiation spectral index, consistent with the observations of GRB 021206 [3]. The ordered field is also expected to induce measurable circular polarization [18].

Two-Component Outflow — The decoupled neutrons in a neutron-rich outflow will undergo β decay into protons at a distance $\sim 4 \times 10^{14} (\gamma_d/15)$ cm. In contrast with the situation in purely hydrodynamic outflow models [19, 20], there may well be *no* interaction between the two decoupled components in the MHD case since their motions are not collinear (see Fig. 1*b*). The latter scenario thus gives rise to a 2-component outflow: an outer (wider) component (comprising the decoupled neutrons) that carries most of the energy and may be responsible (after the neutrons decay) for the bulk of the optical/radio afterglow, and an inner (narrower) component (comprising the original protons) that accounts for the prompt γ-rays and possibly also for much of the X-ray afterglow. A 2-component outflow of this type was inferred in GRB 030329 [21, 22]. A more detailed investigation of this scenario is currently under way. If $E_{K,\text{narrow}} \lesssim E_{K,\text{wide}}$ and $\theta_{j,\text{narrow}}/\theta_{j,\text{wide}} \lesssim 1/3$, this picture would make it possible to reconcile current inferences of the radiated γ-ray energy [e.g., 23] with internal-shock models.

REFERENCES

1. Mészáros, P., ARA&A **40**, 137 (2002).
2. Vlahakis, N., & Königl, A., ApJ **563**, L129 (2001).
3. Coburn, W., & Boggs, S. E., Nature **423**, 415 (2003).
4. Granot, J., ApJ **596**, L17 (2003).
5. Di Matteo, T., Perna, R., & Narayan, R., ApJ **579**, 706 (2002).
6. Daigne, F., & Mochkovitch, R., MNRAS **336**, 1271 (2002).
7. Mészáros, P., Laguna, P., & Rees, M. J., ApJ **415**, 181 (1993).
8. Drenkhahn, G., & Spruit, H. C., A&A **391**, 1141 (2002).
9. Vlahakis, N., & Königl, A., ApJ **596**, 1080 (2003) (VK03a)
10. Vlahakis, N., & Königl, A., ApJ **596**, 1104 (2003) (VK03b)
11. Vlahakis, N., Peng, F., & Königl, A., ApJ **594**, L23 (2003) (VPK03)
12. Levinson, A., & Eichler, D., ApJ **594**, L19 (2003).
13. Fuller, G. M., Pruet, J., & Abazajian, K., Phys. Rev. Lett. **85**, 2673 (2000).
14. Pruet, J., Woosley, S. E., & Hoffman, R. D., ApJ **586**, 1254 (2003).
15. Beloborodov, A. M., ApJ **588**, 931 (2003).
16. Derishev, E. V., Kocharovsky, V. V., & Kocharovsky, Vl. V., ApJ **521**, 640 (1999).
17. Granot, J., & Königl, A., ApJ **594**, L83 (2003).
18. Matsumiya, M., & Ioka, K., ApJ **595**, L25 (2003).
19. Pruet, J., & Dalal, N., ApJ **573**, 770 (2002).
20. Beloborodov, A. M., ApJ **585**, L19 (2003).
21. Berger, E., et al., Nature **426**, 154 (2003).
22. Sheth, K., et al., ApJ **595**, L33 (2003).
23. Bloom, J. S., Frail, D., A., & Kulkarni, S. R., ApJ **594**, 674 (2003).

Polarization Measurements of GRBs with RHESSI

Wayne Coburn* and Steven E. Boggs*[†]

*Space Sciences Laboratory, UCB
[†]Department of Physics, UCB

Abstract. Measuring gamma-ray polarization of the prompt emission from GRBs provides a key piece of the puzzle in understanding the nature of these events. *RHESSI*, while not designed as a polarimeter, can be used to measure polarization for astrophysical sources such as GRBs. In this paper we will discuss using *RHESSI* as a γ-ray polarimeter, including a discussion of the systematic uncertainties and assumptions involved in such a measurement. We also present our results for the polarized GRB021206, as well as our technique applied to GRB030329 and the solar flare of 2003 July 23.

INTRODUCTION

Measurements of polarization from the prompt γ-ray emission of GRBs can help constrain the various models for GRB progenitors. In [1] we presented the results the first such measurement for a singe gamma-ray burst (GRB021206), demonstrating that γ-ray polarization measurements are possible with a satellite currently in orbit. However, the full implications of this measurement will remain uncertain until polarization levels have been measured in many more bursts.

We used the *Reuven Ramaty High Energy Solar Spectroscopic Imager* (*RHESSI*) as our γ-ray polarimeter. *RHESSI* is an array of 9 large volume (300 cm^3) segmented coaxial germanium detectors, and was designed to study solar X-ray and γ-ray emission (3 keV–17 MeV) with high angular resolution (2$'$) in the ∼1° field of view [2]. What makes *RHESSI* useful for GRB studies is the fact that the focal plane detectors are unshielded, and therefore observe GRBs over most of the sky. However, only bright, hard, and long bursts that occur near the focal axis (and therefore near the sun in the sky) are optimal candidates for polarization measurements. Solar proximity, however, makes optical followups of our best candidates difficult. Another advantage of *RHESSI* over other instruments is that the satellite spins with a ∼4 s period. So while only a small number of unique angles can be measured at any given instant, the rotation mean a large number of angles will sampled during a long duration burst. Rotation also averages out the effects of asymmetries in the passive materials that could be mistake for a modulation.

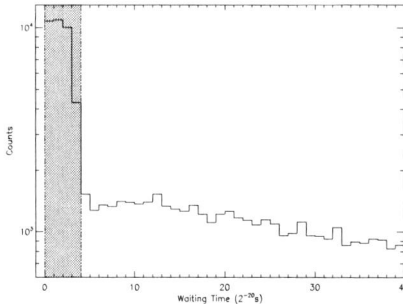

FIGURE 1. Histogram of number of counts versus time between interactions. Note that the vertical axis is logarithmic in scale.

METHOD

From the cross sections for Compton scattering, linearly polarized γ-rays preferentially scatter perpendicular to the polarization vector. This leads to an azimuthal asymmetry in the scattering angles with a 180° period. By plotting the numbers of counts as a function of scattering angle, we can measure (or put an upper limit on) this angular asymmetry. For full reviews of measuring γ-ray polarization, see [3, 4, 5].

For a count rate S and a fractional polarization Π_s, the rate as a function of azimuthal scattering angle θ is given by:

$$\frac{dS}{d\phi} = \left(\frac{S}{2\pi}\right)[1 - \mu_m \Pi_s \cos(2(\phi - \eta))]$$

where η is the angle of polarization and μ_m the average value of the "modulation factor." Converting this formula to discrete binning and identifying the cos() factor with the measured modulation amplitude A_m, gives

$$\Pi_s = \frac{A_m}{\left(\frac{S}{N_{\text{bins}}}\right)\mu_m}$$

where N_{bins} is the number of angular bins.

The modulation factor μ is the modulation amplitude expected for a 100% polarized signal, and determined largely by the details of the instrument design, incidence angle of the GRB, and energy cuts used. It is the calibration required to convert a measured modulation amplitude to a polarization fraction.

RHESSI telemeters all interactions without coincidence information, so determining det/det coincidence has to be done on the ground. The interactions are tagged with 2^{-20} s (a binary microsecond, or bμs) timing resolution. However, the front and rear segments of the *RHESSI* germanium detectors have different electronics chains, so it is expected that interactions in front versus rear detectors will have slightly different time tags. This is what we see when we plot a histogram of counts versus the time interval between counts (Figure 1). In order to maximize our signal-to-noise for our analysis, we chose to use events whose time tags were within 4 bμs.

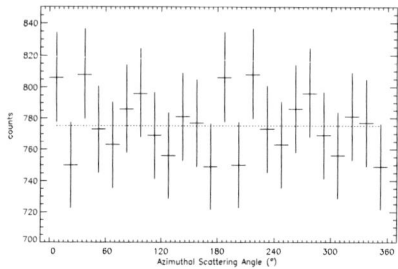

FIGURE 2. ALC of two background intervals taken both before (40 s duration) and after (65 s duration) GRB021206. As we would expect, the background appears unpolarized.

Once all non-coincidence events in the eventlist were removed, we made further energy and detector selections. First, we removed all coincidence events that contained an interaction with a zero or negative channel number. In *RHESSI* eventlists, negative channel numbers indicate things such as ULD triggers and CSA resets, and events containing these should not be included in our analysis. By removing these events *after*, instead of before, identifying coincidences, we are able to avoid the inclusion of bad events in our data.

Next, we looked at the detector numbers for each coincidence event. Given the uncertainty in the timing, we are unable to tell which detector the photon interacted in first. However, if a polarization signal exists, it is 180° symmetric and the interaction order does not matter. We required two unique detectors to be triggered within our time window. We further required that only one or fewer of these detectors had a front/rear coincidence. This is to allow for events that scatter once in the first detector, and then multiple times in the second. Of course events that scatter multiple times in the first detector and are then photoabsorbed in the second are also accepted and will produce an unpolarized background, and we take this factor into account when calculating the final polarization fraction.

Once events with the proper detector combinations were selected, we summed the total energies deposited in each detector. To help remove our chance coincidence background (events where two photons interacted in two detectors within our timing window), we required that at least 30 keV be deposited in each detector. We further required that the *total* energy deposited (summed between the two detectors) be in the 0.15–2.0 MeV range.

Using the locations of the detector centers, and correcting for the known rotation of the spacecraft, we are able to produce an "angular lightcurve" (ALC) for each burst. This is a histogram of the number of counts as a function of scattering angle. Our technique applied to a background interval is given in Figure 2, and to GRB021206 in Figure 4 (crosses).

For a GRB whose lightcurve varies slowly with respect to the rotation of the *RHESSI* spacecraft, we would expect to see the modulation due to polarization (or lack thereof) directly in the data. For many bursts, however, the LC varies quickly and the burst is not long enough to average out over many spacecraft rotations. Since *RHESSI* can only

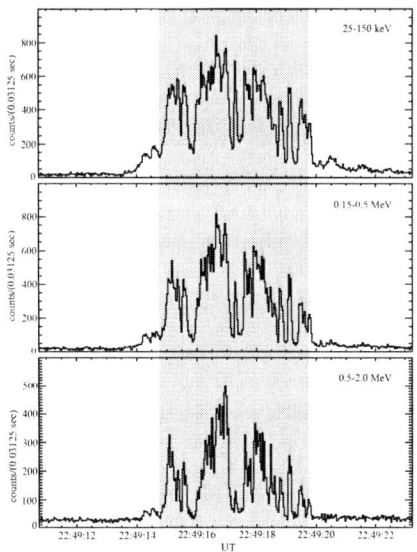

FIGURE 3. LC of GRB021206, in 32 bms and in three energy bands, as observed by *RHESSI*.

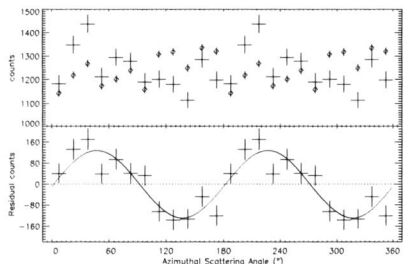

FIGURE 4. ALC for GRB021206. Top: The raw histograms for the data (crosses), and what we would expect if the GRB were unpolarized (diamonds). There is no indication of a modulation in either case. Bottom: The residual counts versus angle once the null ALC has been subtracted from the source ALC. A very clear sinusoidal modulation is detected, which we attribute to polarization of the incident GRB γ-ray photons.

measure a small number of angles at any given instant, for a burst with a complex LC (such as GRB021206, Figure 3) some angles will be sampled more frequently than others. This will lead to a variation in the ALC regardless of the polarization. This variation can be significant, but would be difficult to mistake for a polarization modulation.

To account for any variations that might be due to the source LC and not polarization, we performed a series of Monte Carlo simulations using the *RHESSI* mass model (D. M. Smith, priv. comm.). Using the spectrum measured by *RHESSI* as our input, and varying the relative numbers of counts as a function of azimuthal angle based on the *RHESSI* LC,

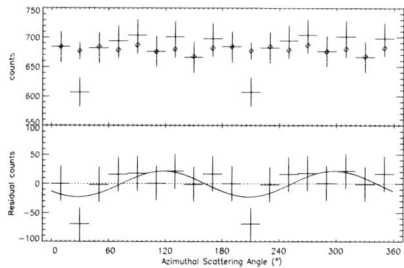

FIGURE 5. ALC for GRB030329. There is no evidence of polarization in the data, although there are not enough counts to exclude moderate polarization to a high significance. The 3σ upper limit on the amount of polarization derived from the data is $\Pi < 80\%$.

we propagated 18-billion photons through the mass model and binned the coincident events in the same way as we binned the data. This "unpolarized" or "null" ALC is plotted in Figure 4 (diamonds). The statistical uncertainties on each point are determined by the uncertainties in the single event counting rates (the template we used as input to the simulations), and not the numbers of coincidence events in the simulations.

When the null ALC is subtracted from the data, a very strong modulation is seen. We note that the existence of the modulation in our data is highly significant, with a chance probability of $< 10^{-8}$. We also note that, given the overall high quality of the *RHESSI* mass model and the large numbers of photons we used to generate the null ALC, the *RHESSI* response to an unpolarized burst is well understood. It is the exact calibration of how sensitive *RHESSI* is to polarization (the average value of the modulation factor μ_m) that leads to the large (25%) error in our estimate of the polarization fraction.

OTHER SOURCES

We have applied our technique to other possible sources of γ-ray polarization. The first is GRB030329 (Figure 5). This nearby burst triggered *HETE*-2 on 2003 March 29 at 11:37:14.67 UT. Since then both the afterglow and underlying supernova associated with GRB030329 have been very well observed, and are the subject of a special section in these proceedings.

Although the GRB was quite bright, because it was located behind *RHESSI* and was a much softer burst than GRB021206, there was severe attenuation due to the spacecraft. Still, we were able generate an ALC in exactly the same way as for GRB021206. We ran Monte Carlo simulations using the *HETE*-2/FREGAT spectrum (K. Hurley, priv. comm.) to estimate our null ALC. As we would expect for a burst of this duration (nearly 8 satellite rotations), the null ALC is nearly constant as a function of angle. We see no evidence of polarization in GRB030329, but are only able to place a weak 3σ upper limit of $\Pi < 80\%$ for the polarization amplitude.

We also applied our analysis to data taken during the γ-ray solar flare observed on 2002 July 23. We used events obtained during a 360 s interval starting at 00:28:00 UT.

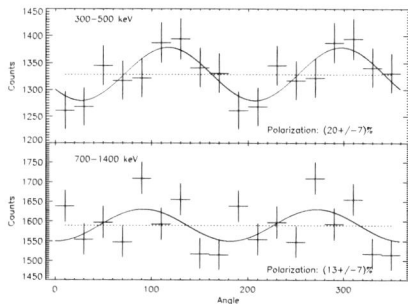

FIGURE 6. ALC for the first 6 minutes of the γ-ray solar flare of 2002 July 23 (00:28:00 UT to 00:34:00 UT) for two energy ranges. Top: In the 300-500 keV band, there is an obvious modulation and polarization at the $\Pi = 20\%$ level. Bottom: There is no evidence for polarization in the 700-1400 keV band, although a fit modulation gives a level of $\Pi = 13\%$.

The flare LC varied slowly with respect to the spacecraft rotation, and due to the long integration time the null ALC averages to a constant. Also, due to the large overall count rate associated with the flare, we were able to use only events that triggered the rear detectors (events that had a simultaneous front detector trigger as well were thrown out).

We once again required that the total energy deposited in a single detector be greater than 30 keV, but again because of the high count rate we were able to separated the events into finer total energy bins. The ALCs in two energy bins are shown in Figure 6. In the lower energy range (0.3-0.5 MeV), we find a polarization amplitude of $\Pi = 20 \pm 7\%$. In the upper energy range (0.7–1.4 MeV), however, we find no evidence of polarization.

CONCLUSIONS

We have developed a simple, physically sound method for searching for polarization in long, bright γ-ray bursts observed with *RHESSI*. We have applied it successfully to GRB021206, and found a polarization fraction of $\Pi = 80 \pm 20\%$. We also applied the technique to GRB030329, however due to the location of the GRB with respect to the spacecraft pointing we are only able to place a weak upper limit on the polarization fraction. We have also searched for polarization in a solar flare, and note that the results differ from what we see in GRBs. We are confident our technique does not produce spurious modulations.

We also note that while this result from GRB021206 is significant, it is a single measurement of one GRB. Further observations are needed to determine the full implications of our result. However, γ-ray polarization **can** be measured with the current generation of γ-ray instruments. It is going to be important for future γ-ray missions to be sensitive to polarization as well.

Addendum Since this conference, [6] has appeared on astro-ph. We are currently awaiting the revised, accepted version so that we may fully address all of the questions raised in their paper, and do so in refereed literature. In the meantime, however, the fact that we see a modulation, therefore that the burst is polarized at some level, is highly significant and not something we consider "controversial."

REFERENCES

1. Coburn, W., and Boggs, S. E., *Nature*, **423**, 415 (2003).
2. Lin, R. P., et al., *Sol. Phys.*, **210**, 1 (2003).
3. Novick, R., *Space Sci.Rev.*, **18**, 389 (1975).
4. Lei, F., Dean, A. J., and Hills, G. L., *Space Sci.Rev.*, **82**, 309 (1997).
5. McConnell, M., Forrest, D., Vestrand, W. T., and Finger, M., "Using BATSE to Measure Gamma-Ray Burst Polarization," in *Proceedings of the 3rd Huntsville Symposium*, edited by C. Kouveliotou, M. F. Briggs, and G. J. Fishman, 1996, p. 851.
6. Rutledge, R. E., and Fox, D. B., Re-analysis of polarization in the gamma-ray flux of grb 021206 (2003), astro-ph/0310385, submitted to MNRAS.

The Polarization Evolution of the Optical Afterglow of GRB 030329

J. Greiner*, S. Klose[†], K. Reinsch**, H.M. Schmid[‡], R. Sari[§], D.H. Hartmann[¶], C. Kouveliotou[‖], A. Rau*, E. Palazzi[††], C. Straubmeier[‡‡], B. Stecklum[†], S. Zharikov[§§], G. Tovmassian[§§], O. Bärnbantner[¶¶], C. Ries[¶¶], E. Jehin***, A. Henden[†††], A.A. Kaas[‡‡‡], T. Grav[§§§], J. Hjorth[¶¶¶¶], H. Pedersen[¶¶¶¶], R. A. M. J. Wijers[♯], A. Kaufer***, H.-S. Park[ℓ], G. Williams[♯♯] and O. Reimer[ℓℓ]

*Max-Planck-Institut für extraterrestrische Physik, 85741 Garching, Germany
[†]Thüringer Landessternwarte, 07778 Tautenburg, Germany
**Universitäts-Sternwarte Göttingen, 37083 Göttingen, Germany
[‡]Institut für Astronomie, ETH Zürich, 8092 Zürich, Switzerland
[§]California Institute of Technology, Theoretical Astrophysics 130-33, Pasadena, CA 91125, USA
[¶]Clemson University, Department of Physics and Astronomy, Clemson, SC 29634, USA
[‖]NSSTC, SD-50, 320 Sparkman Drive, Huntsville, AL 35805, USA
[††]Istituto di Astrofisica Spaziale e Fisica Cosmica, CNR, 40129 Bologna, Italy
[‡‡]Physikalisches Institut, Universität Köln, 50937 Köln, Germany
[§§]Instituto de Astronomia, UNAM, 22860 Ensenada, Mexico
[¶¶]Wendelstein-Observatorium, Universitätssternwarte, 81679 München, Germany
***European Southern Observatory, Alonso de Cordova 3107, Santiago 19, Chile
[†††]USRA, U.S. Naval Observatory, Flagstaff, AZ 86002, USA
[‡‡‡]Nordic Optical Telescope, 38700 Santa Cruz de La Palma, Spain
[§§§]University of Oslo, Institute for Theoretical Astrophysics, 0315 Oslo, Norway
[¶¶¶¶]Astronomical Observatory, NBIfAFG, University of Copenhagen, 2100 Copenhagen, Denmark
[♯]Astronomical Institute Anton Pannekoek, Kruislaan 403, 1098 SJ Amsterdam, The Netherlands
[ℓ]Lawrence Livermore National Laboratory, University of California, Livermore, CA 94551, USA
[♯♯]MMT Observatory, University of Arizona, Tucson, AZ 85721, USA
[ℓℓ]Theoretische Weltraum- und Astrophysik, Ruhr-Universität Bochum, 44780 Bochum, Germany

Abstract. We report 31 polarimetric observations of the afterglow of GRB 030329 with high signal-to-noise and high sampling frequency. The data imply that the afterglow magnetic field has small coherence length and is mostly random, probably generated by turbulence.

INTRODUCTION

The association of a supernova with GRB 030329 [1,2] strongly supports the collapsar model [3] of γ-ray bursts (GRBs), where a relativistic jet [4] forms after the progenitor star collapses. Such jets cannot be spatially resolved because of their cosmological distances. Their existence is conjectured based on breaks in GRB afterglow light curves and the theoretical desire to reduce the GRB energy requirements. Temporal evolution of polarization [5,6,7] may provide independent evidence for the jet structure of the relativistic outflow. Previous single measurements found low-level (1-3 %) polarization

[8-15] in optical afterglows, and the only reports on variable polarization [16,17] were based on few measurements with different instruments and modest signal-to-noise. Here, we report polarimetric observations of the afterglow of GRB 030329 with high signal-to-noise and high sampling frequency [18].

OBSERVATIONS AND RESULTS

GRB 030329 triggered the High Energy Transient Explorer, HETE-2, on March 29, 2003 (11:37:14.67 UT) [19]. The discovery of the burst optical afterglow [20,21] was quickly followed by a redshift measurement [22] for the burster of z=0.1685 ($\sim$800 Mpc). We have obtained 31 polarimetric observations of the afterglow of GRB 030329 with the same instrumentation (plus few more with different instruments) [18] over a time period of 38 days. We performed relative photometry, and derived from each pair of simultaneous measurements at orthogonal angles the Stokes parameters U and Q. In order to obtain the intrinsic polarization of the GRB afterglow, we had to correct for Galactic interstellar polarization (mostly due to dust). We performed imaging polarimetry to derive the polarization parameters of seven stars in the field of GRB 030329, and obtained an interstellar (dust) polarization correction of 0.45% at position angle 155°. Subtraction of the mean foreground polarization was performed in the Q/U plane (Q_{fp}=0.0027±0.0013, U_{fp}=-0.0033±0.0017).

The temporal evolution of the degree and angle of polarization together with the R band photometry is shown in Figure 1, demonstrating the presence of non-zero polarization, $\Pi \sim 0.3 - 2.5\%$ throughout a 38-day period, with significant variability in degree and angle on time scales down to hours. Further, the spectropolarimetric data of the first three nights as well as the simultaneous R and K band imaging polarimetry during the second night show that the relative polarization and the position angle are wavelength independent (within the measurement errors of about 0.1 %) over the entire spectral range. These data imply that polarization due to dust in the host galaxy of GRB 030329 does not exceed $\sim$0.3%.

Figure 1 shows that while the polarization properties show substantial variability (for which no simple empirical relationship is apparent), the R band flux is a sequence of power laws. During each of the power law decay phases the polarization is of order few percent, different from phase to phase, and variable within the phase, but not in tandem with the "bumps and wiggles" in the light curve. We observe a decreasing polarization degree shortly after the light curve break at $\sim$0.4 days (as determined from optical [21,23] and X-ray data [24,25]). Rapid variations of polarization occur $\sim$1.5 days after the burst, and could be related to the end of the transition period towards a new power law phase starting at $\sim$1.7 days. Polarization eventually rises to a level of $\sim$2 %, which remains roughly constant for another two weeks.

Due to relativistic beaming, most of the observed photons arrive from a narrow cone, of opening angle $\theta=1/\Gamma$ around the line of sight. If the magnetic field parallel to the shock front is stronger than the perpendicular component, a resolved afterglow would look like a ring with polarization pointing towards its center [5]. Early on, however, the Lorentz factor is much larger than the inverse opening angle of the jet (1/Γ), and we

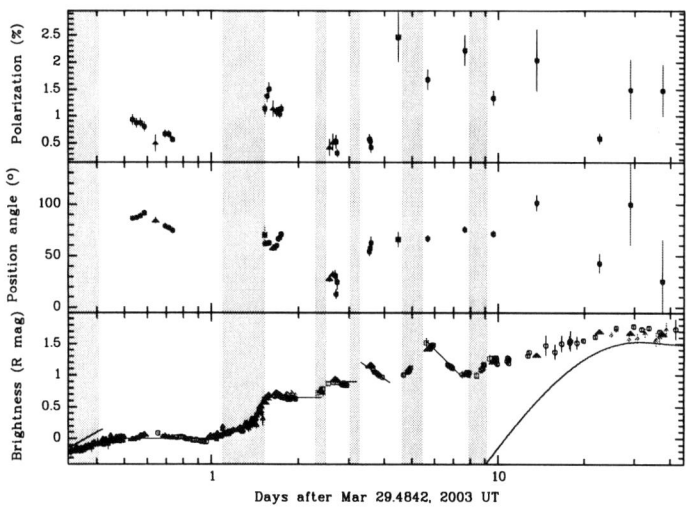

FIGURE 1. Evolution of the polarization during the first 38 days. Top and middle panels show the polarization degree in percent and the position angle in degrees. Spectropolarimetry was performed during the first three nights. The bottom panel shows the residual R band light curve after subtraction of the contribution of a power law $t^{-1.64}$ describing the undisturbed decay during the time interval 0.5-1.2 days after the GRB (i.e., after the early break at 0.4 days), thus leading to a horizontal curve. The symbols correspond to data obtained from either the literature (black), our own observations (gray): the 1m USNO telescope at Flagstaff (circles), the OAN Mexico (open triangles), and FORS1/VLT (filled triangles). Lines indicate phases of power law decay, with the first one from early data [21] (not shown). Vertical gray bars mark re-brightening transitions. Contributions from an underlying supernova (solid curved line) do not become significant until ∼10 days after the GRB.

therefore probe scales that are much narrower than the size of the jet. The jet is uniform over these scales, so the emission pattern (afterglow) has axial symmetry around the line of sight. This symmetry, for an unresolved source, leads to zero polarization. As the ejecta interact with the surrounding medium and decelerate, their Lorentz factor becomes comparable to $1/\Gamma$, so we see most of the emitted photons. As the jet expands laterally, the energy per unit solid angle decreases; the light curve power law decay changes to a steeper slope, marking the so-called "jet break" and, at the same time, axial symmetry is broken (if the jet is not exactly pointed towards us), resulting in non-zero polarization. Since the jet spreads but the offset of the line of sight from the centre of the jet remains constant, axial symmetry is regained and polarization will eventually vanish. Maximal polarization should therefore occur around the jet break time. Some models [5,6,7] suggest the presence of either one or three peaks of polarization, with the most significant peak close to the time of the jet break. If three peaks are present, the position angle of the central peak is rotated by 90° relative to the other two, but remains constant within each peak.

How do our data reflect these properties? A break (change of the power law decay in the light curve) was found at 0.4 days in early optical data [21,23] and confirmed with X-ray data [24,25]. We thus expect maximum polarization near 0.4 days, and a decline thereafter. Our early polarization data (Figure 1) are consistent with this prediction. Though we are missing data before 0.4 days to confirm a peak, the degree of polarization decreases from 0.9% at 0.55 days to about 0.5% at 0.8 days. During this decline the position angle decreases slightly, while the model predicts constancy. The time evolution of the polarization properties during this early phase is thus broadly consistent with the interpretation of the steepening of the light curve at 0.4 days as the jet break, providing independent observational evidence for the crucial assumption of collimated outflows (jets) in GRB explosions.

GRB 030329 is so far the only case where the polarization evolution supports the break as being due to the jet nature. From the jet break time and the isotropic equivalent energy of $9x10^{51}$ erg we calculate the jet opening angle [26] to be about $3°\!.5$, and the actual total energy release during the burst $\sim 2x10^{49}$ erg. This energy is about 25 times smaller than the "standard energy" of GRBs [27] and one order of magnitude larger than the one inferred (assuming isotropic emission) for GRB 980425 (associated with SN 1998bw).

We do not expect the model to apply to the observations after the first re-brightening episode, which started $\sim$1.5 days after the trigger. Indeed the polarization angle changed by 30° with respect to the first night, while the model predicts either no change or a 90° change. What can we still learn from the complex late time behavior? Between 3 and 10 days the magnitude of the polarization changes significantly (a factor of two or more), but the position angle remains fairly constant (fluctuations of less than 10 degrees). This implies that the polarization is not the result of a small number of coherent magnetic field cells with random orientation; such a model would predict that the position angle changes on the same timescale as the magnitude of polarization. Instead, it implies that the position angle is associated with some global geometry.

A change of the position angle by an amount different than ninety degrees, as observed between the first and second day as well as between the second and third day, suggests that the asymmetry of the emission changed direction. Independent of polarization, the steeper decay of the optical light curve after day 10 (once the supernova component is subtracted), as well as a clear break observed at radio frequencies, suggests that the outflow may consist of two components [28]: the first one dominates the light curve and the polarization properties until day 1.5, and has its jet break at day 0.4, and the second one, more mildly relativistic (Lorentz factor of about six), causes the re-brightening at 1.5 days and dominates the light curve and polarization properties thereafter. In this interpretation, the polarization data suggest that these two components do not share the same symmetry axis, which allows for the 30° change in the position angle. Due to its larger angle but similar energy per unit solid angle, the energy content in the second component is comparable to the canonical value of $5x10^{50}$ ergs. Even more than two jet components have been proposed [29] to explain the various wiggles in the lightcurve. If that picture is correct, the polarization data require that the jet axes not be aligned.

Finally, after day ~ 10, there is a hint of some decrease in the polarization degree. At this time, the supernova contributes approximately 60% of the total light in the R band. Radiation transport models of non-symmetric supernovae [30,31] suggest a $\sim 1\%$ degree of polarization. For SN 2003dh we expect an even lower level of polarization since we are oriented towards the SN rotation axis. Thus, it is plausible that the low polarization supernova light dilutes the more highly polarized afterglow.

In summary, our data constitute the most complete and dense sampling of the polarization behaviour of a GRB afterglow to date. The GRB 030329 afterglow polarization probably did not rise above $\sim 2.5\%$, and did not correlate with the flux. The low level of polarization implies that the components of the magnetic field parallel and perpendicular to the shock do not differ by more than $\sim 10\%$, and suggests an entangled magnetic field, probably amplified by turbulence behind shocks, rather than a pre-existing field.

Acknowledgements: This work is primarily based on observations collected at ESO, Chile, with additional data obtained at the German-Spanish Astronomical Centre Calar Alto, operated by the Max-Planck-Institute for Astronomy, Heidelberg, jointly with the Spanish National Commission for Astronomy, the NOT on La Palma, Canary Islands, and the Observatorio Astronomico National, San Pedro, Mexico. We are grateful to the staff at the Paranal, Calar Alto and NOT observatories, in particular A. Aguirre, M. Alises, S. Hubrig, A.O. Jaunsen, C. Ledoux, S. Pedraz, T. Szeifert, L. Vanzi and P. Vreeswijk for obtaining the service mode data reported here.

REFERENCES

1. Stanek, K.Z., et al., ApJ **591**, L17 (2003).
2. Hjorth, J., et al., Nature **423**, 847 (2003).
3. Woosley, S.E., Eastman, R.G., Schmidt, B.P., ApJ **516**, 788 (1999).
4. Meszaros, P., ARAA **40**, 137 (2002).
5. Sari, R., ApJ **524**, L43 (1999).
6. Gruzinov, A., ApJ **525**, L29 (1999).
7. Ghisellini, G., Lazzati, D., MN **309**, L7 (1999).
8. Hjorth, J., et al., Science **283**, 2073 (1999).
9. Wijers, R.A.M.J., et al., ApJ **523**, L33 (1999).
10. Covino, S., et al., A&A **348**, L1 (1999).
11. Rol, E., et al., ApJ **544**, 707 (2000).
12. Björnsson, G., Hjorth, J., Pedersen, K., Fynbo, J.U., ApJ **579**, L59 (2002).
13. Masetti, N., et al., A&A **404**, 465 (2003).
14. Covino, S., et al., A&A **400**, L9 (2003).
15. Bersier, D., et al., ApJ **583**, L63 (2003).
16. Barth, A., et al., ApJ **584**, L47 (2003).
17. Rol, E., et al., A&A **405**, L23 (2003).
18. Greiner, J., et al, Nature 426, 157 (2003).
19. Vanderspek, R., Crew, G., Doty, J., Villasenor, J., Monelly, G., GCN Circ. 1997 (2003).
20. Price, P.A., et al., Nature **423**, 844 (2003).
21. Uemura, M., et al., Nature **423**, 843 (2003).
22. Greiner, J., et al., GCN Circ. 2020 (2003).
23. Burenin, R., et al., Astron. Lett. **29**, 9, 1 (2003).
24. Marshall, F.E., Markwardt, C., Swank, J.H., GCN Circ. 2052 (2003).
25. Tiengo, A., et al., A&A **409**, 983 (2003).
26. Sari, R., Piran, T., Halpern, J., ApJ **519**, L17 (1999).
27. Frail, D.A., et al., ApJ **562**, L55 (2001).
28. Berger, E. et al.,, Nature **426**, 154 (2003).
29. Granot, J., Nakar, E., Piran, T.,, astro-ph/0304563 (2003).
30. Wang, L., Howell, D. A., Höflich, P., Wheeler, J. C., ApJ **550**, 1030 (2001).
31. Leonard, D. C., Filippenko, A. V., Chornock, R., Foley, R. J., PASP **114**, 1333 (2002).

Comparison of Three Afterglow Morphologies

Jay D. Salmonson*, Elena Rossi[†] and Davide Lazzati[†]

*Lawrence Livermore National Laboratory, Liveremore, CA 94551
[†]Institute of Astronomy, University of Cambridge, Madingley Road CB3 OHA, England

Abstract. Herein we compare three functional families for afterglow morphologies: the homogeneous afterglow with constant shock surface energy density, the structured afterglow for which the energy density decays as a power-law as a function of viewer angle, and the gaussian afterglow which has an exponential decay of energy density with viewer angle. We simulate observed lightcurves and polarization curves for each as seen from a variety of observer vantage points. We find that the homogeneous jet is likely inconsistent with observations and suggest that the future debate on the structure of afterglow jets will be between the other two candidates.

The structure of gamma-ray burst afterglows is currently a question of considerable interest. Observations of afterglows have become sufficiently numerous and well sampled that inferences can be made about the morphology and evolution of the external shock, which is thought to generate the afterglow's emission. Specifically, power-law decay slopes, jet-break times and polarization information all can be used to model the structure of the afterglow.

Several clues can be used to infer the structure of afterglow jets. A first clue to the structure of afterglows was the realization that the inferred jet opening angle is inversely proportional to the energy density in a manner suggestive that, in sum, most bursts have similar energies [1]. This realization prompted the postulate that the jet might be structured [2] and that by varying the afterglow external shock energy per solid angle [3] one might encapsulate all observations into a single quasi-universal morphology. Another clue is polarization [4, 5]. With the measurement of polarization in a handful of real afterglows comes the possibility of using this powerful diagnositic as a tool to infer the structure of afterglows [6].

In this paper we model three basic afterglow morphologies in stark simplicity so to compare and discriminate their basic observational differences. All of these models have been previously examined in whole or in part [3, 7, 8, 9, 6], but here we make a comparison of perspective effects on both the lightcurve and polarization curve for each of them. To do this we use a simple spectral model in which peak and cooling frequencies are below the optical band [3, 7] and we neglect lateral expansion.

- **Homogeneous Afterglow** The canonical afterglow, this jet is characterized by a constant afterglow shock surface energy density, ε, within an opening angle, θ_0, and a sharp, Heaviside drop-off at this edge, $\varepsilon(\theta) \propto H(\theta - \theta_0)$. As can be seen in Figure 1, this afterglow exhibits a viewing angle dependent two-stage break in the lightcurve whereby the flux deficit imposed by the near side of the jet coming into view prompts a first steepening, while a second steepening happens when the far

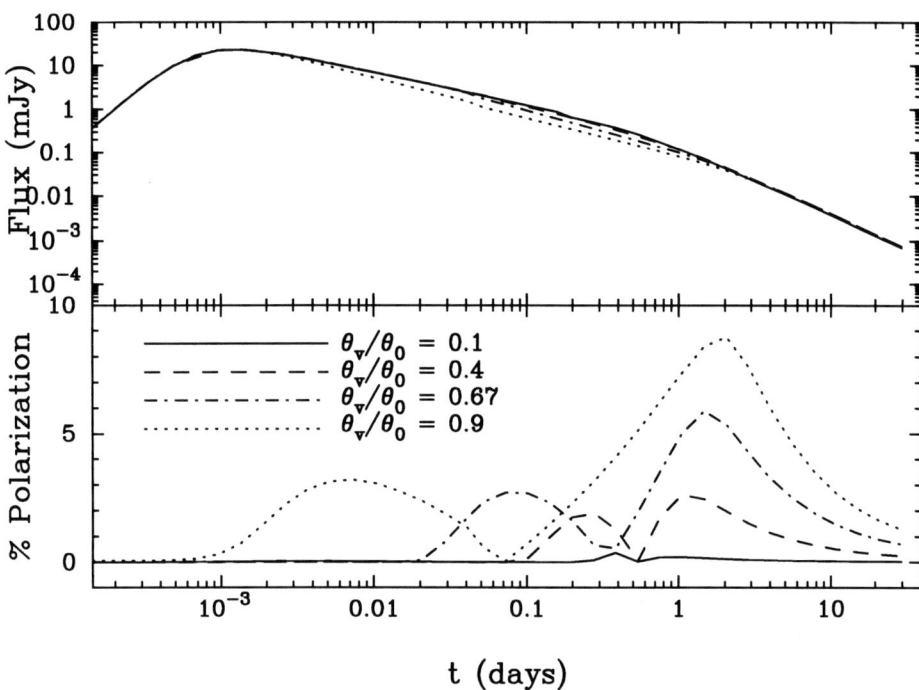

FIGURE 1. Homogeneous jet with $\theta_0 = 5°$ as seen from a range of vantage points, θ_v. The double humped polarization curves indicate a 90° shift in picth angle. Thus this model predicts a null polarization near the jet-break time.

side of the jet comes into view [7]. These two stages roughly correspond to two separate maxima in polarization separated by a 90° shift in the pitch angle. Thus, this model predicts roughly zero polarization at the jet break time, with orthogonal polarization directions before and after the break. To date convincing observations of such behavior have not been reported [6].

- **Structured Afterglow** If the afterglow shock surface energy density is a power-law for angles greater than some narrow core, θ_c, e.g. $\varepsilon(\theta) \propto (1+(\theta/\theta_c)^2)^{-1}$, then the observed lightcurve varies significantly as a function of viewer angle. This raises the intriguing possibility of universality; that all gamma-ray bursts are created roughly equal, but the wide observed variety stems merely from the range of viewer vantage points [3]. As seen in Figure 2, a relatively wide variety of viewing angles produces a variety of lightcurve flux levels and jet-break times. Also, each viewer should see a maximum in polarization coinciding with the jet-break time. For the largest viewing angles, simulations show a flattening in the slope of the lightcurve prior to the jet-break [7]. This is not observed in the data and so provides clues and constraints on universality and structured jets.

- **Gaussian Afterglow** Characterized by an exponential decay of the shock surface energy density, $\varepsilon \propto \exp(-(\theta/\theta_0)^2)$, this morphology might be thought of as a

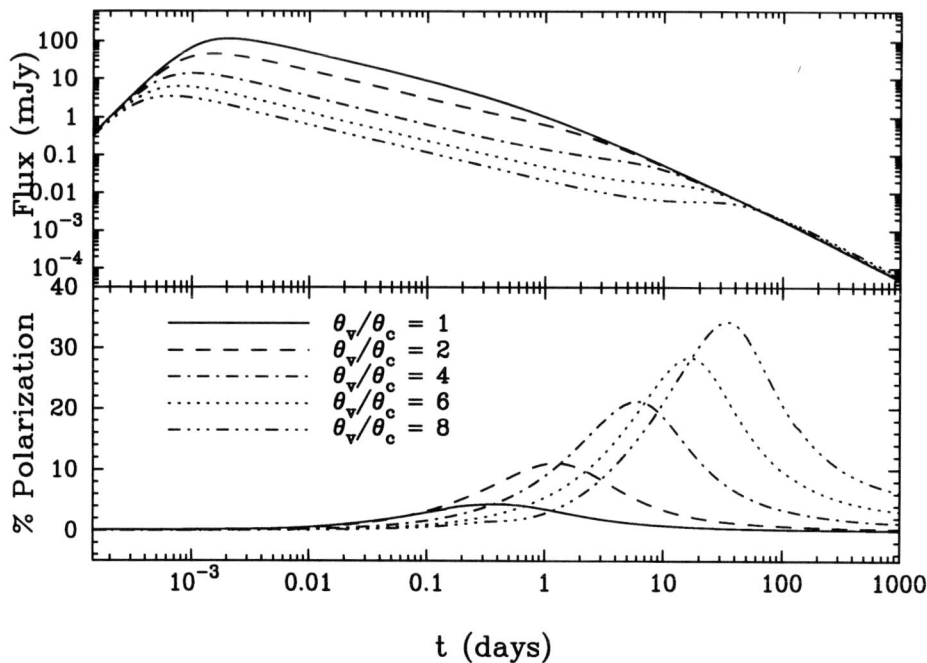

FIGURE 2. Structured jet with core angle $\theta_c = 2°$ as seen from a range of vantage points, θ_v.

more physically realistic version of the homogeneous jet, with blurred edges. The differences between this model and the structured jet are more subtle. From each one can expect a maximum polarization at the jet-break time and at a constant pitch angle (Figure 3).

DISCUSSION

The main point to be made here is that the homogeneous and structured jets are the most distinct from each other, having both divergent lightcurves and polarization curves, while the gausian jet is middling and shares key features from each. In particular, the gausian jet lightcurve, being relatively constant as a function of viewer angle, is more like the homogeneous jet. Thus, the gausian jet does not lend itself to a universal interpretation. Conversely, the gausian jet has a single peaked polarization curve which peaks at the jet-break time similar to the structured jet. This simple breakdown of model behaviors will aid in the job of discriminating between this trichotomy of afterglow morphological families.

The simplistic homogeneous jet, with a predicted 90° shift in polarization pitch angle at the jet-break time, is likely inconsistent with observations. We suggest that the future debate on the structure of afterglow jets will be between the remaining morphological candidates, each with sharp jet breaks and very similar polarization curves. The critical

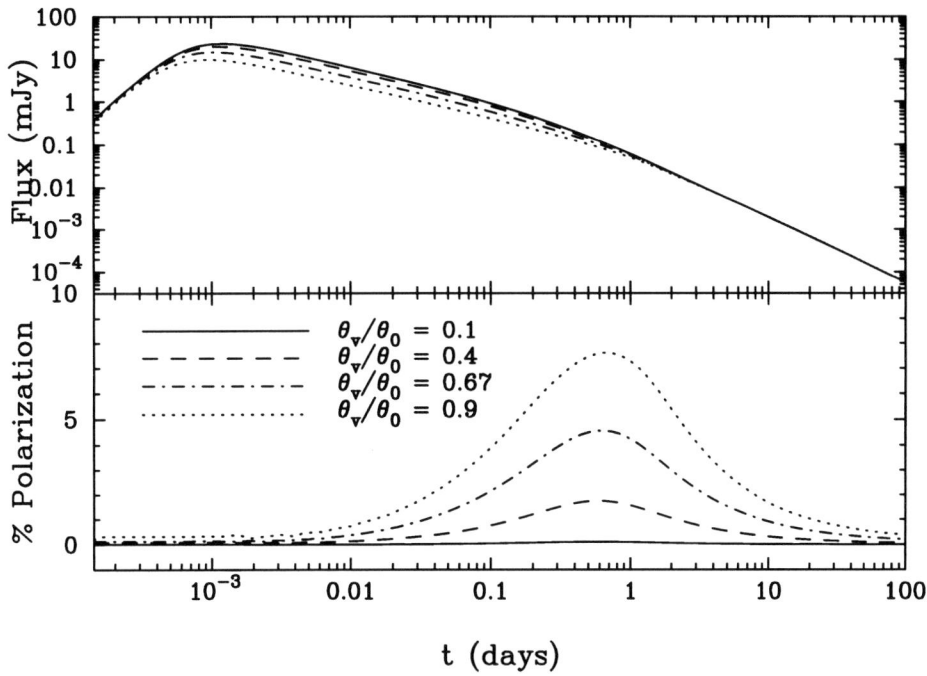

FIGURE 3. Gaussian jet with $\theta_0 = 5°$ as seen from a range of vantage points, θ_v.

discriminator between these two contenders is that the structured jet lends itself to a universal gamma-ray burst paradigm while the gaussian jet requires an innate variety in the progenitor population.

This work was performed under the auspices of the U.S. Department of Energy by University of California Lawrence Livermore National Laboratory under contract W-7405-ENG-48.

REFERENCES

1. Frail, D. A. et al., ApJL **562**, L55–L58 (2001).
2. Salmonson, J. D., and Galama, T. J., ApJ **569**, 682–688 (2002).
3. Rossi, E., Lazzati, D., and Rees, M. J., MNRAS **332**, 945–950 (2002).
4. Ghisellini, G., and Lazzati, D., MNRAS **309**, L7–L11 (1999).
5. Sari, R., ApJL **524**, L43–L46 (1999).
6. Rossi, E., Lazzati, D., Salmonson, J. D., Ghisellini, G., submitted MNRAS (2003).
7. Salmonson, J. D., ApJ **592**, 1002–1017 (2003).
8. Kumar, P., and Granot, J., ApJ **591**, 1075–1085 (2003).
9. Granot, J., and Kumar, P., ApJ **591**, 1086–1096 (2003).

Collapsar Jet Stability at Breakout

Enrique A. Gómez* and Philip E. Hardee*

*206 Gallalee Hall Box 870324, University of Alabama Tuscaloosa, AL 35487-0324

Abstract. The most recent observations of prompt optical afterglows of GRBs like GRB030329 are the strongest evidence so far that jet emission from collapsars produce "long" GRBs associated with supernovae. Recent simulations of collapsar jets indicate that the jet is unstable as it propagates through the atmosphere of the collapsar progenitor up to the breakout point. Recollimation shocks and pinch-body wave modes in the flow are important at breakout for the subsequent jet evolution into a wind. From our stability analysis we predict that pinch-body modes should develop just upstream from the jet head at breakout. Along with the Aloy et al. (2000) observed non-adiabatic density fluctuations, this provides a baseline for the evolution of overrunning relativistic shells in the GRB fireball.

INTRODUCTION

Observations of jet structure in GRB030329 [1, 2, 3] support the prompt (Type I) collapsar model of GRBs [4, 5]. In this model, a rotating He star with mass $> 10\, M_\odot$ fails to eject its envelope after core collapse. Accretion would drive a relativistic gas jet that could break through the He shell if it were powered long enough. Axisymetric, relativistic hydrodynamic simulations by Aloy et. al. (2000) [6] show evidence of pinch-body modes triggered by recollimation shocks. These shocks destroy all upstream perturbations. The jet becomes a fireball after it breaks out from the He layer at 10^{11}cm. The perturbations form just upstream of the last recollimation shock. As these perturbations evolve, they may provide sites for particle acceleration and gamma ray production.

JET ANALYSIS

Aloy et al (2000) [6] made a set of collapsar simulations using the GENESIS code. They simulated two jets with constant energy deposition rates of dE/dt = 10^{50} (C50) and 10^{51} ergs s^{-1} (C51). We analyzed the space-dependent structures in these jets by comparing them to the solutions of the linearized relativistic fluid equations and the corresponding the Kelvin-Helmholtz modes that may be triggered by recollimation shocks within the jet.

Space-dependent perturbations propagate with a dispersion relation given by Hardee, Clarke, & Rosen 1997 [7]. These solutions have real and imaginary components and are related to sound waves propagating with and against the jet flow speed modified by the jet external medium interface.

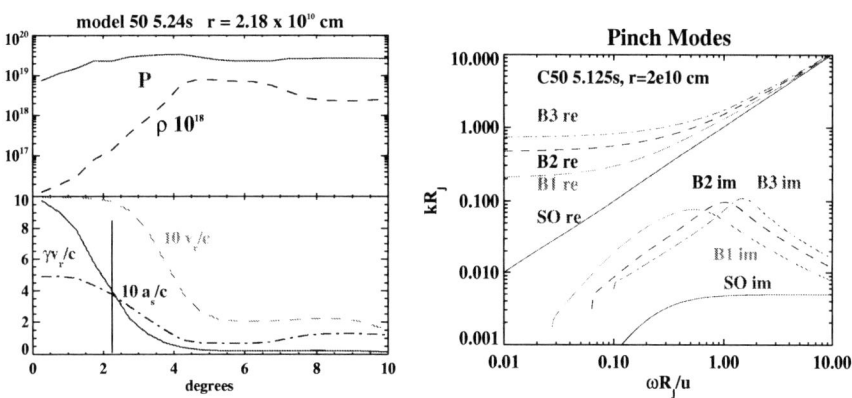

FIGURE 1. Left: Cross section for pressure P, density ρ, velocity v_r/c, Lorentz factor $\gamma v/c$. and sound speed a_s/c for the C50 jet for t=5s at a distance of $r=2.1\ 10^{10}$ cm from the jet engine. The vertical line identifies the edge of the jet. Right: frequency ω and wavenumber k of the real and imaginary solutions of the fundamental (S0), 1st, 2nd, and 3rd pinch body modes (B1, B, B3).

To calculate the value of the relevant parameters in the internal and external medium of the jet, we defined the jet radius R_j where the flow has 0.5 $\gamma^* v_r^*/c$. Here v_r^* is the maximum value for the jet velocity at a given cross section, and γ^* is the corresponding Lorentz factor. We calculated the weighted average for jet speed and sound speed for the regime inside the jet radius, $R < R_j$, and we found the sound speed also as a weighted average for $R_j < R < 3R_j$, which we define as the regime outside the jet radius. The local sound speed is given by the pressure p, the density ρ, a relativistic adiabatic index Γ of 4/3 or 13/9, and the speed of light c. Figure 1 shows a typical jet cross section (in this case at $r=2.1\ 10^{10}$ cm for the C50 jet after 5 seconds) used to derive the solution to the dispersion relation for an adiabatic index $\Gamma =13/9$. From this particular analysis the average jet speed is 0.987 c, the jet sound speed is 0.454 c and the average external sound speed is 0.121 c. The solutions are for the fundamental, first, second, and third pinch body modes.

CONCLUSION

We solved the dispersion relations in Hardee, Clarke, & Rosen 1997 [7] for $k(\omega)$ with the appropriate parameters from each cut of the jets. Using these we found the solutions for the fundamental and the first three body modes for jets C50 and C51. From the solutions we obtain the minimum growth length $\ell^* = (k_I^*)^{-1}$ (where k_I^* is the imaginary wave number) and the wavelengths at the maximum growth rate $\lambda^* = 2\pi/k_R^*$ (where k_R^* is the real wave number). We also found the wavelength for the maximum unstable growth rate λ_m. Given the frequency ω, the wave speed v_w and the growth time $\tau = \ell/v_w$

TABLE 1. Calculated Pinch Body Modes for Model C50

C50 t=3s r=1.3^{10}cm					C50 t=5s r=2^{10}cm			
	λ_m/r	λ^*/r	ℓ^*/r	τ (r/c)		λ_m/r	λ^*/r	ℓ^*/r	τ (r/c)
1B	0.52	0.18	0.22	0.43	1B	0.87	0.29	0.39	0.67
2B	0.24	0.09	0.18	0.35	2B	0.40	0.15	0.31	0.55
3B	0.15	0.06	0.16	0.31	3B	0.26	0.10	0.28	0.49

TABLE 2. Calculated Pinch Body Modes for Model C51

C51 t=2s r=2^9cm					C51 t=4s r=2^{10}cm			
	λ_m/r	λ^*/r	ℓ^*/r	τ (r/c)		λ_m/r	λ^*/r	ℓ^*/r	τ (r/c)
1B	2.35	0.65	1.07	1.65	1B	4.11	0.29	2.61	2.78
2B	1.08	0.34	0.85	1.33	2B	2.24	0.15	2.08	2.22
3B	0.70	0.23	0.76	1.19	3B	1.53	0.10	1.85	1.98

for any given solution are calculated from,

$$v_w = \frac{\omega k_R}{k_R^2 + k_I^2} \quad (1)$$

The solutions for the pinch body modes depend on the distance from the jet engine r because of the change of pressure outside of the jet. We present the calculated theoretical values for the wavelengths of the maximum unstable growth rate λ_m/r, the wavelengths at maximum growth rate λ^*/r, the minimum growth length ℓ^*/r, and the growth times τ in Table 1 for the C50 jet and in Table 2 for the C51 jet. For the maximum growth wavelength λ^* just downstream from the last recollimation shock, the range is $0.06r$ to $0.6r$. The range for the growth length ℓ^* is $0.2r$ to $2r$. Near breakout the growth time τ ranges from $0.3\ r/c$ to $3\ r/c$ or 1 to 10 seconds at the distance of the He shell. In the regime just upstream from breakout, pinch-body modes will have a characteristic wavelength of $\lambda \sim 10^8 - 10^9$ cm. As an example, we compared a segment of the C50 simulation with a first body mode theoretical solution in Figure 2. This segment lies just downstream from the last recollimation shock. We graph the jet evolution from this last shock in units of jet radii R_j. This wavelength spans seven computational zones or 7.8 R_j. For the fastest growing second and third body, the wavelengths would span 4.0 and 2.7 R_j or 4 and 3 computational zones, and this is not sufficient to resolve these modes.

Our numerical analysis of jet stability assumes a sharp lateral boundary between the jet and the external medium. The analytical approach of Aloy et al. (2002) [8] for the same jet assumes an extended shear layer. For Lorentz factors of order 10, they predict a growth time $\tau \sim 0.01r/c$ which is one order of magnitude smaller than our method; however, the characteristic $\lambda^* \sim 0.1r - 0.5r$ is also what we predict. Near the jet head the grid scaling is $\sim R_j$ suppressing the higher order modes, and this should motivate more detailed simulations.

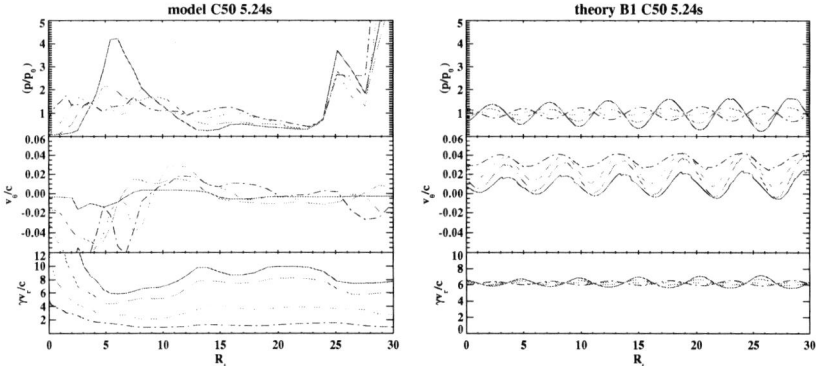

FIGURE 2. Left: Simulation normalized pressure, radial velocity, and Lorentz factor cuts for the C50 jet at t=5 seconds for the jet close to the head (2×10^{10} cm). Cuts are for the angles 0.25°, 1.25°, 2.25° and 3.25° (solid, long dash, short dash and dash-dot correspondingly). Right: Theoretical cuts of the 1st body mode for pressure, radial velocity, and Lorentz factor for the jet. Cuts are at 0.22, 0.44, 0.66 and 0.88 R_j from the jet axis. (solid, long dash, short dash and dash-dot correspondingly). These are plotted in units of jet radii from the last recollimation shock.

ACKNOWLEDGMENTS

E. A. G. wishes to acknowledge the funding of this study through the Alabama Space Grant Consortium and NASA's Graduate Student Research Program. The authors would like to thank Miguel Angel Aloy for his generosity in sharing his simulation results and his comments.

REFERENCES

1. Hjorth, J., et al., *Nature*, **423**, 847 (2003).
2. Price, P. A., et al., *Nature*, **423**, 844 (2003).
3. Uemura, M., et al., *Nature*, **423**, 843 (2003).
4. Woosley, S. E., *Astrophysical Journal*, **405**, 273 (1993).
5. MacFadyen, A. I., and Woosley, S. E., *Astrophysical Journal*, **524**, 262 (1999).
6. Aloy, M. A., et al., *Astrophysical Journal*, **531**, L119 (2000).
7. Hardee, P. E., Clarke, D. A., and Rosen, A., *Astrophysical Journal*, **485**, 533-551 (1997).
8. Aloy, M. A., et al., *Astronomy and Astrophysics*, **396**, 693 (2002).

Large-Scale Magnetic Fields in GRB Outflows: Acceleration, Collimation, and Neutron Decoupling

Nektarios Vlahakis* and Arieh Königl[†]

*Section of Astrophysics, Astronomy & Mechanics, Dept. of Physics, University of Athens, Greece
[†]Dept. of Astronomy & Astrophysics and Enrico Fermi Institute, University of Chicago

Abstract. Using ideal magnetohydrodynamics we examine an outflow from a disk surrounding a stellar-mass compact object. We demonstrate that the magnetic acceleration is efficient ($\gtrsim 50\%$ of the magnetic energy can be transformed into kinetic energy of $\gamma > 10^2$ baryons) and also that the jet becomes collimated to very small opening angles. Observational implications, focusing on the case of an initially neutron-rich outflow, are discussed in Königl's contribution.

IDEAL MAGNETOHYDRODYNAMICS

There is growing evidence in favor of magnetic driving in outflows associated with gamma-ray burst (GRB) sources [e.g., 1, see also Königl's contribution in these Proceedings]. The dynamics of these outflows may be described to zeroth order by the ideal, axisymmetric, hydromagnetic equations, consisting of the Maxwell and momentum equations together with the conservations of baryonic mass and specific entropy. [1] demonstrated that, under the assumptions of a quasi-steady poloidal magnetic field and of a highly relativistic poloidal velocity, these equations become effectively time-independent and the motion can be described as a frozen pulse, generalizing the so-called "frozen pulse" approximation already known in purely hydrodynamic models of GRB outflows [2]. Introducing the magnetic flux function A, the arclength along a poloidal streamline ℓ, and the operator ∇_s that acts while keeping $s \equiv ct - \ell$ constant, the momentum equation can be written as (see [1] for details)

$$\gamma \rho_0 (\mathbf{V} \cdot \nabla_s)(\xi \gamma \mathbf{V}) = \frac{(\nabla_s \cdot \mathbf{E})\mathbf{E} + (\nabla_s \times \mathbf{B}) \times \mathbf{B}}{4\pi} - \nabla P. \quad (1)$$

The large-scale electromagnetic field ($\mathbf{E}, \mathbf{B}$), the bulk flow speed ($\mathbf{V}$), and the total ($e^\pm +$ radiation) pressure can be written as functions of A and the rest baryon density ρ_0[1]

$$\mathbf{B} = \frac{\nabla_s A \times \hat{\phi}}{\varpi} + \mathbf{B}_\phi, \quad \mathbf{E} = -\frac{\Omega}{c}\nabla_s A, \quad \mathbf{V} = \frac{A\Omega^2}{4\pi\gamma\rho_0 c^3 \sigma_M}\mathbf{B} + \varpi\Omega\hat{\phi}, \quad P = Q\rho_0^{4/3}. \quad (2)$$

---

[1] (z, ϖ, ϕ), and (r, θ, ϕ) denote cylindrical and spherical coordinates, whereas subscripts p and ϕ denote poloidal and azimuthal components, respectively.

Faraday's law and the conservations of specific entropy and mass imply that the functions Ω, Q, and σ_M are constants of motion, i.e., functions of A. By integrating equation (1) along $\mathbf{V}_p$ and $\hat{\phi}$ one gets two additional constants of motion,

$$\mu c^2 = \xi \gamma c^2 - \frac{c^3 \sigma_M}{A\Omega} \varpi B_\phi, \quad \frac{\mu c^2 x_A^2}{\Omega} = \xi \gamma \varpi V_\phi - \frac{c^3 \sigma_M}{A\Omega^2} \varpi B_\phi, \quad (3)$$

describing the conservation of the ratio (total energy flux)/(mass flux) and of the total specific angular momentum, respectively. The remaining unknowns are the functions $A(r, \theta)$ and $\rho_0(r, \theta)$. The latter is the solution of Bernoulli's equation [which is obtained after substituting all quantities in the identity $\gamma^2 = 1 + \gamma^2 V_p^2/c^2 + \gamma^2 V_\phi^2/c^2$ using eqs. (2) and (3)], whereas the former controls the shape of the poloidal streamlines and is the solution of the highly nonlinear transfield component of the momentum equation (1).

[1] integrated the last two equations under the r self-similarity assumption $A = r^F f(\theta)$, which makes it possible (with $\Omega \propto A^{-1/F}$, $Q \propto A^{(4-2F)/3F}$, and constant F, σ_M, μ, and x_A) to separate the (r, θ) coordinates. The resulting simplified set of ordinary differential equations can be easily integrated. Two types of boundary conditions at the base of the flow were considered: (1) The case of a strong poloidal magnetic field $(B_p \gtrsim B_\phi)$, which corresponds to a trans-Alfvénic outflow (since the azimuthal field dominates asymptotically and the flow becomes super-fast). (2) The case of a strong azimuthal field $(B_p \ll B_\phi)$, which corresponds to a super-Alfvénic flow.

Trans-Alfvénic flows

A representative solution is shown in Figure 1. Looking at panel (a), which shows the acceleration, one can distinguish three different regimes:
1) $\varpi_1 < \varpi < \varpi_6$ is the fireball phase. The specific enthalpy ξ decreases, resulting in increasing $\gamma \propto \varpi$ ($\xi\gamma \approx$ const, a characteristic of hydrodynamic acceleration), while the specific Poynting flux remains constant (the field is force free). The electromagnetic field only guides the flow, with the bulk of the collimation occurring in this regime.
2) $\varpi_6 < \varpi < \varpi_8$ is the magnetic acceleration regime. The fluid is cold ($\xi \approx 1$), but γ continues to increase (roughly as $\gamma \propto \varpi$) due to the decreasing specific Poynting flux.
3) $\varpi = \varpi_8$ is the asymptotic cylindrical regime. The final Lorentz factor is of the order of the final specific Poynting flux, meaning that $\sim 1/2$ of the total energy (which was mostly electromagnetic initially) is transformed into baryonic kinetic energy ($\gamma_\infty \approx \mu/2$).

The solution presented in Figure 1 describes one shell, corresponding to a particular value of s. By specifying the s dependence in the initial conditions one can examine a multiple-shell outflow and the time dependence of the pulse. For example, the s dependence of the (total energy)/mass flux ratio $\mu(s)c^2$ translates into different final Lorentz factors for distinct shells: $\gamma_\infty(s) \approx \mu(s)/2$. In contrast with Michel's solution [4], in which the classical fast magnetosonic point is located at infinity and $\gamma_\infty \approx \mu^{1/3}$, here this point is encountered at a finite height and most of the magnetic acceleration occurs further out, leading asymptotically to $\gamma_\infty(s) \approx \mu(s)/2 \gg \mu(s)^{1/3}$. Thus, not only is the magnetic acceleration highly efficient, but the stronger dependence of $\gamma_\infty(s)$ on the

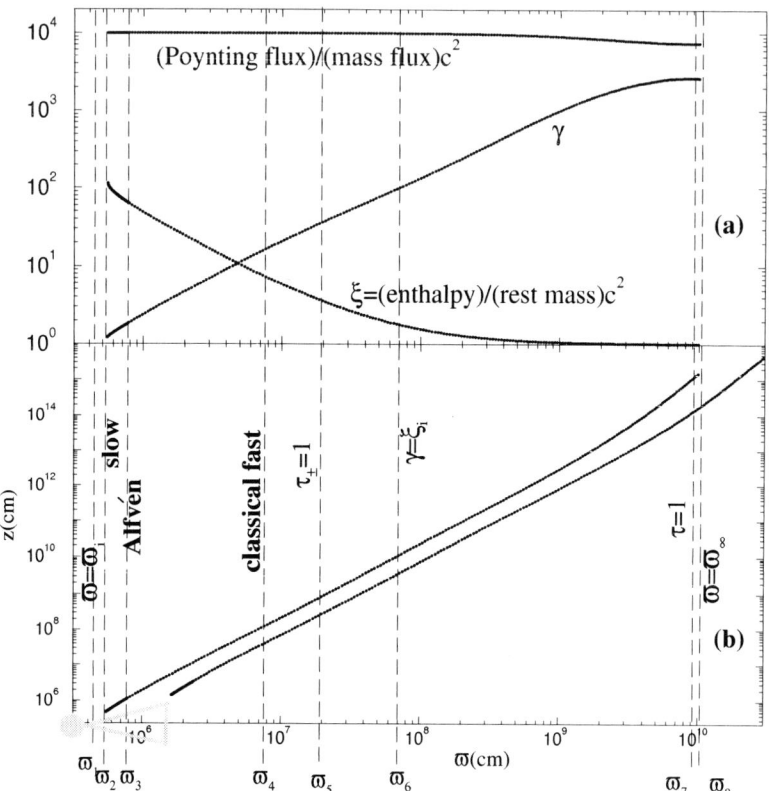

FIGURE 1. Trans-Alfvénic flow solution. (*a*) The Lorentz factor γ, the ratio ξ of the enthalpy to the rest energy, and the ratio of the Poynting flux to the rest-energy flux (*top* curve) are shown as functions of ϖ along the innermost field line. (*b*) The meridional projections of the innermost and outermost field lines are shown on a logarithmic scale, along with a sketch of the central object/disk system. The field lines have a parabolic shape ($z \propto \varpi^2$) for $\varpi \lesssim 10^9$cm and become asymptotically cylindrical. The vertical lines mark the positions of the various transition points along the innermost field line [see text and 3].

initial conditions [through $\mu(s)$] can lead to a larger contrast in γ_∞ between successive shells and hence to a higher efficiency of internal shocks [e.g., 5].

Super-Alfvénic flows

A representative super-Alfvénic solution, corresponding to a base magnetic field ($B_\phi \sim 10^{14}$ G, $B_p \sim 10^{-2} B_\phi$), is shown in Figure 2. The super-Alfvénic solutions are distinguished from the trans-Alfvénic ones in two main respects: (1) During the initial thermal-acceleration phase, some of the internal energy is transformed into electromagnetic energy even as another part is used to increase V_p. (2) During the subsequent magnetic-acceleration phase, the rate of increase of the Lorentz factor with z can be

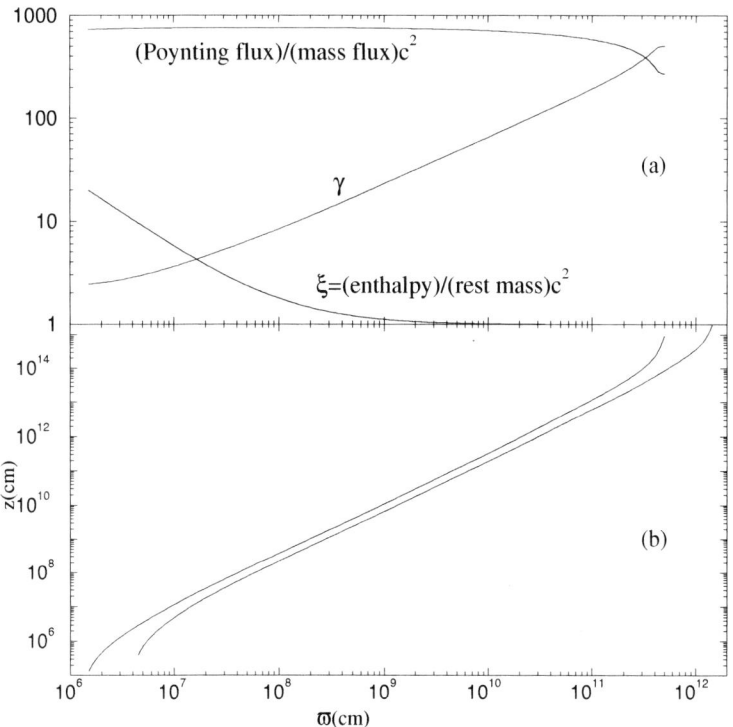

FIGURE 2. Same as Fig. 1, but for a super-Alfvénic solution. Here $\gamma \propto \varpi^{0.46}$, $z \propto \varpi^{1.48}$.

significantly lower than in the trans-Alfvénic case; the rate of increase of the jet radius with z is correspondingly higher. The overall magnetic-to-kinetic energy conversion efficiency is higher. See [6] for further details and analytic scaling relations.

Another potentially important aspect of super-Alfvénic outflows, namely, the possibility that their initial composition is highly neutron-rich, could significantly alleviate the GRB baryon-loading problem. In [7] (see also Königl's contribution) it is shown that, in contrast to the purely hydrodynamic case, the neutrons can decouple at a Lorentz factor that is over an order of magnitude smaller than γ_∞ for the protons.

REFERENCES

1. Vlahakis, N. & Königl, A., ApJ, **596**, 1080 (VK03a) (2003).
2. Piran, T., Shemi, A., & Narayan, R., MNRAS, **263**, 861 (1993).
3. Vlahakis, N., & Königl, A., ApJ, **563**, L129 (2001).
4. Michel, F. C., ApJ, **158**, 727 (1969).
5. Piran, T., Phys. Rep., **314**, 575 (1999).
6. Vlahakis, N. & Königl, A., ApJ, **596**, 1104 (2003).
7. Vlahakis, N., Peng, F., & Königl, A., ApJ, **594**, L23 (2003).

Computational Relativistic Fluids and Jet Formation

G. Richardson*, K.-I. Nishikawa*, S. Koide† and K. Shibata**

National Space Science and Technology Center (NRC/MSFC)
†*Toyama University*
**Kwasan and Hida Observatories, Kyoto University*

Abstract. We present our methodology for numerically solving the fluid equations in general relativistic environments with magnetic fields. Our motivation for the development of such a method is to study the environment around a rotating black hole, specifically the dynamics of the accretion disk and the associated formation of relativistic jets. We present our three-dimensional results confirming previous two-dimensional simulations, which demonstrate the initial stages of jet formation. Without injecting matter, the generated jet evolves into a wind. Consequently, a state change of the jet and accretion disk system is observed.

INTRODUCTION

Relativistic jets have been observed in active galactic nuclei (AGNs), microquasars in our Galaxy, and are believed to originate in regions near accreting (stellar) black holes and neutron stars [1]. To investigate the dynamics of accretion disks and the associated jet formation, we have performed jet formation simulations using a full three-dimensional general relativistic magnetohydrodynamic (MHD) code [2]. The magnetic-acceleration model is a promising model for jet formation [3, 4]. It is believed that the terminal velocity of a jet is comparable to the rotational velocity of the disk at the foot of the jet as seen in nonrelativistic MHD simulations, and that a relativistic jet should be formed near the event horizon [5]. In regions where jet-disk interactions take place near a black hole, the plasma and magnetic fields interact in a complicated manner; therefore three-dimensional general relativistic MHD simulations are required to investigate the dynamics near the horizon.

SIMULATION PARAMETERS AND NUMERICAL METHOD

In the assumed initial state of our simulations, the region is divided into two parts: a background corona surrounding the black hole, and an accretion disk along the horizontal plane of the geometry (see Figure 1a). The coronal plasma is set in a state of transonic free-fall flow (Bondi accretion), with an adiabatic exponent, $\Gamma = 5/3$, specific relativistic enthalpy, $h = 1.3$, and sonic point located at $r = 1.6 r_S$. The Keplerian disk region is located at $r > r_D \equiv 3r_S, |\cos\theta| < \delta$, where $\delta = 1/8$, r_D is the disk radius and r_S the Schwarzschild radius. In this region the density is 100 times that of the background corona, while the orbital velocity, v_ϕ, is relativistic and purely azimuthal: $v_\phi = v_K \equiv c/[2(r/r_S - 1)]^{1/2}$, where v_K is the Keplerian velocity. (Note that this equation reduces to the Newtonian value, $v_\phi = \sqrt{GM/r}$, in the non-relativistic limit $r_S/r \ll 1$). The pressure of both the corona and the disk are assumed to be equal to that of the transonic solution. An initially uniform magnetic field is applied perpendicular to the accretion disk. It is set to the Wald solution [6], which represents the uniform magnetic field around a Schwarzschild black hole; $B_r = B_0 \cos\theta$ and $B_\theta = -\alpha B_0 \sin\theta$ (where α is the lapse function). At the inner edge of the accretion disk, the proper Alfvén velocity is $v_A = 0.015c$ when $B_0 = 0.3\sqrt{\rho_0 c^2}$. The plasma beta of the corona, which is defined as the ratio of gas pressure to magnetic pressure, is: $\beta \equiv p/(B^2/2) = 1.40$ at $r = 3r_S$. The sound speed is calculated from $v_S \equiv c(\Gamma p/h)^{1/2}$.

The 3 + 1 form of the equations used in these simulations for the general relativistic conversion laws governing the plasma and Maxwell equations were derived in [2]. We adapt the Boyer-Lindquist coordinate system and the metric of Schwarzschild space-time. The general-relativistic MHD equations, with the applied initial and boundary conditions, were solved using the simplified total variation diminishing (STVD) method. The STVD method was developed to study violent phenomena such as shocks. This method stems from Lax-Wendoroff's method, with additional diffusion terms. Our simulation domain is, $1.1 r_S \leq r \leq 20 r_S$, $0 \leq \theta \leq \pi$, $0 \leq \phi \leq 2\pi$, with a mesh size of $100 \times 60 \times 120$, The effective linear mesh spacing at $r = 1.1 r_S$ and at $r = 20 r_S$ is $5.38 \times 10^{-3} r_S$ and $0.97 r_S$, respectively. The angular spacing along the polar and azimuthal directions is 5.2×10^{-2} rad. A radiative boundary condition is imposed at $r = 1.1 r_S$ and at $r = 20 r_S$.

SIMULATION RESULTS: JET FORMATION

The numerical results show that matter in the disk loses angular momentum by magnetic braking, then falls into the black hole. The disk falls faster than in the non-relativistic case because of general-relativistic effects that are important below $3 r_S$. A centrifugal shock at $r = 3 r_S$ strongly decelerates the disk. Plasma near the shock is accelerated by the $\mathbf{J} \times \mathbf{B}$ force and forms a two-layered jet structure. Inside this magnetically driven jet, the gradient of gas pressure also generates a jet above the shock region (gas-pressure driven jet).

The evolution of jet formation is shown in Figure 1 where the color gradient shows the proper mass density and the arrows the velocity. The black circle represents the black hole's Schwarzschild radius. Figure 1a presents the initial conditions. Figure 1b ($t = 39.2 \tau_S$) shows a shock forming in the region of $x = 3.5 r_S$. This shock is due to the rapidly infalling material reaching the centrifugal barrier located at $x = 2 r_S$. At this time, the initial signs of jet formation are seen. When the simulation reaches $t = 60.0 \tau_S$ (Figure 1c), the jet structure is clearly seen. The formed jet is identified by high density and pressure (Figure 1c). This phenomena was also seen in previous simulations [7, 8]. In the last frame (Figure 1d) at $t = 128.9 \tau_S$, the accretion disk has thickened and the jet appears to fade into a weak wind that flows near the accretion disk. Matter is not injected into this simulation. A slight asymmetry at the equator is seen and is believed

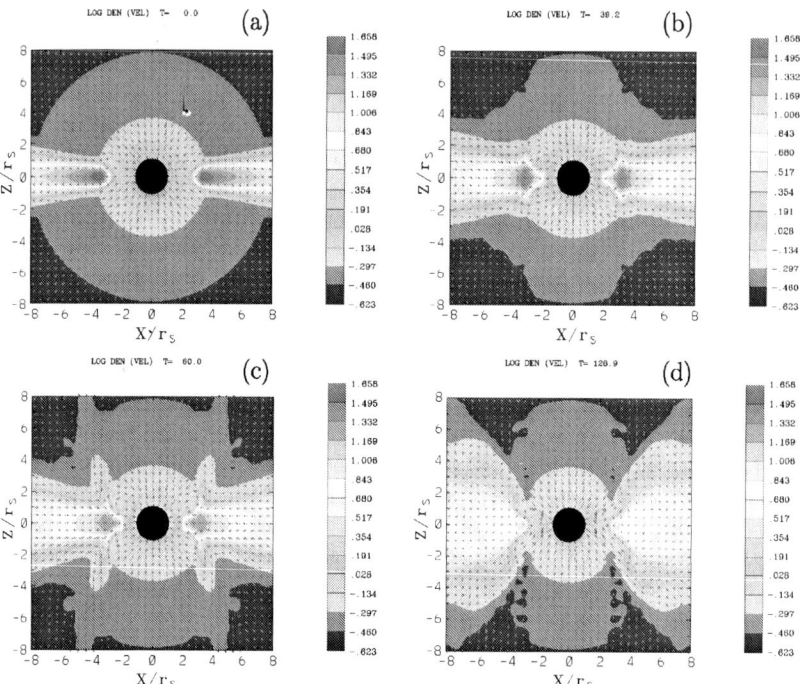

FIGURE 1. The time evolution of the proper mass density with velocity (arrows) in a transonic free-fall corona with an initially uniform magnetic field, at (a) $t = 0.0\tau_S$, (b) $t = 39.2\tau_S$, (c) $t = 60.0\tau_S$, and (d) $t = 128.9\tau_S$. The black hole's Schwarzschild radius is presented by the central black circle in each panel. The jet is fully formed at $t = 60.0\tau_S$ (c). At the later time the wind is formed with a wider angle.

to be numerical and a result of the generated grid. The jet forms in the shape of a hollow cylinder. The pressure in this region is high due to the shock and adiabatic heating at the boundary of the decelerated flow. This high pressure contributes greatly to the jet formation. At the later time the jet is quenched and the pressure decreases, as the shock at $r = 3r_S$ disappears.

The evolution of the jet formation is closely related to the deformed magnetic field. The magnetic field lines, which are initially perpendicular to the accretion disk, are twisted by the accretion disk as the simulation progresses and are pinched by the falling corona and accretion material. Later, the disk drags the magnetic field further in the azimuthal direction, transferring angular momentum outward even as matter falls towards the black hole. The magnetic tension from the pinched field lines increases the magnetic field pressure and generates a jet near the black hole. At $t = 128.9\tau_S$ the magnetic field is relaxed, which is consistent with what is seen in Figure 1d where the jet fades.

SUMMARY AND DISCUSSIONS

We have performed full three-dimensional general relativistic magnetohydrodynamic (GRMHD) simulations of a Schwarzschild black hole with a free-falling corona and thin accretion disk. The initial simulation results show that a bipolar jet (with velocity $\sim 0.3c$) is created as in previous two-dimensional axisymmetric simulations with mirror symmetry at the equator [9, 2]. We show that the jet is initially formed due to the twisting of an initially uniform magnetic field and to shock formation in the region around $r = 3r_S$. At later times, the accretion disk becomes thick and the jet fades resulting in a wind that forms along the disk axis (no streaming from a donor). The wind flows outwards with a wider angle than the initial jet. This evolution of jet-disk coupling suggests that the black hole in a low/hard state with a jet was switched to the high/soft state with a wind. This phenomena will require further investigation for a full understanding [10, 11].

Our current numerical studies include the development of numerical methods that are less dependent on artificial viscosity. Artificial viscosity methods have been shown to produce instabilities at high Lorentz factors. These numerical improvements are important for our goals of modeling jet formation.

ACKNOWLEDGMENTS

This work was partially supported by the NRC Fellowship Program.

REFERENCES

1. Meier, D. L., Koide, S., & Uchida, Y., Science **291**, 84 (2001).
2. Koide, S., Shibata, K., Kudoh, T., ApJ **522**, 727 (1999).
3. Blandford, R. D., & Payne, D. G. MNRAS **199**, 883 (1982).
4. Koide, S., Phys. Rev. D **67**, 104010 (2003).
5. Kudoh, T., Matsumoto, R., & Shibata, K., ApJ **508**, 186 (1998).
6. Wald, R. M., Phys. Rev. D **10**, 1680 (1974).
7. Nishikawa, K.-I., Koide, S., Shibata, K., Kudoh, T., & Sol, H., in Particles and Fields in Radio Galaxies, eds. R. A. Laing & K. B. Blundell, ASPCS **250**, p. 22 (2002).
8. Nishikawa, K.-I., Koide, S., Shibata, K., Kudoh, T., & Sol, H., in New Views on Microquasars, eds. P. Durouchoux, Y. Fuchs, & J. Rodriguez, Center for Space Physics, Kolkata, India, p. 109 (2003).
9. Koide, S., Shibata, K., Kudoh, T., ApJ **495**, L63 (1998).
10. Fender, R., Relativistic flows in Astrophysics, Springer Verlag Lecture Notes in Physics, Ed. A.W. Guthmann, M. Georganopoulos, K. Manolakou and A. Marcowith, **589**, 101 (2002).
11. Fender, R., in Compact Stellar X-Ray Sources', ed. W.H.G. Lewin & M. van der Klis, Cambridge University Press (2003).

Particle Acceleration and Radiation Associated with Magnetic Field Generation from Relativistic Collisionless Shocks

K.-I. Nishikawa*, P. Hardee[†], G. Richardson*, R. Preece**, H. Sol[‡] and G. J. Fishman[§]

*National Space Science and Technology Center, Huntsville, AL 35805
[†]Department of Physics and Astronomy, The University of Alabama, Tuscaloosa, AL 35487
**Department of Physics, University of Alabama in Huntsville, Huntsville, AL 35899 and National Space Science and Technology Center, Huntsville, AL 35805
[‡]LUTH, Observatore de Paris-Meudon, 5 place Jules Jansen 92195 Meudon Cedex, France
[§]NASA-Marshall Space Flight Center, National Space Science and Technology Center, 320 Sparkman Drive, SD 50, Huntsville, AL 35805

Abstract. Shock acceleration is an ubiquitous phenomenon in astrophysical plasmas. Plasma waves and their associated instabilities (e.g., the Buneman instability, two-streaming instability, and the Weibel instability) created in the shocks are responsible for particle (electron, positron, and ion) acceleration. Using a 3-D relativistic electromagnetic particle (REMP) code, we have investigated particle acceleration associated with a relativistic jet front propagating through an ambient plasma with and without initial magnetic fields. We find only small differences in the results between no ambient and weak ambient magnetic fields. Simulations show that the Weibel instability created in the collisionless shock front accelerates particles perpendicular and parallel to the jet propagation direction. The simulation results show that this instability is responsible for generating and amplifying highly nonuniform, small-scale magnetic fields, which contribute to the electron's transverse deflection behind the jet head. The "jitter" radiation from deflected electrons has different properties than synchrotron radiation which is calculated in a uniform magnetic field. This jitter radiation may be important to understanding the complex time evolution and/or spectral structure in gamma-ray bursts, relativistic jets, and supernova remnants.

INTRODUCTION

The most widely known mechanism for the acceleration of particles in astrophysical environments, usually with a power-law spectrum, is Fermi acceleration. This mechanism for particle acceleration relies on the shock jump conditions at relativistic shocks [1]. Most astrophysical shocks are collisionless since dissipation is dominated by wave-particle interactions rather than particle-particle collisions. Diffusive shock acceleration (DSA) relies on repeated scattering of charged particles by magnetic irregularities (Alfvén waves) to confine the particles near the shocks. However, particle acceleration near relativistic shocks is not due to DSA because the propagation of accelerated particles near shocks, in particular ahead of the shock, cannot be described as spatial diffusion. Anisotropies in the angular distribution of the accelerated particles are large, and the diffusion approximation for spatial transport do not apply [2]. Particle-in-cell (PIC) simulations may shed light on the physical mechanism of particle acceleration that involves the complicated dynamics of particles in relativistic shocks [3, 4, 5, 6].

SIMULATION MODEL

The simulations were performed using a $85 \times 85 \times 160$ grid with range of 55 to 85 million particles (27 particles/cell/species for the ambient plasma). Both periodic and radiating boundary conditions are used [7]. The ambient electron and ion plasma has a mass ratio $m_i/m_e = 20$. The electron thermal velocity v_e is $0.1c$, where c is the speed of light. The electron skin depth, $\lambda_{ce} = c/\omega_{pe}$, is 4.8Δ, where $\omega_{pe} = (4\pi e^2 n_e/m_e)^{1/2}$ is the electron plasma frequency (Δ is the grid size).

SIMULATION RESULTS

Flat Magnetized jets

The jet density of the flat jet is nearly $0.741 n_b$. The average jet velocity $v_j = 0.9798c$, and the Lorentz factor is 5 (corresponds to 5 MeV). The time step $\omega_{pe}t = 0.026$, the ratio $\omega_{pe}/\Omega_e = 2.89$, the Alfvén speed $v_A = 0.0775c$, and the Alfvén Mach number $M_A = v_j/v_A = 12.65$. The gyroradii of ambient electrons and ions are 1.389Δ, and 6.211Δ, respectively. In this case, the jet makes contact with the ambient plasma at a 2D interface spanning the computational domain. Therefore, the dynamics of the jet head and the propagation of a shock in the downstream region are studied. The Weibel instability is excited and the electron density is perturbed as shown in Fig. 1a. The electrons are

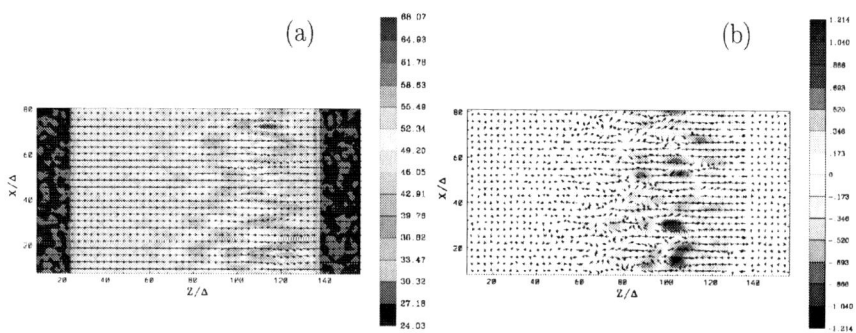

FIGURE 1. The Weibel instability for the flat jet is illustrated in 2D images in the $x-z$ plane ($y = 43\Delta$) in the center of the jet. In (a) the grey scale indicate the electron density with magnetic fields represented by arrows and in (b) the grey scale indicate the y-component of the current density (J_y) with J_z, J_x indicated by the arrows. The Weibel instability perturbs the electron density, leading to nonuniform currents and highly structured magnetic fields.

deflected by the perturbed (small) transverse magnetic fields (B_x, B_y) via the Lorentz force: $-e(\mathbf{v} \times \mathbf{B})$, generating filamented current perturbations (J_z), which enhance the transverse magnetic fields [8, 9]. The complicated current structures due to the Weibel instability are shown in Fig. 1b. The sizes of these structures are nearly the electron skin depth (4.8Δ). This is in good agreement with $\lambda \approx 2^{1/4} c \gamma_{th}^{1/2} / \omega_{pe} \approx 1.188 \lambda_{ce} = 5.7\Delta$ [9]. Here, γ_{th} is a thermal Lorentz factor, and ω_{pe} is the electron plasma frequency. The shapes are elongated along the direction of the jet (the z-direction, horizontal in Fig. 1).

The growth rate of the Weibel instability is calculated to be, $\tau \approx \gamma_{sh}^{1/2} / \omega_{pe} \approx 21.4$ ($\gamma_{sh} = 5$) [9]. This is in good agreement with the simulation results with the jet head located at $z = 136\Delta$. Figure 1 suggests that the "shock" has a thickness from about $z/\Delta = 80 - 130$. Possibly, the "turbulence" assumed for diffusive shock acceleration corresponds to this shock region. The width of the jet head is nearly the electron skin depth (4.8Δ). The size of perturbations along the jet around $z = 120\Delta$ is nearly twice the electron skin depth. This result is consistent with the previous simulations by [4]. The Weibel instability creates elongated shell-type structures which are also shown in counter-streaming jet simulations [3, 5, 6]. The size of these structures transverse to the jet propagation is nearly the electron skin depth (4.8Δ). Note that the size of the perturbations grows larger (see Fig. 1) behind the jet front as smaller scale perturbations merge to larger sizes in the nonlinear stage at the maximum amplitudes [4, 5, 6].

SUMMARY AND DISCUSSIONS

We have performed the first self-consistent, three-dimensional relativistic particle simulations of electron-ion relativistic jets propagating through magnetized and unmagnetized electron-ion ambient plasmas. The Weibel instability is excited in the downstream region behind the jet head, where electron density perturbations and filamented currents are generated. The nonuniform electric field and magnetic field structures slightly decelerate the jet electrons and ions, while accelerating (heating) the jet electrons and ions in the transverse direction, in addition to accelerating the ambient material. The Weibel instability results from the fact that the electrons are deflected by the perturbed (small) transverse magnetic fields (B_x, B_y), and subsequently enhancement of the filamented current is seen [8, 9, 10, 1].

The simulation results show that the initial jet kinetic energy goes to the magnetic fields and transverse acceleration of the jet particles through the Weibel instability. The properties of the synchrotron or "jitter" emission from relativistic shocks are determined by the magnetic field strength, $\mathbf{B}$ and the electron energy distribution behind the shock. The following dimensionless parameters are used to estimate these values; $\varepsilon_B = U_B/e_{th}$ and $\varepsilon_e = U_e/e_{th}$ [9]. Here $U_B = B^2/8\pi$, U_e are the magnetic and electron energy densities, and $e_{th} = nm_i c^2 (\gamma_{th} - 1)$ is the total thermal energy density behind the shock, where m_i is the ion mass, n is the ion number density, and γ_{th} is the mean thermal Lorenz factor of ions. Based on the available diagnostics the following values are estimated; $\varepsilon_B \approx 0.02$ and $\varepsilon_e \approx 0.3$. These estimates are made at the maximum amplitude ($z \approx 112\Delta$).

Our present simulation study has provided the framework of the fundamental dynamics of a relativistic shock generated within a relativistic jet. While some Fermi acceleration may occur at the jet front, the majority of electron acceleration takes place behind the jet front and cannot be characterized as Fermi acceleration. Since the shock dynamics is complex and subtle, further comprehensive study is required for better understanding of the acceleration of electrons and the associated emission as compared with current theory [12]. This further study will provide more insight into basic relativistic collisionless shock characteristics. The fundamental characteristics of such shocks are essential for a proper understanding of the prompt gamma-ray and afterglow emission in gamma-ray bursts, and also to an understanding of the particle reacceleration processes and emission from the shocked regions in relativistic AGN jets.

ACKNOWLEDGMENTS

K.N. is a NRC Senior Research Fellow at NASA Marshall Space Flight Center. This research (K.N.) is partially supported by NSF ATM 9730230, ATM-9870072, ATM-0100997, and INT-9981508. The simulations have been performed on ORIGIN 2000 and IBM p690 (Copper) at NCSA which is supported by NSF.

REFERENCES

1. Gallant, Y. A., Particle Acceleration at Relativistic Shocks, in Relativistic Flows in Astrophysics, eds. A. W. Guthmann, M. Georganopoulos, A. Marcowith, & K. Manolokou, Lecture Notes in Physics, Springer Verlag. astro-ph/0201243 (2003).
2. Achterberg, A., Gallant, Y. A., Kirk, J. G., & Guthmann, A. X., MNRAS **328**, 393 (2001).
3. Nishikawa, K.-I., Hardee, P., Richardson, G., Preece, R., Sol, H., and Fishman, G. J., ApJ **595**, 555 (2003).
4. Silva, L. O., Fonseca, R. A., Tonge, J,.W., Dawson, J. M., Mori, W.B., & Medvedev, M. V., ApJ **596**, L121 (2003).
5. Frederiksen, J. T., Hededal, C. B., Haugbølle, & Nordlund, Å., Proc. From 1st NBSI on Beams and Jets in Gamma Ray Bursts, held at NBIfAFG/NORDITA, Copenhagen, Denmark, August, astro-ph/0303360 (2003).
6. Frederiksen, J. T., Hededal, C. B., Haugbølle, & Nordlund, Å., ApJ, in press (astro-ph/0308104) (2004).
7. Buneman, O., Tristan, in Computer Space Plasma Physics: Simulation Techniques and Software, edited by H. Matsumoto Matsumoto & Y. Omura, p. 67, Terra Scientific Publishing Company, Tokyo (1993).
8. Weibel, E. S., Phys. Rev. Lett. **2**, 83 (1959).
9. Medvedev, M. V. & Loeb, A., ApJ **526**, 697 (1999).
10. Brainerd, J. J., ApJ **538**, 628 (2000).
11. Gruzinov, A., ApJ **563**, L15 (2001).
12. Rossi, E. & Rees, M. J., MNRAS **339**, 881 (2003).

The "Supercritical Pile" Model of GRB: Thresholds, Polarization, Time Lags

Demosthenes Kazanas*, Markos Georganopoulos* and Apostolos Mastichiadis[†]

*NASA/Goddard Space Flight Center, Greenbelt, MD 20771
[†]Dept. of Astronomy, University of Athens, Panepistimiopolis, Athens 15784, Greece

Abstract. The essence of the "Supercritical Pile" model is a process for converting the energy stored in the relativistic protons of a Relativistic Blast Wave (RBW) with Lorentz factor Γ into electron – positron pairs of similar Lorentz factor, while at the same time emitting most of the GRB luminosity at an energy $E_p \simeq 1$ MeV. This is achieved by scattering the synchrotron radiation emitted by the RBW in an upstream located "mirror" and then re-intercepting it by the RBW. The repeated scatterings of radiation between the RBW and the "mirror", along with the threshold of the pair production reaction $p\gamma \to pe^-e^+$, lead to a maximum in the GRB luminosity at an energy $E_p \simeq 1$ MeV, *independent of the value of* Γ. Furthermore, the same threshold implies that the prompt γ–ray emission is only possible for Γ larger than a minimum value, thereby providing a "natural" account for the termination of this stage of the GRB as the RBW slows down. Within this model the γ–ray ($E \sim 100$ keV – 1 MeV) emission process is due to Inverse Compton scattering and it is thus expected to be highly polarized if viewed at angles $\theta \simeq 1/\Gamma$ to the RBW's direction of motion. Finally, the model also predicts lags in the light curves of the lower energy photons with respect to those of higher energy; these are of purely kinematic origin and of magnitude $\Delta t \simeq 10^{-2}$ s, in agreement with observation.

INTRODUCTION

The discovery of GRB afterglows by *BeppoSAX* and the ensuing determination of their redshifts [1], ushered a new era in GRB physics and in our understanding of their time development. While the issue of their distance and energetics was settled by these observations, a number of issues concerning the physics of the RBWs that give rise to the GRB phenomenon still remain open as novel issues have been raised with the advent of observations and accumulation of more GRB of known redshifts [3]. Among the older issues that still remain open is that of the narrow distribution of the GRB peak energy E_p [2], in view of its sensitive dependence on the RBW Lorentz factor ($E_p \propto \Gamma^4$) within the synchrotron model of GRB. Another such issue is that of conversion of the energy stored in relativistic protons in the RBW into electrons. The presence of protons was necessary in the early RBW models [4] for transporting and releasing the energy carried off by the GRB RBW to the distances demanded by observation ($R \sim 10^{16}$ cm). While more recent MHD models [5, 6] are immune from this requirement (though they still need a mechanism for dissipating the magnetic energy), sweeping and accumulating the ISM matter by these MHD flows will store a similar amount of energy into relativistic protons as do more conventional models, still demanding a mechanism for converting this energy to radiation. Models generally resolve this issue by assuming the equipartition of the total energy density between protons and relativistic electrons of arbitrary distributions, as demanded by the need to account for their observed spectral characteristics.

In our view, the most compelling argument in favor our model is that it can answer in a rather straightforward way both the issue of the the limited range in E_p and that of conversion the energy stored into relativistic protons to radiation. At the same time, it has additional implications which seem to be in agreement with the mounting GRB phenomenology; these will be discussed in the following sections.

"SUPERCRITICAL PILE": THE THRESHOLDS

We assume the presence of a population of relativistic protons of form $n(\gamma_p) = n_0 \gamma_p^{-\beta}$ on the frame co-moving with the RBW, that is, moving with Lorentz Factor (LF) Γ with respect to the observer (and also at zero angle to his/her line of sight; see [7] for details). We consider synchrotron photons from $e^- e^+$ pairs of energy γ_e, $\varepsilon_s = b \gamma_e^2$ (b is the value of the magnetic field in units of the critical value $B_c \simeq 4 \cdot 10^{13}$ G; all energies measured in units of the electron mass $m_e c^2$) reflecting off an upstream "mirror" and being re-intercepted by the RBW. Their energy will now be $\varepsilon_s' = \Gamma^2 b \gamma_e^2$. The threshold for $e^- e^+$ pair production of these photons with a proton of Lorentz factor $\gamma_p \simeq \gamma_e b \Gamma^2 \gamma_e^3$ is 2. Assuming that the proton and the electrons are drawn from the relativistic thermal post-shock particle population, $\gamma_e \simeq \gamma_p \Gamma$, leading to the *kinematic threshold* condition

$$b \Gamma_{\text{th}}^5 \simeq 2 \quad \text{or} \quad \Gamma_{\text{th}} \gtrsim \left(\frac{2}{b}\right)^{1/5} \tag{1}$$

Assuming equipartition for the magnetic field for a GRB with total energy $E = 10^{52} E_{52}$ erg, restricted to an angle $\theta \simeq 1/\Gamma$ leads to $\Gamma_{\text{th}} \gtrsim 90 (E_{52}/R_{16})^{-1/12}$.

For this reaction network to be self-sustained, at least one of the reflected synchrotron photons must pair produce with the protons on the RBW (after its reflection by the "mirror") to replace the electron that produced it. Therefore the plasma optical depth τ to the pair producing reaction $p\gamma \to p e^- e^+$ must be at least as large as the inverse of the number of synchrotron photons produced by a given electron. An electron of energy γ produces $\mathcal{N}_\gamma \simeq \gamma/b\gamma^2 = 1/b\gamma$ photons, yielding $\tau = n_{\text{com}} \sigma_{p\gamma} \Delta_{\text{com}} \gtrsim 1/\mathcal{N}_\gamma$, where n_{com}, Δ_{com} are the comoving density and width of the RBW. Considering that the column density is a Lorentz invariant $n_{\text{com}} \Delta_{\text{com}} = nR$ and taking into account the kinematic threshold relation (Eq. 1) the *dynamic threshold* condition reads

$$\sigma_{p\gamma} \Delta_{\text{com}} n_{\text{com}} = \sigma_{p\gamma} R n \gtrsim b\Gamma \quad \text{or} \quad \sigma_{p\gamma} R n \Gamma^4 \gtrsim 2 \ . \tag{2}$$

This latter condition (and the physics behind it) are akin to those of those of a "supercritical" nuclear pile, hence the nomenclature of this model. For the typical values of n and R used in association with GRBs, i.e. $n = 1\, n_0$ cm^{-3} and $R = 10^{16} R_{16}$ cm and considering that $\sigma_{p\gamma} \simeq 5 \cdot 10^{-27}$ cm^2, the criticality condition yields $\Gamma \gtrsim 375 (n_0 R_{16})^{-1/4}$ (Eq. 2 is slightly different from that given in [7]; we would like to thank P. Mészáros for this correction which does not affect the results otherwise).

Assuming the width of the reflecting "mirror" to be thinner than the width of the RBW, blast waves with column densities higher than that implied by Eq. (2) will release the energy stored in relativistic protons explosively on times scales comparable to the RBW light crossing time scale; otherwise, the duration of the burst will be comparable to the time it takes the RBW to cross the width of the "mirror". In this case, prominent emission is halted until the proton column has been built up significantly to conform to the dynamic threshold. For RBW with Γ between those of Eqs. (2) and (1) and assuming the presence of a "mirror" γ-ray emission continues as long as its LF is greater than that implied by the kinematic threshold (Eq. 1). Eventually, when the value of the LF drops below this value, γ-ray emission stops and the RBW enters the stage of afterglow.

THE GRB SPECTRA

Consider a RBW of Lorentz factor Γ. Because of the relativistic focusing of emitted radiation, we need only consider a section of the blast wave of opening half angle $\theta = 1/\Gamma$. The shocked electrons of the ambient medium (and pairs from the $p\gamma \to e^+e^-$ process) produce, as discussed above, synchrotron photons of energy $\varepsilon_s \simeq b\Gamma^2$. These, upon their scattering by the "mirror" and re-interception by the RBW, are boosted to energy $\varepsilon = \varepsilon_s \Gamma^2 = b\Gamma^4$ (in the RBW frame). These photons will then be scattered by the following electron populations: (a) By electrons of $\gamma \simeq 1$, originally contained in the RBW and/or cooled since the explosion. (b) By the hot ($\gamma \simeq \Gamma$), recently shocked electrons to produce inverse Compton (IC) radiation at energies correspondingly $\varepsilon_1 \simeq b\Gamma^4$ and $\varepsilon_2 \simeq b\Gamma^6$ at the RBW frame. (the SSC process will also yield photons at $\varepsilon_{ssc} \simeq b\Gamma^4$, however it turns out that this is not as important and it is ignored here). At the lab frame, the energies of these three components, i.e. $\varepsilon_s, \varepsilon_1, \varepsilon_2$ will be higher by roughly a factor Γ, i.e. they will be respectively at energies $b\Gamma^3$, $b\Gamma^5$ and $b\Gamma^7$. Assuming that the process operates near its kinematic threshold, $b\Gamma^5 \simeq 2$, at the lab frame these components will occur at energies $\varepsilon_s \simeq \Gamma^{-2}$, $\varepsilon_1 \simeq 2 \simeq 1$ MeV and $\varepsilon_2 \simeq \Gamma^2 \simeq 10$ GeV $(\Gamma/100)^2$. This model therefore, produces "naturally" a component in the νF_ν spectral distribution which peaks in the correct energy range. It also predicts the existence of two additional components at an energies $m_e c^2 \Gamma^2$ and $m_e c^2/\Gamma^2$. The high energy emission has been observed from several GRBs [8].

OTHER ISSUES

The simplicity by which this model deals with several outstanding issues of GRB suggest that one should attempt to test its viability by addressing additional GRB systematics and properties.

A. *Time Lags*. It has been observed that, in general, the light curves of soft photons lag with respect to those of the harder ones. A systematic study of these lags in a sample of GRBs with known redshift has determined that these lags range between $0.01 - 0.1$ sec, with their magnitude in inverse correlation to the GRB peak luminosity [9]. The model we presented above can produce lags of the order of magnitude observed: According to the basic premise of our model the emission at $E \sim 50-500$ keV is due to bulk IC scattering of synchrotron photons, by the "cold" electrons of the RBW, after been reflected the "mirror". The eventual energy of these photons depends on the angle of their direction after scattering at the "mirror" with respect to the velocity vector of the RBW (this being highest for a "head-on" collision). Because the distance between the "mirror" and the RBW is of order R/Γ^2, one can easily estimate that the path difference between the soft and hard photons are of the same order of magnitude i.e. $\Delta L \simeq R/\Gamma^2$. Therefore the corresponding time lags (assuming the observer to be along the direction of the velocity of the RBW) should be $\Delta t \simeq \Delta L/c\Gamma^2 \simeq R/c\Gamma^4 \simeq 10^{-2.5} R_{16}/\Gamma_2^4$ sec, in agreement with observations.

B. *Polarization.* The recent results of high $(80 \pm 20\%)$ polarization of GRB 021211 [10] has raised the interest of the community in this particular aspect of GRBs. While models employing synchrotron radiation can produce at best polarization $\lesssim 70\%$ (for totally uniform field geometry), models producing the 100 - 1000 keV radiation by the inverse Compton process can potentially produce polarization approaching 100% for particular orientation of the observer [11, 12, 13]. This is a purely geometric effect: Thomson scattering of unpolarized radiation to an angle $\theta = 90°$ with respect to the incident direction erases all electric field orientations but that perpendicular to the plane defined by the incident and scattering directions leading to 100% polarization in this direction. Since in the lab frame this direction corresponds to an angle $\theta = 1/\Gamma$, the polarization should rise from 0 – 100% in going from $\theta = 0$ to $\theta = 1/\Gamma$ and drop again for larger angles. However, in our model we scatter not unpolarized radiation but the synchrotron radiation from the RBW, which is itself polarized. This leads to a non-zero polarization even for angles close to $\theta = 0°$ thus enhancing the probability that we observe a high polarization signal.

C. *General Considerations* We would like to point out that the arguments presented above have completely ignored the possibility of particle acceleration at the RBW of GRBs (a fundamental requirement of most models). We have dealt with the conversion in pairs of the energy stored only in the thermal population of protons. A non-thermal component will ease the thresholds of Eqs. (1, 2) and allow high energy emission long after the end of the prompt GRB phase, as it appears to be the case in some GRBs [8].

REFERENCES

1. Costa, E. et al., Nature, 387, **783** (1997)
2. Mallozzi, R. S. et al., ApJ, 454, **597** (1995)
3. Lamb, D. Q., these proceedings
4. Rees, M. J. & Mészáros, P., MNRAS, 320, L25 (1992)
5. Vlahakis, N. & Königl, A., ApJ, 563, L129 (2001)
6. Lyutikov, M., these proceedings, also astro-ph/0310040
7. Kazanas, D., Georganopoulos, M. & Mastichiadis, A., ApJ, **587**, L18 (2002)
8. Dingus, B. L., Ap &Sp Sci, **231**, 187 (1995)
9. Norris, J. P., Marani, G. & Bonnell, G. 2002, ApJ, **534**, 248 (2002)
10. Coburn, W., Boggs, S.E., Nature, **423**, 415 (2003)
11. Dar, A., De Rujula, A., astro-ph/0308248 and references therein
12. Lazzati, D., Rossi, E., Ghisellini, G., Rees, M. J., astro-ph/0309038
13. Eichler, D., Levinson, A., astro-ph/0306360

GRB030329

The GRB-SN Connection: GRB 030329 and XRF 030723

J. P. U. Fynbo*, J. Hjorth†, J. Sollerman**, P. Møller‡, J. Gorosabel§, F. Grundahl*, B. L. Jensen†, Michael I. Andersen¶, P. Vreeswijk‖, A. Castro-Tirado§ and the GRACE collaboration††

*Department of Physics and Astronomy, Ny Munkegade, DK-8000 Aarhus C, Denmark
†Astronomical Observatory, Juliane Maries Vej 30, DK–2100 Copenhagen Ø, Denmark
**Stockholm Observatory, Department of Astronomy, AlbaNova, S-106 91 Stockholm, Sweden
‡European Southern Observatory, Karl Schwarzschild-Strasse 2, D-85748 Garching, Germany
§Instituto de Astrofísica de Andalucía, IAA-CSIC, P.O. Box 03004, 18080 Granada, Spain
¶Astrophysikalisches Institut Potsdam, An der Sternwarte 16, D-14482 Potsdam, Germany
‖European Southern Observatory, Casilla 19001, Santiago 19, Chile
††http://zon.wins.uva.nl/ grb/grace/

Abstract. The attempt to secure conclusive, spectroscopic evidence for the GRB/SN connection has been a central theme in most GRB observing time proposals since the discovery of the very unusual GRB 980425 associated with the peculiar type Ib/c SN 1998bw. GRB 030329 provided this evidence to everybody's satisfaction. In this contribution we show the results of a spectroscopic campaign of the supernova associated with GRB 030329 carried out at ESOs Very Large Telescope. We also present preliminary results from a photometric and spectroscopic campaign targeting the X-ray Flash of July 23.

INTRODUCTION

Like GRB 980425, the Gamma Ray Burst (GRB) detected by the HETE-II satellite on March 29, 2003 was born famous. It was so bright in γ-rays that the duty astronomer on the HETE-2 team designated it "monster GRB". The fluence alone places GRB 030329 in the top 0.2% of the 2704 GRBs detected with the Burst And Transient Source Experiment (BATSE) during its nine years of operation. The designation was further justified by the detection of the very bright Optical Afterglow (OA) with an optical magnitude around 12.4 (e.g. [1], [2], [3]). Its redshift was determined with the high resolution UVES spectrograph at the European Southern Observatory (ESO) to be $z = 0.1685$ ([4]), the lowest redshift ever measured for a normal (excluding GRB 980425), long duration GRB. From that point on it was clear to everybody that GRB 030329 offered a unique chance to finally obtain spectroscopic proof of the connection between long duration GRBs and core collapse supernovae (SNe). This connection was believed to exist both from theoretical expectation (e.g. [5], [6] and references therein) and observational hints. The strongest but also most elusive observational hint came from the very unusual GRB 980425 that was associated with the very energetic type Ib/c SN 1998bw ([7], [8]). However, the fact that GRB 980425 had a total equivalent isotropic energy release four orders of magnitude smaller than any other well studied long duration GRB left the reasonable doubt that it possibly was not representative for other GRBs.

Another important set of observations were bumps seen superimposed on the OA lightcurves (e.g. [9], [10]). For GRB 011121 and GRB 021211 there were, in addition to photometric evidence, also tentative but not conclusive spectroscopic evidence for an underlying core collapse SN ([11], [12], [13]). The attempt to secure conclusive, spectroscopic evidence for the GRB/SN connection was therefore a central theme in GRB science, especially from 1999 and onwards.

In this contribution we describe the observations of the afterglow and SN associated with GRB 030329 within the context of GRACE[1]. These results have been published in Hjorth et al. (2003, [18]) and will be more thoroughly described in a paper in preparation. We also briefly review the results obtained by other groups. Finally, we show preliminary results from an extensive GRACE follow-up of the X-Ray Flash (XRF) from July 23 2003. In this case there is photometric evidence for an associated SN.

GRACE OBSERVATIONS OF OA 030329

After the determination of the redshift of GRB 030329, we designed an extensive spectroscopic campaign aimed at detecting and following the supernova expected to be associated with the GRB. Spectra were obtained with the FORS1 or FORS2 spectrographs on six epochs from April 3 through May 1. In the left panel of Fig. 1 we show the six flux-calibrated spectra. As seen, the spectrum evolves from a featureless power-law spectrum to a SN-like spectrum dominated by broad features. The presence of a SN was first reported by Garnavich et al. (2003b, [17]) on April 9 2003 whereby it was designated SN 2003dh. Superimposed on the spectrum are several strong emission lines from the underlying host galaxy (shown in more detail in the right panel of Fig. 1). Shown in the left panel of Fig. 1 with a dashed line is the spectrum of SN 1998bw 33 days after GRB 980425. The similarity between this spectrum and the spectrum of SN 2003dh is striking.

In the left panel of Fig. 2 we show the V-band lightcurve of OA 030329 primarily based on observations from ESO telescopes by GRACE (Guziy et al. 2004, submitted). There is no obvious SN bump seen in the lightcurve.

To investigate the SN component in more detail we performed a spectral decomposition of the afterglow, SN and host galaxy components as follows. While the host galaxy has strong emission lines, its continuum flux upper limit is negligible at early epochs and significantly less than the total flux at the later epochs. The contribution from the host galaxy was therefore accounted for by simply removing the emission lines. Model spectra were constructed as a sum of a power law ($f^{-\beta}$) and a scaled version of one of the SN 1998bw template spectra from Patat et al. ([8]). For each template, or section thereof, a least-squares fit was obtained through fitting of the three parameters: power-law index β, amplitude of afterglow, and amplitude of supernova. In most cases the best fitting index was found to be $\beta = -1.2 \pm 0.05$ which was adopted throughout. We note,

---

[1] Gamma Ray Afterglow Collaboration at ESO, http://zon.wins.uva.nl/~grb/grace/.

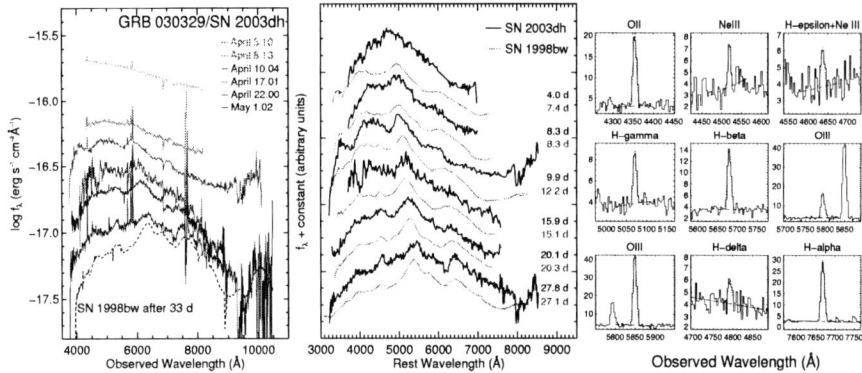

FIGURE 1.

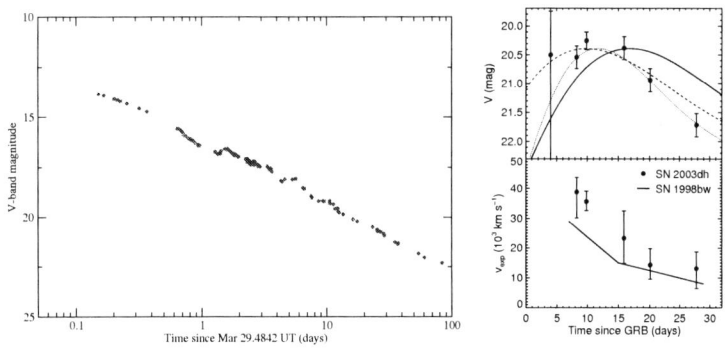

FIGURE 2.

however, that the resulting overall spectral shape of the supernova contribution does not depend on the adopted power-law index or template spectrum. The result of this decomposition is shown in the middle panel of Fig. 1. The striking similarity between the spectra of SN 1998bw and SN 2003dh is clearly seen. The spectral peak wavelength, for both supernovae, is shifting towards the red. The shift is on average 25Å per day for SN 2003dh, which is similar to the evolution of the early spectra of SN 1998bw. The cause of this shift is the growing opacity in the absorption bluewards of 4900Å (rest wavelength).

The spectral decompositions provide the fraction of the total flux in the V-band that is due to the supernova. The resulting SN 2003dh V magnitudes are plotted in the right panel of Fig. 2. The full drawn line in this plot shows the brightness of SN 1998bw as it would have appeared in the V-band at $z = 0.1685$ as a function of time (restframe) since GRB 980425. The dashed line is the same as the solid line, but shifted 7 days earlier. Such an evolution may be expected if the supernova exploded seven days before the GRB. For SN 2003dh, however, this is inconsistent with its spectral evolution (Fig. 2). Dotted line, as for solid line, but evolution speeded up by multiplying time by 0.7.

A faster rise and decay may be expected in asymmetric models in which an oblate supernova is seen pole-on (e.g. Woosley et al. in these proceedings). We assumed 0.20 mag extinction for SN 1998bw (Patat et al., [8]) and none for SN 2003dh. The bottom right plot in Fig. 2 shows the expansion velocities as a function of time in the restframe. Filled circles, SN 2003dh; solid line, SN 1998bw. The SN 2003dh values are our best estimates based on the decomposed spectra (see Hjorth et al., [18], for details). We caution that in some cases these values are very uncertain owing to other features in the spectra around the expected minimum. The solid line shows the trend for SN 1998bw based on the data points in Patat et al. ([8]). The consistent decaying trend in the ejecta velocity of SN 2003dh, together with its very high initial value, indicate that there was no delay between the GRB and the onset of the SN explosion.

Finally we note that the host galaxy is an actively star forming dwarf galaxy with a total luminosity similar to that of the SMC and with a moderate metallicity ([O/H] ≈ -1, [18]). From the OII and Hα emission lines (shown along with the other detected emission lines in the right panel of Figure 1) we infer a star formation rate of a few tenths solar masses per year. In this respect GRB 030329/SN 2003dh is also similar to GRB 980425/SN 1998bw that was hosted by an actively star forming dwarf galaxy of type SBc (Fynbo et al., [15]).

COMPARISON WITH INDEPENDENT STUDIES

A study of SN 2003dh has been presented in the paper by Stanek et al. (2003) and in the thorough and comprehensive paper by Matheson et al. (2003). These authors follow a decomposition strategy very similar to the one described above leading to very similar results for the properties of SN 2003dh. Two spectra during the supernova dominated phase were also secured by Kawabata et al. ([19]). The most significant difference between our and other works is that Matheson et al. ([20]) find a lightcurve for SN 2003dh that is identical to that of SN 1998bw within the erros. This is clearly not the case in the right panel of Figure 2. Bloom et al. ([14]) speculate that there was a significant chromatic slit loss in our April 3 observation due to the fact that we did not observe at the parallactic angle. However, the FORS spectrographs are equipped with atmospheric dispersion correctors (see http://www.eso.org/instruments/fors1/adc.html for details) that should minimize this effect. In a future paper in preparation we will study in more detail if the lightcurve of SN 2003dh is consistent with that of SN 1998bw or not.

XRF 030723

We end this contribution by describing our most recent results from follow-up observations of the X-Ray Flash of July 23, 2003 (Fynbo et al. 2004, [16]). XRF 030723 was located precisely by the SXC detector on HETE-II (Prigozhin et al., [21]). We started observations 4 hrs after the burst and followed the evolution of the lightcurve during two months thereafter. In the left panel of Figure 3 we show the R-band lightcurve from

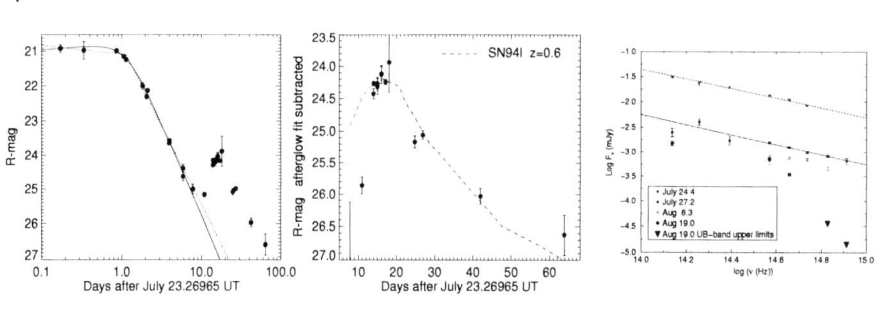

FIGURE 3.

a few hours to about 70 days after the explosion. The decay curve is consistent with being very flat during the first 24 hr after the burst. Around 1 day after the burst the decay slope $dm/d\log t$ steepens to about -2 and it remains so for the following 4-5 days. Around a week after the burst the decay curve starts to deviate from the fast decay and it then quickly rises to a secondary maximum, peaked at around 16 days, followed by a new steep decay. In the middle panel of Figure 3 we show the bump emission with an extrapolation of the afterglow lightcurve subtracted.

Spectroscopic observations were secured on July 26 and on August 8 during the steeply declining afterglow and bump phases respectively. The spectrum of the afterglow from July 26 covers the region from about 3800Å to 8500Å and it shows a featureless continuum with no significant absorption or emission lines. From the lack of Lyα absorption we infer an upper limit to the redshift of about $z = 2.1$. The spectrum taken during the bump on August 8 covers the region 5300Å to 8600Å. It also shows no significant narrow emission or absorption lines.

Multicolor imaging was secured on four epochs. The afterglow spectral energy distribution is well described by a $\beta \approx -1$ power-law over the full range from the U-band to the K-band (right panel of Figure 3). During the bump phase the spectral energy distribution starts to deviate strongly from a power-law shape due to a strong decrease in the flux in the bluest bands.

In a sense the situation for XRF 030723 is opposite to that of GRB 030329. For GRB 030329 the lightcurve did not show a strong bump apparently due to a late break in the afterglow lightcurve that by coincidence balanced out the extra emission from the SN. On the other hand the spectroscopic evidence was unambiguous. For XRF 030723 the photometry shows the most significant late time bump ever detected in an afterglow lightcurve at the time expected for an underlying supernova, but the spectroscopic evidence is more unclear. Nevertheless, it is clear that a SN 1998bw lightcurve is inconsistent with our data at any redshift. So far the best match found is for a SN similar to the type Ic SN 1994I, which had a very early peak time and a rather narrow peak in its lightcurve, at a redshift around $z \approx 0.6$ (Figure 3, middle panel). Interestingly, a SN similar to SN 1994I has also been proposed to be associated to GRB 021211 ([13]). However, even this SN has too slow a rise to match the data.

CONCLUSION

The connection between core collapse SN, more specifically of type Ib/c, and long duration GRBs has been demonstrated to everybody's satisfaction in the case of GRB 030329/SN 2003dh. The results for XRF 030723 show that XRFs are most likely also related to core collapse SN of type Ib/c. However, other potential explanations for the light curve bump should be investigated. A connection between XRFs and SN of type Ib/c would support the hypothesis that XRFs and GRBs are manifestations of the same underlying phenomenon seen either under different viewing angles of with different baryon loadings.

ACKNOWLEDGMENTS

This paper is based on observations collected by the Gamma-Ray Burst Collaboration at ESO (GRACE) at the European Southern Observatory, Paranal, Chile. We acknowledge benefits from collaboration within the EU FP5 Research Training Network "Gamma-Ray Bursts: An Enigma and a Tool". This work was also supported by the Danish Natural Science Research Council (SNF) and by the Carlsberg Foundation.

REFERENCES

1. Price, P., et al., Nature, **423**, 844 (2003)
2. Torii, K., et al., ApJL, **597**, L101 (2003)
3. Sato, R., et al., ApJL, **599**, L9 (2003)
4. Greiner, J. et al., GCN Circ. 2020 (2003a)
5. MacFadyen, A. I. & Woosley, S. E., ApJ, 524, 262 (1999)
6. Dado, S., Dar, A. & De Rújula, A., ApJL, **594**, L89 (2003)
7. Galama, T., et al., Nature, **395**, 670 (1998)
8. Patat, F., et al. ApJ, **555**, 900 (2001)
9. Bloom, J. S., et al., Nature, **401**, 453 (1999)
10. Castro-Tirado, A. J. & Gorosabel, J., A&AS, **138**, 449 (1999)
11. Garnavich, P., et al., ApJ, **582**, 924 (2003a)
12. Greiner, J. et al., ApJ, **599**, 1223 (2003b)
13. Della Valle, M. et al., A&A, **406**, L33 (2003)
14. Bloom, J. S., et al., AJ, submitted (astro-ph/0308034) (2003)
15. Fynbo, J.P.U., et al., ApJL, **542**, L89 (2001)
16. Fynbo, J.P.U., et al., ApJ in press (ApJ preprint doi: 10.1086/421260) (2004)
17. Garnavich, P., et al., IAU Circ. 8114 (2003b)
18. Hjorth, J. et al., Nature, **423**, 847 (2003)
19. Kawabata, K. S., et al., ApJL, **593**, L19 (2003)
20. Matheson, T., et al., ApJ, **599**, 394 (2003)
21. Prigozhin, G., et al., GCN Circ. 2313 (2003)
22. Stanek, K. Z., et al., ApJL, **591**, 17 (2003)

Earliest Detection of the Optical Afterglow of GRB 030329 and its Variability

R. Sato*, N. Kawai*, M. Suzuki*, Y. Yatsu*, J. Kataoka*, R. Takagi*, K. Yanagisawa† and H. Yamaoka**

*Tokyo Institute of Technology, 2-12-1 Ookayama, Meguro-ku, Tokyo 152-8551, Japan.
†Okayama Astrophysical Observatory, Kamogata-cho, Asaguchi-gun, Okayama 712-0232, Japan.
**Kyusyu University, Ropponmatsu, Fukuoka 810-8560, Japan.

Abstract. We report the earliest detection of an extremely bright optical afterglow of the gamma-ray burst GRB 030329 using a 30 cm telescope at the Tokyo Institute of Technology (Tokyo, Japan). Our observation started 67 minutes after the burst and continued for two succeeding nights. Combining our data with those reported in GCN Circulars, we find that the early afterglow light curve of the first half day is described by a broken power-law ($\propto t^{-\alpha}$) function with indices $\alpha_1 = 0.88 \pm 0.01$ (0.047 days $< t < t_{b1}$), $\alpha_2 = 1.18 \pm 0.01$ ($t_{b1} < t < t_{b2}$), and $\alpha_3 = 1.81 \pm 0.04$ ($t_{b2} < t < 1.2$ days), where $t_{b1} \sim 0.26$ days and $t_{b2} \sim 0.54$ days, respectively. The change of the power-law index at the first break at $t \sim 0.26$ days is consistent with that expected from a "cooling break".

INTRODUCTION

GRB 030329 was detected by the HETE-2 satellite on 2003 March 29 at 11:37:14.7 UT. The position was determined by the ground analysis, and the location was reported to the GRB Coordinates Network (GCN) 73 minutes after the burst [1]. A very bright($R \sim$ 13 mag) optical transient (OT) was reported at $\alpha = 10^h44^m50^s.0$, $\delta = +21°31'17.8''$ (J2000.0; [2, 6]) inside the soft X-ray camera (SXC) error circle. This is the brightest GRB ever detected by HETE-2 with a 30–400 keV fluence of 1.2×10^{-4} ergs cm^{-2}, and precise and continuous follow-up observations were carried out by dozens of telescopes located around the world. We report the earliest detection of the optical afterglow of GRB 030329 starting 67 minutes after the burst.

Our observation was performed at the Tokyo Institute of Technology using a 30 cm telescope (Meade LX-200) and an unfiltered CCD camera (Apogee AP6E) equipped with a front-illuminated 1024×1024 CCD chip (Kodak KAF-1001E).

We started observing the preliminary SXC position at 12:44:13 UT on 2003 March 29, 67 minutes (0.047 days) after the burst. [1] The magnitude of GRB afterglow at the very beginning was $R_C \sim 12.4$ mag.

We continued observations for the rest of the night, covering $t \sim 0.05 - 0.30$ days, and we performed observations on the following two nights covering the period of

---

[1] As a HETE-2 Operations graduate student, R. Sato took the initiative and "ran up to the roof to start observing" while the location data were still preliminary.

$t \sim 0.93 - 1.21$ days and $t \sim 2.03 - 2.08$ days, respectively, where t refers to the time since the burst onset.

Our light curve of the GRB 030329 afterglow in the R_C band is shown by combining data from other observations (see Fig 1).

The light curve from the first day cannot be fitted with a single-power-law function. We therefore tried to fit the light curve using two different forms of broken power-law functions. One is given by [4]: $F(t) \propto [(t/t_b)^{\alpha_1' n'} + (t/t_b)^{\alpha_2' n'}]^{-1/n'}$, where t_b is the break time and n' provides a measure of the relative width and the smoothness of the break. The other is a "double-broken power-law" function with two breaks with the following form:

$$F(t) \propto \begin{cases} t^{-\alpha_1} & (t < t_{b1}) \\ [(t/t_{b2})^{\alpha_2 n} + (t/t_{b2})^{\alpha_3 n}]^{-1/n} & (t > t_{b1}) \end{cases} \quad (1)$$

where t_{b1} and t_{b2} are the break times and n provides a measure of the relative width and the smoothness of the break. Here we excluded the "bump" at $t \sim 0.08 - 0.09$ days, which is discussed later.

We found that the former is not acceptable with a reduced χ^2 of 1.72 (285 degrees of freedom [dof]), whereas the latter improves the fit significantly (a reduced χ^2 of 1.06 with 283 dof; see Fig 2). As a result, it is well described by a broken-power-law of the form $\alpha_1 = 0.88 \pm 0.01$ (0.047 days $< t < t_{b1}$), $\alpha_2 = 1.18 \pm 0.01$ ($t_{b1} < t < t_{b2}$), $\alpha_3 = 1.81 \pm 0.04$ ($t_{b2} < t < 1.2$ days), where $t_{b1} \sim 0.26$ days and $t_{b2} \sim 0.54$ days, respectively, and $n = 18.8 \pm 5.1$. Here, α_1 is determined by essentially the full Tokyo Tech data. The parameters α_2 and α_3 are determined by measurements reported by Burenin et al. and the GCN (see the caption to Fig 1).

DISCUSSION

Light curve at $0.05 < t < 0.26$ days

We have presented a light curve of the early phase of the optical afterglow of GRB 030329 starting 67 minutes after the burst.

[7, 5, 8] found that the results of their observations are consistent with the model in which the afterglow emission is generated during the deceleration of the ultra relativistic collimated jet. They found that the break in the power-law light curve, at $t \sim 0.5 - 0.6$ days, can be interpreted as the jet break. Therefore, our major concern is to understand the nature of the first break, t_{b1}, and examine the consistency of the above scenario in the frame work of standard GRB fireball theories ([13]).

Break at $t \sim 0.26$ days

There are two possible break frequencies in the spectra, v_m and v_c. Since v_m and v_c are functions of time, a break in the light curve could be observed when v_c and/or v_m crossed over the observed frequency v_R. Therefore, we examined six possible cases in

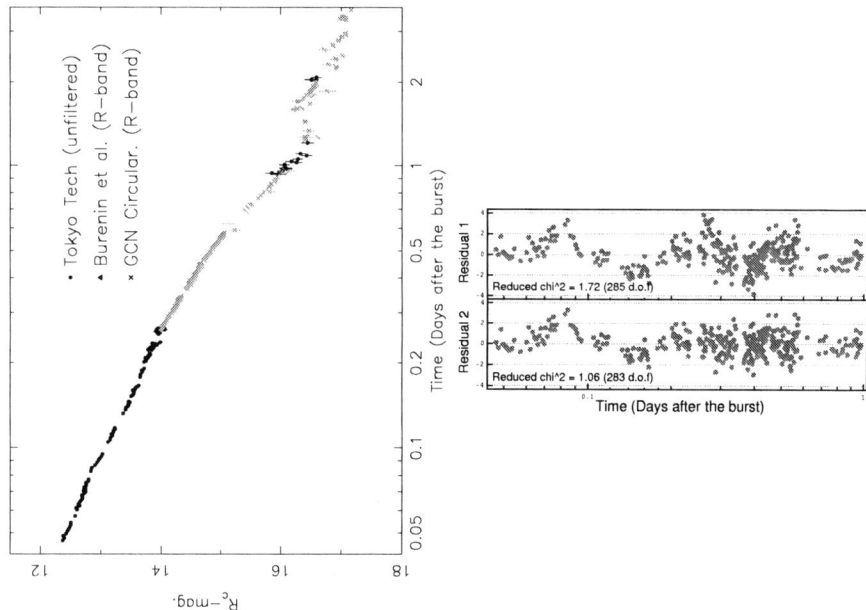

FIGURE 1. *Left*: Light curve of the optical afterglow of GRB 030329. The filled circles are our observations, the filled triangles come from [5], and the rest comes from [16, 17, 18, 19, 20, 21, 22, 23, 24, 25] amd KAIT. These magnitudes were translated using the standard sequence by [26]. *Right*: Residuals of the light curve of the optical afterglow for the two models. (*Top*: Residual from a broken power-law that contains a single, broad break. *Bottom*: Residual from a double-broken power-law (see text).)

order to understand the first break (t_{b1}), according to the relation between v_m, v_c and v_R. We will also extend our discussion to discriminate between "a homogeneous ISM model" [14] and "a pre-existing stellar wind model" [15] for the GRB environment. The relationships between the observed spectral index and model predictions are compared in Table 1.

We first consider the cases 3 and 6 in Table 1. In these two situations, the observed flux at v_R should increase with time, which strongly conflicts with the observed declining light curve. On the contrary, for cases 1 and 4 in Table 1, the predicted optical spectral index would be $\beta = p/2$, where p is the electron spectral index. Since the photon spectral index of this afterglow was $\beta = 0.66$ at $t = 0.25$ days [5], we expect $p = 1.32$, which is unusually flat for an electron population accelerated in a GRB.

Case 5 in Table 1 is also ruled out because the predicted power-law index $\alpha = 0.25$ [10] is too flat. Therefore, we argue that the possible solution is $v_m < v_R < v_c$ for the time region of $t \leq t_{b1}$. In this case, however, if the burst occurred in a pre-existing stellar wind, the optical decay slope is predicted to be $\alpha = 3\beta/2 - \delta/(8 - 2\delta) \sim 1.49$, with $\delta = 2$ for a wind model [9], which is quite steeper than that observed, and hence we can rule out the wind-interaction model. In summary, $v_m < v_R < v_c$, and the ISM model (case 2 in Table 1) is the only possible solution to reproduce both the temporal and spectral index of the optical afterglow of GRB 030329 at 0.05 days $< t <$ 0.26 days.

In such a slow cooling case, the time variation of the afterglow flux is given by $F \propto t^{-3(p-1)/4}$ for $v < v_c$ and $F \propto t^{-(3p-2)/4}$ for $v_c < v$ [10]. By assuming $\alpha_1 = 0.88$, the electron spectral index is estimated as $p = 2.17$. Furthermore, we expect that the power-law slope of the light curve would change from 0.88 to 1.13 for $v_c < v$. Again, this is approximately consistent with the observed spectral index after t_{b1}, where $\alpha_2 = 1.18 \pm 0.01$. Therefore, we conclude that the first break in the optical afterglow light curve at t_{b1} is the most probably cooling break.

Under this assumption, we can determine important physical parameters for the GRB emission. For example, we can estimate ε_B and ε_e, the fractions of the shock energy given to the magnetic field and the electrons at the shock, respectively [10]. In the case of slow cooling, $t_m < t_R < t_c$ would be expected. Since we started our observation 0.047 days after the burst, we can limit the range of t_m as $t_m < t_{obs} = 0.047$ days. For $t_c = t_{b1} = 0.26$ days, $t_m < 0.047$ days, $E = 10^{52}$ ergs, $n = 1$ cm^{-3}, and $v = 0.5 \times 10^{15}$ Hz, we obtain $\varepsilon_B \sim 0.05$ and $\varepsilon_e < 0.20$.

And we also calculated the Lorentz factor which depends on the time and the magnetic field strength using ε_B and the Lorentz factor at two characteristic break times. They are $\gamma = 9.7, B = 0.86$ G at $t_{b1} = 0.26$ days and $\gamma = 7.4, B = 0.64$ G at $t_{b2} = 0.54$ days, respectively.

Bump at $t \sim 0.08 - 0.09$ days

Finally, we comment on a small "bump" of the light curves at $t \sim 0.08 - 0.09$ days (t_{bump}) with an amplitude of ~ 0.1 mag.

Short time variabilities, that is, "bumps and wiggles," may be associated with the forward-/reverse-shock structures along the afterglow-emitting regions [11], the repeated energy injection from the central engine, or the fluctuation in the density of the ISM [12].

First, we can rule out a case with the forward-/reverse-shock structure since it predicts that the light curve should not have the same power-law index before and after the bump. A case with repeated energy injection is also ruled out since after the injection, the light curve after the bump should have the same power-law decay slope, but with a larger normalization. Therefore, we conclude that the bumps in the light curve are likely due to the fluctuation in the external density of the ISM [12].

SUMMARY

We observed an extremely bright optical afterglow of GRB 030329 67 minutes after the burst. Our observational results show that the shocked electrons are in the slow cooling regime with an electron index of 2.17 in this burst, and that the burst occurred in a uniform ISM; that is, GRB 030329 can be understood very well in the predicted "standard" model. We conclude that the first break changes the power-law index by ~ 0.3, consistent with the cooling break in the framework of the standard external shock

TABLE 1. Predicted decay slopes for various theoretical models

Case	Model	Environment	α	Comment
1	$\nu_m < \nu_c < \nu_R$	ISM	0.49	α and β are inconsistent
		Wind	0.49	α and β are inconsistent
2	$\nu_m < \nu_R < \nu_c$	ISM	0.99	OK
		Wind	1.49	α does not fit data
3	$\nu_R < \nu_m < \nu_c$	ISM	-	$\alpha < 0$
		Wind	-	$\alpha < 0$
4	$\nu_c < \nu_m < \nu_R$	ISM	0.49	α and β are inconsistent
		Wind	0.49	α and β are inconsistent
5	$\nu_c < \nu_R < \nu_m$	ISM	0.25	α does not fit data
		Wind	0.25	α does not fit data
6	$\nu_R < \nu_c < \nu_m$	ISM	-	$\alpha < 0$
		Wind	-	$\alpha < 0$

model.

REFERENCES

1. Vanderspek, R. et al., GRB Circ. 1997 (2003).
2. Peterson, B. A. & Price, P. A., GRB Circ. 1985 (2003).
3. Torii, K., GRB Circ. 1986 (2003).
4. Beuermann, K. et al., A&A **352**, L26 (1999).
5. Burenin, R. A. et al., Astronomy Letters **29**, 573 (2003).
6. Price, A., GRB Circ. 2058 (2003).
7. Price, P. A. et al., Nature **423**, 844 (2003).
8. Tiengo, A. et al., A&A **409**, 983 (2003).
9. Panaitescu, A., Meszaros, P., & Rees, M. J., ApJ **503**, 314 (1998).
10. Sari, R., Piran, T., & Narayan, R., ApJ **497**, L17 (2003).
11. Kobayashi, S. & Zhang, B., ApJ **582**, L75 (2003).
12. Nakar, E., Piran, T., & Granot, J., New Astronomy **8**, 495 (2003).
13. Piran, T., Phys. Rep. **314**, 575 (2003).
14. Sari, R. & Piran, T., ApJ **520**, 641 (1999).
15. Chevalier, R. A. & Li, Z., ApJ **520**, L29 (1999).
16. Rumyantsev, V. et al., GRB Circ. 2028 (2003).
17. Klose, S., Hoegner, C., & Greiner, J., GRB Circ. 2029 (2003).
18. Lipkin, Y., Ofek, E. O., & Gal-Yam, A., GRB Circ. 2034 (2003).
19. Stanek, K. Z., Martini, P., & Garnavich, P., GRB Circ. 2041 (2003).
20. Pavlenko, E. et al., GRB Circ. 2050 & 2067 (2003).
21. Fitzgerald, J. B. & Orosz, J. A., GRB Circ. 2056 & 2070 (2003).
22. Li, W., Chornock, R., Jha, S., & Filippenko, A. V., GRB Circ. 2064 (2003).
23. Price, A. & Mattei, J., GRB Circ. 2071 (2003).
24. Cantiello, M. et al., GRB Circ. 2074 (2003).
25. Ibrahimov, M. A. et al., GRB Circ. 2077 (2003).
26. Henden, A., GRB Circ. 2023 (2003).

A New Astrophysical "Triptych": GRB030329/SN2003dh/URCA-2

M.G. Bernardini*, C.L. Bianco*, P. Chardonnet[†], F. Fraschetti**, R. Ruffini* and S.-S. Xue*

*ICRA - International Center for Relativistic Astrophysics and Dipartimento di Fisica, Università di Roma "La Sapienza", Piazzale Aldo Moro 5, I-00185 Roma, Italy.
[†]Université de Savoie, LAPTH - LAPP, BP 110, F-74941 Annecy-le-Vieux Cedex, France.
**Università di Trento, Via Sommarive 14, I-38050 Povo (Trento), Italy.

Abstract. We analyze the data of the Gamma-Ray Burst/Supernova GRB030329/SN2003dh system obtained by HETE-2 (gcn [1]), R-XTE (gcn [2]), XMM (Tiengo et al. [3]) and VLT (Hjorth et al. [4]) within our theory (Ruffini et al. [5] and references therein) for GRB030329. By fitting the only three free parameters of the EMBH theory, we obtain the luminosity in fixed energy bands for the prompt emission and the afterglow (see Fig. 1). Since the Gamma-Ray Burst (GRB) analysis is consistent with a spherically symmetric expansion, the energy of GRB030329 is $E = 2.1 \times 10^{52}$ erg, namely $\sim 2 \times 10^3$ times larger than the Supernova energy. We conclude that either the GRB is triggering an induced-supernova event or both the GRB and the Supernova are triggered by the same relativistic process. In no way the GRB can be originated from the supernova. We also evidence that the XMM observations (Tiengo et al. [3]), much like in the system GRB980425/SN1998bw (Ruffini et al. [6], Pian and et al. [7]), are not part of the GRB afterglow, as interpreted in the literature (Tiengo et al. [3]), but are associated to the Supernova phenomenon. A dedicated campaign of observations is needed to confirm the nature of this XMM source as a newly born neutron star cooling by generalized URCA processes.

A distinctive feature of our model, developed in the framework of the three interpretational paradigms (Ruffini et al. [8, 9, 10]), has been the relation between the photon arrival time at the detector t_a^d and the photon emission time t (see Ruffini et al. [5, 9, 11]):

$$t_a^d = (1+z)\left(t - \frac{\int_0^t v(t')\,dt' + r^\star}{c}\cos\vartheta + \frac{r^\star}{c}\right), \quad (1)$$

where $r(t)$, $v(t)$ and $\gamma(t)$ are the radial coordinate, the velocity and the Lorentz gamma factor of the expanding shell, $r^\star = r(t=0)$, ϑ is the angle between the velocity of the emission point of the photon and the line of sight and z is the cosmological redshift of the source.

In contrast with the relation between t_a^d and t used in the literature, which depends on an instantaneous value of the Lorentz γ factor (see e.g. Rees and Mészáros [12], Eq.(30) in Piran [13]], Eq.(2) in van Paradijs et al. [14], Eq.(2) in Mészáros [15]), Eq.(1) contains an integral which is a function of all previous values of the Lorentz gamma factor along the source world-line since the time $t = 0$. Therefore the knowledge of the Equations Of Motion (EOM) of the source is crucial to the evaluation of Eq.(1). In turns all the quantities which are computed using the EQuiTemporal Surfaces (EQTS, Ruffini et al. [5, 11], Bianco and Ruffini [16]) determined from Eq.(1) become themselves very

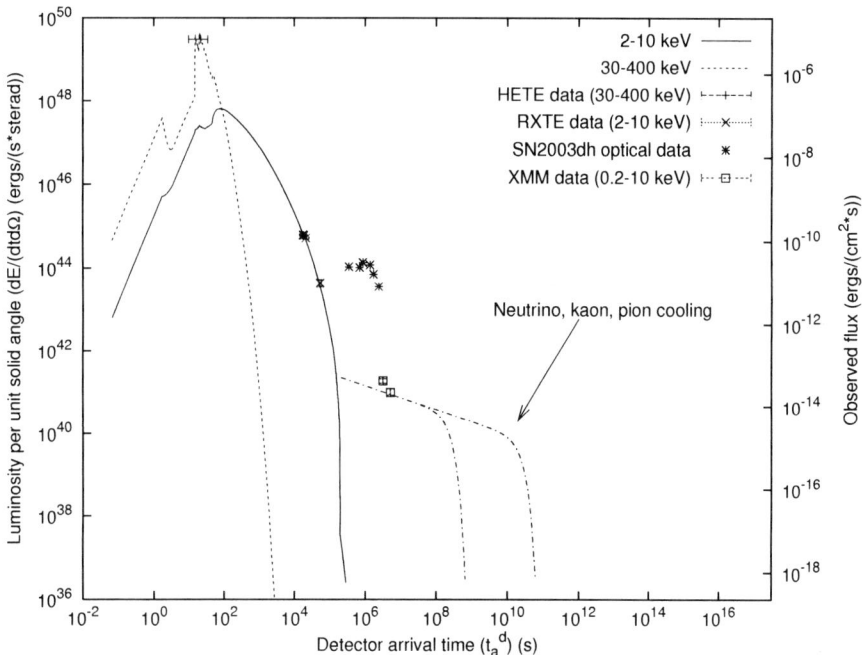

FIGURE 1. The dotted line represents our theoretically predicted GRB030329 light curve in γ-rays (30-400 keV) with the horizontal bar corresponding to the mean peak flux from HETE-2 (gcn [1]). The solid line represents the corresponding one in X-rays (2-10 keV) with the experimental data obtained by R-XTE (gcn [2]). The remaining points refer respectively to the optical VLT data (Hjorth et al. [4]) of SN2003bw and to the X-ray XMM data (Tiengo et al. [3]) of URCA-2. The dash-dotted lines corresponds to cooling theoretical curves of young neutron stars by generalized URCA processes. It is interesting to compare and contrast these results with the ones for GRB980425/SN1998bw (see Fig. 3 in Ruffini et al. [6])

sensitive functions of the EOM. This includes the slope of the afterglow (Ruffini et al. [5]), which is essential in assessing the possible presence of beaming in the source (Ruffini et al. [17]), the luminosity in fixed energy bands and the spectral analysis (Ruffini et al. [18]).

The determination of the EOM leads to a quite complex treatment, which starts from a very special set of initial conditions, proven to be unique. This treatment fits the observed luminosities with a large number of redundancy checks on the EOM (see Fig. 1). It fits as well the time variability in the prompt radiation self-consistently with the determination of the EOM [19].

We have adopted a spherically symmetric distribution for the GRB source and, as initial conditions at $t = 10^{-21}$ s, an e^{+}-e^{-}-photon neutral plasma lying between the radii $r_1 = 2.9 \times 10^6$ cm and $r_2 = 9.0 \times 10^7$ cm. The temperature of such a plasma is 2.1 MeV, the total energy $E_{tot} = 2.1 \times 10^{52}$ erg and the total number of pairs $N_{e^+e^-} = 1.1 \times 10^{57}$. These conditions have been derived evaluating the vacuum polarization processes (Damour and Ruffini [20]) occurring in the dyadosphere of an EMBH (Ruffini

[21], Preparata et al. [22], Cherubini et al. [23], Ruffini and Vitagliano [24, 25], Ruffini et al. [26]). The total energy E_{tot} coincides with the dyadosphere energy E_{dya} which is the first independent parameter of the EMBH theory. The optically thick electron-positron plasma created in the dyadosphere self-propels itself outward reaching ultrarelativistic velocities (Ruffini et al. [27]) and then interacts with the baryonic matter of the remnant of the progenitor star. The baryonic matter component M_B is the second free parameter of the EMBH theory: $B = M_B c^2/E_{dya} = 4.8 \times 10^{-3}$. The $e+$-e^--photon-baryon plasma by further expansion becomes optically thin (Ruffini et al. [28]). As the transparency condition is reached, the Proper-GRB (P-GRB) is emitted with an extremely relativistic shell of Accelerated Baryonic Matter (the ABM pulse) with initial Lorentz gamma factor of $\gamma = 183.6$. It is this ABM pulse which produces the afterglow through its interaction with the ISM, whose average density is best fitted by $<n_{ism}> = 1$ particle/cm^3. In such a collision the "fully radiative condition" is implemented (for details see Ruffini et al. [5]): the internal energy ΔE_{int} which results is instantaneously radiated away.

We have recently assumed that the radiation emitted in the collision between the ABM pulse and the ISM has a thermal spectrum measured in the ABM pulse comoving frame (Ruffini et al. [18]). In our approach the source luminosity is derived from an infinite set of foliations of events on the EQTS, each one characterized by a different thermal spectrum in the comoving frame boosted by a different relativistic transformation obtained from the EOM. The third free parameter of the EMBH theory describes this process of generating the thermal spectrum in the comoving frame. It is given by $1.1 \times 10^{-7} < R = A_{eff}/A_{abm} < 5.0 \times 10^{-11}$, where A_{abm} is the ABM pulse external surface area and A_{eff} is the ABM pulse effective emitting area.

We can then obtain for the GRB030329 the luminosities in given energy bands, computed in the range 2-400 keV with very high accuracy. Fig. 1 shows the results for the luminosities in the 30-400 keV and 2-10 keV bands. Subsequently, the theoretically predicted GRB spectra have been evaluated at selected values of the arrival time [19].

We can now compare these results with those for GRB980425/SN1998bw (Ruffini et al. [6]). We conclude that:

a) The intensity of the GRB versus the Supernova, comparable in the case of GRB980425, becomes 2×10^3 times larger in the case of GRB030329. This crucial fact clearly indicates beyond any doubt the independence of the GRB phenomenon from the Supernova (Ruffini et al. [10]). Moreover, the GRB is generally energetically dominant on the supernova; either the GRB is triggering an induced-supernova event or both the GRB and the Supernova are triggered by the same relativistic process. In no way the GRB can originate from the supernova.

b) In both systems the XMM observations point to the existence of an additional X-ray source, which we consider related to the Supernova phenomenon and not to the GRB. There is the distinct possibility that this source originates from the emission of a newly formed hot neutron star, cooling via generalized URCA processes (Ruffini et al. [6]). It has been recently proposed (Ruffini et al. [29]) to indicate this new physical and astrophysical systems as URCA-1 for GRB980425/SN1998bw and URCA-2 for GRB030329/SN2003dh. A dedicated campaign of observations with XMM is urgently needed in order to explore this unprecedented "triptych" astrophysical systems, formed by a GRB, an induced-supernova and possibly a newly born pulsating hot neutron star.

Details of this results are going to be published in [19].

REFERENCES

1. GCN Circ. 1997 (2003).
2. GCN Circ. 1996 (2003).
3. Tiengo, A., Mereghetti, S., Ghisellini, G., Rossi, E., Ghirlanda, G., and Schartel, N., *A&A*, **409**, 983 (2003).
4. Hjorth, J., Sollerman, J., Møller, P., Fynbo, J. P. U., Woosley, S. E., Kouveliotou, C., Tanvir, N. R., Greiner, J., Andersen, M. I., Castro-Tirado, A. J., Cerón, J. M. C., Fruchter, A. S., Gorosabel, J., Jakobsson, P., Kaper, L., Klose, S., Masetti, N., Pedersen, H., Pedersen, K., Pian, E., Palazzi, E., Rhoads, J. E., Rol, E., van den Heuvel, E. P. J., Vreeswijk, P. M., Watson, D., and Wijers, R. A. M. J., *Nature*, **423**, 847 (2003).
5. Ruffini, R., Bianco, C. L., Chardonnet, P., Fraschetti, F., Vitagliano, L., and Xue, S.-S., "New Perspectives in Physics and Astrophysics from the Theoretical Understanding of Gamma-Ray Bursts," in *COSMOLOGY AND GRAVITATION: X^{th} Brazilian School of Cosmology and Gravitation; 25th Anniversary (1977-2002)*, edited by M. Novello and S. E. P. Bergliaffa, AIP, New York, 2003, vol. 668, p. 16.
6. Ruffini, R., Bernardini, M. G., Bianco, C. L., Chardonnet, P., Fraschetti, F., and Xue, S.-S., "GRB 980425, SN1998bw and the EMBH Model," in *Proceedings of the 34th COSPAR Scientific Assembly*, edited by E. Pian, N. Masetti, and L. Piro, Elsevier, 2003, in press.
7. Pian, E., and et al., ".," in *Proceedings of the 34th COSPAR Scientific Assembly*, edited by E. Pian, N. Masetti, and L. Piro, Elsevier, 2003, in press.
8. Ruffini, R., Bianco, C. L., Chardonnet, P., Fraschetti, F., and Xue, S.-S., *ApJ Lett.*, **555**, L107 (2001).
9. Ruffini, R., Bianco, C. L., Chardonnet, P., Fraschetti, F., and Xue, S.-S., *ApJ Lett.*, **555**, L113 (2001).
10. Ruffini, R., Bianco, C. L., Chardonnet, P., Fraschetti, F., and Xue, S.-S., *ApJ Lett.*, **555**, L117 (2001).
11. Ruffini, R., Bianco, C. L., Chardonnet, P., Fraschetti, F., and Xue, S.-S., *ApJ Lett.*, **581**, L19 (2002).
12. Rees, M. J., and Mészáros, P., *MNRAS*, **258**, 41p (1992).
13. Piran, T., *Phys. Rep.*, **314**, 575–667 (1999).
14. van Paradijs, J., Kouveliotou, C., and Wijers, R. A. M. J., *Ann. Rev. Astron. Astroph.*, **38**, 379 (2000).
15. Mészáros, P., *Ann. Rev. Astron. Astroph.*, **40**, 137 (2002).
16. Bianco, C. L., and Ruffini, R., On the analytic expressions for the equitemporal surfaces in gamma-ray burst afterglows (2004), submitted to A&A.
17. Ruffini, R., Bianco, C. L., Chardonnet, P., Fraschetti, F., and Xue, S.-S., Is there any beaming in GRBs? (2003), submitted to Phys. Rev. Lett.
18. Ruffini, R., Bianco, C. L., Chardonnet, P., Fraschetti, F., Gurzadyan, V. G., and Xue, S.-S., On the instantaneous spectrum of gamma-ray bursts (2004), iJMPD in press.
19. Ruffini, R., Bernardini, M. G., Bianco, C. L., Chardonnet, P., Fraschetti, F., and Xue, S.-S. (2004), in preparation.
20. Damour, T., and Ruffini, R., *Phys. Rev. Lett.*, **35**, 463–466 (1975).
21. Ruffini, R., "Beyond the Critical Mass: The Dyadosphere of Black Holes," in *Black Holes and High Energy Astrophysics, Proceedings of the 49th Yamada Conference*, edited by H. Sato and N. Sugiyama, Universal Ac. Press, Tokyo, 1998.
22. Preparata, G., Ruffini, R., and Xue, S.-S., *A&A*, **338**, L87–L90 (1998).
23. Cherubini, C., Ruffini, R., and Vitagliano, L., *Phys. Lett. B*, **545**, 226 (2002).
24. Ruffini, R., and Vitagliano, L., *Phys. Lett. B*, **545**, 233 (2002).
25. Ruffini, R., and Vitagliano, L., *Int. Journ. Mod. Phys. D*, **12**, 121 (2003).
26. Ruffini, R., Vitagliano, L., and Xue, S.-S., *Phys. Lett. B*, **559**, 12 (2003).
27. Ruffini, R., Salmonson, J. D., Wilson, J. R., and Xue, S.-S., *A&A*, **350**, 334–343 (1999).
28. Ruffini, R., Salmonson, J. D., Wilson, J. R., and Xue, S.-S., *A&A*, **359**, 855 (2000).
29. Ruffini, R., Bernardini, M. G., Bianco, C. L., Chardonnet, P., Fraschetti, F., and Xue, S.-S., ".," in *Proceedings of the Tenth Marcel Grossmann Meeting on General Relativity*, edited by V. G. Gurzadyan, R. T. Jantzen, and R. Ruffini, World Scientific, Singapore, 2003.

The X-ray Afterglow of GRB030329 at Early and Late Times

A. Tiengo[*,†], S. Mereghetti[*], G. Ghisellini[**], E. Rossi[‡], G. Ghirlanda[*] and N. Schartel[§]

[*]*Istituto di Astrofisica Spaziale e Fisica Cosmica – Sezione di Milano "G.Occhialini" (Italy)*
[†]*Università degli Studi di Milano, Dipartimento di Fisica (Italy)*
[**]*INAF-Osservatorio Astronomico di Brera, Merate (Italy)*
[‡]*Institute of Astronomy, Cambridge (UK)*
[§]*XMM-Newton Science Operation Center, ESA, Vilspa (Spain)*

Abstract. Thanks to its extraordinary brightness, the X-ray afterglow of GRB030329 could be studied by *XMM-Newton* up to two months after the prompt Gamma-ray emission. We present the results of two *XMM-Newton* observations performed on May 5 and 29, as well as an analysis of the *Rossi-XTE* data of the early part of the afterglow, discussing in particular the stability of the X-ray spectrum and presenting upper limits on the presence of X-ray emission lines.

INTRODUCTION

GRB030329 is an exceptional Gamma-ray burst for various reasons: it had a very large fluence of $\sim 10^{-4}$ erg cm^{-2} (30–400 keV, [1]), in the top 1% of all observed GRBs; its redshift is z=0.1685 ([2],[3]), which makes it the second nearest GRB; its optical transient was observed at magnitude 13 one hour after the explosion ([4],[5]); it is the first GRB unambiguously associated with a supernova ([6], [7]). The detailed studies in all wavebands made possible by the brightness of this event are yielding an unprecedented understanding on the jet structure, GRB energetics and circumburst environment. In particular, the X-ray afterglow could be studied at late times, with a sensitivity which was not achieved for previous bursts.

The spectral shape and the time evolution of the X-ray afterglow has been already reported and discussed in comparison with preliminary measurements of the optical afterglow by [8].

STABILITY OF THE NON-THERMAL X-RAY SPECTRUM

The first part of the afterglow of GRB030329 was studied with *Rossi-XTE*, which observed it twice in the first 30 hours since the burst explosion. For visibility constraints GRB030329 could not be observed by *XMM-Newton* until May. The first XMM observation was carried out 37 days after the GRB and a second one was done 23 days later.

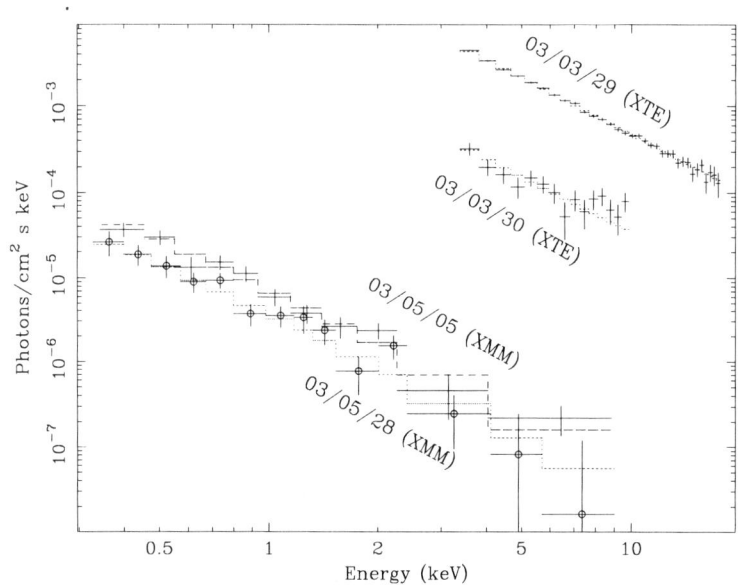

FIGURE 1. The X–ray spectra of the GRB030329 afterglow taken 5 hours, 30 hours (*Rossi-XTE* data), 37 days, and 61 days (*XMM-Newton* data) since the burst explosion. These unfolded spectra are obtained fitting the count spectra with an absorbed power law model with $N_H = 2 \times 10^{20}$ cm^{-2} and $\Gamma = 2.2$.

All the available X–ray spectra of the GRB030329 afterglow are well fitted by an absorbed power law with photon index ~ 2.2 and absorption fixed to the Galactic value in that direction ($N_H = 2 \times 10^{20}$ cm^{-2}). Such a non–thermal model is quite typical for X–ray afterglows, which are usually observed within few days since the GRB explosion. However, it is remarkable that it can fit also the afterglow two months after the GRB, when its flux has already decayed by ~ 4 decades (see Fig.1).

The high quality of the EPIC spectral data allows us to investigate the possible alternative of non–thermal spectra for the late afterglow. We find that a fit to the summed spectra of the two *XMM-Newton* observations with a thermal plasma model (MEKAL) with Solar abundances is not acceptable ($\chi^2/dof = 62.6/30$, see Fig.2). If the abundance is left free to vary, an acceptable fit is obtained (28.5/29) with $kT = 2.4 \pm 0.6$ and a 3σ upper limit on the abundance of $Z < 0.2$.

CONSTRAINTS ON EMISSION LINES

No discrete spectral features have been significantly detected in any of the X–ray spectra of the GRB030329 afterglow.

To derive upper limits on the presence of narrow emission lines in the late X–ray afterglow of GRB030329, we fitted the EPIC MOS and PN spectra of the sum of the two *XMM-Newton* observations with a model consisting of an absorbed (N_H fixed to the Galactic value) power law and a Gaussian emission line with $\sigma = 0$. The line centroid

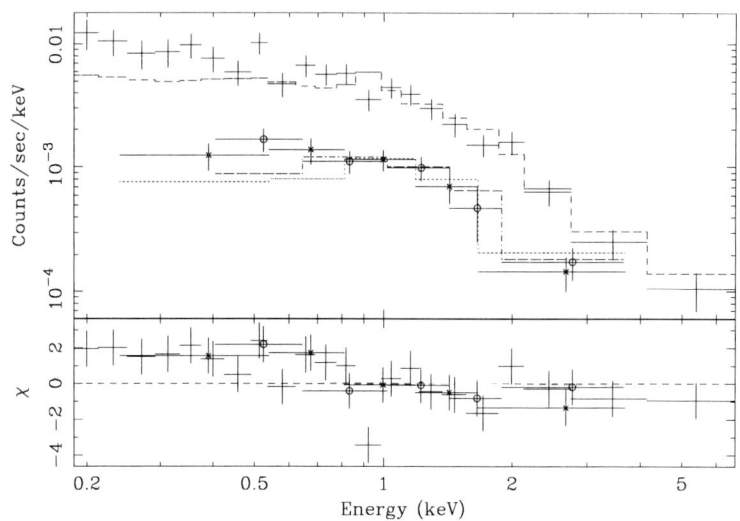

FIGURE 2. EPIC spectrum of the sum of the two *XMM-Newton* observations fitted with a MEKAL model (kT=3.8 keV) with Solar abundances and redshift fixed at z=0.1685 (absorption is fixed at the Galactic value $N_H = 2 \times 10^{20}$ cm^{-2})

was fixed to a grid of values covering the 0.5–6 keV energy band and all the 3σ upper limits on the corresponding normalizations were computed. The maximum values of the corresponding equivalent widths in different energy ranges are reported in Table 1. We consider them a reliable estimate of our sensitivity in detecting narrow emission lines in the EPIC data.

The detection of emission lines in the soft X–ray spectrum has been reported in the afterglows of GRB011211 ([9]), GRB030227 ([10]), GRB020813, and GRB021004 ([11]). In most of these cases the lines could be detected only during short time intervals in the early phases of the afterglow. All the lines were identified with transitions of rest-frame energies between 1.4 and 4.6 keV. For the redshift of GRB030329 their range corresponds to the 1.2–4 keV observed band. The two lines detected with the highest significance in all these cases (Si XIV and S XVI) have energies of 2.22 and 2.77 keV (1.9 and 2.4 keV at z=0.1685) and had equivalent widths smaller than 600 eV. These results can be compared with our upper limits for the afterglow of GRB030329 (Table 1).

Due to the combination of low redshift, small interstellar absorption and high quality spectra, we can put stringent limits to the presence of emission lines with rest–frame energy much lower than in any other GRB afterglow.

On the contrary, no significant information is obtained on the presence of a Fe-K line, which, at z=0.1685, is expected in the 5–6 keV band, where only few photons were collected in the EPIC instrument.

TABLE 1. 3σ upper limits on emission lines

	Equivalent width
0.5–1 keV	<120 eV
1–1.5 keV	<150 eV
1.5–2 keV	<400 eV
2–3 keV	<700 eV
3–5 keV	<2000 eV
5–6 keV	<2800 eV

REFERENCES

1. Ricker, G. R., *IAU Circ.*, **8101** (2003).
2. Greiner, J., Peimbert, M., Estabanet, C., Kaufer, A., Jaunsen, A., Smoke, J., Klose, S., and Reimer, O., *GCN Circ.*, **2020** (2003).
3. Caldwell, N., Garnavich, P., Holland, S., Matheson, T., and Stanek, K. Z., *GCN Circ.*, **2053** (2003).
4. Price, P. A., Fox, D. W., Kulkarni, S. R., Peterson, B. A., Schmidt, B. P., Soderberg, A. M., Yost, S. A., Berger, E., Frail, S. G. D. D. A., Harrison, F. A., Sari, R., Blain, A. W., and Chapman, S. C., *Nature*, **423**, 844 (2003).
5. Torii, K., Kato, T., Yamaoka, H., Kohmura, T., Okamoto, Y., Ohnishi, K., Kadota, K., Yoshida, S., Kinugasa, K., Kohama, M., Oribe, T., and Kawabata, T., *ApJ*, **597**, L101 (2003).
6. Stanek, K. Z., Matheson, T., Garnavich, P. M., Martini, P., Berlind, P., Caldwell, N., Challis, P., Brown, W. R., Schild, R., Krisciunas, K., Calkins, M. L., Lee, J. C., Hathi, N., Jansen, R. A., Windhorst, R., Echevarria, L., Eisenstein, D. J., Pindor, B., Olszewski, E. W., Harding, P., Holland, S. T., and Bersier, D., *ApJ*, **591**, L17 (2003).
7. Hjorth, J., Sollerman, J., Meller, P., Fynbo, J. P. U., Woosley, S. E., Kouveliotou, C., Tanvir, N. R., Greiner, J., Andersen, M. I., Castro-Tirado, A. J., Castro-Cerón, J. M., Fruchter, A. S., Gorosabel, J., Jakobsson, P., Kaper, L., Klose, S., Masetti, N., Pedersen, H., Pedersen, K., Pian, E., Palazzi, E., Rhoads, J. E., Rol, E., van den Heuvel, E. P. J., Vreeswijk, P. M., Watson, D., and Wijers, R. A. M. J., *Nature*, **423**, 847 (2003).
8. Tiengo, A., Mereghetti, S., Ghisellini, G., Rossi, E., Ghirlanda, G., and Schartel, N., *A&A*, **409**, 983 (2003).
9. Reeves, J. N., Watson, D., Osborne, J., Pounds, K. A., O'Brien, P. T., Short, A. D. T., Turner, M. J. L., Watson, M. G., Mason, K. O., Ehle, M., and Schartel, N., *Nature*, **416**, 512 (2002).
10. Watson, D., Reeves, J. N., Hjorth, J., Jakobsson, P., and Pedersen, K., *ApJ*, **595**, L29 (2003).
11. Butler, N. R., Marshall, H. L., Ricker, G. R., Vanderspek, R. K., Ford, P. G., Crew, G. B., Lamb, D. Q., and Jernigan, J. G., *ApJ*, **597**, 1010 (2003).

Structure in Early Afterglow Light Curves: GRB021004 and GRB030329

Makoto Uemura*, Ryoko Ishioka*, Taichi Kato*, Daisaku Nogami[†] and Hitoshi Yamaoka**

*Department of Astronomy, Kyoto University, Sakyou-ku, Kyoto 606-8502, Japan
[†]Hida Observatory, Kyoto University, Kamitakara, Gifu 506-1314, Japan
**Faculty of Science, Kyushu University, Fukuoka 810-8560, Japan

Abstract. Observations by our VSNET team successfully detected early afterglows of GRB021004 and GRB030329. The observational picture of GRB afterglows has been changing from the static fading to the fading with recurring variations thanks to these observations. In GRB021004, we first revealed the afterglow light curve ~ 0.05 day (time after the burst). Our observations revealed the existence of a short plateau phase which lasted for about 2 hours from 0.024 to 0.10 d. In GRB030329, our observation detected complicated structures on the power-law fading even in early times of $\gtrsim 0.05$ d. As well as large bumps (0.3–0.5 mag) observed until 6 d, two sub-bumps appear to be superimposed on an early bump around 0.4 d. These early observations provides new hints to re-construct a model of GRB afterglows: early afterglow emission of $\lesssim 0.1$ d may be affected by the transition through the maximum of forward shock emission. The long-lived, large amplitude variations in GRB030329 indicate that additional energy was supplied for the shock region.

INTRODUCTION

Before the HETE-2 era, it was very difficult for us to receive GRB alerts within a few hours. Because of this, the early behavior of afterglows was poorly studied. GRB990123 is a unique exception. Observations by ROTSE detected a peak in a very early phase and revealed another short-lived and bright component, which is called the "optical flash" [1]. It implies the presence of two components of emission sources, that is, a forward shock region for the ordinary afterglow and a reverse shock region for the optical flash. According to the blast wave model, the peak time of the forward shock emission is predicted at a few tens of minutes after the burst [2]. Observations of early afterglows would hence provide crucial clues for the physical condition of the forward and reverse shock regions.

Owing to prompt localizations by HETE-2, the situation has been improving. The most fantastic examples are GRB021004 and GRB021211, whose positions were reported within a minute after the bursts. In March 2003, furthermore, the "monster", GRB030329, was detected by HETE-2 [3]. The spectral evolution of this burst finally provided evidence for the relationship between GRBs and SNe [4]. Additionally, the above two bright sources, GRB021004 and GRB030329 gave us a unique chance to reveal detailed structure of optical afterglows. Here we report the result of our observation of these objects. The light curves are surprising: they are filled by repeating bumps superimposed on canonical fading trends even in early phases.

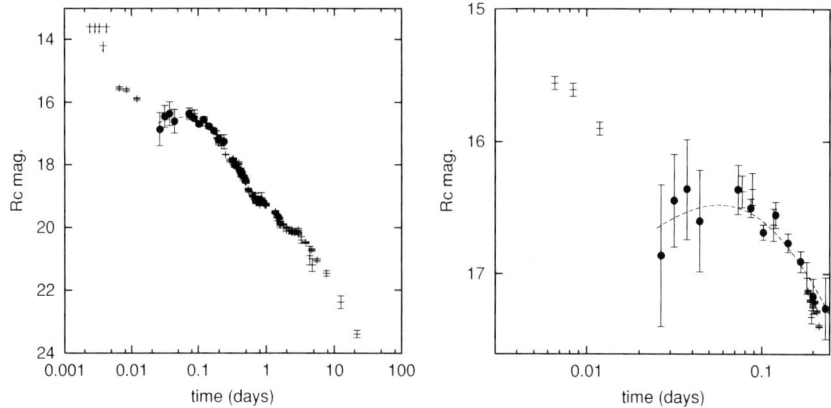

FIGURE 1. Light curves of the optical afterglow of GRB021004. Our observations are indicated by the filled circles. The other symbols are GCN observations (for detailed information, see [5]). The dashed line is the best-fitted model light curve (see the text). Left panel: The whole light curve of the afterglow. Right panel: Enlarged light curve around our observations.

GRB021004

The light curve of the optical afterglow of GRB021004 is shown in Figure 1. Our observations show that, after the initial fading phase, the object experienced a plateau phase, during which it remained at almost constant brightness. The object again drastically changed its fading rate during our observation, and entered an ordinary decay phase. A fading rate after the plateau phase is a standard one among known afterglows [5].

Our observation revealed the presence of a plateau phase preceding the ordinary decay phase. For the nature of this early plateau phase, two types of interpretations have been proposed, that is, i) the natural time evolution of the forward shock emission [6][7][8] and ii) one of the later bumps [9][5].

According to the fireball model, the optical afterglow is predicted to start rapid fading when the observed frequency becomes larger than the typical synchrotron frequency. The time when the typical frequency passes through the optical range has been theoretically estimated to be a few tens of minutes after the burst [2]. We propose that the plateau phase corresponds to a part of such an early evolution phase of the ordinary afterglow from the forward shock. The observed transition time from the early plateau to the decay phase is just what is expected from the theoretical calculation, which favors our scenario. The dashed line shown in Figure 1 is the best-fitted model light curve based on this picture using a smoothly broken power-law model. The transition time was calculated to be 0.10 ± 0.02 d.

Regarding the nature of the plateau phase, an alternative scenario may be possible: that it is one of bumps observed during the later phase [6][7][8] (also see the next sections). The fading trend cannot, however, be described by a simple power-law between the first fading around 0.01 d and the late afterglow after 0.1 d [10].

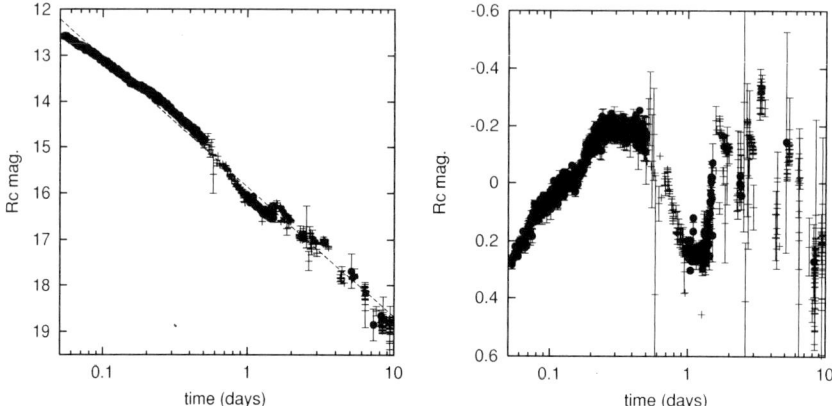

FIGURE 2. Left panel: light curve of the optical afterglow of GRB030329. Symbols are same as Figure1. The dotted line is the best fitted power-law model between 0.05 and 10 days after the burst. Right panel: residual light curve from the power-law decay [11][12].

GRB030329

The light curve of the optical afterglow of GRB030329 is shown in Figure 2. The dotted line is the best fitted power-law model using all points in Figure 2. The light curve exhibits clear and repeating deviations from a canonical fading trend. The residuals from the single power-law model are shown in the right panel of Figure 2. During the first day, the light curve has three breaks at 0.085±0.028, 0.227±0.043, and 0.492±0.029 d. At the end of the first day, the afterglow experienced a rebrightening. After that, observations recorded at least three other rebrightenings. These humps apparently have the same order amplitudes of about 0.3–0.5 mag. According to reports to the GCN, no large color variation was associated with these bumps [11][12].

GRB030329 is the first case in which unambiguous variations were observed before 1 day after the burst. The origin of the modulations can be considered to be a variable density profile of ambient matter or a variable energy in the jet. In the former case, the model predicts the decrease of amplitudes of variations with time [6]. The almost constant amplitudes of bumps in GRB030329 can only be explained with the unrealistic situation that the density increases with radius. On the scenario with the energy variation, two mechanisms have been proposed: one is the patchy shell model, that is, angular variations in the jet. The amplitude of modulations is expected to decrease with time in this model, which however contradicts the observation [6]. The other is the refreshed shock from the collision of two distinct shells. According to this scenario, delayed shells provide additional energy in the shock region, which may be able to explain the constant amplitude of variations in GRB030329 [13]. The slope after bumps, however, contradicts the prediction by the model: in this scenario, the slope after bumps should be constant [6]. On the other hand, observed slopes have different values in the case of GRB030329.

DISCUSSION

Our observations of GRB021004 and GRB030329 revealed that light curves of optical afterglows can be described *not* with a simple power-law fading, but with repeating bumps superimposed on the canonical fading trend. Various scenarios have already been proposed for bumps, and they are summarized as i) the natural time evolution of the afterglow, ii) the variable density of the ambient matter, iii) the patchy shell model, and iv) the refreshed shock model. The first one (i) would appear only in early afterglows. Because it is due to a change of spectral slopes, an apparent bump must be associated with color variations. For the other three models, characteristics expected for observations are summarized in [6].

In the case of the short plateau phase in GRB021004, the information of early color evolution is not available. The most important claim against the scenario (i) is a slow decay during the initial fading phase. Based on the scenario (i), the emission from a reverse shock region must be dominant during this phase. GRB990123 exhibited a much larger fading slope, which is theoretically expected from the reverse shock scenario. On the other hand, the slow fading can be reconciled by the contribution of the forward shock emission [9]. If the scenario (i) is the case, we will detect color variations in other early afterglows.

It is interesting that the break around 0.5 d is believed to be a jet break in GRB030329, while the period of the rapid fading was so short that we cannot find clear evidence for the jet break in Figure 2. The light curve of the X-ray afterglow even supports the jet break at 0.5 d [14], however, the light curve in Figure 2 implies that we need to determine the jet break time of GRB afterglows using multi-wavelengths observations.

This work is partly supported by a grant-in aid from the Japanese Ministry of Education, Culture, Sports, Science and Technology [No. 13640239, 15037205 (TK), 14740131 (HY)]. Part of this work is supported by a Research Fellowship of the Japan Society for the Promotion of Science for Young Scientists (MU and RI).

REFERENCES

1. Kulkarni, S. R. et al., *Nature*, **398**, 389 (1999).
2. Sari, R., and Piran, T., *ApJ*, **520**, 641 (1999).
3. Vanderspek, R. et al., *GRB Circ. Netw.*, **1997** (2003).
4. Hjorth, J. et al., *Nature*, **423**, 847 (2003).
5. Uemura, M., Kato, T., Ishioka, R., and Yamaoka, H., *PASJ*, **55**, L31 (2003).
6. Nakar, E., Piran, T., and Granot, J., *New Astronomy*, **8**, 495 (2003).
7. Li, Z. Y., and Chevalier, R. A., *ApJ*, **589**, L69 (2003).
8. Lazzati, D., Rossi, E., Covino, S., Ghisellini, G., and Malesani, D., *A&A*, **396**, L5 (2002).
9. Kobayashi, S., and Zhang, B., *ApJ*, **582**, L75 (2003).
10. Malesani, D. et al., *GCN*, **1645** (2002).
11. Uemura, M. et al., *Nature*, **423**, 843 (2003).
12. Uemura, M. et al., *PASJ*, **56**, in press (2004).
13. Granot, J., Nakar, E., and Piran, T., <*astro-ph/0304563*> (2003).
14. Tiengo, A. et al., *A&A*, **409**, 983 (2003).

High Resolution Observations of GRB 030329

Greg Taylor*, Dale Frail*, Edo Berger[†] and Shri Kulkarni[†]

*National Radio Astronomy Observatory, Socorro, NM 87801, USA
[†]California Institute of Technology, Pasadena, CA 91125, USA

Abstract. The nearby (z=0.1685) gamma-ray burst of 29 March 2003 has presented us with a unique opportunity to study an event with unprecedented physical resolution. This burst reached flux density levels at centimeter wavelengths more than 50 times brighter than any previously studied event. Here we present the results of VLBI observations that have resolved the radio afterglow, and constrain its proper motion in the sky to <0.3 mas. The size of the afterglow is measured to be ~0.08 mas 24 days after the burst, consistent with expectations of the standard fireball model. In observations taken 51 days after the burst we detect an additional compact, "jet", component at a distance from the main component of 0.28 ± 0.05 mas. The presence of this jet component is not consistent with the standard model.

INTRODUCTION AND RESULTS

GRB 030329, discovered by HETE-2 (GCN 1997), and localized rapidly in the optical bands (GCN 1985) represents a unique opportunity for VLBI observations. At a redshift of $z = 0.1685$ (GCN 2020) this is the nearest cosmological burst detected to date. Observations with the VLA shows the radio afterglow to be the brightest detected so far, with a maximum flux of 55 mJy at 43 GHz one week after the burst [1].

The observations reported on here are based on the first five epochs taken between 3 and 83 days after the burst. All observations employed the VLBA[1]. Other telescopes used in one or more epochs were the Effelsberg 100-m telescope[2] of the MPIfR, the phased VLA, the GBT, the Arecibo telescope (on June 19), and the WSRT tied array (on June 19). The observing runs were typically 5 hours long, with 256 Mbps recording in full polarization with 2 bit sampling. The nearby (1.5°) source J1051+2119 was used for phase-referencing with a 2:1 minute cycle on source:calibrator. The weak calibrator J1048+2115 was observed hourly to check on the quality of the phase referencing.

In the first two epochs (3 and 8 days after the event) the GRB was observed at multiple frequencies within the 5 and 8 GHz bands with the goal of studying the scintillation. The expected scintillation was found indicating a size <0.05 milliarcseconds (mas). A paper combining the flux density measurements from the VLBA and VLA in order to characterize the scintillation and contrain the source size at early times is in preparation.

The third and fourth epochs (at 24 and 51 days after the burst) were carried out at

---

[1] The National Radio Astronomy Observatory is operated by Associated Universities, Inc., under cooperative agreement with the National Science Foundation.
[2] The 100-m telescope at Effelsberg is operated by the Max-Planck-Institut für Radioastronomie in Bonn.

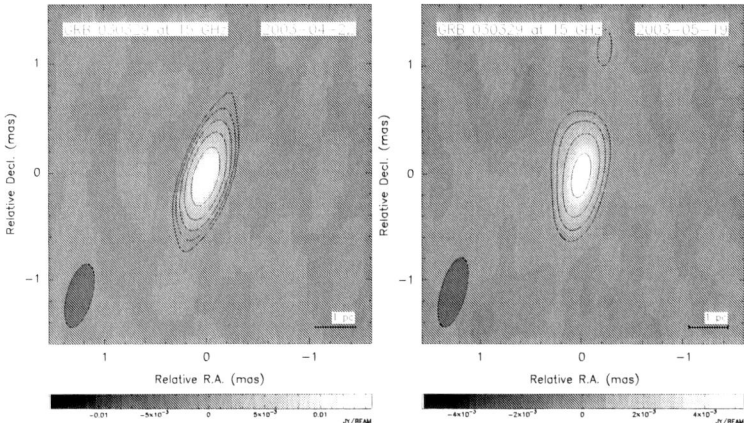

FIGURE 1. Images of GRB 030329 made with the VLBA and Effelsberg telescopes at 15 GHz from 24 and 51 days after the burst respectively. During the May 19 epoch the GBT also participated. Contours are drawn starting at 0.5 mJy/beam and increase by factors of 2. The beam pattern is shown in the lower left corner.

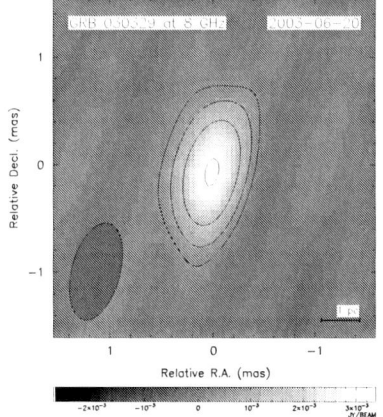

FIGURE 2. An image of GRB 030329 at 8.4 GHz made with the VLBA, Effelsberg, Arecibo, phased VLA, and phased WSRT telescopes taken 83 days after the burst. Contours are drawn starting at 0.4 mJy/beam and increase by factors of 2.

higher frequencies (15 and 22 GHz) in order to attempt to directly resolve the radio afterglow (see Fig. 1). The third epoch was best fit with a resolved Gaussian of size 0.08 mas, although a size of 0 mas cannot be completely ruled out. In the fourth epoch GRB 030329 appears resolved at 15 GHz with a component appearing to the north east at 0.28 ± 0.05 mas from the main component. We tentatively identify this component as a fast jet component. To reach its position would require an average velocity of 19c. This jet component is not detected in the 22 GHz image, though this could be a result of the

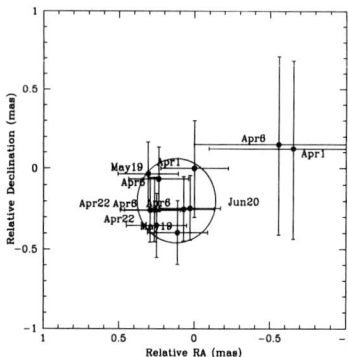

FIGURE 3. The positions derived from the observations to date relative to the first determination on April 1st. Observations at multiple frequencies at a given epoch have been plotted seperately since they are independent measurements. A circle with a radius of 0.3 mas is shown to encompass all current measurements at 8 GHz and above.

reduced sensitivity at 22 GHz and the faintness of the GRB. The reduction in SNR at 22 GHz makes removal of the residual atmospheric errors by self-calibration impossible.

The fifth epoch obtained 83 days after the burst, was carried out at 8.4 GHz on June 19 using an impressive array consisting of the VLBA, GBT, phased VLA, phased WSRT, Effelsberg, and Arecibo. It was hoped that this epoch would confirm the existence of a jet component to the north east, but as shown in Fig. 2, there is only a slight indication of extended emission to the north east. The best fitting Gaussian model has a size of 0.17 ± 0.04 mas. The synthesized beam of the uniformly weighted image shown in Fig. 2 has dimensions 0.92 × 0.47 mas.

The absolute positions derived from all the epochs can be combined in order to constrain the proper motion of the burst. The positions for each frequency and epoch are plotted in Fig. 3.

IMPLICATIONS FOR THE MODELS

The "cannonball" model [4, 3] predicts a motion of 0.05 mas/day, or ∼1 mas between 3 and 24 days after the burst. This motion is clearly excluded by our observations (see Fig. 3), and between 3 and 83 days after the burst the motion is less than 0.004 mas/day (<14c). At the redshift of 0.1685, a motion of 1 mas/year corresponds to 10.9 c. Based on these observations we can rule out the fast cannonball model as currently proposed.

Fireball models of heating by a single relativistic shock front predict that at late times the fireball should look like a ring [5]. Modeling of afterglow light curves in the optical and radio suggest that they are not isotropic events, but are jetted with an initial opening angle of $\theta_0 \sim 0.1$ rad. The size of the fireball depends on the observer's viewing angle, but presumably observers at large angles to the jet could not have seen the burst of gamma-rays. [6] predict an angular size for GRB 030329 that grows linearly with time up to a size of $\sim 0.27(E_{51}/n_0)^{1/3}$ mas some 100 days after the burst and a somewhat slower ($t^{0.4}$) growth after that. VLBI observations at 22 GHz show that GRB 030329 had a size of ∼0.08 mas just 24 days after the burst, roughly consistent with the energies (E_{51}) and exterrnal densities (n_0) derived by [2], but it is not possible to discriminate between a disk and a ring. [6] also predict a shift in the flux centroid from early times by 0.019 mas, but this is well below our current capabilities to measure. In addition to the predictions of the fireball model, in GRB 030329 an extension was detected 50 days after the burst at 0.28 mas from the main component (Fig. 1). This is the first direct evidence for a jet in GRBs, and may indicate that some very fast shocks are produced, though none have been predicted by the standard fireball model.

FUTURE PLANS

Although GRB 030329 has faded considerably, it is still detectable with VLBI techniques. One year after the burst the diameter of the ring is predicted by the standard fireball model to be 3×10^{18} cm or 0.33 mas. This size can be resolved by VLBI observations at 8 GHz. A further epoch is planned for the Global VLBI session in Fall 2003, and another epoch beyond that may be possible if the sensitivity can be improved by an increase in bandwidth.

ACKNOWLEDGMENTS

We are particularly greatful to the schedulers of the VLBA, GBT, Effelsberg, WSRT, and Arecibo telescopes for heroic efforts on behalf of this program.

REFERENCES

1. Berger, E. et al. ApJ **560**, 652 (2001).
2. Berger, E. et al. Nature **426**, 154 (2003).
3. Dado, S., Dar, A., & De Rujula, A., A&A **401**, 243 (2003).
4. Dar, A. et al., GCN 2133 (2003).
5. Granot, J., Piran, T., & Sari, R., ApJ **513**, 679 (1999).
6. Granot, J., & Loeb, A., ApJ **593**, L81 (2003).

The Low-Luminosity Tail of the GRB Distribution: the Case of GRB 980425

Frédéric Daigne* and Robert Mochkovitch*

*Institut d'Astrophysique de Paris, France.

Abstract. The association of GRB 980425 with SN 1998bw at $z = 0.0085$ implies the existence of a population of GRBs with an isotropic-equivalent luminosity which is $\sim 10^4$ times smaller than in the standard cosmological case. We investigate two scenarios to explain them: normal (intrinsically bright) GRBs seen off-axis or intrinsically weak GRBs seen on-axis.

GRB 980425 PROPERTIES

Regarding the properties of its prompt emission[1], GRB 980425 appears as a normal long soft GRB (see Table 1). However if the association with SN 1998bw is correct[2], its redshift is notably lower than all the other measured GRB redshifts. This implies that *GRB 980425 has an unusually weak isotropic-equivalent luminosity.*

TABLE 1. Main observed properties of GRB 980425.

Lightcurve	(single pulse burst)	Spectrum	
Duration	$T_\gamma \simeq 31$ s	Peak energy	$E_p \simeq 68 \pm 40$ keV
Peak flux	$P_\gamma \simeq 2.4 \times 10^{-7}$ erg.cm.$^{-2}$s^{-1}	Low/high-Energy slopes	$\alpha \simeq -0.8 \,/\, \beta \simeq -2.3$
Redshift	$z = 0.008$ ($D_L \simeq 34$ Mpc)	Luminosity	$L_{\gamma,4\pi} \simeq 3 \times 10^{46}$ erg.s^{-1}

PEAK FLUX AND PEAK ENERGY OF A COSMOLOGICAL GRB

We assume that a GRB is produced by relativistic ejecta of Lorentz factor Γ and opening angle $\Delta\theta$ generated by a source at redshift z and observed with a viewing angle θ_0. The intrinsic emission is characterized by the isotropic equivalent luminosity $L_{\rm rad,4\pi}$ and the peak energy $E_{\rm p}$. The observed bolometric peak flux and peak energy are:

– on-axis ($\theta_0 \leq \Delta\theta$):

$$P^{\rm obs} = L_{\rm rad,4\pi}/4\pi D_L^2 \text{ and } E_p^{\rm obs} = E_p/(1+z) \, .$$

– off-axis ($\theta_0 > \Delta\theta$):

$$P^{\rm obs} = \frac{1}{2\left(1 + \Gamma^2 (\theta_0 - \Delta\theta)^2\right)^3} \frac{L_{\rm rad,4\pi}}{4\pi D_L^2} \text{ and } E_p^{\rm obs} = \frac{1}{1 + \Gamma^2 (\theta_0 - \Delta\theta)^2} \frac{E_p}{1+z} \, .$$

Therefore, the peculiar properties of GRB 980425 may have two origins: (i) either it is a normal (intrinsically bright with $L_{\rm rad,4\pi} \gtrsim$ a few 10^{50} erg.s^{-1}) GRB seen off-axis, or (ii) it is an intrinsically weak ($L_{\rm rad,4\pi} \simeq 3 \times 10^{46}$ erg.s^{-1}) GRB seen off-axis.

SCENARIO 1: A NORMAL GRB SEEN OFF-AXIS

Low redshift and large viewing angle: we have indicated in Fig. 1a all *Beppo-SAX* GRBs with a known redshift in an intensity–hardness diagram. *Effect of z:* GRB 980425 is moved from $z = 0.008$ to $z = 1$. The new GRB 980425* is clearly much to weak to be observed. *Effect of θ_0:* GRB 980425* is moved up to $\theta_0 = 0$ (on-axis), assuming that GRB 980425 is seen off-axis with $\theta_0 = \Delta\theta + 4/\Gamma$. The final GRB 980425** is now back in the GRB region. We conclude that *GRB 980425 can be explained as an intrinsically bright GRB observed off-axis, as proposed by Yamazaki et al. [3]*.

Rate of GRB 980425-like events: we assume that GRBs are produced by events characterized by a comoving rate $R(z)$ following the star formation rate, a constant intrinsic luminosity $L_{\rm rad} = 10^{50}$ erg.s$^{-1} = (1-\cos\Delta\theta)L_{\rm rad,4\pi}$ and a constant Lorentz factor $\Gamma = 300$ and that they are detected by a bolometric detector with a threshold $\mathcal{P} = 10^{-8}$ erg.cm.$^{-2}$s^{-1}. The corresponding observed rate $R_{\rm on}(<z)$ (resp. $R_{\rm off}$) of on-axis (resp. off-axis) GRBs up to the redshift z is shown in Fig. 1b-c in two cases: (b) for a distribution of opening angle $p(\Delta\theta) = \sin\Delta\theta$ (large opening angles) and (c) for $p(\Delta\theta) = \delta(\Delta\theta - 2^o)$ (small opening angle). The normalization is $R_{\rm on}(<+\infty) = 1000$ yr^{-1}. This calculation shows that: (i) very low redshift events are extremely rare, just because of a small volume; (ii) at very low redshift, the ratio of the observed rates of on-axis and off-axis GRBs is either ~ 5 (case b) or ~ 1 (case c); (iii) the same ratio for the total rates (up to very large redshifts) is ~ 500 (case b) and ~ 3 (case c). We conclude that *the interpretation of GRB 980425 as an intrinsically bright GRB seen off-axis implies that: (1) either we have observed by chance during the ~ 6 years lifetime of Beppo-SAX a very rare event; (2) or the local GRB rate is much higher than what is assumed gen-*

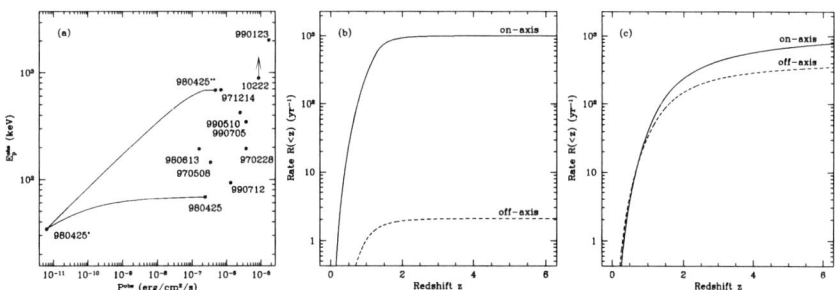

FIGURE 1. GRB 980425 as a normal GRB seen off-axis. Panel (a): location of *Beppo-SAX* GRBs with known redshift in a bolometric peak flux–peak energy diagram. Panels (b) and (c): the expected rate of on-axis and off-axis GRBs in this scenario, assuming a large (panel b) or a small opening angle (panel c).

erally. However in this latter case, we should have already observed a few very bright low redshift GRBs. The proposition[4] that the off-axis events at cosmological distance could explain the recently identified population of XRFs would imply that the total XRF rate is notably smaller than the total GRB rate, which seems to be in contradiction with the data. However, case (c) indicates that these limitations disappear if the distribution of opening angle is clustered around very small values ($\sim 1/\Gamma$).

SCENARIO 2: AN INTRINSICALLY WEAK GRB

Emission of internal shocks: the prompt emission is dominated by the radiation of shock-accelerated electrons. The intrinsic luminosity and peak energy are given by [5] $L_{\text{rad},4\pi} \simeq \alpha_e f_d \dot{E}$ and $E_p = K \dot{E}^x \phi_{xy}(\kappa)/(\bar{\Gamma}^{6x-1}\tau^{2x})$, where $\dot{E}$ is the kinetic energy flux, f_d is the internal shock efficiency, α_e is the fraction of the dissipated energy injected in electrons, $\bar{\Gamma}$ is the mean Lorentz factor, κ is the typical ratio $\Gamma_{\max}/\Gamma_{\min}$ between shells, τ is the variability timescale ($\sim$ observed duration) and K, x and y depend on the dominant radiative process ($\phi_{xy}(\kappa)$ is a steadily increasing function). The standard synchrotron process corresponds to $x = 1/2$ and $y = 5/2$. However smaller values of x and y are necessary to reproduce the hardness-intensity and hardness-fluence correlations observed in GRB pulses[6]. Such values can be expected even for the synchrotron process if the equipartition parameters vary with the shock intensity.

Parameter space: $L_{\text{rad},4\pi}$ and E_p are fixed by 8 parameters: $\dot{E}$, $\bar{\Gamma}$, κ, τ, α_e, K, x and y. The values of α_e, K, x and y depend on the unknown radiative process. We adopt $\alpha_e = 0.5$, $x = y = 0.5$ or $x = y = 0.25$ and we fix K by demanding that a GRB with $\tau = 10$ s, $\kappa = 4$, $\dot{E} = 10^{52}$ erg.s^{-1} and $\bar{\Gamma} = 300$ at $z = 1$ has an observed peak energy $E_p^{\text{obs}} = 200$ keV. The remaining 4 parameters are limited by 3 constraints: (1) the ejecta have to be transparent when the internal shocks appear, (2) the shocked material has

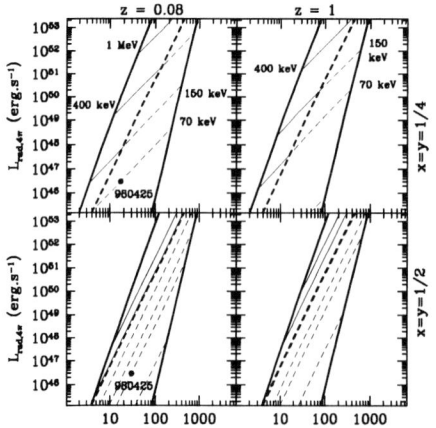

The thick lines show the transparency constraints 1 and 2 on the left side and the external medium constraint 3 on the right side (solid line: uniform medium with $n = 1$ cm^{-3} and dashed line: dense stellar wind with $\rho = A/r^2$ and $A = 5\ 10^{11}$ g.cm^{-1}).
The thin lines show GRBs of constant observed peak energy. In both top panels, the corresponding value of E_p^{obs} is labeled. In both bottom panels, the six lines correspond to $E_p^{\text{obs}} = 10, 70, 150, 400$ and 1000 keV. When GRB 980425 ($L_{\text{rad},4\pi} \simeq 3\ 10^{46}$ erg.s^{-1} and $E_p^{\text{obs}} \simeq 70$ keV) can be reproduced, it is indicated by a big dot.

FIGURE 2. The parameter space of the internal shocks in the Lorentz factor–luminosity plane.

to be transparent to pairs, (3) the internal shocks have to finish before the deceleration radius. In Fig. 2, we have plotted constant E_p^{obs} lines limited by these 3 constraints in a $\bar{\Gamma}$–$L_{rad,4\pi}$ plane for $\tau = 30$ s (closed to the duration of GRB 980425) and $\kappa = 4$.

Results: for a uniform external medium, a large range of luminosities and of peak energies can be obtained and *it is possible to reproduce GRB 980425, which appears as an intrinsically weak and soft GRB with a low injected power in the ejecta ($\dot{E} \sim 3 \times 10^{47}$ erg.s^{-1}) and a low Lorentz factor ($\bar{\Gamma} \sim 20 - 30$). Such parameters could correspond to an event with an unefficient central engine.* In the case of a dense stellar wind, the parameter space is dramatically reduced. For $x = y = 1/2$, it is still possible to produce GRBs with standard luminosity and peak energy but for $x = y = 1/4$ only very hard GRBs can be produced. In both cases, *GRB 980425 is impossible to reproduce*.

Rate of GRB 980425-like events: No "normal" (intrinsically bright) GRB at very low redshift has ever been observed since 1997, whereas GRB 980425 has been observed at $z = 0.008$ in the same period. This shows that the rate of such events in the Universe is much higher than the normal GRB rate. It is only because GRB 980425-like events cannot be observed at large z that the apparent rate is lower. In our scenario, this tells us that: *(1) the parameter space of the GRB-events (especially the distribution of the injected power and the Lorentz factor in the relativistic ejecta) covers a much larger range than only the ultra-relativistic/ultra-energetic region known to produce normal GRBs at cosmological distance; (2) in most cases, the central engine is not able to power such an efficient relativistic ejection, so that GRB 980425-like events are more frequent.* This interpretation is in agreement with the collapsar model: the current estimate of the rate of cosmological GRBs is ~ 1 GRB per 1000 core-collapse SNae, which shows that *most collapses are unable to produce an ultra-relativistic/ultra-energetic ejecta*. Some of them can produce a less energetic ejection, leading to GRB 980425-like GRBs. However the rate of core-collapse SNae within a sphere of 40 Mpc is much higher than the rate of GRB 980425-like events. Then, *a large fraction of collapsing massive stars is even not able to produce a midly relativistic / midly energetic relativistic ejecta. The association of GRB 980425 and GRB 0303029 with a SN Ic suggests that only the collapse of very massive progenitors is able to power a relativistic ejection.*

CONCLUSIONS

We have investigated two scenarios to explain GRB 980425: in the first one, GRB 980425 is a normal (intrinsically bright) GRB seen off-axis[3]. If we exclude the possibility that we observed by chance an extremely rare event, statistical considerations imply that this scenario is possible only under two severe constraints: (1) the local rate of the events responsible for GRBs is much higher than the simple extrapolation of the cosmic rate of the same events; (2) opening angles of the relativistic ejecta leading to GRBs are much narrower than what is usually estimated from the breaks observed in afterglow lightcurves. *The second scenario, where GRB 980425 is an intrinsically low-luminosity GRB seen on-axis, seems to be more realistic, as the parameter space of internal shocks allows such events. The consequences are that (1) in collapsars the central engine in most cases fails to power a highly-relativistic/highly-energetic ejection but can more easily power a midly relativistic/midly-energetic wind; (2) the rate of GRB 980425-like events in the Universe is then much higher than the normal GRB rate but the apparent rate is lower because these events can be detected at low redshift only.*

REFERENCES

1. Frontera, F., Amati, L., Costa, E., Muller, J. M., Pian, E., Piro, L., Soffitta, P., Tavani, M., Castro-Tirado, A., Dal Fiume, D., Feroci, M., Heise, J., Masetti, N., Nicastro, L., Orlandini, M., Palazzi, E., and Sari, R., *ApJS*, **127**, 59–78 (2000).
2. Galama, T. J., Vreeswijk, P. M., van Paradijs, J., Kouveliotou, C., Augusteijn, T., Bohnhardt, H., Brewer, J. P., Doublier, V., Gonzalez, J.-F., Leibundgut, B., Lidman, C., Hainaut, O. R., Patat, F., Heise, J., in 't Zand, J., Hurley, K., Groot, P. J., Strom, R. G., Mazzali, P. A., Iwamoto, K., Nomoto, K., Umeda, H., Nakamura, T., Young, T. R., Suzuki, T., Shigeyama, T., Koshut, T., Kippen, M., Robinson, C., de Wildt, P., Wijers, R. A. M. J., Tanvir, N., Greiner, J., Pian, E., Palazzi, E., Frontera, F., Masetti, N., Nicastro, L., Feroci, M., Costa, E., Piro, L., Peterson, B. A., Tinney, C., Boyle, B., Cannon, R., Stathakis, R., Sadler, E., Begam, M. C., and Ianna, P., *Nat*, **395**, 670–672 (1998).
3. Yamazaki, R., Yonetoku, D., and Nakamura, T., *ApJ*, **594**, L79–L82 (2003).
4. Yamazaki, R., Ioka, K., and Nakamura, T., *ApJ*, **571**, L31–L35 (2002).
5. Daigne, F., and Mochkovitch, R., *MNRAS*, **296**, 275–286 (1998).
6. Daigne, F., and Mochkovitch, R., *MNRAS*, **342**, 587–592 (2003).

GRB 030329 with SARA and TLS

K. Lindsay*, A. Zeh†, D. H. Hartmann*, S. Klose†, S. Shaw**, M. Leake‡,
J. Webb§, B. Stecklum†, M. Williams‡ and E. Howard§

*Clemson University
†Thüringer Landessternwarte Tautenburg, Germany
**University of Georgia
‡Valdosta State University
§Florida International University

Abstract. We present B band observations performed with the SARA 0.9m telescope at KPNO and the Tautenburg 1.34m Schmidt telescope in Tautenburg, Germany, for the optical afterglow of the gamma-ray burst GRB 030329. Observations were carried out from March 30.12 to April 8.29 (UT) with SARA, and from March 30.80 to April 7.00 (UT) with the Tautenburg Schmidt. This burst provides the best covered GRB afterglow light curve to date. The afterglow exhibits a complicated "fine structure" with several episodes of significant re-brightening. These episodes inevitably demand a more detailed numerical fit which introduces a larger number of free parameters. It is difficult to separate the light from the re-brightening phases and the light from the underlying supernova. The photometric data show an average temporal regression in the flux of the optical afterglow, that is typical for GRB afterglow. However, "bumps and wiggles" clearly suggest a degree of complexity resulting from the re-brightening phases that is not easily fitted with standard models. At late time the light curve receives a significant contribution from an underlying supernova component (SN 2002dh). We argue that in cases such as GRB 030329 single site observations are inadequate to properly sample the evolving flux. Consequently, simple broken power law fits of data obtained with just one telescope may lead to rather poor descriptions of the afterglow charcateristics.

OBSERVATIONS

The photometric data for the optical afterglow of GRB 030329 were obtained with the SARA 0.9 m telescope at KPNO, and the TLS 1.34 m Schmidt telescope at the Tautenburg Observatory.

The SARA consortium formed in 1989 with original members Florida Tech, Eastern Tennessee State University, University of Georgia, and Valdosta State University. An association of institutions of higher education in the southeastern United States was formed. The fifth and sixth institutions, Florida International University and Clemson University, joined SARA in 1992 and 1999, respectively. The SARA telescope (Boller & Chivens 0.9 m Cassegrain reflector) was the first major research telescope installed at KPNO. The SARA telescope, CCD canera, and dome are remotely operated [1]. The SARA telescope at KPNO is characterized by: Mirror diameter 0.9 m, Elevation: 2072 m, CCD: Ap7 (512 x 512) pixels, f-number: f/13.5, Field of view: 6' x 6', Limiting magnitude: $\sim$ 20 (V-band, 5 min exposure). Figure 1 shows a SARA snapshot sequence of the decay of the optical afterglow of GRB 030329 in comparison to the DSS image.

The Karl-Schwarzschild-Observatorium, named for astrophysicist Karl Schwarzschild (1873 - 1916), was founded in 1960 as an affiliated institute of the former German

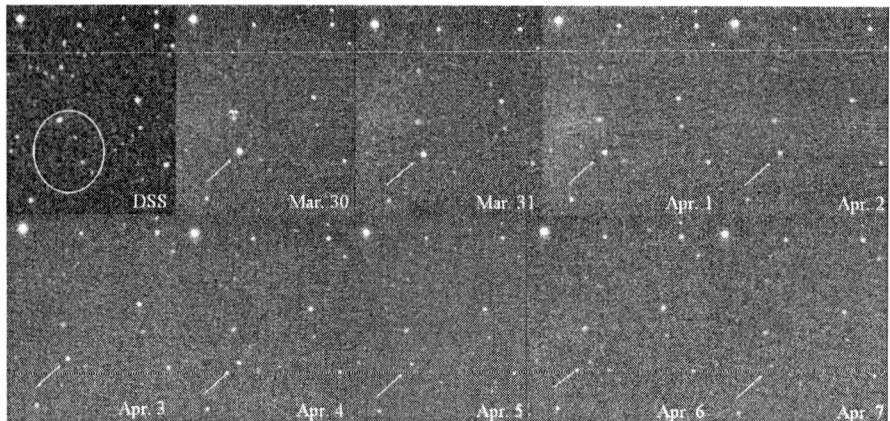

FIGURE 1. (top left-to-right) DSS image of GRB 030329 star field followed by individual 300 second exposure images of the optical afterglow of GRB 030329 taken by the SARA 0.9-m Telescope at KPNO on the evenings of March 30th through April 2nd. (bottom left-to-right) Individual 300 second exposure images of the optical afterglow of GRB 030329 taken by the SARA 0.9-m Telescope at KPNO on the evenings of April 3rd through April 6th, and April 8th.

Academy of Sciences. Since reunification of Germany, it has been renamed as the Thüringer Landessternwarte Karl-Schwarzschild-Observatorium, indicating that it is operated as a State Observatory of the state of Thüringen. The observatory utilizes a ZEISS 2.0 m telescope, capable of three different optical configurations: 1) Schmidt telescope, 2) Cassegrain telescope, and 3) Coudé telescope [2]. The TLS Schmidt telescope is the largest in the world and is characterized by: Mirror diameter: 2.0 m, Corrector plate diameter: 1.34 m, Elevation: 341 m, CCD: 2k x 2k SITe, f-number: f/2.9, Field of view: 36' x 36', Limiting magnitude: 22 (15 minute exposure), and Location: Tautenburg, Germany.

RESULTS & DISCUSSION

We observed the optical afterglow of GRB 030329 with the SARA 0.9m Telescope at KPNO over a period of nine nights between March 30.12 UT and April 8.35 UT, and with the TLS 1.34 m Schmidt Telescope in Tautenburg, Germany over a period of eight nights between March 30.80 and April 7.00. The GRBs proximity to Earth (z = 0.1685) allowed for the afterglow to be detectable in every individual image during every night. All observations described herein were performed in the Johnson-Cousins B-band. All images were dark-subtracted and flat-fielded. Point spread function fitting (with SExtractor) was performed on all data, whether individual or combined images from each night, and the results were calibrated utilizing standards reported by [3]. Based on the magnitude measurements from both telescopes, we constructed the light

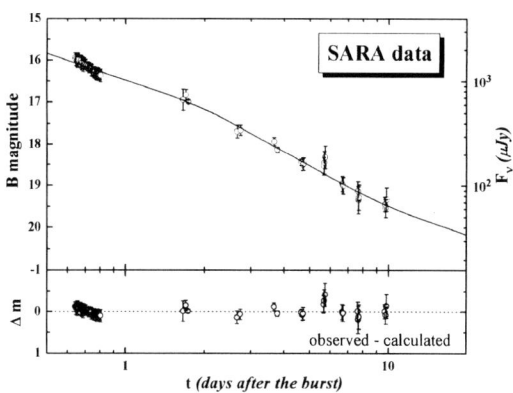

FIGURE 2. Light curve constructed with data collected with the SARA 0.9-m telescope within the first 10 days after the burst.

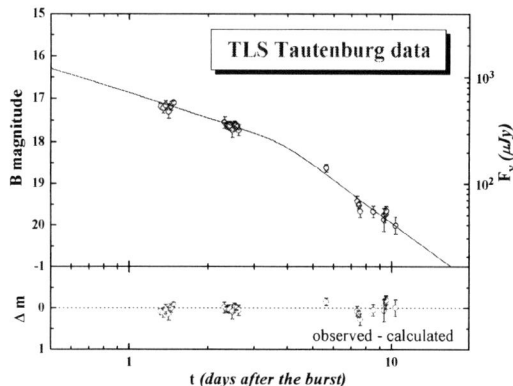

FIGURE 3. Light curve constructed with data collected by the Tautenburg Schmidt 1.34-m telescope in Tautenburg, Germany within the first 10 days after the burst.

curves shown in Figure 1 and 2, in which the afterglow flux is fitted as [4]:

$$F_\nu^{AG}(t) = const.[(t/t_b)^{\alpha_1 n} + (t/t_b)^{\alpha_2 n}]^{-1/n}. \quad (1)$$

The total flux from the OT is the sum of afterglow, host galaxy, and supernova written as,

$$F_\nu^{OT}(t) = F_\nu^{AG}(t) + F_\nu^{host}(t) + kF_\nu^{SN}(t), \quad (2)$$

where k is the scale factor for SN1998bw.

Based on the light curve established from the SARA data, shown in Figure 2, it can be seen that an attempt to interpret a sparsely sampled light curve, is likely to produce misleading results. The fit shown in Figure 2 yields an inaccurate pre-break slope, break time, post-break slope, and provides an unclear picture of the re-brightening phases observed for this afterglow. The same can be said for the light curve established based on the TLS dataset shown in Figure 3. However, based on a comparison made with USNO data, a combined light curve utilizing three B-band data sets establishes a more complete picture for the evolution of the afterglow decay. However, discrepancies remain as even the combined light curve contains gaps, and exhibits measurement inconsistencies due to the use of separate analysis methods. If the characteristics of afterglows are to be fully characterized,it is necessary that essentially gap-free light curves be assembled, uniform analysis be performed, and multi-wavelength data be collected. To improve our understanding of the GRB phenomenon, international coordinated observing campaigns are required to derive consistent data sets with minimal gaps in time, multi-wavelength coverage, and long-term monitoring. This suggests that results from small- and large-aperture telescopes must be combined and coordinated in a global network, to ensure maximal quality of the scientific results. During the Swift era this constitutes a major logistical and resource challenge, as the anticipated high trigger rate could lead to the potential problem of partial coverage (in time, brightness, and wavelength) for many GRBs, depending on the sequence of events. It would be desirable to develop an observing strategy for the global network of follow- up telescopes in which access to resources is optimized. SARA and TLS will be part of this challenge, and we look forward to contributing photometric data to this global effort.

ACKNOWLEDGMENTS

We thank NSF and DAAD for financial support of the joint SARA-TLS project, USNO for B-band comparison data set, and Arne Henden for the GRB 030329 field calibration.

REFERENCES

1. Terry Oswalt, Sara: Southeastern association for research in astronomy, http://www.astro.fit.edu/sara/ (2003).
2. Bernd Fuhrmann, Thüringer Landessternwarte Tautenburg (TLS) Karl-Schwarzschild-Observatorium, http://www.tls-tautenburg.de/tlse1.html (2003).
3. Henden, A., *GRB Circular Network*, **2082**, 1 (2003).
4. Rhoads, J. E., and Fruchter, A. S., *Astrophysical Journal*, **546**, 117–126 (2001).

A Search for Short Time-Scale Optical Variability in the GRB 030329 Afterglow

N. Mirabal*, J. P. Halpern*, M. Bureau* and K. Fathi[†]

Astronomy Department, Columbia University, 550 West 120th Street, New York, NY 10027
[†]*Kapteyn Astronomical Institute, Postbus 800, 9700 Av, Groningen, The Netherlands*

Abstract. We present a densely sampled R-band light curve of the optical afterglow of GRB 030329 obtained during the period 15.5–23 hours after the burst. This dataset allows us to search for short time-scale fluctuations that must be present at some level due to inhomogeneities in the circumburst medium and/or if the afterglow is intrinsically variable.

INTRODUCTION

The recent detection of early fluctuations about the mean power-law decay of GRB afterglows [1, 2] may provide important clues in our understanding of the afterglow evolution. However, since several interpretations have been put forth to explain the fluctuations (e.g. [3]), it is crucial to determine the mechanism ultimately responsible for the variability. Here we present an initial attempt to help constrain the microphysics of GRB afterglows using a rapid optical photometric sequence.

DISCUSSION

High-speed optical photometry of the GRB 030329 afterglow was obtained at the MDM 1.3 m telescope beginning on March 30 03:05 UT, 15.5 hours after the burst. The data consists of a 7.5–hour time series in the R-band with a time resolution of 90 s. Altogether, we obtained a total of 306 points. During this period, the magnitude of the optical afterglow declined from $R = 15.4$ to $R = 16.2$. Figure 1 shows a power-law decay slope $\alpha = -1.931 \pm 0.005$ fitted to the data, where the uncertainties are dominated by faint comparison stars and large airmasses at the beginning and the end of the run.

The computed residuals indicate that the average fluctuation about the power-law fit is smaller than 0.01 mag on time scales of minutes to hours. If the flux variability f_1/f_0 traces the density contrast n_1/n_0 as a function of $f_1/f_0 \propto (n_1/n_0)^{1/2}$ [4], this implies that the enhancements in the vicinity of the GRB are less than a factor of 1.03 in density. In other words, the circumburst medium was close to homogeneous between 1.4×10^{17} cm and 1.7×10^{17} cm from the burst site, assuming a spherical adiabatic blast wave expanding in a stellar wind. Unfortunately this fast sequence by itself cannot place strong limits on intrinsic afterglow variability, but it shows that the afterglow evolution can be highly uniform on short time-scales.

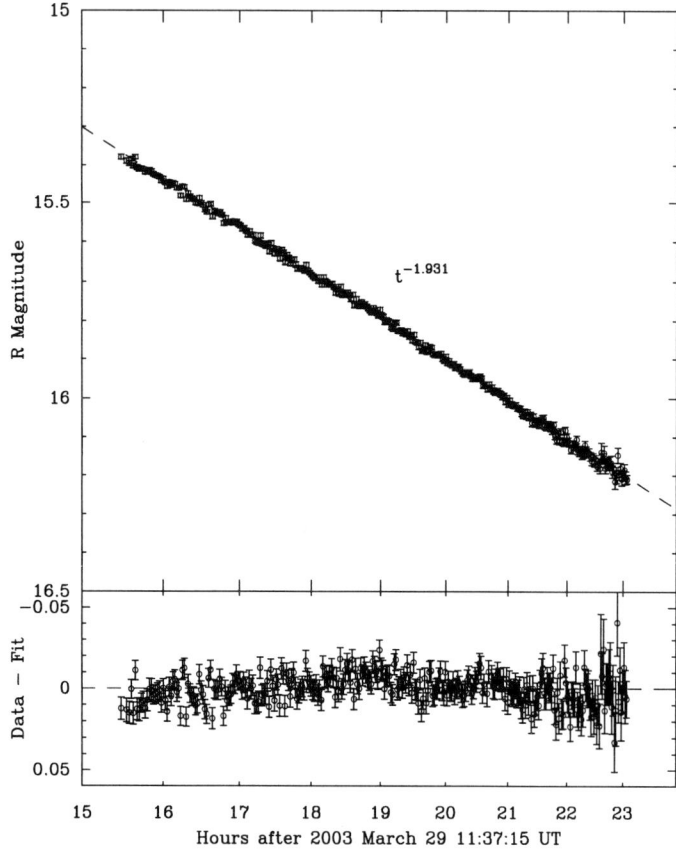

FIGURE 1. *Top*: R-band light curve of the GRB 030329 afterglow. *Bottom*: Difference between data and model.

CONCLUSIONS AND FUTURE WORK

We have shown preliminary results of a high-speed optical photometry program to access short time-scale fluctuations in GRB afterglows. A comprehensive effort to observe and model fluctuations at different time scales might hold the key to understanding the microphysics of GRB afterglows.

REFERENCES

1. Bersier, D., et al., ApJ **584**, L43 (2003).
2. Mirabal, N., et al., ApJ **595**, 935 (2003).
3. Heyl, J. S., & Perna, R., ApJ **586**, L13 (2003).
4. Lazzati, D., et al., A&A **396**, L5 (2002).

Colors of the Optical Afterglow of GRB030329/SN 2003dh

V. Šimon*, R. Hudec* and G. Pizzichini[†]

Astronomical Institute, Academy of Sciences of the Czech Republic, 251 65 Ondřejov, Czech Republic
[†]*IASF/CNR, Sezione di Bologna, via Gobetti 101, 40129 Bologna, Italy*

Abstract. We show that the color indices $(B-V)_0$, $(V-R)_0$, $(R-I)_0$ of the optical afterglow of GRB030329 were consistent with those of the other OAs during $t-T_0 < 10$ days. This suggests that the synchrotron component was initially dominant in the spectrum. Large color variations emerge only for $t-T_0 > 20$ days and suggest a rapid spectral evolution of the underlying supernova SN 2003dh.

DATA ANALYSIS AND RESULTS

Color indices of the optical afterglows (OAs) of GRBs are a powerful and straightforward approach to the study of such events [1]. These indices enable us to resolve small variations of the profile of the spectra and help to determine the emission mechanisms.

This analysis makes use of the data published in the GCN circulars http://gcn.gsfc.nasa.gov/gcn/gcn3_archive.html (ed. S. Barthelmy) and J. Greiner's GRB Web page http://www.mpe.mpg.de/~jcg/grbgen.html. Because of space limitations, the reader is referred to these web pages for full bibliographic references. The Galactic reddening of GRB030329 is only $E_{B-V} = 0.02$ [2] and can be neglected for our purposes. Also the light contribution of the host galaxy was quite small [3] in all the observations considered by us. Details of the determination of the colors of the OAs can be found in [1]. The resulting diagrams are shown in Fig. 1.

We find that the color indices $(B-V)_0 = 0.42 \pm 0.07$, $(V-R)_0 = 0.38 \pm 0.04$, $(R-I)_0 = 0.42 \pm 0.03$, $(I-J)_0 = 0.72 \pm 0.07$ of the OA of GRB030329 (which has redshift $z = 0.168$ [4]) in the observer frame inside the time interval $t-T_0 < 10$ days are consistent with those of most other OAs with $z = 0.43 - 3.5$, determined by [1]. These colors are in agreement with the GRB afterglow fireball emission [5]. This suggests that the synchrotron component, with parameters comparable to the other OAs, was initially dominant in the spectrum of the OA of GRB030329. We also argue that the intrinsic reddening of GRB030329 inside its host galaxy must have been quite low, following the arguments of [1].

We detect large color changes of the OA of GRB030329 only at $t-T_0 > 20$ days (Fig. 1b), giving rise to rapid shifts in the color-color diagram (Fig. 1c). They suggest a rapid evolution of the spectrum of the underlying supernova SN 2003dh (we note that this evolution was prominent only in the color indices, not in the light curves). The

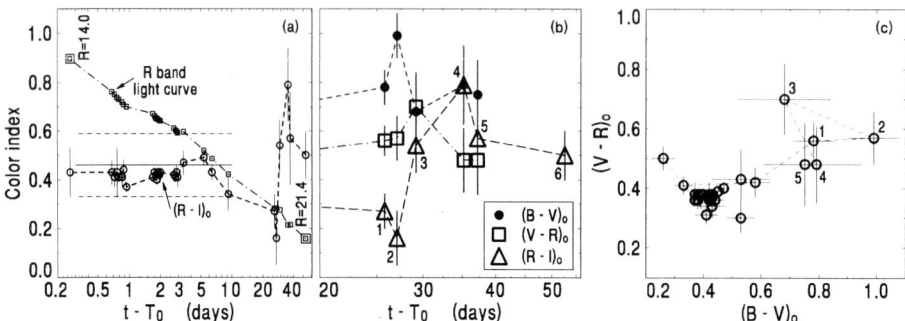

FIGURE 1. (a) Temporal evolution of the color index $(R-I)_0$ of the afterglow of GRB030329 (the course of $(B-V)_0$ and $(V-R)_0$ is similar to that of $(R-I)_0$ for $t-T_0 < 10$ days). The horizontal solid line with the dashed error bars marks the mean $(R-I)_0$ index for the ensemble of the OAs, determined by [1]. The filled circles denote the R band light curve. (b) Color changes during the late phase. Notice that the transient reddening occurs first for $B-V$, then $V-R$, and finally for $R-I$. (c) Color-color diagram of the OA of GRB030329. The numbers at some points allow a comparison between (b) and (c).

features of the Ic-type SN 2003dh appeared in the spectrum of this OA at $t-T_0$ of about a week and strengthened later on [6]. Changes of a large part of the optical and near IR continuum of SN 2003dh, and not only of the lines, must be involved in the complicated color variations. A comparison with the synthetic colors of SN 1998bw, generated from [7] and scaled to $z = 0.168$, reveales large similarities in the time evolution of the colors of these two events at very similar $t-T_0$. It also emerges from this comparison that it is likely that the brightness of SN 2003dh was already declining in $t-T_0 > 20$ days. More details and plots will be given in the forthcoming paper [8].

ACKNOWLEDGMENTS

This study was supported by the grant A3003206 provided by the Grant Agency of the Academy of Sciences of the Czech Republic, the project ESA PRODEX INTEGRAL 14527, and the CNR-AVČR collaborative project Investigation of GRBs (2000/2003).

REFERENCES

1. Šimon, V., Hudec, R., Pizzichini, G., and Masetti, N., *A&A*, **377**, 450 (2001).
2. Schlegel, D., Finkbeiner, D., and Davis, M., *ApJ*, **500**, 525 (1998).
3. Blake, C., and Bloom, J., *GCN*, **2011** (2003).
4. Greiner, J., Peimbert, M., Estaban, C., and et al., *GCN*, **2020** (2003).
5. Sari, R., Piran, T., and Narayan, R., *ApJ*, **497**, L17 (1998).
6. Stanek, K., Matheson, T., Garnavich, P., and et al., *ApJ*, **591**, L17 (2003).
7. Poznanski, D., Avishay, G.-Y., Maoz, D., and et al., *PASP*, **114**, 833 (2002).
8. Šimon, V., Hudec, R., and Pizzichini, G., *submitted to A&A* (2003).

GRB PROGENITORS

The Collapsar Model for Gamma-Ray Bursts

S. E. Woosley*, W. Zhang* and A. Heger†**

*Department of Astronomy and Astrophysics, UCSC, Santa Cruz CA 95064
†Group T6, Los Alamos National Laboratory, Los Alamos, NM 87545
**Department of Astronomy and Astrophysics, University of Chicago Chicago, IL 60637

Abstract. We consider the relation among collapsars, supernovae, X-ray flashes, and other possible off-axis phenomena, e.g., uv-transients. New three-dimensional calculations are presented of relativistic jet propagation and break out in a massive Wolf-Rayet star. The supernovae that accompany GRBs are novel and need not be standard candles. Because of continuing energy input at late times, afterglows need not uniquely reflect either the opening angle or the energy of GRBs. The structure of the jet is neither a top hat nor a (single) Gaussian, and energy and Lorentz factor do not have the same angular dependence. We speculate on the existence of other forms of high energy transients at high redshift.

INTRODUCTION

It is now generally acknowledged that "long-soft" gamma-ray bursts (GRBs) are a phenomenon associated with the deaths of massive stars. At least one burst was accompanied by a nearly simultaneous supernova, SN 2003dh [1, 2], which had an unusual spectrum, copious radio emission, and showed evidence of asymmetric expansion [3]. Bumps in the afterglows of other GRBs, the association of GRB counterparts with regions of active star formation, and growing evidence linking SN 1998bw with GRB 980425, all strongly suggest that SN 2003dh was not exceptional - that many, if not all GRBs have supernova counterparts.[1] This narrows the model space considerably, leaving only theories in which the compact object formed by iron core-collapse is able to energize a relativistic jet. The remaining viable candidates are the millisecond magnetar [4, 5] and the collapsar [6, 7]. We explore here some implications of the collapsar model and some generic predictions associated with relativistic shock break out in a star.

SUPERNOVAE

The supernova produced in the collapsar model is novel in several regards. First, it is aspherical with very high velocities along the polar axis and low velocities in the equatorial plane. Relativistic matter is ejected in varying amounts and energies in a supernova whose total (mostly non-relativistic) kinetic energy is more or less constant. Relativistic flow is channeled along the rotational axis, *but this relativistic jet is not*

---

[1] The converse is not true Most supernovae do not have GRB counterparts. Only about 1% do.

responsible for most of the explosion energy. The jet subtends too small a solid angle and the lateral shock it launches is not that powerful. Most of the energy for the supernova comes from the disk wind [7]. Nucleons flowing off of the accretion disk recombine into iron-group nuclei, and it is this kinetic energy that is responsible for the $\gtrsim 10^{52}$ erg of the explosion. Observations of GRB energies [8] and supernova energies [9, 10, 3] suggest that the energy from this wind - and therefore in the supernova - exceeds by as much as 10, the energy of the GRB-producing jet.

Second, since the ^{56}Ni is made by the wind and not by a spherical shock, its mass is not so limited by shock wave hydrodynamics. Producing 0.5 $M_\odot$ is not so difficult. Naively, one expects that the ^{56}Ni mass will be some fraction of the total mass accreted by the black hole and (naively squared) that the total energy of the GRB will be proportional to this same mass, i.e., brighter, longer GRBs will make brighter supernovae. [2] This simple expectation can be complicated however by electron capture in the disk [11] and by variable efficiency factors relating the GRB luminosity to the accreted mass. There is no clear reason for the supernovae that accompany GRBs to be standard candles, though they might be so, to a factor of several, because the accreted mass is always similar.

Finally, a collapsar continues to provide energy from fallback and accretion a long time after the main burst is over [12]. As the star explodes, the opening angle of the jet increases. The total energy at late times (minutes to a day) could even be greater than the GRB energy. This possibility should be kept in mind when limiting the energies and angles of GRBs by "breaks" in their afterglows observed days after the event [8, 13].

XRFs AND UVFs FOLLOWING JET BREAKOUT

Zhang, Woosley, & Heger [14] recently studied jet break out in collapsars in 2D and 3D. Figs. 1 and 2, from that paper, show some relevant properties of a typical jet as it emerges. The part that makes the GRB, about 3 - 4 degrees in radius with $\Gamma \sim 200$, is surrounded by a lower energy cocoon [15] of "mildly" relativistic material, with Γ up to 20 - 40, which explodes to larger angles. The interaction of this off-axis material with the circumstellar wind of the star seems certain to give rise to high energy transients of some sort. Predicting their properties from first principles is difficult though, because of uncertainty in the emission mechanism and efficiency factors. In contrast to the modulated Lorentz factor found in the central core of the jet, the calculations do not show significant non-monotonic variation in velocity with radius for the off-axis component. External shocks might therefore dominate (though see [21] for a "shockless" internal magnetic dissipation model). Dermer, Chiang, & Böttcher [17] suggested a "dirty fireball" of this sort might produce softer transients. The peak photon energy in the external shock model depends on $n^{1/2}\Gamma^4$ [18], but for reasonable values of $\Gamma \sim 30$ and $n \sim 10^5 r_{15}^{-2}$ cm^{-3}, could be in the keV range. There would be multiple values of Γ in the observer's line of sight, so the spectrum would not be sharply peaked. As in past papers [19, 20, 16, 14], we associate these lower power, possibly softer emissions

---

[2] The brightness, at peak, of a Type I supernova of any subclass is directly proportional to the mass of ^{56}Ni it ejects. This is known as "Arnett's Rule"

with x-ray flashes (XRFs, [22]). Recently several groups [23, 24, 25] have have determined a relation between peak photon energy and GRB equivalent isotropic energy, $E_{\text{peak}} \sim 100\ (E_{\text{iso}}/10^{52}\ \text{erg})^{1/2}$ keV, that fits both GRBs and XRFs. While the core emission in Fig. 2 is close to 10^{54} erg, this is the kinetic energy. The radiative E_{iso} would be less, by perhaps 10, implying a photon energy in the core of ~ 300 keV. In the wings, the scaling relationship would imply peak energies about 10 times less, i.e., hard x-rays. Still softer emission could come from the wings farther out.

If this picture is true, then every GRB is also an X-ray flasher and every GRB light curve will contain, sometimes at a very low level, an XRF coming from the wings of the jet. Given the different emission mechanisms (internal vs. external shocks) and efficiency factors (external shocks are thought to be more efficient), the ratio of the brightnesses is unknown. XRFs could be a large fraction of the observed high energy transient event rate and they are certainly more numerous than GRBs (because of the large angle to which they are visible).

If XRFs are made this way,[3] several predictions emerge. First, since every XRB is just a GRB seen at a different angle, XRFs should share the spatial distribution of GRBs. They should also be accompanied by a supernova (though not necessarily a standard candle). Because they are visible to a larger angle, more XRFs are potentially detectable than GRBs, but because they are less powerful, they are sampled to a shorter distance. It may be that the log N - log S distribution for XRFs will not roll over in the same way as GRBs. We may not have seen to the edge of the distribution yet (hence a more sensitive detector would see a rise in the ratio of XRFs to GRBs).

But why stop with XRFs? For somewhat lower Γ and larger angles, the peak emission could fall into the ultraviolet or even optical range. Then one expects GRBs in the jet core, XRFs in the near wings, and UV flashes in the broad wings. *The most common observable transient produced by jet breakout may not be in gamma-rays or x-rays, but in the ultraviolet* [26]. This important conclusion may have implications for SWIFT.

The total energy in the wings of the jet is not large though. 90% of the relativistic ejecta ($\Gamma > 5$) are within 6 degrees of the axis (integral of Fig. 2). One does not expect this material at large polar angle to contribute appreciably to the afterglow - compared with the decelerated GRB jet itself. However, there may still be a lot of energy radiated at large angle and later times if the central black hole continues to accrete and power a jet after the principal GRB is over. Given the expected fallback in the collapsar, which could easily generate 10^{51} erg, this is likely.

OTHER PREDICTIONS OF THE COLLAPSAR MODEL

This continuing emission could also have important implications for putative x-ray lines in GRBs. During the first day, most of the star is expelled as a supernova which remains

---

[3] An alternate possibility, hard to disprove at the present time, is that XRFs are the *on axis* emissions of collapsars that, for whatever reason, did not develop as high a Lorentz factor as in GRBs. This could reflect instabilities in the jet [14], a more extended progenitor, or simply a jet that, for whatever reason, never attained a very high energy per baryon.

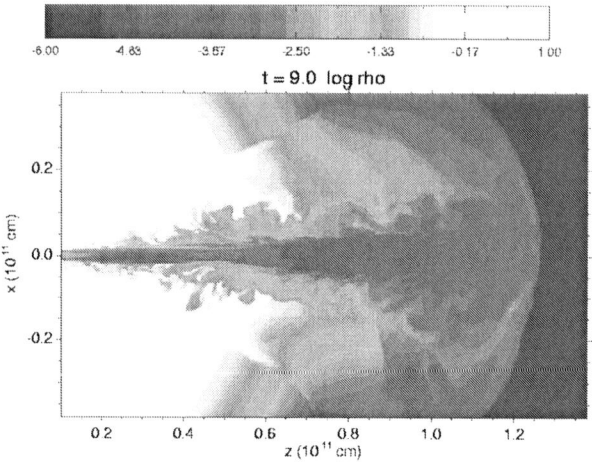

FIGURE 1. Jet breakout in Model 2A of ref. [14]

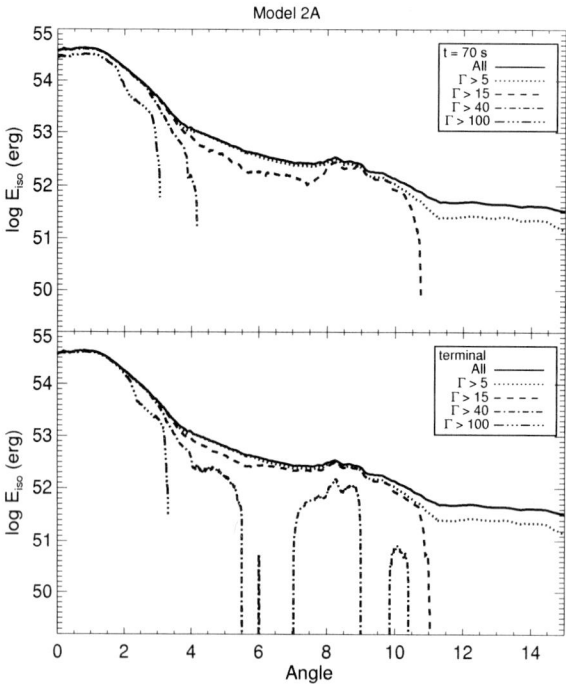

FIGURE 2. Distribution of Lorentz factor and equivalent isotropic energy in Fig. 1. It is assumed that all internal energy converts to kinetic in oredr to calculate a terminal Lorentz factor.

optically thick out to $\sim 10^{15}$ cm for weeks. However, along the polar axis, a jet continues to flow, energized by fallback from the supernova [12] which declines as roughly $t^{-5/3}$. After a day, the power may still be $\sim 10^{46}$ erg s^{-1} (for an accretion rate of 10^{-7} M$_\odot$ s^{-1} with efficiency 5% $\dot{M}c^2$). Internal shocks within this jet could irradiate the slower moving, subrelativistic material along the jet nozzle and make x-ray lines [27]. If this is the origin of the lines, they would be brightest when the star along the rotational axis first becomes optically thin and decline rapidly ($\sim t^{-5/3}$ thereafter).

The collapsar model also predicts that GRBs of duration less than hundreds of seconds will only originate from Type Ib/c supernovae, never from red or blue supergiants (BSGs). Our unpublished calculations show that a jet with typical power for a GRB, left on for 70 s in a blue supergiant (radius 3×10^{12} cm) only reaches 10^{12} cm and that the jet head is advancing subrelativisitically. If the power source at the origin is removed, the jet dissipates quickly, after encountering its rest mass. However, it is still possible that some sort of high energy transient, an XRF or UVF, could originate from a BSG [12]. A very long GRB is also a possibility. Given that most massive stars at high redshift are BSGs when they die, these sorts of transients could be common.

Indeed, at high redshift, several new varieties of high energy transients from collapsars could become visible. Type III collapsars occur when the pair instability leads to core collapse in a rotating star of several hundred solar masses [28]. These objects would produce very energetic transients lasting hundreds of seconds in the rest frame. It is difficult to estimate the jet energy and γ-ray efficiency in such hypothetical objects, but they could resemble ordinary GRBs in terms of Lorentz factor and peak photon energy. If so, one expects powerful bursts of hard x-ray emission from redshift ~ 15 lasting, perhaps, several hours. These would be a distinctive signature of the first stars to form after the "Dark Ages".

But why stop at 300 M$_\odot$? An enduring puzzle in astrophysics has been the origin of the supermassive black holes found in both active and normal galactic nuclei. There may not have been enough time to grow black holes of $\gtrsim 10^6$ M$_\odot$ from scratch before the first quasars are seen. One possibility (e.g.[29, 30]) is that they formed from supermassive stars, also of $\gtrsim 10^6$ M$_\odot$. If these stars rotated anywhere near break up, the new (Kerr) black hole would be surrounded by a disk with mass roughly 10% that of the hole [31].
[4]. The natural hydrodynamic time scale for such a collapse is about a day, and one might expect accretion from the disk to go on for that long [5]. Accretion of $\gtrsim 10^5$ M$_\odot$ in 10^5 s, assuming 1% conversion of the rest mass gives jets with power 10^{52} erg s^{-1} and total energy 10^{57} erg. These factors may be further amplified an additional factor of ~ 100 by beaming. The Lorentz factor and photon energy is unknown, but the accretion rate is not unlike the most energetic GRBs. If emission were in the gamma-ray band, the redshift would give bursts in hard x-rays with GRB-line fluxes lasting several weeks. The greatest uncertainty is the event rate. Including beaming, estimates are in the range

---

[4] We could call such an object a "Collapsar Type IV", the other three types being prompt black hole formation in a massive star [7], black hole formation by fallback [12], and black hole formation in a pair-instability collapse[28].

[5] Shibata & Shapiro [31] estimate a much longer time scale, but ignore cooling by neutrino emission or the disk wind.

a few per year to a few per century [32, 33]. While this is very uncertain, the good news is that a detection one or two would help determine a very interesting, uncertain number - the birth rate of supermassive black holes in our universe.

ACKNOWLEDGMENTS

This research has been supported by NASA (NAG5-12036) and the DOE Program for Scientific Discovery through Advanced Computing (SciDAC; DE-FC02-01ER41176).

REFERENCES

1. Hjorth, J. et al., Nature, **423**, 847 (2003).
2. Stanek, K. Z. et al., ApJL, **591**, L17 (2003).
3. Woosley, S. E., & Heger, A., ApJ, accepted, astroph 0309165 (2003).
4. Usov, V., Nature, **357**, 472 (1992).
5. Wheeler, J. C., Yi, I., Höflich, P., & Wang, L., ApJ, **537**, 810 (2000).
6. Woosley, S. E., ApJ, **405**, 273. (1993).
7. MacFadyen, A., Woosley, S. E., ApJ, **524**, 262 (1999).
8. Frail, D. A., Kulkarni, S. R., Sari, R., Djorgovski, S. G., Bloom, J. S., Galama, T. J., Reichart, D. E., Berger, E., et al., ApJL, **562**, 55 (2001).
9. Woosley, S. E., Eastman, R. G., Schmidt, B. P., ApJ, **516**, 788 (1999).
10. Iwamoto, K.,, Nature, **395**, 672 (1998).
11. Pruet, J., Woosley, S. E., & Hoffman, R. D., ApJ, **586**, 1254 (2003).
12. MacFadyen, A., Woosley, S. E., & Heger, A., ApJ, **550**, 410 (2001).
13. Berger, E., Kulkarni, S. R., Pooley, G., Frail, D., McIntyre, V., Wark, R. M., Sari, R., Soderburg, A. M., Fox, D. W., Yost, S., & Price, P. A., Nature, **426**, 154 (2003).
14. Zhang, W., Woosley,S.E., & Heger, A., ApJ, in press, astroph-0308389 (2003).
15. Ramirez-Ruiz, E., Celotti, A., & Rees, M. J., MNRAS, **337**, 1349 (2002).
16. Zhang, W., Woosley, S. E., & MacFadyen, A., ApJ, **586**, 356 (2003).
17. Dermer, C. D., Chiang, J., & Böttcher, M., ApJ, **513**, 656 (1999).
18. Sari, R. & Piran, T., ApJ, **520**, 641 (1999).
19. Woosley, S. E. 2000, GRBs, 5th Huntsville Symposium, eds. Kippen, Mallozzi, & Fishman, AIP, Vol **526**, 555 (2000).
20. Woosley, S. E. 2001, GRBs in the Afterglow Era, eds. Costa, Frontera, & Hjorh, ESO Astrophysics Symposia, (Springer), **555** (2001).
21. Zhang, B. & Mészáros, P., ApJ, **581**, 1236 (2002).
22. Heise, J., in't Zand, J., Kippen, R. M., Woods, P. M. 2001, GRBs in the Afterglow Era, eds. Costa, Frontera, & Hjorth, ESO Astrophysics Symposia, (Springer), **16** (2001).
23. Lloyd, N. M., Petrosian, V., & Mallozzi, R. S., ApJ, **534**, 227 (2000).
24. Amati, L. et al., A&Ap, **390**, 81 (2002).
25. Lamb, D. Q., Donaghy, T. Q., & Graziani, C., to appear in proc. 2nd VERITAS Symposium on TeV Astrophysics, Chicago, Illinois (2003).
26. Ramirez-Ruiz, E., Woosley, S. E., & Zhang, W., in prreparation for ApJL (2003).
27. McLaughlin, G. C., Wijers, R. A. M. J., & Brown, G. E., & Bethe, H. A., **567**, 454 (2002).
28. Fryer, C. L., Woosley, S. E., & Heger, A., ApJ, **550**, 372 (2001).
29. Rees, M. J., ARAA, **22**, 471 (1984).
30. Bromm, V. & Loeb, A., ApJ, **596**, 34 (2003).
31. Shibata, M., & Shapiro, S. L., APJL, **572**, L39 (2002).
32. Fuller, G. M., & Shi, X., ApJL, **502**, L5 (1998).
33. Woosley, S. E., in preparation for ApJ (2003).

A Field Guide to Collapsars

Enrico Ramirez-Ruiz

School of Natural Sciences, Institute for Advanced Study, Einstein Drive, Princeton, NJ 08540

Abstract. As we have heard at this meeting, it seems likely that GRBs originate in a very small fraction of stars that undergo a catastrophic energy release event toward the end of their evolution. Expressly, the association of some GRBs with supernovae has pointed a finger at deaths of massive stars as the cause of GRBs, or at least a subset thereof. Many of their observed properties can be understood as resulting from outflows, driven by newly formed black holes, which are then subsequently collimated into a pair of anti-parallel jets within a collapsing massive star. The broad features seem to be determined by the mass of the black hole, the amount of matter surrounding it, and the orientation of the observer with respect to its axis of angular momentum. GRB jets, if powered by the black hole itself, may therefore be one of the few observable consequences of how flows close to nuclear density behave under the influence of strong gravitational fields.

INTRODUCTION

The progress in our understanding of GRBs has been a story of consolidation and integration, and there is every indication that this progression will continue. For instance, previously separate phenomena like supernovae and GRBs have now been shown to be different manifestations of the same underlying physical process[1, 2], distinguished mainly by luminosity, and orientation or environment. Though the field is far from being mature, sufficient progress has been made in identifying some essential ingredients. This encourages us to present a preliminary account of the causes and effects of GRB activity. A basic scheme can provide a conceptual framework for describing the observations even when the framework is wrong! The following should be interpreted in this spirit. A distinction should be drawn at this stage. We make no attempt to unify different classes of GRBs, most prominently long and short duration ones, but only to outline some of the physical process that are believed to be most relevant to interpreting GRBs. There could also be more subclasses of classical GRB than just short and long duration – bursts that are mostly γ-rays, bursts that are mostly X-rays [4], bursts with no high energy pulses [3], bursts with afterglows and those without, etc. This interpretation is based upon observations restricted so far to long duration GRBs. Some outstanding theoretical issues are highlighted, along with the types of observation that would help to discriminate between different manifestations. We conclude this article with some brief speculations about assembling these components into a general scheme.

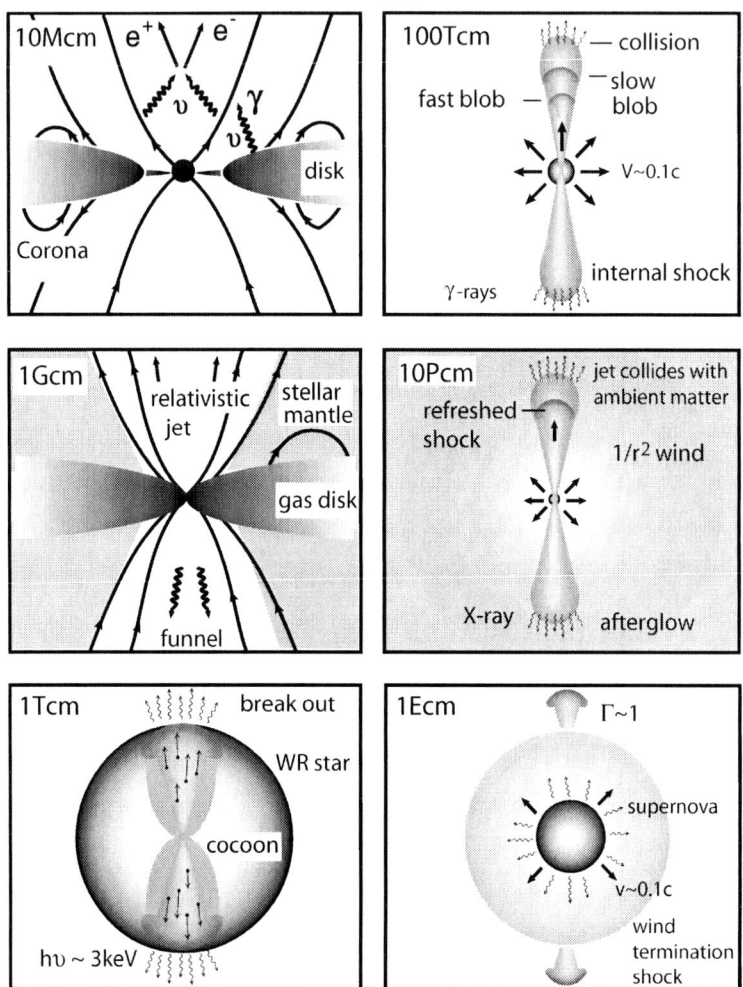

FIGURE 1. Diagram exhibiting GRB activity over various decades in radius ranging from 10^7 cm to 1pc. This is scaled to a black hole of two solar masses. We comment upon the individual frames in the text.

AN ILLUSTRATED TIMELINE

GRB activity manifests itself over a dynamical range of ~ 12 decades in radius. In Figure 1, we show a schematic montage of various decades, exhibiting phenomena which are believed to take place on each of these length scales. The phenomena are not directly observed and the associated frames represent "educated" guesses of their geometrical arrangements. Let us consider these frames in turn, working from the small scales to the large scales.

10Mcm – Black Hole and Debris Torus, 10^2 *km.* The first task in attempting to construct a general scheme of GRBs is to decide which parameters exert a controlling influence upon their properties. It seems unavoidable that the mass of the central black hole is a primary parameter, as it dictates a characteristic length and luminosity scale for GRB activity. The black hole mass is controlled by one key parameter, the mass of the progenitor star[5, 6]. A black hole cannot radiate at all without accretion and the rate at which gas is supplied to the hole, $\dot{M}$, is probably as important as the mass. A GRB is then likely to be triggered if the remaining star has sufficient angular momentum to form a centrifugally supported disc. Yet the actual angular momentum in a presupernova star is unknown. A brief inflow of fallback gas emanating from the stellar mantle supplies mass to the disc at a rate much greater than the Eddington limit[7]. The accretion disc is believed to be neutrino-dominated in its inner regions[8, 9], with thickness proportional to the dimensional mass accretion rate. Thick accretion discs are subject to dynamical instability, though the ultimate consequence of this remains uncertain. Provided that the disc is sufficiently dense to thermalise the radiation, most of the radiation will be emitted roughly with peak frequency somewhat in excess of the local effective temperature, which lies in the MeV range. The relativistic outflow from the black hole is proposed to be focused into two jets[10]. The outflowing hydromagnetic wind can collimate the central relativistic jet, when present, even when it transports a significantly smaller power than the jet. These jets may be further slowed by entrainment.

1Gcm – The Stellar Gas Inflow, 10^4 *km.* This is typically the scale at which $\Gamma \sim \eta$[11]. As the flow accelerates along the jet, the internal energy contained in the radiation and pairs is transformed into kinetic energy of the protons, and by the time the photons can escape, the ions are moving with Lorentz factors $\Gamma \sim 300$. Even if the outflow is not narrowly collimated, some beaming is expected because energy would be channelled preferentially along the rotation axis[12, 13]. Moreover, we would expect baryon contamination to be lowest near the axis, because angular momentum flings material away from the axis, and any gravitationally-bound material with low angular momentum falls into the hole. Beyond this scale, an understanding of the collimation and confinement of a gas-dynamic jet can come about only through a knowledge of the properties of the medium through which it propagates[14]. This is the first extrinsic or environmental effect that comes into play, which may in turn strongly affect what we observe.

1Tcm – Mantle Uncorking, 10^7 *km.* This is the typical size of an evolved massive star. The majority of stellar progenitors, with the exception of some very compact stars, will not collapse entirely during the typical duration of a GRB[15]. A stellar envelope will thus remain to impede the advance of the jet. At this scale, the beam will have evacuated a channel out to some location where it impinges on the stellar envelope[16]. A continuous flow of relativistic fluid emanating from the nucleus supplies this region with mass, momentum, energy, and magnetic flux[17]. Reconversion into random energy occurs at the end of the channel, which is a natural site for particle acceleration[18, 15]. If we balance the jet pressure against that of the surrounding gas, we conclude that these jets must supply a large fraction of their power to the cocoon region[14, 15, 13]. It might not have a ready-made escape route, and yet still have more energy than the envelope binding energy; the cocoon could then expand more or less isotropically through the envelope and violently disrupt it. So in these models the γ-rays would be restricted to a narrow beam, even though outflow with a more moderate Lorentz factor (relevant to the afterglow) could be spread over a wider range of angles[14, 13]. A thermal and a very hard γ-ray signal should precede the canonical, softer γ-rays observed in GRBs[17, 19]. Shock waves and magnetic dissipation in the escaping bubble can also contribute a nonthermal UV/X-ray afterglow, and also excite Fe line emission from thermal gas[16]. For very extended envelopes, the jet maybe unable to break through the envelope. TeV neutrino signals produced by Fermi accelerated relativistic protons within the cork may provide a means of detecting such choked-off, γ-ray dark collapses [18].

100Tcm – The Unsteady Wind, 10^9 km. Velocity differences across the jet profile – maybe imprint as it bores its way through the stellar mantle[13, 20] – provide a source of free energy from particle acceleration through shock waves, hydromagnetic turbulence, and tearing mode magnetic reconnection[21]. In the presence of turbulent magnetic fields built up behind the shocks, the electrons can produce a synchrotron power-law radiation spectrum, whereas the inverse Compton scattering of these synchrotron photons extends the spectrum into the GeV range[22]. Given the precise description of the dynamics, along with the baryon content, magnetic field, and Lorentz factor of the outflow, one is able to predict the gross spectral and temporal features. Yet, we are still a long way from predicting the intensity or intrinsic spectrum of the emitted radiation without also having an adequate theory for particle acceleration in relativistic shocks. Ultrarelativistic outflows, in general, are seen in radio jets, pulsar wind nebulae and GRBs, and there is every reason to undertake comparative studies to understand their general, global behavior. A magnetic field can ensure efficient cooling even if it is not strong enough to be dynamically significant. If, however, the field is dynamically significant in the wind, then its internal motions could lead to dissipation even in a constant η wind (perhaps the truth lies between these two extremes). Instabilities in this magnetized wind may be responsible for particle acceleration[23]. A further effect renders the task of simulating unsteady winds even more challenging. This stems from the likelihood that any entrained matter would be a mixture of protons and neutrons (neutrons, being unconstrained by magnetic fields, could also drift into a jet from the denser walls at its boundary). If a streaming velocity builds up between ions and neutrons (i.e. if they have different Lorentz factors in the outflow) then interactions can lead to dissipation even in a steady jet where there are no shocks [24, 25].

10Pcm – The Stellar Wind Region, 10^{11} km. The external shock becomes important when the inertia of the swept-up external matter starts to produce an appreciable slowing down of the ejecta. As the fireball continues to plough ahead, it sweeps up an increasing amount of external matter, made up of gas which was previously ejected by the progenitor star. This sets a characteristic deceleration length[22]. This deceleration allows slower ejecta to catch up, replenishing and re-energizing the reverse shock and boosting the momentum in the blast wave[26]. The bulk Lorentz factor of the fireball thereafter decreases as an inverse power of the time. As a consequence, the accelerated electron minimum random Lorentz factor and the turbulent magnetic field also decrease as inverse power-laws in time. This implies that the spectrum softens in time, as the synchrotron peak corresponding to the minimum Lorentz factor and field decreases, leading to the possibility of afterglow emission[26]. Even in the simplest case of a wind whose properties do not vary over the life of the star, copious behavior with multiple possible transitions in the observable part of the afterglow lifetime may be seen[27]. The resulting lightcurve depends fairly strongly on the properties of the system, especially the mass-loss rate of the star and the ambient density[28]. The wind history of a WR star during its last few centuries could be quite complicated as the star enters advanced burning stages unlike those in any WR star observed so far. Unlike a red supergiant, the core and surface of a WR star remain in communication at late times. That is, pulsations and flashes in the center during carbon or oxygen burning are communicated to the surface where they are amplified by the steep density gradient.

1Ecm – The Sedov Length, 10^{13} *km*. Finally, we come to the end of the relativistic phase. Obviously, this happens when the mass E/c^2 has been swept-up. This sets a non-relativistic length scale. Beyond this point, the event slowly changes to a classical Sedov-Taylor supernova remnant evolution, leading to a steeper decline in the lightcurve [29]. We have just outlined a particular scheme that we believe to be logically self-consistent but for which there is not a shred of *direct* evidence. For alternative schemes the reader is referred to [30, 23, 31].

PRIME MOVERS

If we were to venture a general classification scheme for GRBs, on the hypothesis that the central engine involves black holes formed in core collapse explosions, we would obviously expect the hole mass, the rate at which the gas is supplied to the hole, the angular momentum of the hole, and the orientation relative to our line of sight to be essential parameters. It is then tempting to identify the high Lorentz factor and GRB energies with the catastrophic formation of a stellar mass black hole of $\sim 5M_\odot$ with $\sim 1\%$ of $M_\odot c^2$ going into a jet outflow. This could be the extreme example of the jet producing supernova discussed by [32], in which instead of halting at the neutron star stage, the collapse continues to the black hole stage, producing an even faster jet in the process. Optimists might hope eventually for a genuine understanding of GRBs, on the level of our present knowledge of stellar structure and evolution. Realists, however, are conscious that there may be more independent parameters than the few (mass, composition, rotation, magnetization) crucial to stellar evolution, and that most observations tell us less about the primary power source than about secondary reprocessing of this power in the stellar environment – from which we can learn about the central engine only by a chain of uncertain inferences.

Our understanding of GRBs has come a long way since their discovery almost 30 years ago, but these enigmatic sources continue to offer major puzzles and challenges. As we have described, our rationalization of the principal physical considerations combines some generally accepted features with some more speculative and controversial ingredients. When confronted with observations, it seems to accommodate their gross features but fails to provide us with a fully predictive theory – but then no such theory exists as yet. What is more valuable, though considerably harder to achieve, is to refine models like the ones advocated here to the point of making quantitative predictions, and to assemble, assess and interpret observations so as to constrain and refute these theories. What we can hope of our present understanding is that it will assist us in this endeavor.

ACKNOWLEDGMENTS

I am very grateful to Ed Fenimore and Mark Galassi for inviting me to take part in a very enjoyable meeting. Special thanks to Martin Rees, Peter Mészáros and Annalisa Celotti for extended collaboration, and to Chryssa Kouveliotou, Andrew MacFadyen, Jay Salmonson, Stan Woosley and Wequin Zhang for discussions. This research has been supported by NASA through a Chandra Postdoctoral Fellowship award PF3-40028.

REFERENCES

1. Hjorth, J. et al., Nature **423**, 847 (2003).
2. Stanek K. Z. et al., ApJ **591**, L17 (2003).
3. Pendleton G. et al., ApJ **464**, 606 (1996).
4. Heise J. et al., in Gamma Ray Bursts in the Afterglow Era, ed. E. Costa, F. Frontera, and J. Hjorth (Springer: Berlin), 16 (2001).
5. MacFadyen A. I., Woosley S. E., Heger A., ApJ **550**, 410 (2001).
6. Fryer C. L., Kalogera V., ApJ **554**, 548 (2001).
7. MacFadyen A. I., Woosley S. E., ApJ **524**, 262 (1999).
8. Woosley S. E., ApJ **405**, 273 (1993).
9. Narayan R., Piran T., Kumar P., ApJ **557**, 949 (2001).
10. Wheeler J. G., Yi I., Hoflich P., Wang L., ApJ **537**, 810 (2000).
11. Shemi A., Piran T., ApJ **365**, L55 (1990).
12. Aloy M. A., Ibanez J. M., Marti J. M., Muller E., MacFadyen A. I., ApJ **531**, L119 (2000).
13. Zhang W., Woosley S. E., MacFadyen A. I., ApJ **586**, 356 (2003).
14. Ramirez-Ruiz E., Celotti A., Rees M. J., MNRAS **337**, 1349 (2002).
15. Matzner C. D., MNRAS **345**, 575 (2003).
16. Mészáros P., Rees M. J., ApJ **556**, L37 (2001).
17. Ramirez-Ruiz E., MacFadyen A. I., Lazzati D., MNRAS **331**, 197 (2002).
18. Mészáros P., Waxman E., Phys. Rev. Lett. **87**, 1102 (2001).
19. Waxman E., Mészáros P., ApJ **584**, 390 (1998).
20. Woosley S. E., these proceedings (2004).
21. Rees M. J., Mészáros P., ApJ **430**, L93 (1994).
22. Piran T., Phys. Rep. **314**, 575 (1999).
23. Lyutikov M., these preoceedings (2004).
24. Derishev E. V., Kocharovsky V. V., Kocharovsky Vl. V., ApJ **521**, 640 (1999).
25. Beloborodov A., ApJ **588**, 931 (2003).
26. Rees M. J., Mészáros P., ApJ **496**, L1 (1998).
27. Wijers R. A. M. J., in Gamma Ray Bursts in the Afterglow Era, ed. E. Costa, Frontera F. and Hjorth J. (Springer: Berlin), 306 (2001).
28. Ramirez-Ruiz E., Dray L., Madau P., Tout C. A., MNRAS **327**, 829 (2001).
29. Waxman E., Frail D., Kulkarni S., ApJ **497**, 288 (1998).
30. Dermer C. D., astro-ph/0202254 (2002).
31. Dar A., De Rújula A., astro-ph/0308248 (2003).
32. Khokhlov A. M. et al., ApJ **529**, L107 (1999).

Circumburst Environments of Gamma-Ray Bursts

Roger A. Chevalier

Department of Astronomy, University of Virginia, P.O. Box 3818, Charlottesville, VA 22903, USA

Abstract. I review current evidence on the surroundings of gamma-ray bursts, in light of the fact that GRB 030329 was associated with the explosion of a massive star (SN 2003dh). Although interaction with either a free wind or the shocked wind from a massive star progenitor might be able to explain GRB observations, rather special circumstances are needed to enable this. In particular, a number of lines of evidence suggest a lower density than would be expected around a Wolf-Rayet star for many bursts. The recent observation of absorption lines in GRB spectra may indicate radiatively accelerated clumps in the circumburst environment, but an outflow from the host galaxy cannot be ruled out.

INTRODUCTION

The finding of SN 2003dh in GRB 030329 has confirmed that massive stars are progenitors of long duration, cosmological GRBs (gamma-ray bursts) with the burst occurring at approximately the same time as the supernova, at least in this case [1, 2, 3]. The Type Ic supernova, which resembled SN 1998bw, can be approximately modeled by an explosion with 8 $M_\odot$ of ejecta [4, 5]. Previous evidence linking GRBs with massive stars has already generated interest in viewing the environment of massive stars as providing the circumburst environment of GRBs; in particular, the steady wind of a massive star has been investigated as a likely environment [6, 7, 8, 9, 10, 11]. Despite this attention to the problem, there has been no general demonstration of consistency of GRB observations with expectations for the surroundings of massive stars.

Both SN 1998bw and SN 2003dh are classified as Type Ic supernovae which are believed to have Wolf-Rayet star progenitors. This type of progenitor is also supported by theoretical arguments on the time for a jet to traverse the progenitor star [12, 13]. If there is a free wind from the progenitor star, the wind density is $\rho = Ar^{-2} = \dot{M}/4\pi r^2 v_w$, where $\dot{M}$ is the mass loss rate and v_w is the wind velocity. Characteristic mass loss parameters for Galactic Wolf-Rayet stars are $\dot{M} = 10^{-5}$ $M_\odot$ yr^{-1} and $v_w = 10^3$ km s^{-1}; the value of the density can be scaled to the corresponding value for these parameters, $A_* = A/(5 \times 10^{11}$ gm cm$^{-1})$. This provides a reference point for mass loss densities. Here, I briefly review the evidence for circumburst environments, based on observations of GRBs, their afterglows, and absorption line features.

GAMMA-RAY BURSTS

In the standard internal shock model for GRBs, the surrounding medium does not play a role in the characteristics of the emission. The radial scale at which the internal shocks occur is given by [14]

$$r_{is} \approx 3 \times 10^{15} t_{var} \gamma_{300}^2 \text{ cm},\qquad(1)$$

where γ_{300} is the mean Lorentz factor of the outflow in units of 300 and it is assumed that there are variations $\Delta\gamma \sim \gamma$ on a timescale t_{var} in s. The radius at which there is substantial deceleration of the relativistic ejecta in a surrounding wind is given by [15]

$$r_{dec} \approx 2 \times 10^{14} E_{53} A_*^{-1} \gamma_{300}^{-2} \text{ cm},\qquad(2)$$

where E_{53} is the energy of the burst in units of 10^{53} ergs. It can be seen that in some circumstances, the interaction with the wind can overlap or precede the generation of internal shocks. Mochkovich & Daigne [15] found that the interaction of a $E_{53} = 0.1$ burst with an $A_* = 0.1$ or 1 wind could have a significant effect on the appearance of the GRB; the input of two pulses with a FRED (Fast Rise Exponential Decay) shape is changed to a different appearance. GRBs like this have not been observed, but an increase in E or reduction in γ would reduce the effect.

The situation is more complicated if the preacceleration by the GRB radiation is included. Beloborodov [16] found that this radiative acceleration could separate the external medium from the relativistic ejecta so that the interaction with the external medium is delayed out to a radius

$$r_{acc} \approx 2 \times 10^{16} E_{53}^{1/2} \text{ cm}.\qquad(3)$$

This effect is especially important for wind interaction and requires self-consistent calculations. The result can be that the interaction with the surrounding medium is delayed and is observed at relatively soft energies. An additional complication is that neutrons in the relativistic outflow can give a delayed interaction at $r_n \approx 8 \times 10^{16} \gamma_{300}$ cm [16]; this effect is not present if the burst is Poynting flux dominated.

Even if most GRBs are produced by internal shocks, there is the possibility that some fraction of the bursts are the result of external shocks. The expectation for this type of burst would be a lack of shells with differing speeds, so that the GRB would have a relatively smooth evolution. This was the case for GRB 021211 and Kumar & Panaitescu [17] have modeled the early smooth evolution of GRB 021211 with the external shock assumption. In their wind model for the interaction, they find $A_* \approx 5 \times 10^{-4}$, which is so low that the model seems implausible. However, the constant density model for the gamma-ray emission requires $n_0 \lesssim 10^{-2}$ cm^{-3}, which is a very low density at the radius of GRB production if the progenitor is a massive star.

AFTERGLOWS

Afterglow radiation results from the interaction of the GRB burst with the surrounding medium, so it should provide a probe of the circumburst medium. The early part of the afterglow may be dominated by the reverse shock wave that goes into the ejecta shell. In the case of interaction with a wind with $A_* \sim 1$, the reverse shock is in the strong cooling regime [9] and a rapid decline from maximum is expected. The rate of decline can be related to the longer time it takes off-axis emission to reach the observer.

The early optical emission from GRB 990123 has been interpreted in terms of an energy conserving reverse shock front [18]. The overall afterglow from this source has been modeled as interaction with a low density ($n_0 \sim 2 \times 10^{-3}$ cm^{-3}), uniform surrounding medium [10, 11]. The early optical afterglow from GRB 021211 was similar to that from GRB 990123, although less luminous, and is likely to be explained by a similar model; again, a low density surroundings appears to be necessary [17]. In the case of GRB 021004, the early optical observations show a relatively flat evolution and are ambiguous. They can be explained by either the decline in the reverse shock phase and subsequent rise from the forward shock in a constant density external medium [19], or by the forward shock in a wind medium [20]. More detailed early information, including colors, is needed to distinguish between these possibilities. Also, the effects of pre-acceleration by the GRB radiation may lead to strong, early optical emission from the external shock wave, which obviates the need for reverse shock emission [16].

Constraints on the surrounding density can also be obtained from early radio observations. A number of bursts have shown relatively strong radio emission at an age of ~ 1 day. This was first seen in GRB 990123, where it was interpreted as emission from the reverse shock wave [21]. The emission requires the reverse shock wave not be in the cooling phase and that synchrotron self-absorption not be strong. Soderberg & Ramirez-Ruiz [22] have examined the early radio emission from 5 bursts and concluded that $A_* < 0.1$ is needed if the bursts are interacting with a wind. Kumar & Panaitescu [23] examined the early optical emission from the external shock in a number of bursts, allowing for pair production in the early external shock interaction, and also concluded that $A_* < 0.1$ for these cases.

A possible problem for the massive star scenario is that in a number of cases, models of afterglows show that interaction with a constant density medium is preferred over interaction with a wind [10, 11, 24]. The most plausible way for such a medium to be produced around a massive star is by a wind that has passed through a termination shock [25]. Such a picture may be able to account for the apparent constant density regions in several observed cases, but only if the surrounding pressure is very high in some sources and A_* is low (< 0.01) in another [26]. The density downstream from a termination shock is higher that what would be present in a free wind, so that a low surrounding density still requires a small value for A_*. The high pressure might occur if the burst is in a starburst region. In this case, there should be a relation of the afterglow properties to the burst position in a galaxy. The results of afterglow modeling show that in general $A_* < 1$.

ABSORPTION LINES

An interesting recent development is the finding of blueshifted absorption lines, including lines of high ionization species, in a number of afterglows. The best studied of these is GRB 021004 [27, 28]. Mirabal et al. [27] find a host redshift of $z = 2.328$ and absorption lines of CIV, SiIV, and Lyman lines at $z = 2.323, 2.317, 2.293$, corresponding to velocities of $-450, -990, -3155$ km s^{-1} relative to the host. In addition to GRB 021004, high excitation, high velocity absorption features have been found in GRB 020813 [29] and GRB 030226 [30, 31, 32]. The absorption lines of CIV in GRB 020813 are at $z = 1.223$ and $z = 1.255$ [29]. The $z = 1.223$ system has a velocity of -4320 km s^{-1} relative to the host. In this case, the blueshifted absorption is also present in a number of lower ionization species (Si II, Al II, Fe II, Mg II, and Mg I); there is no coverage of Lyα. In the case of GRB 030226, strong absorption line systems are present at $z = 1.961$ and $z = 1.984$, with C IV and Si IV present, as well as numerous lower ionization species and Lyα. The velocity separation is 2300 km s^{-1}.

The fact that the systems are always blueshifted and the character of the absorption systems makes it unlikely that these lines are produced in intervening galaxies. Such systems in the spectra of quasars are quite rare. The remaining possibilities are outflows from the host galaxy or from the circumburst medium. There is no evidence for short time variability of the lines, which would imply that the lines are local to the GRB, so the question does not have a clear observational solution. If the flows are from the host galaxies, possible sources are a starburst superwind or an outflow driven by radiation from an AGN (active galactic nucleus). In a thermally driven starburst wind, the high velocity components could not be formed in the diffuse wind because its ionization would be too high. There is the possibility of cooler, denser entrained gas, but a velocity of $4,300$ km s^{-1} is unlikely to be achieved in this gas. Observations of Lyman break galaxies at high redshift show evidence for outflowing gas, but with velocities < 800 km s^{-1} [33, 34]. AGN can show absorption line systems of the observed type, but Mirabal et al. [27] argue against this possibility for GRB 021004 based on the lack of evidence for AGN activity and the low probability that the AGN flow would be aligned with the line of sight. However, low level AGN activity might be difficult to detect at $z = 1.2 - 2.3$ and one picture for AGN outflows is that they occur in conical regions adjacent to a thick accretion flow. A point source near the nucleus might have a non-neglible probability of lining up with the flow.

If the lines are formed in the circumburst region, the observed velocities may relate to those present in the wind and shells around a Wolf-Rayet star progenitor or to clumps that have been radiatively accelerated by the GRB radiation. The high velocity features have velocities that are typical of those present in a freely expanding Wolf-Rayet wind, but the presence of H and low ionization species and the composition of the gas do not support the wind hypothesis [27]. The radiatively accelerated clump hypothesis has the difficulty that the clumps are so close to the GRB that they are completely ionized, unless they are so dense that recombination is important [27, 28]. The presence of the burst in a high pressure starburst region may enable the presence of dense clumps close to the burst [26].

DISCUSSION

The current interpretations of GRBs and their afterglows do not clearly support interaction with the expected surroundings of a Wolf-Rayet star, but they do not rule it out. A number of lines of evidence indicate that the density surrounding bursts is lower than would be expected. Some lowering of the density might occur in a low metallicity region or if the progenitor star has a low mass [26], but this cannot explain the extreme cases of GRB 990123 and GRB 021211. The density surrounding some Wolf-Rayet stars at the end of their lives may be low for unknown reasons. An indication that this is the case comes from the fact that radio observations imply low densities around the Type Ic supernovae SN 1998bw and SN 2002ap, if they do not have very low efficiencies of magnetic field production [26].

GRB 030329 provides an excellent test case for wind interaction because of the evidence that its progenitor was a massive star. The GRB itself showed 2 major pulses; there is no clear distortion from interaction with an external medium. Berger et al. [35] model the afterglow as either interaction with a uniform density medium or a free wind. However, there is a sharp break in the light curve at 0.5 day, which they interpret as a jet break. Such a break may be difficult to produce by interaction with a free wind [36]. Willingale et al. [37] have also modeled the afterglow in terms of interaction with a constant density medium with $n_0 \sim 1$ cm^{-3}. GRB 030329 does not appear to have occurred in a strong starburst region, so it is difficult to attribute these properties to a shocked wind. Even in this case, the expected wind environment does not clearly manifest itself.

ACKNOWLEDGMENTS

I am grateful to Zhi-Yun Li and Claes Fransson for collaboration on these topics. This work was supported in part by NSF grant AST-0307366.

REFERENCES

1. Stanek, K. Z. et al. *ApJ*, **591**, L17 (2003)
2. Hjorth, J. et al., *Nature*, **423**, 847 (2003)
3. Matheson, T., et al. *ApJ*, **599**, 394 (2003)
4. Mazzali, P. A. et al., *ApJ*, **599**, L95 (2003)
5. Woosley, S. E., and Heger, A., *ApJ*, submitted (astro-ph/0309165), (2003)
6. Mészáros, P., Rees, M. J., and Wijers, R. A. M. J., *ApJ*, **499**, 301 (1998)
7. Dai, Z. G., and Lu, T., *MNRAS* **298**, 87 (1998)
8. Chevalier, R. A., and Li, Z.-Y., *ApJ*, **520**, L29 (1999)
9. Chevalier, R. A., and Li, Z.-Y., *ApJ*, **536**, 195 (2000)
10. Panaitescu, A., and Kumar, P., *ApJ*, **554**, 667 (2001)
11. Panaitescu, A., and Kumar, P., *ApJ*, **571**, 779 (2002)
12. MacFadyen, A. I., Woosley, S. E., and Heger, A., *ApJ*, **550**, 410 (2001)
13. Matzner, C. D., *MNRAS*, **345**, 575 (2003)
14. Rees, M. J., and Mészáros, P., *ApJ*, **430**, L93 (1994)

15. Daigne, F., and Mochkovitch, R., in *Gamma-ray Bursts in the Afterglow Era*, Ed. E. Costa, F. Frontera, & J. Hjorth, Berlin: Springer, 2001, p. 324
16. Beloborodov, A. M., in *Beaming and Jets in Gamma Ray Bursts*, in press (astro-ph/0305118), (2003)
17. Kumar, P., and Panaitescu, A., *MNRAS*, **346**, 905 (2003)
18. Sari, R., and Piran,T., *ApJ*, **517**, L109 (1999) (2003)
19. Kobayashi, S., and Zhang, B., *ApJ*, **582**, L75 (2003)
20. Li, Z.-Y., and Chevalier, R. A., *ApJ*, **589**, L69 (2003)
21. Kulkarni, S. R. et al., *ApJ*, **522**, L97 (1999)
22. Soderberg, A. M., and Ramirez-Ruiz, E., *MNRAS*, **345**, 854 (2003)
23. Kumar, P., and Panaitescu, A., *MNRAS*, submitted (astro-ph/0309161), (2003)
24. Yost, S. A., Harrison, F. A., Sari, R., & Frail, D. A., *ApJ*, **597**, 459 (2003)
25. Wijers, R. A. M. J., in *Gamma-ray Bursts in the Afterglow Era*, Ed. E. Costa, F. Frontera, & J. Hjorth, Berlin: Springer, 2001, p. 306
26. Chevalier, R. A., Li, Z.-Y., and Fransson, C., *ApJ*, **606**, 369 (2004)
27. Mirabal, N. et al., *ApJ*, **595**, 935 (2003)
28. Schaefer, B. et al., *ApJ*, **588**, 387 (2003)
29. Barth, A. et al., *ApJ*, **584**, L47 (2003)
30. Greiner, J., Guenther, E., Klose, S., and Schwarz, R., GCN 1886 (2003)
31. Price, P. A., Fox, D. W., Djorgovski, S. G., Cote, P., and Jordan, A., GCN 1889 (2003)
32. Chornock, R., and Filippenko, A. V., GCN 1897 (2003)
33. Pettini, M., Shapley, A. E., Steidel, C. C., Cuby, J., Dickinson, M., Moorwood, A. F. M., Adelberger, K. L., and Giavalisco, M., *ApJ*, **554**, 981 (2001)
34. Frye, B., Broadhurst, T., and Benítez, N., *ApJ*, **568**, 558 (2002)
35. Berger, E. et al., *Nature*, **426**, 154 (2003)
36. Kumar, P., and Panaitescu, A., *ApJ*, **541**, L9 (2000)
37. R. Willingale, R., Osborne, J.P., O'Brien, P.T., Ward, M.J., Levan, A., and Page, K.L., *MNRAS*, **349**, 31 (2004)

Dynamos, Super-pulsars and Gamma-ray Bursts

Stephan Rosswog* and Enrico Ramirez-Ruiz[†]

*International University Bremen, Germany
[†]Institute for Advanced Study, Princeton, USA

Abstract. The remnant of a neutron star binary coalescence is expected to be temporarily stabilised against gravitational collapse by its differential rotation. We explore the possibility of dynamo activity in this remnant and assess the potential for powering a short-duration gamma-ray burst (GRB). We analyse our three-dimensional hydrodynamic simulations of neutron star mergers with respect to the flow pattern inside the remnant. If the central, newly formed super-massive neutron star remains stable for a good fraction of a second an efficient low-Rossby number $\alpha - \Omega$-dynamo will amplify the initial seed magnetic fields exponentially. We expect that values close to equipartition field strength will be reached within several tens of milliseconds. Such a super-pulsar could power a GRB via a relativistic wind, with an associated spin-down time scale close to the typical duration of a short GRB. Similar mechanisms are expected to be operational in the surrounding torus formed from neutron star debris.

INTRODUCTION

While there is mounting evidence that the long-soft variety GRBs are related directly to the death of massive stars and go along with supernova explosions, there is so far little evidence about the progenitor of short-hard GRBs. The most popular candidates are binary coalescences of either a double neutron star or a stellar mass black hole with a neutron star. Most often a "unified picture" for the GRB central engine, a new-born black hole plus a debris disk, is invoked. We will explore here the possibility that the central object of the merger remnant produces a GRB via magnetic processes *before* collapsing to a black hole. Further scenarios with ultra-magnetised neutron stars have been suggested, for example, by Usov (1992, [1]; 1994, [2]), Duncan and Thompson (1992, [3]), Thompson and Duncan (1993, [22]), Meszaroz and Rees (1997, [4]), Katz (1997, [5]) and Kluzniak and Ruderman (1998, [6]).

The merger of two neutron stars results in a massive central object, a thick, hot and dense torus of neutron star debris and some material on highly eccentric/unbound orbits [7, 8, 9]. The central object of the remnant is rapidly differentially rotating [10, 9, 11, 12] with rotational periods ranging from ~ 0.4 to ~ 2 ms [12]. Differential rotation is known to be very efficient in stabilising stars that are substantially more massive than their non-rotating maximum mass. For example, Ostriker and Bodenheimer (1968, [13]) constructed differentially rotating white dwarfs of 4.1 $M_\odot$. A recent investigation analysing differentially rotating polytropic neutron stars [14] finds it possible to stabilise systems even beyond twice the typical neutron star mass of 2.8 $M_\odot$. The exact time scale of this stabilisation is difficult to determine, as all the poorly known high-density nuclear physics ("exotic" condensates etc.) could influence the results, but estimates of

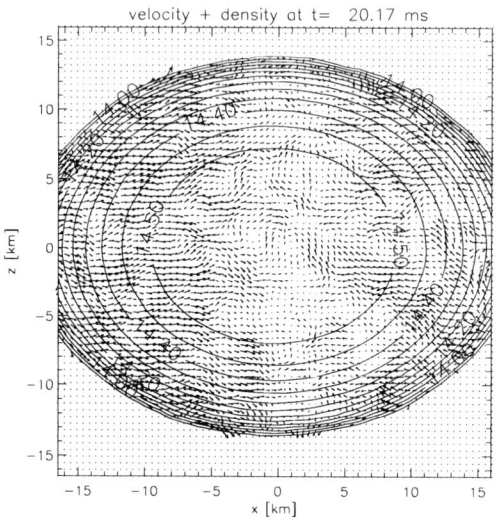

FIGURE 1. Velocity field (space-fixed frame) inside the central object of the remnant of a neutron star coalescence. The labels at the contour lines refer to $\log(\rho)$, typical fluid velocities are $\sim 10^8$ cm/s.

up to many seconds are not unrealistic.

SIMULATIONS

We have performed 3D simulations of the last inspiral stages and the subsequent coalescence for about 20 ms. We use a temperature and composition dependent nuclear equation of state that covers the whole relevant parameter space in density, temperature and composition [15, 12]. In addition, a detailed, multi-flavour neutrino treatment has been applied to account for energy losses and compositional changes due to neutrino processes. The neutrino treatment and the results concerning the neutrino emission have been described in detail in [16]. To solve the hydrodynamic equations we use the smoothed particle hydrodynamics method (SPH), the simulations are performed with up to more than a million SPH particles. We use Newtonian self-gravity plus extra forces emerging from the emission of gravitational waves. The details of the production runs as well as those of several test runs can be found in [12, 16, 18]. Results focusing particularly on gamma-ray bursts have been presented in [17, 18, 19].

DYNAMO ACTION IN MERGER REMNANTS

Before proceeding further with the argumentation, it is worth pointing out that the fluid flow never becomes axisymmetric during the simulation and therefore Cowlings anti-dynamo theorem does not apply here.

The central object of the merger remnant is differentially rapidly rotating with rotation periods below 1 ms over a large fraction of the central object's radius, examples of rotation profiles can be found in [12]. When the stellar surfaces come into contact, a vortex sheet forms between them across which the tangential velocities exhibit a discontinuity. This vortex sheet is Kelvin-Helmholtz-unstable with the shortest modes growing fastest. These fluid instabilities lead complex flow patterns inside the central object of the merger remnant. In the orbital plane they manifest themselves as strings of vortex rolls that may merge (see Fig. 8 in [12]). An example of the flow pattern perpendicular to the orbital plane is shown in Fig. 1. This pattern caused by fluid instabilities exhibits "cells" of size $l_c \sim 1$ km and velocities of $v_c \sim 10^8$ cm/s.

Moreover, the neutrino optical depth drops very steeply from $\sim 10^4$ at the centre to the edge of the central object (see Fig. 11 in [16]). For this reason the outer layers loose neutrino energy, entropy and lepton number at a much higher rate than the interior, this leads to a gradual build-up of a negative entropy and lepton number gradient which will drive vigorous convection [20, 21]. We expect this to set in after a substantial fraction of the neutrino cooling time (i.e. on time scales longer than our simulated time) when a lot of the thermal energy of the remnant has been radiated away. The situation is comparable to the convection in a new-born protoneutron star, but here we have around twice the mass and the matter is much more deleptonized ($Y_e \sim 0.1$). As both the fluid instabilities and the neutrino-driven convection have very similar properties, we will not further distinguish between them in this context. Assuming neutrinos to be the dominant source of viscosity [22] we estimate a viscous damping time scale of the order $\tau_c \sim l_c^2/\nu_\nu \sim 60$ s, where ν_ν is the neutrino viscosity. In other words the fluid pattern will be damped out only on a time scale that is much longer that the time scales of interest here.

We expect an efficient $\alpha - \Omega$-dynamo to be at work in the merger remnant. The differential rotation will wind up initial poloidal into a strong toroidal field ("Ω-effect"), the fluid instabilities/convection will transform toroidal fields into poloidal ones and vice versa ("α-effect"). Usually, the Rossby number, $Ro \equiv \frac{\tau_{rot}}{\tau_{conv}}$ is adopted as a measure of the efficiency of dynamo action in a star. In the central object we find Rossby numbers well below unity, ~ 0.4, and therefore expect an efficient amplification of initial seed magnetic fields. A convective dynamo amplifies initial fields exponentially with an e-folding time given approximately by the convective overturn time, $\tau_c \approx 3$ ms; the saturation field strength is thereby independent of the initial seed field (Nordlund et al. 1992, [23]).

Adopting the kinematic dynamo approximation we find that, if we start with a typical neutron star magnetic field, $B_0 = 10^{12}$ G, as seed, equipartition field strength in the central object will be reached (provided enough kinetic energy is available, see below) in only ≈ 40 ms. The equipartition field strengths in the remnant are a few times 10^{17} G for the central object and around $\sim 10^{15}$ G for the surrounding torus (see Fig. 8 in [18]). To estimate the maximum obtainable magnetic field strength (averaged over the central object) we assume that all of the available kinetic energy can be transformed into magnetic field energy. Using the kinetic energy stored in the rotation of the central object, $E_{kin} = 8 \cdot 10^{52}$ erg for our generic simulation, we find $\langle B_{co} \rangle = \sqrt{3 \cdot E_{kin}/R_{co}^3} \approx 3 \cdot 10^{17}$ G (note that if only a fraction of 0.1 of the equipartition pressure should be reached this would still correspond to $\sim 10^{17}$ G).

GAMMA-RAY BURSTS

There are various ways how this huge field strength could be used to produce a GRB. The fields in the vortex rolls (see Fig. 8 in [12]) will wind up the magnetic field fastest. Once they reach field strengths close to the local equipartition value they will become buoyant, float up, break through the surface and possibly reconnect in an ultra-relativistic blast [6]. The time structure imprinted on the sequence of such blasts would then reflect the activity of the fluid instabilities inside the central object. The expected lightcurve of the GRB would therefore be an erratic sequence of sub-bursts with variations on millisecond time scales.

Simultaneously such an object can act as a scaled-up "super-pulsar" and drive out an ultra-relativistic wind. A similar configuration, a millisecond pulsar with a magnetic field of a few times 10^{15} G, formed for example in an accretion-induced collapse, has been suggested as a GRB-model by Usov (1992, [1]; 1994, [2]). The kinetic energy from the braking of the central object is mainly transformed into magnetic field energy that is frozen in the outflowing plasma. At some stage the plasma becomes transparent to its own photons producing a blackbody component. Further out from the remnant the MHD-approximation breaks down and intense electromagnetic waves of the rotation frequency of the central engine are produced. These will transfer their energy partly into accelerating outflowing particles to Lorentz-factors in excess of 10^6 that can produce an afterglow via interaction with the external medium. The other part goes into non-thermal synchro-Compton radiation with typical energies of ~ 1 MeV [2].

SUMMARY

We have discussed the possibility of dynamo action in the central object created in a neutron star merger, which is expected to be stabilised against gravitational collapse via differential rotation. If it remains so for a good fraction of a second then the initial neutron star magnetic fields are expected to be amplified by a low-Rossby number $\alpha - \Omega$-dynamo. In principle enough rotational energy is available to attain an average field strength in the central object of $3 \cdot 10^{17}$ G. Locally the equipartition field strength (ranging from 10^{16} to a few times 10^{17} G depending on the exact position in the remnant) may be reached. This will cause the corresponding fluid parcels to float up and produce via reconnection an erratic sequence of ultra-relativistic blasts. In addition the central object can act as a "super-pulsar" of $\sim 10^{17}$ G that transforms most of its rotational energy into an ultra-relativistic wind with frozen-in magnetic field. As shown in [2] such a wind will result in a black-body component plus synchro-Compton radiation. Such a super-pulsar will spin-down in ~ 0.2 s, just the typical duration of a short GRB.

We have only discussed magnetic processes in the central object of the remnant, but very similar processes are expected from the surrounding torus [24]. Here, however, longer time scales and lower magnetic field strengths are expected, the equipartition fields being around 10^{15} G.

ACKNOWLEDGMENTS

The reported simulations have been performed using the UK Astrophysical Fluids Facility (UKAFF) and the supercomputer of the Mathematical Modelling Centre of the University of Leicester (HEX).

REFERENCES

1. Usov, V.V., Nature, 357, 472 (1992)
2. Usov, V.V., MNRAS, 267, 1035 (1994)
3. Duncan, R.C. and Thompson, C., ApJ, 392, L9 (1992)
4. Meszaros, P. and Rees, M.J., ApJ, 482, L29 (1997)
5. Katz, J.I., ApJ, 490, 633 (1997)
6. Kluzniak, W.; Ruderman, M., ApJ, 505, L113 (1998)
7. Ruffert M., Janka H.-T., Schäfer G., 1996, A & A, 311, 532
8. Ruffert M., Janka H.-T., Takahashi K., Schäfer G., 1997, A & A, 319, 122
9. Rosswog, S.; Liebendörfer, M.; Thielemann, F.-K.; Davies, M. B.; Benz, W.; Piran, T., A&A, 341, 400 (1999)
10. Rasio, F.A. and Shapiro, S.L., Class.Quant.Grav. 16, R1-R29 (1999)
11. Faber, J.A., Rasio, F.A. and Manor, J.B.,Phys.Rev. D63, 044012(2001)
12. S. Rosswog, M.B. Davies, MNRAS, **334**, 481 (2002)
13. Ostriker,P. and Bodenheimer, J.P., ApJ, 151, 1089 (1968)
14. Lyford, N. D., Baumgarte, T.W. and Shapiro, Stuart L., ApJ, 583, L410 (2002)
15. Shen H., Toki H., Oyamatsu K., Sumiyoshi K., Nuclear Physics, A **637**, 435 (1998); Shen H., Toki H., Oyamatsu K., Sumiyoshi K., Prog. Theor. Phys., **100**, 1013 (1998)
16. S. Rosswog, M. Liebendörfer, MNRAS, 342, 673 (2003)
17. S. Rosswog, E. Ramirez-Ruiz, MNRAS, **336**, L7 (2002)
18. S. Rosswog, E. Ramirez-Ruiz, MNRAS, 343, L36 (2003)
19. S. Rosswog, E. Ramirez-Ruiz, M.B. Davies, MNRAS, 345, 1077 (2003)
20. Epstein, R.I., MNRAS, 188, 305 (1979)
21. Burrows,A. and Lattimer, J.M., Phys. Rep., 163, 51 (1988)
22. Thompson, C. & Duncan, R.C., ApJ, 408, 194 (1993)
23. Nordlund et al., ApJ, 392, 647 (1992)
24. Narayan, R., Paczynski, B. and Piran, T., ApJ, 395, L83 (1992)

GRB 021004: A Possible Shell Nebula around a Wolf-Rayet Star Gamma-Ray Burst Progenitor

N. Mirabal*, J. P. Halpern*, R. Chornock[†], A. V. Filippenko[†] and D. M. Terndrup**

Astronomy Department, Columbia University, 550 West 120th Street, New York, NY 10027
[†]*Department of Astronomy, 601 Campbell Hall, University of California, Berkeley, CA 94720-3411*
**Department of Astronomy, Ohio State University, Columbus, OH 43210*

Abstract. The rapid localization of GRB 021004 by the HETE-2 satellite allowed nearly continuous monitoring of its early optical afterglow decay, as well as high-quality optical spectra that determined a redshift of $z = 2.328$ for its host, an active starburst galaxy with strong Lyman-α emission and several absorption lines. Spectral observations show multiple absorbers blueshifted by up to 3,155 km s^{-1} relative to the host galaxy Lyman-α emission. We argue that these correspond to a fragmented shell nebula, gradually enriched by a Wolf-Rayet wind over the lifetime of a massive progenitor bubble. In this scenario, the absorbers can be explained by circumstellar material that have been radiatively accelerated by the GRB emission. Dynamical and photoionization models are used to provide constraints on the radiative acceleration from the early afterglow.

INTRODUCTION

Gamma-ray bursts (GRBs) have been a challenge for astronomers ever since their serendipitous discovery by the Vela satellites in the early 1970s [1]. However, evidence collected over the past six years now links "long-duration" (> 2 s) GRBs to the deaths of massive stars. Some of the clearest information about the nature of GRBs comes from the coincidence of the unusual GRB 980425 with SN 1998bw [2], and the discovery of the Type-Ic supernova, SN 2003dh, nearly simultaneous with GRB 030329 [3]. The temporal coincidence of these events proves that long-duration GRBs are associated with peculiar Type-Ic supernovae (SNe), and thus are a consequence of the evolution of massive stars [4]. The evidence also strongly supports the collapsar model for GRBs where a rotating massive star (typically a Wolf-Rayet star) undergoes core collapse to a black hole surrounded by an accretion disk wind [5].

While it is now generally accepted that some long-duration GRBs are associated with the deaths of massive stars, considerable uncertainty remains as to what the precise nature of the progenitor star is. Theories of stellar evolution suggest that massive stars lose a large fraction of their mass through strong stellar winds [6]. As a result, one expects to find a substantial amount of circumstellar material in the vicinity of long-duration GRBs. Indeed, observations of some GRB environments show compatibility with a wind-like medium [7]. There have also been reports of strong UV absorption lines in GRB afterglows with outflow velocities of up to 4,260 km s^{-1} [8, 9, 10]. We discuss how these observations can provide information on the GRB progenitor.

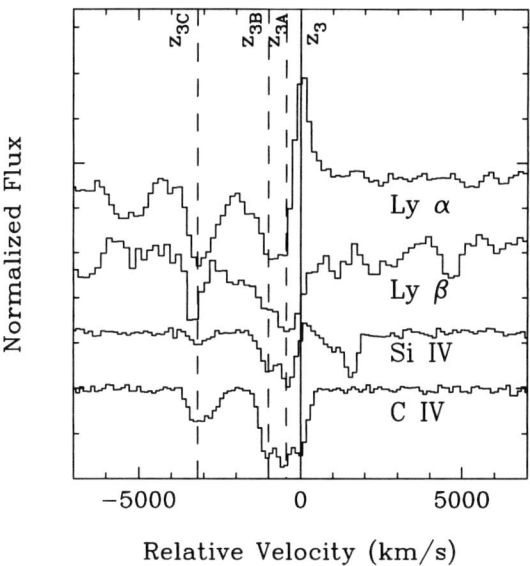

FIGURE 1. Outflowing Ly α, Ly β, C IV, and Si IV absorbers in the GRB 021004 afterglow spectrum plotted in velocity space. As zero velocity we use the systemic redshift of the host galaxy $z = 2.328$. The dashed lines indicate blueshifted absorbers at 450, 990, and 3,155 km s^{-1}.

SPECTRAL OBSERVATIONS OF THE GRB 021004 AFTERGLOW

Possibly the best example of velocity shifts in GRB afterglows is the optical spectrum of GRB 021004, an active starburst galaxy with strong Ly-α emission and several absorption lines. Spectral observations of its afterglow revealed multiple blueshifted kinematic components at $z_{3A} = 2.323$, $z_{3B} = 2.317$, and $z_{3C} = 2.293$ with radial velocities of ~ 450, ~ 990, and $\sim 3,155$ km s^{-1} relative to the systemic velocity of the host galaxy (Figure 1). The absorption components also show velocity widths broader than the expected thermal widths at the instrumental resolution, indicating internal motions within each component. Such velocity structure is highly unusual for large-scale absorbing material near or around the GRB host galaxy.

One is thus led to consider scenarios where the absorbers are closer to the GRB progenitor system (i.e., associated). Hot, massive stars generally have expanding material characterized by velocities of up to 3,000 km s^{-1}, which originates through the scattering of stellar radiation in the stellar wind. Radial outflows seen via blueshifted resonance lines are also a common trait of a large fraction of Seyfert galaxies [11]. These are thought to represent massive outflows of highly ionized gas from their active galactic nuclei. Although there are marked differences between Seyferts and GRBs, the similarities in their spectra may shed light on the physical conditions of the blueshifted components. Building off this premise, we explore a scenario where the absorbers in the GRB 021004 spectrum are the result of outflowing stellar material associated with the GRB progenitor.

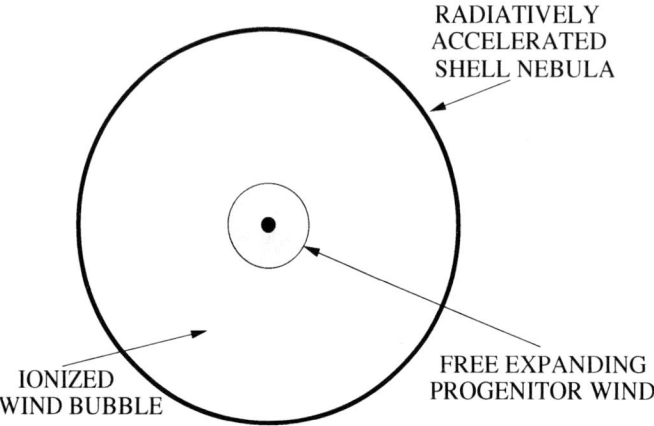

FIGURE 2. Schematic cross-section of a stellar-wind bubble model with various features including the termination of the wind and shell nebula. The model cannot reproduce the great wealth of structure observed around massive stars.

SHELLS, FILAMENTS AND CLUMPS

One of the most interesting consequences of stellar evolution is that stellar-wind bubbles are intimately connected to the mass-loss history of their central star. On its way to the Wolf-Rayet phase, a main-sequence star is thought to evolve through a luminous blue variable (LBV) or red supergiant (RSG) stage. The slow winds (10–50 km s^{-1}) generated during the LBV or RSG phase expand into the interior of the main-sequence bubble until the mass driven by the wind is comparable to the mass of circumstellar material. This condition sets the characteristic radius of expansion R_s roughly given by

$$R_s = \sqrt{\frac{\dot{M}\tau}{n_0}} \text{ pc.} \qquad (1)$$

where τ is wind lifetime in units of 10^6 years, $\dot{M}$ is the mass-loss rate in units of $10^{-6} M_\odot$ yr^{-1}, and n_0 corresponds to the density of the surroundings in units of cm^{-3}.

Interestingly, stellar winds carry not only mass but kinetic energy into the ambient medium. Such injection of energy leads naturally to the formation of overdense shell nebulae (with expansion velocities v $\approx$ 40 km s^{-1}) along the wind profile. Figure 2 shows the predicted physical structure of a shell nebula formed at the termination of a massive stellar wind. Soon after entering its Wolf-Rayet phase, a fast wind (1000–3000 km s^{-1}) starts sweeping the LBV or RSG material, eventually overtaking the main-sequence gas residing around the progenitor star. The combination of streaming winds and internal instabilities within the wind results in a complex morphology characterized by fragmented shells, filaments and clumps. Additional structure is likely to be introduced by wind-wind collisions in close-binary systems. But the basic picture is confirmed through the variety of morphologies observed in the surroundings of isolated Wolf-Rayet stars [12].

Apart from providing a complex circumstellar environment, a wind bubble configuration either isolated or in a close-binary system will contain a significant amount of mass from all past stellar phases. If GRBs are indeed formed by the death of massive stars, overdense stellar regions within the bubble may produce detectable spectral features in the optical afterglow. From the absence of N V and the presence of Ly α, Si IV and C IV, we inferred that the absorbing material in the GRB 021004 spectrum was dominated by He-burning and core nucleosynthesis products. Such composition is in good agreement with a massive late-type carbon-rich Wolf-Rayet star embedded inside an interstellar bubble, in which the Wolf-Rayet wind has gradually enriched the bubble interior. So far the argument is consistent with the abundances, but let us explore the kinematics.

PHOTOIONIZATION AND DYNAMICAL MODELS

As part of the analysis, we modeled the radiation environment in a wind bubble system following a GRB explosion using detailed photoionization models. The models are especially constructed to provide constraints on the physical conditions of blueshifted absorbers. For these particular simulations, we used the photoionization code IONIZEIT described in [13], which includes time-dependent photoionization processes taking place under a predetermined GRB afterglow ionizing flux. Perhaps the most dramatic outcome of the simulations is that material intrinsic to the GRB progenitor should be subjected to increasing velocity and gradual ionization.

In order to explain the absence of significant variability following the first afterglow spectrum, we concluded that radiation acceleration by bound-free transitions had to be most efficient in the early stages of the GRB [9, 14]. Figure 3 shows the predicted velocity profile for one set of initial conditions. Faster outflowing absorbers can be explained by nearby circumstellar material, while more distant absorbers will evolve slower. After final velocity is achieved the absorbers simply coast along. In particular, the simulations are able to reproduce the kinematics of GRB 021004 if high-density clouds are placed at a distance $0.3 < d < 30$ pc from the GRB. The constraints from these models are consistent with the typical sizes of stellar winds around Wolf-Rayet stars in our Galaxy, and provide quite possibly the first direct spectral signature of circumstellar material around a GRB progenitor. They also lend support to the collapsar model for GRBs where a rotating massive star undergoes core collapse to a black hole [5].

CONCLUSIONS AND FUTURE WORK

Observations of the early afterglow of GRB021004 show multiple absorption features blueshifted by up to 3,155 km s^{-1} relative to the host galaxy of the GRB. The features may indeed have been caused by circumstellar material from a Wolf-Rayet progenitor wind located at a distance $0.3 < d < 30$ pc from the GRB site that has been radiatively accelerated by the GRB afterglow emission. While at this stage we cannot distinguish between an isolated Wolf-Rayet star and a close-binary system, the observational data on GRB 021004 could be the first direct spectral signature of material in the surroundings

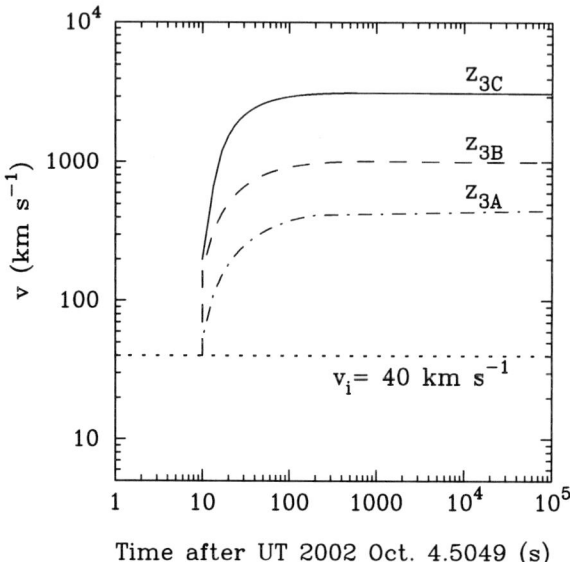

FIGURE 3. Simulated velocity profiles for radiatively-accelerated outflows. The dotted line corresponds to an initial outflowing velocity of $v \approx 40$ km s^{-1}.

of a GRB. Our findings motivate the need to undertake intensive surveys for variable and accelerating resonance lines (C IV, Si IV, N V, and O VI), along with Ly α and lower ionization species. In addition to the observations, intensive numerical modeling is required to deal with the overionization that takes place when the progenitor gas is exposed to strong GRB emission. Lastly, we note that the advent of the *Swift* GRB mission (see http://swift.gsfc.nasa.gov/) should bring unique access to early multiwavelength observations of GRBs, which will provide a critical diagnostic tool for GRB progenitors.

REFERENCES

1. Klebesabel, R. W., Strong, I. B., & Olson, R. A., ApJ **182**, L85 (1973).
2. Galama, T. J., et al., Nature **395**, 670 (1998).
3. Stanek, K. Z., et al., ApJ **591**, L17 (2003).
4. Hjorth, J., et al., Nature **423**, 847 (2003).
5. Woosley, S. E., ApJ **405**, 273 (2003).
6. Cassinelli, J. P., ARA&A **17**, 275 (1979).
7. Chevalier, R. A., & Li, Z.-Y., ApJ, **520**, L29 (1999).
8. Barth, A. J., et al., ApJ **584**, L47 (2003).
9. Mirabal, N., et al., ApJ **595**, 935 (2003).
10. Fox, D. W., these proceedings (2003).
11. Crenshaw, D. M., et al., ApJ **516**, 750 (1999).
12. Marston, A. P., ApJ **475**, 188 (1997).
13. Mirabal, N., et al., ApJ **578**, 818 (2002).
14. Schaefer, B., et al., ApJ **588**, 387 (2003).

Stellar Collapse and the Formation of Black Holes

Chris L. Fryer* and Rejean Dupuis[†]

*T-6, MS B227, Los Alamos National Laboratory, Los Alamos, NM 87545
[†]Physics & Astronomy, University of Glasgow, Glasgow G12 8QQ, UK

Abstract. We review the engines behind neutrino-driven supernovae and gamma-ray bursts. Combined with our understanding of the convection-enhanced, neutrino-driven supernova mechanism, the stellar collapse can explain all of the supernova-like explosions observed from normal supernovae, to weak explosions and jet-like hypernovae. Combining this theoretical understanding with observations suggests that the collapsar rate is roughly 1/1000th that of normal supernovae.

INTRODUCTION

The collapsar engine is increasingly becoming the favored power source for long-duration gamma-ray bursts and a new class of jet-driven explosions, also known as hypernovae. Indeed, astronomers seem to be following the fashion of invoking jet-like explosions from stellar collapse to explain a variety of observed explosions (Wang & Wheeler, [1]). However, both physics and observational constraints limit the number of jet-like explosions in astrophysics. Theoretical and observational evidence all suggests that >99% of all stellar collapses do not produce collapsar explosions. In this paper, we review this evidence.

Before we dive into the details of stellar collapse, let's review the constraints on collapsars:

- The star must collapse to a black hole. This can occur in stars that produce weak or no supernova explosion.
- The star must have sufficient angular momentum to form a disk around that black hole. The ideal angular momentum (j) range in the core lies between $10^{16} < j < 10^{18} \mathrm{cm}^2 \mathrm{s}^{-1}$.
- **GRBs only** - The star must lose its hydrogen envelope. A massive star that retains its hydrogen envelope will go into a giant phase prior to collapse. If the engine turns off before the jet breaks through this giant, the jet will not be relativistic enough to produce GRBs. It is difficult to maintain the engine for such long times (> 1000 s).

In this paper, we will focus our attention on the first criterion and how our understanding of core-collapse supernovae constrains the number of collapsars. To do this, we must understand the how the neutrino-driven mechanism behind normal supernovae and the conditions that cause it to fail. We present current simulation results of black-hole forming stars and conclude with a comparison of the theoretical results with current

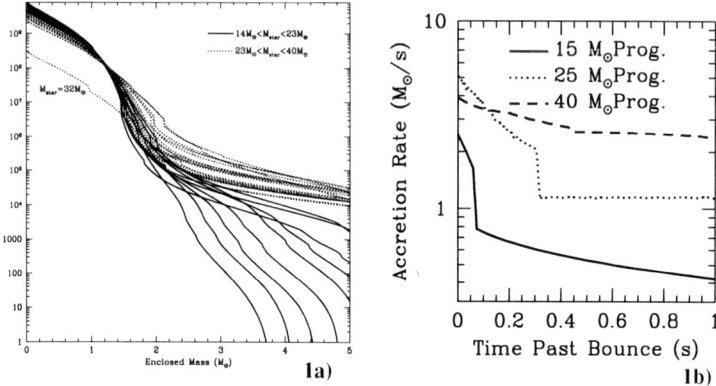

FIGURE 1. Fig. 1a) Density versus enclosed mass for stellar models just prior to collapse with initial stellar masses ranging from $14\,M_\odot$ up to $40\,M_\odot$ (Woosley, Heger, Weaver, [2]). Out to 1.2-$1.4\,M_\odot$, there is very little difference in the stellar densities. Fig 1b) Accretion rates onto the convective region in the supernova engine as a function of time past bounce. The pressure from this infall must be overcome to drive a supernova explosion. The lower accretion rates of lower mass stars make these stars easier to explode. See Fryer ([3]), [4]) for more details.

observations.

WHAT MAKES THE SUPERNOVA ENGINE FAIL

Massive stars collapse when the iron core in their center becomes so hot and dense that the iron in the core dissociates (removing all the nuclear binding energy it gained throughout its life) and the electrons in the core capture onto protons. This core, supported by thermal and electron degeneracy pressure, loses its support almost instantly as its initial contraction drives further iron dissociation and electron capture. This nearly free-fall collapse halts only when the core reaches nuclear densities where nuclear forces and neutron degeneracy pressure halt the collapse. Because the inner $1.4\,M_\odot$ of all massive stars have nearly identical structures, this collapse and bounce phase is nearly the same for all stars above $\sim 12-15\,M_\odot$ (Figure 1a).

All stars collapse, bounce, and form a very similar proto-neutron star. And for all stars, the bounce shock stalls when its density decreases sufficiently for neutrinos to escape and sap the shock of its energy. The stalled shock leaves behind a convectively unstable

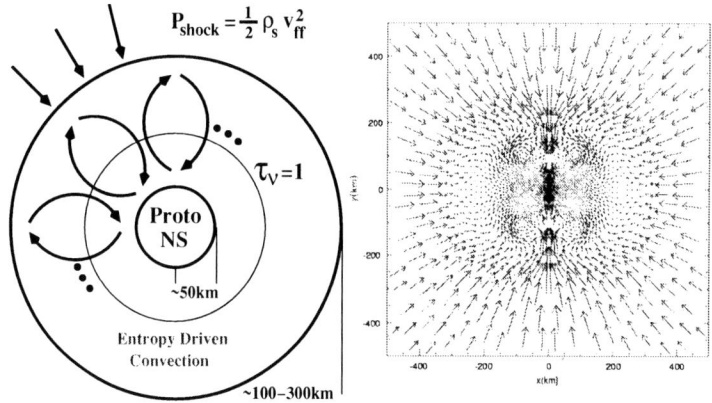

FIGURE 2. Left: Diagram of the neutrino-driven convective engine. At its center is a hot proto-neutron star that emits neutrinos and heats material just above its surface. This material is convectively unstable and rises, pushing against the rest of the infalling star. The infalling star makes a lid for this neutrino-heated pressure cooker. When this lid is blown off, an explosion occurs. On the right is a simulation plot, showing the downflows and upflows of this convective engine ([5]).

layer above the proto-neutron star that is confined by the rest of the infalling star (Figure 2). If the convective motions can blow off the infalling matter, a supernova explosion is born. If not, the proto-neutron star collapses to from a black hole: a "collapsar" ([6]). The fate of all these stars would also all be the same except for the fact that the density beyond $1.4\,M_\odot$ does differ considerably. The accretion of this matter forms the lid that the star must blow off to explode, and the accretion rates onto the convective region can vary dramatically (Fig. 1b).

These differences, although just at the edge of the supernova engine, cause the different fates of massive stars, from strong explosions and neutron star remnants to weak explosions with enough fallback to form black holes to stars that collapse and do not explode until, for sufficiently fast rotators, a disk forms around the black hole, driving a collapsar explosion (Fig. 3). Fryer (1999) not only predicted a range of fates, but also a range of neutron star and black hole masses. He also predicted that stars above $20\,M_\odot$ should form black holes and either weak explosions via the supernova mechanism or strong collapsar outbursts.

For a reasonable mass distribution of stars, roughly 10-20% of all massive stars would collapse to form black holes (Fryer & Kalogera 2001). This black hole fraction is an upper limit of the total number of jet-like (collapsar) explosions from stellar collapsar.

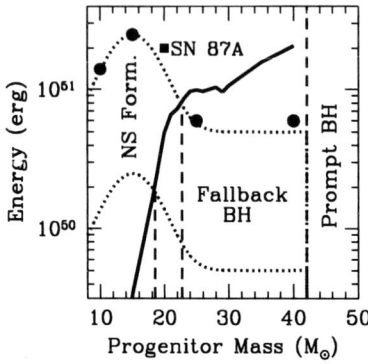

FIGURE 3. Explosion energy from simulations (solid circles) and binding energy of the stellar envelope ($\equiv M_{\text{star}}$ - 3 M$_\odot$: solid lines) of stars as a function of initial progenitor mass (a rough a fit to the solid circles). The dotted lines denote rough estimates of the supernova explosion energy as a function of mass. This energy must drive the explosion and eject the outer layers of the star. Where the dotted and solid lines cross marks the transition between neutron star and black hole formation. If 100% of this energy goes to ejecting the stellar envelope Assuming 3 M$_\odot$ is the maximum neutron star mass, the progenitor mass for neutron star/black hole transition is at 23 M$_\odot$. If only 10% of this energy goes into ejecting the mass (or the explosion energies are weaker than predicted), the transition mass is at 18 M$_\odot$. Above $\sim 40-45$ M$_\odot$, Fryer (1999) predicts the star produces no explosion whatsoever with the standard neutrino-driven mechanism and collapses directly into a black hole.

But what fraction of these black holes actually produce collapsars? Observations suggest that both collapsars and weak supernovae occur for stars more massive than 20 M$_\odot$ (Nomoto et al. 2004), confirming the results of supernova simulations. The fact that hypernovae occur nearly 10 times more frequently than weak supernovae is balanced by the fact that weak supernovae are 100 times fainter. For these luminosity limited sample, we find that the collapsar engine works for 1 in 100 of all black hole forming stars, or roughly 1 in every 1000 stellar collapses.

GRBS FROM STELLAR COLLAPSE

When the supernova fails and the remnant collapses to a black hole, the rest of the star continues to fall down onto the black hole. To form a collapsar, we must turn around this infall along the polar axis. Just as with the supernova engine, an explosion occurs when the energy injected into the polar region can blow off the pressure of the infalling material. As the density in the polar region decreases, it becomes easier and easier to drive an explosion. If magnetic fields are the driving force, we can predict very little as we do not know the energy injection rate to even a few orders of magnitude. But if neutrino annihilation is the source of power, we can quantitatively predict when the polar lid will be driven back and an explosive jet will occur. For very massive stars (progenitors of the Collapsar type I explosion - MacFadyen & Woosley 1999), a large

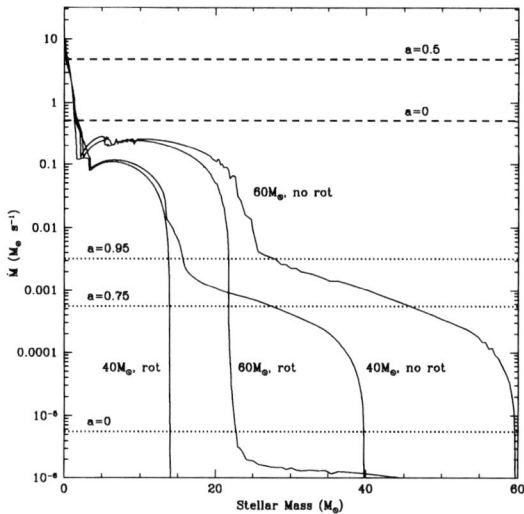

FIGURE 4. Accretion rates along the polar axis versus mass for rotating and non-rotating 40,60 M$_\odot$ stars. The GRB jet can not be launched until these rates fall below a critical value based on the energy injection. For the neutrino-annihilation power source, we can caluclate this critical rate for a range of black hole accretion disks based on the black hole spin (a) and the disk accretion rate (dotted lines refer to 0.1 M$_\odot s^{-1}$, dashed lines refer to 1.0 M$_\odot s^{-1}$. When the accretion rates fall below the critical value (the mass with higher accretion rates must accrete), an explosion is launched. The corresponding black hole masses for rotating stars range are $\sim$15-20 M$_\odot$. See Fryer & Meszaros ([7]) for details.

amount of material is accreted prior to launching a jet. The final black hole masses as a function of disk structure are shown in Figure 4.

ACKNOWLEDGMENTS

This work was funded under the auspices of the U.S. Dept. of Energy, and supported by its contract W-7405-ENG-36 to Los Alamos National Laboratory, the University of Arizona, and by a DOE SciDAC grant #DE-FC02-01ER41176.

REFERENCES

1. Wang, L., and Wheeler, J. C., *ApJ*, **504**, L87–L90 (1998).
2. Woosley, S. E., Heger, A., and Weaver, T. A., *RvMP*, **74**, 1015–1071 (2002).
3. Fryer, C. L., *ApJ*, **522**, 413–418 (1999).
4. Fryer, C. L., *International Journal of Modern Physics D*, **12**, 1795–1835 (2003).
5. Dupuis, F. C. L., R., and Heger, A., *ApJ*, **in preparation** (2004).
6. Woosley, S. E., *ApJ*, **405**, 273–277 (1993).
7. Fryer, C. L., and Meszaros, P., *ApJ*, **588**, L25–28 (2003).

Numerical Simulations of Relativistic Jets in Collapsars

Weiqun Zhang*, S. E. Woosley* and A. Heger[†][**]

*Department of Astronomy and Astrophysics, UCSC, Santa Cruz CA 95064
[†]Group T6, Los Alamos National Laboratory, Los Alamos, NM 87545
[**]Department of Astronomy and Astrophysics, University of Chicago Chicago, IL 60637

Abstract. Relativistic jets in collapsars are studied by numerical simulations. Such jets are believed to give rise to outbursts of high-energy emission known as gamma-ray bursts. The propagation and break out of the jets are examined in multi-dimensional numerical simulations using a special relativistic hydrodynamics code. During its propagation, the jet is collimated by the passage through the stellar mantle. As it erupts, the highly relativistic jet core is surrounded by a cocoon of less energetic, but still moderately relativistic ejecta that expands and becomes visible at larger polar angles. We predict a distribution of energy and Lorentz factor with viewing angle in the jet beam and its cocoon. These imply that what is seen may vary greatly with viewing angle. In particular, we predict the existence of a large number of low energy GRBs with mild Lorentz factors that may be related to GRB 980425/SN 1998bw and to the recently recognized cosmological X-ray flashes.

INTRODUCTION

It is widely believed that at least some long soft Gamma-ray bursts (GRBs) are made from relativistic jets produced in massive stars. The recent discovery of the GRB-SN association [1, 2] favors the collapsar model [3, 4]. Numerical simulations of relativistic jets from collapsars has shown that the collapsar model is able to explain many of the observed characteristics of GRBs [5, 6].

In our previous 2-dimensional simulations of relativistic jet propagation in the star and its wind, we found that a relativistic jet born deep inside the star can penetrate the stellar mantle. Then in the stellar wind, the jet can acquire a Lorentz factor high enough to make a GRB after its internal energy is converted to kinetic energy. However, The emergence of the jet and its immediate interaction with the circumstellar medium needs further examination, especially in a calculation that finally resolves the stellar surface at breakout and includes the jet cocoon as well as its core.

In this paper, we present some recent two- and three-dimensional numerical studies on the breakout of relativistic jets from massive Wolf-Rayet stars. More details will be presented elsewhere [7]. We follow the emergence of the jets for a sufficient length of time to ascertain the properties of the cocoon explosion that surrounds the GRB and we study the dependence of the results on the dimensionality of the calculation. We also study precessing jets and find that the properties of the emergent jet are quite sensitive to whether the jet maintains its orientation (to within a critical angle) for durations of 10 s or so, the time it takes the jet to traverse the star.

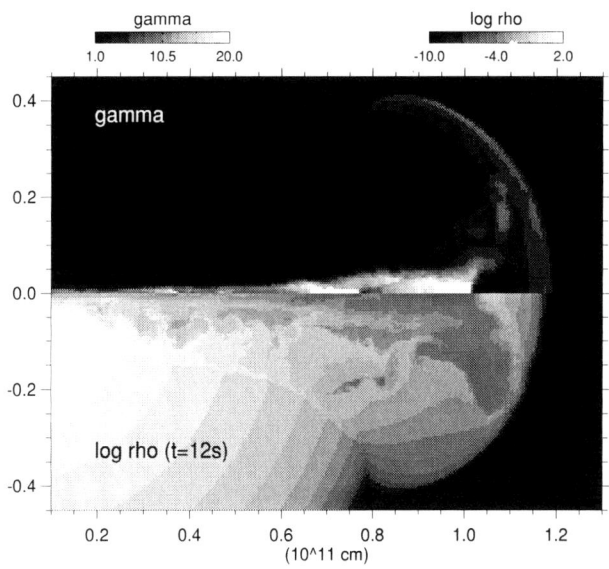

FIGURE 1. The structure of the density (bottom panel) and Lorentz factor (top panel) for Model 2A. The density is on a logarithmic scale, the Lorentz factor is on a linear scale. The jet emerges from the star with a cocoon surrounding the jet beam and a dense "plug" at the head of the jet.

RESULTS

Two-Dimensional Models of Jet Breakout

The initial stellar model is taken directly from a presupernova star with a very finely zoned surface calculated by [8]. We continue to employ our multi-dimensional relativistic hydrodynamics code that has been used previously to study relativistic jets in the collapsar environment. The mass interior to 1.0×10^{10} cm is removed from the presupernova star and replaced by a point mass. No self gravity is included. This should suffice since we are studying phenomena that happen on a relativistic time scale and the speed of sound is very sub-luminal. While jets presumably go out both axes, we follow here only one, assuming symmetry in the other hemisphere. Jets are injected along the rotation axis (the center of the cylindrical axis of the grid) through the inner boundary. Each jet is defined by its power (excluding rest mass energy), $\dot{E}$, its initial Lorentz factor, Γ_0, and the ratio of its total energy (excluding rest mass energy) to its kinetic energy, f_0.

The relativistic jet begins to propagate along the z-axis shortly after its initiation. The jet consists of a supersonic beam, a shocked cocoon, a bow shock, and is narrowly collimated. A snapshot of Model 2A at 12 s is given in Figure 1.

In all cases studied the high Lorentz factor characteristic of common GRBs is confined to a narrow angle of about 3 to 5 degrees with a maximum equivalent isotropic energy in highly relativistic matter along the axis of $\sim 3 \times 10^{53}$ to 3×10^{54} erg. Considerable energy is also at larger angles with Lorentz factors, $\Gamma \sim 10$ to 20. At an angle of 10

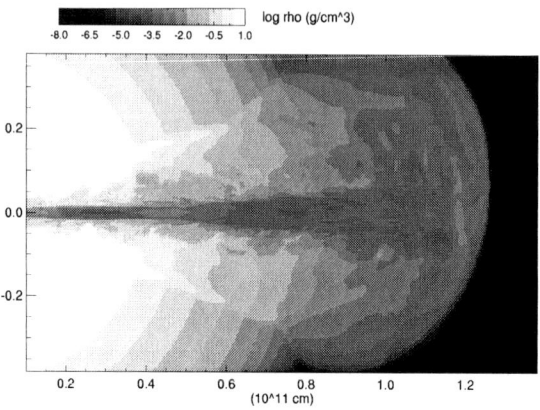

FIGURE 2. The Structure of the density for Model 3BL at 8 s. The density is on a logarithmic scale. The density of the "plug" is greatly diminished in these asymmetric jets.

degrees for example, the equivalent isotropic energy in matter with $\Gamma > 20$ is $\sim 10^{52}$ erg in Model 2A. The possibility that the off-axis emission from material with $\Gamma \sim 10$ to 20 corresponds to X-ray Flashes (XRFs) will be discussed later.

Three-Dimensional Models of Jet Breakout

It is very important to repeat these calculations in 3D to ensure that our 2D results are valid and to examine three-dimensional jet instabilities. For the three-dimensional models, the same helium star was remapped into a three-dimensional Cartesian grid. The grid employed in all 3D models was Cartesian with 256 zones each along the x- and y-axes and 512 along the z-axis (jet axis).

As expected, our Model 3A with perfect initial cylindrical symmetry closely resembles 2D models. The bulk properties of jets calculated in 2D - including collimation and modulation - would be the same in 3D. However, assuming a jet that initially has perfect cylindrical symmetry does not fully exercise the 3D code. In our 3D models 3BS and 3BL, the jets were initiated asymmetrically. In particular, Model 3BS differed from 3A in a 1% asymmetry in input energy from one side of the jet to the other, yet the structure of the emergent jet and cocoon is strikingly different. In Model 3BL with a 10% asymmetry the difference is even more striking (Figure 2). More dramatic is the difference in the high density "plug" among Models 3A, 3BS, and 3BL. In the latter two where the 2D symmetry was mildly broken, the plug has a much lower density and is not prominent. A movie of these runs shows the plug forming, slipping off to the side, then forming again.

Several other models were calculated to explore the collimation properties of non-radial jets that precessed. With a precession angle of 3 degrees, the jet escapes the star with its relativistic flow at least partly intact. However, models with larger precession angles show the break up of the jet. Because of the "baryon-poisoning" in these jets, there will be no GRB of the common variety.

DISCUSSIONS

Our simulations have shown that there is a lot of energy in the material at large angles ($\sim$ 10 degrees) after the jet breaks out of the star. We argue that ordinary bursts seen off axis might appear as hard x-ray transients of one sort or another [9, 10]. We have identified them with GRB 980425 and with the recently recognized XRFs. If this is valid, XRFs and GRBs should be continuous classes of the same basic phenomenon sharing many properties. They should have a similar spatial distribution to GRBs because they are essentially the same sources. However, because they are much less luminous, their log N -log S distribution would not exhibit the same roll over attributed in GRBs to seeing the "edge" of the distribution. Their median redshift should be considerably smaller. XRFs may, most frequently, be seen in isolation and will be characterized by softer spectra, but there would also be an underlying XRF in every GRB since the emission of the mildly relativistic cocoon material is beamed to a larger angle that includes the poles. In some cases these XRFs might be seen as precursors or extended hard X-ray emission following a common GRB. They should be associated with supernovae. Indeed XRFs may more frequently serve as guideposts to jet-powered supernovae than GRBs, especially the nearby ones.

ACKNOWLEDGMENTS

This research has been supported by NASA (NAG5-12036) and the DOE Program for Scientific Discovery through Advanced Computing (SciDAC; DE-FC02-01ER41176). This research used resources of the National Energy Research Scientific Computing Center, which is supported by the Office of Science of the U.S. Department of Energy under Contract No. DE-AC03-76SF00098.

REFERENCES

1. Stanek, K. Z. et al., ApJL, **591**, L17 (2003).
2. Hjorth, J. et al., Nature, **423**, 847 (2003).
3. Woosley, S. E. , ApJ, **405**, 273. (1993).
4. MacFadyen, A., Woosley, S. E. , ApJ, **524**, 262 (1999).
5. Aloy, M. A., Müller, E., Ibáñez, J. M., Martí, J. M., & MacFadyen, A., ApJL, **531**, L119 (2000).
6. Zhang, W., Woosley, S. E., & MacFadyen, A. , ApJ, **586**, 356 (2003).
7. Zhang, W., Woosley, S. E., & Heger, A., ApJ, submitted, astro-ph/0308389 (2003).
8. Heger, A. & Woosley, S. E., American Institute of Physics Conference Series, **662**, 214 (2003).
9. Woosley, S. E., *GRBs, 5th Huntsville Symposium*, eds. Kippen, Mallozzi, & Fishman, AIP, Vol 526, 555 (2000).
10. Woosley, S. E., *GRBs in the Afterglow Era*, eds. Costa, Frontera, & Hjorh, ESO Astrophysics Symposia, (Springer), 555 (2001).

The First Steps in the Life of a Short GRB

M.-A. Aloy*, H.-T. Janka* and E. Müller*

*Max-Planck-Institut für Astrophysik. Karl-Schwarzschild-Str. 1. 85741 Garching, Germany
maa@mpa-garching.mpg.de

Abstract. We present some results of relativistic hydrodynamic simulations of post-neutron star merger disks as potential candidates to be progenitors of short gamma-ray bursts. We discuss some of the generic conditions under which a gamma-ray burst can be initiated in this kind of progenitor and the characteristics of the resulting outflow.

INTRODUCTION

Due to their different duration and spectral properties GRBs are commonly divided in two classes: short (≤ 2 s) and long (≥ 2 s) GRBs [1]. Observations of long GRBs have shed some light on the type of progenitors and environments in which such progenitors reside. Observations of short GRBs are less numerous and it has not been possible to detect them in multi frequency searches. Thus, we know very little about the progenitors of the subclass of short GRBs.

The scenario arising after the merging process of a compact binary system consists of a central black hole (BH) of a few solar masses girded by a thick accretion torus whose mass is of $0.05 - 0.3 M_\odot$ [2, 3]. Once the thick disk is formed, up to $\sim 10^{51}$ ergs can be released above the poles of the BH in a region that contains less than $10^{-5} M_\odot$ of baryonic matter due to the release of energy via $\nu-\bar{\nu}$ annihilations. In principle, this may lead to the acceleration of this matter to ultrarelativistic speeds accounting for a successful GRB. If the duration of the event is related to the lifetime of the system [4] this kind of events can only belong to the class of short GRBs because the expected time scale on which the BH engulfs the disk is fractions of a second [2].

In this work we address the question of whether a local deposition of energy around the remnant left over from the merger of two compact objects (two neutron stars or a neutron star and a BH) can yield the formation of relativistic, collimated plasma outflows that may account for short GRBs. We employ 2D general relativistic hydrodynamic (GRHD) numerical simulations to study the properties of the outflows generated when pure thermal energy is released in a wide angle cone around the rotation axis of the system consisting of a BH surrounded by a thick accretion torus. We use the code GENESIS [5] to integrate the GRHD equations in spherical (r, θ) coordinates assuming axisymmetry. Among the issues that we want to study are the viability of the scenario of compact object mergers for producing ultrarelativistic outflows; which is the mechanism that can account for collimation of the outflowing plasma; which are the expected durations of the GRB events generated in this framework and whether they are related to the time during which the source of energy is active.

INITIAL MODEL AND NUMERICAL SET UP

We have constructed two initial models in which the gravitational field is provided by a Schwarzschild BH of $3M_\odot$ (models type-A) and $2.44M_\odot$ (models type-B) located at the center of the system (effects on the dynamics due to the self-gravity of the accretion torus or the external environment are neglected). These black holes are surrounded by thick accretion disks for which the initial configurations are built either by guidance through the data of Ruffert & Janka [2] and then letting the model relax until a torus of $0.1 M_\odot$ is obtained (type-A), or analytically following a prescription very close to that of Font & Daigne [6] in order to build an equilibrium torus of $0.07 M_\odot$ around a Schwarzschild BH (type-B). The initial models include an environment which is of high density and non uniform in type-A models. In type-B models it is spherically symmetric, with low density which decreases with radius ($\rho \sim r^{-3.4}$) and having a total mass of $2.52 \cdot 10^{-7} M_\odot$. These initial configurations mimic the expected state of a remnant of the merger of a compact binary system sufficiently well for our purposes. We assume equatorial symmetry, and we cover $90°$ in the angular θ-direction with 200 uniform zones. In the r-direction the computational grid consists of 500 zones spaced logarithmically between the inner boundary and an outermost radius of $R_{\max} = 2 \cdot 10^{10}$ cm (thus, we can study the evolution of any outflow up to ~ 0.65 s). The equation of state assumes nuclei to be disintegrated to free non-relativistic nucleons, treated as a mixture of Boltzmann gases and includes the contribution from radiation, and an approximate correction due to $e^+ e^-$–pairs (as in refs. [7, 8]). Complete ionization is assumed, and the effects due to degeneracy are neglected.

In a consistent post neutron star merger model an outflow will be powered by any process which gives rise to a local deposition of energy and/or momentum, as e.g., $\nu \bar{\nu}$–annihilation, or magneto-hydrodynamic processes. We mimic such a process by releasing pure thermal energy in a prescribed cone around the rotational axis of our system. In the radial direction the deposition region extends from the inner grid boundary located at 2 gravitational radii ($R_g = GM/c^2$; G, M and c being the gravitational constant, the mass of the BH and the speed of light in vacuum, respectively) to the outer radial boundary. In the angular direction, the opening half–angle of the deposition cone (θ_0) was chosen to be in the range $30° - 75°$. From the annihilation rate distribution computed in [2] and [3], we infer a power law distribution for the energy deposited per unit of volume in the surrounding of the system whose explicit form was approximated as $\dot{q} = \dot{q}_0 z^{-n}$, where z is the distance along the rotation axis, n is the power law index ($n = 5$ hereafter) and $\dot{q}_0$ is a normalization factor that we use to fix the total energy deposition rate ($\dot{E}$).

RESULTS

We have done a parameter study addressing different aspects of the dynamics of outflows resulting from neutron star merger remnants. We concentrate here on the dependence of the resulting outflow on the increase of $\dot{E}$ from $5 \cdot 10^{49}$ erg s^{-1} to $5 \cdot 10^{51}$ erg s^{-1}, using a fixed value of the opening half–angle (θ_0) of the deposition region.

For energy deposition rates larger than a certain threshold $\dot{E}_{\text{th}}$, all the models lead to either relativistic jets or ultrarelativistic winds (i.e., fireballs). The threshold is due to the need of overcoming the ram pressure p_{ram} that is exerted by the infalling external medium onto the new born fireball close to its initiation site. For type-A models we find $\dot{E}_{\mathrm{th}} \sim 10^{49}\,\mathrm{erg\,s^{-1}}$, while for type-B models $\dot{E}_{\mathrm{th}} \lesssim 10^{48}\,\mathrm{erg\,s^{-1}}$. The smaller value in type-B models is due to their smaller ambient density.

Depending on the energy deposition rate and on the ambient density we find that the outflows are either jets (i.e., outflows where the lateral boundaries are causally connected) having a very small opening angle ($\lesssim 8°$) or relatively wide opening angle ($\lesssim 25°$) winds (i.e., the lateral boundaries are not causally connected). Models close to the thresholds of the energy deposition rate or with a high density environment tend to form relativistic ($\Gamma \sim 10$), low density, knotty jets whose head propagates at mildly relativistic speeds($\sim 0.6c$). In contrast, models well above the threshold with dense environments or, independent of the deposition rate, in case of diluted environments either tend to form conical, ultrarelativistic ($\Gamma \gtrsim 400$) winds which are smooth, propagate at relativistic speeds ($\sim 0.97c$) and can be fitted by analytic power laws in case of models of type-A, or they propagate at ultrarelativistic speeds ($\gtrsim 0.9999c$) being rather irregular due to the effect of large Kelvin-Helmholtz (KH) instabilities originating from their interaction either with the torus, or with the environment in case of type-B. Indeed, the growth of KH modes determines whether the profiles of the physical variables are smooth and monotonically decreasing in the θ-direction (type-A), or non-smooth and non monotonic (type-B). An effect of the KH instabilities is to entrain mass into the relativistic outflows of type-B models through the side edges of the fireballs. The amount of entrained mass is comparable with that of models of type-A. However, in type-B models, there is much less mass swept by the front radial edge of the outflows (because there is much less mass in the ambient) leading to a highly relativistic propagation velocity of the fireballs in type-B models, while only allowing for mildly relativistic speeds in type-A models.

Increasing the energy deposition rate yields an increase of the average Lorentz factor, and a decrease of the average density of the outflow at any given time. In models of type-A, the increase of $\dot{E}$ results in a transition in the outflow morphology from relativistic jets ($\dot{E} < 10^{51}\,\mathrm{erg\,s^{-1}}$) to ultrarelativistic wind-like outflows ($\dot{E} > 10^{51}\,\mathrm{erg\,s^{-1}}$). In models of type-B, we find ultrarelativistic winds for all energy deposition rates considered ($\dot{E} > 5 \cdot 10^{48}\,\mathrm{erg\,s^{-1}}$). For energy deposition rates producing conical wind structures, the opening angle of the outflow (θ_w) is quite insensitive to the exact value of $\dot{E}$ (although it slightly increases with increasing $\dot{E}$), being its value $\sim 20° - 30°$.

The opening angle of the resulting outflow, provided it is a jet, is set by the environmental conditions (mainly, the density). If the generated outflow is a wind, then the complete collimation process happens in less than 1 ms (approximately, the light crossing time of the torus) and θ_w is set by the opening angle of the torus, and neither by the external medium (which has much less inertia in type-B models because it is much more rarefied), nor by the angular size of the deposition region.

We have checked the evolution of two models after the shut down of the energy deposition. It turns out that outflows propagating in high density environments (type-A) will not yield successful GRBs while models with diluted environments (type-B) can do so. The reason being that, in type-B models, the speed of propagation of the leading radial edge of the fireball is highly ultrarelativistic (with a Lorentz factor of ≈ 100) by

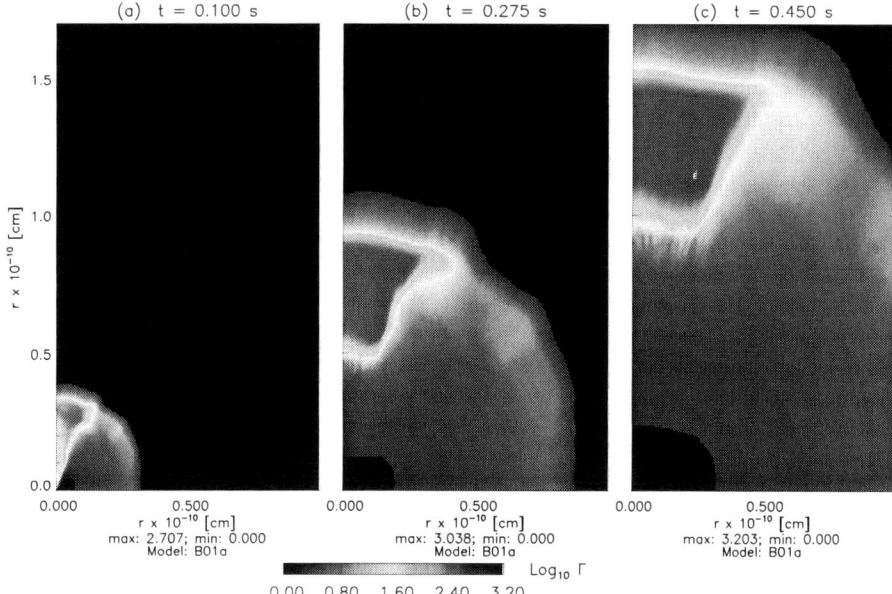

FIGURE 1. Snapshots of the evolution of the Lorentz factor (in logarithmic scale) for a type-B model with $\dot{E} = 2 \times 10^{50}\,\mathrm{erg\,s^{-1}}$ and $\theta_0 = 45°$. Note the stretching of the fireball (blue region corresponding to Lorentz factor > 250) as it moves away from the central source.

the time that the energy release is shut down. This is not the case in type-A models because they sweep up more ambient mass and the leading front of the outflow slows down. We find that in type-B models the fireball even stretches substantially in radial direction, because the propagation velocity of its leading front is larger than its rear edge (Fig. 1). This points to the possibility that the duration of the GRB emission can be much larger than the time of the activity (i.e., of release of energy) of the central source.

REFERENCES

1. Kouveliotou, C., Megan, C. A., Fishman, G. J., and et al., *ApJ*, **413**, L101 (1993).
2. Ruffert, M., and Janka, H. T., *A&A*, **344**, 573 (1999).
3. Janka, H. T., Eberl, T., Ruffert, M., and Fryer, C. L., *ApJ*, **527**, L39 (1999).
4. Sari, R., and Piran, T., *ApJ*, **485**, 270 (1997).
5. Aloy, M. A., Ibáñez, J. M., Martí, J. M., and Müller, E., *ApJ Suppl. Series*, **122**, 151 (1999).
6. Font, J. A., and Daigne, F., *MNRAS*, **334**, 383 (2002).
7. Aloy, M. A., Müller, E., Ibáñez, J. M., Martí, J. M., and MacFadyen, A. I., *ApJ*, **531**, L119 (2000).
8. Aloy, M. A., Ibáñez, J. M., Miralles, J. A., and Urpin, V., *A&A*, **396**, 693 (2002).

MHD Simulations of the Collapsar Model for GRBs

Daniel Proga*, Andrew I. MacFadyen†, Philip J. Armitage* and Mitchell C. Begelman*

*JILA, University of Colorado, Boulder, CO 80309-0440, USA
†California Institute of Technology, Mail Code 130-33, Pasadena, CA 9112

Abstract. We present results from axisymmetric, time-dependent magnetohydrodynamic (MHD) simulations of the collapsar model for gamma-ray bursts. Our main conclusion is that, within the collapsar model, MHD effects alone are able to launch, accelerate and sustain a strong polar outflow. We also find that the outflow is Poynting flux-dominated, and note that this provides favorable initial conditions for the subsequent production of a baryon-poor fireball.

INTRODUCTION

The collapsar model is one of most promising scenarios to explain the huge release of energy in a matter of seconds, associated with gamma-ray bursts (GRBs; [1, 2, 3, 4, 5]). In this model, the collapsed iron core of a massive star accretes gas at a high rate ($\sim 1 M_\odot \, s^{-1}$) producing a large neutrino flux, a powerful outflow, and a GRB. Although the association of long duration GRBs with stellar collapse is now secure ([6, 7]), basic properties of the central engine are uncertain. In part, this is because previous numerical studies of the collapsar model did not explicitly include magnetic fields, although they are commonly accepted as a key element of accretion flows and outflows.

We present a study of the time evolution of 2.5-dimensional, magnetohydrodynamic (MHD) flows in the collapsar model. This study is an extension of existing models of MHD accretion flows onto a black hole (BH; [8]). In particular, we include a realistic equation of state, photodisintegration of bound nuclei and cooling due to neutrino emission. Our study is also an extension of collapsar simulations by [3], as we consider very similar neutrino physics and initial conditions but solve MHD instead of hydrodynamical equations.

MODELS

We begin the simulations after the 1.7 $M_\odot$ iron core of a 25 $M_\odot$ presupernova star has collapsed and study the ensuing accretion of the 7 $M_\odot$ helium envelope onto the central black hole formed by the collapsed iron core. We consider a spherically symmetric progenitor model, but with spherical symmetry broken by the introduction of a small, latitude-dependent angular momentum and a weak radial magnetic field. For more details, see [9]).

RESULTS

We find that after a transient episode of infall, lasting 0.13 s, the gas with $l \gtrsim 2R_{SC}$ piles up outside the black hole and forms a thick torus bounded by a centrifugal barrier near the rotation axis. Soon after the torus forms (i.e., within a couple of orbital times at the inner edge), the magnetic field is amplified by the magnetorotational instability (MRI, e.g., [10]) and shear. We have verified that most of the inner torus is unstable to MRI, and that our simulations have enough resolution to resolve, albeit marginally, the fastest growing MRI mode. The torus starts evolving rapidly and accretes onto the black hole. Another important effect of magnetic fields is that the torus produces a magnetized corona and an outflow. The presence of the corona and outflow is essential to the evolution of the inner flow at all times and the entire flow close to the rotational axis during the latter phase of the evolution. We find that the outflow very quickly becomes sufficiently strong to overcome supersonically infalling gas (the radial Mach number in the polar funnel near the inner radius is ~ 5) and makes its way outward, reaching the outer boundary at $t = 0.25$ s. Due to limited computing time, our simulations were stopped at $t = 0.28215$ s, which corresponds to 6705 orbits of the flow near the inner boundary. We expect the accretion to continue much longer, roughly the collapse timescale of the Helium core (~ 10 s), as in [3].

Figure 1 shows the flow pattern of the inner part of the flow at $t = 0.2437$ s. The left and middle panels show density and $|B_\phi|$ maps, respectively. The two maps are overlaid with the direction of the poloidal velocity. The right panel shows the flow domains of different polarity of B_ϕ overlaid with the direction of the poloidal magnetic field. The polar regions of low density and high B_ϕ coincide with the region of an outflow. We note that during the latter phase of the evolution not all of the material in the outflow originated in the innermost part of the torus – a part of the outflow is "peeled off" the infalling gas at large radii by the magnetic pressure (see Fig. 2 in [9]). We imposed a weak poloidal field of a given polarity at the outer boundary. This means that *initially* B_ϕ changes sign only across the equator, which is a relatively unfavorable configuration for subsequently producing B_ϕ reversals in the jet. Nevertheless, we find that the polarity of B_ϕ changes with time. This is because the flow loses memory of the initial polarity by the time it reaches the inner MRI-dominated regions where the jet is formed.

Sikora et al. (2003) [11] argue that Poynting flux-dominated jets with reversing B-fields provide a natural and efficient way to dissipate energy via the reconnection process. Poynting flux dominated jets have been found in previous numerical simulations whereas jets with reversing B-fields appear as a relatively new result. Therefore, we have reviewed results from [8] to check whether polar outflows generated during adiabatic accretion onto BHs also exhibit reversing B_ϕ. Simulations reported by [8] are suitable to study the flow pattern on relatively large length scales because they have been continued for a few dynamical time scales at a distance of $\sim 10^3$ BH radii.

Figure 2 shows the flow domains of different polarity of B_ϕ overlaid with the direction of the poloidal velocity and magnetic field (left and right panel, respectively) for run D

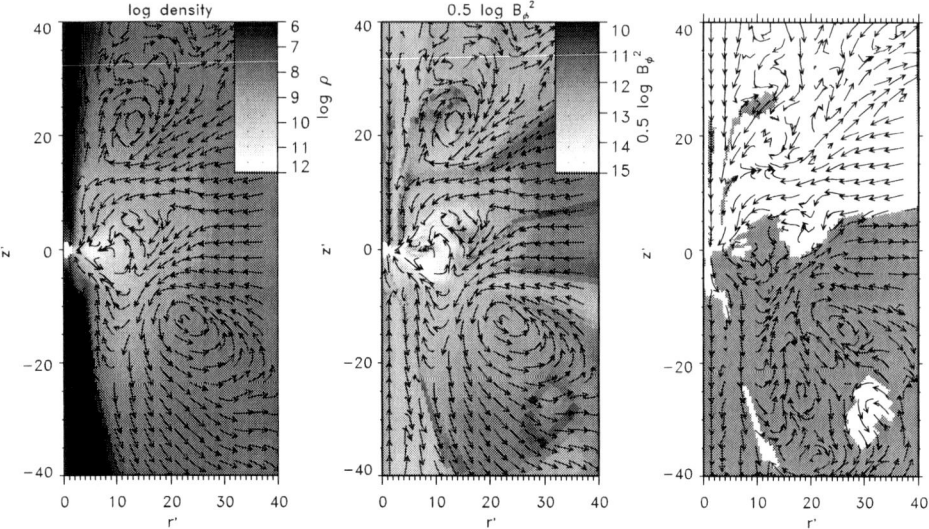

FIGURE 1. Maps of logarithmic density, toroidal magnetic field, and toroidal magnetic field domains with different polarity (white and grey regions) for Proga et al.'s [9] GRB simulation at $t = 0.2437$ s (left, middle, and right panel, respectively). The arrows in the left and middle panels indicate the direction of the poloidal velocity while the arrows in right panel indicate the direction of the poloidal magnetic field. The length scale is in units of the BH radius (i.e., $r' = r/R_S$ and $z' = z/R_S$

in [8]. It appears then that reversing B-fields are not unique to our GRB simulations and that domains with different polarity can be relatively large and long-lived.

CONCLUSIONS

We have performed time-dependent two-dimensional MHD simulations of the collapsar model. Our simulations show that: 1) soon after the rotationally supported torus forms, the magnetic field very quickly starts deviating from purely radial due to MRI and shear. This leads to fast growth of the toroidal magnetic field as field lines wind up due to the torus rotation; 2) The toroidal field dominates over the poloidal field and the gradient of the former drives a polar outflow against supersonically accreting gas through the polar funnel; 3) The polar outflow is Poynting flux-dominated; 4) The polarity of the toroidal field can change with time; 5) The polar outflow reaches the outer boundary of the computational domain (5×10^8 cm) with an expansion velocity of 0.2 c; 6) The polar outflow is in a form of a relatively narrow jet (when the jet breaks through the outer boundary its half opening angle is $5°$); 7) Most of the energy released during the accretion is in neutrinos, $L_v = 2 \times 10^{52}$ erg s^{-1}. Neutrino driving will increase the outflow energy (e.g., [12] and references therein), but could also increase the mass loading of the outflow if the energy is deposited in the torus.

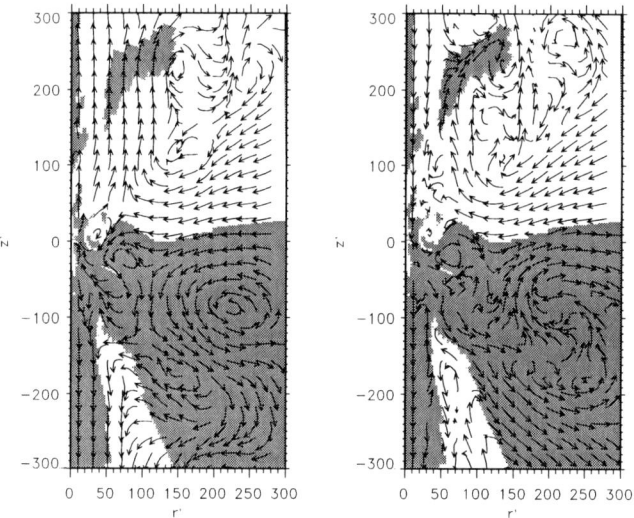

FIGURE 2. Maps of toroidal magnetic field domains with different polarity (white and grey regions) overplotted with the direction of the poloidal velocity (left panel) and the direction of the poloidal magnetic field (right panel) at the end of Proga & Begelman's simulation of adiabatic accretion onto a BH (run D in [8]). The length scale is also in units of the BH radius; the same as in Figure 1. Note however the difference in the r' and z' ranges.

ACKNOWLEDGMENTS

DP acknowledges support from NASA under LTSA grants NAG5-11736 and NAG5-12867. MCB acknowledges support from NSF grants AST-9876887 and AST-0307502.

REFERENCES

1. Woosley, S. E., ApJ **405**, 273 (1993).
2. Paczyński, B., ApJ **494**, L45 (1998).
3. MacFadyen, A., Woosley, S. E., ApJ **524**, 262 (1999).
4. Popham, R., Woosley, S. E., Fryer, C., ApJ **518**, 356 (1999).
5. MacFadyen, A., Woosley, S. E., Heger, A., ApJ **550**, 410 (2001).
6. Hjorth, J., et al., Nature **423**, 847 (2003).
7. Stanek, K. Z., et al., ApJ **591**, L17 (2003).
8. Proga, D., Begelman, M. C., ApJ **592**, 767 (2003).
9. Proga, D., MacFadyen, A. I., Armitage, P. J., Begelman M. C., ApJ **599**, L5 (2003).
10. Balbus, S. A., Hawley, J. F., ApJ **376**, 214 (1991).
11. Sikora M., Begelman M. C., Coppi P., Proga D., submitted to ApJ Letters (astro-ph/0309504)(2003).
12. Fryer, C. L., Mészáros, P., ApJ **588**, L25 (2003).

Searching for GRB Remnants in Nearby Galaxies

S. G. Bhargavi*, J. Rhoads[†], R. Perna**, J. Feldmeier[‡] and J. Greiner[§]

Indian Institute of Astrophysics, Sarjapur Road, Bangalore 560 034 India
[†]*Space Telescope Science Institute, 3700 San Martin Drive, Baltimore, MD 21218, USA*
**Dept. of Astrophysical Sciences, Princeton University, 4 Ivy Lane, Princeton, NJ, 08544, USA*
[‡]*Case Western Reserve University, Cleveland, OH 44106-1712, USA*
[§]*Max-Planck Institute for Extraterrestrial Physics, Munich, Germany*

Abstract. Gamma Ray Bursts (GRBs) are expected to leave behind GRB remnants, similar to how "standard" supernovae (SN) leave behind SN remnants. The identification of these remnants in our own and in nearby galaxies would allow a much closer look at GRB birth sites, and possibly lead to the discovery of the compact object left behind. It would also provide independent constraints on GRB rates and energetics. We have initiated an observational program (2002, [1]) to search for GRB remnants in nearby galaxies. The identification is based on specific line ratios, such as $OIII/H_\beta$ and $HeII/H_\beta$, which are expected to be unusually high in case of GRB remnants according to the theoretical predictions of Perna et al. (2000, [2]). The observing strategies and preliminary studies from a test run at 2.34 m VBT as well as archival data from planetary nebulae surveys of spiral galaxies are discussed.

INTRODUCTION

The intense X-ray/UV radiation accompanying GRBs has dramatic effects on their environment: it can photoionize regions of ~ 100 pc size (Loeb & Perna 1998, [3]), and destroy dust on scales of tens of pc (e.g. Waxman & Draine 2000, [4]). Moreover, similar to what happens for SN remnants, a powerful blast wave is driven into the medium. While it takes a very short time (compared to the duration of the most intense X-ray UV radiation from the burst) to alter the equilibrium state of the medium, it takes a very long time for the medium to recover its original state, as detailed in the next section. This means that, while it is highly unlikely to observe a GRB in a nearby galaxy, there is a significant probability to find a GRB remnant.

Identification of GRB remnants in our own and nearby galaxies would allow a close study of GRB birth sites, and therefore provide independent diagnostic of their progenitors. Similar to how pulsars are found in association with SN remnants, the compact remnant objects left over from the GRB explosions can then be found in association with GRB remnants. Moreover, an estimate of the number of GRB remnants in the local universe would allow independent constraints on GRB rates and energetics.

GRB REMNANTS

Two phases can be distinguished in the life of a GRB remnant:

Cooling remnants (Perna, Raymond & Loeb 2000, [2]): due to the radiation flux of the GRB and its afterglow.
The X-ray/UV radiation accompanying a GRB heats and ionizes the surrounding medium. An emission spectrum is expected to be produced from the cooling ionized gas, the cooling time being of the order of $\sim 10^5 (T/10^5 K)/(n/cm^{-3})$ yr.

Slowing remnants (Loeb & Perna 1998, [3]; Efremov et al. 1998, [5]): due to the slowing blast-wave.
The relativistically expanding blast wave resulting from a GRB explosion takes $\sim 10^7$ years to slow down and merge with the ISM.

Combining these time scales with the present GRB rate of $\sim (10^6 - 10^7 f_b \, yr)^{-1}$ per galaxy (f_b is the beaming fraction) (e.g. Wijers et al. 1998, [6]), it can be seen that there is a substantial probability of finding GRB remnants in any galaxy at any given time.

Identifying GRB Remnants

In this paper, we report on our initial search strategy for cooling GRB remnants. Although the duration of the cooling phase is much shorter than the lifetime of the blastwave, these cooling remnants are much easier to identify due to their unique spectral signatures that allow one to distinguish them from other sources such as shock heated gas in SN remnants, HII regions and planetary nebulae (Perna, Raymond & Loeb 2000, [2]).

- High value for the line ratio [OIII]$\lambda 5007 / H_\beta$ (~ 100 for most of the cooling phase).
- Unusually high value for He II $\lambda 4686 / H_\beta$ ratio (up to ~ 100 at the beginning of the cooling phase).
- Time-dependent increase in the ratio [OIII]/[OII] indicating cooling of the gas.
- High SII$\lambda 6717/H_\beta$ as compared to HII regions.

HII regions are also photo-ionized like GRB remnants and can sometimes have high OIII$\lambda 5007/H_\beta$ ratios (≈ 3), but are characterized by lower temperatures in comparison to GRB remnants. Therefore it is useful to measure the [OIII]λ 5007/[OIII]λ 4363 line ratio which is temperature sensitive and increases with time in a cooling gas. The oxygen-rich SNR might show high OIII/H_β occationally but only during a brief period of incomplete cooling. The He II $\lambda 4686 /H_\beta$ ratio is weak in both HII regions as well as SNRs. Further, GRB remnants have physical sizes of ~ 100 pc, and can be distinguished from planetary nebulae (PNe) which look like point sources in external galaxies.

TABLE 1. Telescope Parameters

	VBO	IOA	KPNO
Size	2.34 m	2.01 m	4 m
Longitude:	78°49'36"E	78°57'51"E	111°36'59"W
latitude:	12°34'36"N	32°46'46"N	31°57'12"N
Altitude:	725 m	4500 m	2100 m
Seeing (typ.):	2".5	< 1"	1"
F-ratio:	f/3.25 prime	f/9 cassegrain	f/3.1 prime
Image scale:	0".6/pix	0".17/pix	0".42/pix
FOV:	10' × 10'	7' × 7'	14' × 14'

OBSERVATIONS

VBO data

We have initiated an observational search for cooling GRB remnants using the 2.34 m Vainu Bappu Telescope (VBT) at Kavalur, India, (2002, [1]) and plan to use the new 2 meter Himalayan Chandra Telescope of the Indian Astronomical Observatory (IOA) at Hanle, featured by cloudless skies and low atmospheric water vapour. Nearby galaxies will be observed in narrow-band filters [OIII]λ 5007, [OIII]λ 4363, He II λ4686 and H_β to measure the various line ratios and to identify the candidate GRB remnants for further investigations.

In a test run of observations this summer, NGC 3627 and NGC 3351 were imaged at the 2.34 m VBT using the narrow-band filters [OIII]λ 5007, [OIII]λ 4363 and H_β.

KPNO archival data

In addition to the narrow-band data taken with the VBT, we are searching for GRB remnants using archival data originally taken to search for planetary nebulae in spiral galaxies (Feldmeier et al. 1997 [7]; Ciardullo et al. 2002 [8]). The data consists of narrow band [O III] λ5007 and Hα + [NII] exposures, along with a λ5300 continuum image. In some cases, there is additional R data as well. The seven galaxies (NGC 891, 2403, 3627, 3351, 3368, 5194/5, 5457) are all luminous spiral galaxies, and should provide reasonable targets for search.

While the [OIII]λ5007/H_α ratios are typically 5 in PNe, we expect it to be $\sim$ 30 for a candidate GRB remnant (using H_α/H_β of 2.8 for emission nebulae). Table 1 shows the telescope parameters.

SEARCH STRATEGIES

The images are reduced in the standard manner using the IRAF software. All images are aligned and positionally registered before continuum subtraction. Our goal is to identify

candidate GRB remnants, and follow them up with additional imaging and spectroscopic observations. Since we have just begun our search, our results are still preliminary. In order to separate the potential GRB remnant candidates from other sources (HII regions, SN remnants, and PNe), we are using the following selection criteria:
1. Objects must have a signal-to-noise greater than 9.
2. Objects must appear non-stellar, and have a SExtractor star/galaxy classifier value less than 0.95.
3. Objects must have an [OIII]$\lambda 5007$/ Hα ratio that is 2σ larger than the mean of the distribution. This removes almost all H II regions from the sample, as they have low [O III]$\lambda 5007$ / Hα ratios.

Each candidate is then visually inspected to confirm their candidature against artifacts. Currently, we are finding tens of candidates in each galaxy, though there is significant scatter from galaxy to galaxy (NGC 891 having none, most likely due to its edge-on orientation). We require further observations to confirm the list of candidates.

CONCLUSIONS

In this paper we report the preliminary studies we carried out to search for the GRB remnants in nearby galaxies from a test run at VBT as well as archival data from KPNO. In the present investigations we find about 20-30 candidates in each galaxy. Additional observations are required to measure other line ratios and to check the validity of candidates. We also plan to use data from other existing surveys to find preliminary candidates in nearby galaxies. A photoionized remnant of radius ~ 100 pc would subtend an angle of $2''$ on the sky at the distance of Virgo cluster (~ 20 Mpc), where a typical galaxy would subtend 2 arcmin. We require a multiple-fibre spectrograph which can take simultaneous spectra across the entire image of a nearby galaxy.

ACKNOWLEDGMENTS

We thank Prof. A. Saha (KPNO) for lending the narrow-band filters for the observations at VBT. SGB acknowledges the science visit hosted by STScI and the guest user facility extended by Prof. D. Lamb (University of Chicago) in preparing this poster paper.

REFERENCES

1. Bhargavi, S. G., Cowsik, R. & Perna, R. In 'GRBs in the afterglow era', Proceedings of the 3rd Rome workshop (2002).
2. Perna, R., Raymond, J. & Loeb, A.,ApJ **533**, 658 (2000).
3. Loeb, A. & Perna, R., ApJ, **503L**, 35 (1998).
4. Waxman, E. & Draine, B. ApJ, 537, 796 (2000).
5. Efremov, Y. N., Elmegreen, B. G., hodge, P. W., ApJ, **501L**, 163 (1998).
6. Wijers, R. A. M. J., Bloom, J. S., Bagla, J. S. & Natarajan, P. 1998, MNRAS **294**, L13 (1998).
7. Feldmeier, J. J., Ciardullo, R. & Jacoby, G. H., ApJ **479**, 231 (1997).
8. Ciardullo, R., Feldmeier, J. J., Jacoby, G. H., Kuzio de Naray, R., Laychak, M. B. & Durrell, P. R., ApJ **577**, 31 (2002).

General Relativistic MHD Simulations of the Gravitational Collapse of a Rotating Star with Magnetic Field as a Model of Gamma-Ray Bursts

Y. Mizuno*, S. Yamada[†], S. Koide** and K. Shibata[‡]

Department of Astronomy, Kyoto University, Kyoto 606-8502
[†]*Department of Science and Engineering, Waseda University, Tokyo 169-8555*
**Department of Engineering, Toyama University, Toyama 930-8555*
[‡]*Kwasan and Hida Observatory, Kyoto University, Kyoto 607-8471*

Abstract. We have performed 2.5-dimensional general relativistic magnetohydrodynamic (MHD) simulations of the gravitational collapse of a magnetized rotating massive star as a model of gamma ray bursts (GRBs). This simulation showed the formation of a disk-like structure and the generation of a jet-like outflow inside the shock wave launched at the core bounce. We have found the jet is accelerated by the magnetic pressure and the centrifugal force and is collimated by the pinching force of the toroidal magnetic field amplified by the rotation and the effect of geometry of the poloidal magnetic field. The maximum velocity of the jet is mildly relativistic (~ 0.3 c).

INTRODUCTION

GRBs and the afterglows are well described by the fireball model, in which a relativistic outflow is generated from a compact central engine. Rapid temporal decay of several afterglows is consistent with the evolution of a highly relativistic jet with bulk Lorentz factors $\sim 10^2 - 10^3$. The formation of relativistic jets from a compact central engine remains one of the major unsolved problems in GRB models.

From recent observations, some evidence was found for a connection between GRBs and the death of massive stars. This evidence includes a correlation between star forming regions and the position of GRBs inside the host galaxy [1], a "bump" resembling the light curves of Type Ic supernovae in the optical afterglow of several GRBs [2, 3], and association of GRB980425-SN1998bw [4, 5] and GRB030329-SN2003df [6, 7]. Several authors [8, 9, 10] have studied beaming angles and energies of a number of GRBs. They have found that central engines of GRBs release supernova-like energies ($\sim 10^{51}$ erg). It is thus probable that a major subclass of GRBs is a consequence of the collapse of a massive star.

In this study, we perform 2.5-dimensional general relativistic MHD simulations of the gravitational collapse of a rotating star with magnetic field as a model for a collapsar. The collapsar is in some sense an anisotropic supernova, and it is considered that relativistic jets from collapsars are launched by MHD processes in accreting matter and/or by neutrino annihilation [11, 12].

NUMERICAL METHOD

In order to study the formation of relativistic jets from a collapsar we use a 2.5-dimensional general relativistic magnetohydrodynamics (GRMHD) code [13] .We assume that the gravitational potential is constant in time. We neglect nuclear burnings and neutrinos in this simulation.

We consider the following situation as the initial condition for the simulations: A few $M_\odot$ black hole is produced at the center of the stellar remnant with a weak shock standing at a radius of a few hundred km and the post-shock gas falling onto the central black hole. We consider a non-rotating black hole as the central black hole. We employ 1-dimensional supernova simulation data of Bruenn [14] to obtain the initial density, pressure and radial velocity distribution. We add the effect of stellar rotation and intrinsic magnetic field. The initial rotational velocity distribution is assumed to be a function of the distance from the rotation axis only. The initial magnetic field is assumed to be uniform and parallel to the rotational axis. See Mizuno et al. [15] for details.

RESULTS

Formation of Jet

The stellar matter falls onto the central black hole at first. This collapse is anisotropic due to the effects of rotation and magnetic field. The accreting matter falls more slowly on the equatorial plane than on the rotational axis. The matter piles up on the equatorial plane, and a disk-like structure is formed near the central black hole. Since the magnetic field is frozen into the plasma, it is dragged by the accreting matter and amplified. The amplified magnetic field expands outwards as Alfvén waves and launches an outgoing shock wave.

The jet-like outflow is generated behind the shock wave. The shock wave has the high magnetic pressure gradient force and high centrifugal force. These forces push back the accreting matter and construct the jet-like outflow. The jet-like outflow formed in the simulation is thus magnetically driven. The jet has a mildly relativistic velocity, ~ 0.3 c (the poloidal velocity is ~ 0.1 c). It exceeds the escape velocity. Thus, the jet is likely to get out of the stellar remnant. The magnetic field plays an important role in the collimation. Not only the pinching force of the toroidal magnetic field but also the geometry of the poloidal magnetic field (i.e. poloidal magnetic pressure) plays a crucial role.

Dependence on the Initial Magnetic Field Strength and Initial Rotatinal Velocity

As the initial magnetic field strength increases, the jet velocity increases and the magnetic twist decreases. However, for stronger magnetic field the jet velocity decreases with increasing initial magnetic field strength and the magnetic twist still continues

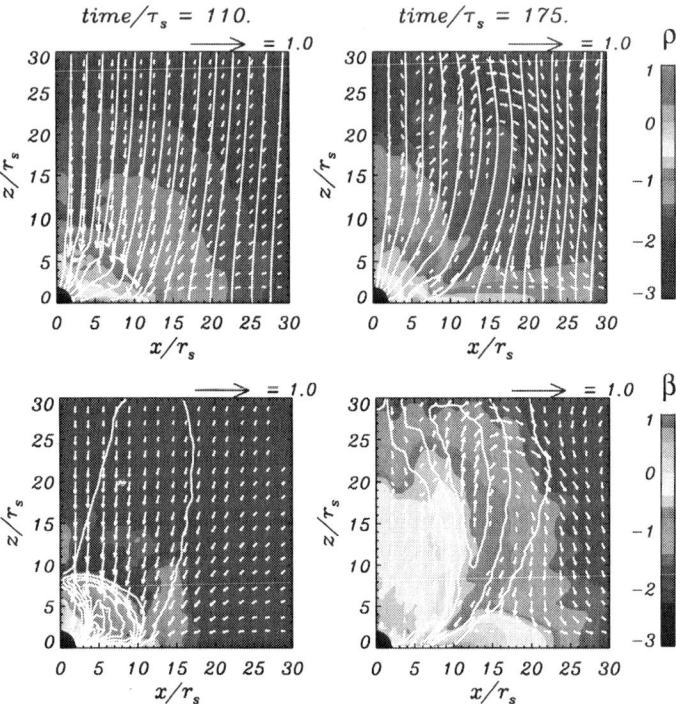

FIGURE 1. The time evolution of the density (upper panel) and the plasma beta (lower panel). The grey scale shows the value of the logarithm of density and the plasma beta. The white curves depict magnetic field lines (upper panel) and contour of the toroidal magnetic field (lower panel). The arrows represent the poloidal velocities normalized by the light velocity. Accreting matter forms a disk-like structure. The shock wave is generated near the central black bole. The jet-like outflow is produced inside the shock wave.

decreasing. As the initial rotational velocity becomes faster, the jet velocity becomes faster and the magnetic twist becomes stronger up to a certain value. For faster initial rotational velocity, the vertical component of the jet velocity and the magnetic twist are almost constant.

In order to produce a strong jet, the magnetic field has to be twisted significantly so that it can store enough magnetic energy. If the initial magnetic field is strong, the magnetic field cannot be twisted significantly because Alfvén waves propagate as soon as the magnetic field is slightly twisted. As a result, the jet velocity does not rise up and the magnetic twist remains weak. Therefore, weaker initial magnetic fields are favorable for a stronger jet. The dependence on the initial rotational velocity can also be understood by as the dependence on the initial magnetic field strength. If the initial rotation is sufficiently fast, the magnetic field is twisted significantly and stores enough energy to produce a fast jet. However, the stored magnetic energy has a limit. It is determined by the competition between the propagation time of Alfvén waves and the rotation time of the disk (or the twisting time).

DISCUSSION

We have studied the generation of a jet from gravitational collapse of a rotating star with magnetic field by using the 2.5D general relativistic MHD simulation code. The maximum velocity of the jet is mildly relativistic ($\sim 0.3c$). This result is consistent with Newtonian case [16]. The jet is too slow for the jet of GRBs. We have to consider other acceleration mechanism. The break out of the jet through the stellar surface is the most applicable acceleration mechanism. When the jet goes through the stellar surface, the strong density gradient may accelerate the jet. In fact, some authors [17, 18] have showed numerically that a significant acceleration of the jet occur and the terminal Lorentz factor becomes as high as $\Gamma \sim 50$. We think it is important to simulate the propagation of the jet outside the stellar surface by a GRMHD code properly evaluating the importance of magnetic fields for the dynamics and further propagation of the jet.

ACKNOWLEDGMENTS

This work was partially supported by a Grant-in-Aid for the 21st Century COE "Center for Diversity and Universality in Physics" and Grants-in-Aid for the scientific resarch from Ministry of Education, Science, Sports, Technology and Culture of Japan through No.14079202, No.14540226, and No.14740166.

REFERENCES

1. Bloom, J. S., Kulkarni, S. R., & Djorgovski, G., AJ **123**, 1111 (2002).
2. Bloom, J. S. et al., ApJ **572**, L45 (2002).
3. Garnavich, P. M. et al., ApJ **582**, 924 (2002).
4. Iwamoto, K. et al., Nature **395**, 672 (1998).
5. Woosley, S. E., Eastman, R. G., & Schmidt, B., ApJ **516**, 788 (1999).
6. Stanek, K. Z. et al., ApJ **591**, L111 (2003).
7. Hjorth, J. et al., Nature **423**, 847 (2003).
8. Frail, D. et al., ApJ **562**, L55 (2001).
9. Panaitescu, A. & Kumar, P., ApJ **560**, L49 (2001).
10. Bloom, J. S., Frail, D. A., & Kulkarni, S. R., ApJ **594**, 674 (2003).
11. Woosley, S. E., ApJ **405**, 273 (1993).
12. MacFadyen, A. I. & Woosley, S. E., ApJ **524**, 262 (1999).
13. Koide, S., Phys. Rev. D **67**, 104010 (2003).
14. Bruenn, S. W., in Nuclear Physics in the Universe, ed. M.W. Guidry & M.R. Strayer (Philadelphia: Institute of Physics Pub.), 31 (1992).
15. Mizuno, Y., Yamada, S., Koide, S., & Shibata, K., astro-ph/0310017 (2003).
16. Proga, D., MacFadyen, A. I., Armitage, P. J., & Begelman, M. C., astro-ph/0310002 (2003).
17. Aloy, M. A., et al., ApJ **531**, L119 (2000).
18. Zhang, W., Woosley, S. E., & MacFadyen, A. I., ApJ **586**, 356 (2003).

GRB CONNECTION TO SUPERNOVAE

Previously Claimed(/Unclaimed) X-ray Emission Lines in High Resolution Afterglow Spectra

N. Butler*, A. Dullighan*, P. Ford*, G. Ricker*, R. Vanderspek*, K. Hurley†, J. Jernigan† and D. Lamb**

*Center for Space Research, Massachusetts Institute of Technology, MA
†Space Sciences Laboratory, Berkeley, CA
**Department of Astronomy, University of Chicago, IL

Abstract. We review the significance determination for emission lines in the *Chandra* HETGS spectrum for GRB 020813, and we report on a search for additional lines in high resolution Chandra spectra. No previously unclaimed features are found. We also discuss the significance of lines sets reportedly discovered using XMM data for GRB 011211 and GRB 030227. We find that these features are likely of modest, though not negligible, significance.

INTRODUCTION

Multiple luminous X-ray lines have been claimed in spectra taken with *XMM* of 2 GRB afterglows. (GRB 011211, Reeves et al. [1]; GRB 030227, Watson et al. [2]) The statistical significance of the GRB 011211 lines has been called into question by Rutledge & Sako [3], and we address this question. In Butler et al. [4], we discuss high resolution spectra from *Chandra* for the X-ray afterglows to GRB 020813 and GRB 021004, and we report the discovery of moderately low significance spectral lines in the case of GRB 020813. Our detection (with an independent instrument) supports the claimed multiple$-\alpha$ line detections in *XMM* data, and the high spectral resolution facilitates a clearer determination of the line significances. There are 3 additional bursts with high resolution *Chandra* spectra (GRB 991216, GRB 020405, and GRB 030328), which we have analyzed to search for discrete spectral features. Only the data for GRB 991216 yielded a claimed line detection in the literature [5]. My collaborators and I analyzed the GRB 030328 data [6, 7, 8] and reported no line detections. Null detections for GRB 020405 are reported by Mirabal et al. [9].

S AND SI LINES FOR GRB 020813?

In Butler et al. [4], we describe our data reduction and continuum fits for the *Chandra* HETGS observation of GRB 020813 (and GRB 021004). We also describe our emission line search method which involves successively binning the spectral data by factors of 2 in order uncover deviations from the continuum fit. This procedure is sensitive to resolved emission lines. It allows for quick and easy determination of line significances, easily verified by Monte Carlo. One prominent emission line was found for GRB 020813. Considering the number of wavelength bins searched, we estimate a multiple-trial (i.e. blind search) significance for the line of 3.3σ. If the line is identified with the $K\alpha$ transition in H-like S, a low significance line possibly due to the $K\alpha$ transi-

TABLE 1. Lines detected in the spectra of GRB 991216, GRB 020405, and GRB 030328 are shown along with the implied redshift (z_{x-ray}). This z can be compared to that measured in the optical ($z_{optical}$). For GRB 991216, the data do not constrain the ionization state of the candidate Fe line.

Afterglow Source	Lines Detected	z_{x-ray}	$z_{optical}$	Significance
GRB 991216	Fe (I–XXVI)	~1	≥ 1.02	> 3.7σ
GRB 020405	Ar XVIII, Mg XI, Mg XII	0.63 ± 0.03	0.695	< 2σ
GRB 030328	Mg XI	1.52	≥ 1.52	< 1σ

tion in H-like Si can be identified. The significance of the pair turns out to be modestly better (3.5σ) than the significance of the S line alone [4].

Our S line significance determination agrees with the estimate made using the deprecated likelihood ratio test [see, e.g., 10]. To check whether this fact is statistically meaningful, we apply Monte Carlo integration to establish the true distribution for the log-likelihood. We form 10^4 simulated data sets using power-law model parameters (with Galactic absorption) drawn from the posteriori distribution (the distribution of model parameters given the observed data). Each data set is then fitted with this model, then with this model plus a Gaussian emission line. The number of iterations yielding larger improvements in $\Delta\chi^2$ than the observed value ($\Delta\chi^2$ =15.5 for 3 additional degrees of freedom) is recorded. To ensure that the parameter space for the emission line is adequately explored in each Monte Carlo iteration, we use FFTs to determine the most significant line-like residual on a fine line centroid wavelength grid ($\delta\lambda = 0.01$Å) and line width grid (dyadic intervals from $\sigma_\lambda = 0.02$Å to $\sigma_\lambda = 10.24$Å). We find that 13 of 10^4 runs yield a larger $\Delta\chi^2$ than the observed value, consistent with the significance estimates quoted above. We find consistent results independent of whether uniform or delta function priors are assumed on the power-law model parameters, indicating that the parameters are well constrained by the data.

LINES IN OTHER HIGH RESOLUTION CHANDRA SPECTRA

We apply the methods which led to the discovery of the lines for GRB 020813 to the 4 other high resolution spectra taken with *Chandra* of GRB X-ray afterglows (GRB 991216, GRB 020405, GRB 021004, and GRB 030328). We reduce the data as described in Butler et al. [4], with the exception that we jointly fit the gratings data along with the 0th-order data, instead of fitting the gratings data alone, in order to best determine the continuum fits. The continuum fits and the line searches in the gratings data are described in detail in Nat Butler's Ph.D. thesis [11]. The results are shown in Table 1. No highly significant, previously unreported lines are detected, and the Fe line reported by Piro et al. [5] is robustly detected with a multiple trials significance (conservatively) better than 3.7σ. This line is apparent in each of the 4 independent 1st-order HETGS spectra, and it is also seen in the 0th-order data. It is, in our opinion, the best case for an emission line in a GRB X-ray afterglow to date.

THE SIGNIFICANCE OF THE XMM MULTIPLE-α LINES

We reduce the EPIC-pn data for GRB 011211 following Reeves et al. [12], finding 537 net counts in the first 5 ksec of the observation. To search for emission lines, we employ the "matched filter" technique described in Rutledge & Sako [3]. We correct a number of minor errors in that work: (1) Rutledge & Sako [3] base their significance estimates on model continua determined using χ^2 fits of sparsely binned ($\sim$ 12 counts/bin) data. We find that the grouped data bin boundaries are not robust and that modest shifting can arise due to minor changes in the source and background selection regions. This shifting is sufficient to occasionally wash out the indication of line emission. We choose to fit continuum models to the unbinned spectrum. (2) Rutledge & Sako [3] uniformly sample from a number of possible continuum models rather than sampling according to the posteriori distribution of possible models given the observed data. For example, in the case of χ^2 fitting, our approach would suggest sampling from a model A in frequency relative to a model B as $\exp\{-0.5\chi_A^2/\chi_B^2\}$, assuming uniform priors on the model parameters. (3) Rutledge & Sako [3] also unjustifiably increase the normalizations on their continuum models (relative to the best fit values) in order to force each to yield the total number of observed source counts.

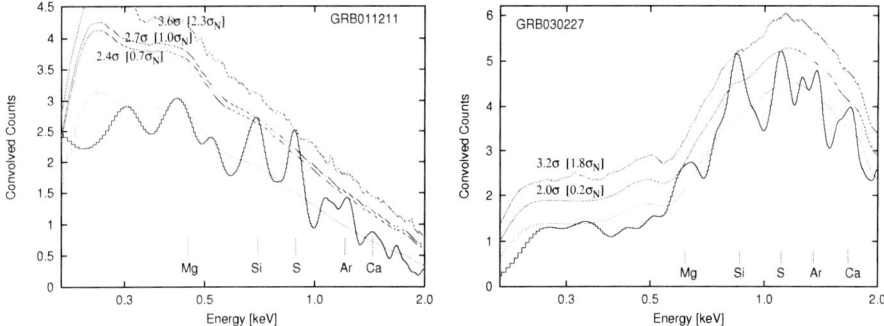

FIGURE 1. The raw PI spectra for 2 *XMM* X-ray afterglows are convolved with the line response function (i.e. "the matched filter" of Rutledge & Sako [3], dark curve). Significance contours are calculated using an MCMC method, which does not rely on binning the data. The locations of emission lines claimed by Reeves et al. [1] and Watson et al. [2] are indicated.

We model the continuum emission as an absorbed power-law, with the absorption fixed at the Galactic value as in Reeves et al. [12]. We apply the Markov Chain Monte Carlo (MCMC) algorithm described in Appendix B of van Dyke et al. [13] to determine the distribution of null model parameters given the data in raw PI bins. The background is modelled using a broken power-law with a break at 1.35 keV as in Rutledge & Sako [3]. We consider a uniform (i.e. totally non-informative) prior distribution on each model parameter for the source and background counts. From 10^4 models generated in the fashion, we simulate 10^4 spectra and apply the matched filter of Rutledge & Sako [3] to each.

The first panel in Figure 1 shows the result of applying this filter to the observed data (dark line). The mean result from the simulated spectra is plotted as a dotted line. At the locations of the reported S, Si, and Ar emission lines [see, 1], the solid dark curve in the first panel of Figure 1 deviates from the mean at $> 2\sigma$ significance. The significance contour reached by each feature is plotted, and the individual (single-trial) significances of the lines are 3.6σ, 2.7σ, and 2.4σ. Generating an additional 10^4 model sample parameters and reapplying the matched filter to the simulated spectra for each model, we count the number of times that a simulated model breaks through each significance contour between 0.2 and 2.0 keV. This allows us to determine that the multiple trials significances (σ_N values in Figure 1) for the emission lines are $2.3\sigma_N$, $1.0\sigma_N$, and $0.7\sigma_N$. The significance of the line set can then be estimated as 5 times the product of these probabilities (i.e. 2.4σ), where the factor of 5 takes into account that the data have been analyzed in 5 time regions. This is higher than the $\sim 1.0\sigma$ value determined by Rutledge & Sako [3], and it is lower than the 3.9σ initially suggested by Reeves et al. [1].

Recently, Watson et al. [2] have claimed the detection of a line set in GRB 030227, remarkably similar to the detections claimed for GRB 011211 by Reeves et al. [1]. Watson et al. [2] quote a significance determined solely via the likelihood ratio test ($2.7-4.4\sigma$, depending on the number of degrees of freedom). We reduce the EPIC-pn data for GRB 030227 following Watson et al. [2], finding 1593 source counts in the final 10 ksec of the observation. In our MCMC analysis, we describe the background counts as a power-law. The source counts are modelled as an absorbed power-law as in Watson et al. [2]. All model parameter prior distributions are taken as uniform. As displayed in the second panel of Figure 1, the claimed Mg, Si, and S lines have significances $\gtrsim 2.0\sigma$ (single-trial). This data set is the last of 4 time slices within the observation. Performing the multi-trial significance calculation as above, we find the the significance of the line set is 1.3σ. Thus, we find that the line emission for GRB 030227 is less significant than that for GRB 011211. We stress that this significance estimate, like that made for GRB 011211 above, is a lower bound on the significance of the line emission. If the line centroids can be argued to be constrained by the physics, as argued in Watson et al. [2], then the significance would increase.

CONCLUSIONS

We conservatively estimate the significances for line emission in GRB 991216 (*Chandra*), GRB 020813 (*Chandra*), GRB 011211 (*XMM*), and GRB 030227 (*XMM*) as $> 3.7\sigma$, 3.5σ, 2.4σ, and 1.3σ, respectively. We do not find the line emission for GRB 011211 (which has galvanized many GRB researchers) to be entirely insignificant, as claimed by Rutledge & Sako [3]. However, we find none of the observations to be highly significant. Hopefully, early observations with Swift in the coming few years will decide conclusively whether this emission is real.

REFERENCES

1. Reeves, J. N., et al. 2002, Nature, 415, 512
2. Watson, D., et al. 2003, *ApJ*, 595, L29
3. Rutledge, S., & Sako, M. 2003, MNRAS, 339, 600
4. Butler, N., et al., *ApJ* **597** (2003).
5. Piro, L., et al., Science, 290, 955 (2000).
6. Butler, N. R. et al., *GCN*, 2007 (2003).
7. Ford, P. G., et al., *GCN*, 2027 (2003).
8. Butler, N. R. et al., *GCN*, 2076 (2003).
9. Mirabal, N., et al., *ApJ* **587**, 128 (2002).
10. Protassov, R., et al., *ApJ* **571**, 545 (2002).
11. Butler, N., Ph.D. Thesis, MIT (2003).
12. Reeves, J. N., et al., A&A, 403, 463 (2003).
13. van Dyke, D., et al., *ApJ* **548**, 224 (2001).

SN 2002lt and GRB 021211: a SN/GRB Connection at $z = 1$

M. Della Valle[*], D. Malesani[†], S. Benetti[**], V. Testa[‡], M. Hamuy[§], L.A. Antonelli[‡], G. Chincarini[¶ ‖], G. Cocozza[‡ ††], S. Covino[¶], P. D'Avanzo[¶], D. Fugazza[¶ ‡‡], G. Ghisellini[¶], R. Gilmozzi[§§], D. Lazzati[¶¶], E. Mason[§§], P. Mazzali[***] and L. Stella[‡]

[*]*INAF, Osservatorio Astrofisico di Arcetri, largo E. Fermi 5, 50125 Firenze, Italy*
[†]*International School for Advanced Studies (SISSA/ISAS), via Beirut 2-4, 34014 Trieste, Italy*
[**]*INAF, Osservatorio Astronomico di Padova, vicolo dell'Osservatorio 5, 35122 Padova, Italy*
[‡]*INAF, Osservatorio Astronomico di Roma, via Frascati 33, 00040 Monteporzio (Roma), Italy*
[§]*Carnegie Observatories, 813 Santa Barbara Street, Pasadena, California 91101, USA*
[¶]*INAF, Osservatorio Astronomico di Brera, via E. Bianchi 46, 23807 Merate (Lc), Italy*
[‖]*University of Milano–Bicocca, Department of Physics, Piazza delle Scienze, 20126 Milano, Italy*
[††]*University of Roma Tor Vergata, Dept. of Physics, via d. Ricerca Scientifica, 00133 Roma, Italy.*
[‡‡]*INAF, TNG, Roque de Los Muchachos, PO box 565, 38700 Santa Cruz de La Palma, Spain*
[§§]*ESO, Alonso de Cordova 3107, Casilla 19001, Vitacura, Santiago, Chile*
[¶¶]*Institute of Astrophysics, University of Cambridge, Madingley Road, CB3 0HA Cambridge, UK*
[***]*INAF, Osservatorio Astronomico di Trieste, via Tiepolo 11, 34131 Trieste, Italy*

Abstract. We present spectroscopic and photometric observations of the afterglow of GRB 021211 and the discovery of its associated supernova, SN 2002lt. The spectrum shows a broad feature (FWHM = 150 Å), around 3770 Å (in the rest-frame of the GRB) which we interpret as Ca H+K blueshifted by 14 400 km/s. Overall, the spectrum shows a suggestive resemblance with the one of the prototypical type-Ic SN 1994I. This might indicate that GRBs are produced also by standard type-Ic supernovæ.

INTRODUCTION.

After long years of study, we have now some convincing evidence that long-duration gamma-ray bursts (GRBs) are produced by the death of massive stars. The earliest hint was the spatial and temporal coincidence between SN 1998bw and GRB 980425 (at $z = 0.0085$; [1]). However, this association was hardly representative of the whole class of GRBs: GRB 980425 was indeed a very dim event, its gamma-ray energy being lower than that of classical GRBs by ~ 4 orders of magnitude. SN 1998bw was also a peculiar event, belonging to the class of the so-called 'hypernovæ': it showed unusual expansion velocities and a very high luminosity, the latter being the effect of copious Nickel production (see e.g. [2]).

Very recently, however, SN 2003dh was discovered associated with GRB 030329 ([3, 4, 5]). An intensive spectral monitoring showed a strict similarity between SN 2003dh and SN 1998bw, thereby conclusively proving that hypernovæ can generate classical GRBs. SN 2003dh was however somewhat fainter than SN 1998bw ([6]).

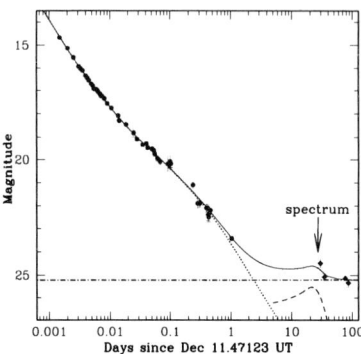 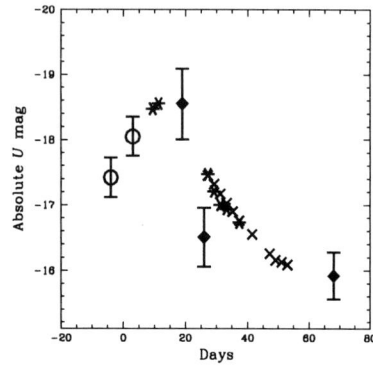

FIGURE 1. **Left panel.** R-band light-curve of the optical afterglow of GRB 021211. Data come from our observations (diamonds), the literature (filled circles; [10, 11, 18]), and HST measurements (open circles; [13]). The dotted and dot-dashed lines represent the afterglow and host contribution respectively. The dashed line shows the light curve of SN 1994I reported at $z = 1.006$ and dereddened with $A_V = 2$ (from [19]). The solid line shows the sum of the three contributions. **Right panel.** Comparison of the U-band light curves of the rebrightening of GRB 021211, after subtracting the host galaxy contribution (filled diamonds and open circles) and of SN 1994I, dereddened with $A_V = 1.8$ (asterisks; crosses have been obtained from the V-band lightcurve after assuming $U - V =$ constant).

We present here photometric and spectroscopic observations highlightening the association between GRB 021211 and SN 2002lt ([7]). GRB 021211 ([8]) was a rather dim event, its total (isotropic) gamma-ray energy being $E_{\mathrm{iso}} = (6 \pm 0.5) \times 10^{51}$ erg, at the low end of the energy distribution of GRBs ([9]). Prompt optical observations allowed an early discovery of the optical afterglow, just few minutes after the GRB onset (e.g. [10, 11]). The afterglow also turned out to be quite faint, about 3 magnitudes in the R optical band. The redshift was determined to be $z = 1.006$ ([12]). Optical/NIR colors showed little or no extinction local to the host galaxy. The intrinsic faintness of the afterglow made this event a good candidate for searching a supernova component.

DATA AND ANALYSIS

We observed the optical afterglow of GRB 021211 with the ESO VLT–UT4 equipped with the FORS 2 instrument, during the period January – March 2003 (20 – 100 days after the GRB). Low-resolution spectroscopy was performed on Jan 8.27 UT.

Photometric data show a rebrightening of the afterglow starting ~ 15 days after the burst ([13]) and reaching the maximum, $R \sim 24.5$, during the first week of January. The contribution of the host galaxy, estimated from our late-epoch images, is $R = 25.22 \pm 0.10$. Therefore, the intrinsic magnitude of the bump was $R = 25.24 \pm 0.38$. The significance of the rebrightening is at the 4-σ level. At the time of the maximum, the afterglow contribution, extrapolated from earlier epochs, is smaller than 5% under the most conservative assumptions. This suggests that the bump is powered by some different component.

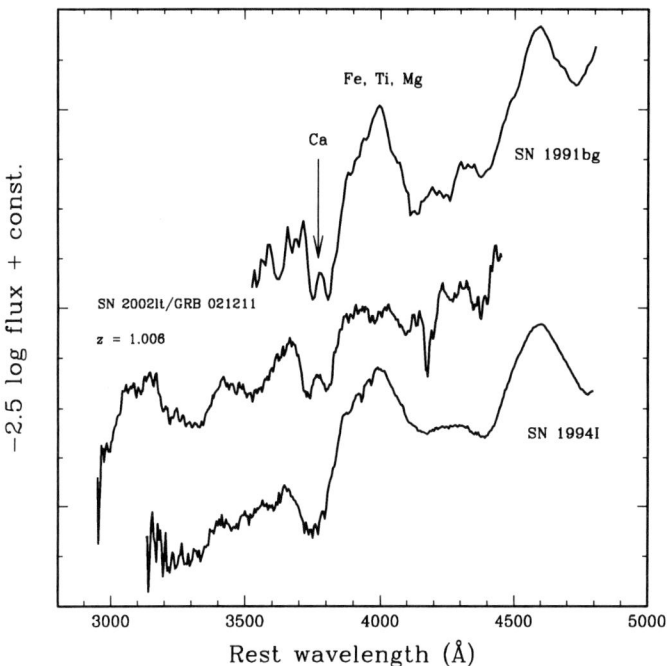

FIGURE 2. Spectrum of the afterglow+host galaxy of GRB 021211 (middle line), taken on Jan 8.27 UT (27 days after the burst). For comparison, the spectra of SN 1994I (type Ic, bottom) and SN 1994I (type Ia, top) are shown, both showing the Ca absorption.

To investigate the nature of the rebrightening, we obtained a spectrum with VLT+FORS 2. The reduced spectrum covered the range of wavelengths (6000 – 9000) Å at an acceptable S/N (> 3). The resolution was about 19 Å, and the integration time was 4×1 h with a seeing of $0''.6 - 1''.4$. We confirm the detection of the emission line at $\lambda = 7472.9$ Å ([12]), which may be interpreted as [O II] 3727 Å in the rest frame, leading to a determination of the redshift $z = 1.006$.

Figure 2 shows our spectrum smoothed and cleaned from the emission line [O II]. The spectrum of the afterglow is characterized by broad low-amplitude undulations blueward and redward of a broad absorption, the minimum of which is measured at ~ 3770 Å (in the rest frame of the GRB), whereas its blue wing extends up to ~ 3650 Å. The absorption feature in our spectrum is a characteristic signature of the SN ejecta (see Figure 2) and it is due to Ca II H+K absorption lines. The blueshifts corresponding to the minimum of the absorption and to the edge of the blue wing imply velocities of $v \sim 14\,400$ km/s and $v \sim 23\,000$ km/s respectively. The more convincing resemblance is found with SN 1994I, a prototypical type-Ic event, 9 days after its B-band maximum ([14]). It is difficult to explain such broad absorption feature in terms of other

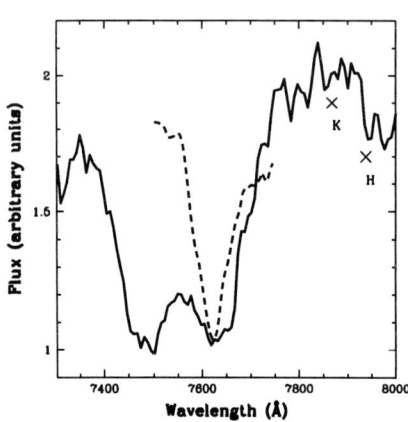 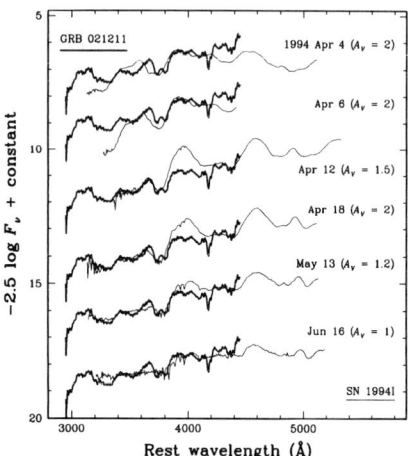

FIGURE 3. Left panel. Comparison of the absorption deep (solid line) in GRB 021211 and the telluric feature (dashed line), smoothed with the same boxcar filter 55 Å wide. The two patterns do not match, thereby confirming that the feature is intrinsic to the afterglow. **Right panel.** Comparison of our spectrum with the ones of SN 1994I at various epochs (spectra taken from [14]).

components. It was argued that the telluric absorption at ~ 7600 Å can contaminate, by chance, the absorption feature. Figure 3 (left panel) shows that even in the case that the subtraction of the telluric line was not effective (which is not the case), its profile cannot reproduce the broad and double-structured deep in our spectrum. Also in Figure 3 the position of the rest-frame Ca H+K edges is marked (crosses), showing that no contamination comes from the host galaxy.

IMPLICATIONS

The spectroscopic features observed during the rebrightening of the afterglow of GRB 021211 and the similarity of its lightcurve with the one of SN 1994I indicate that the bump was indeed powered by a supernova. This is therefore the third GRB (second in chronological order) for which a SN association was spectroscopically confirmed. The IAU dubbed this event SN 2002lt ([15]).

Supernova dating. Using SN 1994I as a template, our photometric and spectroscopic data allow us to estimate the time at which the SN exploded, and to compare it with the GRB onset time. Due to the limited wavelength coverage and to the lack of multi-time spectroscopic observations, our spectrum yields only a shallow contrain, suggesting that the SN went off between ~ 50 and 0 days before the GRB (see Figure 3, right panel).

Information from the photometry yields more stringent limits. Again, after assuming SN 1994I as a template, the best match is achieved if the SN and the GRB exploded simultaneously or separated, at most, by a few days. We stress however that also in this case the dataset is not rich enough to set firm bounds. Moreover, this result also depends upon the assumed rising time of the template SN, which was quite short (~ 12 days) for

SN 1994I ([16]).

GRB progenitor. It is also interesting to note that SN 1994I, the spectrum of which provides the best match to that observed in GRB 021211, is a typical type-Ic event rather than an exceptional 1998bw-like object, as the one proposed for association with GRB 980425 and GRB 030329 ([1, 3, 4]). If the SN associated with GRB 021211 indeed shared the properties of SN 1994I, this would open the interesting possibility that GRBs may be associated with standard type-Ic SNe, and not only with the more powerful events known as 'hypernovæ'. However, we should note that the recently studied SN 2002ap ([17]) showed significantly broader lines than our case and this difference vanished after maximum, such that it may be not easy to distinguish between the two types of SNe.

REFERENCES

1. Galama T.J., Vreeswijk P.M., van Paradijs J., et al. Nature, **395**, 670 (1998).
2. Nomoto K., et al., in *Gamma ray bursts, 5th Huntsville Symposium*, AIP Conf. Ser., Vol 526, ed. R.M. Kippen, R.S. Mallozzi, & G.J. Fishman (New York: Melville), 622 (2000).
3. Stanek K.Z., Matheson T., Garnavich P.M.; et al. ApJ, **591**, L17 (2003).
4. Hjorth J., Sollerman J., Møller P., et al., Nature, **423**, 847 (2003).
5. Matheson T., Garnavich P.M., Stanek K.Z., et al., ApJ, submitted (astro-ph/0307435) (2003).
6. Mazzali P.A., Deng J., Tominaga N., et al., ApJL, in press (astro-ph/0309555) (2003).
7. Della Valle M., Malesani D., Benetti S., et al., A&A, **406**, L33 (2003).
8. Crew G.B., Lamb D.Q., Ricker G.R., et al., ApJ, submitted (astro-ph/0303470) (2003).
9. Frail D.A., Kulkarni S.R., Sari R., et al., ApJ, **562**, L55 (2001).
10. Li W., Filippenko A.V., Chornock R., & Jha S., ApJ, **586**, L9 (2003).
11. Fox D.W., Price P.A., Soderberg A.M., et al., ApJ, **586**, L5 (2003).
12. Vreeswijk P.M., Fruchter A., Hjorth J., & Kouveliotou C., GCN Circ **1785** (2002).
13. Fruchter A.S., Levan A., Vreeswijk P.M., Holland S.T., & Kouveliotou C., GCN Circ **1781** (2002).
14. Filippenko A.V., Barth A.J., Matheson J., et al., ApJ, **450**, L11 (1995).
15. Della Valle M., Malesani D., Benetti S., Testa V., & Stella L. IAUC **8197**, (2003).
16. Iwamoto K., Nomoto K., Höflich P., et al., ApJ, **437**, L115 (1994).
17. Mazzali P.A., Deng J., Maeda K., et al., ApJ, **572**, 61 (2002).
18. Pandey S.B., Anupama G.C., Sagar R., et al., A&A, **408**, L21 (2003).
19. Lee M.G., Kim E., Kim S.C., et al., Journ. Korean Astr. Soc., **28**, 31 (1995).

Search for Correlations Between BATSE Gamma-Ray Bursts and Supernovae

Jiří Polcar*[†], Martin Topinka**[†], René Hudec[†], Věra Hudcová[†], Nicola Masetti[‡], Graziella Pizzichini[‡] and Eliana Palazzi[‡]

*Masaryk University Brno, Czech Republic
[†]Astronomical Institute of the Academy of Sciences of Czech Republic, Ondřejov, Czech Republic
**Charles University Prague, Faculty of Mathematics and Physics, Czech Republic
[‡]CNR IASF/TESRE Sezione di Bologna, Italy

Abstract. We report on the search for positional and temporal correlations between gamma-ray bursts (GRBs) in the CGRO BATSE catalog and the supernovae (SNe) detected during the BATSE era (1991-2000). By making use of innovative procedures we have carried out a statistical analysis of the results. We have checked if the possibly correlated objects differ in physical properties from the rest of the GRB/SN sample. Although at least two GRB-SN coincidences are known, our preliminary results do not indicate significant correlations between GRBs and supernovae which are presently detected by SN searches, but provide new insights into correlations based on physical properties of GRB and SN samples.

INTRODUCTION

The origin and the source of gamma-ray bursts (GRBs) still remain a puzzle. This tremendous energy could be released on a short time-scale during a collapse of a massive star in a supernova-like explosion. There are several pieces of observational evidence supporting a connection between GRBs a supernovae (SNe): some GRBs reveal an underlying SN in an optical afterglow lightcurve (e.g. GRB980326), sometimes supported by spectral and color signs (e.g. GRB030329 and SN2003dh) and there is a space-time coincidence known (e.g. GRB980425 and SN1998bw).

We used current BATSE GRB catalog (up to date April 2003) [1] and combined data from Harvard [2], SAI [3], and Asiago Padova [4] catalogs of SNe.

SEARCH FOR SPACE-TIME CORRELATION

We developed a fast matching engine for matching databased data from two chosen catalogs written in `perl` under Linux OS. We looked for space and time coincidences. A SN and a GRB make a pair if a position of a SN falls into the GRB errorbox and if they fit into the same time window. In time domain it is not theoretically clear what is the time delay between eventual gamma-ray emission of a SN (which could be observed as a GRB) and its optical manifestation. We simplified our problem and divided SNe into type Ia SN associated with white-dwarf explosion and core-collapse SN which covers

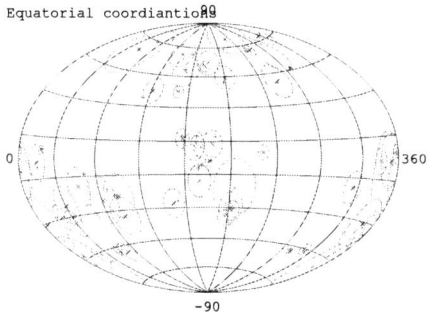

FIGURE 1. Final Match - GRB/SN pairs

the rest of well-defined SN types. The time delay between the time of explosion and the time of maximum of a SN lightcurve and the width of the time interval correspond to the type of a SN and to the level of uncertainty we have. We used 0 ± 30 days for core-collapsed SNe and 20 ± 7 days for type Ia SNe [5] in our analysis. Unfortunately only for a fraction of SNe information about the date of maximum is available and not all SNe have their type defined or have it defined with doubts. We approach this problem statistically and we assume the time delay between the time of maximum and the time of discovery to be the median (-4 days) of all known time delays. We used the weighted average of all types and time delays for SNe of an unknown type. Larger uncertainty we had longer time interval window we used to cover the possible SN types.

There are 2799 SNe in all known catalogs and 1014 of them in BATSE era. Current BATSE catalog collects 2707 GRBs. We found 55 possibly connected GRB/SN pairs, containing 49 GRBs and 50 SNe (see Fig. 1) using this data. Note that we, a priori, do not know how many of these coincidences are due to a real correlation and how many are there just by chance.

SEARCH FOR PHYSICAL PROPERTIES CORRELATIONS

The number of all events and the number of pairs is too small to give any significant conclusion. We auume that only a fraction of GRBs and SNe are correlated and also that this fraction is intrinsically physically different from the rest of the sample. We analyzed all available physical quantities of both GRBs and SNe. None of GRBs detected by BATSE has a redshift measured. In the analysis below, we assume that if there is a redshift of the paired SN, it is also the redshift of the corresponding GRB.

From a variety of plots we pinpoint only the ones we might find interesting or suspicious. The absolute magnitude of SN vs isotropic energy equivalent of GRB shows a surprising correlation not only affected by strong dependence on redshift, and duration of GRB vs hard to soft ratio of GRB which shows excess in longer and softer GRBs (see Fig. 2 and Fig. 3).

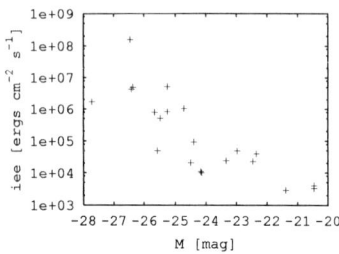

FIGURE 2. Abs. magnitude of SNe vs isotropic energy equiv. of GRBs

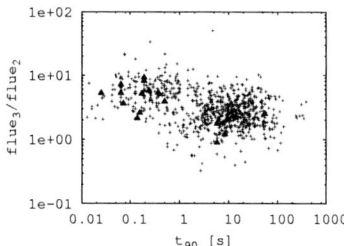

FIGURE 3. Hard to soft ratio of GRBs vs duration T90 of GRBs

An interesting feature is that a ratio between the core-collapse SNe and the white-dwarf (type Ia) SNe is much higher for the subsample of matched SNe than in the whole sample by a factor of 5.97.

CAVEATS

There are several crucial caveats in our analysis which put huge uncertainty and lowers the confidence level of our result. Surprisingly, more fatal problems are at the side of SNe:

a) Errorboxes of BATSE GRBs from in 1991-2000 era are incredibly large in comparison with precise astronomical measurement in optical bands.

b) On the other hand the BATSE catalog is an excellent source of calibrated and reduced data belonging to one detector with no need to recalculate the data.

c) There is a limiting flux in CGRO BATSE detector and, more seriously, there is a telescope-to-telescope, survey-to-survey and day-to-day changing magnitude limit for observing SNe. Some faint events could remain undetected in both samples (GRBs and SNe).

d) Neither GRBs nor SNe were not being detected by all-sky and 24 hours a day monitors. The sky covering at one time is only 33% for the BATSE GRB detector. The situation for SN sky coverage is much worse, the selection effect is very high and the probability function of observing a particular location at the sky is impossible to reconstruct. Due to this (and partially due to point c) there is still a non-negligible chance that we have not detected GRB (which lasts generally for few seconds only) of a particular SN and via versa. This is the main problem of our search and of determining the validity of our results.

e) There is an uncertainty in guessing the time of maximum of SN if there is one or two observations only.

f) There is an uncertainty in guessing the time delay between gamma and optical SN emission.

g) There are several pieces of evidence that the majority of GRB explosions are beamed into a narrow cone. Thus, the total number of GRBs and possible related SNe is much larger than observed number of these events and, because the size of the beaming angle is a function of emission frequency, there could be a SN connected to an invisible GRB if we assume larger opening angle for optical emission.

h) Dark bursts: a fraction of quite well-localized GRBs were investigated very carefully for an occurrence of possible optical afterglow with no success even when the observational conditions were good and the observations followed within few minutes. These dark bursts should be counted in statistically. Unfortunately, it is very difficult to estimate the probability of an optical detection for each one of them.

CONCLUSIONS

Even though there is a long list of caveats, there is a chance to get a meaningful result from such analysis. At this moment we are not able to conclude if there is any positive or negative or no correlation between GRBs and SNe. In the future better and proper observations of GRB afterglows should solve this problem. Meanwhile we shall try to improve our correlation method for the past events and to add rotation test, Monte-Carlo simulation could provide a more complete statistical background and, therefore, maximize the amount of information one could get from BATSE and SNe catalogs. Also filtering according to particular SN types will be done.

ACKNOWLEDGMENTS

We acknowledge the support by the Grant Agency of the Academy of Sciences of the Czech Republic, grant A3003206 and the CNR/AV CR collaborative project Investigation of GRBs.

REFERENCES

1. http://cossc.gsfc.nasa.gov/batse/index.html
2. http://cfa-www.harvard.edu/iau/lists/Supernovae.html
3. http://www.sai.msu.su/sn/
4. http://merlino.pd.astro.it/~supern/snean.txt
5. Petschek, A.G. *Supernovae*, Springer Verlag New York 1996

How Can The SN-GRB Time Delay Be Measured?

J. P. Norris* and J. T. Bonnell*

*Laboratory for High Energy Astrophysics,
NASA/Goddard Space Flight Center, Greenbelt, MD, USA 20771

Abstract. The connection between SNe and GRBs, launched by SN 1998bw / GRB 980425 and clinched by SN 2003dh / GRB 030329—with the two GRBs differing by a factor of ~ 50000 in luminosity—so far suggests a rough upper limit of ~ 1–2 days for the delay between SN and GRB. Only four SNe have had nonnegligible coverage in close coincidence with the initial explosion, near the UV shock breakout: two Type II, and two Type Ic, SN 1999ex and SN 1998bw. For the latter, only a hint of the minimum between the UV maximum and the radioactivity bump served to help constrain the interval between SN and GRB. Swift GRB alerts may provide the opportunity to study many SNe through the UV breakout phase: GRB 980425 "look alikes"—apparently nearby, low-luminosity, soft-spectrum, long-lag GRBs—accounted for half of BATSE bursts near threshold, and may dominate the Swift yield near threshold, since it has sensitivity to lower energies than did BATSE. The SN to GRB delay timescale should be better constrained by prompt UV/optical observations alerted by these bursts. Definitive delay measurements may be obtained if long-lag bursters are truly nearby: The SNe/GRBs could emit gravitational radiation detectable by LIGO-II if robust non-axisymmetric bar instabilities develop during core collapse, and/or neutrino emission may be detectable as suggested by Meszaros et al. [1]

NATURE OF PROBLEM

The timescale between gamma-ray burst (GRB) onset and supernova collapse, $T_{0,GRB}$ - $T_{0,SN} = \Delta T$, is characteristic of the progenitor's collapse. Knowledge of ΔT would constrain the physics relating GRB jet dynamics and SN core collapse. However, most GRBs with known redshifts are distant ($z_{median} \sim 1+$) and luminous, and thus SN onsets are swamped by the GRB afterglow. In fact without the GRB alert, the SN would remain undetected. The light curves of SNe are usually not well observed prior to the "radioactivity bump"—during the UV breakout phase and interjacent minimum. Since SN classification began in earnest in ~ 1992, more than 1700 SNe have been logged [2], but only four SNe have been observed for which any coverage of the shock breakout phase was obtained, two Type II(b) and two Type 1c (1987A, 1993J, 1998bw, and 1999ex [4, 5, 6, 7]). The circumstances were unusual in each case. Both Type II's were bright; SN 1998bw was alerted by the GRB; and SN 1999ex was fortuitously discovered in the same galaxy being monitored for the progress of a Type II, SN 1999ee. Some details concerning the possible degree of constraint on ΔT derivable from observations of SNe onset are discussed in the next section. The prospects for constraints on $T_{0,SN}$ are of the order of ± 0.5 day.

The best chance for empirical constraints on ΔT may come from long-lag GRBs whose distance scale is still ill-defined as a group. Long-lag GRBs may be relatively nearby compared to many luminous GRBs [8], and so as in the case of GRB 980425, detection and study of the SN onset may be much easier. But the Berger et al. [9] study of radio emission from Type Ic SNe suggests that most are less energetic than 1998bw, and may not be accompanied by GRBs. Thus important questions are, how frequently will we be able to study nearby SNe/GRBs—possibly like 1998bw/980425—and what channels will yield constraining information about ΔT and jet dynamics?

Modeling by Iwamoto et al. [10] yielded a predicted core collapse time for SN 1998bw within +0.7/-2 days of GRB 980425. In their Figure 1 illustrating light curves of three Type Ic SNe, the full-width half maxima span a dynamic range > 3. Thus, while SN 1998bw has often been quoted as a template, Type 1c SN light curves are highly variable one to the next, and seemingly unreliable for prophesying the extent of SN-like bumps in afterglows of distant GRBs, much less for predicting SN onset.

Predictions for ΔT vary over timescales of a few to $\sim 10^6$ seconds, with timescales longer than 1–2 days excluded for one case so far (1998bw/980425 Iwamoto et al.). The central issue is whether the SN happens first (core-collapse) followed by the GRB (residual torus accretion)—or if the events are (nearly) simultaneous—and this relates to strength of the two, or one, gravitational wave (GW) signals. Some current scenarios involve such a two-stage process with the timescale between collapse and explosion governed by accretion and other torques, and by black hole spin rate [11, 12].

BACKGROUND ON SUPERNOVA ONSETS

Observations of UV shock breakout in SNe are rarely obtained, due to a combination of factors. This phase is most prominent in the UV, whereas discovery proceeds from optical searches; the timescale from SN onset to end of breakout is short; and searches may require an increase in brightness (beginning of radioactive decay phase) to signal a SN and thence to proceed with observations to measure the light curve. Yet observing the rise of the UV breakout would best constraint $T_{0,SN}$ in the electromagnetic channel.

From examination of the light curves of the four SNe that have any observations of the onset and breakout phase, three trends are apparent: (1) Progressing through Type II (1987A), IIb (1993J), Ic (1999ex), and extreme Ic (1998bw), the hydrogen envelopes are increasingly depleted (then absent for Ic's), probably related to the duration of minima becoming shorter; this trend takes some study to discern due to the differing abscissa scales in the several references. (2) The minima tend to come earlier at redder colors (barely so for 1993J). (3) The minima are deeper with bluer color, to the extent that, in two of the cases, only an inflection is seen in R and redder colors. For 1998bw, coverage in (I, V, B, U) was less complete at early times than in V and R. Woosley et al. [13] discuss model UBV light curves for SN 1987A and Ib models, the first SN for which UV breakout was observed.

The closest in appearance of the four light curves are the two Type Ic, 1998bw and 1999ex. The pair have nearly equal radioactive decay widths, but different shapes especially near the minima, whose depths differ by two magnitudes. These shape variations near minimum emphasize the difficulty of using one Type 1c's light curve to infer an accurate onset time for another, even when the widths of the radioactive decay bumps are quite comparable.

The general conclusion must be that measuring SN onsets via the electromagnetic channel with sufficient accuracy required to constrain the interval between core collapse and GRB explosion to a small fraction of a day—comparable to some predictions [12]—may be at best difficult. Promising alternative approaches will probably still involve relatively nearby GRB sources and observations in neutrino or gravitational wave channels.

LONG-LAG GAMMA-RAY BURSTS

The closest GRB sources may be those with relatively smooth emission, the individual pulse appearing as the canonical fast-rise exponential decay. These "long-lag" GRBs tend to have very few, wide pulses and spectral lags between BATSE energy bands of order 1 s or greater. This group dominates the BATSE sample of long bursts approaching trigger threshold, below a peak flux of ~ 0.7 photons cm^{-2} s^{-1}. GRB 980425 was the famous example, for which the redshift is presumed to be z = 0.0085, associated with SN 1998bw. Whether a one or two-branch lag-luminosity relationship is more appropriate [14, 15, 16], may determine the average distance scale to these sources.

Several groups predicted subclasses similar to the observed long-lag bursts—with ultra-low luminosities compared to most GRBs, soft spectra, and possibly long spectral lags, where these properties may be attributed to a combination of low Lorentz factor, large jet opening angle, and/or large viewing angle [17, 18, 19]. However, the Berger et al. [8] survey of 32 Type 1c SNe found radio luminosities below that of 1998bw, suggesting that GRB 980425 may be anomalously and uniquely nearby. Regardless, long-lag bursts may be the closest subset given their tendency to occur near BATSE threshold, and thus may offer the best opportunities to explore ΔT in any channel, neutrino, gravitational or electromagnetic.

Swift's BAT should also detect many long-lag bursts. Their peak in $\nu F(\nu)$ clusters below 100 keV, compared to the median value for bright BATSE bursts of ~ 230 keV [20], and their lower power-law indices are steep. Calculations by Band [21] indicate that Swift's sensitivity to such bursts will be a factor of several better than BATSE's, as expected given the effective area vs. energy curve for Swift.

Fryer et al. [22] and Blondin et al. [2] discuss the possibility of bar instabilities developing during core collapse, giving rise to GWs. The ratio of rotational kinetic energy to gravitational potential must be sufficiently large. In an optimistic case [8], alerted by nearby long-lag GRBs, the GW emission would be detectable by LIGO II. Assuming 10 cycles, f $\sim$ 200-800 Hz, source < 50 Mpc, then h/Hz$^{-1/2} \sim 4 \times 10^{-24}$. This strain falls near the predicted LIGO threshold on the high frequency end. If the 2-branch lag-luminosity relationship obtained, then we would expect ~ 4 long-lag GRBs yr^{-1} (< 50 Mpc), with a rate within 100 Mpc ~ 30 yr^{-1}, with a few yr^{-1} detectable by LIGO II. However, SN 1998bw may be anomalously nearby in this picture (its low luminosity partially attributable to viewing angle?), and most long-lag GRBs could be considerably more distant and undetectable in GWs.

SUMMARY

From SN 1998bw/GRB 980425 and SN 2003dh/GRB 030329 we have conclusive evidence that highly energetic core-collapse SNe are associated with GRBs. To understand the SN/GRB relationship, we need to constrain $T_{0,GRB} - T_{0,SN} = \Delta T$ which is characteristic of the progenitor's collapse sequence. ΔT would also constrain the physics of GRB jet dynamics. For only four SNe, and only two Type Ic's, the UV shock breakout phase was observed. With such a small sample, and the inhomogeneity of Type Ic SNe light curves, it will be difficult to make progress. Moreover, GRB afterglows may be sufficiently bright to overwhelm the UV breakout rise to maximum, thus inhibiting investigation of timescales much shorter than currently probed for ΔT. Regardless of their mean redshift ($z \sim 0.01$–0.1 ?), long-lag, soft-spectrum, low-brightness and thus apparently low-luminosity GRBs—possibly the closest subset of GRBs—may represent the best chance for constraining ΔT. Long-lag bursts dominate near BATSE threshold, and their spectra are commensurate with Swift's BAT energy response. Thus we should detect a higher fraction of long-lag bursts with Swift. Eventually, the best channels for probing short ΔTs may be gravitational waves or neutrino emission [1]. However, depending on the mean distance to long-lag bursts and the particulars of GW and neutrino radiation, summation of several events over a timescale of a few years may be necessary to detect these signals.

REFERENCES

1. Meszaros, P., Kobayashi, S., Razzaque, S. & Zhang, B., astro-ph/0305066 (2003).
2. Blondin, J.M., Mezzacappa, A., & DeMarino, C., ApJ **584**, 971 (2003).
3. "List of Supernovae," http://cfa-www.harvard.edu/iau/lists/Supernovae.html
4. Hamuy, M., et al., AJ **95**, 1 (1988).
5. Lewis, J.R., et al., MNRAS **266**, L27 (1994).
6. Galama, T.J., et al., Nature **395** 670 (1998).
7. Stritzinger, M., et al., AJ **124**, 2100 (2002).
8. Berger, E., Kulkarni, S.R., & Frail, D.A., & Soderberg, A.M., astro-ph/0307228 (2003).
9. Norris, J.P., in The Astrophysics of Gravitational Wave Sources, AIP 686, ed. J.M. Centrella (New York: AIP), p. 74 (2003).
10. Iwamoto, K., et al., Nature **395**, 672 (1998).
11. MacFayden, A. I., Woosley, S. E., & Heger, A., ApJ **550**, 410 (2001).
12. Fryer, C.L., & Meszaros, P., ApJ **588**, L25 (2003).
13. Woosley, S.E., et al., ApJ **318**, 664 (1987).
14. Norris, J.P., Marani, G.F., & Bonnell, J.T., ApJ **534**, 238 (2000).
15. Norris, J.P., ApJ **579**, 386 (2002).
16. Band, D.L., Norris, J.P., Bonnell, J.T., these proceedings
17. Woosley, S.E., MacFayden, A.I., A&AS **138**, 499 (1999).
18. Ioka, K., Nakamura, T., ApJ |bf 554, L163 (2001).
19. Salmonson, J.D., ApJ **546**, L29 (2001).
20. Preece, R.D., et al., ApJS **126**, 19 (2000).
21. Band, D.L., ApJ **588**, 945 (2003).
22. Fryer, C.L., Holz, D.E., & Hughes, S.A., ApJ **565**, 430, (2002).

GRB 980425 in the Off-Axis Jet Model of the Standard GRBs

Ryo Yamazaki*, Daisuke Yonetoku† and Takashi Nakamura*

*Department of Physics, Kyoto University, Kyoto 606-8502, Japan
†Department of Physics, Kanazawa University, Kakuma, Kanazawa, Ishikawa 920-1192, Japan

Abstract. Using a simple off-axis jet model of GRBs, we can reproduce the observed unusual properties of the prompt emission of GRB 980425, such as the extremely low isotropic equivalent γ-ray energy, the low peak energy, the high fluence ratio, and the long spectral lag when the jet with the standard energy of $\sim 10^{51}$ ergs and the opening half-angle of $10° \leq \Delta\theta \leq 30°$ is seen from the off-axis viewing angle $\theta_v \sim \Delta\theta + 10\gamma^{-1}$, where γ is a Lorentz factor of the jet. For our adopted fiducial parameters, if the jet that caused GRB 980425 is viewed from the on-axis direction, the intrinsic peak energy $E_p(1+z)$ is ~ 2.0–4.0 MeV, which corresponds to those of GRB 990123 and GRB 021004. Our model might be able to explain the other unusual properties of this event. We also discuss the connection of GRB 980425 in our model with the X-ray flash, and the origin of a class of GRBs with small E_γ such as GRB 030329.

INTRODUCTION

There are some GRBs that were thought to be associated with SNe [1, 2]. GRB 980425 / SN 1998bw, located at $z = 0.0085$ (36 Mpc), was the first event of such class [3, 4, 5, 6]. It is important to investigate whether GRB 980425 is similar to more or less typical long duration GRBs. However, GRB 980425 showed unusual observational properties. The isotropic equivalent γ-ray energy is $E_{iso} \sim 6 \times 10^{47}$ ergs and the geometrically corrected energy is $E_\gamma = (\Delta\theta)^2 E_{iso}/2 \sim 3 \times 10^{46}$ ergs $(\Delta\theta/0.3)^2$, where $\Delta\theta$ is the unknown jet opening half-angle. These energies are much smaller than the typical values of GRBs. The other properties of GRB 980425 are also unusual; the large low-energy flux [7], the low variability [8], the long spectral lag [9], and the slowly decaying X-ray afterglow [5, 6].

Previous works suggest that the above peculiar observed properties may be explained if the standard jet is seen from the off-axis viewing angle (e.g.[10, 11]). Following this scenario, the relativistic beaming effect reduces E_{iso} and hence E_γ. In this paper, in order to explain all of the observed properties of GRB 980425, we reconsider the prompt emission of this event using our simple jet model [12, 13, 14, 15].

SPECTRAL ANALYSIS OF GRB 980425 USING BATSE DATA

We analysis the time-averaged observed spectral properties of GRB 980425. Using the BATSE data of GRB 980425, we analyze the spectrum within the time of FWHM of the peak flux in the light curve of BATSE channel 2. We fit the observed spectrum with the Band function. The best-fit values are $\alpha = -1.0 \pm 0.3$, $\beta = -2.1 \pm 0.1$, and $E_p = 54.6 \pm 20.9$ keV, which are consistent with those derived by previous papers [7, 3]. This spectral property is similar to one of the recently identified class of the X-ray flash (XRF) [16, 17]. The observed fluence of the entire emission is $S(20–2000 \text{keV}) = (4.0 \pm 0.74) \times 10^{-6}$ erg cm^{-2}, thus we find $E_{iso} = (6.4 \pm 1.2) \times 10^{47}$ ergs. The fluence ratio is $R_s = S(20–50 \text{keV})/S(50–320 \text{keV}) = 0.34 \pm 0.036$.

MODEL OF PROMPT EMISSION OF GRB 980425

We use a simple jet model of prompt emission of GRBs, where an instantaneous emission of infinitesimally thin shell is adopted [10, 12, 13, 14, 15]. See Yamazaki et al. [15] for details. We fix model parameters as $\alpha_B = -1$, $\beta_B = -2.1$, $\gamma v_0' = 2600$ keV, and $\gamma = 100$. Normalization of emitted luminosity is determined so that E_γ be observationally preferred value of $1.15 \times 10^{51 \pm 0.35} (h/0.7)^{-2}$ ergs [21] when we see the jet from the on-axis viewing angle $\theta_v = 0$. Our calculations show that on-axis intrinsic peak energy becomes $E_p^{(\theta_v=0)}(1+z) \sim 1.54 \gamma v_0' \sim 4.0$ MeV, in order to reproduce the observed quantities of GRB 980425. Indeed, there are some GRBs with higher intrinsic E_p; for example, $E_p(1+z) \sim 2.0$ MeV for GRB 990123 and 3.6 MeV for GRB 021004 [18, 19].

The left panel of Figure. 1 shows E_{iso} as a function of the viewing angle θ_v. When $\theta_v \leq \Delta\theta$, E_{iso} is constant, while for $\theta_v \geq \Delta\theta$, E_{iso} is considerably smaller than the typical value of $\sim 10^{51-53}$ ergs because of the relativistic beaming effect.

We next calculated E_p and R_s for the set of $\Delta\theta$ and θ_v^* that reproduces the observed E_{iso} of GRB 980425. For our parameters, $\Delta\theta$ should be between $\sim 18°$ and $\sim 31°$, and then θ_v^* ranges between $\sim 24°$ and $\sim 35°$ in order to reproduce the observation results. Thorough discussions on the right panel of Figure 1 is found in [15].

DISCUSSION

We have found that when the jet of opening half-angle of $\Delta\theta \sim 10–30°$ is seen from the off-axis viewing angle of $\theta_v \sim \Delta\theta + 6°$, observed quantities can be well explained. Observed low variability can be explained since only subjets at the edge of the cone contribute to the observed quantities [12]. If the time unit parameter $r_0/c\beta\gamma^2$ is about 3 sec, which is in the reasonable parameter range, the spectral-lag of GRB 980425 can be also explained.

Our result might be able to explain the slowly decaying X-ray afterglow of GRB 980425. If we assume the density profile of ambient matter as $n = n_0(r/r_{ext})^{-2}$ with $n_0 r_{ext}^2 = 4 \times 10^{17}$ cm^{-2}, the break in the afterglow light curve should occur at $t_b = 3.1 \times 10^2$ days $E_{51}(\Theta/0.4\text{rad})^2$, where Θ is defined by $\Theta^2 = (\Delta\theta)^2 + \theta_v^2$, and E is the total energy in the collimated jet [20, 11]. Since our calculation suggests Θ should range between 0.4 and 0.67 rad, t_b is consistent with the observation [11]. Up to the break time, one can estimate the flux in the X-ray band as $F(2-10\text{keV}) \propto t^{-0.2}$, where we assume $\theta_v \gg \Delta\theta$ and the spectral index of accelerated electrons as $p = 2.2$ [20, 11]. This result is also consistent with the observation [5, 6]. Furthermore,

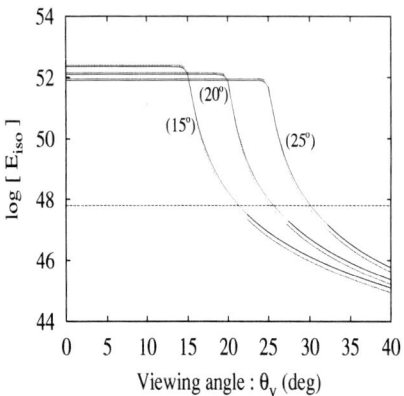 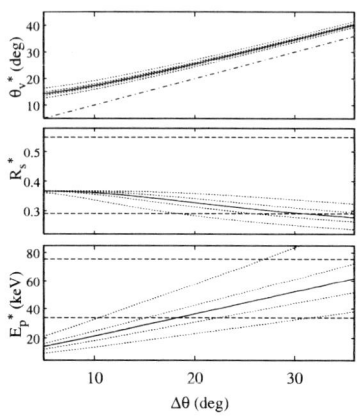

FIGURE 1. (Left panel): the isotropic equivalent γ-ray energy E_{iso} is shown as a function of the viewing angle θ_v for a fixed jet opening half-angle $\Delta\theta$. The source is located at $z = 0.0085$. The values of $\Delta\theta$ are shown in parentheses. Solid lines correspond to the case of $\gamma v_0' = 2600$ keV, while dotted lines $\gamma v_0' = 1300$ keV. Horizontal dashed line represents the observed value of GRB 980425. (Right panel): the upper panel shows θ_v^* for which E_{iso} is the observed value of GRB 980425, while the middle and the lower panels represent the fluence ratio $R_s^* = R_s^{(\theta_v=\theta_v^*)}$ and the peak energy $E_p^* = E_p^{(\theta_v=\theta_v^*)}$, respectively. Solid lines correspond to the fiducial case. The dotted lines represent regions where E_{iso} becomes $(6.4 \pm 1.2) \times 10^{47}$ ergs when E_γ is in 1 σ and 5 σ level around the fiducial value, respectively. The dot-dashed line in the upper panel represents $\theta_v^* = \Delta\theta$. Horizontal dashed lines in the middle and the lower panels represent the observational bounds.

the adopted value of $n_0 r_{ext}^2$ corresponds to the mass loss rate of the progenitor star $\dot{M} = 1.3 \times 10^{-6} M_\odot \text{ yr}^{-1} (v_W/10^3 \text{ km s}^{-1})$, which might be able to explain the radio data (see [22]).

The observed quantities of small E_p and large fluence ratio R_s are the typical values of the XRF [7, 16, 17]. The operational definition of the *BeppoSAX*-XRF is a fast X-ray transient with duration less than $\sim 10^3$ s which is detected by WFCs and not detected by the GRBM. If the distance to the source of GRB 980425 were larger than ~ 90 Mpc, the observed flux in the γ-ray band would have been less than the limiting sensitivity of GRBM, so that the event would have been detected as an XRF.

We might be able to explain the origin of a class with low E_γ such as GRB 980326 and GRB 981226 [21], and GRB 030329 whose E_γ is about $\sim 5 \times 10^{49}$ ergs if the jet break time of ~ 0.48 days is assumed [23, 24]. Let us consider the jet seen from a viewing angle $\theta_v \sim \Delta\theta + \gamma_i^{-1}$, where γ_i is the Lorentz factor of a prompt γ-ray emitting shell. Due to the relativistic beaming effect, the observed E_γ of such a jet becomes an order of magnitude smaller than the standard energy. At the same time, the observed peak energy E_p is small because of the relativistic Doppler effect. In fact, the observed E_p of the above three bursts are less than ~ 70 keV. In our model the fraction of low-E_γ GRBs becomes $2/(\gamma_i\Delta\theta) \sim 0.1$ since the mean value of $\Delta\theta \sim 0.2$, while a few of them are observed in ~ 30 samples [21]. In later phase, the Lorentz factor of afterglow emitting shock γ_f is smaller than γ_i, so that $\theta_v < \Delta\theta + \gamma_f^{-1}$. Then, the observed properties of afterglows may be similar to the on-axis case $\theta_v \ll \Delta\theta$; hence the observational estimation of the jet break time and the jet opening angle remains the same.

ACKNOWLEDGMENTS

This work was supported in part by Grant-in-Aid for Scientific Research of the Japanese Ministry of Education, Culture, Sports, Science and Technology, No.05008 (R.Y.), No.14047212 (T.N.), and No.14204024 (T.N.).

REFERENCES

1. Della Valle, M. et al., A&A **406**, L33 (2003).
2. Stanek, K.Z. et al., ApJ **591**, L71 (2003).
3. Galama, T.J., et al., Nature **395**, 670 (1998).
4. Kulkarni, S.R., et al., Nature **395**, 663 (1998).
5. Pian, E. et al,, ApJ **536**, 778 (2000).
6. Pian, E. et al,, astro-ph/0304521 (2003).
7. Frontera, F. et al., ApJS **127**, 59 (2000).
8. Fenimore, E. E. & Ramirez-Ruiz. E., astro-ph/0004176 (2000)
9. Norris, J.P., Marani, G.F., & Bonnell, J.T., ApJ **534**, 248 (2000).
10. Ioka, K., & Nakamura, T., ApJ **554**, L163 (2001).
11. Nakamura, T., Prog. Theor. Phys. Suppl **143**, 50 (2001).
12. Yamazaki, R., Ioka, K., & Nakamura, T., ApJ **571**, L31 (2002).
13. Yamazaki, R., Ioka, K., & Nakamura, T., ApJ **591**, 283 (2003).
14. Yamazaki, R., Ioka, K., & Nakamura, T., ApJ **593**, 941 (2003).
15. Yamazaki, R., Yonetoku, D., & Nakamura, T., ApJ **594**, L79 (2003).
16. Heise, J., et al., in Proc. 2nd Rome Workshop: GRBs in the Afterglow Era, astro-ph/0111246 (2001).
17. Kippen, R. M., et al., in Proc. Woods Hole Gamma-Ray Burst Workshop, astro-ph/0203114 (2002).
18. Amati, L., et al., A&A **390**, 81 (2002).
19. Barraud, C., et al., A&A **400**, 1021 (2003).
20. Nakamura, T., ApJ **522**, L101 (1999).
21. Bloom, J.S., et al., ApJ **594**, 674 (2003).
22. Waxman, E., astro-ph/0310320 (2003).
23. Tamagawa, T. et al., in this proceeding (2004).
24. Vanderspek, R. et al., GCN circ. 1997 (2003).

Color Superconductivity in Compact Stars and Gamma Ray Bursts

A. Drago*, A. Lavagno† and G. Pagliara*

*Dipartimento di Fisica, Università di Ferrara and INFN, Sezione di Ferrara, 44100 Ferrara, Italy
†Dipartimento di Fisica, Politecnico di Torino and INFN, Sezione di Torino, 10129 Torino, Italy

Abstract. We study the effects of color superconductivity on the structure and formation of compact stars. We find that a huge amount of energy, of the order of 10^{53} erg, can be released in the conversion from a (metastable) hadronic star into a (stable) hybrid or quark star. If the conversion takes place immediately after the deleptonization of the proto-neutron star, the released energy can help Supernovae to explode. If the conversion is delayed the energy released can power a Gamma Ray Burst. A delay between the Supernova and the subsequent Gamma Ray Burst is possible, in agreement with the delays proposed in recent analysis of astrophysical data.

INTRODUCTION

The increasing quantity of new data from X-ray satellites provide important informations on the structure of compact stellar objects and on the observations of Gamma-Ray Bursts (GRBs), indicating the possibility that some of the GRBs are associated with a Supernova (SN) explosion. It has not yet been clarified if the two explosions are always simultaneous or if, at least in a few cases, a time delay can exist, with the SN preceding the GRB [1, 2, 3, 4, 5].

The effect of the transition to deconfined Quark Matter (QM) on explosive processes like SNs and GRBs has been discussed by many authors. The possibility that deconfinement takes place during the core-collapse of massive stars at the moment of the bounce could help the SN to explode by increasing the mechanical energy associated with the bounce [6, 7]. However, it seems more plausible that deconfinement takes place only when the proto-neutron star has deleptonized and cooled down to a temperature of a few MeV [8, 9]. In Refs. [10, 11], we discuss the scenario in which neutron stars having a small enough mass can exist as metastable Hadronic Star (HS) if a non-vanishing surface tension is present at the interface between Hadronic Matter (HM) and QM. The process of quark deconfinement can then be a powerful source for GRBs and it can also explain the possible delay between a SN explosion and the subsequent GRB.

In recent years, many theoretical works have investigated the possible formation of a diquark condensate in QM, at densities reachable in the core of a compact star [12]. The formation of this condensate can deeply modify the structure of the star as shown in Refs.[13]. The aim of this paper is to show that the formation of diquark condensate can significantly increase the energy released in the conversion from a purely HS into a more stable star containing deconfined QM. The newly formed Hybrid Star (HyS) or Quark Star (QS) cools down emitting neutrinos and the subsequent neutrino-antineutrino annihilation can power a GRB.

EQUATION OF STATE OF BETA-STABLE MATTER

The EOS appropriate to the description of a compact star has to satisfy beta-stability conditions. Moreover two charges are conserved, the baryonic and the electric one. These conditions need to be satisfied in the hadronic, in the quark and in the mixed phase by imposing the Gibbs condition. Concerning the hadronic phase we use the relativistic non-linear Glendenning-Moszkowski model (GM1-GM3) [14]. For the quark matter phase we adopt a MIT-bag like model in which the formation of a diquark condensate is taken into account in a simple and effective way, as proposed in Refs.[13, 15], but by considering a μ dependent gap resulting from the solution of the gap equation [12]. The shapes of the gaps Δ_i are shown in Fig. 1. As we will see below, both the magnitude and the position of the gap deeply affects the energy released in the conversion from HS to HyS or QS.

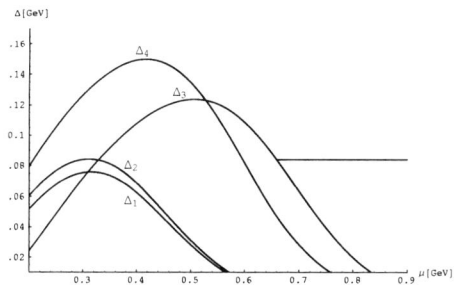

FIGURE 1. Gap Δ_i as function of the chemical potential, for four different parameter sets.

NUCLEATION TIME AND ENERGY RELEASED

The possibility of a relation between GRBs and SN explosions has been extensively discussed in the literature. Concerning the existence of a time delay between the two explosions, very important information can be obtained from the analysis of the optical afterglow, since in two cases (GRB980425/SN1998bw, GRB030329/SN2003dh) the spectrum of a type Ic SN emerged once the spectrum of the GRB afterglow has been subtracted. It is therefore possible, at least in principle, to estimate the date of the SN explosion and to estimate the delay (if any) between the two explosions.

In the model we are proposing, the central density of a pure HS increases, due to spin down or mass accretion, until its value approaches the deconfinement critical density. At this point a spherical virtual drop of QM can form. To compute the time needed to form a bubble of quarks having a radius larger than the critical one, we use the technique of quantum tunneling nucleation in the WKB approximation [10, 16, 17]. We can assume that the temperature has no effect in our scheme: for values of $B^{1/4} \sim 160 - 180$ MeV,

TABLE 1. Energy released ΔE in the conversion to hybrid or quark star, for various sets of model parameters, assuming the hadronic star mean life-time $\tau = 1$ yr. M_{cr} is the gravitational mass of the hadronic star at which the transition takes place, for fixed values of the surface tension σ and of the mean life-time τ. BH indicates that the hadronic star collapses to a Black Hole. A dash (–) indicates situations in which the Gibbs construction does not provide a mechanically stable EOS.

Hadronic Model	$B^{1/4}$ [MeV]	σ [MeV/fm^2]	$M_{cr}/M_\odot$	ΔE $\Delta=0$	Δ_1	Δ_2	Δ_3	Δ_4
GM3	160	20	0.69	20	65•	69•	76•	148•
GM3	160	30	0.91	32	90•	95•	106•	196•
GM3	160	40	1.00	38	100•	105•	119•	216•
GM3	170	10	1.12	0	34	40	68	162•
GM3	170	20	1.26	4	44	50	86	185•
GM3	170	30	1.39	11	53	60	104	207•
GM3	170	40	1.49	BH	62	68	120	224•
GM3	180	10	1.55	BH	11	13	BH	–
GM3	180	20	1.61	BH	BH	22	BH	–
GM3	180	30	1.67	BH	BH	BH	BH	–
GM1	160	10	0.45	11	41•	44•	47•	96•
GM1	160	20	0.72	28	75•	79•	86•	160•
GM1	160	30	0.96	48	108•	114•	127•	220•
GM1	160	40	1.18	72	142•	148•	166•	276•
GM1	170	10	1.17	18	59	65	96	191•
GM1	170	20	1.33	33	79	85	124	226•
GM1	170	30	1.45	50	96	103	150	254•
GM1	170	40	1.60	BH	122	128	BH	290•
GM1	180	10	1.63	BH	BH	72	BH	–
GM1	180	20	1.72	BH	BH	BH	BH	–
GM1	180	30	1.79	BH	BH	BH	BH	–

which we use in this paper, the critical density ρ_1 separating pure HM from mixed phase is larger than $4\rho_0$ for $Z/A \sim 0.3$, i.e. for an isospin fraction typical of a newly formed and hot proto-neutron star. This critical density typically exceeds the central density of hot and not too massive stars. Therefore, the mixed phase can form only when the star has deleptonized and its temperature has dropped down to a few MeV [8, 9]. When the temperature is so low, only quantum tunneling is a practicable mechanism.

In Table 1, we show the value of M_{cr} for various sets of model parameters. In the conversion process from a metastable HS into an HyS or a QS a huge amount of energy ΔE is released. ΔE is the difference between the gravitational mass of the metastable HS and that of the final HyS or QS having the same baryonic mass. We see in the Table that the formation of a CFL phase allows to obtain values for ΔE which are one order of magnitude larger than the corresponding ΔE of the unpaired QM case ($\Delta = 0$). Moreover, we can observe that ΔE depends both on magnitude and position of the gap.

In the model we are presenting, the GRB is due to the cooling of the just-formed HyS or QS via neutrino - antineutrino emission. The subsequent neutrino-antineutrino annihilation generates the GRB. The duration of the prompt emission of the GRB is therefore regulated by two mechanisms: 1) the time needed for the conversion of the HS into a HyS or QS, once a critical-size droplet is formed and 2) the cooling time of the justly formed HyS or QS. Concerning the time needed for the conversion into QM of at least a fraction of the star, the seminal work by [18] has been reconsidered in Ref.[19], where it has been shown that the stellar conversion is a very fast process, having a duration much shorter than 1s. On the other hand, the neutrino trapping time, which provides the cooling time of a compact object, is of the order of a few ten seconds [20], and it gives the typical duration of the GRB in our model.

CONCLUSIONS

We have found that the superconducting gap deeply affects the energy released in the conversion from hadronic star into hybrid or quark star. We assume that the deconfinement transition only takes place when the star has deleptonized and cooled down, in agreement with the results of Ref.[8, 9]. If deconfinement occurs immediately after deleptonization, the energy released can help the Supernova to explode. If, at variance, the transition is delayed, a metastable hadronic star can form. Its subsequent transition to a stable configuration, containing deconfined quark matter, can power a GRB via the annihilation of neutrinos and antineutrinos emitted during the cooling of the newly formed compact star. The energy released is significantly increased by the effect of the chemical-potential dependent superconducting gap and it can reach a value of the order of 10^{53} erg. The proposed mechanism could explain recent observations indicating a possible delay between a Supernova and the subsequent Gamma Ray Burst [1, 2, 3].

REFERENCES

1. L. Amati et al., Science **290**, 953 (2000).
2. J.N. Reeves et al., Nature **416**, 512 (2002).
3. J. Hjorth et al., Nature **423**, 847 (2003).
4. R.E. Rutledge, M. Sako, Mon. Not. R. Astron. Soc. **339**, 600 (2003).
5. J.N. Reeves et al., Astron.Astrophys. **403**, 463 (2003).
6. A. Drago, U. Tambini, J. Phys. G. **25**, 971 (1999).
7. D.K. Hong, S.D.H. Hsu, F. Sannino, Phys. Lett. B **516**, 362 (2001) and references therin.
8. O.G. Benvenuto, G. Lugones, Mon. Not. R. Astron. Soc. **304**, L25 (1999).
9. J.A. Pons et al., Phys. Rev. Lett. **86**, 5223 (2001).
10. Z. Berezhiani et al., Astrophys. J. **586**, 1250 (2003).
11. A. Drago, G. Pagliara, A. Lavagno, Phys. Rev. D **69**, 057505 (2004).
12. M.G. Alford, K. Rajagopal and F. Wilczek, Nucl. Phys. B **537**, 443 (1999).
13. M.G. Alford, S. Reddy, Phys. Rev. D **67**, 074024 (2003).
14. N.K. Glendenning, S.A. Moszkowski, Phys. Rev. Lett. **67**, 2414 (1991).
15. G. Lugones, J. E. Horvath, Phys. Rev. D **66**, 074017 (2002).
16. I.M. Lifshitz, Yu. Kagan, Zh. Eksp. Teor. Fiz. **62**, 385 (1972) [Sov. Phys. JETP, **35**, 206 (1972)].
17. K. Iida and K. Sato, Phys. Rev. C **58**, 2538 (1998).
18. A. Olinto, Phys. Lett. B **192**, 71 (1987).
19. J.E. Horvath, O.G. Benvenuto, Phys. Lett. B **213**, 516 (1988).
20. M. Prakash et al., Phys. Rept. **280**, 1 (1997).

The GRB 980425-SN1998bw Association in the EMBH Model

F. Fraschetti*, M. G. Bernardini†, C. L. Bianco†, P. Chardonnet**, R. Ruffini† and S. S. Xue†

Università di Trento, Via Sommarive 14, I-38050 Povo (Trento), Italy
†*ICRA - International Centre for Relativistic Astrophysics and Dipartimento di Fisica, Università di Roma "La Sapienza", Piazzale Aldo Moro 5, I-00185 Roma, Italy*
**Université de Savoie, LAPTH - LAPP, BP 110, F-74941 Annecy-le-Vieux Cedex, France*

Abstract. Our GRB theory, previously developed using GRB 991216 as a prototype, is here applied to GRB 980425. We fit the luminosity observed in the 40–700 keV, 2–26 keV and 2–10 keV bands by the BeppoSAX satellite. In addition the supernova SN1998bw is the outcome of an "induced gravitational collapse" triggered by GRB 980425, in agreement with the GRB-Supernova Time Sequence (GSTS) paradigm (Ruffini et al. [1]). A further outcome of this astrophysically exceptional sequence of events is the formation of a young neutron star generated by the SN1998bw event (Ruffini et al. [2]). A coordinated observational activity is recommended to further enlighten the underlying scenario of this most unique astrophysical system.

Our GRB theory (Ruffini et al. [3, 4, 1, 5, 6] and references therein), previously successfully applied to GRB 991216 used as a prototype, is applied to GRB 980425 (Pian et al. [7]) and SN1998bw (Galama and et al. [8]). This event allows to test the validity of the theory over a range of energies of 6 orders of magnitude: both sources appear to be spherically symmetric and the respective total energies are $E_{tot} \simeq 5 \times 10^{53}$ ergs and $E_{tot} \simeq 10^{48}$ ergs.

The theory, therefore, explains all the observed features of the bolometric intensity variations of the afterglow as well as the spectral properties of the source and, in the specific case of GRB 980425 (Ruffini et al. [2]), it also allows to clarify the general astrophysical scenario in which the GRB actually occurs. In this system, in fact, we propose that GRB 980425 has been the trigger of a phenomenon of "induced gravitational collapse" (Ruffini et al. [1]) originating the supernova explosion and we also witness the birth of a young neutron star out of the supernova event. This extraordinary coincidence of these three astrophysical events represents an unprecedented scenario of fundamental importance in the field of relativistic astrophysics.

The observational situation of this system is quite complex. In addition to the source GRB 980425 and the supernova SN1998bw, two X-ray sources have been found by BeppoSAX in the error box for the location of GRB 980425: a source *S1* and a source *S2* (Pian et al. [7]). Our approach is the following. We first interpret the GRB 980425 within the EMBH theory. This allows the computation of the luminosity, spectra, Lorentz gamma factors, and more generally all the dynamical aspects of the source. Having characterized the features of GRB 980425, we can gradually approach the remaining part of the scenario, disentangling the GRB observations from the supernova ones and from the sources *S1* and *S2*. This leads to a natural time sequence of events and to their autonomous astrophysical characterization.

Our approach has focused on identifying the energy extraction process from the black hole (Christodoulou and Ruffini [9]) as the basic energy source for the GRB phenomenon. The distinguishing feature is a theoretically predicted source energetics all the way up to $1.8 \times 10^{54} (M_{BH}/M_\odot)$ ergs for $3.2 M_\odot \leq M_{BH} \leq 7.2 \times 10^6 M_\odot$ (Damour and Ruffini [10]). In particular, the formation of a "dyadosphere", during the gravitational collapse leading to a black hole endowed with electromagnetic structure (EMBH) has been indicated as the initial boundary conditions of the GRB process (Ruffini [11], Preparata et al. [12]).

The equations of motion in our theory depend only on two free parameters: the total energy E_{tot}, which coincides with the dyadosphere energy E_{dya}, and the amount M_B of baryonic matter left over from the gravitational collapse of the progenitor star, which is determined by the dimensionless parameter $B = M_B c^2 / E_{dya}$. Our best fit corresponds to $E_{dya} = 1.1 \times 10^{48}$ ergs, $B = 7 \times 10^{-3}$ and the ISM average density is found to be $\langle n_{ism} \rangle = 0.02$ particle/cm^3. The plasma temperature and the total number of pairs in the dyadosphere are respectively $T = 1.028$ MeV and $N_{e\pm} = 5.3274 \times 10^{53}$.

Recently, within the EMBH theory, we have developed an attempt to theoretically derive the GRB spectra out of first principles as well as the GRB luminosity in fixed energy bands (Ruffini et al. [13]). We have adopted three basic assumptions: **a)** the resulting radiation as viewed in the comoving frame during the afterglow phase has a thermal spectrum and **b)** the ISM swept up by the front of the shock wave, with a Lorentz gamma factor between 300 and 2, is responsible for this thermal emission. **c)** We also assume, like in our previous papers (Ruffini et al. [3, 4, 5, 6]), that the expansion occurs with spherical symmetry.

The temperature T of the black body in the comoving frame is then

$$T = \left(\frac{\Delta E_{\text{int}}}{4\pi r^2 \Delta \tau \sigma \mathscr{R}} \right)^{1/4}, \qquad (1)$$

where $\mathscr{R} = A_{eff}/A_{abm}$ is the ratio between the "effective emitting area" and the ABM pulse surface A_{abm} (in this case the best fit value of $\mathscr{R}$ is monotonically decreasing from 4.81×10^{-10} to 2.65×10^{-12}), σ is the Stefan-Boltzmann constant and ΔE_{int} is the proper internal energy developed in the collision between the ABM pulse and the ISM in the proper time interval $\Delta \tau$ (see Ruffini et al. [6, 13]). The ratio $\mathscr{R}$, which is a priori a function that varies as the system evolves, is evaluated at every given value of the laboratory time t.

All the subsequent steps are now uniquely determined by the equations of motion of the system. The basic tool in this calculation involves the definition of the EQuiTemporal Surfaces (EQTS) for the relativistic expanding ABM pulse as seen by an asymptotic observer. The key to determining such EQTS (see Fig. 1 in Ruffini et al. [5]) is the relation between the time t in the laboratory frame at which a photon is emitted from the ABM pulse external surface and the arrival time t_a^d at which it reaches the detector.

The results are given in Fig. 1 where the luminosity is computed as a function of the arrival time for three selected energy bands.

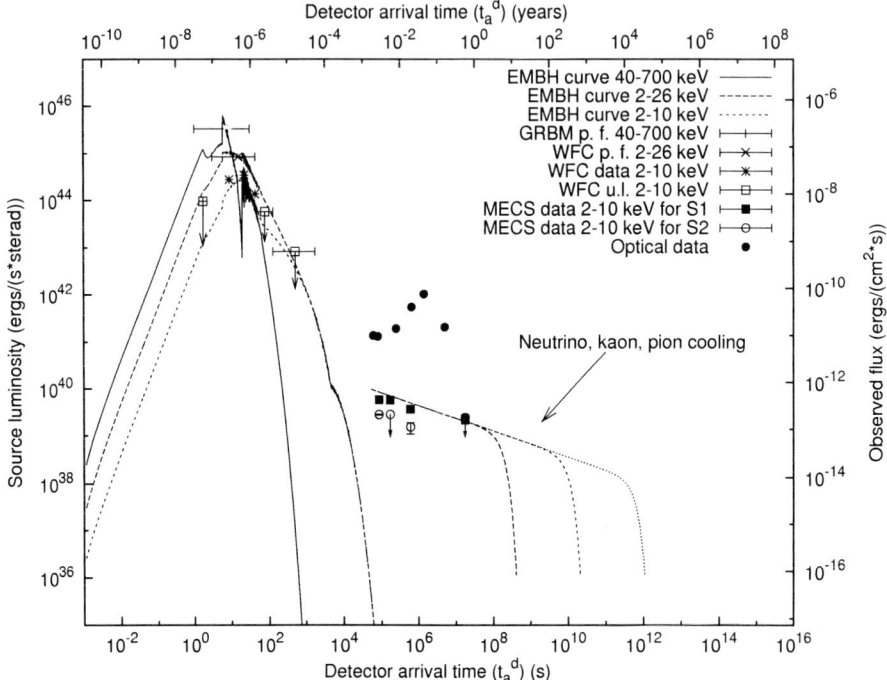

FIGURE 1. The light curves in selected bands are reported as well as the MECS light curves in the 2-10 keV band of S1 and S2 (Pian et al. [7]) as well as the optical data (Iwamoto [14]). Here are also reported theoretical models of neutron star cooling (Canuto [15]).

In Fig. 1 the luminosities in the three bands are represented together with the optical data of SN1998bw (black dots), the source S1 (black squares) and the source S2 (open circles). It is then clear that GRB 980425 is separated both from the supernova data and from the sources *S1* and *S2*.

While the occurrence of the supernova in relation to the GRB has already been discussed with the GRB-Supernova Time Sequence (GSTS) paradigm (Ruffini et al. [1]), we like to address here a different fundamental issue: the possibility of observing the birth of a newly formed neutron star, possibly pulsating, out of the supernova event, which in turn has been triggered by the GRB 980425.

In the early days of neutron star physics it was clearly shown by (Gamow and Schoenberg [16]) that the URCA processes are at the very heart of the supernova explosions. The neutrino-antineutrino emission described in the URCA process is the essential cooling mechanism necessary for the occurrence of the process of gravitational collapse of the imploding core. Since then, it has become clear that the newly formed neutron star can be still significantly hot and in its early stages will be associated to three major radiating processes (Tsuruta [17, 18], Tsuruta et al. [19], Canuto [15]): **a)** the thermal radiation from the surface, **b)** the radiation due to neutrino, kaon, pion cooling, and **c)** the possible influence in both these processes of the superfluid nature of the supra-nuclear density neutron gas. Qualitative representative curves for these cooling processes, which are still today very undetermined due to the lack of observational data, are shown in Fig. 1.

It is of paramount importance to follow the further time history of the two sources *S1* and *S2*. If, as we propose, *S2* is a background source, its flux should be practically constant in time and this source has nothing to do with the GRB 980425 / SN1998bw system. If S1 is indeed the cooling radiation emitted by the newly born neutron star, it should be possible to notice a very drastic behavior in its luminosity as qualitatively expresses in Fig. 1.

The complete details on the source with all numerical values and explicit relations is going to appear in (Ruffini et al. [20]).

REFERENCES

1. Ruffini, R., Bianco, C. L., Chardonnet, P., Fraschetti, F., and Xue, S.-S., *ApJ Lett.*, **555**, L117 (2001).
2. Ruffini, R., Bernardini, M. G., Bianco, C. L., Chardonnet, P., Fraschetti, F., and Xue, S.-S., "GRB 980425, SN1998bw and the EMBH Model," in *Proceedings of the 34th COSPAR Scientific Assembly*, edited by E. Pian, N. Masetti, and L. Piro, Elsevier, 2003, in press.
3. Ruffini, R., Bianco, C. L., Chardonnet, P., Fraschetti, F., and Xue, S.-S., *ApJ Lett.*, **555**, L107 (2001).
4. Ruffini, R., Bianco, C. L., Chardonnet, P., Fraschetti, F., and Xue, S.-S., *ApJ Lett.*, **555**, L113 (2001).
5. Ruffini, R., Bianco, C. L., Chardonnet, P., Fraschetti, F., and Xue, S.-S., *ApJ Lett.*, **581**, L19 (2002).
6. Ruffini, R., Bianco, C. L., Chardonnet, P., Fraschetti, F., Vitagliano, L., and Xue, S.-S., "New Perspectives in Physics and Astrophysics from the Theoretical Understanding of Gamma-Ray Bursts," in *COSMOLOGY AND GRAVITATION: X^{th} Brazilian School of Cosmology and Gravitation; 25th Anniversary (1977-2002)*, edited by M. Novello and S. E. P. Bergliaffa, AIP, New York, 2003, vol. 668, p. 16, Allegato 8.
7. Pian, E., Amati, L., Antonelli, L., Butler, R., Costa, E., Cusumano, G., Danziger, J., Feroci, M., Fiore, F., Frontera, F., Giommi, P., Masetti, N., Muller, J., Nicastro, L., Oosterbroek, T., Orlandini, M., Owens, A., Palazzi, E., Parmar, A., Piro, L., in 't Zand, J., Castro-Tirado, A., Coletta, A., Fiume, D. D., Sordo, S. D., Heise, J., Soffitta, P., and Torroni, V., *ApJ*, **536**, 778 (2000).
8. Galama, T., and et al., *Nature*, **59**, 395 (1998).
9. Christodoulou, D., and Ruffini, R., *Phys. Rev. D*, **4**, 3552 (1971).
10. Damour, T., and Ruffini, R., *Phys. Rev. Lett.*, **35**, 463 (1975).
11. Ruffini, R., "Beyond the Critical Mass: The Dyadosphere of Black Holes," in *Black Holes and High Energy Astrophysics, Proceedings of the 49th Yamada Conference*, edited by H. Sato and N. Sugiyama, Universal Ac. Press, Tokyo, 1998.
12. Preparata, G., Ruffini, R., and Xue, S.-S., *A&A*, **338**, L87 (1998).
13. Ruffini, R., Bianco, C. L., Chardonnet, P., Fraschetti, F., Gurzadyan, V., and Xue, S.-S., On the instantaneous spectrum of gamma-ray bursts (2004), iJMPD, in press.
14. Iwamoto, K., *ApJ Lett.*, **512**, L47 (1999).
15. Canuto, V., "Neutron Stars, Physics and astrophysics of neutron stars and black holes," in *Proceedings of the International School of Physics "Enrico Fermi"*, edited by R. Giacconi and R. Ruffini, Amsterdam, North Holland Publishing Co., 1978.
16. Gamow, G., and Schoenberg, M., *Phys. Rev.*, **59**, 539 (1941).
17. Tsuruta, S., Ph.D. thesis, Columbia University (1964).
18. Tsuruta, S., *Phys. Rep.*, **56**, 237 (1979).
19. Tsuruta, S., Teter, M. A., Takatsuka, T., Tatsumi, T., and Tamagaki, R., *ApJ*, **571**, L143 (2002).
20. Ruffini, R., Bernardini, M. G., Bianco, C. L., Chardonnet, P., Fraschetti, F., and Xue, S.-S. (2004), in preparation.

GRB 970228 Within the EMBH Model

A. Corsi*, M. G. Bernardini*, C. L. Bianco*, P. Chardonnet[†], F. Fraschetti**, R. Ruffini* and S.-S. Xue*

*ICRA - International Center for Relativistic Astrophysics and Dipartimento di Fisica, Università di Roma "La Sapienza", Piazzale Aldo Moro 5, I-00185 Roma, Italy.
[†]Université de Savoie, LAPTH - LAPP, BP 110, F-74941 Annecy-le-Vieux Cedex, France.
**Università di Trento, Via Sommarive 14, I-38050 Povo (Trento), Italy.

Abstract. We consider the gamma-ray burst of 1997 February 28 (GRB 970228) within the ElectroMagnetic Black Hole (EMBH) model. We first determine the value of the two free parameters that characterize energetically the GRB phenomenon in the EMBH model, that is to say the dyadosphere energy, $E_{dya} = 5.1 \times 10^{52}$ ergs, and the baryonic remnant mass M_B in units of E_{dya}, $B = M_B c^2 / E_{dya} = 3.0 \times 10^{-3}$. Having in this way estimated the energy emitted during the beam-target phase, we evaluate the role of the InterStellar Medium (ISM) number density (n_{ISM}) and of the ratio $\mathscr{R}$ between the effective emitting area and the total surface area of the GRB source, in reproducing the observed profiles of the GRB 970228 prompt emission and X-ray (2-10 keV energy band) afterglow. The importance of the ISM distribution three-dimensional treatment around the central black hole is also stressed in this analysis.

The GRB 970228 [1] had an important role in solving the origin of GRBs through the first detection of counterparts at other wavelengths: the afterglow phenomenon, long-lived multi-wavelength emission, was discovered following GRB 970228 at X-ray ([2], Costa et al. [3]) and optical ([4], van Paradijs et al. [5]) wavelengths.

We consider of great interest to compare the predictions of the ElectroMagnetic Black Hole (EMBH) theory (see Ruffini et al. [6] and references therein) with the first afterglow observed by the Beppo-SAX satellite.

We are also interested in testing the efficiency of the model in reproducing the GRB 970228 prompt emission: in the 40-700 keV energy band the burst was characterized by an initial 5 s strong pulse followed, after about 30 s, by three additional pulses of decreasing intensity (Frontera et al. [7]). The InterStellar Medium (ISM) number density (n_{ISM}) inhomogeneities have an important role in interpreting this profile within the EMBH model.

Our analysis starts establishing the value of the two free parameters that determine energetically the GRB phenomenon in the EMBH model: the total energy deposited in the dyadosphere E_{dya} (Ruffini et al. [8]) and the amount of the baryonic matter left over in the collapse process of the EMBH progenitor star (Ruffini et al. [8]), that can be parametrized by the dimensionless parameter $B = M_B c^2 / E_{dya}$.

With the choice of $E_{dya} = 5.1 \times 10^{52}$ ergs and $B = 3.0 \times 10^{-3}$, the EMBH model predicts that a 98% of the total energy E_{dya} is emitted during the so-called beam-target phase (Ruffini et al. [9]), that is to say during the collision of the Accelerated Baryonic Matter-pulse (ABM-pulse) with the ISM (Ruffini et al. [6]). During this phase, the internal energy developed in the collision is instantaneously radiated away (fully radiative condition) and, as a consequence, the resulting shape of the light curve is strictly linked to the ISM distribution and number density (Ruffini et al. [10]). We use a one-dimensional treatment of the ISM, where the n_{ISM} is a function of the radial distance from the central black hole (Ruffini et al. [10]).

In order to reproduce the observed profile of the GRB 970228, n_{ISM} has to range between the values of 10^{-2} particles/cm^3 and 200 particles/cm^3 in the region of space within 2.00×10^{15} cm and 4.95×10^{16} cm from the central black hole. Since 2.00×10^{15} cm and beyond 4.95×10^{16} cm, the ISM number density has a constant value of 1 particle/cm^3 (details are given in Ruffini et al. [11], Ruffini et al. [12]).

The correct spectral distribution of the energy emitted during the the beam-target phase depends on the $\mathscr{R}$ parameter (Ruffini et al. [13]). As a consequence, the theoretical curves in selected energy bands are strictly related to this parameter. $\mathscr{R}$ is a function of the radial distance from the EMBH and it represents the ratio between the effective emitting area of the ABM-pulse and its total surface area:

$$\mathscr{R} = A_{eff}/A_{ABM} \qquad (1)$$

According to Ruffini et al. [13], by assuming a black-body spectrum in the co-moving frame for the radiation emitted during the collision with the ISM, the spectral distribution of the energy emitted results to be dependent on the temperature of the emitting black body (Ruffini et al. [13], Ruffini et al. [14]):

$$T = \left(\frac{\Delta E_{int}}{4\pi r^2 \Delta \tau \sigma \mathscr{R}}\right)^{1/4} \qquad (2)$$

where ΔE_{int} is the proper internal energy developed in the collision of the ABM-pulse with the ISM in the proper time interval $\Delta \tau$, r is the radial coordinate of the ABM-pulse, σ is the Stefan-Boltzmann constant.

In the case of GRB 970228 we find $\mathscr{R}$ monotonically varying from 3.7×10^{-12} to 8.8×10^{-11} when the radial coordinate r goes from 7.0×10^{14} cm to 5.0×10^{17} cm. With this result, the first peak in the 40-700 keV observed light curve is correctly reproduced by the model (details are given in Ruffini et al. [11], Ruffini et al. [12]). The three additional pulses, that follow the first one after a gap in the emission, are reproduced by the model in terms of the mean luminosity. The Fast Rise Exponential Decay (FRED) shape that emerges in the theoretical light curve is a consequence of the one-dimensional treatment of the ISM. To solve this problem, a three-dimensional treatment of the ISM distribution is required (details are given in Ruffini et al. [11], Ruffini et al. [12]).

About the X-ray afterglow, in Fig.1 we present the theoretical curve in the 2-10 keV energy band compared with the observed data by Beppo-SAX (Costa et al. [3]) and ASCA [15]. The afterglow phase corresponds to the ABM-pulse expansion in the region beyond 4.95×10^{16} cm, where the number density of the ISM has a constant value, $n_{ISM} = 1$ particle/cm^3. We can see that there is a good agreement ($\chi^2=0.5$) between the theoretical light curve in the 2-10 keV energy band and the observed data by Beppo-SAX and ASCA.

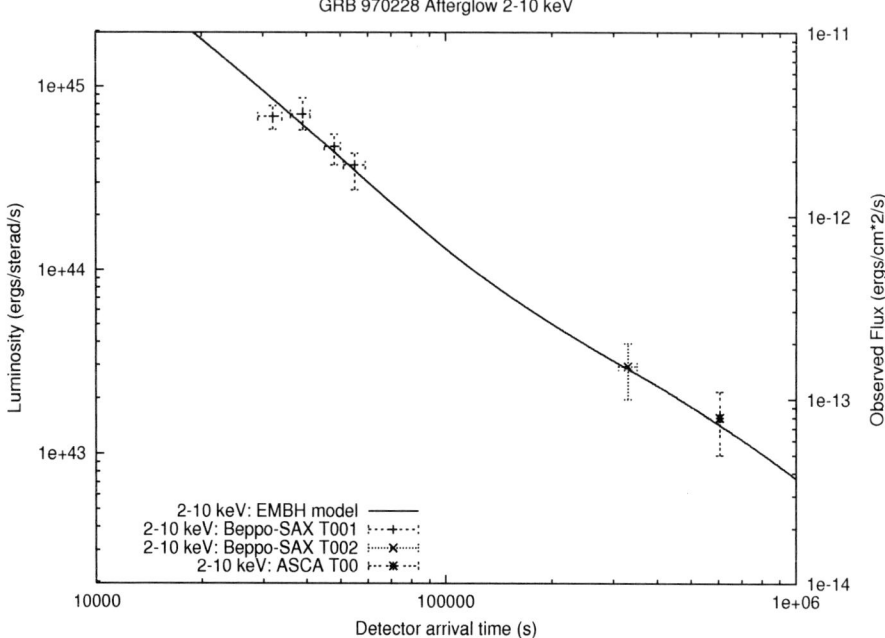

FIGURE 1. Afterglow 2-10 keV: the solid line represents the theoretical light curve for the 2-10 keV emission in the EMBH model. The points are the GRB 970228 2-10 keV afterglow data observed by Beppo-SAX (Costa et al. [3]) and ASCA ([15])

From this analysis we conclude that:

- a mask of density inhomogeneities of the ISM is needed in the region of space between 2.00×10^{15} cm and 4.95×10^{16} cm from the black hole, in order to reproduce the structure of the GRB 970228 prompt emission;
- a three-dimensional treatment of the ISM is required in order to improve the theoretical predictions of the model (details are given in Ruffini et al. [11], Ruffini et al. [12]);
- finally, a good result is obtained with a constant value of the $n_{ISM} = 1 \text{particle/cm}^3$ for the 2-10 keV afterglow emission.

REFERENCES

1. IAU Circ. 6572 (1997).
2. IAU Circ. 6576 (1997).
3. Costa, E., Frontera, F., Heise, J., Feroci, M., in't Zand, J., Fiore, F., Cinti, M. N., Dal Fiume, D., Nicastro, L., Orlandini, M., Palazzi, E., Rapisarda, M., Zavattini, G., Jager, R., Parmar, A., Owens, A., Molendi, S., Cusumano, G., Maccarone, M. C., Giarrusso, S., Coletta, A., Antonelli, L. A., Giommi, P., Muller, J. M., Piro, L., and Butler, R. C., *Nature*, **387**, 783 (1997).

4. IAU Circ. 6584 (1997).
5. van Paradijs, J., Groot, P. J., Galama, T., Kouvelioutou, C., Strom, R. G., Telting, J., Rutten, R. G. M., Fishman, G. J., Meegan, C. A., Pettini, M., Tanvir, N., Bloom, J., Pedersen, H., Nørdgaard-Nielsen, H. U., Linden-Vørnle, M., Melnick, J., van der Steene, G., Bremer, M., Naber, R., Heise, J., in't Zand, J., Costa, E., Feroci, M., Piro, L., Frontera, F., Zavattini, G., Nicastro, L., Palazzi, E., Bennet, K., Hanlon, L., and Parmar, A., *Nature*, **386**, 686 (1997).
6. Ruffini, R., Bianco, C. L., Chardonnet, P., Fraschetti, F., Vitagliano, L., and Xue, S.-S., "New Perspectives in Physics and Astrophysics from the Theoretical Understanding of Gamma-Ray Bursts," in *COSMOLOGY AND GRAVITATION: X^{th} Brazilian School of Cosmology and Gravitation; 25th Anniversary (1977-2002)*, edited by M. Novello and S. E. P. Bergliaffa, AIP, New York, 2003, vol. 668, p. 16.
7. Frontera, F., Costa, E., Piro, L., Muller, J. M., Amati, L., Feroci, M., Fiore, F., Pizzichini, G., Tavani, M., Castro-Tirado, A., Cusumano, G., Dal Fiume, D., Heise, J., Hurley, K., Nicastro, L., Orlandini, M., Owens, A., Palazzi, E., Parmar, A. N., in't Zand, J., and Zavattini, G., *ApJ*, **493**, L67 (1998).
8. Ruffini, R., Salmonson, J. D., Wilson, J., and Xue, S.-S., *A&A*, **855**, 359 (2000).
9. Ruffini, R., Bianco, C. L., Chardonnet, P., Fraschetti, F., and Xue, S.-S., *ApJ Lett.*, **555**, L113 (2001).
10. Ruffini, R., Bianco, C. L., Chardonnet, P., Fraschetti, F., and Xue, S.-S., *ApJ Lett.*, **581**, L19 (2002).
11. Ruffini, R., Bianco, C. L., Bernardini, M. G., Corsi, A., Chardonnet, P., Fraschetti, F., and Xue, S.-S. (2003a), in preparation.
12. Ruffini, R., Bernardini, M. G., Bianco, C. L., Bernardini, M. G., Corsi, A., Chardonnet, P., Fraschetti, F., and Xue, S.-S., "," in *Proceedings of the 10th Marcell Grossmann Meeting*, 2003b, in preparation.
13. Ruffini, R., Bianco, C. L., Chardonnet, P., Fraschetti, F., Gurzadyan, V., and Xue, S.-S., *IJMPD* (2004), in press.
14. Ruffini, R., Bernardini, M. G., Bianco, C. L., Chardonnet, P., Fraschetti, F., and Xue, S.-S., "GRB 980425, SN1998bw and the EMBH Model," in *Proceedings of the 34th COSPAR Scientific Assembly*, edited by E. Pian, N. Masetti, and L. Piro, Elsevier, 2003, in press.
15. IAU Circ. 6593 (1997).

DARK VERSUS BRIGHT GRBs

Chandra Observations of the Optically Dark GRB 030528

N. Butler*, A. Dullighan*, P. Ford*, G. Ricker*, R. Vanderspek*, K. Hurley†, J. Jernigan† and D. Lamb**

*Center for Space Research, Massachusetts Institute of Technology, MA
†Space Sciences Laboratory, Berkeley, CA
**Department of Astronomy, University of Chicago, IL

Abstract. The X-ray-rich GRB 030528 was detected by the HETE satellite and its localization was rapidly disseminated. However, early optical observations failed to detect a counterpart source. In a 2-epoch ToO observation with Chandra, we discovered a fading X-ray source the likely counterpart to GRB 030528. The source brightness was typical of X-ray afterglows observed at similar epochs. Other observers detected an IR source at a location consistent with the X-ray source. The X-ray spectrum is not consistent with a large absorbing column.

OBSERVATIONS

The X-ray-rich (i.e. for the fluence S, $\log[S_X(2-30 \text{ kev})/S_\gamma(30-400 \text{ kev})] > -0.5$) GRB 030528 was detected by the *HETE* satellite [1] with a $2'$ radius (90% confidence) SXC localization. The initial SXC error region was later revised [2] after the discovery of an unaccounted-for systematic effect, resulting in a shift in position center and an expansion of the error region to $2.5'$ radius. Early R-band observations reaching $R \approx 18.7$ roughly 140 minutes after the burst [3] and unfiltered observations reaching 20.5 magnitude roughly 14 hours after the burst [4] failed to detect a counterpart.

On 3 June, the *Chandra Observatory* targeted the field of GRB 030528 as part of a series of GTO target-of-opportunity observations focusing on optically-dark GRBs discovered by *HETE*. The 25 ksec observation spanned the interval 12:22-20:08 UT, 5.97 - 6.29 days after the burst. The revised SXC error region from Villasenor et al. [2] was completely contained within the field-of-view of the *Chandra* ACIS-S3 chip. From 9 June 8:14 UT to 9 June 14:19 UT, 11.8 to 12.1 days post-burst, *Chandra* again targeted the field of GRB 030528 for a 20 ksec second epoch (E2) observation with ACIS-S3.

CHANDRA E1 SOURCES

As reported in Butler et al. [5], four candidate sources were detected within the revised SXC error region. Seven additional nonstellar point sources were detected within the entire ACIS-S3 field-of-view. Positions and other data for these sources are shown in Table 1. None of the sources were anomalously bright relative to objects in *Chandra* deep field observations [see, e.g., 6]. We had performed deep observations with Magellan prior to the *Chandra* observation, but none of the *Chandra* sources were in our field of view. However, near-IR observations of a portion of the SXC error region containing two of the E1 *Chandra* sources revealed a fading Ks-band source [7]. Between 0.7 and 3.6 days after the burst a fade by 0.9 magnitudes was observed for a source spatially coincident with the brightest *Chandra* source. After the E1 *Chandra* observation, deep observations in the radio (6.8 days after the burst) [8] and in I-band (8.7 days after the burst, I>21.5) [9] failed to detect a counterpart source.

TABLE 1. Four point sources are detected in the 0.5-8.0 keV band in the *Chandra* E1 observation lying within the revised SXC error region. Eight additional, non-stellar point sources are detected in the ACIS-S3 field-of-view. From the E1 net counts, we calculate E2$_{90\%}$, the 90% confidence region for the expected net counts in E2, following Kraft et al. [10]. The columns labeled "ΔC" and "P_C" are explained in the text. Small values of P_C indicate sources likely to have faded between E1 and E2. Source #17 was situated on a chip gap in E2.

#	Chandra Name	Epoch 1 Net (Bg)	E2$_{90\%}$	Epoch 2 Net (Bg)	ΔC	$P_C^{(\%)}$
1	CXOU J170400.3-223710	39.5 (1.5)	24.1,41.1	8.5 (2.5)	6.97	0.01
4	CXOU J170348.4-223826	30.1 (2.9)	16.4,30.2	20.3 (2.7)	0.01	37.7
9	CXOU J170400.1-223548	10.8 (2.2)	4.1,12.6	5.4 (3.6)	0.17	28.8
10	CXOU J170354.0-223654	9.2 (2.8)	3.4,12.6	8.3 (2.7)	0.00	44.7
...						
2	CXOU J170358.7-224237	30.6 (3.4)	14.1,28.0	23.9 (2.1)	0.00	36.1
3	CXOU J170355.7-223503	23.1 (2.9)	11.8,24.6	15.1 (3.9)	0.03	2.3
5	CXOU J170342.8-223548	23.7 (4.3)	12.4,26.0	10.7 (6.3)	0.85	9.8
8	CXOU J170403.9-223543	12.7 (2.3)	5.7,16.0	4.4 (2.6)	0.94	8.9
14	CXOU J170341.4-223646	6.1 (4.9)	1.1,9.0	9.7 (5.3)	0.00	48.8
15	CXOU J170411.2-224032	11.0 (4.0)	4.3,14.4	5.5 (3.5)	0.20	24.1
17	CXOU J170345.8-224133	10.5 (4.5)	1.1,4.0	...		

AFTERGLOW CONFIRMED IN E2

Table 1 shows the number of counts detected in E1, along with the 90% confidence interval for E2 based on the E1 values. The E2 observations were reported in Butler et al. [11]. We have used a circular extraction region for each source, with radius set to 2 times the 95% encircled energy radius r. This varies over the chip and is approximated via $r = 2.05 - 0.55*d + 0.18*d^2$ arcsec, with d measured in arcminutes from the center of the ACIS-S3 chip. We use an annular background region ten times larger than the signal region, centered on and surrounding the signal region. The exposure is calculated separately for each source extraction region in each epoch.

GRB X-ray afterglows typically fade in brightness with time as $t^{-1.3}$, with t measured from the GRB [12]. Assuming no spectral evolution, this implies a count-rate fade factor of approximately 2.5 between E1 and E2. We can test whether the data for each source prefers a fade versus a constant count rate by fitting the data for each source first with a single-rate model (Model A), then fitting the data with a model allowing the E2 rate to be lower than the E1 rate (Model B). We do the fits by maximizing the logarithm of the Poisson likelihood (i.e. the Cash [13] statistic C). We then simulate 10^4 data sets for each source using the count rate determined from Model A, and we count the number of these which yield a larger ΔC than the observed value when fit with Model B. These fractions are expressed as probabilities (P_C) in Table 1.

Of the four sources in the revised SXC error region (#'s 1,4,9,10), source #1 has faded far below the 90% confidence range established in E1. The significance of the fade is approximately 3σ, and we estimate the temporal index (assuming a power-law fade) to be $\alpha = 2.0 \pm 0.8$. This is somewhat steeper than the typical $t^{-1.3}$ fade for X-ray afterglows [12], though it is characteristic of afterglows which have undergone a so-called "jet-break" [14]. None of the other *Chandra* sources were observed to fade at a high level of significance, and source #1 (also a fading IR source as discussed above) is extremely likely to be counterpart to GRB 030528.

COLUMN DENSITY CONSTRAINTS

We reduce the spectral data using the standard CIAO[1] processing tools. We use "contamarf"[2] to correct for the quantum efficiency degradation due to contamination in the ACIS chips, important for energies below ~ 1 keV. There are not enough source counts for detailed spectral fitting with the *Chandra* data. However, we can use the instrumental response determined in the steps above in combination with the total number of detected source and background counts in the 0.5-8 keV band (Table 1) and the number of source plus background counts detected in the 0.5-0.6 keV band (2 counts) to constrain the model column density. Assuming a power-law spectrum, Figure 1 shows how the two low-energy counts become increasingly improbable as the column density increases. Except for the case of high redshift ($z > 1$), Figure 1 implies a column density $\lesssim 10^{22}$ cm^{-2}. This implies an extinction in R-band of $A \lesssim 3$ mag. At high redshift ($z \sim 3$), the column is likely $\lesssim 10^{23}$ cm^{-2}.

CONCLUSIONS

An X-ray observation ~ 6 days after GRB 030528 detected the afterglow at a flux level (1.4×10^{-14} erg cm^{-2} s^{-1}) typical for GRB X-ray afterglows at that epoch [see, 12]. Here we have assumed a typical power-law spectrum with photon index $\Gamma = 1.9$ and the Galactic $N_H = 1.6 \times 10^{21}$ cm^{-2}. A second epoch observations decisively revealed that source #1 had faded, establishing securely that this was the counterpart X-ray afterglow to GRB 030528. The X-ray spectrum appears to imply a fairly low column density, which in turn implies a fairly low amount of reddening in the source frame. Thus, although the detection of a near-IR counterpart with no detection of an optical counterpart for this burst perhaps points toward dust extinction, we find no supporting

---

[1] http://cxc.harvard.edu/ciao/
[2] http://space.mit.edu/CXC/analysis/ACIS_Contam/script.html

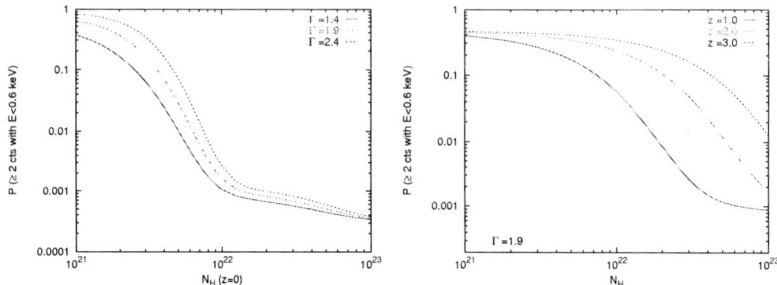

FIGURE 1. The probability is very low that a power-law spectrum with the indices shown (left plot) and $N_H \gtrsim 10^{22}$ cm^{-2} could have yielded two source counts below 0.6 keV, as observed. The possibility that the counts could have come from the background is accounted for, using an annular background region approximately 100 times larger than the source region and surrounding the source region. The expected number of background counts in the 3.3″ radius source extraction region is 0.02 counts. Because the host redshift is unknown, any local contribution in excess of the Galactic column (right plot) may be less well constrained. For $z = 1$, it is likely that the local column is $\lesssim 10^{22}$ cm^{-2}.

evidence in the X-ray data. The publication of additional photometric data in various passbands for this burst, if available, would help to constrain any possible extinction by dust in the GRB host galaxy. We will perhaps learn that the afterglow to GRB 030528 was intrinsically faint rather than heavily extinguished, as appears to be common in may GRBs [see, e.g., 15, 16].

REFERENCES

1. Atteia, J-L., et al., *GCN*, 2256 (2003).
2. Villasenor, J., et al., *GCN*, 2261 (2003).
3. Ayani, K., & Yamaoka, H., *GCN*, 2257 (2003).
4. Valentini, G., et al., *GCN*, 2258 (2003).
5. Butler, N. R., et al., *GCN*, 2269 (2003).
6. Rosati, P., et al., *ApJ* **566**, 667 (2002).
7. Greiner, J., Rau, A., & Klose, S., *GCN*, 2271 (2003).
8. Frail, D. A., & Berger, E., et al., *GCN*, 2270 (2003).
9. Mirabel, N., & Halpern, J., *GCN*, 2273 (2003).
10. Kraft, R., Burrows, D., & Nousek, J., *ApJ* **374**, 344 (1999).
11. Butler, N. R., et al., *GCN*, 2279 (2003).
12. Costa, E., et al., *A&AS* **138**, 425 (1999).
13. Cash, W., *ApJ* **228**, 939 (1979).
14. Frail, D. A., et al., *ApJ* **562**, L55 (2001).
15. Berger, E., et al., *ApJ* **581**, 981 (2002).
16. Ricker, G. R., et al., *in these proceedings* (2003).

Discovery of the Faint Near-IR Afterglow of GRB 030528[1]

A. Rau*, J. Greiner*, S. Klose[†], J. Castro Cerón**, A. Fruchter**, A. Küpcü Yoldaş*, J. Gorosabel**, A. Levan**, J. Rhoads** and N. Tanvir[‡]

*Max-Planck Institute for extraterrestrial Physics, Garching, Germany
[†]Thüringer Landessternwarte, Tautenburg, Germany
**Space Telescope Science Institute, Baltimore, USA
[‡]Department of Physical Science, Univ. of Hertfordshire, Hatfield Herts, UK

Abstract. We report on the discovery of the near-IR transient of the long-duration gamma-ray burst GRB 030528 and its underlying host galaxy. The near-IR transient was first observed in $JHKs$ with SofI at the 3.6 m ESO-NTT 16 hrs after the burst and later observations revealed a fading Ks-band afterglow. The afterglow nature was confirmed by *Chandra* observations which found the source to be a fading X-ray emitter. The lack of an optical afterglow and the early faintness in the near-IR ($Ks > 18.5$ mag) place GRB 030528 in a parameter space usually populated by dark bursts. We find the host to be an elongated blue galaxy.

OBSERVATIONS

On May 28 2003 *HETE-2* localized a moderately bright (peak flux of 4.8×10^{-8} erg/cm^2/s at 30-400 keV) long-duration Gamma-ray burst (GRB) [1] 107 min after the observed prompt emission (13:03 UT). Shortly after, ToO observations of the circulated $2'$ radius error circle were initiated with the the 3.6 m ESO-New Technology Telescope (NTT) equipped with the Son of ISAAC (SofI) infrared spectrograph and imaging camera at La Silla/Chile. Here we report on $JHKs$ imaging data of the first three nights supplemented by later I-, Js- and K-band imaging with the Mosaic2 imager at the 4 m Blanco Telescope at the Cerro Tololo Inter-American Observatory (CTIO), the Infrared Spectrometer And Array Camera (ISAAC) at the 8.2 m Very Large Telescope (VLT) in Paranal and the 3.8 m United Kingdom Infra-Red Telescope Fast-Track Imager (UKIRT UFTI) on Mauna Kea, respectively (see Table 1 for an observing log).

NTT-SofI observations were accomplished 16, 41 and 87 hrs after the GRB. Unfortunately, the *HETE-2* localization was revised to a $2\rlap{.}'5$ radius error circle displaced by $1\rlap{.}'3$ from the earlier position [2] after these images were taken. Given the $5\rlap{.}'5$ field of view of NTT-SofI, our data were not covering the entire revised error circle. This, and the non-detection of a transient in optical ($R > 18.7$ mag 2.3 hrs after the burst [3]) and radio[4] observations of the crowded field would have made the discovery of the counterpart nearly impossible without *Chandra* observations. Following the detection of four

---

[1] for the GRACE collaboration

TABLE 1. Observation log.

Instrument	Date (Start UT)	Filter	Exp. [min]	Seeing	Magnitude
NTT-SofI	29.5. 04:58	J	15	0.8″	20.9±0.9
NTT-SofI	29.5. 05:16	H	15	0.8″	19.9±0.9
NTT-SofI	29.5. 05:32	Ks	15	0.8″	18.5±0.4
NTT-SofI	30.5. 04:54	J	20	1.6″	>20.1±0.7
NTT-SofI	30.5. 05:16	H	20	1.1″	>18.9±1.6
NTT-SofI	30.5. 05:40	Ks	20	1.1″	18.8±0.3
NTT-SofI	1.6. 04:07	Ks	60	0.8″	19.5±0.6
Blanco-Mosaic2	4.6. 02:10	I	40	1.2″	21.4±0.5
UKIRT-UFTI	12.6. 08:55	K	116	0.6″	19.3±0.2
Blanco-Mosaic2	30.6. 03:00	I	40	1.1″	21.4±0.5
VLT-ISAAC	17.9. 00:12	Js	50	0.6″	21.3±0.4
VLT-ISAAC	27.9. 00:27	Js	60	0.9″	21.4±0.4
VLT-ISAAC	29.9. 23:38	Js	94	0.7″	20.9±0.4
VLT-ISAAC	1.10. 00:04	Js	60	0.6″	21.0±0.6

X-ray sources inside the revised error circle on June 3 [5] one of these sources (CXOU J170400.3–223710) was found to be fading in our SofI-Ks-band images [6]. A second *Chandra* observation on June 9 showed a significant fading of the counterpart candidate while the other sources inside the error circle did not reveal any brightness decline [7].

The NTT-SofI and VLT-ISAAC data were reduced using the *Eclipse* package[8]. The reduction of the Blanco-Mosaic2 data was performed with *bbpipe*, a script based on the *IRAF/MSCRED*. The photometry was performed with *IRAF/DAOPHOT*. We calibrated the $JJsHK$ and Ks images against several stars contained in the 2MASS All-Sky Point Source Catalog[2] from the neighbourhood of the GRB. The Mosaic2 I-band images were calibrated using USNOFS field photometry[9]. The magnitudes are corrected for foreground extinction[10] (A_K=0.22, A_H=0.35, A_J=0.54 and A_I=1.17) for the given galactic coordinates.

THE NEAR-IR AFTERGLOW

Observations from the first night (∼16 hrs after the burst) showed the afterglow to be near the detection limit of the JH-band SofI images. The source is well visible in Ks at an extinction corrected magnitude of 18.5±0.4. It was thus ∼1.5 mag brighter than the faintest K-band[3] afterglow observed at that time after the burst, that of GRB 971214.

The afterglow shows a probable fading in Ks from 16 to 80 hrs after the burst by ∼1 mag (Fig. 1). This corresponds to a decay with a slope of α∼0.6 Ks/K-band data from June 1 (NTT) and June 12 (UKIRT) shows constant magnitudes. Therefore, the host galaxy seems to dominate the K-band emission already after ∼3 days. Unfortunately, the observing conditions from the second night (41 hrs after the burst) make the

---

[2] http://irsa.ipac.caltech.edu/applications/Gator/
[3] We assume here $K \sim Ks$, which is justified within the error of the estimated magnitudes.

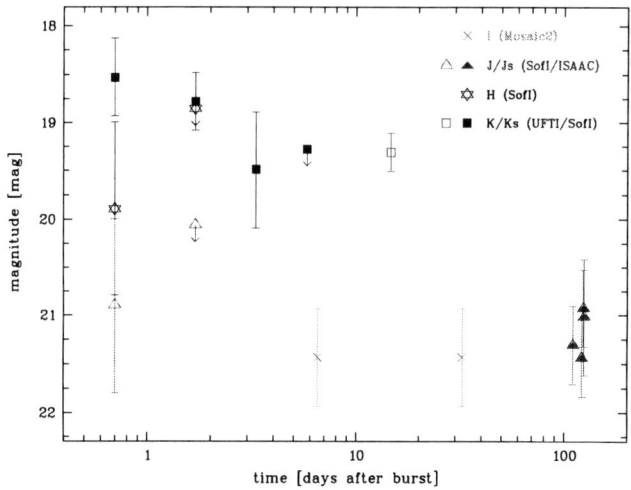

FIGURE 1. Light curves in I (marked by a cross), J/Js (triangle/filled triangle), H (star) and K/Ks (square/filled square). For J and H only upper limits exist for t=1.7 days. The Ks-band upper limit at t=5.8 days is from [11]. All magnitudes are corrected for foreground extinction in the Galaxy.

estimation of an early fading in J and H not feasible. No decay is seen in the J/Js-band[4] when comparing the SofI observations and the ISAAC observations from >100 days. This suggests, that the afterglow emission was strongly extinguished in the J-band (in addition to the galactic extinction) and the emission dominated by the host galaxy already at the time of our first SofI observation.

It is unlikely that we would have found this IR counterpart without *Chandra*, and the non-detection in the optical would have led to the classification of GRB as a dark burst. While ~60% of all well localized GRBs have no detected optical/NIR afterglow, the fast reaction time and deep detection limit of ~19.5 in Ks were important for the discovery of the near-IR transient of GRB 030528.

THE HOST GALAXY

Late time K-band imaging with UKIRT revealed the potential underlying host galaxy (see Fig. 2). The data show an East-West elongated object at the position of the transient with a size of $\sim 1.5'' \times 0.8''$. This shape suggests the host to be either an elliptical or edge-on disk galaxy. The apparent brightness of $K \sim 19.5$ mag and I-$K \sim 2$ mag places the host of GRB 030528 well in the mean of the sample of GRB host galaxies detected in the K-band [12] and is more typical for a star-forming rather than for an elliptical galaxy.

---

[4] The J and Js filters differ slightly in width and efficiency. For a flat spectrum a first order assumption of $J \sim Js + 0.4$ mag can be used.

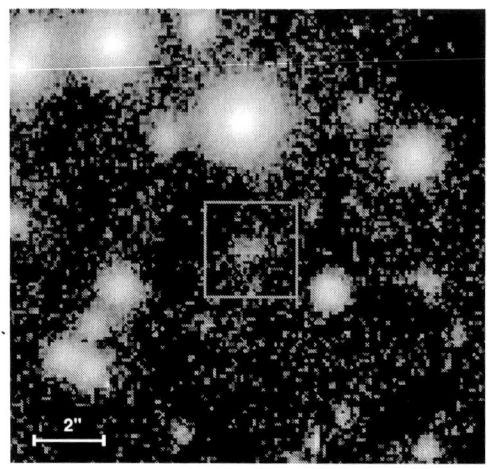

FIGURE 2. K-band image taken with the 3.8m UKIRT equipped with the UFTI on June 12, 15 days after the prompt emission. North is up and East to the right. The potential host galaxy of GRB 030528 (square) shows an elongation in East-West direction.

ACKNOWLEDGMENTS

This work is primarily based on observations collected at the European Southern Observatory, Chile, under the proposal 71.D-0355 (PI: E.v.d.Heuvel) with aditional data obtained at the Cerro Tololo Inter-American Observatory and the United Kingdom Infra-Red Telescope. This publication makes use of data products from the Two Micron All Sky Survey, which is a joint project of the University of Massachusetts and the Infrared Processing and Analysis Center/California Institute of Technology, funded by the National Aeronautics and Space Administration and the National Science Foundation.

REFERENCES

1. Atteia, J-L., Kawai, N., Lamb, D., Ricker, G., Woosley, S., et al., GCN, **2256**, (2003)
2. Villasenor, J., Butler, N., Crew, G., Doty, J., Prigozhin, G., et al., GCN, **2261**, (2003)
3. Ayani, K., and Yamaoka, H., GCN, **2257**, (2003)
4. Frail, D.A. and Berger, E., GCN, **2270**, (2003)
5. Butler, N., Dullighan, A., Ford, P., Ricker, G., Vanderspek, R., et al., GCN, **2269**, (2003)
6. Greiner, J., Rau, A. and Klose, S., GCN, **2271**, (2003)
7. Butler, N., Dullighan, A., Ford, P., Ricker, G., Vanderspek, et al., GCN, **2279**, (2003)
8. Devillard, N., The Messenger, **87**, 19, (1997)
9. Henden, A., GCN, **2267**, (2003)
10. Schlegel, D.J., Finkbeiner, D.P., Davis, M., ApJ **500** 525, (1998)
11. Bogosavljevic, M., Mahabal, A. and Djorgovski, S.G., GCN, **2275**, (2003)
12. le Floc'h, E., Duc, P-A., Mirabel, I.F., Sanders, D.B., Bosch, G., et al., A&A, **400**, 499, (2003)

Dust and Gamma-Ray Bursts: Mutual Implications

S.D. Vergani*, E. Molinari*, F.M. Zerbi* and G. Chincarini[†]

INAF-OAB, via Bianchi 46, 23807 Merate, Italy
[†]*Università Milano Bicocca*

Abstract. In a cosmological context dust has been always poorly understood. This is true also for the statistics of Gamma-Ray Bursts (GRBs). We therefore started a program to understand the role of dust both in GRBs and as function of z. We present a composite model that considers a rather generic distribution of dust in a spiral galaxy and considers the effect of changing some of the parameters characterizing the dust grains, size in particular. Computations from this model of GRBs missed by dust absorption agree with the hypothesis of host galaxies with an extinction curve similar to that of the Small Magellanic Cloud, but the host cloud could be characterized also by dust with larger grains. Unfortunately, the present statistics lack significance, being based on incompatible observations, at different times from the burst and with different limiting magnitudes. To confirm our findings we need a set of homogeneous infrared observations. The use of forthcoming dedicated infrared telescopes, like REM [1], will provide a wealth of new afterglow observations.

THE MODEL

The GRB afterglow radiation reaches the observer after interacting with circumburst material, host ISM, IGM and the ISM of our galaxy. In this work we consider only the interaction with the dust in the ISM of the host galaxy.

Our dust model is based on the Mathis, Rumpl & Nordsiek [2] model, improved by Mathis [3]. The size of the grains plays a strong role – in particular, large grains cause the extinction curve to flatten (Figure 1).

We model host galaxy similar to our Milky Way, where the dust is distributed as neutral hydrogen, both molecular (H_2) and atomic (HI). The H_2 clouds are present in two distinct morphologies, which we catalogue as giant molecular clouds (GMC) and dense clouds (DC). Neutral atomic hydrogen is present in a diffuse form throughout the disk system.

We suppose that the GRBs are distributed as the luminosity (barionic mass) of the host Milky Way-like galaxy. The burst occurs randomly using as a weighting function the distribution of mass of the galactic model and the observer is located randomly over a 4π solid angle. The amount of NH is integrated along the line of sight. The amount of dark GRBs due to dust absorption can now be estimated, once we have fixed the limiting magnitude of our telescopes in the various passbands and the typical apparent magnitude of GRB afterglows. We randomly associate to each GRB a jet opening angle θ. Considering the luminosity L proportional to θ^{-2}, we calculate the magnitudes R and K of GRBs at 100s and 5000s, taking as reference GRB990123 shifted at z=1 and the color data by [4]. To compute the limiting magnitudes we assume use of the REM

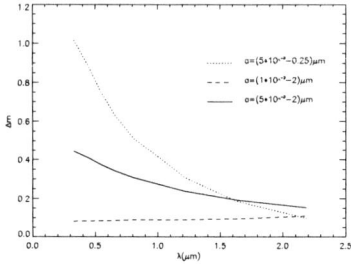

FIGURE 1. Extinction curves produced by varying the dust grain size in case of hydrogen column density of $10^{21} cm^{-2}$. The dotted steeper curve represents the extinction considering our Galaxy-like dust model (STD, see Table 2). Dashed and solid lines describe the extinction curve in the case of a dust composed of larger grains (OLG and LRG, respectively).

TABLE 1. Extinction limits computed on the basis of the limiting magnitudes of different instruments and of GRB990123 afterglow light curve. NH values are computed using our dust model.

	(1)	(2)	(3)*	(4)	(5)†	(6)	(7)	(8)	(9)**	(10)
	sec	m_R	m_K	$R_{\lim}$	$K_{\lim}$	A_R	A_K	NH_R	NH_K
prompt	100	10	7.5	19	15.5	9	8	0.86	2.13
late	5000	16	13.5	24	20.5	8	7	0.76	1.87

* (3)(4) In these columns we report GRB990123-like light curve values.
† (5)(6) Prompt observation simulated with REM ([5], Table 2.3), late observation simulated with ISAAC and FORS (see http://www.eso.org/oserving/etc/ for its ETC time estimates)
** (9)(10) in units of $10^{22} cm^{-2}$

telescope for the fast response to the GRB alert in R and K bands, and the FORS and ISAAC camera of ESO/VLT for the long term observations. For each GRB we calculate the extinction in R and K caused by the traversed NH column density. We are then able to compute the percentage of lost afterglows due to dust absorption under the assumption that the host galaxy at z=1 has dust properties similar to Milky Way. Results are shown in Table 1, where the values of NH quoted in columns (9) and (10) represent the computed amount of NH needed, according to our dust model, to obtain the extinction values of column (7) and (8). The percentage of dark GRBs due to dust obscuring is rather low (~8%). There is no relevant difference between R and K observations and between percentages relative to observations taken at 100s and 5000s. It also happens that no burst, out of the 500 considered, occurs statistically inside a molecular cloud. In this scenario the majority of dark bursts could be due to high redshift Ly-α absorption.

GRBS IN MOLECULAR CLOUDS

We then consider the case where GRBs follow the distribution of giant molecular clouds placing 5000 GRBs inside our GMC.

We consider the results of [6] and [7] that compute the radius up to which the dust is

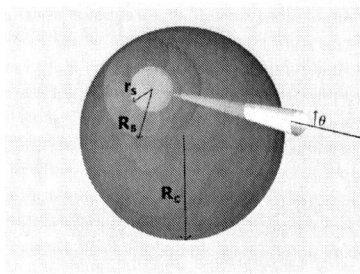

FIGURE 2. Geometry around the GRB location explaining the shapes of the four regions whose content in terms of dust is parameterized in our models.

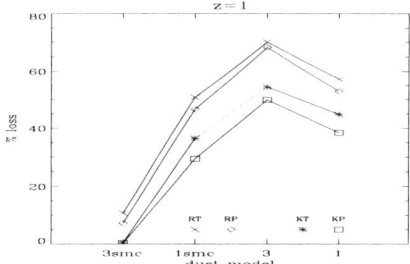

FIGURE 3. Trends for the most relevant models of the percentage for prompt (RP, KP) and late (RT, KT) observations in R and K bands of dark bursts caused by dust extinction.

sublimated, which is also a function of dust grain size. Inside the host cloud we consider both the case of standard galactic dust and the case of dust, already present before the sublimation took place, with larger grain size with $a_{min} = 0.005\mu$m, $a_{max} = 2\mu$m. The larger size is because of the connection of GRBs and intense star formation regions ([8] 2001), as described in §2. Moreover, in the case of large grains, we have to consider two different sublimation radii: an inner one (r_s) up to which all grains are destroyed and a larger one (R_s) up to which only the grains with a radius smaller than 1μm can be sublimated ([9]). R_s and r_s vary with the intensity of the peak luminosity of the optical flash, that we can suppose depends on the GRB jet opening angle θ.

To each of the 5000 simulated GRBs is associated a random line of sight passing through 4 regions: $\Re 1$ is the inner region from the GRB to r_s (the radius up to which all dust grains are destroyed); $\Re 2$ goes from r_s to R_s, where large grains (radii from 1μm to 2μm) are allowed to survive; $\Re 3$ is the rest of the undisturbed host cloud dust; $\Re 4$ is the host galaxy outside the host cloud, consisting of all other molecular clouds and of the diffuse medium (see Figure2). In these regions we place 5 kinds of dust yielding ten different models. Table 2 explains all the cases.

In Figure 3 we plot percentages for relevant cases.

TABLE 2. Regions around GRBs and their contents in term of dust type as adopted in our computations. Dust model used are: STD (our model of standard Galactic dust, with grain radii from 0.005μm to 0.25μm); LRG (with added larger grains, from 0.005μm to 2μm); OLG (only large grains, left from partial sublimation, with radii from 1μm to 2μm); GRA (only large graphite grains, left from partial sublimation) SMC (dust with low extinction similar to Small Magellanic Clouds); SUB (no dust, completely sublimated by the burst).

(1)	(2)	(3)	(4)	(5)
	$\Re 1$	$\Re 2$	$\Re 3$	$\Re 4$
model 1	SUB	OLG	LRG	STD
model 1b	SUB	GRA	LRG	STD
model 2	SUB	SUB	STD	STD
model 3	SUB	OLG	STD	STD
model 3b	SUB	GRA	LRG	STD
model 1smc	SUB	OLG	LRG	SMC
model 1smcb	SUB	GRA	LRG	SMC
model 2smc	SUB	SUB	SMC	SMC
model 3smc	SUB	OLG	SMC	SMC
model 3smcb	SUB	GRA	SMC	SMC

CONCLUSIONS

In the case of GRBs occurring inside giant molecular clouds, our simulations show that dust in GRB host galaxies does not have the same properties as galactic dust, otherwise dark GRBs would happen more often. Furthermore our results agree with the hypothesis of a dust extinction curve of GRB host galaxies similar to that of the SMC. Moreover, if the host molecular cloud is characterized by large dust grains, high redshift plays a minimal role in the causes of dark GRBs. The opposite holds if the dust inside the host molecular cloud also has SMC properties.

ACKNOWLEDGMENTS

The authors want to thank Stefano Covino and Daniele Malesani for useful discussions.

REFERENCES

1. Chincarini, G., Zerbi, F.M., Antonelli, A. et al., ESO Messenger, **113**, 40 (2003).
2. Mathis, J.S., Rumple W. & Nordsieck K.H., Astrophys. J., **217**, 425 (1977).
3. Mathis, J.S., Astrophys. J., **308**, 281 (1986).
4. Simon, V., Hudec, R., Pizzichini, G. & Masetti, N., Astron. Astrophys., **377**, 450 (2001).
5. Vergani, S.D., "laurea" thesis (2002).
6. Waxman, E. & Draine, B.T., Astrophys. J., **537**, 796 (2000).
7. Reichart, D.E., Astrophys. J., submitted (astro-ph 0107546) (2001).
8. Galama, T. & Wijers, R.A.M.J., Astrophys. J., **549**, L209 (2001).
9. Venemans, B.P. & Blain, A.W., MNRAS, **325**, 1477 (2001).

Four Years of Observations of GRB Localizations with TAROT

M. Boër*[†], A. Klotz*, C. Thiébaud*, J.-L. Atteia**, R. Malina[‡], J. de Freitas Pacheco[§] and H. Pedersen[¶]

*Centre d'Etude Spatiale des Rayonnements (CNRS-UPS), BP 4346, 31028 Toulouse Cedex 4
[†]Present address: Observatoire de Haute Provence (CNRS-OAMP), 04870 Saint Michel l'Observatoire, France
**Laboratoire d'Astrophysique de Toulouse (CNRS-UPS), 14, ave. E. Belin, 31400 Toulouse, France
[‡]Laboratoire d'Astrophysique de Marseille (CNRS-OAMP), Traverse du Siphon, 13376 Marseille, France
[§]Observatoire de la Côte d'Azur (CNRS-OCA), 06304 Nice, France
[¶]Copenhagen University Observatory, Copenhagen, Denmark

Abstract. We present a summary of the observations performed with the Télescope à Action Rapide pour les Objets Transitoires (TAROT - Rapid Action Telescope for Transient Objects) performed over the period 1999 - 2003. Seventeen GRB localization observations where performed shortly after the burst (10s - 90min.), and in at least one case, even while the source was still active in gamma-rays. During this period CGRO, HETE-2 and INTEGRAL were in operation. Though no alert was missed, no source was detected, to a magnitude limit between $R = 15$ and $R = 20$. Future plans are also presented, featuring the duplication of TAROT at ESO - La Silla.

INTRODUCTION

In 1999, the Télescope à Action Rapide pour les Objets Transitoires (TAROT – Rapid Action Telescope for Transient Objects, see web site at the following url: `http://tarot.cesr.fr`) became routinely operational at the Calern Observatory in the French Alps. This 25 cm aperture telescope reaches a limiting magnitude of about $R < 20$ (17 in 10s) and is able to point a source and start acquisition within 1-2 seconds upon the receipt of an alert from the GCN [1]. All operations and processing of images are completely autonomous. In the event of a GRB alert observed by TAROT a duty astronomer is awaked and can get the images from the web. The main goal of TAROT is the observation of prompt optical counterparts from GRB sources and their early afterglow. TAROT has reacted to alerts from CGRO/BATSE, HETE-II and INTEGRAL. We are currently duplicating TAROT for installation at the European Southern Observatorty (La Silla) by mid-2004 [2]. In this paper we present the main results from TAROT, as well as its perspective of evolution including its duplication at ESO.

CURRENT SETUP OF THE INSTRUMENT

At present we use a prototype camera [3] designed and build by CESR (CNRS-UPS), CEMES (CNRS), and INSU/DT (CNRS), using a Thomson THX 7899M 2k x 2k chip. At present we are able to read this camera in 2s for a readout noise of 15e- (20e- for 1s). In its final version the "regular" readout time will be 1s for a rms noise of about 9e-, with "rapid modes" at 0.5 and 0.25s. However, we plan to replace this camera with a commercial ANDOR DW436 camera, based on a thin EEV 42-40, 2k x 2k back illuminated CCD. Though the readout time will be longer (5 seconds in non binned mode), the use of a less noisy, back illuminated camera will result in a substantial gain in sensitivity.

All TAROT operations are fully robotic. The MAJORDOME software [4] efficiently schedules observations. Should an alert occurs, the current observation is interrupted and the telescope slews to the designated position and the observation starts within 1-2s. The Data Processing Software analyzes within a minute the last acquired frame, and produces as an output a list of sources with their identification in the USNO A2.0 catalogue. If the telescope perform alert observations, the duty astronomer is awaked and can get the results at home through a dedicated web page. Sources from the USNO catalogue are superimposed on the TAROT image, allowing quick identification of potential "candidate" events. The user can interactively get a map of asteroids in the TAROT field of view (FOV), and extract a 10x10 arcmin archival image from the DSS, centered on any position within the FOV. In addition a web page and an email is sent with the position of variable sources in the field.

OBSERVATION AND RESULTS

We refer the reader to [5] for a discussion of the data taken with CGRO-BATSE. We concentrate here on the data acquired with HETE-2 and INTEGRAL.

HETE

TAROT has been in operation since the beginning of the HETE mission. The limiting magnitude we reached is between R = 16 and 20 depending on the event. As an example, for the "dark" GRB 030115 [6] we co-added the 10 first frames to reach a limiting magnitude of 19. Alert GRB 030324 was sent in 18 seconds, allowing a prompt start of TAROT observations before the end of the gamma-ray activity of the burst. Unfortunately, the source was located into the galactic plane, with more than 14 magnitude of absorption in the R band. The alert of Aug. 17, 2003 was in fact due to a real GRB outside the field of view but it occurred while an X-ray source (GRS 1915+105, visible by the HETE SXC detectors), lead to a wrong online localization. However, we note that there was a strong shortening of the time taken by HETE to propagate localizations.

TABLE 1. Number of alerts per year since 1999, until December 1st, 2003

Year	Number of alerts
1999	6
2000	5
2001	0
2002	1
2003	5

INTEGRAL

TAROT has received only one valid INTEGRAL alert (GRB 030501). The alert delay was 16s, and observations started 2s later, while the burst was still active (it had a duration of 40s). No counterpart has been found, probably because of the position of this burst in the galactic plane.

DISCUSSION

Table 1 summarizes the number of GRB localizations observed so far. For BATSE, we quote only the alerts for which the sources were within the field of view at any moment of the observation. False alerts from HETE and INTEGRAL have been removed as well. Figure 1 gives the time when the TAROT observation starts. The shaded area represents the TAROT pointing time, underlining that the dominant delay come from the computation of the position from the spacecraft and/or SC team, and the transmission time. The example of HETE 030115 and INTEGRAL 030501 show that provided these times are short, TAROT may well observe a GRB while it is still active at gamma-ray wavelengths. This is important for the study of the shocks, (internal, reverse) and of the prompt GRB/Afterglow transition.

CONCLUSION AND PERSPECTIVES

The experiment is currently running smoothly. The data processing software has been enhanced, and is more rapid and robust. The supervising software has been now replaced by a major revision, in the framework of ESO duplication. All TAROT parameters are now accessible through the web, and the reliability has been enhanced.

Within the FROST consortium we are duplicating TAROT for installation in the former GMS building at La Silla. TAROT-S will be installed this year [2]. To ensure proper operations and reduce on-site maintenance, the whole telescope will be extensively tested in France in operational conditions prior shipping to La Silla. Remote software maintenance will be possible.

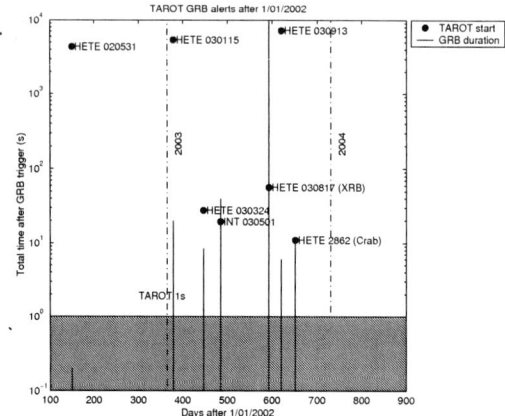

FIGURE 1. Delay between the start of the GRB and the TAROT start observations (dots). This delay is dominated by the on-board computation of the position and the time taken to send it to the observatories. The lines represent the GRB duration, while the shaded area is the time taken by TAROT to close the current observation, to slew and start the new one.

For a given observatory, the SWIFT burst rate is expected to be about 2-3 per month. With the addition of HETE and INTEGRAL bursts, we expect to be able to derive quantitative data for dark and bright GRBs, as well as short events. For the two instruments of the TAROT network, this means that as of the launch of SWIFT by mid-2004, about 50 event will be studied each year.

ACKNOWLEDGMENTS

The *Télescope à Action Rapide pour les Objets Transitoires* (TAROT) has been funded by the *Centre National de la Recherche Scientifique* (CNRS), *Institut National des Sciences de l'Univers* (INSU) and the *Carlsberg Fundation*. It has been built with the support of the *Division Technique* of INSU (INSU/DT). The TAROT CCD camera was built by a collaboration between CESR and CEMES with support from the CNRS technological fund.

REFERENCES

1. Barthelmy, S., Proceedings of the 4th Huntsville Symposium, AIP conf. proc. 428, edts. C.A. Meegan, R.D. Preece, and T.M. Koshut, p. 99 (1997).
2. Boër, M., et al., The ESO Messenger, **113**, 45 (2003).
3. Pinna, H., et al., in *Optical Detectors for Astronomy II: State-of-the-Art at the Turn of the Millenium*, ESO, p. 339 (2000).
4. Bringer, M., Boër, M., Peignot, C., Fontan, G., Merce, C., Exper. Astrophys **12**, 34 (2001).
5. Boër, M., *et al.*, A&A **378**, 76 (2001).
6. Klotz, A., Boër, M., and Atteia, J.L., A&A, **404**, 815 (2003).

LATE AFTERGLOWS

Damped Lyα Systems in GRB Afterglows

Paul Vreeswijk*, Sara Ellison*†, Cédric Ledoux*, Ralph Wijers**, Jens Hjorth‡, Johan Fynbo§, Andrew Fruchter¶ and the GRACE collaboration‖

*European Southern Observatory, Santiago, Chile
†P. Universidad Católica de Chile, Santiago, Chile
**Astronomical Institute 'Anton Pannekoek', University of Amsterdam & Center for High Energy Astrophysics, Amsterdam, The Netherlands
‡Niels Bohr Institute, Astronomical Observatory, Copenhagen University, København, Denmark
§Department of Physics and Astronomy, Århus University, Århus, Denmark
¶Space Telescope Science Institute, Baltimore, USA
‖

Abstract. We present VLT spectroscopy of the afterglow of GRB 030323, which shows a damped Lyα (DLA) absorption line at a redshift $z=3.372$ with an inferred neutral hydrogen column density of log $N(HI)=21.90\pm0.07$, larger than any (GRB- or QSO-) DLA HI column density inferred directly from Lyα in absorption. We discuss several other properties of the GRB host, such as the metallicity ([S/H]=-1.3 ± 0.2), the H_2 content, and the detection of SiII*, which has never been clearly detected in QSO-DLAs. From the latter we infer an order of magnitude estimate for the particle density: $n_{H^0} \sim 100$ cm^{-3}. Comparison of the HI column densities and metallicities in the available sample of GRB-DLAs (which is still very small) with those in the sample of QSO-DLAs suggests that both of these quantities are larger in GRB-DLAs.

INTRODUCTION

Damped Lyα (DLA) systems are conventionally found in absorption toward QSOs; they are characterized by their neutral hydrogen column density: log $N(HI) \geq 20.3$. The HI column density distribution of QSO absorption-line systems, ranging from the low-density Lyα forest, through the Lyman limit systems to the higher-density DLA systems, follows a power law with index of about −1.5, which means that the bulk of the neutral hydrogen in the universe is contained within the highest HI column density systems: the DLAs. Despite intense searches, only few counterparts of DLA absorbers have been detected [see 1]; the persistent light of the background QSO makes deep imaging of the DLAs very difficult. Linking DLAs with galaxy type has therefore proven to be difficult; some advocate large, fast-rotating, thick-disk galaxies [e.g. 2], others prefer faint, gas-rich dwarfs [3, 4].

GRB-DLAS

Gamma-ray bursts (GRBs) are, just as QSOs, distant and bright sources. In several GRB spectra, evidence for the presence of a DLA line has been found, as described in more detail below. In all these cases, the DLA system is located at or near the redshift of the

GRB itself, i.e. located in the GRB host galaxy. The first GRB afterglow to be associated with a DLA system was GRB 000301C at $z=2.040$ [5]. The Lyα line is just at the blue edge of the spectrum, where the error spectrum rises very rapidly. The resulting signal-to-noise ratio of the Lyα absorption line is not so high. The resulting column density is log N(HI)=21.2±0.5, i.e. with a very large uncertainty. Another DLA system was detected in the line-of-sight towards GRB 000926 at $z=2.038$ [6]. Its column density is also very high: log N(HI)=21.3. Note that both GRB hosts are a factor of about 10 above the classical DLA definition limit. Along the GRB 020124 ($z=3.198$) sight line, Hjorth et al. [7] find evidence for a DLA with log N(HI)=21.7. From the afterglow lightcurve, they also obtain an upper limit on the line-of-sight reddening, suggestive of a low dust content. We note that all these three DLAs have been detected by the group of Jens Hjorth and colleagues, which is the Copenhagen node of the GRACE collaboration. The fourth GRB DLA, with a neutral hydrogen column density larger than any previously detected DLA, is the one that we discuss in more detail below: GRB 030323 at $z=3.372$. More recently, a DLA line was also found in the afterglow of GRB 030429, with an HI column density of log N(HI)=21.5 (Jakobsson, Fynbo et al. 2004, in prep.).

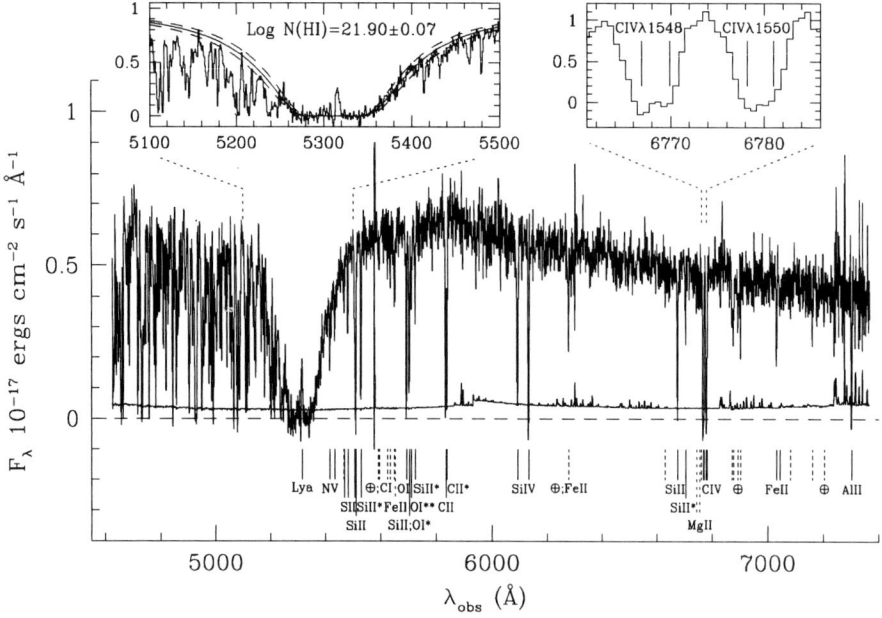

FIGURE 1. Combined VLT/FORS2 spectrum of GRB 030323, including the Poisson error spectrum. The left inset shows the normalized spectrum with the model fitting to the damped Lyα line, including the 1σ errors. This column density is currently the highest of any DLA line measured through Lyα in absorption (see Figure 2).

THE HIGHEST HI COLUMN DENSITY DLA TO DATE

Following the HETE-II detection of GRB 030323 [8] and the discovery of its optical counterpart by Gilmore et al. [9], we performed follow-up spectroscopic observations. The combined, dereddened VLT/FORS2 spectrum (with a resolution R~2000) of GRB 030323 is shown in Figure 1. We identify many metal absorption lines from different ions: NV, SII, SiII, SiII*, SiIV, OI, CII*, CII, CIV, FeII and AlII, strong Lyα absorption, and the intervening Lyα forest blueward of this. All the identified lines are indicated in Figure 1. The average redshift of the metal absorption lines is $z=3.372\pm0.001$.

A fit to the strong Lyα absorption line yields log N(HI)=21.90±0.07, which is the highest column density measured so far for a (QSO- or GRB-) DLA system. Figure 2 shows a comparison of the HI column density distribution of QSO-DLAs (taken from the compilation of Curran et al. [10]) and GRB-DLAs [5, 6, 7]. For completeness, we also show the two GRBs for which Lyα was detected but which do not qualify as a DLA: GRB 011211 (Vreeswijk et al. 2004, in prep.) and GRB 021004 [11]. It is quite striking that out of 7 GRB afterglows for which Lyα was redshifted into the observable spectrum, 5 show evidence for a high column density DLA.

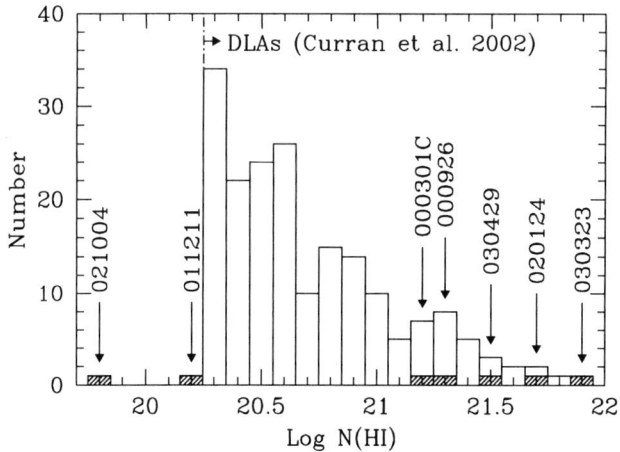

FIGURE 2. Histogram of the column densities of DLA systems measured through the damping wings of Lyα discovered against a background QSO, taken from Curran et al. [10]. We do not include the sub-DLA systems on this plot, which are located at column densities lower than log N(HI)=20.3; i.e. the apparent cut-off is not real. The shaded histogram shows GRBs for which the redshift was large enough to detect Lyα. Out of 7 GRBs, 5 show neutral hydrogen column densities above the DLA threshold $N(HI) \geq 2 \times 10^{20}$ atoms cm^{-2}.

It is generally assumed that the apparent HI column density limit of log N(HI)~22 for QSO-DLAs is due to an observational bias against the detection of such high-column density systems, as these would obscure the background QSO if they contain some dust [e.g. 12, 13]. However, a radio-selected QSO survey for DLAs by Ellison et al. [14] did not uncover a previously unrecognized population of N(HI)> 10^{21} cm^{-2} DLAs in front of faint QSOs. An alternative scenario was proposed by Schaye [15]: the lack of high HI column density systems would be due to the conversion of HI to H$_2$ molecules as the neutral gas density increases. In GRB 030323, we do not see any evidence for the presence of H$_2$ to support this scenario. Future GRBs with possible larger HI column densities will provide further constraints on the existence of such a conversion of the neutral gas to H$_2$ at high HI column densities.

METALLICITY, H$_2$ CONTENT AND SILICON II* DETECTION

From the few unsaturated metal lines detected in the GRB 030323 spectrum, we measure iron and sulphur metallicities of [Fe/H]=−1.5±0.2 and [S/H]=−1.3±0.2. In Figure 3, we compare the GRB-DLAs for which the Zn, S or Si metallicity has been measured (GRB 000926 and the GRB 030323 measurement above) with a sample of QSO-DLAs taken from Prochaska et al. [16]. Although the GRB-DLA sample is still small, there is a hint that GRB-DLAs are more metal-rich than QSO-DLAs, which one would expect if GRBs originate in massive-star forming regions. We note that Savaglio and colleagues [17] also pointed out large Zn column densities in three GRB host galaxies (one of which is GRB 000926) with respect to QSO-DLAs, while they found the Fe column densities to be similar to those of QSO-DLAs.

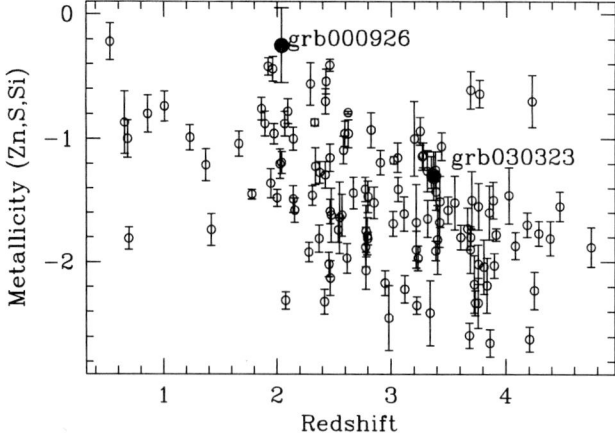

FIGURE 3. Comparison of the metallicities of a sample of QSO-DLAs, taken from Prochaska et al. [18] (open circles), with the two GRBs for which a metallicity has been determined (solid circles): GRB 000926 and GRB 030323. Although the GRB-DLA sample is too small to draw any conclusions, the GRB hosts are located at the metal-rich end of the QSO-DLA distribution.

From the detection of the fine-structure line SiII* λ1533 (see Figure 1), we very crudely estimate the particle density: $n_{H^0} \sim 100$ cm^{-3}, under the assumption that these fine-structure levels are populated by collisions, and not through direct excitation by infra-red photons (which is not an important excitation mechanism in the case of SiII) or fluorescence [see 19]. This particle density is higher than typically inferred for QSO-DLA environments [19]. As this line has never been clearly detected up to now in QSO-DLAs, the detection of this and other SiII* lines in the GRB 030323 spectrum suggests an origin in the vicinity of the GRB place of birth (e.g. the star-forming region in which it exploded). From the measured H I column density and the rough estimate of the particle density, we infer a size of 26 pc for the Si II* absorbing region; as the total H I column density that we measured may not be all associated with the Si II* absorbing region, this size estimate is actually an upper limit. With the particle density so high, one would expect hydrogen molecules to be present, which we do not detect. We obtain a rather strong upper limit on the molecular fraction: $f \equiv 2N(H_2)/(2N(H_2)+N(HI)) \lesssim 10^{-6}$, for both the GRB environment and the host galaxy of GRB 030323. This can be explained by the low metallicity of the gas [see 20], but it may also be that the molecules in the vicinity of the GRB explosion have been dissociated by the strong GRB UV/X-ray emission [e.g. 21].

The spectrum of GRB 030323 shown in Figure 1 was taken when the afterglow brightness was around R=21.5. In the Swift era, the burst alert delay will be shorter than it is at present, and the error circle will be considerably smaller, allowing afterglows to be discovered when they are much brighter. This should allow high-resolution spectroscopy of a large sample of afterglows, which can provide statistical information about the distribution of the gas in high-redshift star-forming regions, in addition to the evolution of the metallicity, dust and H_2 contents of GRB host galaxies.

REFERENCES

1. Møller, P., Warren, S. J., Fall, S. M., Fynbo, J. P. U., and Jakobsen, P., *ApJ*, **574**, 51–58 (2002).
2. Wolfe, A. M., Lanzetta, K. M., Foltz, C. B., and Chaffee, F. H., *ApJ*, **454**, 698 (1995).
3. Haehnelt, M. G., Steinmetz, M., and Rauch, M., *ApJ*, **495**, 647 (1998).
4. Ledoux, C., Petitjean, P., Bergeron, J., Wampler, E. J., and Srianand, R., *A&A*, **337**, 51–63 (1998).
5. Jensen, B. L., Fynbo, J. P. U., Gorosabel, J., Hjorth, J., and et al., *A&A*, **370**, 909–922 (2001).
6. Fynbo, J. P. U., Gorosabel, J., Møller, P., Hjorth, J., and et al., "The optical afterglow and host galaxy of GRB 000926," in *Lighthouses of the Universe: The Most Luminous Celestial Objects and Their Use for Cosmology, Proc. of the MPA/ESO/MPE/USM Joint Astronomy Conf.*, eds. M. Gilfanov, R. Sunyaev, & E. Churazov (Garching: Springer), p.187; preprint astro-ph/0110603, 2001.
7. Hjorth, J., Møller, P., Gorosabel, J., Fynbo, J. P. U., and et al., *ApJ*, **597**, 699–705 (2003).
8. Graziani, C., Shirasaki, Y., Matsuoka, M., Tamagawa, T., and et al., *GRB Circular Network*, **1956** (2003).
9. Gilmore, A., Kilmartin, P., and Henden, A., *GRB Circular Network*, **1949** (2003).
10. Curran, S. J., Webb, J. K., Murphy, M. T., Bandiera, R., and et al., *Publications of the Astronomical Society of Australia*, **19**, 455–474 (2002).
11. Møller, P., Fynbo, J. P. U., Hjorth, J., Thomsen, B., and et al., *A&A*, **396**, L21–L24 (2002).
12. Ostriker, J. P., and Heisler, J., *ApJ*, **278**, 1–10 (1984).
13. Fall, S. M., and Pei, Y. C., *ApJ*, **402**, 479–492 (1993).
14. Ellison, S. L., Yan, L., Hook, I. M., Pettini, M., and et al., *A&A*, **379**, 393–406 (2001).
15. Schaye, J., *ApJ*, **562**, L95–L98 (2001).
16. Prochaska, J. X., Gawiser, E., Wolfe, A. M., Cooke, J., and Gelino, D., *ApJS*, **147**, 227–264 (2003).
17. Savaglio, S., Fall, S. M., and Fiore, F., *ApJ*, **585**, 638–646 (2003).
18. Prochaska, J. X., Gawiser, E., Wolfe, A. M., Castro, S., and Djorgovski, S. G., *ApJ*, **595**, L9–L12 (2003).
19. Silva, A. I., and Viegas, S. M., *MNRAS*, **329**, 135–148 (2002).
20. Ledoux, C., Petitjean, P., and Srianand, R., *MNRAS*, **346**, 209–228 (2003).
21. Draine, B. T., and Hao, L., *ApJ*, **569**, 780–791 (2002).

On the Shallow Decay of Some GRB Afterglows

A. Panaitescu* and P. Kumar*

Department of Astronomy, University of Texas at Austin

Abstract. Half of the radio afterglows for which there is a good temporal coverage exhibit, after 10 days from the burst, a decay which is shallower than at optical frequencies, contrary to what is expected within the simplest form of the standard model of relativistic fireballs or jets. We investigate possible ways to decouple the radio and optical decays. First, the radio and optical emissions are assumed to arise from the same electron population, and we allow for either a time-varying slope of the power-law distribution of electron energy or for time-varying microphysical parameters. Then we consider two scenarios where the radio and optical emissions arise in distinct parts of the GRB outflow, either because the outflow has an angular structure or because there is a long-lived reverse shock. We find that only the last scenario is compatible with the observations.

The anomalous radio afterglows. The radio emission of all well-observed GRB afterglows decays after about day 10. This is consistent with what is expected in the standard fireball model for GRB afterglows. In this model, the characteristic synchrotron frequency v_i at which electrons with the typical post-shock energy ($e_i = \varepsilon_i m_p c^2 \Gamma$ where Γ is the fireball Lorentz factor) radiate is

$$v_i \sim 30 \left(\frac{\mathscr{E}}{10^{53}\,\text{ergs}}\right)^{1/2} \left(\frac{\varepsilon_i}{0.03}\right)^2 \left(\frac{\varepsilon_B}{10^{-3}}\right)^{1/2} \left(\frac{t}{10\text{d}}\right)^{-3/2} \text{GHz}. \tag{1}$$

Here, $\mathscr{E}$ is the fireball's kinetic energy per solid angle and the parameters ε_i and ε_B quantify the fraction of the post-shock energy imparted to electrons[1] and to the magnetic field. If the outflow is collimated, the evolution of v_i becomes faster, $v_i \propto t^{-2}$, after the "jet-break" time when the jet starts to expand laterally.

Equation (1) indicates that, for reasonable afterglow parameters, the injection frequency v_i is expected to cross the radio domain around 10 days, after which the radio afterglow should decay as $F_r \propto t^{-\alpha_r}$ with $\alpha_r = (3p-1)/4$ for a fireball interacting with a wind medium, $\alpha_r = (3p-3)/4$ for a fireball decelerated by a uniform medium, and $\alpha_r = p$ for jet spreading laterally. In the standard afterglow model, the optical light-curve decay is expected to be the same, apart from a difference $|\alpha_o - \alpha_r| = 1/4$ arising when the cooling frequency (v_c) is below the optical domain.

For five radio afterglows (970508, 980329, 980703, 000418, 021004), the radio and optical decay indices are consistent with this basic "prediction" of the fireball model, but it is not so for the other five well-monitored radio afterglows (991208, 991216, 000301, 000926, 010222), which are shown in Figure 1. For these cases, it is natural

---

[1] For a power-law distribution of electron energies $dN/de \propto e^{-p}$ at $e > e_i$, the total electron energy is $[(p-1)/(p-2)] \times \varepsilon_i$ for $p > 2$.

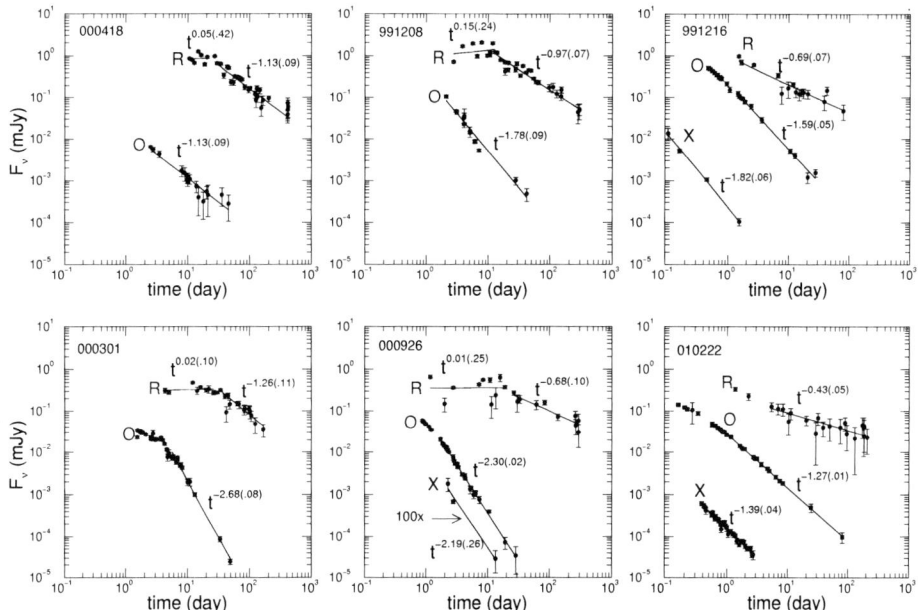

FIGURE 1. Radio (8 GHz), Optical (5×10^{14} Hz), and X-ray (5 keV) light-curves for five anomalous GRB afterglows whose radio light-curves decay slower than in the optical and for the afterglow 000418, for which the decays seen in these domains have the same indices, as expected in the simplest version of the standard fireball/jet afterglow model. Power-law fits are indicated for each frequency range, errors are given in parentheses.

to investigate first if the difference between the radio and optical decay indices could be caused by that the injection frequency, v_i, remains above the radio domain (of typical frequency $v_r \sim 10$ GHz) until the last radio measurements[2], usually around 100 days after the burst. According to equation (1), this requires electron and/or magnetic field parameters close to equipartition ($\varepsilon_i, \varepsilon_B \gtrsim 0.1$), particularly if the jet-break time occurs at around 1 day, as is the case for the afterglows 991216, 000301, 000926 and 010222 (see the break exhibited by their optical light-curves).

For $v_r < v_i$ there are three relevant cases which yield a decaying radio emission. If the GRB ejecta is spherical (or a sufficiently wide jet), the radio light-curve decays only if the circumburst medium is a wind (r^{-2} density profile) and if $v_r < v_c < v_i$, in which case $\alpha_r = 2/3$, as observed for the afterglows 991216 and 000926. If the GRB ejecta is collimated, a decaying radio light-curve can be obtained either for $v_i < v_c$ or for $v_c < v_i$. In the former case $\alpha_r = 1/3$, close to the value observed for the afterglow 010222, while the latter case yields $\alpha_r = 1$, as seen in the afterglow 991208. Despite this

---

[2] That radio afterglows decay after 10 days indicate that the self-absorption frequency is below the radio domain, thus it cannot account for this difference

nice agreement between the expected and observed radio decay indices for $v_r < v_i$, none of the above cases actually work well. The first and third case require dense external media (homogeneous or wind-like) to maintain $v_c < v_i$ until about 100 days, leading to self-absorbed sources at radio frequencies even at day 10, which is inconsistent with the decaying radio light-curves, and to a synchrotron flux at v_i above 100 mJy at 100 days, which is 1,000 times larger than the radio flux observed at that time. The second case leads to very tenuous external media, which are rather inconsistent with those expected for a massive star GRB progenitor.

Thus we have to investigate departures from the standard afterglow model in its simplest form that could account for the shallow radio decays observed for the above five anomalous afterglows. Since the $v_r < v_i$ case cannot explain all the properties of the radio afterglow emission, we will consider that the injection frequency, v_i, is below the radio domain when the radio decay is observed. In fact, the passage of v_i through this domain is the most natural cause for the onset of the radio decay seen at about 10 days in the afterglows 991208, 000301, 000926 and 010222, as well as in other afterglows whose radio and optical decay indices are consistent with the fireball model expectations.

These departures fall in two categories. In the first, we assume that the radio and optical emissions arise from the same region of the GRB fireball. If the same electron population gives both the radio and optical afterglows, then different decay indices can be obtained either if the slope p of the electron energy distribution is time-varying or if there is a spectral break between these two domains. In the second category, we assume that the radio and optical emissions arise from different fireball regions, as could be the case with a structured fireball or a long-lived reverse shock. We note that, in the scenarios that we describe below, involving time-varying afterglow parameters, the break seen in the optical light-curve of most afterglows shown in Figure 1 may arise from a rapid variation of the parameters under investigation and not necessarily from the tight collimation of the GRB outflow.

Electron distribution with time-varying slope. We consider that, at all times of interest (1-100 days after the burst), the electron distribution injected by the forward shock is a power-law of exponent $-p$ that varies in time[3]. It is evident that, in this case, the radio and optical light-curves cannot both be power-laws in time. Given that optical measurements have smaller uncertainties, we can determine the evolution of the electron index p that yields a power-law optical light-curve and test the consequences of that evolution on the afterglow emission at other frequencies.

It can be shown that, in order to obtain $\alpha_r < \alpha_o$, the index p must increase in time, asymptotically approaching the value that the observed α_r required if the index p were constant in time. Thus the optical afterglow spectrum should soften in time. For such a behavior of p, the radio light-curve should steepen in time, while the X-ray light-curve should decay faster that in the optical and should flatten in time. In general, these features are not quantitatively consistent with the multiwavelength observations of the

---

[3] An alternative scenario for the $\alpha_o - \alpha_r$ index difference of the anomalous afterglows is an electron distribution that is not a power-law and whose shape does not change in time. This scenario is largely unconstrainable with current observations.

anomalous afterglows: the optical spectrum of 991208 appears to harden in time while the decays of the X-ray light-curve of 991216 (during day 1) and of the radio light-curves of 000926 and 010222 (at 100 days) are much shallower than expected. Only for 000301 the radio light-curve and optical spectrum are consistent with the consequences of an evolving electron distribution index.[4]

Variable electron and magnetic field parameters. As the self-absorption and injection break frequencies must be below the radio domain when the radio light-curves decay, the only remaining break that could decouple the radio and optical decay indices must be the cooling frequency v_c. However its evolution must be much faster than that for a constant magnetic parameter ε_B, to account for the magnitude of the observed $\alpha_o - \alpha_r$. Since we want, in fact, to explain not just the difference $\alpha_o - \alpha_r$, but the observed values of α_r and α_o, we must also allow for the electron energy parameter ε_i to be time-varying. We shall also consider that the fireball's kinetic energy per solid angle $\mathscr{E}$ (or the kinetic energy for a jet) is time-varying, either because of radiative losses or an energy injection in the fireball through some less relativistic ejecta which catch-up with the leading edge of the ejecta (which is decelerated by the interaction with the circumburst medium).

Because the quantities pertaining to the fireball dynamics are power-laws in the observer time t, it is then natural to restrict our attention to time-varying ε_i, ε_B, and $\mathscr{E}$ that evolve as power-laws with t: $\mathscr{E} \propto t^e$, $\varepsilon_i \propto t^i$, $\varepsilon_B \propto t^b$. Radiative losses yield $e \geq -3/7$ for a fireball interacting with a homogeneous medium, $e \geq -1/3$ if the medium is wind-like, and $e \geq -3/5$ for a spreading jet and any type of external medium. Using standard equations for the afterglows spectral characteristics, we can calculate the decay indices for the radio and optical light-curves as functions of the parameters e, i and b and use the observed α_r and α_o to constraint them.

In this way it can be shown that, if there is no energy injection ($e \leq 0$), then ε_i must decrease in time while ε_B must increase so fast that the magnetic field strength $B \propto \varepsilon_B^{1/2}$ is constant or increases in time. This rather extreme requirement is somewhat alleviated if the external medium density increases with radius, as such a density profile leads to a faster evolution of the cooling frequency. If there is an energy injection, leading to $e \geq 0$, then there are solutions with constant ε_B, however the peak flux and injection frequency evolutions implied by this scenario are in strong conflict with those inferred from the radio emission of the afterglow 991208. These conclusions apply to both a spherical fireball and a collimated outflow.

Structured outflow. If the outflow has a non-uniform angular distribution of the kinetic energy per solid angle $\mathscr{E}$, it is possible that the optical emission arises predominantly from a core of higher $\mathscr{E}$ while the radio afterglow is emitted by a surrounding envelope of lower $\mathscr{E}$. This possibility is suggested by the dependence of the injection frequency v_i on $\mathscr{E}$ (eq. [1]). For simplicity and maximal effect of the outflow structure, we consider that both the core and envelope have an uniform distribution of $\mathscr{E}$. In this scenario, the break exhibited by the afterglow optical light-curve is caused by the collimation of the core, while the onset of the radio decay is due to the passage through the

---

[4] There are no X-ray measurements for the afterglow 000301 to further test this scenario

radio domain of the v_i for the envelope emission. To explain the observed α_r and α_o, we must allow for different electron indices p in the outflow core and envelope.

The test that this scenario must pass is that the radio and optical emissions are decoupled. More specifically, the softening emission from the optical core must not overshine in the radio the emission from the envelope and the emission from the envelope must not be brighter in the optical than the faster decaying emission from the core. The first condition is equivalent to a lower limit on the v_i frequency for the optical core at the time when the optical light-curve break is observed (which is the time when the core edge becomes visible to the observer). With the aid of equation (1), this leads to a lower limit on the injection frequency for the radio envelope at the time when the radio light-curve begins to decay. For the parameters of the radio and optical emission of the five anomalous afterglows, the latter lower limit falls invariably above 10 GHz, contrary to what is implied by the observed onset of the radio decay. The second condition leads to an upper limit on the cooling frequency for the envelope emission, which implies very dense external media and a self-absorbed radio emission, inconsistent with the observed radio decay.

Reverse shock. If there is a continuous inflow of slower ejecta from the GRB progenitor into the leading edge of the GRB fireball, then the emission from the long-lived reverse shock crossing the incoming ejecta could dominate the radio afterglow emission, while the optical afterglow arises as usually from the forward shock. Given that a collimated outflow undergoing lateral spreading and delayed energy injection loses angular uniformity, we restrict our attention to a spherical outflow, which maintains its isotropy during the injection.

For simplicity, we parameterize the distribution of ejecta mass with Lorentz factor as a power-law, $d\mathcal{M}_i/d\Gamma \propto \Gamma^{-(q+1)}$. The fireball energy can also vary in time (e.g. $\mathcal{E} \propto t^e$), as described for a structured outflow[5]. Just as for the scenario involving variable microphysical parameters, one can calculate the radio and optical decay indices as function of the parameters q and e, and then determine these parameters with the aid of observations. We find that the anomalous radio afterglows require $q > 3$ for a homogeneous medium, $q > 4$ for a wind medium, and $e < 1/3$, indicating that the incoming ejecta can carry at most the same energy as the initial outflow energy.

The reverse-forward shock scenario must also pass the test discussed above for a structured outflow, leading to a lower limit on the forward shock v_i frequency and an upper limit on the reverse shock cooling frequency. Two other constraints can be obtained by requiring that the forward shock yields the flux normalization seen in the optical and that the v_i frequency for the reverse shock crosses the radio domain when the onset of the radio fall-off is seen. These four requirements can be converted into constraints on the fundamental afterglow parameters $\mathcal{E}$, ε_i, ε_B, and n (external medium density). For the anomalous afterglows we find that reasonable values of these parameters satisfy the observational requirements.

Note: More details on the features of the scenarios described here and the calculations behind the results presented can be found at *astro-ph/0308273*.

---

[5] If the energy of the incoming ejecta is dominant, then the exponents q and e are not independent, but they are so for a negligible energy injection, i.e. $e \lesssim 0$

Relativistic Wind Bubbles

Z. G. Dai

Department of Astronomy, Nanjing University, Nanjing 210093, China; dzg@nju.edu.cn

Abstract. Gamma-ray bursts (GRBs) are currently believed to originate from highly magnetized, rapidly rotating compact objects. After the GRB, such an object may directly lose its rotational energy through some magnetically-driven processes, which produce an ultrarelativistic wind dominated possibly by the energy flux of electron-positron pairs. The interaction of this wind with an outward-expanding fireball leads to a relativistic wind bubble. Here we discuss the effects of this wind bubble. We find that when the wind energy significantly exceeds the initial energy of the fireball, the bulk Lorentz factor of the wind bubble decays more slowly than before, and more importantly, the reverse-shock emission could dominate the afterglow emission, which yields a bump in afterglow light curves.

INTRODUCTION

The recent observation of high linear polarization during the prompt γ-ray emission of GRB 021206 ([1]) suggests that GRBs be driven by highly magnetized, rapidly rotating compact objects. Two currently popular scenarios for GRBs are the merger of a compact binary or the collapse of a massive star (for a recent review see [2]). In both scenarios, a remaining object is a rapidly rotating black hole surrounded by an accretion disk or a millisecond magnetar. After the GRB (i.e., during an afterglow), this central object may lose directly its rotational energy to produce an ultrarelativistic wind by some magnetically-driven processes (e.g., the magnetic dipole radiation or the Blandford-Znajek mechanism). The interaction of this outflow with an outward-expanding fireball implies a continuous injection of the stellar rotational energy into the fireball. Dai and Lu ([3]) and Zhang and Mészáros ([4]) discussed the evolution of a relativistic fireball by assuming a pure electromagnetic-wave energy outflow, while Rees & Mészáros ([5]) and Sari & Mészáros ([6]) took into account a variable and baryon-dominated injection.

However, based on the successful models of the well-observed Crab Nebula (Rees & Gunn, [7]; Kennel & Coroniti, [8]), a realistic, continuous outflow during the afterglow is expected to be ultra-relativistic and dominated by the energy flux of electron-positron pairs. In the case of an afterglow, therefore, it is natural to expect that the central object still produces an ultra-relativistic e^+e^--pair wind, whose interaction with the fireball leads to a relativistic wind bubble. This can be regarded as *a relativistic version of the Crab Nebula*. In this paper, we discuss the dynamics of such a wind bubble and its emission signatures. More details can be seen in Dai ([9]).

THE BUBBLE DYNAMICS AND LIGHT CURVES

We assume that a burst itself arises from a series of explosive reconnection events. After the GRB, we are left with a highly magnetized, rapidly rotating compact object. No matter whether this object is a Kerr black hole or a millisecond magnetar, the luminosity of a resulting relativistic wind will evolve with time as $L_w \propto (1+t/T_0)^{-2} \sim$ constant for $t < T_0$ and $\propto t^{-2}$ for $t > T_0$, where t is the observer time in day, and T_0 is the "initial" spin-down timescale at the onset of the afterglow.

As in the Crab Nebula, a highly magnetic, rapidly rotating object at the center of an afterglow generates a highly relativistic wind dominated by the energy flux of e^+e^- pairs, with bulk Lorentz factor of γ_w. Because γ_w is much larger than the Lorentz factor of the medium swept up by a fireball, this wind passes through a shock front and decelerates to match the expansion velocity of the swept-up medium. As a result, a relativistic wind bubble should include two shocks: a reverse shock that propagates into the cold wind and a forward shock that propagates into the ambient medium. Thus, there are four regions separated in the bubble by these shocks: (1) the unshocked medium, (2) the forward-shocked medium, (3) the reverse-shocked wind gas, and (4) the unshocked cold wind, where regions 2 and 3 are separated by a contact discontinuity.

We denote n_i and P_i' as the baryon number density and pressure of region "i" in its own rest frame respectively, and γ_i is the Lorentz factor of region "i" measured in the local medium's rest frame. Based on the jump conditions for a relativistic shock (Blandford & McKee, [10]), the pressure of region 3 is calculated by $P_3' \simeq L_w/(12\pi r^2 \gamma_3^2 c)$. Neglecting the presence of the reverse shock and the radiative energy loss of region 2, and assuming an ambient interstellar medium with constant density of n_1, the properties of the shocked medium in region 2 should satisfy the Blandford-McKee adiabatic self-similarity solution with the similarity variable χ at any radius r. According to Blandford & McKee ([10]), the pressure and Lorentz factor of the shocked medium at radius r are given by

$$P_2'(r) = \frac{4}{3} n_1 m_p c^2 \gamma_2^2 \chi^{-17/12}, \qquad (1)$$

$$\gamma_2(r) = \gamma_2 \chi^{-1/2}, \qquad (2)$$

where m_p is the proton mass and γ_2 is the Lorentz factor of the fluid just behind the forward shock. Along the contact discontinuity, $P_3' = P_2'(r)$ and $\gamma_3 = \gamma_2(r)$, which yield

$$\chi = (16\pi \gamma_2^4 R^2 n_1 m_p c^3 / L_w)^{12/29} (r/R)^{24/29}$$

$$\simeq 1.95 (L_{w,47} t/E_{52})^{-12/29}, \qquad (3)$$

where $L_{w,47} = L_w/10^{47}\,\mathrm{erg\,s}^{-1}$, R is the forward shock radius, and $E_0 = 10^{52} E_{52}$ ergs is the "initial" isotropic-equivalent kinetic energy of the fireball. To obtain the second equality of equation (3), we have considered the thin-shell approximation and the Blandford-McKee Lorentz factor

$$\gamma_2 = \left(\frac{17 E_0}{1024\pi n_1 m_p c^5 t^3}\right)^{1/8}, \qquad (4)$$

where the shock radius $R \simeq 4\gamma_2^2 ct$ is used. Letting $\chi = 1$, we define a critical time $t_{\text{cr}} = 5.0 E_{52} L_{w,47}^{-1}$ days. At this time, the injection energy to the fireball significantly exceeds its initial energy. For $t < t_{\text{cr}}$, the similarity variable $\chi > 1$ and the Lorentz factor of region 3 decays as

$$\gamma_3 = \gamma_2 \chi^{-1/2} \simeq 4.4 E_{52}^{-19/232} n_1^{-1/8} L_{w,47}^{6/29} t^{-39/232}, \tag{5}$$

where n_1 is in units of $1\,\text{cm}^{-3}$. It should be emphasized that the dynamics denoted by equations (4) and (5) are simply calculated by equating the pressures of the two-sided shocked fluids at the contact discontinuity. In this derivation, we have neglected any work done on region 2 by region 3 because the pressure of region 3 is much less than $P_2'(R)$ at $t < t_{\text{cr}}$.

Once the observer's time exceeds t_{cr}, the similarity variable $\chi = 1$. At this stage, the total kinetic energy of region 2 is approximated as $E_{\text{kin},2} = (\gamma_2^2 - 1) M_{\text{sw}} c^2$, where $M_{\text{sw}} = (4\pi/3) R^3 n_1 m_p$ is the swept-up medium mass. Energy conservation requires that any increase of kinetic energy of region 2 should be equal to work done by region 3,

$$dE_{\text{kin},2} = \gamma_3 P_3' dV_3', \tag{6}$$

where $dV_3' = 4\pi R^2 dR' = 4\pi R^2 (dR/\gamma_3)$ is the volume change of region 3 in its own rest frame. Since regions 2 and 3 should keep velocity equality along the contact discontinuity (viz., $\gamma_2 = \gamma_3$), we rewrite equation (6) as

$$\frac{d\gamma_2}{dR} = \frac{4\pi R^2 [P_3' - (\gamma_2^2 - 1) n_1 m_p c^2]}{2\gamma_2 M_{\text{sw}} c^2}, \tag{7}$$

whose solution is

$$\gamma_2 = \gamma_3 = \left(\frac{L_w}{128 \pi n_1 m_p c^5 t^2} \right)^{1/8}, \tag{8}$$

where the dependence of γ_2 on t is consistent with the one derived for a pure electromagnetic energy injection by Dai & Lu ([3]).

Therefore, we find the dynamics of a relativistic wind bubble: In the case of $t_{\text{cr}} > T_0$, the wind bubble should evolve based on equations (4) and (5); for $t_{\text{cr}} < T_0$, however, the Lorentz factors of the wind bubble decays initially as $\gamma_2 \propto t^{-3/8}$ and $\gamma_3 \propto t^{-39/232}$ at $t < t_{\text{cr}}$ (stage I), subsequently as $\gamma_2 = \gamma_3 \propto t^{-1/4}$ at $t \in (t_{\text{cr}}, T_0)$ (stage II), and finally again as $\gamma_2 \propto t^{-3/8}$ at $t > T_0$ (stage III).

Dai ([9]) has discussed in detail light curves of the emission from a relativistic wind bubble at three stages based on the standard afterglow fireball scenario of Sari, Piran & Narayan ([11]). Figure 1 presents an example. We can see that the emission flux from region 2 decays rapidly at time $< t_{\text{cr}}$, subsequently fades more slowly at time $\in (t_{\text{cr}}, T_0)$, and finally declines based on the initial evolution law (Dai & Lu, [3]). More importantly, the emission from region 3 dominates the afterglow emission, which leads to a bump in afterglow light curves.

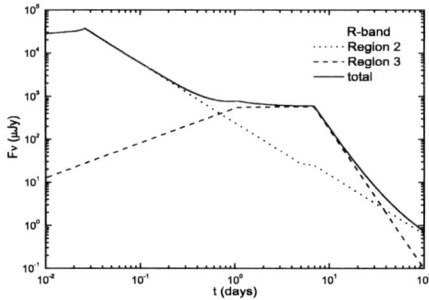

FIGURE 1. R-band light curves of the emissions from regions 2 (dotted line) and 3 (dashed line). The solid line corresponds to the total flux.

CONCLUSIONS

Based on the successful models of the Crab Nebula, we have discussed the dynamics of a relativistic wind bubble and its emission signatures, and found that that when the injection energy significantly exceeds the initial energy of the fireball, the bulk Lorentz factor of the wind bubble declines more slowly than before. In addition, the reverse-shock emission could dominate the afterglow emission, which leads to a bump in afterglow light curves. It should be pointed out that the magnetic field in the reversely-shocked region (i.e., region 3) of the wind bubble includes two components: one is originally toroidal and another is turbulently generated by the reverse shock. Depending on the ratio of both components, the afterglow polarization might be highly variable.

This work was supported by the National Natural Science Foundation of China (grants 10233010 and 10221001) and the National 973 Project (NKBRSF G19990754).

REFERENCES

1. Coburn, W., & Boggs, S. E., Nature, **423**, 415 (2003).
2. Mészáros, P., ARA&A, **40**, 137 (2002).
3. Dai, Z. G., & Lu, T., Phys. Rev. Lett., **81**, 4301 (1998).
4. Zhang, B., & Mészáros, P., ApJ, **552**, L35 (2001).
5. Rees, M. J., & Mészáros, P., ApJ, **496**, L1 (1998).
6. Sari, R., & Mészáros, P., ApJ, **535**, L33 (2000).
7. Rees, M. J., & Gunn, J. E., MNRAS, **167**, 1 (1974).
8. Kennel, C. F., & Coroniti, F. V., ApJ, **283**, 694 (1984).
9. Dai, Z. G., astro-ph/0308468, submitted to ApJ (2003).
10. Blandford, R., & McKee, C., Phys. Fluids, **19**, 1130 (1976).
11. Sari, R., Piran, T., & Narayan, R., ApJ, **497**, L17 (1998).

IR and Optical Observations of GRB 030115

Allyn Dullighan*, George Ricker*, Nathaniel Butler* and Roland Vanderspek*

Center for Space Research, Massachusetts Institute of Technology, Cambridge, MA

Abstract. We present an upper limit on the brightness of the afterglow of the long GRB 030115 measured from Infrared (IR) images taken with the Magellan Classic Cam instrument of Ks > 22 at 6.2 days after the burst. We also present measurements of the host galaxy of GRB 030115 from archival optical and IR HST images taken with the Advanced Camera for Surveys and the Near Infrared Camera and Multi Object Spectrometer 25+ days after the burst. GRB 030115 is classified as an Optically Dark Burst, as its afterglow was found in the J, H, and K IR bands after a null result was reported in the optical. It is the first HETE GRB to have its afterglow found initially in the IR.

INTRODUCTION

The High Energy Transient Explorer (HETE) detected Gamma Ray Burst (GRB) 030115 (=H2533) [1] on January 15, 2003, at 03:22:34.28 UT. The Wide Field Monitor (WXM) position was reported to the community over the GRB Coordinate Network (GCN) at 04:33:07 UT, and a 2 arcminute error radius, Soft X-ray Camera (SXC) position was reported at 04:46:34 UT on January 15, 2003. Followup optical observations were begun early, with observations as deep as the Digital Sky Survey beginning between 1-2 hours after the burst, [2, 3, 4], but no conclusive evidence for a fading counterpart was found. Infrared (IR) observations were begun at ∼5 hours after the burst by Levan et al. [5] and a fading IR source was reported to the GCN about 15 hours after the burst.

We made further IR observations of this source with the Magellan 6.5 meter Baade telescope using the Classic Cam instrument on January 23, 2003, but we were not able to analyze the data because of difficulties with the astrometry of the images. The recent release of archived[1] Hubble Space Telescope (HST) Near Infrared Camera and Multi Object Spectrometer (NICMOS) and Advanced Camera for Surveys (ACS) observations of GRB 030115 allowed for more accurate astrometric calibration of our images. Our analysis of the Classic Cam images and the public HST archived data is presented below.

INFRARED DATA AND ANALYSIS

Our IR observations were taken with the Classic Cam IR instrument at the Magellan telescopes at Las Campanas Observatory in Chile. Classic Cam is a Rockwell NICMOS-

---

[1] http://archive.stsci.edu/hst/

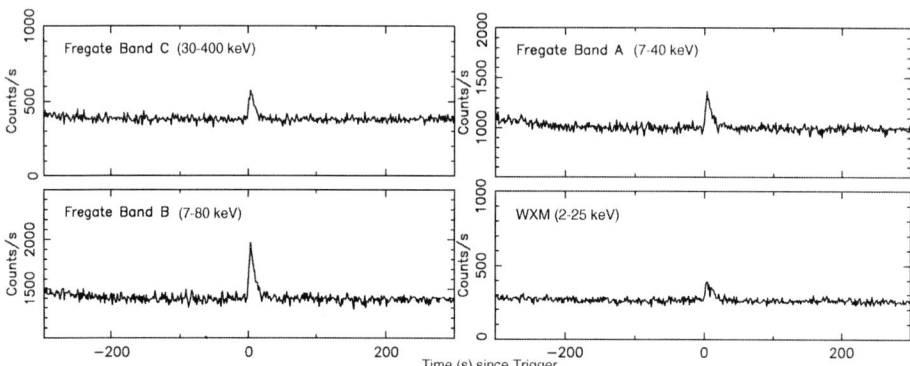

FIGURE 1. The HETE Fregate and WXM lightcurves of GRB 030115. The burst lasted ~20sec placing it well within the 'Long/Soft' GRB catagory, and had a waveband evolution typical of these bursts.

3 HgCdTe array with a 256x256 array of 40 μm pixels. It was positioned at one of the f/11 foci of the 6.5m Baade telescope, but is no longer in use. The Classic Cam imager has a maximum field of view of ~30x30 arcseconds, which we used with the K-short filter. The USNO catalog star ~14" to the West of the GRB was used to align the telescope with the field. At the time of our observations, however, engineering issues with the telescope meant that the camera orientation was not well known. Therefore, with only one bright star in the field of view, it was almost impossible to determine the orientation of the field. Stacking of all 45 good images from the data set revealed three other weak sources in the field, but the finder charts available at the time did not clearly resolve other sources in the extremely small ~30x30" field around the USNO star.

The release of three HST NICMOS F160W filtered images from February 10, 2003, covering the field of GRB 030115 gave an opportunity to orient these images, as the HST images were much deeper, and so clearly resolved the faint sources not seen in the finding charts. Stacking the three images allowed for a clean removal of cosmic rays in the field in IRAF. A comparison between the three stacked NICMOS and our Classic Cam image (below) shows the corrected orientation. Unfortunately, from the image below, you can see that the GRB was below our sensitivity limit for the observation. A limiting magnitude has been calculated at Ks~22 for our Stacked Classic Cam image. IRAF aperture photometry of the stacked NICMOS image gives a magnitude of 24.8 for the host galaxy in the F160W filter (~H band), when calibrated using the zero point photometry keywords in the HST header files of the images[2].

---

[2] see http://www.stsci.edu/hst/nicmos/documents/handbooks/DataHandbookv5/

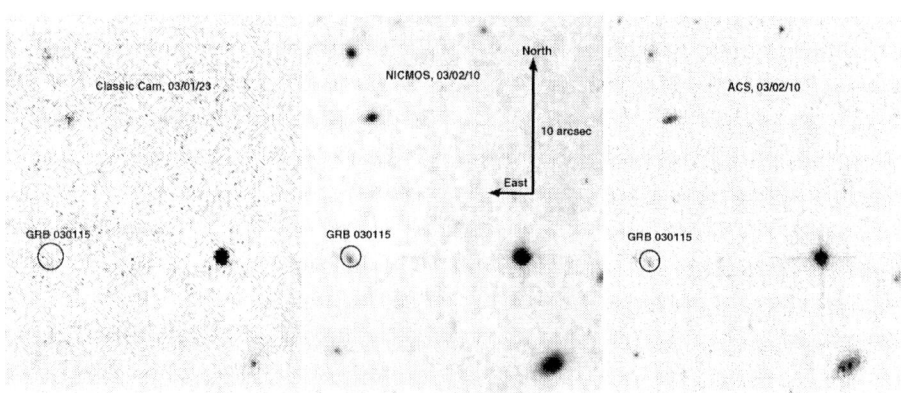

FIGURE 2. A comparison of the ClassiCAM stacked image (left), the NICMOS stacked image (center), and the ACS two-color image (right) of GRB 030115 at the same astrometric scale.

OPTICAL DATA AND ANALYSIS

The IR transient was reported to the GCN at $\sim$16 hours after the burst [5]. At that time earlier observers were able to re-examine their images, and the transient was found in the optical bands at R$\sim$21.5 on January 15, 2003, at 05:25 UT [6] with respect to a nearby USNO-A2 catalog star, or R$\sim$21.9 when recalibrated against the USNO-B1 catalog. Late time optical observations were made by Garnavich [7] on Jan. 29.4 UT, 2003, and placed the magnitude of the GRB and host at R=25.2, when recalibrated to the USNO-B1 catalog magnitudes.

The public archived ACS images we obtained were taken on February 10, 2003. There were images available in two filters, F606W and F814W, which are approximately equal to the broad band V and I filters, respectively. Only one image of good quality was available in each filter, so science quality cosmic ray rejection was not available, but an attempt was made to remove the worst of the cosmic rays by combining the two filters to make the image above. It clearly shows the small irregular host galaxy of the GRB with no obvious point source remaining, as well as the bright comparison USNO catalog star used for photometric calibration. The small companion to the host galaxy at 1.3" to the Northeast, reported by Garnavich [7], is also resolved in these images. It is comparatively more blue than the host, only $\sim$0.3 magnitudes dimmer than the host in the F606W filter, and then progressively fainter in the longer wavebands.

IRAF aperture photometry was performed on each of the two ACS images. A small 5 pixel (0.25") radius aperture was used to avoid contamination by cosmic rays, which unfortunately led to not all of the galaxy being contained in the aperture. Our photometry was calibrated against the USNO-B1 catalog star 14 arcseconds to the west of the host galaxy (B=20.640, R=19.450, I=19.690). The errors in moving between the different filters have not been accurately calibrated, and so the magnitudes reported below should be taken as preliminary. We find a magnitude of $\sim$26 for the central region of the host galaxy in both filters, F606W and F814W. This gives a surface brightness of $\sim$24 magnitudes per square arcsecond.

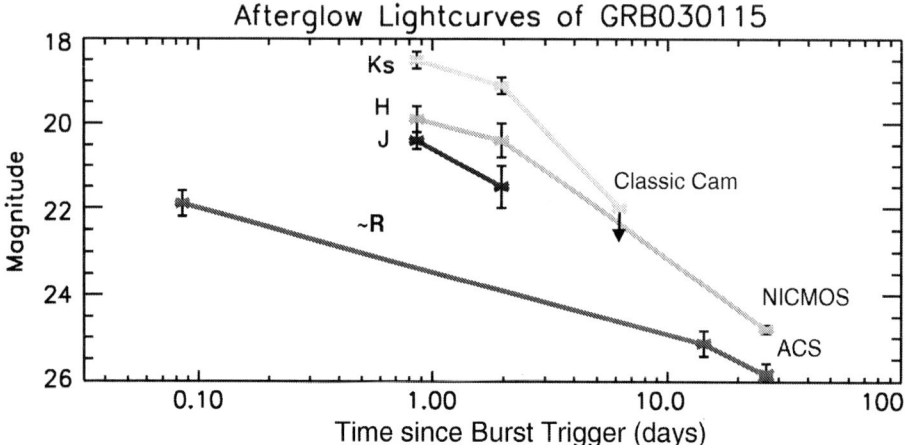

FIGURE 3. An updated lightcurve for GRB 030115, plotted using data from GCNs referenced below and including the new points from Classic Cam and HST.

CONCLUSIONS

We can place a new upper limit in the afterglow of GRB 030115 of Ks > 22 on January 23, 2003. This upper limit is too faint to be consistent with the afterglow decay of ∼0.7 between the two earlier Ks measurements reported by Kato et al. [8], which implies a steepening of the light curve at some time between ∼2 and 6 days after the burst. We also place the magnitude of the afterglow plus host galaxy in F160W (∼H band) at 24.8 at 25 days after the burst. Optically, the host galaxy of GRB 030115 is seen to be a small, irregular galaxy, perhaps interacting with its companion to the Northeast. A slight count excess at the position of the GRB in the Digital Sky Survey (DSS) image of the region was reported by Masetti et al. [6], but this is most likely noise, given the faintness of the host galaxy in the ACS images.

REFERENCES

1. Kawai, N., et al., GCNC 1816 (2003).
2. Atteia, J.L., et al., GCNC 1810 (2003).
3. Castro-Tirado, A., et al., GCNC 1807 (2003).
4. Masetti, N., et al., GCNC 1811 (2003).
5. Levan, A., et al., GCNC 1818 (2003).
6. Masetti, N., et al., GCNC 1823 (2003).
7. Garnavich, P., GCNC 1848 (2003).
8. Kato, D., et al., GCNC 1830 (2003).
9. Kato, D., et al., GCNC 1825 (2003).
10. Šimon, V., et al., A&A **377**, 450-461 (2001).

Observations of Optical Afterglows of Gamma-Ray Bursts from Loiano

C. Bartolini*, A. Guarnieri*, A. Piccioni*, G. Pizzichini[†] and P. Ferrero[†]

*Astronomy Department, University of Bologna, via Ranzani 1, 40127, Bologna, Italy
[†]IASF/CNR, Sezione di Bologna, via Gobetti 101, 40129 Bologna, Italy

Abstract. We report on our program for multiband photometry observations of Optical Transients of Gamma-Ray Bursts at the 152 cm telescope in Loiano, 40 Km south of Bologna, Italy. Observations in the four semesters starting on July 1st, 2001, were performed for 13 events and allowed positive detections for six of them.

THE TELESCOPE

The 152 cm Cassini telescope of the Bologna Astronomical Observatory is located in Loiano, circa 40 Km. south of Bologna, Italy. The location is Lat. 44° 15' 30" N Long 11° 20' 12" E, altitude 785 m. a.s.l. Additional information can be found at: http://www.bo.astro.it/loiano/index.htm. Observations are limited to declinations higher than -10 degrees. The telescope, equipped with an EEV CCD and Johnson-Cousins Filters, has been used for Target of Opportunity observations of Optical Afterglows of Gamma-Ray Bursts. We report on results for the four semesters starting on July 1st, 2001.

RESULTS

In the mentioned time interval, from mid-2001 to mid-2003, it was possible to observe 13 Gamma-Ray Burst locations for which either the error box was small enough for the field of view of the telescope or the OT had already been detected. Observations are not always possible, either, of course, because of bad weather, or because the telescope is scheduled for coordinated campaigns with other observatories which cannot be cancelled, or because the instrument at the focus of the telescope at that time is incompatible with contemporaneous mounting of the CCD imager. Alerts for new observations were obtained from the GRB Coordinates Network(GCN) at: http://gcn.gsfc.nasa.gov/gcn/gcn_main.html. We had negative detections for the following GRBs: 020317, 020812, 020819, 030217, 030227, 030324 and 030416. Positive detections were obtained for GRBs 020813, 021004, 030226, 030329 and 030418.

For GRB021004 (Figure 1) the observations in Loiano revealed that a new "plateau" in the afterglow was present (GCN1603). We show in figures 2, 3 and 4 (right panel) our

TABLE 1. GCNs by our group, or which include our observations

	GCN			
GRB 020317	1287			
GRB 020812	1486			
GRB 020819	1509	1511		
GRB 021004	1603			
GRB 030217	1873			
GRB 030226	1892	1940		
GRB 030324	1963			
GRB 030328	2008			
GRB 030324	2003	2030	2136	2228*

* correction to GCN 2136

detections for GRBs 030226, 030328 and 030418. Figure 4 (left panel) shows our limit to the magnitude for the OT of GRB030416, which is lower, in magnitude, than those of other observers, but partially covers a time gap between them.

A log of our observations can be retrieved by sftp using hostname: ermione.bo.astro.it, username: publicGRB, password: GRB_bo.

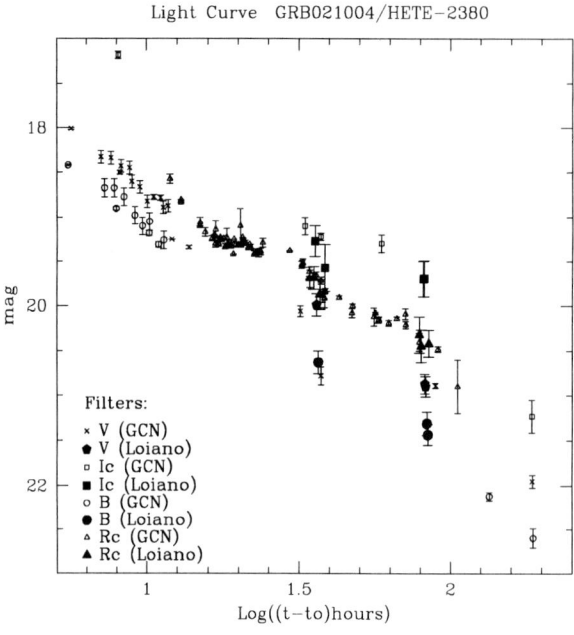

FIGURE 1. Light curve for the OT of GRB021004: data from Loiano and GCN

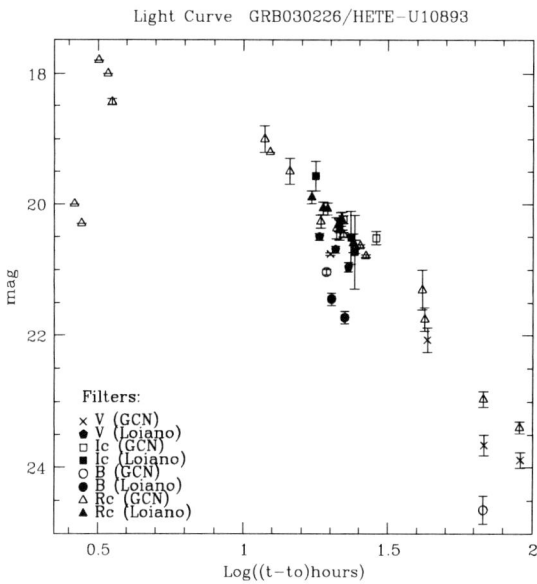

FIGURE 2. Light curve for GRB021004, data from Loiano and GCNs.

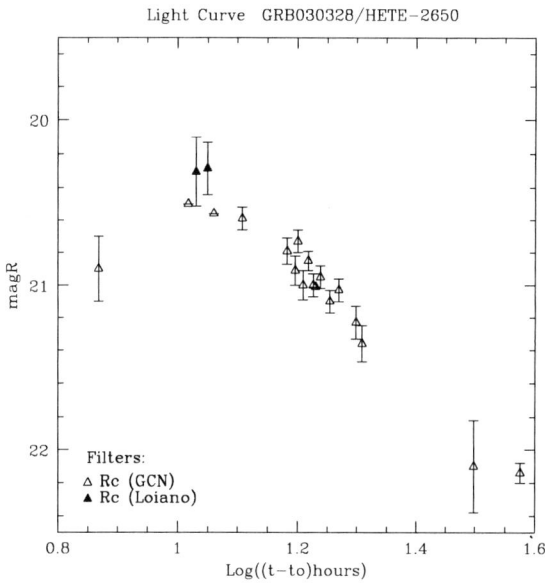

FIGURE 3. Light curve for GRB030226, data from Loiano and GCNs.

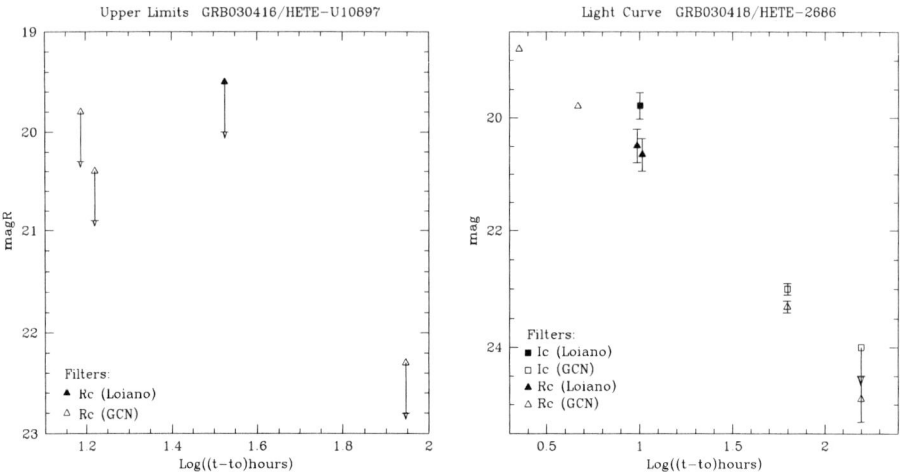

FIGURE 4. GRB0030416: upper limits; GRB030418: lightcurve, data from Loiano and GCNs.

CONCLUSIONS

Our data are an example of what can be achieved by ToO observations at an optical, nonrobotic telescope of this size, 152 cm, at a moderate altitude. We can expect to be able to observe reasonably soon after the dissemination of a small enough error box, ($\sim$15'), possibly discover the OT, contribute to continuous coverage of the OT decay and confirm observations by other telescopes. We make an effort, when possible, to use several Johnson-Cousin filters. OTs have been found to have colors which are fairly typical and independent of the OT magnitude, at least during the first ten days after the GRB [1] a property which we hope to test for as many OTs as possible.

ACKNOWLEDGMENTS

We are indebted to the Bologna Observatory for the use of the Cassini telescope. We are also very much indebted to the night observers at the telescope in Loiano, Stefano Bernabei, Ivan Bruni, Antonio De Blasi and Roberto Gualandi for their invaluable help in performing the observations together with us and to Scott Barthelmy for establishing and maintaining the GRB Coordinates Network. Dr. Nicola Masetti was for some months involved in this collaboration.

REFERENCES

1. Simon, V., Hudec, R., Pizzichini, G. & Masetti, N., Astron. & Astrophys. **377**, 450 (2001).

GRB Afterglows in the Deep Newtonian Phase

Y. F. Huang*, K. S. Cheng†, Z. G. Dai* and T. Lu**

Department of Astronomy, Nanjing University, Nanjing 210093, China
†*Department of Physics, the University of Hong Kong, Hong Kong, China*
**Purple Mountain Observatory, Chinese Academy of Sciences, Nanjing 210008, China*

Abstract. In many GRBs, afterglows have been observed for months or even years. It deserves noting that at such late stages, the remnants should have entered the deep Newtonian phase, during which the majority of shock-accelerated electrons will no longer be highly relativistic. However, a small portion of electrons are still ultra-relativistic and capable of emitting synchrotron radiation. Under the assumption that the electrons obey a power-law distribution according to their kinetic energy (not simply the Lorentz factor), we calculate optical afterglows from both isotropic fireballs and beamed ejecta, paying special attention to the late stages. In the beamed cases, it is found that the light curves are universally characterized by a flattening during the deep Newtonian phase. Implication of our results on orphan afterglows is also addressed.

IMPORTANCE OF NEWTONIAN PHASE

GRBs have been recognized as the most relativistic phenomena in the Universe. In 1997, Wijers et al. once discussed GRB afterglows of the non-relativistic phase [1]. However, for quite a long period, many authors were obviously beclouded by the energetics of GRBs and emission in the non-relativistic phase was generally omitted. In 1998, Huang et al. stressed the importance of the Newtonian phase for the first time [2]. In fact, the Lorentz factor of GRB blastwave evolves as

$$\gamma \approx (200 - 400) E_{51}^{1/8} n_0^{-1/8} t_s^{-3/8}, \quad (1)$$

in the ultra-relativistic phase. It is clear that the shock will enter the trans-relativistic phase within several months, and will become non-relativistic soon after that. Fig. 1a illustrates the condition clearly. Today, this point has been realized by more and more authors [3, 4, 5, 6, 7, 8, 9, 10].

MODEL

A refined generic dynamical model has been proposed by Huang et al. [12], which is mainly characterized by

$$\frac{d\gamma}{dm} = -\frac{\gamma^2 - 1}{M_{\text{ej}} + \varepsilon m + 2(1 - \varepsilon)\gamma m}. \quad (2)$$

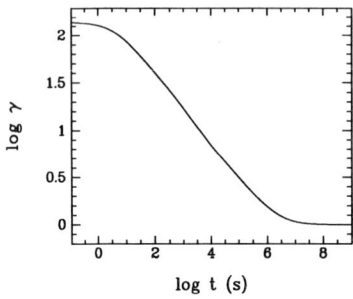
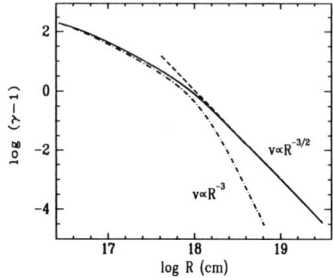

FIGURE 1. (a) Left panel, evolution of an adiabatic fireball ($E_0 = 10^{51}$ ergs, and $n = 1$ cm^{-3}) [2], which becomes non-relativistic in a few months; (b) Right panel, Eq. 2 (solid line) is correct in both the relativistic and the Newtonian phases [11].

Fig. 1b shows clearly that this equation is applicable in both the ultra-relativistic phase and the Newtonian phase. For a realistic description of the overall dynamical evolution of isotropic fireballs and collimated jets, we refer to Huang et al. [12, 13, 14].

GRB afterglows mainly come from synchrotron emission of shock-accelerated electrons. In the ultra-relativistic case, these electrons are generally assumed to distribute as $dN'_e/d\gamma_e \propto \gamma_e^{-p}$, with $\gamma_{e,\min} \sim \xi_e(\gamma-1)m_p/m_e$. However, we noticed that $\gamma_{e,\min}$ will typically be less than 2.0 when $t \geq$ a few months. This means most electrons will no longer be ultra-relativistic in the deep Newtonian phase [15].

We have suggested that the correct distribution function that is also applicable in the deep Newtonian phase should be [15],

$$\frac{dN'_e}{d\gamma_e} \propto (\gamma_e - 1)^{-p}, \quad (\gamma_{e,\min} \leq \gamma_e \leq \gamma_{e,\max}). \tag{3}$$

In the deep Newtonian phase, most electrons are now non-relativistic and their cyclotron radiation cannot be observed in the optical bands. But there are still many relativistic electrons capable of emitting synchrotron radiation. With the help of Eq. (3), optical afterglows can be calculated conveniently by integrating synchrotron emission from those electrons with Lorentz factors above a critical value ($\gamma_{e,\text{syn}}$) [15].

NUMERICAL RESULTS

We present our numerical results in Fig 2. Note that the light curves in Fig. 2a steepen slightly in the deep Newtonian phase. It is consistent with the analytical solution of,

$$S_R \propto \begin{cases} t^{(3-3p)/4}, & (\gamma \gg 1), \\ t^{(21-15p)/10}, & (\beta \ll 1). \end{cases} \tag{4}$$

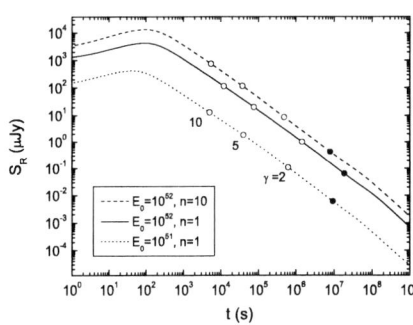 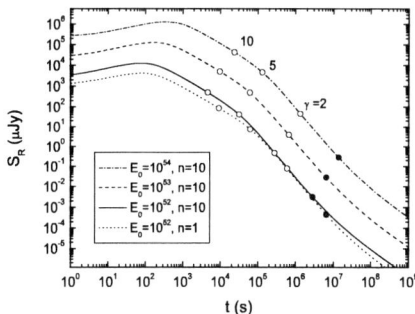

FIGURE 2. R-band optical afterglows from isotropic fireballs (**a**, left panel) and conical jets (**b**, right panel) [15]. The black dot on each light curve indicates the moment when $\gamma_{e,\min} = \gamma_{e,\text{syn}} \equiv 5$, and open circles mark the time when the bulk Lorentz factor $\gamma = 2$, 5 and 10 respectively.

The light curve in Fig. 2b flattens in the deep Newtonian phase, which is also consistent with the analytical solution [3],

$$S_R \propto \begin{cases} t^{(3-3p)/4}, & (\gamma > 1/\theta_0), \\ t^{-p}, & (\gamma < 1/\theta_0 \text{ and } \beta \sim 1), \\ t^{(21-15p)/10}, & (\beta \ll 1). \end{cases} \quad (5)$$

IMPLICATIONS ON ORPHAN AFTERGLOWS

Orphan afterglows are regarded as a useful tool for measuring the beaming angle of GRBs [16, 17, 18, 19, 20, 21, 22, 23, 24, 25, 26]. However, Huang et al. pointed out that there may exist large numbers of failed GRBs [27], i.e., fireballs with initial Lorentz factors $1 \ll \gamma_0 \ll 100$, which fail to produce GRBs but are likely to give birth to X-ray flashes. The simple discovery of orphan afterglows then does not necessarily mean that GRBs are highly collimated.

To judge whether an orphan afterglow comes from a failed GRB or a jetted but off-axis GRB, a $\log S_R - \log t$ light curve will be helpful. However, such a log-log light curve is usually not available for orphan afterglows, since the trigger time is unknown. Fig. 3 illustrates the effect of the uncertainty of the trigger time on the light curve.

To overcome the problem, Huang et al. suggested that the most important thing is to monitor the orphan for a relatively long period [27]. Obviously, the calculation of afterglows in the deep Newtonian phase is necessary in the studies of orphan afterglows.

This research was supported by the National Natural Science Foundation of China (10003001, 10221001 and 10233010), the FANEDD (Project No: 200125), the National 973 Project, and a RGC grant of Hong Kong SAR.

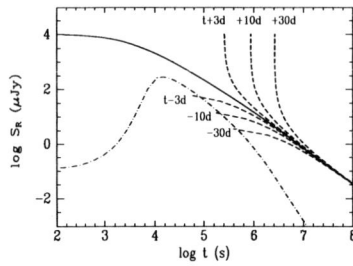 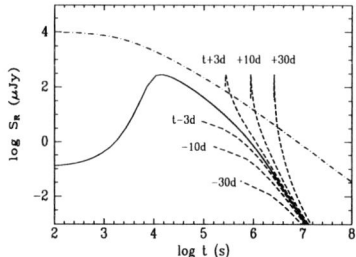

FIGURE 3. Direct comparison of the two kinds of orphan afterglows [27]. In the left panel, a failed GRB orphan is shifted by $t \pm 3$ d, $t \pm 10$ d, and $t \pm 30$ d to show the effect of the uncertainty of trigger time. In the right panel, a jetted GRB orphan is shifted similarly.

REFERENCES

1. Wijers, R., Rees, M. J., and Mészáros, P., *MNRAS*, **288**, L51 (1997).
2. Huang, Y. F., Dai, Z. G., and Lu, T., *A&A*, **336**, L69 (1998).
3. Livio, M., and Waxman, E., *ApJ*, **538**, 187 (2000).
4. Frail, D., Waxman, E., and Kulkarni, S. R., *ApJ*, **537**, 191 (2000).
5. Dermer, C. D., Böttcher, M., and Chiang, J., *ApJ*, **537**, 255 (2000).
6. Dermer, C. D., and Humi, M., *ApJ*, **556**, 479 (2001).
7. Piro, L., and et al., *ApJ*, **558**, 442 (2001).
8. in't Zand, J. J. M., and et al., *ApJ*, **559**, 710 (2001).
9. Panaitescu, A., and Kumar, P., *MNRAS, submitted, astro-ph/0308273* (2003).
10. Zhang, B., and Mészáros, P., *Int. J. Mod. Phys. A, in press (astro-ph/0311321)* (2003).
11. Huang, Y. F., *astro-ph/0008177* (2000).
12. Huang, Y. F., Dai, Z. G., and Lu, T., *MNRAS*, **309**, 513 (1999).
13. Huang, Y. F., Gou, L. J., Dai, Z. G., and Lu, T., *ApJ*, **543**, 90 (2000).
14. Huang, Y. F., Dai, Z. G., and Lu, T., *MNRAS*, **316**, 943 (2000).
15. Huang, Y. F., and Cheng, K. S., *MNRAS*, **341**, 263 (2003).
16. Rhoads, J. E., *ApJ*, **487**, L1 (1997).
17. Perna, R., and Loeb, A., *ApJ*, **509**, L85 (1998).
18. Mészáros, P., Rees, M. J., and Wijers, R., *New Astron.*, **4**, 303 (1999).
19. Grindlay, J. E., *ApJ*, **510**, 710 (1999).
20. Lamb, D. Q., *Phys. Report*, **333**, 505 (2000).
21. Totani, T., and Panaitescu, A., *ApJ*, **576**, 120 (2002).
22. Levinson, A., and et al., *ApJ*, **576**, 923 (2002).
23. Nakar, E., Piran, T., and Granot, J., *ApJ*, **579**, 699 (2002).
24. Granot, J., Panaitescu, A., Kumar, P., and Woosley, S. E., *ApJ*, **570**, L61 (2002).
25. Rhoads, J. E., *ApJ*, **591**, 1097 (2003).
26. Yamazaki, R., Ioka, K., and Nakamura, T., *ApJ*, **593**, 941 (2003).
27. Huang, Y. F., Dai, Z. G., and Lu, T., *MNRAS*, **332**, 735 (2002).

Optical Orphan Afterglows: Observational Aspects

R. Hudec

Astronomical Institute, Academy of Sciences of the Czech Republic, CZ–251 65 Ondřejov, Czech Republic

Abstract. GRBs may be detected and investigated not only by its high-energy radiation, but also by its optical emission. We discuss the recent independent possibilities to detect, to identify, and to investigate orphan optical afterglows and optical transients of GRBs.

INTRODUCTION

An orphan afterglow (OA) is an (yet hypothetical) optical afterglow without detectable Gamma ray emission (due to different beaming). These events are predicted by theory but not yet confirmed by observation. The rate of Orphan Optical Afterglows (OOAs) may exceed the GRB rate; hence the improved gamma-ray burst (GRB) statistics is expected with numerous consequences such as improved statistics of host galaxies, redshift distribution, cosmological conclusions, etc. The searches for OOAs usually assume independent optical searches. So far, no optical orphan afterglow has been confirmed. However, this does not necessarily means that such triggers do not exist, since their detection is difficult and heavily affected by high background rate. The recent detection of optical afterglows and optical transients (OTs) of gamma ray bursts allows considering independent optical ground-based detection of these phenomena. Taking into account the range of typical magnitudes of OAs detected so far, one can conclude that the optical surveys achieving limiting magnitudes better than 19 to 23 for stars and/or 10 for 1 min exposures may detect OAs and OTs of GRBs. This is feasible with recent telescopes and observing techniques, so this opens the possibility of independent optical searches. However, due to low expected rate of the phenomena, these searches must be of large field of view, that is, CCD surveys and/or deep patrol plates are suitable.

INDEPENDENT OPTICAL SEARCHES

The possibility to record the GRBs by their optical emissions is challenging. The optical surveys may provide a larger sample of detected triggers (due to different beaming) and a much better localization accuracy (1 arcmin or better) than provided by gamma ray satellite detectors. The larger sample of OAs and their host galaxies may be crucial for understanding the nature of GRBs as well as for related cosmological implications. Many statistical studies are still quite affected by a relatively small size of the sample.

The actual rate of OAs can also place additional constraints on the afterglow appearance fraction and, perhaps, the initial beaming angle of GRB sources. The UV flashes predicted by some theories such as [1] could be detected and studied. The corresponding delays regarding GRBs could serve to study the nature of the sources. This can be addressed only by surveys, not by follow-up devices since the flashes may precede the GRBs ([1], [2]). The optical flashes preceding GRB expected by theory ([2]) can be also detected and analyzed. If compared with satellite projects, the optical surveys are surely much more cost-effective. It seems to be feasible that both flaring (OTs) as well as fading (OAs) optical emission related to GRBs may be detected by optical sky patrols. Although the true rate of these triggers remains unknown, it is sure that their rate is substantially (by several orders of magnitude) below the background rate; hence the good knowledge of all background triggers must be available as well as a reliable technique for their classification and climination. Hence, it is clear that the main problem of independent searches for OAs and OTs is the background.

THE BACKGROUND PROBLEM

As background we consider the events not related to GRBs but showing similar transient behaviour. The most important background sources are SNe, AGN flares/brightening, stellar flares, variable stars, luminous extragalactic blue variable stars (LBVs), optical transients of unknown nature and origin, as well as various types of nonastrophysical triggers ([3]).

Supernovae. The supernovae, especially those of Ia type, may represent an important source of confusion due to their occurrence rates, rise and decay timescales, and magnitudes. The expected rate of occurrence for faint events (down to mag R23) is roughly 2 deg^{-2} or 0.0015 arcmin^{-2} ([4]; [5]; [6]). But, at least some SNe are related to GRBs (e.g. SN 1998bw and GRB 980425).

AGN flares. The flares and brightening of AGN may achieve quite large amplitudes. The spread of particular AGN flare amplitudes is large (0.1 ... 6.7 mag). Also, there is growing evidence for large amplitude (more than 10 mag) flares on AGN ([7], [8]). Before, the largest known QSO amplitude for 3C 279 was more than 6.7 mag, but with under sampled light curve ([9]) . The estimated AGN/QSOs surface densities are (10-37) deg^{-2} with lim mag B/V 20.5 ([10]; [11]) and 111 deg^{-2} with limiting magnitude B 22.6 ([12]). Most of the QSOs are variable: the typical QSO variability in a sample of 149 optically selected QSOs is 0.26-0.33 in B and 0.22-0.30 in R ([13]). There are $\sim$100 variable (by more than 0.1 mag) QSOs deg^{-2}, that is, 1 variable QSO in 6×6 arcmin^2 brighter than B 22.6 ([12]). 97% QSOs below B 22.5 are variable ([14]). The QSO variability seems to increase with decreasing luminosity ([13]).

Stellar flares. Also the flares on stars may achieve very large amplitudes. There is growing evidence for such large amplitude (5 mag and more) stellar flares. Flare stars detected by extended plate surveys can serve as an example: the two pairs of OTs-flare stars detected by [15] and [8] which seem to represent large amplitude flares (5-9 mag) from otherwise typical dMe flare stars. The exact statistics of such events is however unknown but the recently operated and developed wide field CCD sky monitors may be able to address this question soon.

Variable stars. Also other types of variable stars may exhibit light behaviour similar to OAs and OTs. The Y Dra Mira type variable star located inside the error box of the GRB 910709 detected and positioned by COMPTEL can serve as an example. The star exhibits light variations between 6 and 15 mag, with gradual light decrease after maximum (period 322 d). Although the physical relation would be difficult to explain, the object shows a maximum almost exactly coincident with the GRB date. The estimated rates are available only for variable stars brighter than 20 mag: 80 deg^{-2} for /bII/ < 20 deg and 4 deg^{-2} for /bII/ > 40 deg ([16], however the discovery probability is ~ 0.1 (blinkmicroscope). No statistics for variable stars below 20 mag is available (no systematic surveys). Variable stars are observed more commonly in decline than in increase since the declines are typically more slowly, for example, in delta Cep stars, U Gem stars, flare stars, novae etc. Again, the recently operated and developed sensitive sky CCD surveys may soon be able to address this question and provide a reliable statistics of faint (below mag 20) variable stars.

Luminous Blue Variable Stars. The Luminous Blue Variable Stars (LBVs) are both galactic and extragalactic objects, poorly physically understood, with sporadic violent flares, which can occasionally mimic the SNe and the OAs behaviour. The LBVs in NGC 3432 (SN2000ch) can serve as an example ([17]). In some extreme cases, they can show analogous behaviour as OAs of GRBs on the time scale of few days after the brightness maximum.

OTs of unknown nature and origin. There are real OTs of unknown nature but of astrophysical origin detected both on emulsions and CCDs. Few examples (real CCD detections) are listed below: **(i) OT 970215:** real CCD detection, V 13 mag, nothing down 20 mag on the position, amplitude more than 7 mag ([18]), **(ii) OT 950806:** real object: detected on 20 CCD frames, peak magnitude I 7.5, amplitude more than 10 mag, nothing down mag 21 48 hrs after detection ([19]), **(iii) OT triggers found by SNe searches:** mystery events found at a rate of about 0.15 deg^{-2}/per time scale (between 10 min and 3 days) limiting mag R 23.5 ([20]). In three cases, no host galaxy seen down mag R 24, in one case, host galaxy clearly visible. Two events were at a low galactic latitude, and hence can be flare stars. Note that SN searches reject events with timescales less than 3 min (as cosmic rays) hence can detect OAs but not all OTs. These searches are very limited so far (6 SN runs done, 2-6 sq. deg. per run, [20]). This means that the OAs may be among detected SNe.

Recently, we have some powerful tools how to distinguish OAs and SNe (and other background events): **(i) Light curve**, **(ii) Peak luminosity** (only for objects with known redshift), and **(iii) Color information**: most of OAs have R–I = 0.46, V–R= 0.40, B–V=0.47 ([21]).

CONCLUSIONS

The GRBs can be detected not only from spacecrafts by their high-energy radiation, but also from the ground by their optical afterglow and optical transient emission. We have suitable tools for that. Numerous ground-based experiments, such as CCD-based devices and telescopes, digitized plate surveys and digitized archival plates, SNe and microlensing surveys, etc., can be used for these purposes. The independent optical searches for GRBs are however heavily affected by the presence of background. Even the recent high localization accuracy of GRBs cannot exclude positional coinciding OTs as just random unrelated coincidences. The exact background rate of unrelated triggers such as variable stars and extragalactic objects is unknown but may be rather high. The expected rate of faint (below B 23) variable AGNs is unknown. The rate of faint (below mag 20) variable stars is unknown. For brighter variable stars, the expected rate is 0.03 - 0.5 (depending on galactic latitude) in a 5×5 arcmin box (lim mag 20). The expected rate of faint (lim mag R 23) SNe is 0.4 in a 5×5 arcmin box. The estimated total number of optically variable sources of astrophysical origin (SNe+AGN+VS) is 1–2 in a 5×5 arcmin box, lim mag 23. It is obvious that we need a better statistics of faint variable optical objects and that the rate of background triggers exceeds the expected rate of OOAs at least by several orders.

ACKNOWLEDGMENTS

This study was supported by the grant A3003206 provided by the Grant Agency of the Academy of Sciences of the Czech Republic.

REFERENCES

1. Protheroe, R., and Bednarek, W., *astro-ph/9904279* (1999).
2. Paczynski, B., *astro-ph/0108522* (2001).
3. Hudec, R., *Astroph. Letters and Comm.*, **28**, 359 (1993).
4. Evans, R., and et al., *ApJ*, **345**, 752 (1989).
5. Pain, R., *ApJ*, **473**, 356 (1996).
6. Brainerd, J., *AIP Conf. Proc.*, **428**, 545 (1998).
7. Hudec, R., and et al., ",", in *Blazar continuum variability*, edited by H. Miller, J. Webb, and J. Noble, ASP Conf. Series 110, ASP, New York, 1996, p. 129.
8. Hudec, R., Luginbuhl, C., Vrba, F., and et al., ",", in *The Transparent Universe, Proceedings of the 2nd INTEGRAL Workshop held 16-20 September 1996*, edited by C. Winkler, T. Courvoisier, and P. Durouchoux, ESA, ESA, Paris, 1997, p. 481.
9. Eachus, L. J., and Liller, W., *ApJ*, **200**, L61 (1975).
10. Iovino, A., and et al., *A&AS*, **119**, 165 (1996).
11. Hartwick, F., and Scade, D., *ARA&A*, **28**, 437 (1990).
12. Trevese, D., and et al., *AJ*, **98**, 1 (1989).
13. Cristiani, S., and et al., *A&A*, **321**, 123 (1997).
14. Trevese, D., and Kron, R., ",", in *Multi-Wavelength Continuum Emission of AGNs*, edited by T. Courvoisier and A. Blecha, IAU, IAU, 1994.
15. Greiner, J., and Motch, C., *A&A*, **294**, 177 (1995).
16. Hudec, R., and Wenzel, W., *A&AS*, **120**, 707 (1996).
17. Wagner, R., and et al., *ApJ* (2003).
18. Vidal-Saiz, J., and et al., *IBVS Budapest*, **4324**, 1 (1996).
19. Toth, I., and et al., *A&A*, **315**, 153 (1996).
20. Schmidt, B., *private communication* (1999).
21. Šimon, V., Hudec, R., Pizzichini, G., and Masetti, N., *A&A*, **377**, 450 (2001).

The Optical Afterglow of GRB 030226[1]

S. Klose*, J. Greiner†, A. Zeh*, A. Rau†, A. A. Henden**, D. H. Hartmann‡, N. Masetti§, A. J. Castro-Tirado¶, J. Hjorth‖, E. Pian††, N. R. Tanvir‡‡, R. A. M. J. Wijers§§ and E. van den Heuvel§§

*Thüringer Landessternwarte Tautenburg, 07778 Tautenburg, Germany
†Max-Planck-Institut für extraterrestrische Physik, 85741 Garching, Germany
**USNO/USRA, U. S. Naval Observatory, Flagstaff station, Flagstaff, AZ 86002
‡Clemson University, Department of Physics and Astronomy, Clemson, SC 29634
§Istituto di Astrofisica Spaziale e Fisica Cosmica, CNR, Sez. di Bologna, Via Gobetti 101, 40129 Bologna, Italy
¶Instituto de Astrofísica de Andalucía (IAA-CSIC), P.O. Box 03004, 18080 Granada, Spain
‖Astronomical Observatory, University of Copenhagen, Juliane Maries Vej 30, 2100 Copenhagen, Denmark
††INAF, Osservatorio Astronomico di Trieste, Via Tiepolo 11, 34131 Trieste, Italy
‡‡Department of Physical Sciences, Univ. of Hertfordshire, College Lane, Hatfield Herts, AL10 9AB, UK
§§University of Amsterdam, Kruislaan 403, 1098 SJ Amsterdam, The Netherlands

Abstract. We report on optical and near-infrared follow-up observations of the afterglow of GRB 030226 performed with the telescopes at ESO La Silla and Paranal. Our observations started 0.2 days after the burst when the afterglow was at a magnitude of $R \approx 19$. One week later the magnitude of the afterglow had fallen to $R=25$, and at two weeks after the burst it could no longer be detected ($R > 26$). VLT blue-band spectra show two absorption line systems at redshifts $z = 1.962 \pm 0.001$ and 1.986 ± 0.001, indicating a metal-rich environment of the burster, and placing the burster very likely at $z = 1.986 \pm 0.001$.

INTRODUCTION

GRB 030226 was discovered by the *HETE-2* satellite on 2003 February 26.15731 UT (3:46 UT). Its duration was longer than 100 seconds, its peak flux and fluence in the 30-400 keV band were $\sim 1.2 \times 10^{-7}$ erg cm^{-2} s^{-1} and 5.7×10^{-6} erg cm^{-2}, respectively [1]. The optical counterpart was detected ~ 2.5 hours later [2] at a magnitude $R \sim 18.5$ [3] at coordinates (J2000) RA = 11:33:04.9; Dec = +25:53:55.6 [4], well within the *HETE-2* Wide Field X-ray monitor (WXM) and Soft X-ray Camera (SXC) error boxes.

---

[1] on behalf of the GRACE collaboration

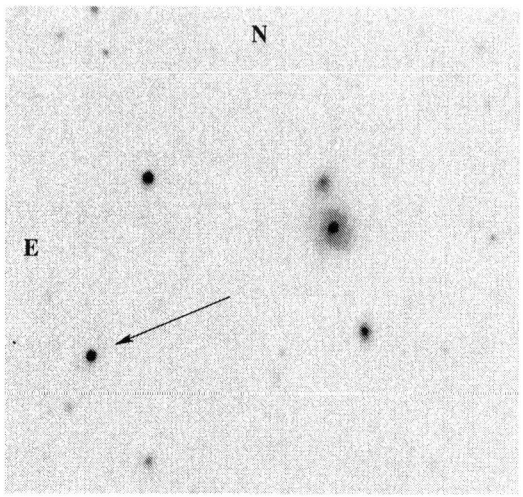

FIGURE 1. *J*-band image of the afterglow of GRB 030226 obtained 5 hrs after the burst with VLT/ISAAC. The field size is approximately $1' \times 1'$. The optical transient is marked with an arrow.

OBSERVATIONS

Imaging in the near-infrared (NIR) bands started ~ 4.5 hrs after the burst using VLT/ISAAC at ESO Paranal (Fig. 1). Simultaneously, imaging in the optical was performed using the ESO NTT telescope at La Silla equipped with the multi-purpose instrument EMMI. Further data were obtained in the following nights with the NTT equipped with the SUperb Seeing Imager (SuSI2) and VLT/FORS1.

The early report of the discovery of the GRB afterglow by Fox et al. [2] allowed for a rapid spectroscopic observation of the optical transient [5]. These observations were performed 5 hrs after the burst with VLT Yepun/FORS2. A 900 s exposure using the 300V grism was obtained when the afterglow was at a magnitude of $B \approx 19.8$. A second spectroscopic run was performed one night later using VLT Antu/FORS1 equipped with the 600B grism.

RESULTS AND DISCUSSION

In Fig. 2 we display the results of the fits based on our ESO *R*-band data. The best fit using the equation suggested by [6] is $\alpha_1 = 0.70 \pm 0.25$, $\alpha_2 = 2.66 \pm 0.32$, $t_b = 1.04 \pm 0.12$ days, and $n = 1.2 \pm 1.1$ (χ^2/d.o.f.= 3.8), where the symbols have their usual meaning: α_1 is the pre-break slope, α_2 is the post-break slope, t_b is the break time and n characterizes the smoothness of the fit. Potential fluctuations in the light curve are apparent. Kulkarni et al. [7] also noticed an unusual behavior of the afterglow light curve between 2.3 and 3.3 days after the burst.

GRB 030226 joins the set of bursts with well-observed light curves which allow for

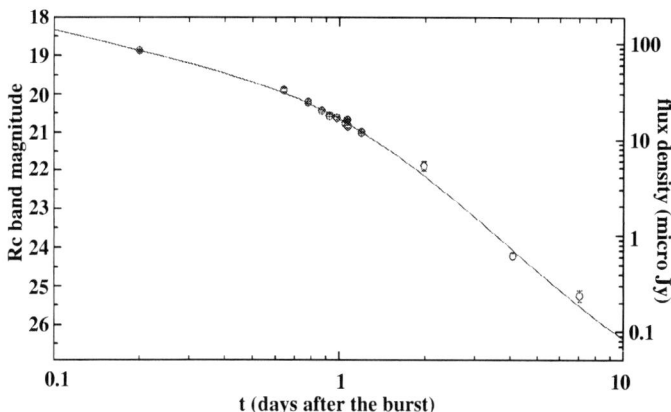

FIGURE 2. The R-band light curve of the afterglow of GRB 030226 based on our data obtained at the telescopes at ESO. Note the rapid fading of the afterglow. No host galaxy is apparent in our data down to $R > 26$.

a check of their color evolution and temporal behavior. Our multi-band data allowed us to search for color changes before, during, and after the break in the light curve. Before the break time, based on our multi-color data set on day 0.2, we obtain for the spectral slope β across the optical/NIR bands $\beta = 0.81 \pm 0.06$. Based on the observed spectral energy distribution of the GRB afterglow no evidence for dust in the GRB host galaxy was found. After the break time, on day 4.1, we obtain $\beta = 0.98 \pm 0.05$. In other words, there is no evidence for a change in the spectral slope of the afterglow in the optical/NIR bands before and after the break time. This is in agreement with the prediction of the simplest jet model, according to which the break should be achromatic if it is due the increasing relativistic opening angle when the fireball decelerates [8].

The host of GRB 030226 represents another impressive example of a high-z galaxy for which we obtained information about the metal content of its interstellar medium, although the potentially underlying host remained undetectable on our VLT images. The VLT spectra of the optical transient show several absorption lines, but no prominent emission lines. From the absorption lines two redshift systems at $z=1.962\pm0.001$ and 1.986 ± 0.001 can be determined (Fig. 3). We find systems of Fe II, Al II, C IV, O I, C II, Si IV, and Si II. Their rest-frame equivalent widths are comparable to those found in other GRB afterglows (e.g., [9, 10, 11]). We conclude that the interstellar medium close to the burster is possibly enriched with α-group elements. We tentatively identify this as being either a feature of the wind from a Wolf-Rayet star, which was the GRB progenitor, or an interstellar medium shaped by the nucleosynthesis of type II supernovae.

ACKNOWLEDGMENTS

This paper is based on observations collected at the European Southern Observatory, La Silla and Paranal, Chile (ESO Programme 70.D-0523, PI: Ed van den Heuvel).

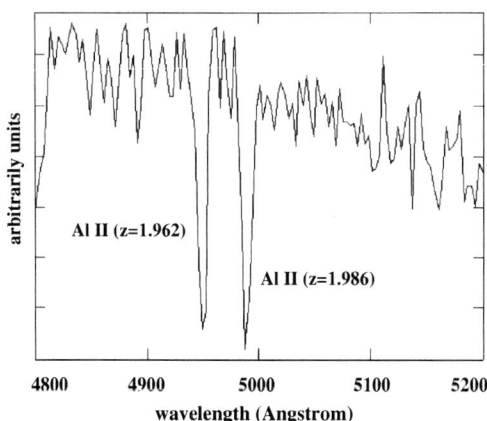

FIGURE 3. Zoom-in of the VLT/FORS2 spectrum of the GRB afterglow obtained 5 hrs after the burst. Shown here are the two pronounced aluminium absorption lines at redshifts z=1.962 and 1.986.

We are highly indebted to the ESO staff at La Silla and at Paranal for performing the observations.

REFERENCES

1. Suzuki, M. et al., GCN Circ. 1888 (2003).
2. Fox, D. W., Chen, H. W., & Price, P. A., GCN Circ. 1879 (2003).
3. Garnavich, P.M., von Braun K., & Stanek, K.Z., GCN Circ. 1885 (2003).
4. Price, P. A., Fox, D.W., & Chen, H.W., GCN Circ. 1880 (2003).
5. Greiner, J., Guenther, E., Klose, S., & Schwarz, R., GCN Circ. 1886 (2003).
6. Beuermann, K. et al., A&A **352**, L26 (2002).
7. Kulkarni, S. et al., GCN Circ. 1911 (2003).
8. Mészáros, P., & Rees, M., MNRAS **306**, L39 (1999).
9. Castro, S. et al., ApJ **586**, 128 (2003).
10. Masetti, N. et al., A&A **374**, 382 (2001).
11. Matheson, T. et al., ApJ **582**, L5 (2003).

Colors of Optical Afterglows of GRBs and Their Time Evolution

V. Šimon*, R. Hudec*, G. Pizzichini† and N. Masetti†

Astronomical Institute, Academy of Sciences of the Czech Republic, 251 65 Ondřejov, Czech Republic
†*IASF/CNR, Sezione di Bologna, via Gobetti 101, 40129 Bologna, Italy*

Abstract. The study of the color indices of 23 optical afterglows (OAs) of GRBs shows that the color variations during the decline of the OAs are very small during $t - T_0 < 10$ days. The colors in the observer frame concentrate at $(B-V)_0 = 0.45 \pm 0.15$, $(V-R)_0 = 0.41 \pm 0.09$, $(R-I)_0 = 0.46 \pm 0.13$, $(I-J)_0 = 0.88 \pm 0.15$, but are more scattered in $(U-B)_0$. These findings entirely confirm our previous ones [1], determined from a smaller ensemble of the OAs with a smaller range of the redshifts z. The strong concentration of the color indices of most OAs also suggests that the intrinsic reddening (i.e. inside their host galaxies) must be quite similar and relatively small for these events. The maximum reddening which can be allowed by the scatter of the colors of the individual OAs is about $E_{B-V} = 0.2$. This suggests that they are not deep inside the star-forming regions or, alternatively, that the dust is destroyed by the energetic initial flash.

INTRODUCTION

Color indices of the optical afterglows (OAs) of GRBs are a powerful, straightforward and innovative approach to the study of such events. This method has been applied to other phenomena in astrophysics for many decades but is only very rarely applied in the analysis of the OAs. The color indices enable us to resolve small variations of the profile of the spectra. The method of the color indices makes use of the photometric observations, which can be obtained by small and medium sized telescopes, and helps us to: *(a)* search for the common properties of the afterglows; *(b)* understand the related physical processes; *(c)* find out whether an optical event is related to a GRB even without available gamma-ray detection by using the color indices of the OAs; *(d)* search for the orphan afterglows; *(e)* search for the relations among colors, luminosities and decay rates of the OAs (if the redshift z is known); *(f)* constrain the properties of the local interstellar medium of GRBs.

COLLECTION AND ANALYSIS OF THE DATA

This analysis makes use of the data published in the GCN circulars [2], in J. Greiner's Web page [3], and in the journals. Because of space limitations, the reader is referred to these web pages for full bibliographic references on each OA. The photometry of the OAs often comprises unorganized observations. The light curves of the OAs in the individual passbands were therefore plotted and critically examined. They were usually

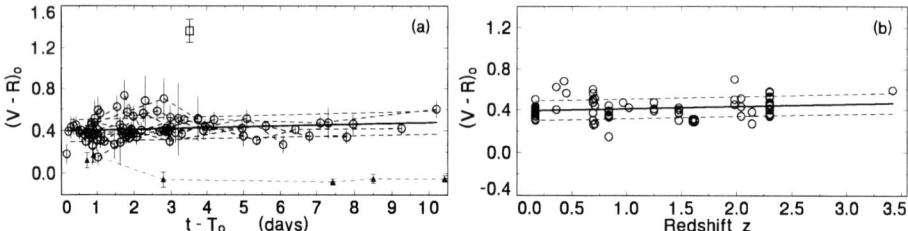

FIGURE 1. (a) Time evolution of the color index of an ensemble of the OAs. Color indices of a given OA are connected by the lines for convenience. Their error bars are also shown. The fit (solid straight line) with its standard deviation (dashed stright lines) represents the whole ensemble of 23 OAs (except for GRB000131 (square)). The brightness of the OAs falls by several magnitudes over this time interval. SN 1998bw (triangles) is not included in the fit. (b) Colors of all OAs with known redshift z plotted as a function of z (only OAs with $z < 3.5$ and $t - T_0 < 10.2$ days). The weight of each OA, used for the fit, was proportional to the number of available color indices. Notice also that the color of the OA of GRB030329 ($z = 0.17$) is in very good agreement with those of the events at higher z.

found to be free of complicated rapid changes. A meaningful interpolation between the neighbouring measurements was thus possible, at least within $t - T_0 < 10$ days, where T_0 refers to the time of the GRB. The typical standard deviations of the indices lie within $0.04 - 0.2$ mag. The indices were corrected for the Galactic reddening according to [4]. The light contribution of the host galaxies was quite small for $t - T_0 < 10$ days and could be neglected. More details can be found in [1].

The time evolution of the color index $(V - R)_0$ is shown in Fig. 1a. The course of $(B - V)_0$, $(R - I)_0$ and $(I - J)_0$ is very similar to $(V - R)_0$ but a scatter is observed in $(U - B)_0$. Fig. 1b shows that $(V - R)_0$ is largely independent on z. This is true also for $(B - V)_0$, $(R - I)_0$ and $(I - J)_0$ (except for $(U - B)_0$) as it was shown in [1], but now with more cases. The color-color diagrams in Fig. 2 include the representative reddening paths for $E_{B-V} = 0.5$. Because the observer in a given passband will detect radiation at progressively shorter wavelengths in the source rest frame with the increasing z, the reddening paths, appropriate for $U - B$, $B - V$, $V - R$ and $R - I$, are plotted [1].

RESULTS

We found that the typical color indices of the OAs in the observer frame inside the time interval $t - T_0 < 10$ days are $(B - V)_0 = 0.45 \pm 0.15$, $(V - R)_0 = 0.41 \pm 0.09$, $(R - I)_0 = 0.46 \pm 0.13$, $(I - J)_0 = 0.88 \pm 0.15$, independent on redshifts between $z = 0.17$ and 3.5. This implies a very smooth spectral shape, with no bumps or strong lines, between the observed B to I passbands (2000 − 5600 Å in the rest frame). There is often a break in the OA at $t - T_0 < 10$ days; our findings hold also beyond this break and imply that it is achromatic. The average $(R - I)_0$, $(V - R)_0$ and $(B - V)_0$ colors are consistent with a power-law spectral distribution with $\beta \sim 1$ (in accordance with the GRB afterglow 'fireball' emission model [5]). Color changes of the OAs are negligible although the brightness declines by several magnitudes. This implies that the shape of

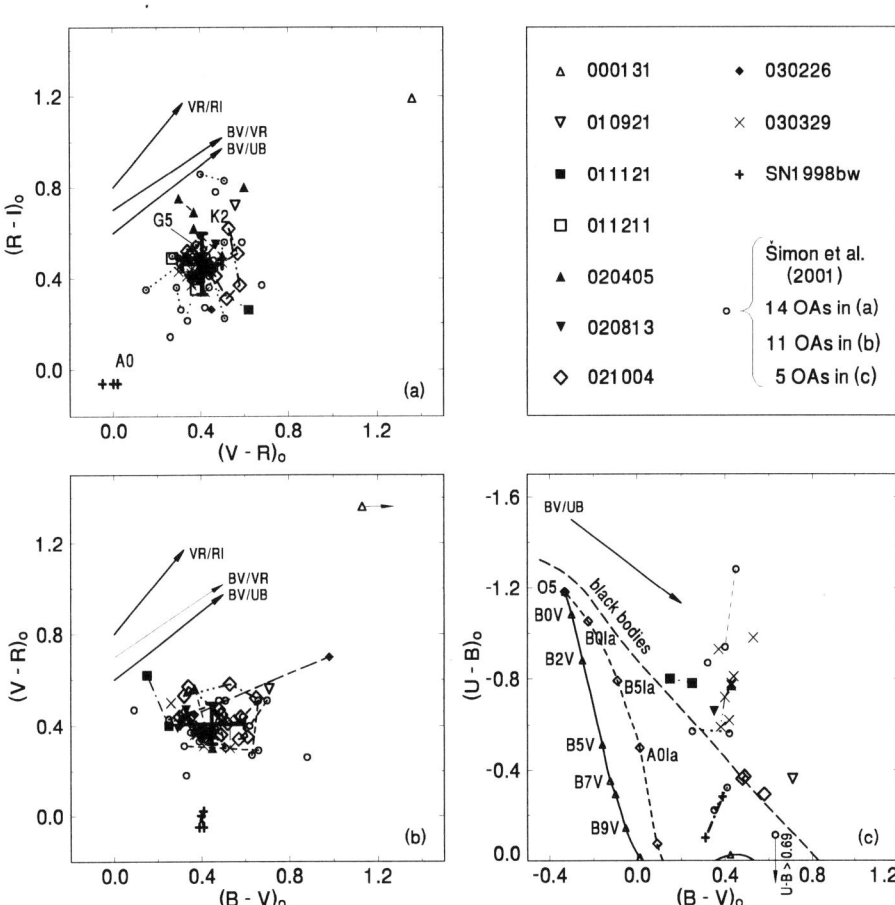

FIGURE 2. (a) $V - R$ vs. $R - I$ diagram of the OAs ($t - T_0 < 10.2$ days). (b) $B - V$ vs. $V - R$ diagram of the OAs. (c) $U - B$ vs. $B - V$ diagram of the OAs. The colors are corrected for the Galactic reddening. Multiple indices of the same OA are connected by the lines to denote the time evolution. The indices of the older OAs, determined by [1] (14 OAs in (a); 11 OAs in (b); 5 OAs in (c)), are plotted as the empty circles. Notice the strong concentration of most OAs in a small region of the color-color diagram in (a) and (b) and a noticeable scatter of the OAs in $U - B$ in (c). The mean colors (centroid) of the whole ensemble of the OAs (except for GRB000131 and SN 1998bw) are marked by the large cross. The representative reddening paths for $E_{B-V} = 0.5$ mag are also shown. Positions of the main-sequence stars are included for comparison.

the spectrum of the OA does not change significantly while the luminosity decreases by a large amount. The outlying position of GRB000131 can be explained because, due to its high redshift $z = 4.5$, the Lyman α spectral break lies in the optical region ($\lambda \approx 6700$ Å).

As we also showed before in [1], the spectra of most OAs are quite similar while their luminosity in a given $t - T_0$ differs: the spectral shape in the 'fireball' model [5] does not depend on the input energy, while the luminosity of the OA at a particular epoch does depend on it. GRBs and OA are not standard candles.

The strong concentration of most colors of the OAs in the color-color diagrams implies that the *intrinsic reddening (inside their host galaxies) must be quite similar for all of them and relatively small*. Otherwise, if the reddening were large it would be quite unlikely to obtain such similar values of absorption in all cases. The maximum reddening which can be allowed by the scatter of the colors of the individual OAs is about $E_{B-V} = 0.2$. There are two possibilities: *(a)* GRBs with detectable OA lie only on the Earth-watching side of a star-forming region; *(b)* density and dust abundance of the local interstellar medium is substantially reduced by energetic initial flash (models by [6]). We are, however, aware that in some cases the OAs appear to have steeper optical spectra with $\beta \sim 2$ or higher (and thus redder colors, see [1] for more). Nevertheless, the strong concentration of the color indices of the OAs in Fig. 2 allows one to infer that maybe there is not a smooth transition between the events considered in our analysis and these OAs with steep optical spectra.

The colors of the OA of GRB030329 are in good agreement with those of most other OAs for $t - T_0 < 10$ days and speak in favour of the dominant synchrotron component in the spectrum (this event is analyzed in detail in a companion paper [7]).

The evolution of the observed color indices also puts constraints on the models for the synchrotron radiation of the OAs. E.g. some theoretical light curves [8] suggest a reddening of the OA within the observed spectral region (near the frequency 10^{15} Hz) during $t - T_0 < 1$ day, which is not observed. Any real reddening must be quite small to be in accordance with the observations.

It can be seen that the $B-V$, $V-R$ and $R-I$ colors are a powerful tool to determine whether an event is related to a GRB even from the optical observations alone (for $z < 3.5$) and are a promising way to identify orphan afterglows (of course, provided that the orphan afterglows have the same colors as the OAs with the detected GRB trigger).

ACKNOWLEDGMENTS

This study was supported by the grant A3003206 provided by the Grant Agency of the Academy of Sciences of the Czech Republic, the project ESA PRODEX INTEGRAL 14527, and the CNR-AVČR collaborative project Investigation of GRBs (2000/2003).

REFERENCES

1. Šimon, V., Hudec, R., Pizzichini, G., and Masetti, N., *A&A*, **377**, 450 (2001).
2. Barthelmy, S. (2004), URL http://gcn.gsfc.nasa.gov/gcn/gcn3_archive.html.
3. Greiner, J. (2004), URL http://www.mpe.mpg.de/~jcg/grbgen.html.
4. Schlegel, D., Finkbeiner, D., and Davis, M., *ApJ*, **500**, 525 (1998).
5. Sari, R., Piran, T., and Narayan, R., *ApJ*, **497**, L17 (1998).
6. Waxman, E., and Draine, B., *ApJ*, **537**, 796 (2000).
7. Šimon, V., Hudec, R., and Pizzichini, G., ",", in *these proceedings*, edited by E. E. Fenimore and M. Galassi, 2004, p. xxx.
8. Downes, T., Duffy, P., and Komissarov, S., *MNRAS*, **332**, 144 (2002).

The GRB 030227 Detected by INTEGRAL: Another Sign of Compton Scattering in X-rays

A.J. Castro-Tirado[*], J. Gorosabel[†,**], S. Guziy[*,‡], D. Reverte[*], J. M. Castro Cerón[**], A. de Ugarte Postigo[*], N. Tanvir[§], S. Mereghetti[¶], A. Tiengo[¶], S. B. Pandey[||], N. Masetti[††], H. Pedersen[‡‡], M. D. Pérez Ramírez[§§] and the GRACE Collaboration[¶¶]

[*]*Instituto de Astrofísica de Andalucía (IAA-CSIC), P.O. Box 3.004, 18.080 Granada Spain.*
[†]*Instituto de Astrofísica de Andalucía (IAA-CSIC), P.O. Box, 3.004, 18.080 Granada Spain.*
[**]*Space Telescope Science Institute, 3700 San Martín Drive, Baltimore MD 21218 USA.*
[‡]*Nikolaev State University, Nikolaev, Ukraine.*
[§]*Department of Physical Sciences, Univ. of Hertfordshire, College Lane, Hatfield, UK.*
[¶]*Istituto di Astrofisica Spaziale e Fisica Cosmica, Milano, Italy.*
[||]*State Observatory, Manora Peak, Naini Tal 263129, Uttaranchal, India.*
[††]*Istituto di Astrofisica Spaziale e Fisica Cosmica, Bologna, Italy.*
[‡‡]*Astronomical Observatory, Univ. of Copenhagen, Copenhagen, Denmark.*
[§§]*European Space Agency, Noordwijk, The Netherlands.*
[¶¶]

Abstract. Multiwavelengthp observations of a GRB detected by INTEGRAL (GRB 030227) revealed a dim optical afterglow (OA) that would not have been detected by many previous searches due to its faintess (R~ 23). This OA was seen to decline following a power law decay with index $\alpha = -0.95 \pm 0.16$. The spectral index β of the OA yields -1.32 ± 0.15, with the intrinsec absorption consistent with zero. These values may be explained by a relativistic expansion of a fireball in an homogeneous medium. We also find evidence for inverse Compton scattering in X-rays. A possible break is detected at $\sim$ 1.5 days.

INTRODUCTION

GRB 030227 was discovered in the INTEGRAL IBIS data by the automatic IBAS software [1] on 27 February 2003. The burst started at 08:42:03 UT and lasted for $\approx$ 18 s, putting it in the "long-duration" class of GRBs. It had a peak flux of 1.1 photons cm^{-2} s^{-1} and a fluence of 7.5×10^{-7} erg cm^{-2} in the 20-200 keV range [2].

The prompt dissemination (50 min) of the GRB position enabled triggering of a target of opportunity (ToO) observation with ESA's XMM-Newton satellite, starting $\sim$8 hr after the event. This observation revealed a fading X-ray source consistent with the IBIS error circle which was identified as the X-ray afterglow of GRB 030227 [3, 4]. Optical follow-up observations led to the discovery of the optical afterglow, as has been reported elsewhere [5].

RESULTS AND DISCUSSION

The lightcurve of the GRB 030227 OA

The R band lightcurve shows that the source was declining in brightness as it can be seen in Figure 3 of Castro-Tirado et al. (2003, [5]). Most GRB optical counterparts appear to be well characterised by a power law decay plus a constant flux component, $F(t) \propto (t-t_0)^\alpha + F_{host}$ where, $F(t)$ is the total measured flux of the counterpart at time $(t-t_0)$ after the onset of the event at t_0, α is the temporal decay index and F_{host} is the flux of the underlying host galaxy. F_{host} is negligible in this case, as indicated by the lack of flattening of the optical light curve at the time of our last observations (we estimate $R_{host} > 26$). From a least squares linear regression to the observed R band fluxes, a power law decline with $\alpha_R = -1.10 \pm 0.14$ (χ^2/dof = 1.53) provides an acceptable fit but it is ruled out by the early upper limits. However, if we exclude the 3.6ESO data point, a power law decline with $\alpha_R = -0.95 \pm 0.16$ (χ^2/dof = 0.64) provides a better fit, which is also just consistent with the earlier upper limits. This flux decay of the GRB 030227 OA agrees with the decay in X-rays $\alpha_X \simeq -1.0 \pm 0.1$ ([2, 6]) and is comparable to the slow decline rates observed in other GRB OAs. There are OAs for which a break or a smooth, gradual transition does occur within 1-2 days of the burst. The 3.6ESO measurement might give indication of a break around ~ 1.5 days but the lack of further data at later epochs precludes confirmation.

An upper limit to the redshift of $z \leq 3.5$ can be estimated from the absence of the onset of the Lyman forest blanketing in the optical data, consistent with the estimates from the X-ray spectra ($z \sim$ 3-4, [2], $z \sim 1.6$, [6]).

Evidence for inverse Compton scattering

We have determined the spectral flux distribution of the GRB 030227 OA on 28.24 UT Feb 2003 (mean epoch of the *HK* band images) by means of our *BRHK* broad band photometric measurements obtained with the different telescopes. We interpolated the B & R band magnitudes to that epoch, and fitted the observed flux distribution with a power law $F_\nu \propto \nu^\beta$, where F_ν is the flux density at frequency ν, and β is the spectral index. The optical flux densities at the wavelengths of *BRHK* bands have been derived without subtracting the contribution of any host galaxy, assuming a reddening $E(B-V)$ = 0.46 from the DIRBE/IRAS dust maps [7]. In converting the magnitude into flux, the effective wavelengths and normalisations given in [8] were used. The flux densities are 2.4, 2.7, 10.2 and 15.0 μJy at the *BRHK* bands, corrected for Galactic reddening (but not for possible intrinsic absorption in the host galaxy, which is consistent with zero). The fit to the NIR/optical data $F_\nu \propto \nu^\beta$ gives $\beta_{opt/NIR} = -1.25 \pm 0.14$ (χ^2/dof = 0.08).

Figure 1 shows the broadband spectrum (from near-IR/optical photometry to X-rays) for the GRB 030227 afterglow. The NIR/optical spectral index ($\beta_{opt/NIR} = -1.25 \pm 0.14$) is consistent with the spectral index for the unabsorbed X-ray spectrum ($\beta_X = -0.94\pm0.05$, [2]) but they do not match each others' extrapolations, similarly to GRB 000926 [9] and GRB 010222 [10]. We have investigated whether considerable extinction

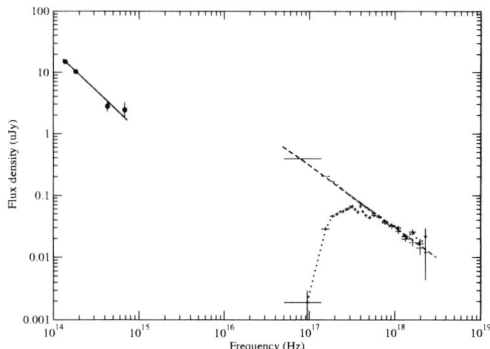

FIGURE 1. The broadband spectrum for the GRB 030227 afterglow at $t_0 + 0.87$ days. The NIR/optical spectrum (solid line) has been dereddened assuming $E(B-V) = 0.46$ from the DIRBE/IRAS dust maps. A fit $F_\nu \propto \nu^\beta$ to the NIR/optical data gives $\beta_{\rm opt/NIR} = -1.25 \pm 0.14$. The absorbed X-ray spectrum from XMM-Newton (dotted line) can be unabsorbed and represented by a power law (dashed line) with photon index $\Gamma = 1.94 \pm 0.05$ (i.e. $\beta_X = -0.94 \pm 0.05$), considering $N_H = 6.8 \times 10^{22}$ cm^{-2} at a redshift $z \sim 4$, following Mereghetti et al. ([2]). This is equivalent to consider a value of $N_H = 1.1 \times 10^{22}$ cm^{-2} at $z \sim 1.6$, as derived by and Watson et al. ([6]). Adapted from Castro-Tirado et al. ([5]).

in the host galaxy (considering different extinction laws) could produce such effect, in order to reproduce an optical-IR spectrum as an extrapolation of the X-ray spectrum, but found this not feasible. This results in no spectral break between the NIR/optical and X-ray bands, which will be also consistent with the similar decay indexes in both bands ($\alpha_R = -0.95$ and $\alpha_X = -1.0$). Thus, we suggest that in contrast to the NIR/optical band, where synchrotron processes dominate, there is an important contribution of inverse Compton scattering to the X-ray spectrum besides line emission ([6]), as it has been proposed for GRB 000926 [9]. This implies a lower limit on the density of the external medium, $n \geq 10$ cm^{-3} [11].

Adiabatic expansion or cooling regime ?

Many afterglows exhibit a single power law decay index. Generally this index is $\alpha \sim -1.3$, a reasonable value for the spherical expansion of a relativistic blast wave in a constant density interstellar medium, according to the standard fireball model [12]. In fact, the value of α for GRB 030227 falls within the boundaries defined by the observations made to date, from -0.67 ± 0.1 in the GRB 020331 [13] to -1.73 ± 0.04 in the GRB 980519 [14]. β only depends on p (the exponent of the power-law distribution of the Lorentz factor for the relativistic electrons) and it does not depend on the geometry of the expansion. Hereafter we will assume $\alpha \simeq -1.0$ and $\beta \simeq -1.0$ for the NIR/optical/X-ray bands. Several models have been explored in order to reproduce the observed values of α and β.

For an adiabatic expansion ($\nu_m < \nu < \nu_c$) in a constant density insterstellar medium (ISM), $\beta = (1-p)/2$ and $\alpha = -3(p-1)/4$, where ν is the observing frequency, ν_c is the cooling break frequency and ν_m is the synchrotron peak frequency [15]. For a spherical adiabatic expansion with the density $n \propto r^{-s}$ with $s = 2$ (inhomogeneous medium due to a stellar wind, [16]), $\beta = (1-p)/2$ and $\alpha = -(3p-1)/4$. In both cases, we have considered values of p in the range 1.8 to 3, as observed in most afterglows detected to date [17], but we cannot reproduce the observed values of α and β.

For the cooling regime ($\nu_c < \nu$) in both the ISM ($s = 0$) and wind ($s = 2$) cases, the evolution is similar at NIR/optical and X-ray wavelengths, with $\beta = -p/2$ and $\alpha = -(3p-2)/4$. The best results are obtained for $p \simeq 2.0$, from which we derive $\alpha \simeq -1.0$ and $\beta \simeq -1.0$, consistent with our measurements.

In light of the previous arguments, we propose that both the observed slow decay in the NIR/opt/X-ray lightcurves and the intrinsic spectrum, are consistent with a fireball in the cooling regime with $p = 2.0$, but we cannot distinguish between the $s = 0$ and $s = 2$ cases. Only a detection in radio ($\nu < \nu_c$ at $t_0 + 0.87$ days) would have allowed us to discriminate both models.

CONCLUSIONS

We presented multiwavelength observations of the afterglow associated with the GRB 030227. This would be one of the few OAs detected to date to an X-ray rich GRB. The decay index in the R-band lightcurve is $\alpha_R = -0.95 \pm 0.16$, with a possible break detection at $t_0 + \sim 1.5$ days. The optical-NIR spectrum at $t_0 + 0.87$ days allowed us to determine a spectral index $\beta_{opt/NIR} = -1.25 \pm 0.14$. The multiwavelength spectrum can be modeled by the expansion of a fireball (with $p = 2.0$) in the cooling regime. We also found evidence for inverse Compton scattering in X-rays.

REFERENCES

1. Mereghetti, S., Götz, D., Borkowski, J. et al., A&A, **411**, L291 (2003).
2. Mereghetti, S., Götz, D., Tiengo, A. et al., ApJ **590**, L73 (2003).
3. Loiseau, N., Gilomo, M., González-Riestra, R. et al., GCN Circ. 1901 (2003).
4. González-Riestra, R., Guainazzi, M., Loiseau, N. et al., IAUC 8087 (2003).
5. Castro-Tirado, A. J. et al. 2003, A&A **411**, L315 (2003).
6. Watson, D., Reeves, J.N., Hjorth, J., Jakobsson, P. & Pedersen, K., ApJ **595**, L29 (2003).
7. Schlegel, D.J., Finkbeiner, D.P. & Davis, M., ApJ, **500**, 525 (1998).
8. Fukugita, M., Shimisaku, K. & Ichikawa, T., PASP **107**, 945 (1995).
9. Harrison, F. A., Yost, S. R., Sari, R. et al., ApJ, **559**, 123 (2001).
10. in´t Zand, J.J.M., Kuiper, L., Amati, L. et al., ApJ, **559**, 710 (2001).
11. Panaitescu, A. & Kumar, P., ApJ, **543**, 66 (2000).
12. Mészáros, P. & Rees, M.J., ApJ, **476**, 232 (1997).
13. Dullighan, A., Monnelly, G., Butler, N. et al., GCN Circ. 1382 (2002).
14. Jaunsen, A.O., Hjorth, J., Bjornsson, G. et al., ApJ, **546**, 127 (2001).
15. Sari, R., Piran, T., & Narayan, R., ApJ, **497**, L17 (1998).
16. Chevalier, R. A. & Li, Z.-Y., ApJ, **536**, 195 (2000).
17. van Paradijs, J., Kouveliotou, C. & Wijers, R.A.M.J., ARA&A, **38**, 379 (2000).

GRBs AND COSMOLOGY

Energetics and the GRB Hubble Diagram

J. S. Bloom[†]

Harvard-Smithsonian Center for Astrophysics, 60 Garden Street, Cambridge, MA 02138, USA
Harvard Society of Fellows, 78 Mount Auburn Street, Cambridge, MA 02138 USA

Abstract. The energy release in prompt γ- and X-rays is a fundamental parameter in the study of the explosion mechanism and the central engines of GRBs. Though the apparent standardizability of GRB energetics have lead some to suggest that GRBs might be useful for cosmography, it is unlikely that GRBs can competitive with other high-precision cosmography experiments. Instead, by fixing a cosmology, there is a great deal more to be learned about long duration GRBs themselves.

INTRODUCTION

Before the afterglow revolution led to the discovery of burst distances, the energies of cosmological GRBs were estimated by simple order-of-magnitude. Assuming a Hubble scale for the luminosity distance (c/H_0) and a fluence of $S_\gamma = 10^{-6}$ erg s^{-1}, Paczyński [1] reckoned that the typical cosmological GRB releases $4\pi S_\gamma c^2 H_0^{-2} (1+z)^{-1} \approx 1.5 \times 10^{51}$ erg (adopting a Hubble constant of $H_0 = 71$ km s^{-1} Mpc^{-1})[1]. To date, there have been 33 luminosity distances measured by way of spectroscopic redshifts. The importance of obtaining new redshifts to constrain burst energetics continues unabated but the resource-expense for large aperture spectroscopy remains a considerable obstacle. One alternative that was discussed in a number of talks at this meeting, is the use of so-called "distance indicators"—metrics based upon prompt burst properties used to infer the intrinsic luminosity or energy. Calibrated against a sample of bursts with known redshift, distance indicators can yield estimates of burst redshifts. In addition to the variability [3] and spectral lag [4] indicators, Atteia [5] recently described the construction of a "psuedo-redshift" metric based upon the intensity, spectrum, and burst duration. While promising, the rms error on the distribution of the logarithm of the ratio of true redshift to pseudo redshift is $\sigma_{\log z'/z} = 0.18$, comparable to the rms of the true redshift distribution (see figure 2). If the use of distance indicators is ever to be vindicated, proponents should endeavor to make redshift *predictions* for new *Swift* bursts; thus far, all redshifts from distance indicators have been post-dictions.

Aside from redshifts to measure luminosity distances, precision energetics require the measurement of burst fluences, spectra, and geometry-corrections. The geometry-

---

[1] That this order–of–magnitude estimation is so close to the current value of GRB energies (1.3×10^{51} erg [2]) is an apparent conspiracy of three effects: 1) the luminosity distances to the 33 GRBs with redshifts span more than a factor of 50, 2) the fluences of those GRBs with redshifts span a range of 500, and 3) bursts appear to be collimated with a dispersion of over 150 in geometry corrections to the isotropic-equivalent energy (see [2] and references therein).

corrections are derived principally from measurements of temporal breaks in afterglow light curves (see [6] for a recent review).

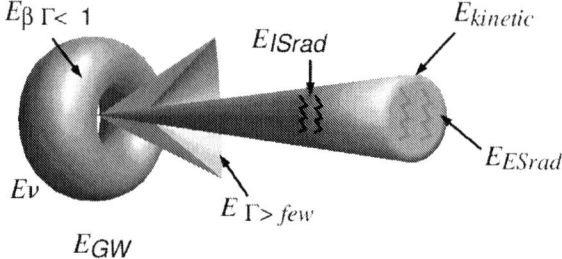

FIGURE 1. Schematic of the various sources of energy radiation in a GRB. The energy liberated in the prompt emission from the (supposed) internal shock ($E_{ISrad} = E_\gamma$) is discussed herein. Afterglow radiation from the external shock (E_{ESrad}) has been used to measure the kinetic energy contained in the outflowing blastwave ($E_{kinetic}$) [7]. Berger et al. have argued that energy radiated in a wide-angle jet of mildly relativistic material ($E_{\Gamma > few}$) was more than the energy contained in $E_{kinetic}$ and E_γ for GRB 030329, although this claim is controversial [8, 9]. Energy contained in the associated SN of 030329, as I suggest here, may be comparable to $E_{kinetic} + E_\gamma$. The energy radiated in neutrinos (E_ν) and gravitational waves (E_{GW}) may also be considerable but emission from either channel has not yet been detected. All of these channels are critical line-items to the complete GRB energy budget.

The first assemblage of geometry-corrected energies revealed an apparent constant release in the sample energy E_γ [10]. This result, using 17 GRBs, was based on the assumptions that the circumburst environments are homogeneous (all with density $n = 0.1$ cm^{-3}) and the energy per steradian was constant across the jet head (a so-called "top-hat" jet). Revisiting the energetics distribution, we recently relaxed this first assumption, making use of measured densities from individual afterglows [2]. Despite the reasonable claim that top-hat jet models are too simplistic to characterize realistic jets from collapsars (see figs. 8–9 in [11]), our new geometry-corrected energy distribution under the simple top-hat assumption is remarkably narrow (with a few exceptions addressed below). Of course, the prompt energy release in high-energy photons is only part of the total GRB energetics story. Figure 1 serves as a graphical admonition of this point. Given the rather humbling notion that some (if not most) of the radiated energy is not electromagnetic, the apparent constancy of energy released in prompt γ-rays is even more surprising.

(MIS)USE OF THE APPARENT STANDARD ENERGY AND GRB HUBBLE DIAGRAM

Jet properties and Unified Models

If the structure of the jet is not uniform, then the interpretation of the temporal break as being directly related to the jet collimation angle is incorrect. As noted [12, 13], only when the energy per unit solid angle drops as the inverse square of the angle off the center axis, will the true energy be equivalent as under the top-hat assumption.

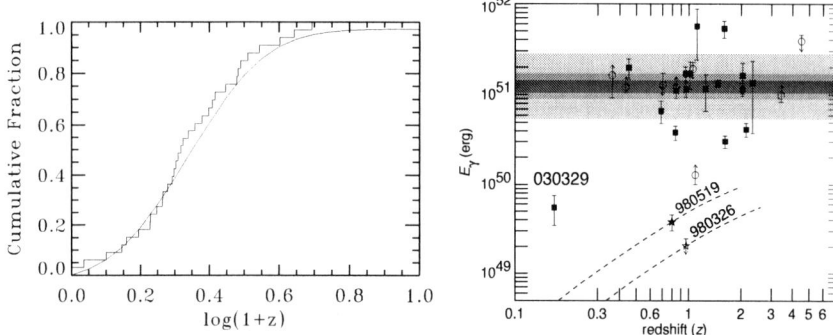

FIGURE 2. *left panel:* Cumulative distribution of GRBs with spectroscopic redshift measurements. The solid curve represents a log Gaussian fit to the distribution with a mean of log $(1 + \bar{z}) = 0.34$ ($\bar{z} = 1.19$) and $\sigma = 0.18$. This rms is comparable to the current precision of calibrated distance indicators, implying that the predictive power of such indicators is limited. Compiled by Andy Friedman. *right panel:* Inferred geometry-corrected prompt emission energy release in 27 GRBs in the co-moving bandpass 20–2000 keV (adapted from Bloom et al. 2003; [2]). The dark bands represent 1, 2 and 5 σ of a log Gaussian fit to the distribution after one sigma-clipping iteration. Almost all bursts fall within a factor of two of 10^{51} erg (the *WMAP* cosmology is assumed). However, there are three significant energy outliers. Though no spectroscopic redshift is known for GRBs 980326 and 980519, at any possible redshift the bursts are underluminous by a factor of more than 10. Although there is still some controversy about jet opening angle, GRB 030329 also appears underluminous.

There appears to be little physical motivation for the existence of such a universal jet structure. In fact, at this meeting S. Woosley suggested a two-component Gaussian for jet structures might be a better approximation to the results from collapsar models (see also [11]).

Since the structure of GRB jets is far from settled (see a recent review by Frail; [6]) several groups have begun to test the observational ramifications of structured jets versus top-hat jets (e.g., [14]). In flushing out the resulting brightness distributions from various jet scenarios, more recent work (e.g. D. Lamb, this meeting) have fixed $E(\theta) \propto \theta^{-2}$ to reproduce the standard energy result. However, it is important to emphasize that the standard energy result is based on the assumption of a homogeneous circumburst medium and a top-hat functional form for the jet. Placing this standard energy constraint on the universal jet model is unwarranted: again, aside from aesthetics, there is little reason to require that GRB energies are constant once the top-hat assumption is relaxed.

Cosmography

One may wonder whether GRBs would provide a meaningful independent estimate of Ω... Perhaps GRBs will not be able to contribute meaningfully to the myriad of measurements before other methods become more precise, but it is likely that they provide a useful consistency check based on independent objects.   Cohen & Piran (1997) [15]

Years before the standard energy result, Cohen and Piran [15] suggested that if the spread in the observed GRB luminosity function was small ($\sigma_L/L < 1$) then GRBs might be used for *cosmography* (measuring the fundamental parameters of the universe). They also recognized, as evident from the statement above, that the use of GRBs for such a purpose was unlikely to be competitive in the age of high-precision measurements of Type Ia supernovae and the microwave background. We now know that the observed luminosity (to be sure: energy) function is quite large ($\sigma_{E_{iso}(\gamma)}/E_{iso}(\gamma) \gg 1$) until the geometry-corrections are taken into account.

The apparent standardizability of E_γ have led some to (re)propose that the ensemble of GRBs energies might be useful for cosmography. At the "Lighthouses of the Universe" conference in Garching (2001), using a GRB Hubble Diagram (redshift versus apparent distance), we pointed out that the Frail et al. dispersion about a constant energy was smaller than the distribution of Type Ia supernovae peak B-band luminosities in 1992. Schaefer [16], using two independent luminosity distance indicators calibrated on nine GRBs with known redshift, produced a GRB Hubble diagram using only information from the prompt burst of more than 100 GRBs. Since the distances were derived under the assumption of a given cosmology, Schaefer cautioned the need to avoid circular reasoning when constraining cosmological parameters using the ensemble. Using the expanded sample of GRBs with known redshift, we [2] produced a self-consistent Hubble Diagram for a number of different cosmologies and found that the current sample was not a sensitive measure of Ω_m or Ω_Λ. Figure 3 illustrates this point.

The measurement of E_γ is non-trivial since it requires an accurate measurement of the fluence, burst spectrum, ambient density, and jet opening angle. Even if the geometry correction is found accurately through afterglow modeling, the fluences and the parameters of burst spectra are rarely measured to better than 10%. Takahashi et al. [17] found in a (self-consistent) simulation of 500 GRBs that the equation of state parameter w (let alone Ω_m and Ω_Λ!) could be measured with an accuracy of 20%. They required that the log rms uncertainty in the redshifts (found from distance indicators) to be $\sigma(\log z) = 0.02$ [recall that Atteia's pseudo-redshifts are currently uncertain with $\sigma(\log z) = 0.18$]. Even if the distance indicator is perfectly calibrated (from a theoretical framework, for example), this essentially requires both the fluence and k corrections to be measured to an accuracy of $\leqslant 6\%$ (for a burst at $z = 1$). Given the limited bandpass size of the BAT on *Swift*, k-corrections are likely to be uncertain by more than 10% for all new bursts in the near future.

Another difficulty with GRB energetics for the purpose of cosmography is the lack of local calibrators. Without a local calibrator to pin down the true energy, sidestepping the lower rungs of the distance ladder, GRBs only probe the *shape* of the distance versus redshift curve. Using different cosmologies, the standard energy value changes, but the dispersion about that energy remains relatively similar. Good local calibrators might one day be discovered with a wide-field GRB survey, but it is amusing to note that the two lowest redshift bursts, GRB 030329 ($z = 0.17$) and GRB 980425 ($z = 0.008$), appear to be underenergetic by more than a factor of 10 and 10^3, respectively.

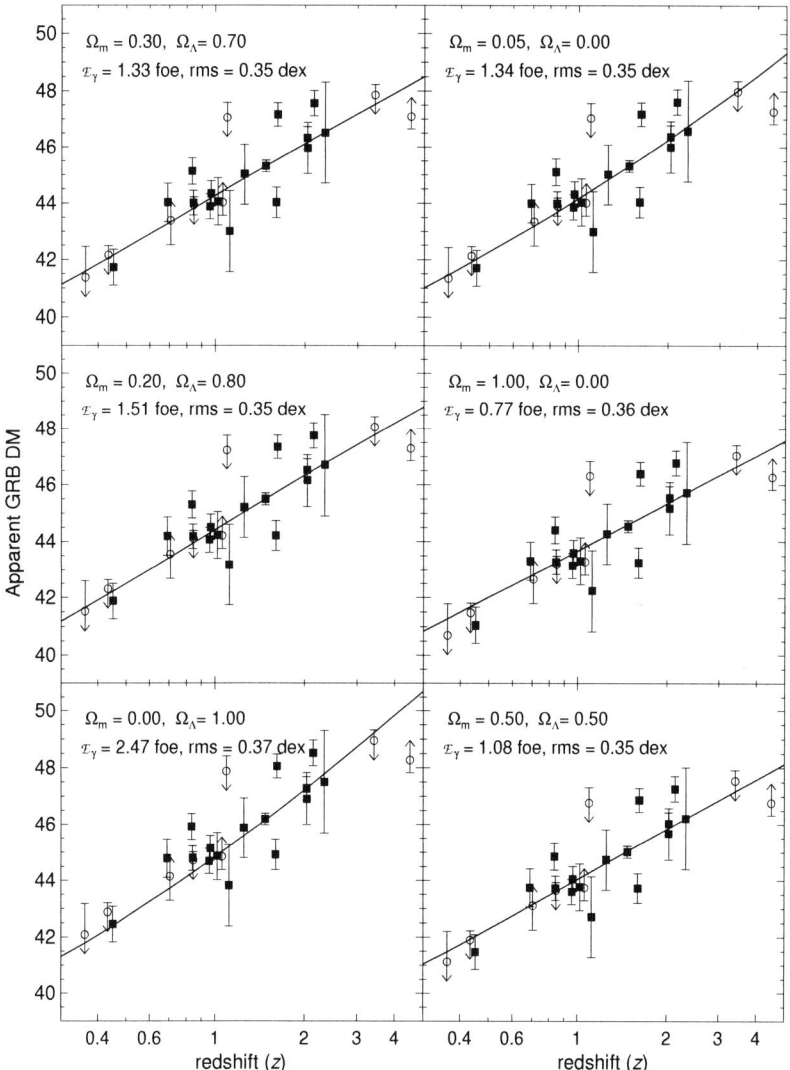

FIGURE 3. The GRB Hubble Diagram illustrating the difficulty in making cosmographic measurements with GRBs. Shown are Hubble Diagrams for six different cosmologies for the 24 GRBs (pre-030329) with known redshift and a measurement or constraint on $t_{\rm jet}$. The solid squares represent the apparent GRB distance modulus (DM) in the given cosmology with associated 1 σ errors. Open circles show those sources with upper or lower limits on the DM measurement. The solid curves are the theoretical distance moduli for the given cosmology. Though the values of standard energy, $\mathcal{E}_\gamma$, vary by more than a factor of 3 over the cosmologies shown, the observed variance about the theoretical curves is almost the same (≈ 0.35 dex). From [2].

PRACTICAL USES OF GRB ENERGETICS

Does the apparent underluminous nature of the low-redshift 980425 and 030329 imply that the standard *WMAP* cosmology is vastly incorrect? No. Instead, under the assumption of the standard cosmology, these two bursts appear to be genuine outliers in the energetics distribution. Indeed, GRB 030329 may be the first GRB with a known redshift that belongs to the sub-class of cosmological GRBs called fast-faders [2], recognized as bursts whose afterglows fade rapidly ($f_\nu \propto t^{-2}$) less than 0.5 days from trigger. Two other bursts (without redshifts)—GRB 980326 and GRB 980519—also belong to this category. Interestingly, from the perspective of energetics in the kinetic energy of the blastwave (as measured by X-ray afterglows), these three bursts also appear underenergetic (see Figure 2; [7]). Curiously, two out of three known fast-faders (030329 and 980326) also show evidence for an associated energetic supernova. The nature of the fast-fader subclass is not known: these bursts could be truly underenergetic explosions or occur in a inhomogeneous environment with a complex density profile. We estimate that about 10–15% of long duration GRBs are fast-faders, and so *Swift* will discover and study about 10 such bursts per year.

With a given cosmology, we can also begin to constrain the evolution of GRB energies as a function of redshift. Fitting for $E_\gamma \propto E_0(1+z)^m$, A. Friedman found that the current sample is consistent with no evolution ($m = 0$) to less than 1 σ. This apparent non-evolution of GRB energies with redshift will be rigorously tested with the order–of–magnitude more GRBs in the *Swift* era.

REFERENCES

1. Paczyński, B., *ApJ*, **308**, L43–L46 (1986).
2. Bloom, J. S., Frail, D. A., and Kulkarni, S. R., *ApJ*, **594**, 674–683 (2003).
3. Reichart, D. E., Lamb, D. Q., Fenimore, E. E., Ramirez-Ruiz, E., Cline, T. L., and Hurley, K., *ApJ*, **552**, 57–71 (2001).
4. Norris, J. P., *ApJ*, **579**, 386–403 (2002).
5. Atteia, J.-L., *A&A*, **407**, L1–L4 (2003).
6. Frail, D., Gamma-ray bursts: Jets and energetics (2003), astro-ph/0311301.
7. Berger, E., Kulkarni, S. R., and Frail, D. A., *ApJ*, **590**, 379–385 (2003).
8. Granot, J., Nakar, E., and Piran, T., The variable light curve of GRB 030329: The case for refreshed shocks (2003), submitted; astro-ph/0304563.
9. Lipkin, Y. M., Ofek, E. O., Gal-Yam, A., et al., Detailed optical light curve of GRB 030329 (2003), in prep.
10. Frail, D. A., Kulkarni, S. R., Sari, R., Djorgovski, S. G., Bloom, J. S., Galama, T. J., Reichart, D. E., Berger, E., Harrison, F. A., Price, P. A., Yost, S. A., Diercks, A., Goodrich, R. W., and Chaffee, F., *ApJ (Letters)*, **562**, L55–L58 (2001).
11. Zhang, W., Woosley, S. E., and MacFadyen, A. I., *ApJ*, **586**, 356–371 (2003).
12. Zhang, B., and Mészáros, P., *ApJ*, **581**, 1236–1247 (2002).
13. Kumar, P., and Granot, J., The evolution of a structured relativistic jet and GRB afterglow light-curves (2003), ApJ, in press.
14. Perna, R., Sari, R., and Frail, D., *ApJ*, **594**, 379–384 (2003).
15. Cohen, E., and Piran, T., *ApJ (Letters)*, **488**, L7 (1997).
16. Schaefer, B. E., *ApJ (Letters)*, **583**, L67–L70 (2003).
17. Takahashi, K., Oguri, M., Kotake, K., and Ohno, H., Probing dark energy with gamma-ray bursts (2003), astro-ph/0305260.

Towards Measuring the Cosmic Gamma-Ray Burst Rate

Paul A. Price* and Brian P. Schmidt*

Research School of Astronomy & Astrophysics, Mount Stromlo Observatory via Cotter Road, Weston, ACT, 2611, Australia.

Abstract. If the progenitors of gamma-ray bursts (GRBs) are massive stars, then we expect the GRB rate to follow the cosmic star formation rate at least to moderate redshifts. Here we attempt to quantify selection effects in the measurement of GRB redshifts as a function of the GRB energy release, afterglow luminosity and redshift. We apply these to a sample of 24 GRBs with optical afterglows and measured redshifts, and present the resultant measurement of the GRB rate over $0.2 < z < 5$. The GRB rate density evolution is consistent with measurements of the UV luminosity density. Our simulations predict that approximately 50% of searches for optical afterglows result in a nondetection, which is broadly consistent with the observed rate of "dark bursts". We finally apply our simulations to the case of the Swift satellite mission and show that small robotic telescope projects may be able to identify ~ 1 GRB beyond $z > 7$ during the 3 year Swift mission. This rate is largely limited by the relatively poor sensitivity in the near-infrared, but may be increased if small telescopes are coupled with more sensitive NIR imaging in order to uncover a larger fraction of GRBs beyond the most distant known objects at $z > 6.5$.

INTRODUCTION

Both indirect and direct evidence points to massive stars as the progenitors of the long/soft class of gamma-ray bursts (GRBs). Theoretically, if the progenitors of GRBs are massive stars, then the GRB rate should follow the cosmic star formation rate (to first order) since the lifetimes of massive stars are short compared to the age of the Universe. In practise, the GRB rate may not perfectly track the star formation rate for a variety of reasons, such as metallicity, binarity, and initial mass functions changing with redshift. However, other methods of measuring the cosmic star formation rate are also susceptible to some of these effects.

SAMPLE AND SELECTION EFFECTS

There are three steps to measuring a redshift for a particular GRB. Firstly, the GRB must be detected and localised by satellite missions (e.g. BeppoSAX, IPN, HETE-2). Secondly the afterglow must be identified in the radio, x-ray or optical, and finally, the redshift may be measured from absorption lines in the optical afterglow, or emission lines from the underlying host galaxy. Our sample consists of 24 GRBs with optical afterglows and measured redshifts prior to 2003, an increase of more than 50% over previous attempts to measure the GRB rate.

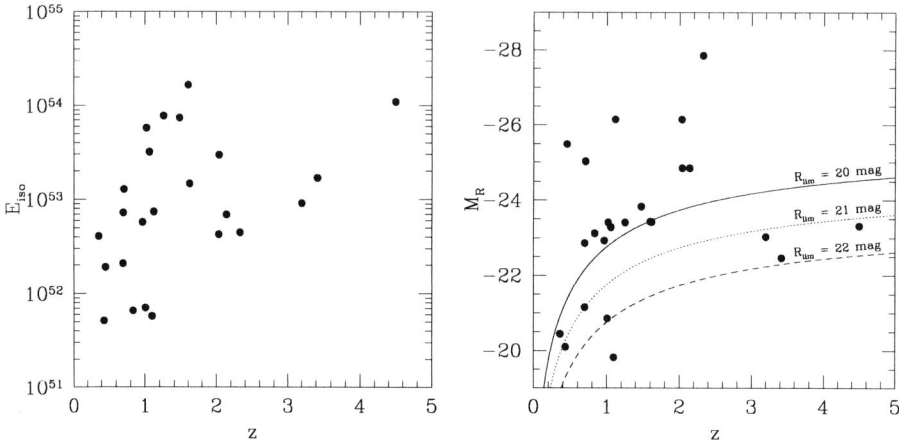

FIGURE 1. Left: Isotropic-equivalent gamma-ray energy releases (a proxy for the isotropic peak luminosity) as a function of redshift for GRBs in our sample. Right: Absolute R-band magnitudes for GRBs afterglows in our sample, plotted as a function of redshift. The solid, dotted and dashed lines demonstrate the sensitivity to a "typical" afterglow as a function of redshift for imaging observations one day after the GRB, with limiting magnitudes of $R = 20$, 21, and 22 mag respectively. In both plots, redshift-dependent selection effects are evident as a sparsity of GRBs with the faintest quantities at higher redshifts.

Figure 1 demonstrates that selection effects are not negligible. We attempt to determine a realistic selection function using a Monte Carlo code to trace observations of a GRB from detection by spacecraft, identification of the afterglow and spectroscopy of the afterglow and host galaxy.

GRB RATE FROM CURRENT MEASUREMENTS

The calculated selection effects in E_{iso}, M_R and z are applied iteratively to produce self-consistent distributions of each quantity. In Figure 2 we plot the GRB rate density (solid circles), corrected for the selection effects according to our simulations. The errors are statistical (Poisson) only and do not include systematics. As a comparison, we also plot the star formation rate measurements based on the ultraviolet luminosity density, corrected for extinction ([1]; open symbols).

The evolution of our measured GRB rate is consistent with the measured star formation rate measured from the UV luminosity over the covered redshift range, but is not yet particularly constraining. We find that the measured GRB rate is not strongly dependent upon most of our input assumptions. By far the most sensitive parameter is the assumed extinction distribution. Adding extinction causes the slope of the GRB rate density to increase markedly, with more GRBs at higher redshifts. Clearly, if the true GRB rate density is to be measured accurately, an accurate extinction distribution must be used. Since GRB optical afterglows generally have power-law spectra, they appear to

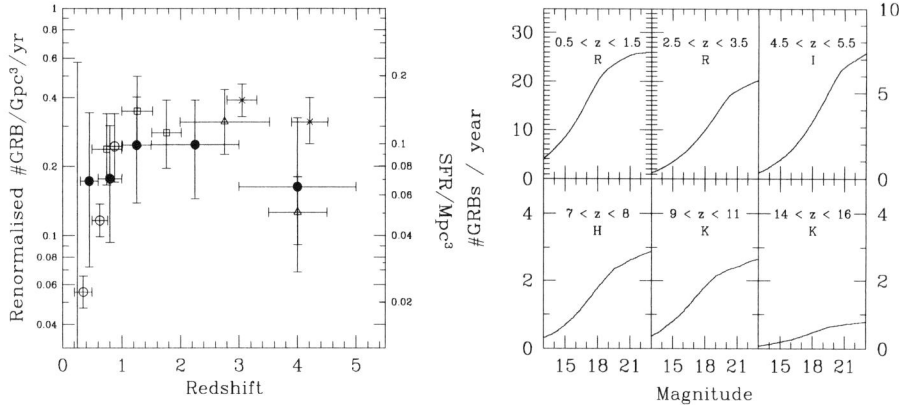

FIGURE 2. Left: The measured GRB rate density as a function of redshift (solid circles), normalised to a total of 500 GRBs per year out to $z = 5$. For comparison we also plot measurements of the extinction-corrected star formation rate density [1]. Right: The expected number of afterglows that should be detected as a function of limiting magnitude at half an hour after the GRB, in various redshift bins. Imaging at 3 minutes instead of half an hour (as is possible for robotic telescopes) should buy an additional equivalent sensitivity of ~ 2.5 mag for these plots. This assumes 150 Swift GRBs per year.

be well-suited for the measurement of extinction curves (e.g. [2, 3]).

Our simulations also have relevance to the nature of "dark bursts" — GRBs which do not appear to have an associated optical afterglow. Using our measured redshift distribution, we predict that an afterglow should have been identified in 50% of cases that a GRB localisation was observed. The observed rate, of $\sim 30\%$ [4, 5] was measured mostly in the IPN era (year 2000), which provided slow localisations and probably resulted in an over-estimate of the rate of "dark bursts". Hence our predicted value is broadly consistent with the observations. If we remove extinction completely from our simulations, we predict a dark burst rate of 40%. This indicates that approximately 40% of "dark bursts" are simply intrinsically faint, and that 10–30% of dark bursts are due to extinction at the source.

GRBS IN THE SWIFT ERA

In order to assist planning of future afterglow observations, we next applied our code to simulating Swift GRBs. Swift is expected to localise GRBs beyond $z = 5$ [6], but due to Lyman absorption, the R-band flux will be suppressed and so response in the near in-

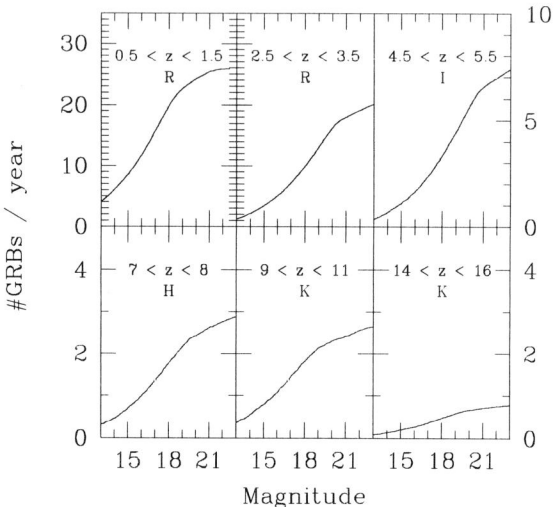

FIGURE 3. Expected number of GRBs as a function of limiting magnitude half an hour after the GRB trigger for different redshift bins.

frared (NIR) is crucial. Telescopes such as REM [7] and PROMPT (P. I.: D. E. Reichart) should be able to identify such afterglows, and direct larger telescopes to the position for spectroscopy. Our simulations predict that such systems should detect more than half of GRBs out to $z = 4$. The identification efficiency drops rapidly at $z = 5$ due to the Lyα absorption in the optical and less sensitivity in the NIR.

We next assumed that the GRB rate density increases monotonically out to $z = 1.25$, after which it is flat out to $z = 10$. We do not consider GRBs beyond this limit in this simulation. Using the measured distributions of $E_{\rm iso}$ and M_R, we calculated the redshift distribution of GRBs for which the above assumed experiment would obtain redshifts. The distribution peaks at $z \approx 1.5$ and contains approximately 75% of Swift GRBs (i.e. the "dark burst" fraction is 25%). 1% of these GRBs are beyond $z = 6$ (allowing study of the epoch of reionisation) and 0.5% are beyond $z = 7$ (which will be the most distant known objects in the Universe). Therefore, over a 3-year Swift mission, a single ground-based experiment with the above parameters may have the opportunity to study ~ 1 very-high redshift GRB. Plans to pounce on any GRB without an optical afterglow using more sensitive NIR imagers will increase the discovery rate of GRBs at very high redshift.

To demonstrate this, we show in Figure 3 the expected number of GRBs as a function of limiting magnitude half an hour after the GRB trigger for different redshift bins, extending our toy model for the star-formation rate density to $z = 16$. The limiting magnitudes are appropriate for the band shown (R for $z \sim 1$ and $z \sim 3$, I for $z \sim 5$, H for $z \sim 7.5$, and K for $z \sim 10$ and $z \sim 15$).

CONCLUSION

We have demonstrated the measurement of the cosmic GRB rate density from a sample of 24 GRBs with redshifts in the Beppo-SAX, IPN and HETE-2 eras. The measured GRB rate density is consistent with the star formation rate density measurements from the UV luminosity density, and not strongly dependent on assumptions of response times and sensitivities. We expect that Swift will provide a more homogenous sample, significantly reducing the uncertainty in GRB measurements. The extinction distribution is the largest uncertainty in the measurement of the GRB rate, but we are optimistic that future work will be able to measure this crucial function directly from the afterglow observations.

Using our Monte Carlo simulations, we find that a significant number of GRBs ($\sim 40\%$) do not have afterglows simply because they are too intrinsically-faint for current searches to detect, while about one third evade detection because they are significantly extinguished by dust.

We have also applied our simulations to the case of Swift combined with a prompt robotic optical/NIR multicolour photometry telescope to identify afterglows and a larger telescope for absorption spectroscopy. We find that the small telescopes with a limiting magnitude $H \sim 14$ mag will have difficulty finding a large number of GRBs at $z > 6$, and need to be combined with more sensitive NIR imaging in order to stand a significant chance of identifying and obtaining spectra for GRBs beyond the currently most distant known objects. This is especially the case if the light curves of GRB afterglows are flat at early times, as has been observed for three of five GRBs.

ACKNOWLEDGEMENTS

BPS and PAP thank the Australian Research Council for supporting Australian GRB research.

REFERENCES

1. Steidel, C. C., Adelberger, K. L., Giavalisco, M., Dickinson, M., and Pettini, M., *Ap.J.*, **519**, 1–17 (1999).
2. Price, P. A. et al., *Ap.J.Lett.*, **572**, L51–LL55 (2002).
3. Galama, T. J. et al., *Ap.J.Lett.*, **587**, 135–142 (2003).
4. Djorgovski, S. G., Frail, D. A., Kulkarni, S. R., Bloom, J. S., Odewahn, S. C., and Diercks, A., *Ap.J.*, **562**, 654–663 (2001).
5. Fynbo, J. P. U. et al., **369**, 373–379 (2001).
6. Lamb, D. Q., and Reichart, D. E., *Ap.J.*, **536**, 1–18 (2000).
7. Zerbi, F. M. et al., *Gamma-ray Bursts in the Afterglow Era*, 434, (2001).

SCUBA Observations of the Host Galaxies of Gamma-Ray Bursts

V. E. Barnard[*], N. R. Tanvir[†], A. W. Blain[**], A. Fruchter[‡], C. Kouveliotou[§], P. Natarajan[¶], E. Ramirez-Ruiz[‖], E. Rol[††], I. A. Smith[‡‡], R. P. J. Tilanus[*] and R. A. M. J. Wijers[††]

[*]*Joint Astronomy Centre, Hilo, USA*
[†]*University of Hertfordshire, Hatfield, UK*
[**]*Caltech, Pasadena, USA*
[‡]*STSI, Baltimore, USA*
[§]*MSFC, Huntsville, USA*
[¶]*Yale University, New Haven, USA*
[‖]*IoA, Cambridge, UK*
[††]*University of Amsterdam, Netherlands*
[‡‡]*Rice University, Houston, USA*

Abstract. In recent years, a population of galaxies with huge infrared luminosities and dust masses has been discovered in the submillimetre. Observations suggest that the AGN contribution to the luminosities of these submillimetre-selected galaxies is low; instead their luminosities are thought to be mainly due to strong episodes of star formation following merger events. Our current understanding of GRBs as the endpoints in the life of massive stars suggest that they will be located in such galaxies.

We have observed a sample of well-located GRB host galaxies in the submillimetre. Comparing the results with the general submillimetre-selected galaxy population, we find that at low fluxes ($S_{850} \leq 4$ mJy), the two agree well. However, there is a lack of bright GRB hosts in the submillimetre. This finding is reinforced when the results of other groups are included. Possible explanations are discussed. These results help us assess the roles of both GRB host galaxies and submillimetre-selected galaxies in the evolution of the Universe.

STAR FORMATION IN THE SUBMILLIMETRE

Our understanding of galaxy evolution was revolutionised by the observations of the IRAS satellite [1], which revealed large populations of dust-enshrouded, actively star-forming galaxies in the local Universe. This picture was extended to high redshifts by several instruments, predominantly the Submillimetre Common-User Bolometer Array (SCUBA, [2]) on the James Clerk Maxwell Telescope (JCMT). SCUBA observes simultaneously at 450 and 850 μm and hence is sensitive to emission from dust, heated to $\sim$ 40 K by optical and UV light form stars. SCUBA's unique sensitivity to submillimetre (submm) radiation is enhanced by the negative k-correction suffered by the greybody radiation, which makes SCUBA almost equally able to see galaxies at z of 1 and 10.

SCUBA has now discovered in excess of 200 galaxies; however progress in understanding these objects has been relatively slow. Identification of the SCUBA galaxies at optical wavelengths is hampered by two factors. Firstly, SCUBA galaxies are inherently faint in the optical (typically $I > 26$ [3]). Secondly, at SCUBA's primary wavelength of 850 μm, the angular resolution is only $\sim 14"$. The surface density of suitably optically faint galaxies is too high to then allow secure identifications. An alternative, but extremely time-consuming route, is to make interferometric detections of CO molecular lines in the millimetre [4, 5, 6], providing a position at which to search with optical telescopes. Only five redshifts have thus been obtained so far for SCUBA-selected galaxies.

However recent studies by Chapman et al. [7] have provided some progress, using a pre-selection in the radio to obtain redshifts for several tens of SCUBA galaxies. This technique has shown that SCUBA galaxies are at $z = 0.7 - 3.7$, with a median of 2.4. Furthermore, optical observations of these radio-selected SCUBA galaxies reveal that they are large systems, with irregular morphologies suggesting that they are in the earliest stages of mergers with enhanced star formation as a result [8]. These results confirm that much, perhaps the majority, of high-redshift star formation takes place in massive dusty systems with luminosities $L > 10^{12} L_\odot$ [9].

Inclusion of the SCUBA galaxy population is thus essential for any study of the star formation history of the Universe, but with SCUBA's confusion limit at around 2 mJy, the population seen by SCUBA accounts for only about 30% of the submm background detected by *COBE*-FIRAS. Furthermore the role of AGN in the huge luminosities of SCUBA galaxies is unclear; a recent estimate suggests hard X-ray AGN are found in only $\sim 10\%$ of SCUBA galaxies [10] but the evolution of SCUBA galaxies is not yet understood; several studies suggest that they are the high-redshift precursors of the local Universe massive spheroids [11, 12, 13] since their star formation rates and volume density match that predicted by evolutionary scenarios. To make progress in understanding both SCUBA galaxies and their role in the history of the Universe, we must look to other observations.

GRBS AS TRACERS OF STAR FORMATION

Several recent pieces of evidence link GRB events with star forming regions in their host galaxies. The recent coeval detection of supernova SN2003dh with GRB 030329 [14, 15] in particular has firmly established the association between GRBs and massive stars. Since high-mass stars are short-lived, the regions of galaxies surrounding GRBs are likely to be associated with ongoing star formation. Hence GRBs may be an alternative way to measure star formation in the Universe. They have several advantages in this regard. Their extreme luminosity means that they can be seen, if they exist, at extremely high redshifts, even through high intervening column densities. This ties in well with SCUBA since as explained above, SCUBA has a similar redshift-independent sensitivity. GRBs can pinpoint star formation in previously undetected objects; and hence if GRB rates can be shown to correlate with star formation it will not even be necessary to observe the host galaxy to include its contribution to the history of the Universe. Furthermore a great deal of other information can be obtained from the afterglow observations alone, such as metallicities, column densities etc, and this technique will not suffer from the impact of AGN.

However GRBs suffer from certain disadvantages too. GRBs can suffer from large selection effects; in order to know more about a GRB it is essential that its afterglow be discovered. This biases studies to those GRBs which take place in a sufficiently dense ISM for the afterglow to be detected, as well as to low-z and low-dust environments, although the GRB will clear out the dust in its very local environment [16]. We do not yet know whether the variety seen in GRBs is due to the progenitors or the ISM environments, but the effect of varying observing strategies must also be considered.

OBSERVATIONS OF GRB HOST GALAXIES WITH SCUBA

To make progress in understanding the relationship that both GRBs and SCUBA galaxies have with global star formation, we have undertaken two studies to observe GRB host galaxies with SCUBA. In our first study [17], we observed a small sample of so-called *'dark bursts'*, GRBs for which optical afterglows were not found. A simple explanation of these events may be that dust obscures the optical afterglow, which would make the host galaxies prime candidates for SCUBA galaxies. Recently the afterglow of GRB 030115 (see article in this proceedings) was detected in the infrared but not the optical, demonstrating that some dark bursts may be due to dust obscuration. We observed the hosts of five GRBs[1] whose afterglows were seen at either radio or X-ray wavelengths, but not at optical wavelengths despite deep, rapid searches.

Of this sample, only one source, GRB 000210, gave a marginal detection, which was confirmed when combined with further data by Berger et al. [19]. This rate of detection matches that predicted by a model which assumes that SCUBA galaxy incidence and GRB rates are both directly correlated with the global star formation rate [20]. Whilst this was only a small sample, the result supports the idea that GRBs and SCUBA galaxies provide complementary measures of star formation in the Universe. However it also indicates that 'dark bursts' are not always reliable signposts to dusty host galaxies; classifying all dark bursts as one type of object is unlikely to be physically accurate.

In our second study (Tanvir et al. in prep.) we observed all remaining accurate GRB locations with SCUBA to measure whether the apparent relation between SCUBA galaxies and GRB hosts continued with a larger unbiased sample. Combining our results with those of a similar study by Berger et al. [19], we have a sample of 20 hosts with < 1.4 mJy errors. The results are plotted in Fig. 1. The histogram represents the binned observational results, whereas the data points represent the predictions of the model by Ramirez-Ruiz et al. ([20]) mentioned above. The error bars indicate the counting statistics in each observational bin.

This figure indicates that the observations agree reasonably well with the model. However, the observational results are skewed to lower fluxes than the model predicts. Of 20 GRBs, only three hosts were detected, and the maximum flux was 3.74 mJy for GRB 010222 [21]. No host was found to have a flux above 4 mJy, whereas the model predicts that $> 10\%$ of GRB hosts should be in this bin. Whilst the statistics are not

---

[1] Note that in Barnard et al. [17], the observations of GRB 001025 were thought to have used an unreliable position. However later analysis [18] suggests that the observed position was the correct position of the GRB after all, and hence the results for this GRB are included here.

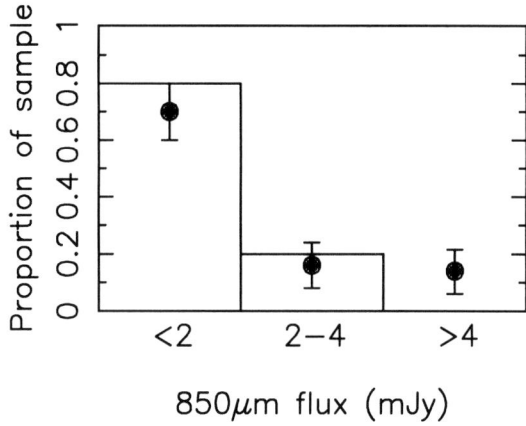

FIGURE 1. Results of the wide sample of all well-located GRB locations with SCUBA. Histogram is observational results binned into three (different-width) bins; points with error bars show prediction of a model by Ramirez-Ruiz et al. ([20]) convolved with observational errors and adjusted for confusion effects. The error bars represent the counting statistics of the proportion expected in each bin.

yet strong enough to confirm this apparent departure from the model, it is interesting to speculate on possible causes.

Firstly, we note that there is likely to be a small population of truly dark bursts – such as GRB 030115 – where dust obscures the site of the burst so that its afterglow is not discovered. Further infrared and submm observations of GRB afterglows are necessary to quantify this population. A similar selection effect may be towards detecting GRBs preferentially at low redshifts - whilst the GRBs themselves can be detected at high redshifts, the afterglow observations are limited. Since the recent work of Chapman et al. ([7]) confirms that SCUBA galaxies are a high-redshift population this may cause an important bias.

Alternatively, it may be the case that the submm-luminous phase and the GRB epoch of the host galaxies do not coincide. Chapman et al. ([8]) used their recent HST observations of the radio-selected sample to model the submm evolution of the SCUBA galaxies; for one example they concluded that the submm-luminous phase may be as brief as 5 Myr. Hence the GRB events may occur in these galaxies after the submm flux has dwindled, and we only see in the submm those few whose flux has a slower decline.

Another explanation may be that GRBs occur preferentially in low-metallicity systems, which is predicted as a consequence of the popular collapsar model [22]. Whilst low metallicities have been inferred for several GRB hosts (for example [23]) the fact that three submm hosts have been confirmed by our studies and those of Berger et al. [19] demonstrates that there is not a complete anti-correlation. It is likely in any case that starbursting galaxies have a very patchy distribution of dust and gas.

Another issue is the extent to which SCUBA galaxy fluxes are boosted by AGN dust heating. If AGN in fact have a larger impact on SCUBA galaxy fluxes then the model points in Fig. 1 are inappropriate. If SCUBA galaxies are, as is thought, proto-elliptical galaxies, then all models predict a QSO phase at some point after the onset of the merger event [11, 12, 13]. However the impact of the QSO on the submm flux, and its epoch of activity relative to star formation, has not been strongly quantified yet.

Finally we note that the three detected submm hosts show some differences from the general (submm brighter) SCUBA galaxy population. Whilst SCUBA galaxies show a surprising variety in the optical [3], the GRB hosts seem to be bluer and perhaps more compact [19]. They are also found at lower redshifts than the general SCUBA galaxy population, all being at $z < 1$. However this is for only a very small sample which, as noted above, may already be biased towards low redshifts.

CONCLUSIONS

We have completed an initial survey of the hosts galaxies of GRBs in the submm, in an attempt to understand how both populations relate to star formation rates in the Universe as a whole. We find that whilst there is a general overlap between the host population and the known SCUBA galaxy population, there is a dearth of bright GRB hosts in the submm. Suggestions to explain this disparity include both sample selection effects and physical factors such as timescales of the various phases of galaxy luminosity, metallicity in GRB hosts and the unknown AGN contribution to SCUBA galaxy fluxes.

Further studies with more complete samples of GRB hosts will overcome some of the selection effects discussed above. If the discrepancy between the model and the observations continues to increase in significance then models of SCUBA galaxy evolution will need to include the implications of this result.

REFERENCES

1. Sanders, D. B., and Mirabel, I. F., *ARA&A*, **34**, 749–+ (1996).
2. Holland, W. S., Robson, E. I., Gear, W. K., and et al., *MNRAS*, **303**, 659–672 (1999).
3. Smail, I., Ivison, R. J., Blain, A. W., and Kneib, J.-P., *MNRAS*, **331**, 495–520 (2002).
4. Frayer, D. T., Ivison, R. J., Scoville, N. Z., Yun, M., Evans, A. S., Smail, I., Blain, A. W., and Kneib, J.-P., *ApJL*, **506**, L7–L10 (1998).
5. Frayer, D. T., Ivison, R. J., Scoville, N. Z., Evans, A. S., Yun, M. S., Smail, I., Barger, A. J., Blain, A. W., and Kneib, J.-P., *ApJL*, **514**, L13–L16 (1999).
6. Neri, R., Genzel, R., Ivison, R. J., Bertoldi, F., Blain, A. W., Chapman, S. C., Cox, P., Greve, T. R., Omont, A., and Frayer, D. T., *ApJL*, **597**, L113–L116 (2003).
7. Chapman, S. C., Blain, A. W., Ivison, R. J., and Smail, I. R., *Nat.*, **422**, 695–698 (2003).
8. Chapman, S. C., Windhorst, R., Odewahn, S., Yan, H., and Conselice, C., *astro-ph/0308197* (2003).
9. Blain, A. W., Smail, I., Ivison, R. J., Kneib, J.-P., and Frayer, D. T., *Phys. Rep.*, **369**, 111–176 (2002).
10. Almaini, O., Scott, S. E., Dunlop, J. S., Manners, J. C., Willott, C. J., Lawrence, A., Ivison, R. J., Johnson, O., Blain, A. W., Peacock, J. A., Oliver, S. J., Fox, M. J., Mann, R. G., Pérez-Fournon, I., González-Solares, E., Rowan-Robinson, M., Serjeant, S., Cabrera-Guerra, F., and Hughes, D. H., *MNRAS*, **338**, 303–311 (2003).
11. Lilly, S. J., Eales, S. A., Gear, W. K. P., Hammer, F., Le Fèvre, O., Crampton, D., Bond, J. R., and Dunne, L., *ApJ*, **518**, 641–655 (1999).
12. Archibald, E. N., Dunlop, J. S., Jimenez, R., Friaça, A. C. S., McLure, R. J., and Hughes, D. H., *MNRAS*, **336**, 353–362 (2002).
13. Granato, G. L., DeZotti, G., Silva, L., Danese, L., and Magliocchetti, M., *ApSS*, **281**, 497–500 (2002).
14. Hjorth, J., Sollerman, J., Møller, P., Fynbo, J. P. U., Woosley, S. E., Kouveliotou, C., Tanvir, N. R., Greiner, J., Andersen, M. I., Castro-Tirado, A. J., Castro Cerón, J. M., Fruchter, A. S., Gorosabel,

J., Jakobsson, P., Kaper, L., Klose, S., Masetti, N., Pedersen, H., Pedersen, K., Pian, E., Palazzi, E., Rhoads, J. E., Rol, E., van den Heuvel, E. P. J., Vreeswijk, P. M., Watson, D., and Wijers, R. A. M. J., *Nat.*, **423**, 847–850 (2003).
15. Stanek, K. Z., Matheson, T., Garnavich, P. M., Martini, P., Berlind, P., Caldwell, N., Challis, P., Brown, W. R., Schild, R., Krisciunas, K., Calkins, M. L., Lee, J. C., Hathi, N., Jansen, R. A., Windhorst, R., Echevarria, L., Eisenstein, D. J., Pindor, B., Olszewski, E. W., Harding, P., Holland, S. T., and Bersier, D., *ApJL*, **591**, L17–L20 (2003).
16. Waxman, E., and Draine, B. T., *ApJ*, **537**, 796–802 (2000).
17. Barnard, V. E., Blain, A. W., Tanvir, N. R., Natarajan, P., Smith, I. A., Wijers, R. A. M. J., Kouveliotou, C., Rol, E., Tilanus, R. P. J., and Vreeswijk, P., *MNRAS*, **338**, 1–6 (2003).
18. Watson, D., Reeves, J. N., Osborne, J., O'Brien, P. T., Pounds, K. A., Tedds, J. A., Santos-Lleó, M., and Ehle, M., *A&A*, **393**, L1–L5 (2002).
19. Berger, E., Cowie, L. L., Kulkarni, S. R., Frail, D. A., Aussel, H., and Barger, A. J., *ApJ*, **588**, 99–112 (2003).
20. Ramirez-Ruiz, E., Trentham, N., and Blain, A. W., *MNRAS*, **329**, 465–474 (2002).
21. Frail, D. A., Bertoldi, F., Moriarty-Schieven, G. H., Berger, E., Price, P. A., Bloom, J. S., Sari, R., Kulkarni, S. R., Gerardy, C. L., Reichart, D. E., Djorgovski, S. G., Galama, T. J., Harrison, F. A., Walter, F., Shepherd, D. S., Halpern, J., Peck, A. B., Menten, K. M., Yost, S. A., and Fox, D. W., *ApJ*, **565**, 829–835 (2002).
22. MacFadyen, A. I., and Woosley, S. E., *ApJ*, **524**, 262–289 (1999).
23. Vreeswijk, P. M., Moller, P., and Fynbo, J. P. U., *astro-ph/0308164* (2003).

Near-Infrared Colors of Gamma-Ray Burst Afterglows and Cosmic Reionization History

Akio K. Inoue*, Ryo Yamazaki* and Takashi Nakamura*

Department of Physics, Kyoto University, Kyoto 606-8502, Japan

Abstract. We propose a method for examining the cosmic reionization history by using near-infrared (NIR) colors of afterglows of high redshift ($5 < z < 25$) gamma-ray bursts (GRBs) that will be detected by the *Swift* satellite. The broad-band photometry has a much higher sensitivity than that of the spectroscopy. A prompt NIR photometric follow-up of the high-z GRB afterglows will reveal how many times the reionization occurred in the universe.

INTRODUCTION

The Gunn–Peterson trough shortward of the Lyα resonance [1] in the spectrum of quasars with redshift $z \sim 6$ indicates that the end of the reionization epoch is at $z \sim 6$ [2]. The recent observation of the polarization of the cosmic microwave background (CMB) by *WMAP* suggests that the beginning of the reionization is $z \sim 20$ [3]. Now we should investigate how the reionization proceeded.

In this paper, we show that the near-infrared (NIR) photometric follow-ups of the GRB afterglows are very useful to investigate how many times the cosmic reionization occurred. Although many techniques to reveal the reionization history have been proposed so far, the NIR photometric colors may be the most promising technique using the *current* facilities.

In order to adopt our photometric method, we need to determine redshifts of GRBs by other ways, for example, detections of iron lines in X-ray afterglow spectra [4] and of the Lyα break in NIR spectra, or empirical methods by using only the γ-ray data [e.g., 5, 6]. Even if redshifts of GRBs are unknown prior to follow-up observations, it is worth performing NIR photometry in as early phase as possible since $\sim 10\%$ of GRBs are expected to be located at $z > 10$ [7]. In practice, such follow-up observations against every GRB are possible. Even after the early NIR follow-up, could be able to determine the redshifts by other observations.

We adopt a standard set of the ΛCDM cosmology throughout the paper: $H_0 = 70$ km s^{-1} Mpc^{-1}, $\Omega_M = 0.3$, $\Omega_\Lambda = 0.7$, and $\Omega_b = 0.04$.

WAS THE UNIVERSE REIONIZED TWICE?

Recently, Cen [8] proposed a new scenario of the reionization; the universe was reionized twice. At $z \sim 20$, the first reionization was made by the Population III (Pop III) stars

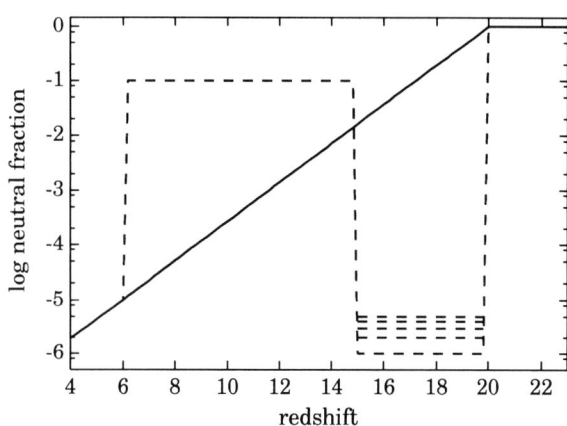

FIGURE 1. Schematic reionization histories. The solid curve is a single gradual reionization case. The dashed curve is a double reionizations case as suggested by Cen [8].

with a top-heavy initial mass function (IMF). Then, the universe was partially recombined at $z \sim 15$ when the transition from the Pop III to II occurred and the UV emissivity was suddenly suppressed because of the different IMF. Finally, the UV photons from the Pop II stars increased gradually and ionized the universe again at $z \sim 6$. We show these two reionization histories schematically in Figure 1. The solid line is a single gradual reionization case which is considered usually, and the dashed lines are Cen's double reionizations. We show some cases of the neutral fraction at the first reionization epoch in Cen's scenario.

Let us discuss what observational difference appear between Cen's scenario and usual single reionization. We have examined the expected NIR colors as a function of the source redshift for the afterglow spectrum $f_\nu \propto \nu^{-1/2}$ case (the observing time less than several hours [9]). Differences between the two reionization histories appear in $I - J$ and $K - L$ colors. In Figure 2, we show such differences in the $K - L$ color.

The afterglows beyond the drop-out redshift of the K-band ($z^K_{\text{Ly}\alpha,\text{out}} = 19.4$) really drop out of the filter in the single reionization case, whereas we can see still such afterglows through the K-band in the twice reionized universe. On the other hand, the afterglows with $z > 23$ drop out of the K-band even in the double reionization case because the Lyβ break goes out of the filter transmission width, i.e. $\lambda_\beta(1 + z_{\text{drop}}) = \lambda^K_{\text{max}}[= \lambda_\alpha(1 + z^K_{\text{Ly}\alpha,\text{out}})]$, where λ^K_{max} is the maximum wavelength of the K-band filter.

In $I - J$ color, we find differences due to the difference of the increasing rate of the neutral fraction around $z \sim 6$ in two reionization histories although we do not show it here. Any difference between the single and double reionizations cannot be found in the $J - H$ and the $H - K$ colors because the neutral fractions in both cases are high enough ($> 10^{-4}$) to extinguish the continuum bluer than the Lyα break completely. That is, the GRB afterglows beyond the drop-out redshift cannot be seen in the J and H-bands for both of reionization histories.

Let us summarize how to confirm or refute Cen's scenario: We can conclude that

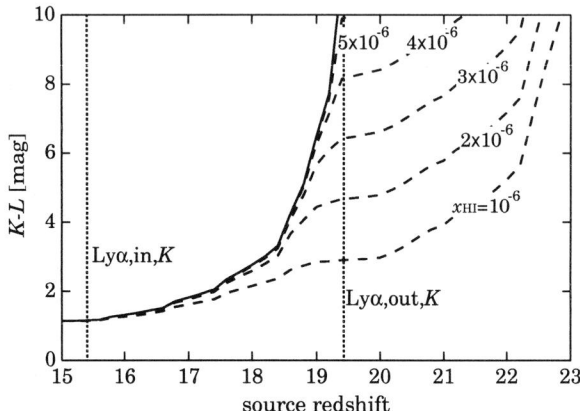

FIGURE 2. Expected $K-L$ color of GRB afterglows. The solid and dashed curves correspond to the cases of the single and double reionizations depicted in Figure 1, respectively. Some cases of the neutral hydrogen fraction in the first reionization are shown for dashed curves. Two dotted vertical straight lines indicate the source redshifts at which the Lyα break enters and goes out of the K filter.

the universe was reionized twice if (1) the GRB afterglows with $z > z^{K}_{\text{Ly}\alpha,\text{out}}(=19.4)$ is detected in the K-band and (2) the afterglows with $z > z^{J}_{\text{Ly}\alpha,\text{out}}(=10.8)$ or $z > z^{H}_{\text{Ly}\alpha,\text{out}}(=14.0)$ drop out of the J or H-bands. However, the null detection of the $z > z^{K}_{\text{Ly}\alpha,\text{out}}$ GRB afterglows in the K-band does not reject Cen's scenario at once. The null detection only shows the neutral fraction at $z \sim 20$ is larger than $\sim 10^{-5}$. In any case, the deep and prompt K-band photometry of the high-z GRB afterglows is useful to examine the ionization state at $z \sim 20$ very much.

ADVANTAGE OF NIR COLORS METHOD

The largest advantage of the NIR color method is its easier availability and higher sensitivity than other methods. The limiting magnitude with a high signal-to-noise ratio of the broad-band NIR photometry reaches much deeper than 20 mag with only a few minute exposure with a 8-m class ground based telescope. Even if the apparent dispersion of the afterglow luminosities is very large, the uncertainty can be controlled in a low level if we use colors, which are independent of the absolute luminosity. This point is also an important advantage of the NIR color method. Therefore, the NIR multi-color follow-ups of the GRB afterglows are strongly encouraged.

The detailed spectroscopy of the red damping wing of the Lyα break provides us with the neutral hydrogen column density to the source [10]. However, the spectral resolution required is $\lambda/\Delta\lambda \sim 5000(10^5/\tau_{\text{Ly}\alpha})$. For a lower opacity, a much higher resolving power is needed, so that the limiting magnitude becomes significantly shallow. That is, the method of damping wing measurement do not have sensitivity for a low opacity case. On the other hand, the NIR colors method is practically sensitive against $\tau_{\text{Ly}\alpha} < 10$.

Therefore, these two methods are complementary to each other.

The dispersion measure in GRB radio afterglows may be promising in near future [11, 12]. If we observe the radio afterglow at about 100 MHz within about 1000 s after the burst occurrence, the delay of the arrival time of the low frequency photons may be detectable by the Square Kilometer Array. As well as the measurement of the red damping wing of the Lyα line, this technique is sensitive to $x_{HI} > 0.1$ (i.e., $\tau_{Ly\alpha} > 10^5$). Thus, this technique and our NIR color method are also complementary each other.

Metal absorption lines like O I λ1302 can be useful [13]. However, the expected equivalent width of the lines is very small as < 5 Å. We require the spectroscopy with a resolving power ~ 5000. Unfortunately, observations with a ground-based telescope may be difficult because a huge number of Earth's atmospheric OH lines conceal the metal lines. Only JWST will have such a high spectral resolution among the future space telescopes having a NIR spectrograph. Although measurement of the CMB polarization anisotropy is also useful, we must await the launch of Planck because the sensitivity of *WMAP* is not enough [14]. The technique using the hydrogen 21 cm absorption line may be also good if there are enough bright background radio source. However, even GRB afterglows are too faint [15].

Details of calculations and further discussions can be found in Inoue, Yamazaki, & Nakamura [16].

ACKNOWLEDGMENTS

This work is supported in part by the Grant-in-Aid for Scientific Research of the Japanese Ministry of Education, Culture, Sports, Science and Technology, No.14047212 (TN) and No.14204024 (TN), and also supported by a Grant-in-Aid for the 21st Century COE "Center for Diversity and Universality in Physics". AKI and RY thank the Research Fellowship of the Japan Society for the Promotion of Science for Young Scientists.

REFERENCES

1. Gunn, J. E., & Peterson, B. A., ApJ **142**, 1633 (1965).
2. Becker, R. H., et al., AJ **122**, 2850 (2001).
3. Kogut, A., et al., ApJS **148**, 161 (2003).
4. Mészáros, P. & Rees, M.J., ApJ **591**, L91 (2003).
5. Fenimore, E. E., & Ramirez-Ruiz, E., astro-ph/0004176 (2003).
6. Yonetoku, D., Murakami, T., Nakamura, T., Yamazaki, R., Inoue, A. K., & Ioka, K., ApJL, submitted, astro-ph/0309217 (2003).
7. Bromm, V., & Loeb, A., ApJ **575**, 111 (2003).
8. Cen, R., ApJ **591**, 12 (2003).
9. Sari, R., Piran, T., & Narayan, R., ApJ **497**, L17 (1998).
10. Miralda-Escudé, J., ApJ **501**, 15 (1998).
11. Ioka, K., ApJL, in press, astro-ph/0309200 (2003).
12. Inoue, S., MNRAS, in press, astro-ph/0309364 (2003).
13. Oh, S. P., MNRAS **336**, 1021 (2002).
14. Haiman, Z., & Holder, G. P., ApJ **595**, 1 (2003).
15. Furlanetto, S. R., & Loeb, A., ApJ **579**, 1 (2003).
16. Inoue, A. K., Yamazaki, R., & Nakamura, T., ApJ, in press, astro-ph/0308206 (2004).

Detectability of Long GRB Afterglows From Very High Redshifts

Lijun Gou*, P. Mészáros*†, Tom Abel* and Bing Zhang*

*Dept. Astronomy & Astrophysics Pennsylvania State University
†Institute for Advanced Study, Princeton,

Abstract. Detectability of high redshift GRB afterglows are carefully discussed. Using standard assumptions, we find that Chandra, XMM and Swift XRT can potentially detect GRBs in the X-ray band out to very high redshifts $z \gtrsim 30$. In the K and M bands, the JWST and ground-based telescopes are potentially able to detect GRBs even one day after the trigger out to $z \sim 16$ and 33.

INTRODUCTION

Gamma-ray bursts are thought to be associated with the formation of massive stars, possibly with first generation star expected to exist at $z \gtrsim 18$. The natural question which needs to be quantified is the degree of detectability of GRBs with current or future detectors, if they occur at much higher redshifts than those currently sampled. Lamb & Reichart ([1]) used specific templates such as GRB 970228 observed at one day to estimate the highest redshifts at which such bursts could be observed using Swift. Ciardi & Loeb ([2]) calculated the flux evolution with redshift of common GRBs and discussed the flux change with redshift at several epochs in the infrared bands. These papers considered only forward shock radiation as known before 2000 and some effects of the galactic mean density evolution, but they did not consider the primeval star-formation environment.

In our simulation we have calculated the flux evolution of typical GRBs based on current knowledge of GRB physics in a more realistic way ([3]). Among the refinements introduced are: (1) The contribution from reverse shocks is considered as a crucial element. (2) We have taken up to date GRB parameters, e.g. incorporating new estimates of the typical magnetic equipartition parameter ε_B about one order of magnitude or more smaller than the electron parameter ε_e in the forward shock, and a possibly higher ε_B in the reverse shock. This has a significant effect on the GRB evolution. (3) We consider GRB external densities motivated both by views on the typical protogalaxy density evolution with redshift, and by views on the conditions around the first stars to form in the universe in the pre-galactic era. (4) We consider both the Lyman-α and photoionization absorption as well as our own galactic extinction. (5) We compare the expected fluxes in the X-ray and near IR bands to the sensitivity of various detectors such as Chandra, XMM, Swift XRT and JWST.

AFTERGLOW CHARACTERISTICS

Spectrum. For the forward shock, we have taken the standard broken power law spectrum of GRBs ([4]). For reverse shock, it is also described by a similar broken power law but with different critical frequencies and peak flux. The critical frequencies and peak flux between forward shock and reverse shock at crossing time are connected by the relations below ([5]; [6]):

$$\frac{v_{m,r}(t_\times)}{v_{m,f}(t_\times)} = (\gamma_\times^2/\eta)^{-2}\mathcal{R}_\mathcal{B} \;,\; \frac{v_{c,r}(t_\times)}{v_{c,f}(t_\times)} = \mathcal{R}_\mathcal{B}^{-3} \;,\; \frac{F_{v,m,r}(t_\times)}{F_{v,m,f}(t_\times)} = (\gamma_\times^2/\eta)\mathcal{R}_\mathcal{B} \quad (1)$$

where

$$\gamma_\times = \min(\eta, \eta_c) \;,\; \text{and} \;\; \mathcal{R}_\mathcal{B} \equiv B_r/B_f = (\varepsilon_{B,r}/\varepsilon_{B,f})^{1/2}. \quad (2)$$

Here $\mathcal{R}_\mathcal{B}$ reflects a stronger B field in reverse shock, as inferred from the analyses of the GRB 990123 and 021200 data ([6]), and subscripts 'r' and 'f' stand for reverse and forward shock, respectively. In our calculations we set $\mathcal{R}_\mathcal{B} = 1$ as the standard case. Scalings for those reverse shock quantities are given by Kobayashi ([7]).

GRB density environment. In our calculation, for simplicity we consider only $n \sim$ constant case, which also favors most of the observed cases that have been analyzed (Frail et al., [8]). We can assume a typical average value n_0 for n at redshift $z = 1$. Next we consider the density evolution with redshift around GRBs. We concentrate on two very different types of dependencies. (1) Based on hierarchical models of galaxy formation ([9]), for a fixed host galaxy mass, this yields $n(z) = n_0(1+z)^4$ ([2]), this is called the density evolution model. (2) Recent numerical simulations of primordial star formation indicate that the particle number density around the first stars at very high redshift could be approximately independent of redshift because of strong radiation pressure from the central massive star, which dominates and smooths any variations in the original galactic number density around the stars ([10]). Here we assume that, for this case (2), this stellar dominance applies to all GRBs originating from massive stars, so $n(z) = n$.

Intergalactic and galactic absorption. As it propagates through the intergalactic medium (IGM), the afterglow radiation from a burst occurring at some redshift z is subject to several absorption processes. The most important are Lyman-α absorption, photoionization of neutral hydrogen, and photoionization of He II as well as the extinction of our own galaxy in UV and x-ray. Thus, one expects that between the Lyman-α frequency corresponding to the source frame and approximately 5×10^{16} Hz (below which the galactic extinction for the above column density becomes large), the flux observed from a high redshift GRB will be totally suppressed. Outside this range, the observed flux is much less affected by the intergalactic and galactic absorption.

NUMERICAL RESULTS

Infra-red redshift dependence. Looking at Figure 1, which is the standard case of $\mathscr{R}_{\mathscr{B}} = 1$ in our calculation, several features can be noted: (1) At early times, e.g. $t = 10$ mins and $t = 2$ hours, we can differentiate the constant density profile from the evolving density profile in both K and M bands. However, at late times it becomes difficult to do so in both bands, although the total flux in the M band is somewhat different for both density profiles at relatively low redshifts. (2) The break in the light curve for the $n = $ constant case at $t = 10$ minutes is caused by the transition from $t > t_{\text{dec,s}}$ to $t < t_{\text{dec,s}}$. (3) There is a sharp decline in the emitted flux in light curves at redshift $z \sim 17$ for K band and at $z \sim 36$, which are caused by the lyman-α and photoionization absorption of neutral hydrogen in IGM. Sensitites on the figures are all for JWST. We can see that in can detect GRBs up to redshift 16 in the JWST K band and 36 in the M band.

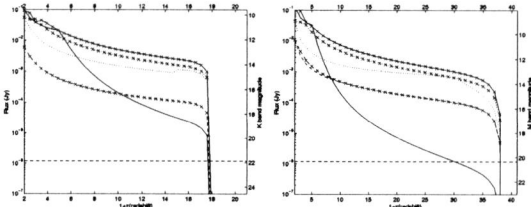

FIGURE 1. Combined forward and reverse shock observed flux as a function of redshift for $\varepsilon_{Bf} = 0.025$ and $\mathscr{R}_B = B_r/B_f = 1$. Forward shock (symbols), reverse shock (without symbols). Solid, dashed and dotted lines indicate emission at different observer times t=10 mins, t=2 hours and t=1 day respectively. a): K-band ($v = 1.36 \times 10^{14}$ Hz); b): M-band ($v = 6.3 \times 10^{13}$ Hz). Straight lines: in K and M bands sensitivities for JWST K & M bands are estimated for a resolution R=1000, S/N=10 and integration time of 1 hour. Parameters: $n = 1$ cm$^{-3}, \varepsilon_e = 0.1, E_{52} = 10, p = 2.5, \eta = 120$.

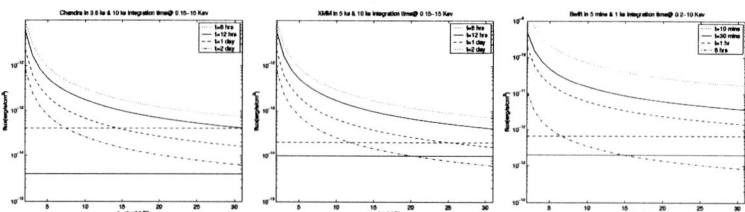

FIGURE 2. Light curves as a function of redshifts. Flux has been integrated over the observing energy ranges of 0.4-6 keV for Chandra, 0.15-15 Kev for XMM and 0.2-10 keV for Swift, respectively. a): for Chandra, the fluxes for the observer times $t_{obs} = 8$ hour, 12 hours, 1 day and 2 days as compared to its sensitivity horizontal lines for integration times of 3.6 ks (dashed) and 10 ks (solid). b): for XMM, same observer times as Chandra's. The sensitivity horizontal lines are for integration times of 5 ks (dashed) and 10 ks (solid). c): for Swift XRT, the fluxes are for observer time $t_{obs} = 10$ mins, 20 mins, 1 hour. The sensitivity horizontal lines are for integration times 300 s (dashed) and 1 ks (solid). Parameters: $\varepsilon_{Bf} = 0.001, \mathscr{R}_B = B_r/B_f = 5, \varepsilon_e = 0.1, E_{52} = 10, p = 2.5, \eta = 120$.

X-ray flux redshift dependence. The X-ray band flux evolution and its redshift dependence is simpler than in the O/IR bands because the reverse shock emission is generally negligible, and we need only consider the forward shock emission. One obvious characteristic of Figure 2 is that the flux from the two different density profiles are the same at all the redshifts for a given time, because the emission in both cases is in the density-independent regime. On Figure 2 (a) and (b), the X-ray flux is calculated for observer times $t = 8$ hours, 12 hours, 1 day and 2 days for Chandra and XMM, respectively. The flux is integrated over their own energy band. As we can see based on the figures that if GRBs exist at very high redshift, we can expect these detectors to be able to measure them in x-ray band.

DISCUSSION

We have considered the spectral time evolution and the flux in the near-IR K- and M-bands, as well as in the X-ray band, from GRBs at very high redshifts and different times. Because most of the radiation in the optical and ultraviolet bands from high-z GRBs, including GRBs, is absorbed by the IGM and the diffuse gas in our own galaxy. The X-ray and infrared bands are therefore of major importance for detecting and tracking high-z GRBs.

In the X-ray band, the Chandra and XMM sensitivities are substantially higher than those of Swift XRT, but their slewing time ($\lesssim 1$ day) limitations make Swift XRT a unique instrument for X-ray follow-up during the first day after a GRB trigger, when the burst is brighter. In spite of this, all three spacecraft should be able to detect very distant GRBs, if they exist, e.g. at $z \gtrsim 30$. In the K and M bands (2.2 and 4.8 μm) the JWST and other telescopes should be able to detect afterglows out to $z \lesssim 16$ and 33 within observer times 1 day for integrations times (with JWST) of 1 hour (see Figure 1).

ACKNOWLEDGMENTS

This research is supported through NASA NAG5-9153, NAG5-9192, NAG5-13286, and the Monell Foundation.

REFERENCES

1. Lamb, D.Q., & Reichart, D.E., ApJ, **536**, 1 (2000).
2. Ciardi, V, & Loeb, A., ApJ, **540**, 687 (2000).
3. Gou, L.J., Mészáros, P., Abel, T., & Zhang, B., ApJ, 604, 508 (astro-ph/0307489) (2004).
4. Sari, R., Piran, T., & Narayan, R., ApJ, **497**, L17 (1998).
5. Kobayashi, S., & Zhang, B., ApJ, **582**, L75 (2003).
6. Zhang, B., Kobayashi, S., & Mészáros, P., ApJ, **595**, 950 (2003).
7. Kobayashi, S., ApJ, **545**, 807 (2000).
8. Frail, D.A. et al., ApJ, **562**, L55 (2001).
9. Kauffmann, G., White, S.D.M., & Guiderdoni, B., MNRAS, **264**, 201 (1993).
10. Whalen, D., Abel, T., & Norman, M.L., ApJ, submitted (astro-ph/0310284) (2003).

Probing Cosmological Parameters with GRBs

T. Di Girolamo*, M. Vietri[†] and G. Di Sciascio*

INFN, Sezione di Napoli, Italy
[†]*Scuola Normale Superiore, Pisa, Italy*

Abstract. In light of the recent finding of the narrow clustering of the geometrically-corrected gamma-ray energies emitted by Gamma Ray Bursts (GRBs), we investigate the possibility of using these sources as standard candles to probe cosmological parameters such as the matter density Ω_m and the cosmological constant energy density Ω_Λ. By simulating different samples of gamma-ray bursts, based on recent observational results, we find that Ω_m (with the prior $\Omega_m + \Omega_\Lambda = 1$) can be determined with accuracy $\sim 7\%$ with data from 300 GRBs, provided a local calibration of the standard candles be achieved.

INTRODUCTION

Recent studies have pointed out that Gamma-Ray Bursts (GRBs) may be considered as standard cosmological candles. The prompt γ-ray energies of GRBs, after correction for the conical geometry of the jet, result clustered around a mean value of a few 10^{50} erg [1].

Since the discovery that GRBs lie at cosmological distances, about 30 redshifts have been measured. Apart from the controversial case of GRB 980425, possibly associated with the nearby supernova SN1998bw (at $z = 0.0085$), all other redshifts are spread within the wide 0.17–4.5 range. Therefore GRBs could be good candles to probe cosmological parameters [2] [3].

GRBs are thought to be associated with the death of massive (and short lived) stellar progenitors. Therefore the rate of GRB events per unit cosmological volume should be a tracer of the global history of star formation.

Hence, we have now all the information necessary to perform simulations of GRB distributions in a given cosmological model. Universal parameters such as the matter density fraction Ω_m and the cosmological constant energy fraction Ω_Λ, can be constrained by fitting the Hubble diagrams corresponding to such simulated distributions. It is the aim of this paper to simulate different GRB distributions and investigate their ability to determine the cosmological parameters Ω_m and Ω_Λ. Both universes with and without a cosmological constant Λ will be considered.

TO START: A KS TEST

First, in order to show what we are aiming at, we performed a Kolmogorov-Smirnov (KS) test on two data sets made of 300 GRBs simulated in two different cosmological

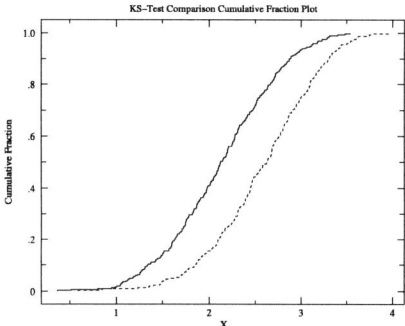

FIGURE 1. Comparison of the cumulative fractions obtained with the parameter $X \equiv \log d_L^2(Gpc)$ calculated for two data sets of 300 GRBs simulated in an universe with $\Omega_m = 1$ and $\Omega_\Lambda = 0$ (full line) and in one with $\Omega_m = 0.3$ and $\Omega_\Lambda = 0.7$ (dashed line). The corresponding probability that the two data sets are drawn from the same distribution is $Q_{KS} = 2.48 \cdot 10^{-14}$.

models, one with $\Omega_m = 1$ and $\Omega_\Lambda = 0$ and the other with $\Omega_m = 0.3$ and $\Omega_\Lambda = 0.7$, but both with a Hubble constant $H_0 = 65$ km s^{-1} Mpc^{-1} (as it will be assumed throughout the paper). We assume that GRBs are indeed standard candles with true prompt γ-ray energy released, E_γ, following a Gaussian distribution in its logarithm with mean $\mu = 50.7$ (if E_γ is expressed in *erg* units) and $\sigma = 0.3$ (corresponding to a multiplicative factor of 2) [1], and that they are distributed in the universe according to the model of star formation rate $R_{SF1}(z)$ reported in [4], which matches the $\log N - \log P$ relation (GRB number counts vs. peak photon flux) obtained with BATSE data. Applying the KS test on the redshift distributions, we found that the probability that the two data sets are drawn from the same distributions is $Q_{KS} = 0.031$, a "no man's land" value for this test. On the other hand, the application of the KS test on the parameter $\log d_L^2(z)$, where $d_L(z)$ is the luminosity distance, resulted in a significant probability $Q_{KS} \sim 10^{-14}$, which tells us that it is possible to discriminate between the two different cosmological models if a set of 300 GRB luminosity distances is known (see Figure 1).

DATA SET SIMULATIONS IN A $\Lambda = 0$ COSMOLOGY

We consider now a $\Lambda = 0$ cosmology, in which the only contribution to the density parameter is given by Ω_m. We assume for GRBs the same energy distribution as for the KS test. However, the assumed mean value is not relevant for our investigation, since it is the dispersion value that constrains the cosmological density parameter.

The standard candle energy is related to the fluence of the burst $f_\gamma = E_\gamma(1+z)/(4\pi d_L^2(z))$ via the luminosity distance $d_L(z)$. In order to have a linear propagation of errors throughout our simulations, we choose to construct with GRBs a Hubble diagram $\log d_L^2 - z$, since the distribution of the parameter $\log d_L^2$ is the same of that of $\log E_\gamma$, and therefore it is Gaussian.

In order to study the ability of GRBs in probing the cosmological parameters as a function of their number, we have simulated different samples with $N_{GRB} = 10, 30,$

TABLE 1. Mean values of the fitted cosmological density parameters Ω_m and Ω_Λ, of their error $\Delta\Omega$ and their dispersion S_Ω obtained by fitting 10^2 GRB sample realizations with N_{GRB} distributed according to function $R_{SF1}(z)$ of [4] in an Einstein-de Sitter universe ($\Omega_m = 1$, left) and in a flat universe with input values $\Omega_m = 0.3$ and $\Omega_\Lambda = 0.7$ (right).

N_{GRB}	$<\Omega_m>$	$<\Delta\Omega_m>$	S_{Ω_m}	N_{GRB}	$<\Omega_m>$	$<\Omega_\Lambda>$	$<\Delta\Omega>$	S_Ω
10	0.9983	0.2997	0.3097	10	0.3195	0.6805	0.1004	0.1307
30	1.0158	0.1895	0.1993	30	0.2973	0.7027	0.0763	0.0700
100	0.9937	0.0993	0.1108	100	0.3002	0.6998	0.0363	0.0351
300	0.9959	0.0599	0.0629	300	0.3023	0.6977	0.0219	0.0222
1000	1.0009	0.0332	0.0351	1000	0.3001	0.6999	0.0120	0.0125

100, 300 and 1000. Moreover, in order to be free from statistical fluctuations, we have performed 10^2 realizations of each of these samples.

The simulation of a GRB consists of the random sampling of both the redshift z and the true γ-ray energy released E_γ, according to the respective adopted distributions. Given a cosmological model, from these coupled values we obtain the corresponding value for the parameter $\log d_L^2$, which we plot on the Hubble diagram as a function of z. At this point we perform a χ^2 minimization of the simulated data to see with which accuracy the fit reproduces the input cosmology. The measurement error on $\log d_L^2$ is assumed to be $\sigma = 0.3$. The mean results of our repeated fits in an Einstein-de Sitter universe ($\Omega_m = 1$) are reported in the left side of Table 1.

DATA SET SIMULATIONS IN A Λ-DOMINATED COSMOLOGY

We move now to a Λ-dominated cosmology, in which the contributions to the density parameter are given by the mass density, Ω_m, and by the cosmological constant energy density, Ω_Λ. In light of the recent observations of the cosmic microwave background anisotropy [5], we put the prior of a flat universe $\Omega_m + \Omega_\Lambda = 1$.

Again, in order to study the ability of GRBs in probing the cosmological parameters in a Λ-dominated universe, we have simulated 10^2 realizations of GRB samples with N_{GRB} = 10, 30, 100, 300 and 1000. The χ^2 minimization of the resulting Hubble diagrams has been performed considering $\log d_L^2$ depending only on the fit parameter Ω_m, i.e., using the relation $\Omega_\Lambda = 1 - \Omega_m$. The right side of Table 1 reports the general results of our repeated fits for a flat cosmology with input values $\Omega_m = 0.3$ and $\Omega_\Lambda = 0.7$ (which are those adopted in [1]).

Focussing on the samples with $N_{GRB} = 300$, which represent the future data set expected from the *Swift* satellite experiment, Figure 2 shows one of the Hubble diagrams $\log d_L^2 - z$ obtained with the simulations (left), together with the distribution of the best fit values of the matter density fraction Ω_m for 10^3 sample realizations (right).

Finally, we must remark that the analysis in [1] assumes of course a particular set of cosmological parameters to derive the standard γ-ray energy of GRBs. To avoid a circular logic we should assume a candle calibration with a local sample of sources, a prospect which can now be considered possible in light of the discovery of the near GRB

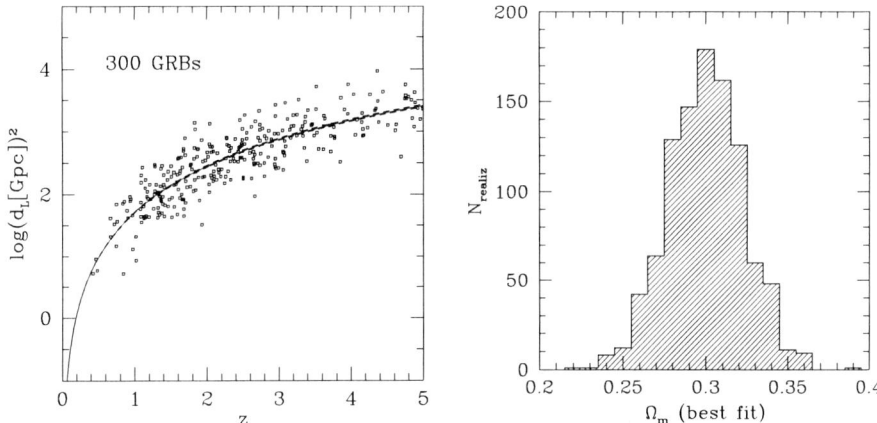

FIGURE 2. Left: Hubble diagram $\log d_L^2 - z$ with data simulated for a sample of 300 GRBs in a flat universe with density parameters $\Omega_m = 0.3$ and $\Omega_\Lambda = 0.7$. The solid curve shows the function $\log d_L^2(z)$ in the assumed cosmology, while the dashed curves give the dispersion about the best fit parameter (upper curve corresponds to lower Ω_m). Right: Histogram with the distribution of the best fit values of the matter density Ω_m for 10^3 realizations of a sample of 300 GRBs in a flat universe with density parameters $\Omega_m = 0.3$ and $\Omega_\Lambda = 0.7$. The distribution has a mean $<\Omega_m> = 0.3001$, a median $\Omega_m(med) = 0.3002$, a dispersion $S_{\Omega_m} = 0.0228$, and a kurtosis $k_{\Omega_m} = 3.0993$, to be compared with the value of a Gaussian distribution, *i.e.*, 3.

030329, with redshift as low as $z = 0.1685$.

CONCLUSIONS

We have simulated different samples of GRBs adopting γ-ray energy and redshift distributions consistent with recent observational results, in order to investigate their ability to probe cosmological parameters such as the density fractions Ω_m and Ω_Λ. Our result is that in a Λ-dominated flat universe the accuracy in the determination of the matter density Ω_m is $\sim 40\%$ for a sample with $N_{GRB} = 10$ and an excellent $\sim 4\%$ for $N_{GRB} = 1000$.

REFERENCES

1. Frail, D. A., et al., ApJ, **562**, L55 (2001).
2. Schaefer, B.E., ApJ, **583**, L67 (2003).
3. Takahashi, K., et al., astro-ph/0305260 (2003).
4. Porciani, C., and Madau, P., ApJ, **548**, 522 (2001).
5. Bennett, C. L., et al., ApJS, **148**, 1 (2003).

GENERAL OBSERVATIONS

GRblog: A Database for Gamma-Ray Bursts

Robert Quimby*, Erin McMahon* and Jeremy Murphy*

*McDonald Observatory, University of Texas at Austin, Austin, TX 78712

Abstract. GRBlog is an on-line database providing researchers with quick access to all information reported in the GCN Circulars. Users of the GRBlog web site (grad40.as.utexas.edu/grblog.php) can search the circulars and produce afterglow light curve plots, or compile data tables. The site also offers advanced search capabilities to aide in statistical studies or comparative research. Most of the GCNs have already been entered into the GRBlog database, with the remainder to follow shortly.

MOTIVATION

The GRB Coordinates Network (GCN), created by Scott Barthelmy[1], has proven to be an invaluable tool for the rapid dissemination of gamma-ray burst observations, and vital to the success of afterglow studies. The Notices portion of the GCN conveys burst locations and types automatically, in a format easily parsed at robotic observatories. The details of any ensuing observations are then sent out to the GRB community via the GCN Circulars. Although the Circulars are automatically distributed, the free format nature of these messages can make it difficult for the community to sort through all the information and find the data they need to address their research questions.

The first important step to organize the GCN Circulars was made by Jochen Greiner, who created a web site with the Circulars sorted by bursts and maintains a master table listing key observational data for each burst[2]. We have taken the further step of inserting all the data from the GCN Circulars into a relational database with a web interface called GRBlog[1]. In this format, the data can easily be retrieved by the entire community. Users of GRBlog can freely download observational data compiled from a number of Circulars in a tabular format, or produce plots of afterglow light curves.

THE DATABASE

We have implemented GRBlog using a PostgreSQL database, with a PHP web interface. To insert a new Circular into the database, we first download the ASCII file from the GCN archives. We then mark up that text with a custom formatting language, similar to Latex. For example, we would convert the passage "R = 18.5 mag" to "\filter{R} = \mag{18.5} mag". The marked up message is then passed through a PHP interpreter which converts it to HTML, and saves the marked up data to

---

[1] See http://grad40.as.utexas.edu/grblog.php

various tables (e.g. a table for optical observations). There are some common patterns as to how one might express a given observation in the GCN Circulars. We have identified several of these and have in place some routines to automatically mark up the Circulars. We then manually check the messages and make adjustments as necessary.

Once the messages have been marked up and their data appropriately stored in the database, PHP routines sort through the data and create HTML web pages on the fly.

Message View

The Message View is used to display the actual text of a Circular. The GRBlog default home page for example displays the 20 most recent messages. When more than one message is selected, only the first few lines of the text are displayed with a link to the full message. Each message begins with a title bar giving both the Circular subject line, and a link to the original Circular in the GCN archive. Data collected during the markup process is used to automatically classify the message, and a message category icon (e.g. eye glasses for an optical observation report) is shown on the right. Some of the marked up data will be converted to a link to another page. For example, the burst name GRB030329 will serve as a link to the 030329 burst page.

When a single message is selected, it will begin with a formated version of the Circular header followed by the list of authors (each name is a link to a list of messages the author is credited with). The message text will follow with embedded links to bursts, message references, and websites. Any tables will be presented as an in-line HTML table. At the end of the message text, the number of citations to the current message will be shown with a link to a list of those messages. Below the citation count are tables summarizing any observational data presented in the message.

Burst View

The Burst View is used to quickly convey all the available observational data for a given burst, or list of bursts. Shown in a table for each burst are the fluence, RA and DEC, number of optical observations, etc., as available. For HETE bursts, the trigger number will be determined on the fly by looking for messages reporting both the HETE trigger number and the GRB name, and a link will be provided to the appropriate HETE Burst Page[3]. When afterglow observations are available, detection of the transient is given as "yes" or "no" which serves as a link to the full light curve (it will be "yes" only if a message in the GRBlog database gives a magnitude for the OT which is not an upper limit).

When a single burst is selected, all of the messages related to that burst are displayed below the data summary table. An example of the burst table is shown in Fig. 1.

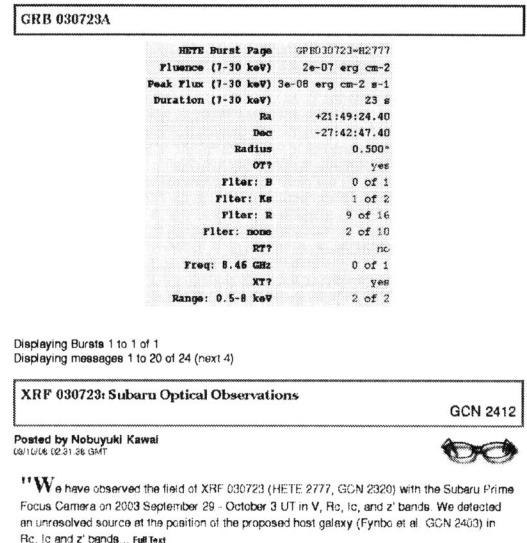

FIGURE 1. Detail from a screen shot of the GRB030723 burst page. Values in the table serve as links to other pages and light curve plots. Below the burst data summary are the introductions to the GCN circulars related to this burst (only one is shown). View the full page at http://grad40.as.utexas.edu/grblog.php?view=burst&GRB=030723.

Data View

Data can be extracted from a collection of messages and displayed either in a table, or as a plot. The source for each data point is given in the tables, and can be viewed from plots by clicking on a given data point. As an example, to view the optical light curve for GRB030723, first go to the burst page, then click on "yes" in the table next to "OT?" (Fig. 1). The data will be plotted using the magnitudes and observation times given in the Circulars (see Fig. 2); they will not be converted to the same reference system. To see what reference system a given observation employs, look for the "refsys" column in the data table. Light curves can also be made from radio or X-ray observations. GRB data for collections of bursts can also be displayed using the search form.

SEARCHING

The web site includes a search form to allow users to find messages, bursts, or data which meet specific criteria. One can search for all messages that contain the word "ROTSE" for example, or to find all bursts with afterglows discovered within one hour of the burst. The search form has fields to limit the search to specific burst names, author names, fluence ranges, durations, optical magnitudes, spectra, etc., and all of the fields can be combined to produce complex searches.

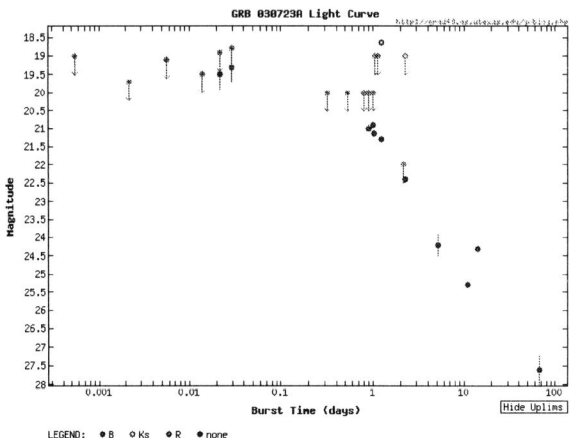

FIGURE 2. Lightcurve of GRB030723 generated automatically by GRBlog.

FUTURE

We will soon have all of the GCN Circulars entered into our database, and we will continue to add new Circulars as they become available. Once the database is complete with respect to the GCNs, we will consider expanding the database to include IAU Circulars, and possibly papers from astro-ph.

Although difficult to implement, using a markup language to convert prose into a machine readable format has proven quite successful. We note here that in the Swift era when the volume of GCN Circulars increases significantly, it may be important to agree on a common format for presenting observations to the community. If the GCNs reports were marked up to clearly identify data in a manner similar to that used by GRBlog, then authors would continue to have great freedom in how they express their results, but the results would be easier for the community to parse.

ACKNOWLEDGMENTS

We thank Bradley E. Schaefer, J. Craig Wheeler, Pawan Kumar, Peter Höflich, Chris Gerardy, and Alin Panaitescu for valuable suggestions on the layout and functionality of the GRBlog website. This work is supported in part by NSF AST 0098644.

REFERENCES

1. Barthelmy, S. D., The gcn web site (2003), URL http://gcn.gsfc.nasa.gov/gcn.
2. Greiner, J., Grb web page (2003), URL http://www.mpe.mpg.de/~jcg/grbgen.html.
3. Vanderspek, R., *GCN Circ.*, **2421** (2003).

Was the X-ray Afterglow of GRB 970815 Detected?

N. Mirabal*, J. P. Halpern*, E. V. Gotthelf* and R. Mukherjee[†]

Astronomy Department, Columbia University, 550 West 120th Street, New York, NY 10027
[†]*Dept. of Physics & Astronomy, Barnard College, New York, NY 10027*

Abstract. GRB 970815 was a well-localized gamma-ray burst (GRB) detected by the All-Sky Monitor (ASM) on the Rossi X-Ray Timing Explorer (RXTE) for which no afterglow was identified despite follow-up *ASCA* and *ROSAT* pointings and optical imaging to limiting magnitude $R > 23$. While an X-ray source, AX/RX J1606.8+8130, was detected just outside the ASM error box, it was never associated with the GRB because it was not clearly fading and because no optical afterglow was ever discovered. We recently made deep optical observations of the AX/RX J1606.8+8130 position, which is blank to a limit of $V > 24.3$ and $I > 24.0$, implying an X-ray–to–optical flux ratio $f_X/f_V > 500$. In view of this extreme limit, we analyze and reevaluate the *ASCA* and *ROSAT* data and conclude that the X-ray source AX/RX J1606.8+8130 was indeed the afterglow of GRB 970815, which corresponds to an optically "dark" GRB. Alternatively, if AX/RX J1608+8130 is discovered to be a persistent source, then it could be associated with EGRET source 3EG J1621+8203, whose error box includes this position.

INTRODUCTION

One of the most intriguing results from six years of GRB follow-ups at optical wavelengths is that roughly 60% of well-localized GRBs lack an optical transient despite intensive ground-based searches (e.g. [1]). Some of these "dark" GRBs could simply be due to a failure to image deeply or quickly enough. However, in certain cases the afterglow may have been missed in the optical either because it is obscured by dust in the host galaxy, or because it is located at high-redshift ($z > 5$). We discuss here X-ray and optical observations of GRB 970815, which support an interpretation consistent with quite possibly the first detection of a "dark" GRB in the afterglow era, preceding GRB 970828 in that category [2].

X-RAY OBSERVATIONS

GRB 970815 was localized by the ASM aboard RXTE on UT 1997 Aug. 15.50623, with a duration of ≈ 130 s [3]. Simultaneous detection with two of the ASM scanning cameras refined the position of GRB 970815 to the small error box shown in Figure 1. The superposed annulus based on the BATSE and Ulysses triangulation confirms the ASM position. Following the prompt localization by *RXTE*, two X-ray observations were made that cover the entire *RXTE* error box, one by *ASCA* [4] and one by the *ROSAT* High Resolution Imager (HRI) [5]. Analysis of the data revealed

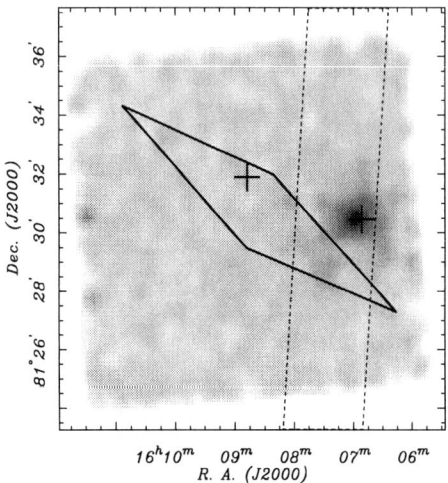

FIGURE 1. *ASCA* CCD SIS image of the field of GRB 970815, with the *RXTE* ASM error box (*solid line*) and Ulysses/BATSE annulus (*dashed lines*) superposed. *ROSAT* HRI point sources are indicated by *crosses*.

no source brighter than 1×10^{-13} erg cm^{-2} s^{-1} within the *RXTE* error box. There was, however, a source AX/RX J1606.8+8130 just outside the *RXTE* error box with an average flux $F_x(2-10\ \text{keV}) = 4.2 \times 10^{-13}$ erg cm^{-2} s^{-1}. Figure 1 shows the combined *ASCA* SIS image and the location of AX/RX J1606.8+8130 with respect to the burst error box. While AX/RX J1606.8+8130 lies just outside the *RXTE* error box, it is within the BATSE/Ulysses annulus. Another marginally significant *ROSAT* source RX J1608.8+8131 lies inside the *RXTE* error box, but it was not detected in the earlier *ASCA* observation. Hereafter we concentrate our discussion on AX/RX J1606.8+8130.

X-RAY LIGHT CURVE

Figure 2 shows the combined X-ray light curve of AX/RX J1606.8+8130. The *ASCA* light curve includes the sum of counts from all four detectors. The *ROSAT* points correspond to an extrapolated flux in the 2–10 keV band assuming the power-law spectral parameters derived from the *ASCA* spectra ($\Gamma = 1.64 \pm 0.35$ and $N_H < 1.3 \times 10^{21}$ cm^{-2}), which might not be entirely valid if an additional spectral component contributes significantly in the HRI soft band. The individual *ASCA* and *ROSAT* components of the light curve show no obvious evidence for variability. However, AX/RX J1606.8+8130 is consistent with a $F_x \propto t^{-1.4}$ flux decay between the *ASCA* and *ROSAT* observations, easily within the range of well-studied GRB X-ray afterglows. Moreover, the integrated 2–10 keV X-ray fluence corresponds to $\approx 10\%$ of the GRB fluence, in agreement with the properties of other GRBs [6].

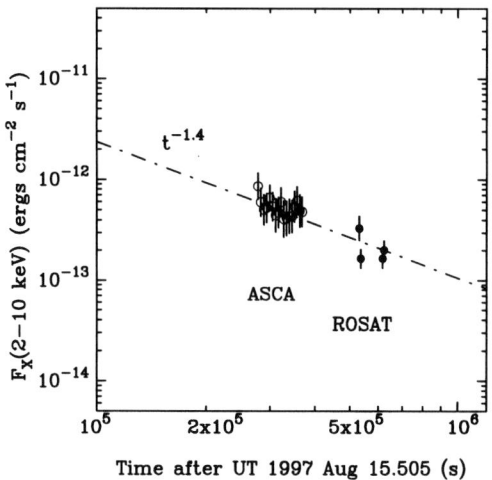

FIGURE 2. The X-ray light curve of AX/RX J1606.8+8130, derived by assuming the simultaneous power-law fit to the *ASCA* spectra. The *dash-dot* line shows a simple power-law decay $F_x \propto t^{-1.4}$, although the variation in the *ASCA* points are also consistent with no overall decay.

OPTICAL OBSERVATIONS OF AX/RX J1606.8+8130

Following the rapid dissemination of the *RXTE* position for GRB 970815, a number of groups conducted optical imaging of its error box including the position of AX/RX J1606.8+8130 as early as 17 hr after the burst [7]. At the time, no significant variable sources were found at the X-ray position or within the *RXTE* error box to an upper limit $R > 23$ [7]. Years later while conducting a search for the γ-ray source 3EG J1621+8203 [8], we reexamined the X-ray position of AX/RX J1606.8+8130 in several optical filters. Figure 3 shows the adopted $10''$ radius *ROSAT* error circle around the X-ray position, which is still optically blank to a 3σ limit of $V > 24.3$. In other filters, AX/RX J1606.8+8130 shows no evidence of a host galaxy or any other optical counterpart to limits of $B > 21.5, R > 22.0$, and $I > 24.0$.

WAS AX/RX J1606.8+8130 THE AFTERGLOW OF GRB 970815?

Starting with the observed X-ray flux density f_X, we can extrapolate a broad-band spectrum of the form $f_R = f_X(\nu_R/\nu_X)^{-\beta}$ where f_R is the R-band optical flux density at a frequency ν_R and β is the X-ray spectral index. From the *ASCA* spectra we have $f_X \approx 0.10\,\mu$Jy ($\nu_X = 4.84 \times 10^{17}$ Hz) at a time $t \sim 3.74$ days after the burst, and $\beta \approx 0.64$. The optical flux density evolution would then correspond to $f_R(t_d) \approx 55 t_d^{-1.4}\,\mu$Jy where t_d is days elapsed since the BATSE detection of GRB 970815. This translates into $R \approx 19.0$ on UT 1997 Aug. 16.31. Therefore, the predicted magnitude is brighter than the $R > 23$ upper limit reported at that time [7]. Such difference would tentatively support a

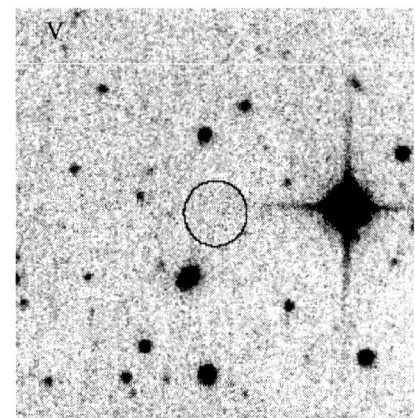

FIGURE 3. Zoom-in on a deep V image from the MDM Observatory at the location of the unidentified X-ray source AX/RX J1606.8+8130. The field is $2'$ across, and the *ROSAT* HRI error circle is $10''$ in radius. The 3σ upper limit is $V > 24.3$. North is up and east is to the left.

"dark" GRB classification. Nonetheless, the chance superposition of 3EG J1621+8203 and GRB 970815 introduces a slight doubt about the nature of AX/RX J1606.8+8130. This is because "dark" GRBs can temporarily mimic the characteristics of a plausible counterpart for unidentified EGRET sources, namely, rotation-powered pulsars [9].

CONCLUSIONS AND FUTURE WORK

In summary, the properties of AX/RX J1606.8+8130 support the idea that GRB 970815 corresponds to an optically "dark" GRB, quite possibly the first detection of a "dark" GRB in the afterglow era, preceding GRB 970828 in that category [2]. However, because of the chance superposition between 3EG J1621+8203 and GRB 970815, a slight doubt remains about the nature of AX/RX J1606.8+8130. *Chandra* observations are planned that should resolve the ambiguity of the possible connection between AX/RX J1606.8+8130 and either GRB 970815 or 3EG J1621+8203.

REFERENCES

1. Fynbo, J. U., et al., A&A **369**, 373 (2001).
2. Djorgovski, S. G., et al., ApJ **562**, 654 (2001).
3. Smith, D. A., et al., ApJ **526**, 683 (1999).
4. Murakami, T., et al., IAU Circular **6732**, 1 (1997).
5. Greiner, J., IAU Circular **6742**, 2 (1997).
6. Frontera, F., et al., ApJS **127**, 59 (2000).
7. Harrison, T. E., et al., IAU Circular **6721**, 1 (1997).
8. Mukherjee, R., Halpern, J. P., Mirabal, N., & Gotthelf, E., ApJ **574**, 693 (2002).
9. Mirabal, N., & Halpern, J. P., ApJ **547**, L137 (2001).

The Optical Afterglow of GRB 020305

J. Gorosabel[*,†], J. P. U. Fynbo[**], A. S. Fruchter[*], P. Nugent[‡], J. M. Castro Cerón[*], A. Levan[§], J. Rhoads[*], D. Bersier[*], I. Burud[*], A. J. Castro-Tirado[†] and J. Hjorth[¶,||]

[*]*3700 San Martin Drive, Baltimore, MD 21218, USA*
[†]*Instituto de Astrofísica de Andalucía (IAA-CSIC), Camino Bajo de Huétor, 24, E-18008 Granada, Spain*
[**]*Department of physics and Astronomy, University of Aarhus, Ny Munkegade, DK-8000 Århus C, Denmark*
[‡]*Lawrence Berkeley National Laboratory, MS 50-F, 1 Cyclotron Road, Berkeley, CA 94720, USA*
[§]*Department of Physics and Astronomy, University of Leicester, University Road, Leicester, LE1 7Rh, UK.*
[¶]*Niels Bohr Institute, Astronomical Observatory, University of Copenhagen, Juliane Maries Vej 30, DK-2100 Copenhagen Ø, Denmark*
[||]*on behalf of GOSH (GRB Optical Studies with HST).*

Abstract. We report ground-based and HST(+STIS) imaging of the GRB 020305 optical afterglow. The light curve and the spectral energy distribution of the afterglow might show evidence of a supernova (SN) component present $\sim$ 15 days after the GRB, which could be explained by a nearby I_c SN. However, a SN1998bw-like template does not yield a completely satisfactory simultaneous fit to the SED and the lightcurve. We discuss some possible reasons for this behaviour.

INTRODUCTION

GRB 020305 was localised by the HETE-II satellite on March 5.496818 UT (Ricker et al., [1]). The error box was later reduced with the triangulation annulus given by the Ulysses spacecraft. The high-energy emission as seen by Ulysses consisted of two broad pulses of gamma-ray radiation separated by $\sim$ 200s, with the total duration of the GRB $\sim$ 280s (Hurley et al., [2]). Price et al. ([3]) reported the presence of an object consistent with the HETE-II/IPN error box in images taken 20 hours after the GRB. Further imaging confirmed the fading behaviour of the candidate (Lee et al. [9]; Ohyama et al. [4]). In the present paper we report ground and space based optical observations carried out for the optical afterglow (OA) of GRB 020305 from 12 to 321 days after the burst.

OBSERVATIONS AND DATA REDUCTION

We monitored the GRB 020305 OA in the *R*-band with the 2.56-m Nordic Optical Telescope (NOT) on 2002 March 16-22. The optical calibration was carried out observing the Landolt field PG1047+003 (Landolt, [5]) with the 3.58-m New Technology Telescope (NTT). Additionally, the OA was observed during three epochs (mean epochs;

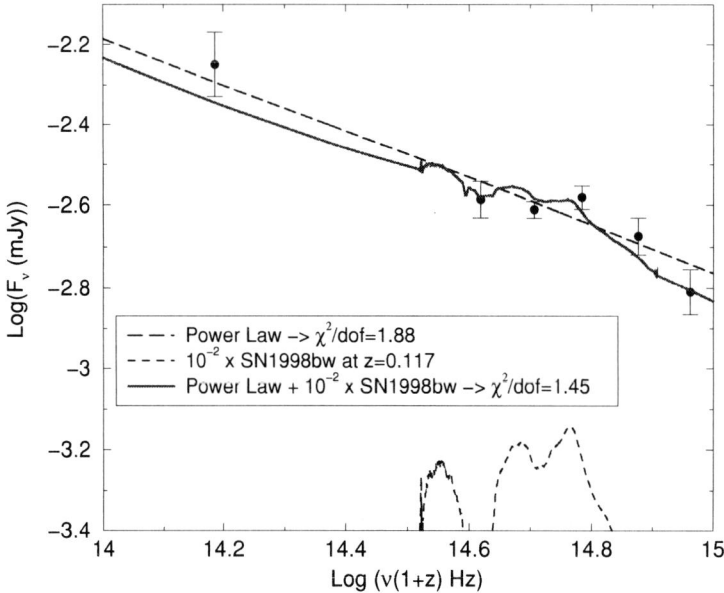

FIGURE 1. The plot shows the $UBVRIK$-band SED. The SED fit is improved with respect to the power law when a nearby ($z \sim 0.1$) dimmed (amplitude 1%) SN1998bw-like SN is included. The frequencies are given in the rest frame.

April 13, June 16 and December/January 2002/2003) using the Space Telescope Imaging Spectrograph (STIS) aboard the *HST*. For each of the three epochs *HST* observed the GRB field through the broad STIS Long-Pass and 50CCD filters, thus providing limited colour information.

AFTERGLOW SED

The optical multicolour imaging carried out during the first NOT observing night (March 16.9498-17.2258 UT) allowed us to construct the SED of the OA. Furthermore, making use of the K'-band detection on March 14.3 UT by Burud et al. ([6]) the optical SED was extended to the near-IR. The mentioned optical/near-IR data where extrapolated in time to a common date at March 17.0 UT, assuming an optical decay α ($F_\nu \sim t^{-\alpha}$), value of 1.3 as reported by (Lee et al., [9]) and consistent with our results. In any case, considering that the data points were taken close to March 17.0 UT, any reasonable value of α (i.e., $0.5 < \alpha < 2.5$) yields a similar SED. The optical and near-IR magnitudes have been transformed to flux densities following the conversion factors given by Fukugita et al. ([7]) and Allen ([8]), respectively. Then, the fluxes were corrected by the foreground Galactic extinction ($E(B-V) = 0.053$; Schlegel et al., [9]). The SED can be roughly approximated by a pure power law, but the result is

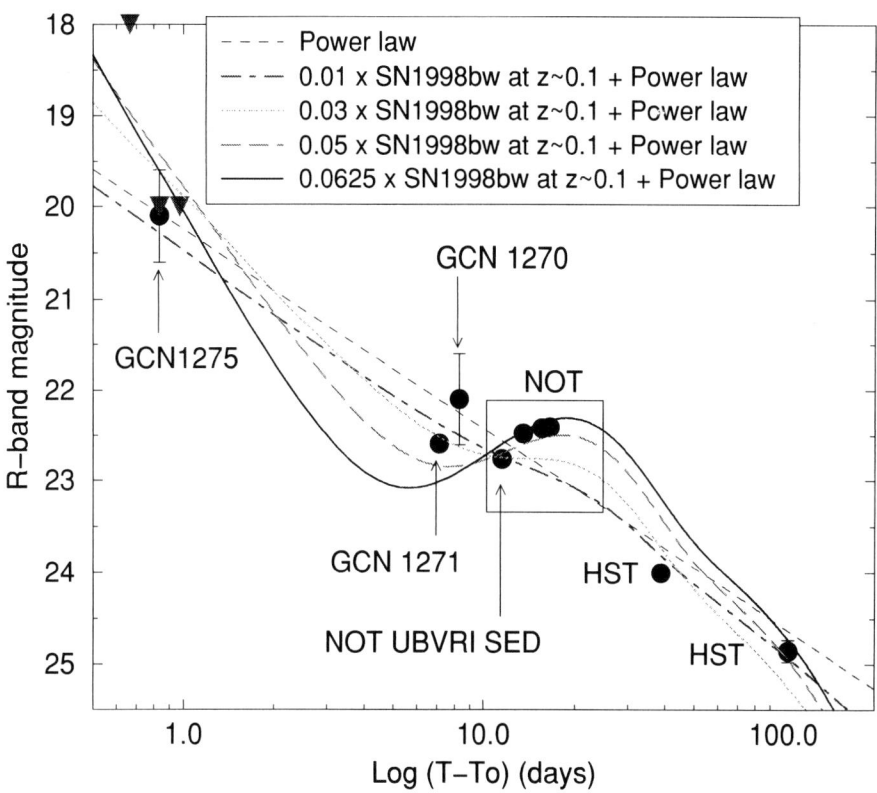

FIGURE 2. The plot shows R-band lightcurve of the GRB 020305 OA, which shows a prominent bump at ~15 days after the GRB. The different dotted lines show the R-band lightcurve resultant of fitting a power law plus a SN component for different amplitudes respect to SN1998bw. The solid line shows a power law fit with a decay index $\alpha = 1.3$

not completely satisfactory ($\chi^2/dof = 1.88$). As it is shown in Figure 1 the SED shows a potential bump at the V-band and also a small depression at the I-band. The detections in the UB-bands make difficult to reproduce the OA SED using SN1998bw templates (Patat et al. 2001) unless the redshift is $z \leq 0.4$. Formally the best fit is obtained introducing a SN1998bw spectrum at $z \sim 0.117$ having a dimming factor of ~ 100. For this redshift the SED fit is slightly improved with respect to a power law fit ($\chi^2/dof = 1.45$ vs. $\chi^2/dof = 1.88$, see Figure 1).

OPTICAL LIGHTCURVE

Figure 2 shows the R-band lightcurve of GRB 020305, being the triangles upper limits and the circles detections. The NOT data points are the ones included in the square.

The last two points were obtained subtracting the last HST visit images (Dec 2002 - Jan 2003) to the previous two ones (~ April 13, June 16 2002). The transformation of the STIS magnitudes to the R-band was performed with PHOTCAL, assuming a spectral index value of $\beta = -1$. As it is shown the lightcurve shows a prominent bump at ~15 days after the GRB. The plot also shows the fit when a fraction (1%, dot-dashed; 3%, dotted; 5%, long dashed; 6.25% continous thick line) of the SN1998bw light curve is added. As it is shown the best results are obtained when the SN1998bw peak is dimmed by a factor of ~20 (5% dotted line of Figure 2), corresponding to a $M_V \sim -16$ peak magnitude.

CONCLUSIONS

Both the SED and the lightcurve show a potential SN bump at ~ 12 days after the GRB. A SN1998bw template can reproduce both the SED and the lightcurve if strong dimming factors are assumed (1% for the SED and 5% for the lightcurve). However, both dimming factors do not agree, so a simultaneous SED+lightcurve solution is not possible based on SN1998bw. The main discrepancy comes from the K-band point (Burud et al., [6]), which makes the SED attenuation factor too high. Currently we are reexamining the original K-band data in order to double check the $UBVRIK$-band SED. Another explanation for this disagreement could be the different SN associated to GRB 020305 in comparison to SN1998bw. Other scenarios (dust echos, Esin & Blandford, [10]; refreshed shocks, Panaitescu et al., [11]) predict also bumps in the lightcurve/SED, so they should be explored. The mere NOT U-band detection gives an upper limit for the redshift of $z < 2.5$. This is inconsistent with the redshift ($z = 4.9$) predicted by Atteia et al. ([12]). A paper on the GRB 020305 OA is in progress (Gorosabel et al., [13]).

REFERENCES

1. Ricker, G., Atteia, J.-L., Kawai, N., et al., GCN 1262 (2002).
2. Hurley, K., Cline, T., Ricker, G., et al., GCN 1263 (2002).
3. Price, P., Fox, D.W., et al.,, GCN 1267 (2002).
9. Lee, B.C., Lamb, D.Q., Tucker, D.L., et al., GCN 1275 (2002).
4. Ohyama, Y., Yoshida, M., Kawabata, K.S., et al., GCN 1271 (2002).
5. Landolt, A.U., AJ **104**, 340. (1992)
6. Burud, I., Rhoads, J., Fruchter, A., & Griep, A., GCN 1283 (2002).
7. Fukugita, M., Shimasaku, K. & Ichikawa, T., PASP, **107**, 945 (1995).
8. Allen, C.W., *Allen's Astrophysical Quantities*, 4th edition, ed. A. N. Cox (2000).
9. Schlegel, D.J., Finkbeiner, D.P., & Davis, M., ApJ **500**, 525 (1998).
10. Esin, A.A., & Blandford, R., ApJ **534**, L151 (2000).
11. Panaitescu, A., Meszaros, P., & Rees, M.J., ApJ **503**, 314 (1998).
12. Atteia, J.-L., A&A 407, L1 (2003).
13. Gorosabel, J., Fynbo, J.P.U., Fruchter, A., et al., A&A in preparation (2004).
14. Patat, F., Cappellaro, E., Danziger, J., et al., ApJ **555**, 900 (2001).

DMSP 14 Observations of GRB011121 and the Giant SGR1900+14 Flare of 98/08/27

James Terrell* and Ray W. Klebesadel*

*Los Alamos National Laboratory, ISR-2, MS B244, Los Alamos, NM 87545, USA

Abstract. The bright gamma-ray burst GRB011121 was observed by DMSP 13 and DMSP 14 at a time resolution of 2s. Event data also obtained by DMSP 14 covered 13.1s of the burst at a time resolution of 12.8ms, and an energy range of 53-3000 keV. Fourier analysis gives evidence, at 95% confidence, of a 1.7s oscillation in the event data. DMSP 14 data for the giant 98/08/27 flare of SGR1900+14 are also presented, giving high-time-resolution data, not previously available, on the initial outburst.

INTRODUCTION

The launch of DMSP 14 on 4 April 1997 considerably increased the capabilities of the DMSP gamma-ray burst detection system. DMSP 14 carries two gamma-ray detectors, each with $\sim$100 cm^2 of NaI (the SSBX2 model). This is more sensitive than the two detectors aboard DMSP 13, each with 40 cm^2 of CSI. Both spacecraft observed the very bright gamma-ray burst GRB011121, and high-time-resolution data were obtained by DMSP 14, with the triggering of the event data system. A preliminary report has been published [1]. Similar results have been obtained for many other gamma-ray bursts [2], as well as for the extremely bright outburst of SGR1900+14 on 27 August 1998. Of particular interest is the time structure of the beginning of the flare.

OBSERVATIONS OF GRB011121

The gamma-ray burst GRB011121 was detected by BeppoSAX at 67641 UT [3]; its position was found to be (173.6059,−76.027)(J2000) [4]. It was also observed by a number of other spacecraft [5]. It was found to have an optical afterglow [6], with a redshift of z=0.36 [7], and an underlying supernova signature [5]. The time history of this GRB, as observed by DMSP 14, is shown in Figure 1 (left), for five different energy thresholds. The data for DMSP 13 (not shown) were similar. The burst lasted for about 20s. The event data from DMSP 14 are shown in Figure 1 (right), summed to 0.2048s to reduce counting rate fluctuations.

There is an appearance of a 1.7s fluctuation in the event data. Fourier analysis gives more evidence of this, as shown in Figure 2 (left). The highest peak in this figure, not counting the very low-frequency peak due to the 13.1s rise and fall of the event data, is at a period of 1.7±0.1s, the peak having a formal significance of 95% (i.e., there is

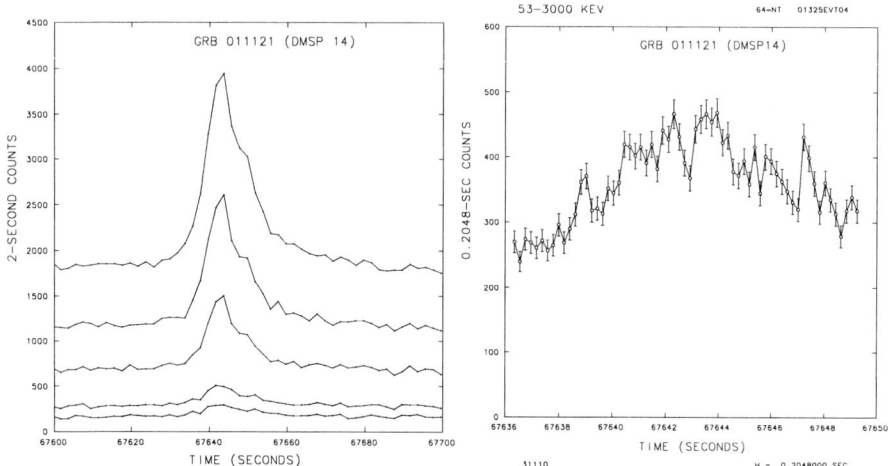

FIGURE 1. **Left:** DMSP 14 data for GRB011121 are shown here as 2s totals. The detection thresholds are for 53, 129, 228, 462, and 580 keV; the upper limit on the channels was 1000 keV. **Right:** DMSP 14 event data (time resolution 12.8ms) are shown as summed to 204.8ms to reduce counting-rate fluctuations. There is evidence of a 1.7s oscillation in the 13.1s record of GRB011121.

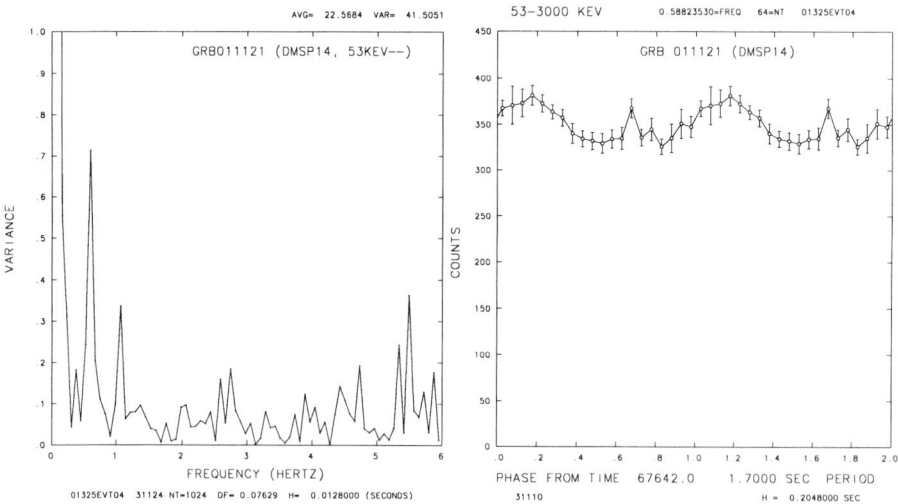

FIGURE 2. **Left:** Fourier analysis of GRB011121 event data shows a prominent peak at 0.6Hz, plus several smaller peaks which are not statistically significant. **Right:** A phase plot of the GRB011121 event data shows some evidence of a second harmonic.

a 5% probability of a peak of this height appearing by chance somewhere in the power spectrum). A phase plot of the same data, seen in Figure 2 (right), gives some evidence of a second harmonic.

The usually assumed models of gamma-ray bursts do not include rotational or other periodicities, except for soft gamma-ray (SGR) bursts, which are often very intense but are believed not to be produced by the same mechanisms as GRB. Thus it would seem unlikely for such a periodicity to be found in this case. However, many GRB's do give evidence of a number of counting-rate peaks at intervals of a few seconds. There are usually not enough peaks in the relatively short bursts to yield convincing evidence of periodicity. This particular event, GRB011121, lasts long enough for a 1.7s periodicity to be evident, although not with overwhelming statistical significance.

OBSERVATIONS OF THE GIANT FLARE (98/08/27) FROM SGR1900+14

The soft gamma-ray repeaters have produced the most powerful outbursts of gamma—rays, such as the notable burst from SGR0526−66 on 5 March 1979 [8] It may be noted that only the soft gamma-ray output has repeated, not (so far) the powerful high-energy gamma-ray outbursts. The other very strong SGR outburst occurred on 27 August 1998, from SGR1900+14 [9]. Figure 3 (left) shows the first 250s of the burst, as observed by DMSP 14. The 5.16s periodicity is apparent, although only in the lowest-energy data (53-129 keV). The data from DMSP 13 (not shown) give much less evidence of periodicity, probably due to a slightly higher energy threshold, $\sim$60keV.

The high-time-resolution event data obtained from DMSP 14 on 98/08/27, seen in Figure 3 (right), show the time structure of the initial peak in detail. The initial burst, lasting about 0.2s, rose to a maximum counting rate of $\sim$200,000 c/s. Such a high rate led to considerable counting losses, amounting to more than 50% at the highest rates, and $\sim$11% at the counting rate of the second major peak. The second burst, about 0.5s after the initial peak, rose to a lower but still considerable counting rate level, and also lasted about 0.2s. There are also numerous shorter peaks in the count rate, at roughly 0.1s intervals. The cause of such structure in the initial bursts from SGR1900+14 is not clear at this time.

CONCLUSIONS

The data from GRB01112 give evidence, at 95% confidence, of a 1.7s periodicity. Data from the 98/08/27 outburst from SGR1900+14 are given here as an example of periodic emission from an SGR, as well as for the interest of this new data. There may be more similarity in the sources of such periodic gamma-ray outbursts than is commonly accepted.

ACKNOWLEDGMENTS

The design and execution of these DMSP gamma-ray experiments owes much to the work of J. W. Griffee, whose continuing advice has been very helpful. This work has

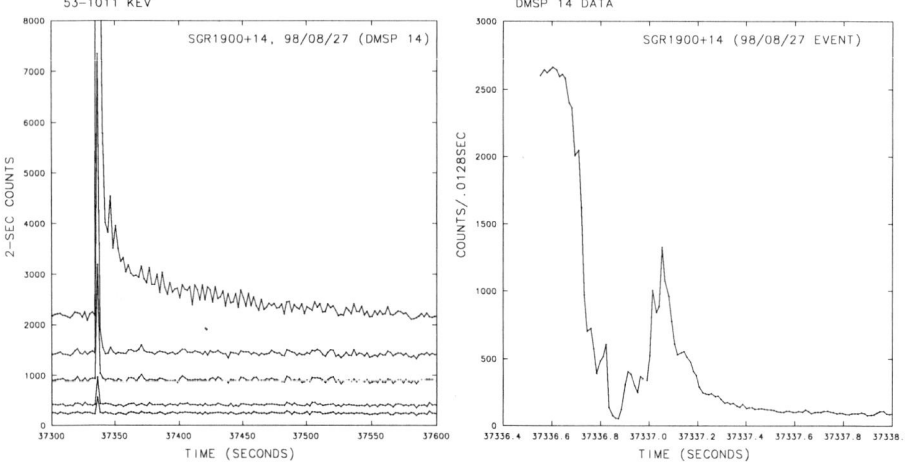

FIGURE 3. Left: DMSP 14 data for the strong outburst of SGR1900+14 on 27 August 1998. The 2s data show the 5.16s periodicity, present only at the lowest energy (53keV threshold). Other data thresholds are 129, 228, 462, and 580 keV. **Right:** DMSP 14 event data for the initial part of the SGR1900+14 outburst, showing two prominent peaks and numerous smaller peaks. The data have not been corrected for counting-rate losses.

been supported by the U. S. Department of Defense, and by the U. S. Department of Energy.

REFERENCES

1. Terrell, J., and Klebesadel, R. W., BullAPS **47**, 140 (2002).
2. (e.g.) Terrell, J., Lee, P., Klebesadel, R. W., and Griffee, J. W., 4th Huntsville GRB Symposium, AIP **428**, 54 (1998).
3. Piro, L., GCN1147 (2001).
4. Piro, L., GCN1149 (2001).
5. Price, P. A., et al., ApJ **572**, L51 (2002).
6. Wyrzykowski, L., Stanek, K. Z., and Garnavich, P. M., GCN1150 (2001).
7. Infante, L., Garnavich, P. M., Stanek, K. Z., and Wyrzykowski, L., GCN1152 (2001)
8. Mazets, E. P., et al., Nature **282**, 587 (1979).
9. Hurley, K., et al., Nature **397**, 41 (1999).

GENERAL THEORY

An Integrated Universal Collapsar Gamma-ray Burst Model

Jay D. Salmonson

Lawrence Livermore National Laboratory, Liveremore, CA 94551

Abstract. Starting with two assumptions: (1) gamma-ray bursts originate from stellar death phenomena or so called "collapsars" and (2) that these bursts are quasi-universal, whereby the majority of the observed variation is due to our perspective of the jet, an integrated gamma-ray burst model is proposed. It is found that several of the key correlations in the data can be naturally explained with this simple picture and another possible correlation is predicted.

INTRODUCTION

A wealth of data is continuing to be accrued that indicates gamma-ray bursts derive from stellar death events. To date there are two certain directly observed associations of a GRB with a type Ib/c supernova: GRB980425 with sn1998bw [1] and GRB030329 with sn2003dh[2]. In several other cases a ruddy bump[e.g. 3, 4], ostensibly the underlying supernova, has been observed in the decaying GRB afterglow lightcurves. These direct indications, coupled with the fact that GRBs tend to originate in star forming regions[5, 6], seem to indicate that GRB explosions are the product of a stellar death event which also produces a supernova - so called "collapsars"[7].

At the same time evidence is emerging to indicate that cosmological gamma-ray bursts might be quasi-universal in that the innate variation from burst to burst might be of order a factor of two or so, while the observed variation of fluxes and timescales can be as much as two orders of magnitude. In the same spirit of the unified models for active galactic nuclei, the wide range of observed GRB quantities is dependent on the observer viewing angle with respect to a single, universal structured jet.

Herein I simply combine these two proposed features of GRBs and attempt to formulate an integrated model. It is found that by literally interpreting the hydrodynamic nature of the emerging jet from a collapsar, as shown from numerical simulations, it is relatively natural to be able to broadcast material into the range of angles and energies inferred from the correlations in the data. A model is sketched out which nominally satisfies the key correlations in the data. As such, this model provides an integrated picture of what a universal collapsar GRB might look like.

THE MODEL

Begin with a collapsar. Numerical simulations[8, 9] demonstrate that introducing an energy source at the center of a collapsing star can produce an energetic jet that bores along the rotation axis of the star and erupts from the surface intact, thereby blowing up the star in the process. They also demonstrate modest Lorentz factors, $\gamma \sim 10-20$ of the jet as it emerges from the star, with a similar internal energy per rest mass $\eta \sim 10-20$.

If GRBs derive from collapsars, I argue that, regardless of the nature of the source of energy at the core of the star, these simulations must be qualitatively correct. Therefore, the first assumption of this model is that a relativistic jet emerging from a star is hydrodynamic in nature; the second law of thermodynamics virtually ensures that shocks, instabilities, magnetic reconnections will render the emerging material hot and with high entropy. Furthermore, I make the plausible assumption, inspired by said jet simulations, that the energy of the emerging jet material will be roughly equally partitioned between kinetic energy and internal energy: $\gamma \approx \eta$.

As such, when a parcel of jet material emerges from the star, no longer being confined, it will explode. This explosion must be isotropic in the co-moving frame of the material and will blow it into a thin spherical shell with terminal Lorentz factor equal to η. In the lab frame the parcel emerged from the star with Lorentz factor, γ, therefore the bulk of the energy in the shell will be beamed into an angle $1/\gamma$ (see Fig. 1). Still in the lab frame, the terminal Lorentz factor of the exploded shell moving along the jet axis will be $\Gamma \sim \gamma\eta$ and, since $\gamma \approx \eta$, will vary from $\Gamma \approx 2\gamma^2$ along the jet axis, to $\Gamma \approx \gamma^2$ at an angle $1/\gamma$ from the jet axis. So in this model $1/\gamma$ is angular scale and $\Gamma \propto \gamma^2$ is the specific energy (i.e. Lorentz factor) scale. This secondary acceleration scenario has also been proposed in the "firework" model of GRBs from highly magnetized black holes [10].

An immediate consequence of this model is a simple correspondence between opening angle, θ, and Lorentz factor, Γ: $\theta \propto 1/\gamma \propto 1/\sqrt{\Gamma}$. This provides a direct correspondence between viewing angles, as inferred from afterglow jet-break times, and the Lorentz factors of the GRB prompt emission. These viewing angles range from $3°$ to $20°$, implying a range of γ from 20 to 3 [11]. Assuming $\eta \approx \gamma$, the final Lorentz factor Γ will range from $2\gamma\eta$ to $\gamma\eta$, or from 800 to 9.

It is an intriguing correspondence that this range of Lorentz factors, from ~ 1000 for the brightest bursts to ~ 10 for the dimmest, is consistent with that found by using kinematic arguments [12] and assuming GRB980425 was only mildly relativistic, $\Gamma_{980425} \sim 2$. Furthermore, several workers[13, 14] have independently given $\Gamma_{990123} \sim 1000$, one of the brightest burst ever observed. In fact it might be telling that the lower end of this range, $\Gamma \sim 10$, is about the lower limit that compactness of the source can allow for optically thin, non-thermal emission [15]. Such a range of Lorentz factors, ranging over a factor of ~ 100 is perhaps unorthodox, however, given that luminosities, timescales and energies span this range, I argue that such variation in Lorentz factor is, at least, plausible.

So in this model the jet that emerges from the collapsar has a variable Lorentz factor, $3 \lesssim \gamma(t) \approx \eta(t) \lesssim 20$ which, by secondary expansion, generates a sequence of concentric expanding shells, each with its own terminal Lorentz factor, $10 \lesssim \Gamma \lesssim 1000$, and opening

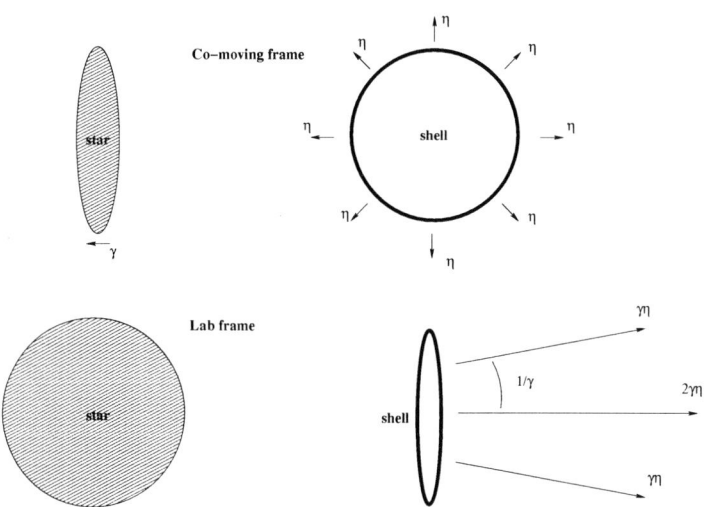

FIGURE 1. A schematic representation of a blob of material emitted from a dying star with Lorentz factor γ and specific internal energy η. No longer confined by the star, the blob expands isotropically, in its co-moving frame, into a shell moving with terminal Lorentz factor η. In the lab frame this material is boosted to a Lorentz factor $\sim \gamma\eta$, but with a characteristic beaming angle $1/\gamma$.

angle, $20° \gtrsim \theta \sim 1/\gamma \gtrsim 3°$. The subsequent collision of these shells creates a viewing angle dependent variety of GRBs and the interaction with the interstellar medium creates a viewer angle dependent afterglow.

COMPARISON WITH THE CORRELATIONS

An intriguing property of GRBs and their afterglows that has emerged in the last few years is that, save a few interesting exceptions, they tend to align along several simple correlations. The existence of simple monotonic relationships of several key observable energy and timescales strongly implies a single variable that governs the observed variation. With this realization it becomes an intriguing possibility that all bursts are, to first order, quite similar and that the dominant contributor to the observed diversity is observer angle with respect to the burst jet axis. This paradigm has the added benefit of identifying the expected variation with viewing angle as being the source of the observed breadth of the correlations. As described in the following, the integrated universal collapsar model outlined here can describe, or consistently draw inferences from, most of the existing correlations.

- ✔ $E_\gamma \propto \text{constant}$ The fact that the majority of cosmological GRBs seem to have a constant total energy in gamma-rays[11, 16], despite the wide variation in their luminosities and timescales, is an intriguing and mysterious clue about the progenitor. However, this mysterious clue becomes a prosaic and even trivial matter of perspective if the energy per steradian of the afterglow shock varies as $\varepsilon \propto \theta^{-2}$

with respect to the jet axis[17, 18]. Thus the observed variety stems merely from the variety of observer angles on a universal jet. If one takes this afterglow energy distribution as given, then one can derive the implied energetics of the emerging fireball. This gives that the energy and mass distributions of the emitted blobs are

$$\frac{d\varepsilon}{d\gamma} \propto \gamma, \quad \frac{dM}{d\gamma} \propto 1/\gamma. \tag{1}$$

The implication that the most energetic material is the least mass loaded is very plausible; the lightest material moves the fastest. There is no obvious physical reason that these scalings be exact and they are likely approximate.

✔ $L \propto \Delta t^{-1}$ One of the earliest discovered correlations is that of peak luminosity, L, with the reciprocal temporal lag, Δt, of lower energy gamma-rays behind their higher energy counterparts within a prompt emission pulse [19]. A kinematic interpretation[20] simply states that luminosity will scale with Lorentz factor and temporal lag will vary inversely with Lorentz factor[see also Sec. 3 of ref. 21]. In this mode observers nearer the jet axis will see higher velocity material and thus larger luminosities and shorter lags.

✔ $L \propto$ variability   The trend that more luminous bursts tend to be more variable, with more, and narrower, spikes[22, 23] has a natural explanation in this model. As with the lag-luminosity relationship discussed above, the highest energy shells will have their energy confined more closely along the jet axis, and thus the more energetic collisions will radiate more closely along the jet axis. Thus observers near the jet axis will observe more collisions, therefore more variability, than those at higher viewing angles.

✔ $\Delta t \propto t_j$ A surprisingly tight connection between the prompt gamma-ray phase and subsequent afterglow phase[21] is seen by this linear correlation between the temporal lag, Δt, of lower energy gamma-rays behind their higher energy counterparts within a prompt emission pulse, and the jet-break time, t_j, at which the afterglow temporal power-law decay index steepens. This correlation is demonstrable by noting $t_j \propto \varepsilon^{1/3} \theta^{8/3} \propto \theta^2 \propto \gamma^{-2} \propto \Gamma^{-1}$, where $\varepsilon \propto \theta^{-2}$ from the structured jet, and $\Delta t \propto \Gamma^{-1}$ from the kinematics of the lag-luminosity relationship, so $t_j \propto \Delta t$.

✗ $E_{pk} \propto \sqrt{E_{iso}}$ A clue to the gamma-ray emission mechanism is that the peak of the spectral energy distribution, E_{pk}, varies like the square root of the inferred isotropic burst energy, E_{iso} [24, 25]. This relationship is quite enigmatic. It is the opinion of the author that until the nature of the emission mechanism is better understood, this correlation remains unexplained within the context of this model or any other. Future work will address this issue.

☞ Prediction: $t_{dec} \propto \theta^{14/3}$ A prediction of this model is that there should be a rather strong dependence of the deceleration time, t_{dec}, which roughly corresponds to the commencement of the forward shock afterglow emission, on viewing angle, θ. Specifically, $t_{dec} \propto (\varepsilon \Gamma^{-8})^{1/3} \propto \Gamma^{-7/3} \propto \theta^{14/3}$. Thus the expected range of afterglow onsets should be $(20°/3°)^{14/3} \sim$ seconds per day. Therefore, given the very prompt, ~ 1 second, onset of the afterglow and attendant reverse shock of bright burst such as 990123, one would expect an onset of the external shock afterglow for weaker bursts to be seen at ~ 1 day, which is precisely what is observed [e.g. 970508, 26].

The model introduced here is very coarse. However, it does describe several of the key correlations encompassing a broad range of physical observables, and suggests another, in the context of a collapsar model by simply and naturally employing the result of numerical simulations that the jet eruption is hydrodynamic. This logically and directly leads to the idea of secondary acceleration and expansion of the equipartition heated gas emerging from the star. Indeed, if GRBs both derive from collapsars and are quasi-universal, then it is highly plausible that at least some of the key features described here must come into play. It is interesting that the picture of the gamma-ray burst jet described here is qualitatively different than what has been described in the literature to date.

ACKNOWLEDGMENTS

I wish to thank E. Ramirez-Ruiz, S. Woosley and A. MacFadyen for useful discussions regarding this work. This work was performed under the auspices of the U.S. Department of Energy by University of California Lawrence Livermore National Laboratory under contract W-7405-ENG-48.

REFERENCES

1. Galama, T. J. et al. Nature **395**, 670–672 (1998).
2. Stanek, K. Z. et al. ApJL **591**, L17–L20 (2003).
3. Bloom, J. S. et al. Nature **401**, 453–456 (1999).
4. Reichart, D. E., ApJ **554**, 643–659 (2001).
5. Hogg, D. W., and Fruchter, A. S., ApJ **520**, 54–58 (1999).
6. Bloom, J. S., Kulkarni, S. R., and Djorgovski, S. G., AJ **123**, 1111–1148 (2002).
7. MacFadyen, A. I., and Woosley, S. E., ApJ **524**, 262–289 (1999).
8. Zhang, W., Woosley, S. E., and MacFadyen, A. I., ApJ **586**, 356–371 (2003).
9. Aloy, M. A., Müller, E., Ibáñez, J. M., Martí, J. M., & MacFadyen, A., ApJL **531**, L119 (2000).
10. Barbiellini, G., Celotti, A., and Longo, F., MNRAS **339**, L17–L21 (2003).
11. Frail, D. A. et al. ApJL **562**, L55–L58 (2001).
12. Salmonson, J. D., ApJL **546**, L29–L31 (2001).
13. Panaitescu, A., and Kumar, P., ApJ **554**, 667–677 (2001).
14. Soderberg, A. M., and Ramirez-Ruiz, E., *AIP Conference Series 662*, 2003, pp. 172–175.
15. Piran, T., Phys. Rep. **333**, 529–553 (2000).
16. Bloom, J. S., Frail, D. A., and Kulkarni, S. R., ApJ **594**, 674–683 (2003).
17. Rossi, E., Lazzati, D., and Rees, M. J., MNRAS **332**, 945–950 (2002).
18. Zhang, B., and Mészáros, P., ApJ **571**, 876–879 (2002).
19. Norris, J. P., Marani, G. F., and Bonnell, J. T., ApJ **534**, 248–257 (2000).
20. Salmonson, J. D., ApJL **544**, L115–L117 (2000).
21. Salmonson, J. D., and Galama, T. J., ApJ **569**, 682–688 (2002).
22. Ramirez-Ruiz, E., and Fenimore, E., *astro-ph/0004176* (2000).
23. Reichart, D. E. et al. ApJ **552**, 57–71 (2001).
24. Lloyd, N. M., Petrosian, V., and Mallozzi, R. S., ApJ **534**, 227–238 (2000).
25. Amati, L. et al. AAP **390**, 81–89 (2002).
26. Pian, E. et al. ApJL **492**, L103 (1998).

Electromagnetic (versus Fireball) Model of GRBs

M. Lyutikov

Physics Department, McGill University, Montreal, QC, H3A 2T8I Canada

Abstract. We briefly review the electromagnetic model of Gamma Ray Bursts and then discuss how various models account for high prompt polarization. We argue that if polarization is confirmed at a level $\Pi \geq 10\%$ the internal shock model is excluded.

ELECTROMAGNETIC MODEL

The electromagnetic model interprets Gamma Ray Bursts (GRBs) as relativistic, electromagnetic explosions [1], (also Lyutikov & Blandford, in preparation), see Fig. 1. It is assumed that a rotating, relativistic, stellar-mass progenitor loses much of its rotational energy in the form of a Poynting flux during the active period lasting ~ 100 sec. The energy to power the GRBs comes eventually from the rotational energy of the progenitor, converted into magnetic energy by the dynamo action of the unipolar inductor, so that the central source acts as a power-supply generating a current flow (along the axis, the surface of the bubble and the equator). Initially a non-spherically symmetric, electromagnetically dominated bubble expands non-relativistically inside the star, most rapidly along the rotational axis of the progenitor. The velocity of expansion of the bubble is determined by the pressure balance on the contact between magnetic pressure in the bubble and the ram pressure of the stellar material. After the bubble breaks out from the stellar surface and most of the electron-positron pairs necessarily present in the initial outflow quickly annihilate, the bubble expansion becomes highly relativistic. After the end of the source activity most of the magnetic energy is concentrated in a thin shell inside the contact discontinuity between the ejecta and the shocked circumstellar material. The electromagnetic shell pushes ahead of it a relativistic blast wave into the circumstellar medium. Current-driven instabilities develop in this shell at a radius $\sim 3 \times 10^{16}$ cm and lead to acceleration of pairs which are responsible for the γ-ray burst. At larger radii the energy contained in the electromagnetic shell is mostly transferred to the preceding blast wave. Particles accelerated at the fluid shock may combine with electromagnetic field from the electromagnetic shell to produce the afterglow emission.

The electromagnetic model produces a "structured jet" with energy $E_\Omega \propto \sin^{-2}\theta$ in a natural way (in fact, there is no proper "jet", but rather a non-spherical outflow and non-spherical shock wave); there is no problem with "orphan afterglows" since GRBs are produced over large solid angles; X-ray flashes are interpreted as GRBs seen "from the side", but their total energetics should be comparable to proper GRBs; the model can qualitatively reproduce hard-to-soft spectral evolution as synchrotron emission in an ever decreasing magnetic field $B \propto \sqrt{L}/r$ (L is luminosity, r is emission radius), akin to "radius-to-frequency mapping" in radio pulsars; similarly, the correlation

$E_{peak} \sim \sqrt{L}$ is also a natural consequence. Finally, high polarization of prompt emission may also be produced [2] (it should correlate with the spectral index; if there is a mixing between circumstellar material and ejecta, *e.g.* due to Richtmyer-Meshkov instability, and if optical polarization is seen, then the position angles of the prompt emission and afterglow should coincide and be constant over time; fractional polarization should be independent of the "jet break" time, but may show variations due to turbulent mixing).

POLARIZATION OF PROMPT EMISSION

A very high linear polarization (nominally 80%) has been reported in RHESSI observations of GRB021206 [3]. Although the uncertainty in the measured polarization is large and a degree of reservation about the result is necessary, the observation, **if correct**, puts strong constraints on the GRB models. In particular, it is inconsistent with the internal shock model, as we argue below.

High polarization fraction cannot be naturally produced in a fireball model, since the magnetic fields expected in this model are produced on small microscopic scales. In order to account for the high polarization fraction of the prompt emission, the fireball model needs to make four limiting assumptions three of which are made specifically in order to explain polarization (some, but not all are listed in [4]).

First, the turbulent magnetic fields, generated presumably by the Weibel instability, are assumed to be exactly two-dimensional in the whole shocked region. Weibel instability [5] indeed produces one-dimensional current filaments oriented normal to the surface of the shock front, but the typical coherence size of the magnetic field is the ion skin depth $\delta_i = c/\omega_{p,i} = (m_i c^2/e)\Gamma r \sqrt{c/L} \sim 5\,\mathrm{cm}\, L_{50}^{1/2} \Gamma_2 r_{12}$ (when the shock scale is $\sim r/\Gamma \sim 10^{10}$ cm, nine orders larger). Two dimensional inverse cascade that sets in after the field generation does increase field coherence to tens or hundreds of ion skin depths, but it still remains microscopically small. Numerical simulations of the Weibel turbulence cannot run for long enough times to describe the late evolution of currents, still we consider it unreasonable to expect that current will support alignment on such large scales. One of the reasons is that the postshock material is expected to be strongly MHD turbulent. In fact, in the fireball model MHD turbulence is *needed* in order to accelerate particles by the Fermi mechanism. This turbulence will easily randomize subtle current structures. Thus, the finely-aligned, one-dimensional currents and the presence of turbulence necessary to accelerate particles are in contradiction in the fireball model. In addition, one may expect that oppositely directed currents created by the Weibel instability will eventually close-up creating three-dimensional magnetic structures.

Secondly, in order to produce high polarization in the fireball model the two-dimensional turbulent magnetic field should be viewed "from the side", so that the line of sight lies in the plane of the field. This imposes a constraint on the position of the observer: the viewing angle with respect to the jet axis should be $\theta_{ob} \sim 1/\Gamma$. Thirdly, in order to make the emitting layer quasiplanar, the jet opening angle should be very small: $\Delta\theta \leq 1/\Gamma$. Both these assumptions are not generic to the fireball model and are made exclusively in order to maximize polarization. (Generally, if one needs to assume $\Delta\theta, \theta_{ob} \leq 1/\Gamma$, then the emission mechanism had better be inverse Compton.)

Fourthly, since the observed burst was multi-peaked, many emitting shells are required. In order to reproduce high polarization it is assumed that all shells move with the same Lorentz factor [4]. This assumption runs contrary to the very basic postulate of the fireball model that the emitting shells are due to collision of material moving with *different Lorentz factors*, so that the velocity of the resulting shocks must be different.

Thus, the fireball model needs to fine-tune several parameters to explain polarization. An argument of exclusivity has been invoked: the burst was not like any other burst, so it tells nothing about the other bursts. This is virtually equivalent to neglecting the results altogether.

Though one can possibly argue in favor of chance coincidence of $\Delta\theta$ and θ_{ob} (but the GRB rate also goes up by $\Gamma^2 \sim 10^4$), one cannot bypass the problem with the spread in Lorentz factors and turbulent randomization of fields. In order to maximize polarization the spread in the Lorentz factors of the emitting shells must be small. This, in principle, can be achieved by a careful arrangement of shells so that collisions occur only between the two blobs that are moving with the same Lorentz factor (if the source emits shells with $\Gamma_1, \Gamma_2, \Gamma_1$ etc). This is an extremely contrived situation. A more generic case is that the blobs' Lorentz factors are randomly distributed. In this case the fireball model needs to walk a thin line: larger polarization would require smaller spread in Lorentz factors, but then the energy available for dissipation is small, so that the total energy of the burst will be very large (and the burst GRB 021206 was unusually luminous to start with), aggravating even further the efficiency problem of internal shocks.

In addition to the problems specific to the fireball model, there is a kinematic depolarization of synchrotron radiation due to the inhomogeneous expansion velocity [2] (it was taken into account by [4]). Electrical vectors of waves emitted by different parts of the flow are rotated by different amounts during a boost into observer's frame, so that the *observed electric fields are generally not orthogonal to the observed magnetic field*. Averaging over emitting volume reduces total polarization.

Thus, even if turbulence downstream and current closure are completely neglected, the effects of relativistic kinematics, randomness of magnetic field, spread in Lorentz factors and (presumably) not a perfect fine-tuning of $\Delta\theta$ and θ_{ob} will all contribute to reduce polarization. Generically, each of these effects will contribute a factor of two, so that the resulting polarization will not exceed $\sim 10\%$. (Mathematically, when all depolarization effects are minimized at once, polarization may be somewhat higher, but still $\leq 20\%$).

In the case of electromagnetic model the magnetic field has a large coherence scale, $\sim r$, so that within a visible patch of linear size r/Γ the field is quasihomogeneous. There are also depolarization effects. First, there is kinematic depolarization discussed above [2]. Secondly, possible presence of random component of magnetic field would lead to further decrease of polarization. But generally *a random component is not needed in the electromagnetic models*. What is needed is presence of currents, so that the magnetic field is inhomogeneous, but it still can be ordered. The field structure of the Sweet-Parker reconnection layer gives an excellent example (see Fig. 2.a). Random components of the magnetic field are naturally expected and one should account for it. In fact, the very amount of the dissipated magnetic energy may be related to the random component of the field (*e.g.* MacFadyen, these proceedings). For efficient accelerations of electrons one then would need $\delta B/B \sim 1$. In this case the total polarization decreases (Fig. 2.b), remaining reasonably high for $\delta B/B \leq 1$.

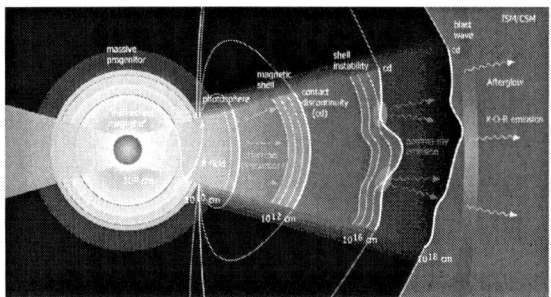

FIGURE 1. Overview of the electromagnetic model

The three competing models of GRBs are compared in Fig. 3. The electromagnetic model is the best contender. It does not require fine tuning of parameters, all observers should see large polarization regardless of the viewing angle, Lorentz factor etc. The only constraint is that the random component of the field is not dominant.

The cannonball model [6] (and other models invoking inverse Compton scattering [7, 8]) can in principle produce very high polarization. It does make an unphysical (in our opinion) assumptions $\Delta\theta$, $\theta_{ob} \leq 1/\Gamma$, but this was inherent to the model before polarization results came out. There is a large degree of fine-tuning: it is assumed that all "cannonballs" are flying within an extremely narrow cone. If their directions were to have a scatter $\Delta\theta > 1/\Gamma \sim 10^{-3}$ polarization would drop to zero. Still, it has the advantage that in the best case it can produce up to 100% polarization.

The internal shock model is the weakest player in the field. It needs to make very contrived and contradictory assumptions to reproduce observations.

Is RHESSI polarization real?

High polarization of prompt emission, if confirmed, would put strong constraint on the emission mechanisms and GRB models. Only inverse Compton may produce polarization as high as 80%. Electromagnetic models can get to $\sim 50\%$, but with a random component a comfortable range is $\leq 30-40\%$. For internal shock anything above $\sim 10\%$ is unreasonable. Obviously, the RHESSI polarization result is highly doubtful, so that an independent confirmation or refutal is a must. If the results are not confirmed, i.e. polarization is $\Pi \leq 20\%$, that won't exclude any model since a multitude of depolarization effects may intervene.

ACKNOWLEDGMENTS

I would like to thank R. Blandford, M. Medvedev, V. Pariev and D. Lazzati.

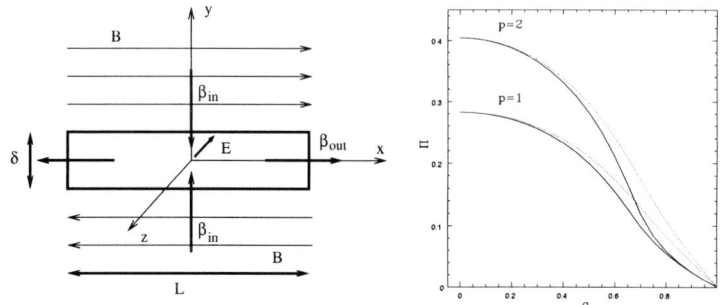

FIGURE 2. (a) An example of ordered inhomogeneous magnetic field (relativistic Sweet-Parker model of reconnection, after [9]). (b) Polarization fraction in the electromagnetic model as a function of q, the ratio of the rms fluctuations to the total field $q = \sqrt{<B_{rn}^2>}/B$ for different values of the particle power-law index p. Solid lines are for two-dimensional random magnetic field confined to the $\mathbf{e}_\theta - \mathbf{e}_\phi$ plane, dashed lines - for the three-dimensional random magnetic field. The large scale magnetic field is B_ϕ. For details see [2].

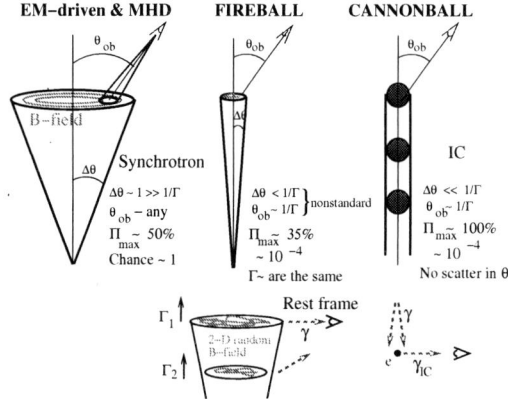

FIGURE 3. Comparison of different models for prompt polarization. For the fireball and cannonball models the maximum polarization given is for a single emitting shell (or ball).

REFERENCES

1. Lyutikov M., Blandford R., 'Electromagnetic Outflows and GRBs", in "Beaming and Jets in Gamma Ray Bursts", R. Ouyed, J. Hjorth and A. Nordlund, eds., astro-ph/0210671 (2002).
2. Lyutikov, M., Pariev, V., Blandford, R., accepted by ApJ, astro-ph/0305410 (2003).
3. Coburn, W., Boggs, S.E., Nature, **423**, 415 (2003).
4. Nakar, E., Piran, T., Waxman, E., astro-ph/0307290 (2003).
5. Medvedev, M. V., Loeb, A., ApJ, **526**, 697 (1999).
6. Dar, A., De Rujula, A., astro-ph/0308248 (2003).
7. Eichler, D., Levinson, A., astro-ph/0306360 (2003).
8. Lazzati, D., Rossi, E., Ghisellini, G., Rees, M. J. , astro-ph/0309038 (2003).
9. Lyutikov, M., Uzdensky D., ApJ, **589**, 893 (2003).

On Hadronic Models for the Anomalous γ-ray Emission Component in GRB 941017

Charles D. Dermer* and Armen Atoyan[†]

*Code 7653, Naval Research Laboratory, Washington, DC 20375-5352 USA
[†]Centre de Recherche Mathématiques, Universite dé Montréal, Montréal, Canada H3C 3J7

Abstract. González et al. (2003, [1]) have reported the discovery of an anomalous radiation component from $\approx 1 - 200$ MeV in GRB 941017. This component varies independently of and contains $\gtrsim 3\times$ the energy found in the prompt ~ 50 keV – 1 MeV radiation component that is well described by the relativistic synchrotron-shock model. Acceleration of hadrons to very high energies by GRBs could give rise to a separate emission component. Two models, both involving acceleration of ultra-high energy cosmic rays with subsequent photomeson interactions, are considered. The first involves a pair-photon cascade initiated by photohadronic processes in the GRB blast wave. Calculations indicate that the cascade produces a spectrum that is too soft to explain the observations. A second model is proposed where photopion interactions in the GRB blast-wave shell give rise to an escaping collimated neutron beam. The outflowing neutrons undergo further photopion interactions to produce a beam of hyper-relativistic electrons that can lose most of their energy during a fraction of a gyroperiod in the Gauss-strength magnetic fields found in the circumburst medium. This secondary electron beam produces a hard synchrotron radiation spectrum that could explain the anomalous component in GRB 941017.

INTRODUCTION

Based on joint analysis of BATSE LAD and EGRET TASC data, González et al. (2003, [1]) recently reported the detection of an anomalous MeV emission component in the spectrum of GRB 941017 that decays more slowly than the prompt emission detected with the BATSE LAD in the $\approx 50 - 300$ keV range. The multi-MeV component lasts for $\gtrsim 200$ seconds (the t_{90} duration of the lower-energy component is 77 sec), and is detected with the BASTE LAD near 1 MeV and with the EGRET TASC between ≈ 1 and 200 MeV. The spectrum is very hard, with a photon number flux $\phi(\varepsilon) \propto \varepsilon^{-1}$, where $\varepsilon = h\nu/m_e c^2$ is the observed dimensionless photon energy.

This component is not predicted or easily explained within the standard leptonic model for GRB blast waves, though it possibly could be related to Comptonization of reverse-shock emission by the forward shock electrons [2], including self-absorbed reverse-shock optical synchrotron radiation [3]. Another possibility is that hadronic acceleration in GRB blast waves could produce this component.

We consider two models involving acceleration of hadrons at the relativistic shocks of GRBs. In the first model, the anomalous radiation component would be explained by the pair-photon cascade radiation initiated by photohadronic processes between high-energy hadrons accelerated in the GRB blast wave and the internal synchrotron radiation field. We find, however, that the cascade radiation spectrum is too soft to be able to explain

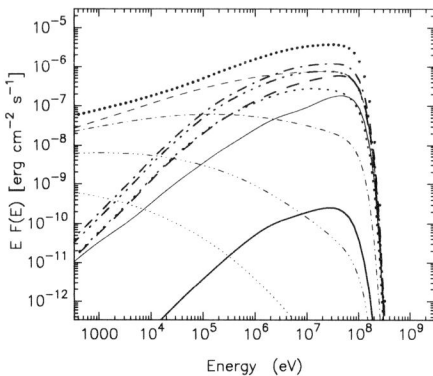

FIGURE 1. Photon energy fluence from an electromagnetic cascade initiated by photopion secondaries in a model GRB, with parameters given in the text. Five generations of Compton (heavy curves) and synchrotron (light curves) are shown. The first through fifth generations are given by solid, dashed, dot-dashed, dot-triple–dashed, and dotted curves, respectively. The total cascade radiation spectrum is given by the upper bold dotted curve.

the anomalous emission component of GRB 941017. In the second model, photomeson interactions in the relativistic blast wave produce a beam of ultra-high energy neutrons, as proposed for blazar jets [4]. Photopion production of these neutrons with photons outside the blast wave produce a directed hyper-relativistic electron-positron beam in the process of charged pion decay and the conversion of high-energy photons from π^0 decay. These energetic leptons produce a synchrotron spectrum in the radiation reaction-limited regime extending to $\gtrsim$ GeV energies, with properties in the 1 – 200 MeV range similar to that measured from GRB 941017. If this model is correct, detection of this component therefore gives important indirect evidence for the acceleration of ultrahigh energy cosmic rays in GRB blast waves.

HADRONIC MODELS FOR GRB 941017

We assume that efficient proton acceleration to ultra-high energies takes place in GRB blast waves, as required in models where GRBs accelerate high-energy cosmic rays [5, 6, 7]. Photopion interactions of these protons with internal synchrotron photons and with photons from an external radiation field create neutral particles and charged pions that decay and initiate an electromagnetic cascade within the GRB blastwave.

One possible model is to attribute the anomalous component to radiation of the cascade produced within the blast wave. Fig. 1 shows the hadron-initiated cascade radiation for a model GRB at redshift $z = 1$, with hard X-ray fluence $\Phi_{tot} = 3 \times 10^{-4}$ erg cm^{-2}, a light curve of 100 second duration divided into 50 pulses of 1 second each, and Doppler factor $\delta = 100$ (see [8, 4] for more details about the model). The total amount of accelerated proton energy $E' = 4\pi d_L^2 \Phi_{tot} \delta^{-3}(1+z)^{-1}$ is injected into the comoving frame of the GRB blast wave.

The various generations of synchrotron and Compton radiation initiated by the cascade are shown in Fig. 1, along with the total radiation spectrum. As can be seen, the isotropic cascade radiation approaches the spectrum of an electron distribution cooling by synchrotron losses, that is, a spectral index between -1.5 and -2. This is too soft to explain the observations of GRB 941017 with a spectral index $= -1$. Moreover, the isotropic cascade radiation should decay at the same rate as the synchrotron radiation. Thus the radiation from an electromagnetic cascade in the GRB blast wave cannot explain the anomalous component in GRB 941017.

We propose a second possibility based on our neutral beam model [4]. Ultrarelativistic neutrons formed in the reaction $p + \gamma \to n + \pi^0$ are not confined by the magnetic field in the GRB blastwave shell and flow out to create an energetic neutron beam. These neutrons are subject to further photopion processes with photons in the surrounding medium to form charged and neutral pions. In the Gauss-type magnetic fields surrounding GRB sources that we assume here, charged π and μ at energies $\lesssim 10^{18}$ eV decay rather than lose energy through synchrotron emission [9]. The charged pions decay into ultrarelativistic electrons and neutrinos, whereas the decay of π^0 produces two γ rays that are promptly converted into electron-positron pairs on this same external radiation field. These energetic electrons (including positrons) are initially produced in the direction of the GRB jet.

The spectra of secondary electrons created by the neutron beam displays a sharp cutoff at energies $\lesssim 10^{14}$ eV as a consequence of the high threshold for photomeson interactions (see Ref. [8] for more details). Electrons with Lorentz factor γ lose energy through synchrotron radiation in an ordered magnetic field with strength B at the rate $-d\gamma/dt = \sigma_T B^2 \gamma^2 \sin^2\psi/(4\pi m_e c)$, where ψ is the electron pitch angle. The corresponding synchrotron energy-loss time scale $t_{syn} = \gamma/|d\gamma/dt|$. The gyration frequency $\omega_B = eB/m_e c\gamma$, and is independent of pitch angle. When $\omega_B t_{syn} \ll 1$, the electron loses almost all of its energy into synchrotron radiation in a time less than the gyroperiod. We use the term "hyper-relativistic" to refer to electrons in this radiation-reaction regime of synchrotron emission [10, 11]. The mean energy of synchrotron photons from electrons that enter the hyper-relativistic regime is independent of magnetic field (e.g., [12]).

Electrons which cool before rotating by an angle equal to the jet opening angle θ_j will produce synchrotron photons that are primarily directed within the jet opening angle. This condition is defined by $\omega_B t_{syn} \lesssim \theta_j$, which applies to electrons with $\gamma \gtrsim \gamma_{hr}(\theta_j) = \sqrt{4\pi e/(\theta_j \sigma_T B \sin^2\psi)} \cong 3 \times 10^7/[\sin\psi\sqrt{(\theta_j/0.1)B(G)}]$. Here we have taken a typical jet opening half-angle $\theta_j = 0.1$ because this is the average value implied by the analysis leading to the standard energy reservoir result of GRBs [13]. Lower-energy electrons with $\gamma < \gamma_{hr}(\theta_j)$ radiate their energy over a much larger solid angle and longer time.

The characteristic synchrotron photon energy $\mathscr{E}_\gamma = m_e c^2 \varepsilon$ radiated by electrons which lose their energy within the jet opening angle θ_j is

$$\mathscr{E}_j \cong \frac{\hbar eB \sin\psi}{m_e c} \frac{\gamma_{hr}^2(\theta_j)}{(1+z)} \cong \frac{500}{(\theta_j/0.1)[(1+z)/2]\sin\psi} \text{ MeV}. \quad (1)$$

Hyper-relativistic electrons with $\gamma > \gamma_{hr}(\theta_j)$ rapidly lose energy through synchrotron losses and deposit all of their energy along the direction of the jet. Electrons at lower energies are deflected to angles $\theta > \theta_j$, and their emission is not seen by an on-axis observer. Hence the distribution of electrons along the jet direction always has an effective low-energy cutoff at $\gamma_{hr}(\theta_j)$. The production spectrum of the electrons can have an intrinsic cutoff γ_{co} due either to the low-energy cutoff in the escaping neutron spectrum, or to the neutron-induced photomeson secondary spectrum outside the GRB blast wave. If $\rho \equiv \gamma_{co}/\gamma_{hr}(\theta_j) \geq 1$, then the observed synchrotron spectrum is a power law with -1.5 index for $\mathscr{E}_j \lesssim \mathscr{E}_\gamma \lesssim \rho^2 \mathscr{E}_j$, and a photon spectrum with the same spectral index as the accelerated protons and escaping neutrons at photon energies $\mathscr{E}_\gamma \gtrsim \rho^2 \mathscr{E}_j$ [10]. If $\rho < 1$, then the observed photon spectrum at $\mathscr{E}_\gamma \gtrsim \mathscr{E}_j$ has the same spectral index as the primary neutrons.

At photon energies $\mathscr{E}_\gamma \ll \mathscr{E}_j$, the observed spectrum is produced by the same hyper-relativistic electrons with $\gamma \gtrsim \gamma_{hr}(\theta_j)$, but at energies ε well below the peak energy $3\gamma^2 \varepsilon_B$, where $\varepsilon_B \equiv B/B_{cr}$ and $B_{cr} = 4.41 \times 10^{13}$ G is the critical magnetic field. We now derive this spectrum.

The differential energy radiated per dimensionless energy interval $d\varepsilon$ per differential solid angle element $d\Omega$ in the direction θ with respect to the direction of an electron moving with Lorentz factor γ is given by

$$\frac{dE}{d\varepsilon d\Omega} = \frac{e^2}{3\pi^2 \bar{\lambda}_C} \left(\frac{\varepsilon}{\gamma \varepsilon_B}\right)^2 (1 + \gamma^2 \theta^2)^2 (\Lambda_\parallel + \Lambda_\perp), \qquad (2)$$

where $\bar{\lambda}_C = \hbar/m_e c = 3.86 \times 10^{-11}$ cm is the electron Compton wavelength, and $\Lambda_\parallel = K_{2/3}^2(\xi)$ and $\Lambda_\perp = (\gamma\theta)^2 K_{1/3}^2(\xi)/[1 + (\gamma\theta)^2]$ are factors for radiation polarized parallel and perpendicular to the projection of the magnetic field direction on the plane of the sky defined by the observer's direction [14]. The factor $\xi = \varepsilon/\hat{\varepsilon}$, where $\hat{\varepsilon} = 3\varepsilon_B \gamma^2/(1+\gamma^2\theta^2)^{3/2}$, and $K_n(x)$ is a modified Bessel function of the second kind, with asymptotes $K_n(x) \to \frac{1}{2}\Gamma(n)(2/x)^n$ in the limit $x \ll 1$, and $K_n(x) \to \sqrt{\pi/2x} \exp(-x)$ in the limit $x \gg 1$. The condition $\xi \ll 1$ corresponds to $\varepsilon \ll \hat{\varepsilon}$ where $K_n(\xi)$ are in their power-law asymptotes, and $\xi \gtrsim 1$ or $\varepsilon \gtrsim \hat{\varepsilon}$ is where $K_n(\xi)$ are in exponential decline. The characteristic energy $\hat{\varepsilon}$ approaches $3\varepsilon_B \gamma^2$ when $\gamma\theta \ll 1$, and $\hat{\varepsilon}$ declines with θ according to the relation $\hat{\varepsilon} \cong 3\varepsilon_B \gamma^2/(\gamma\theta)^3$ when $\gamma\theta \gg 1$. When $\varepsilon \ll \hat{\varepsilon}$, then $\Lambda_\parallel \gg \Lambda_\perp$ and $dE/d\varepsilon d\Omega = (dE_{syn}/d\varepsilon d\Omega) \simeq 3^{1/3}(1.07e/\pi)^2(\gamma\varepsilon/\varepsilon_B)^{2/3}/\bar{\lambda}_C \propto \varepsilon^{2/3}$. For a fixed value of ε, this emissivity exponentially cuts off when $\varepsilon \gtrsim 3\varepsilon_B/\gamma\theta^3$, or when $\theta \gtrsim \theta_{max} = (3\varepsilon_B/\gamma\varepsilon)^{1/3}$.

The synchrotron emission spectrum in the limit $\mathscr{E}_\gamma \ll \mathscr{E}_j$, integrated over solid angle, is simply given by

$$\frac{dE}{d\varepsilon} \simeq 2\pi \int_0^{(3\varepsilon_B/\gamma\varepsilon)^{1/3}} d\theta\, \theta \left(\frac{dE_{syn}}{d\varepsilon d\Omega}\right) \simeq \frac{3e^2}{\pi \bar{\lambda}_C} \propto \varepsilon^0. \qquad (3)$$

This differs from the energy index $+1/3$ for the electrons in the classical regime averaged over a complete orbit because, in this case, $d\theta \sin\theta \to d\theta \sin\psi$ in the integration in eq. (3) [15].

NEUTRON BEAMS FROM GRBS

In our model, ultrarelativistic protons undergo photomeson interactions with the internal and external radiation photons [8], producing a beam of outflowing neutrons. Subsequent interactions of these neutrons with the external radiation field generate a beam of hyper-relativistic electrons. (The external photons could be due to pre-supernova thermal or GRB synchrotron radiations scattered by the stellar wind in the collapsar model, or by plerionic emission in the SA model.) The synchrotron radiation has a low-energy cutoff at $\mathscr{E}_\gamma < \mathscr{E}_j \sim 500$ MeV – 1 GeV, with a specific characteristic power-law number spectral index equal to -1, as observed for GRB 941017 [1].

The ≈ 200 s decay time of the anomalous emission can be explained by the emission of hyper-relativistic electrons from the edges of a jet blastwave at distances $R \approx 10^{15}(\theta_j/0.1)^{-2}[(1+z)/2]^{-1}$ cm, implying significant opacity to photomeson processes due to external photons at these distances. The differing GRB external radiation and density environments which determine the intensity of target photons could account for the unusual spectrum of GRB 941017. The neutron beam can carry up to $\sim 50\%$ of the entire energy of injected protons [4], which can be further reprocessed about equally into neutrinos and hyper-relativistic electrons. This nevertheless requires a large nonthermal hadron-to-lepton load in the GRB blast wave in order for the fluence of the anomalous component to be comparable with that of the normal component. This large hadron load is also consistent with predictions linking cosmic rays to GRB sources [16].

Our model predicts that most of the energy of hyper-relativistic electrons will be observed at $\gtrsim 500$ MeV – GeV energies as a delayed radiation component, well-suited for observations with *GLAST*. GRBs with anomalous γ-ray emission components should also be bright neutrino sources detectable with *IceCube*.

Discussions with M. González, B. Dingus, A. Königl and D. Lazzati are gratefully acknowledged. This work is supported by the Office of Naval Research and the NASA *GLAST* program.

REFERENCES

1. González, M. M., Dingus, B. L., Kaneko, Y., et al., *Nature*, **424**, 749 (2003).
2. Granot, J., and Guetta, D., *Astrophys. J. Lett.*, **598**, L11 (2003).
3. Pe'er, A., and Waxman, E. 2003, *Astrophys. J. Lett.*, submitted (astro-ph/0310836)
4. Atoyan, A., and Dermer, C. D., *Astrophys. J.*, **586**, 79 (2003).
5. Vietri, M., *Astrophys. J.*, **453**, 883 (1995).
6. Waxman, E., *Phys. Rev. Lett.*, **75**, 386 (1995).
7. Dermer, C. D., *Astrophys. J.*, **574**, 65 (2002).
8. Dermer, C. D., and Atoyan, A., *Phys. Rev. Lett.*, **91**, 071102 (2003).
9. Rachen, J. P., and Mészáros, P. 1998, *Phys. Rev. D*, **58**, 123005.
10. Nelson, R. W., and Wasserman, I., *Astrophys. J.*, **371**, 265 (1991).
11. Aloisio, R., and Blasi, P., *Astropar. Phys.*, **18**, 195 (2002).
12. de Jager, O. C., Harding, A. K., Michelson, P. F., et al. 1996, *Astrophys. J.*, **457**, 253.
13. Frail, D., et al., *Astrophys. J.*, **562**, L55 (2001).
14. Jackson, J. D., *Classical Electrodynamics* (Wiley, New York) (1975).
15. Rybicki, G. B., and Lightman, A. P., *Radiative Processes in Astrophysics* (Wiley, New York) (1979).
16. Wick, S. D., Dermer, C. D., and Atoyan, A., submitted to *Astropar. Phys.* (astro-ph/0310667) (2003).

Evidence for GRB Induced Extinctions in the Fossil Record?

Thomas G. Kaye

804 Seton Ct. Wheeling, IL 60090

Abstract. Particular problems in the fossil record are placed in context with a GRB event causing death by radiation exposure, followed by comet showers from a disruption in the Oort Cloud. The absence of pollen below the iridium layer at the KT and Triassic-Jurassic boundaries, the pattern of extinction on land and in the oceans and divergence times of molecular DNA studies may all have a common root in ionizing radiation.

INTRODUCTION

Increased ionizing radiation at the surface of the Earth can be caused by gamma rays destroying the ozone layer, muons from cosmic rays and neutrinos. The production mechanisms are GRB's, supernovas and collapsing stars. While arguments remain for and against these events having catastrophic consequences on Earth, this paper attempts to correlate a proposed terrestrial radiation event with enigmas in the fossil record.

Geologic Problems

Detailed investigation of the KT boundary clays shows that there are actually two distinct, approximately centimeter thick, layers associated with the event. The upper smectitic layer contains iridium and shocked quartz associated with the bolide impact. The lower kaolinitic layer lacks iridium and shocked minerals [1] and is distinct from the Cretaceous sediments below. This lower kaolinitic layer is missing all pollen except for ferns [1] suggesting that the plants were eradicated before the impact occurred. The Triassic-Jurassic boundary also shows an abrupt end to pollen diversity before the Ir spike [2]. Fern pollen again dominates the ecosystem through the transition suggesting a similarity between the TJ and KT boundaries.

The case for a bolide impact at the KT is strong. Iridium has been found at other extinction boundaries but not enough to validate an impactor as the culprit in all the major extinctions. Brecher [3] and others [4] proposed that a GRB would cause perturbation of the Oort Cloud, sending comet showers toward the inner solar system. Comet showers have been detected at the Eocene-Oligocene boundary using helium^3 ratios [5]. For the GRB scenario, this suggests that bolide impacts would be associated with some, but not all extinctions, a pattern seen in the fossil record. The fact that the bolide impact is easily incorporated into the GRB hypothesis, allows all the current data to remain in place.

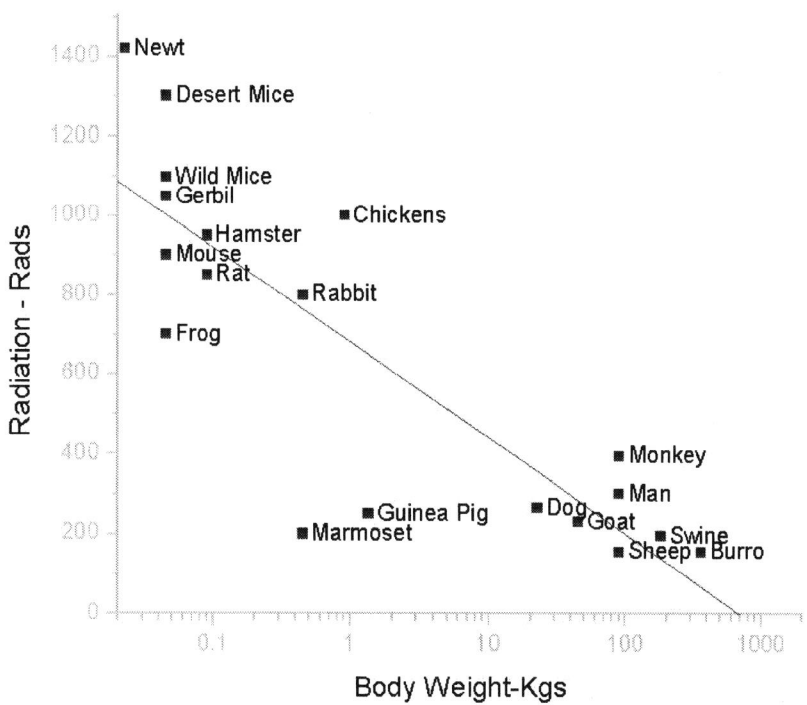

FIGURE 1. Radiation lethal dose scales with size in land animals.

A radiation event that killed off the plant and animal life, and was then followed by a bolide impact, would explain the missing pollen in the ejecta blanket as well as the lack of bolide impacts at other major extinctions.

High dose radiation has two effects, death and mutation [6]. Following the Cambrian Explosion, primitive multicellular life forms suddenly diversified into all the major groups we see today. Since there were no larger life forms to kill off, a radiation event at this stage in Earth's history would only be expected to cause mutations in the existing smaller organisms. We also see a possible link to GRB induced comet showers from Culler et al. [7] who suggest that a major cratering episode took place on the Moon near the time of the Cambrian Explosion.

Biologic Effects

The KT extinction preferentially killed all land animals over 23 kg. This pattern of extinction is one of the main arguments against the bolide impact as the lone cause of extinctions. The bolide scenario predicts devastating effects on both large and small animals, which does not match the observed extinction pattern. Cold War experiments determined lethal doses of radiation for various animals. Figure 1 shows that radiation sensitivity scales with size. This suggests that exposure to high levels of radiation would preferentially kill the larger animals while allowing survival of smaller organisms [8], in agreement with the paleontological record.

Oxygen content is highly effective in changing radiation resistance [6] [9]. This fact is used effectively in treating cancer patients with radiation therapy where increased blood flow to tumors causes the cancerous tissues to become highly radiosensitive. This would allow certain larger animals with lower blood oxygen ratios, such as crocodiles, to survive radiation exposure in spite of their body mass.

The effects of the KT extinction in the ocean were different from those on land. Whereas smaller terrestrial life forms survived, the plankton were killed off at the oceans' surface, while larger animals faired better below [10]. A similar plankton die-off occurred at the Triassic-Jurassic boundary [11]. Plankton has low radiation resistance and receives a lethal dose at about 300 rads. The water column provides excellent shielding against radiation [9], therefore massive radiation exposure would be expected to have its greatest effect at the surface, a pattern supported by the fossil record.

Molecular Problems

There is an ongoing debate between paleontologists and molecular biologists on the divergence times of major fossil groups [12] [13]. In general terms, DNA accumulates mutations at a constant rate over time. The accumulated differences between two gene pools gives an indication of how far in the past they became independent species. The argument here centers on the fact that the DNA data always seems to indicate older divergence times than found in the fossil record [14]. A radiation event could spontaneously add mutations above the background rate [6] to the entire living gene pool. This would add artificial "clock ticks" putting the molecular divergence times in conflict with the fossil record.

SUMMARY

The enigmas of missing pollen, large animal extinction, reverse effects in the ocean, changing molecular clocks and explosive diversification all seem to share a common connection with radiation overdose. Multiple astronomical scenarios exist that could cause such events on Earth. While arguments remain on the likelihood of these occurrences, there is no argument that close proximity of any such events would have disastrous effects on Earth's environment.

ACKNOWLEDGMENTS

The author would like to thank John Innis, Darin Croft and Mark Trueblood for their contributions to this paper.

REFERENCES

1. Bohor, B. F., *Geology*, **15(10)**, 896 (1987).
2. Olsen, P. E. e. a., *Science*, **296(5571)**, 1305–1307 (2002).
3. Brecher, K. (1997), 191st AAS Meeting, Washington, DC.
4. Dar, A., and Rujula, A. D., *Frascati Physics Series*, **XXVI**, 513–523 (2002).
5. Farley, K. A. e. a., *Science*, **280**, 1250–1253 (1998).
6. Lawrence, C. W., Cellular radiobiology (1971).
7. Culler, T. S., *Science*, **287(5459)**, 1785 (2000).
8. Conway, D. J., pers. com. (1999), head of Radiology, Childrens Memorial Hospital. Chicago.
9. Frigerio, N. A., Your body and radiation (1966).
10. Raup, D. M., The nemesis affair, New York (1986).
11. Ward, P. D. e. a., *Science*, **292(5519)**, 1148–1150 (2001).
12. Archibald, J. D., *Science*, **285(5436)**, 2031 (1999).
13. Normile, D., *Science*, **281(5378)**, 774–775 (1998).
14. Fortey, R., *Science*, **293(5529)**, 438–439 (2001).

Observations of X-ray Bursts by HETE-2

Y. E. Nakagawa*, T. Yamazaki*, M. Suzuki[†], A. Yoshida*, N. Kawai[†], D. Takahashi*, M. Matsuoka**, Y. Shirasaki[‡], T. Tamagawa[§], K. Torii[§], T. Sakamoto[†], Y. Urata[†], R. Sato[†], Y. Yamamoto[†], E. E. Fenimore[¶], M. Galassi[¶], D. Q. Lamb[||], C. Graziani[||] and G. Ricker[††]

*Department of Physics, Aoyama Gakuin University, Sagamihara, Kanagawa 229-8558, Japan
[†]Department of Physics, Tokyo Institute of Technology, Meguro-ku, Tokyo 152-8551, Japan
**JAXA, Tsukuba, Ibaraki 304-8505, Japan
[‡]National Astronomical Observatory of Japan, Osawa, Mitaka, Tokyo 181-8588, Japan
[§]RIKEN, Hirosawa, Wako, Saitama 351-0198, Japan
[¶]Los Alamos National Laboratory, Los Alamos, NM 87545, USA email
[||]Department of Astronomy and Astrophysics, University of Chicago, Chicago, IL 60637, USA
[††]Center for Space Research, Massachusetts Institute of Technology, Cambridge, MA 02139, USA

Abstract. The scientific instruments aboard HETE-2 are pointed to the Galactic center region each summer. The WXM has detected more than 407 X-ray busrts from 17 known X-ray burst sources such as GS 1826−238, 4U 1850−087 and 4U 1820−30. In particular, 173 events were localized to GS 1826−238 in 2001, 2002 and 2003.

We find that the X-ray bursts from GS 1826−238 were produced periodically with recurrence intervals of 4.20, 3.57 and 3.73 hours in 2001, 2002 and 2003 respectively. These recurrence intervals until 2002 are shorter than that observed with Beppo-SAX/WFC from 1996 to 1998 [3]. In contrast, the interval in 2003 is slightly increasing.

INTRODUCTION

HETE-2 (High Energy Transient Explorer 2) is a small satellite for cosmic high-energy transient phenomena [1]. It was flown into an equatorial orbit with an inclination of about 2 degrees at the altitude of about 600 km. Although HETE-2 was mainly designed and built to research gamma-ray bursts, it is able to observe X-ray bursts because HETE-2 points to the galactic center region each summer. The WXM instrument on-board HETE-2 plays an important role for localizing the burster position [2]. The list of X-ray bursts observed by HETE-2 for the period from 1 January 2001 through 31 July 2003 is given in Table 1. The total number of events is 407, and the most frequently observed burster is GS 1826−238. We present the results of HETE-2 observations of X-ray bursts with a central focus on GS 1826−238.

Figure 1 shows an example of WXM light curves and spectra for GS 1826−238, 4U 1850−087 and 4U 1812−121. The energy band of these light curves and spectra is from 2 to 25 keV. These spectra are well fitted by black body model, and the best fit parameters are given in Table 2.

TABLE 1. The X-ray burst list for the period from 1 January 2001 through 31 July 2003.

Counterpart	Event Number
GS 1826−238	173
4U 1728−34	62
4U 1820−30 (NGC6624)	45
4U 1850−087 (NGC6617)	28
Aql X−1	17
4U 1916−053	17
4U 1812−121	13
SAXJ 1750−29	11
4U 1832−330	10
Terzan 5	7
Total	407

TABLE 2. The best fit parameters for the spectra in Figure 1. We use the black body model for the fitting. The errors are evaluated using 90% confidence level.

parameter	GS 1826−238	4U 1850−087	4U 1812−121
kT (keV)	2.02±0.2	2.45±0.09	1.87±0.07
burst luminosity (10^{38} erg/sec)	1.41±0.1	1.86±0.07	1.70±0.065
χ_ν^2 (d.o.f)	1.18 (24)	1.14 (27)	0.75 (24)

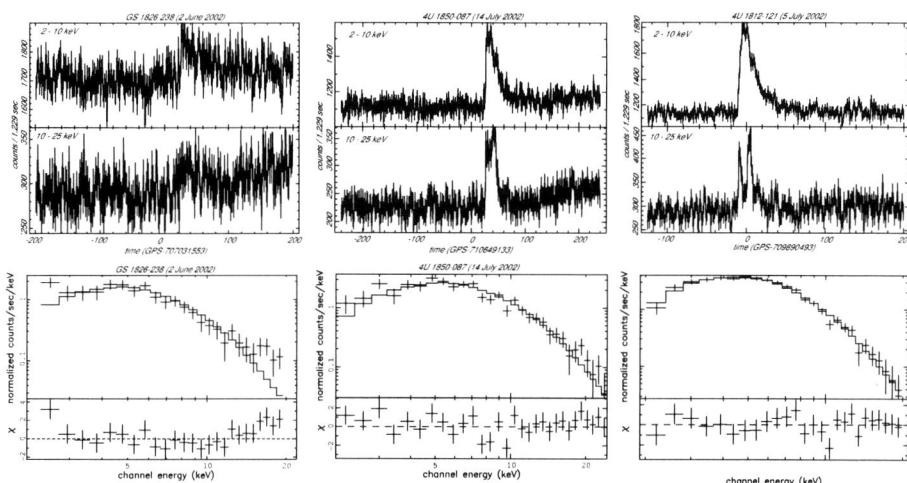

FIGURE 1. These figures are example of the light curves and the spectra. (a) and (b) are for GS 1826−238, (c) and (d) are for 4U 1850−087 and, (e) and (f) are for 4U 1812−121. The best fit parameters are given in Table 2.

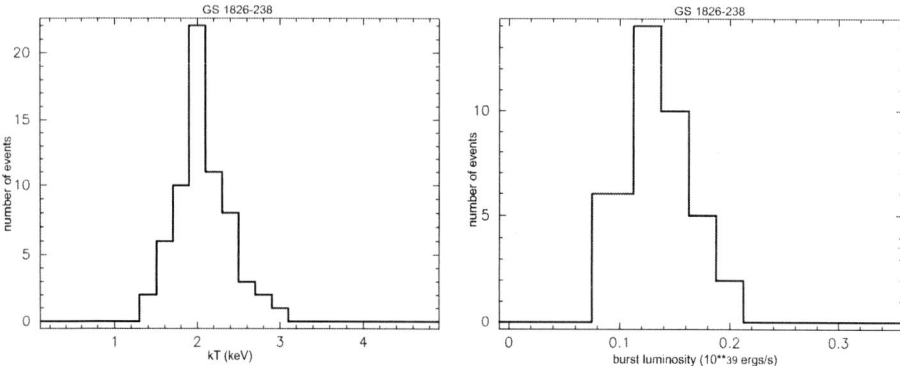

FIGURE 2. The derived temperature and luminosity distributions. The kT and burst luminosity distribution peaked at about 2 keV and 0.13×10^{39} erg s^{-1} respectively.

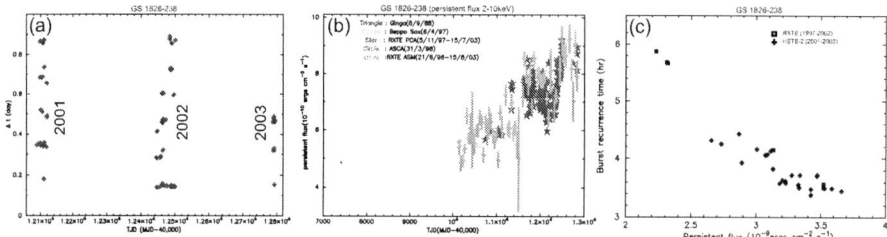

FIGURE 3. (a)The recurrence interval of GS 1826−238. (b)The flucutuation of the persistent flux with the error bar which is evaluated using 90% confidence level. (c)The relationship between persistent flux and recurrence interval.

ANALYSIS OF GS 1826−238

Here we report on mainly GS 1826−238 which was the most frequently observed X-ray burster by HETE-2. The derived temperature and burst luminosity distributions are summarized in Figure 2. The kT and burst luminosity distribution peaked at about 2 keV and 0.13×10^{39} erg s^{-1} respectively and display rather narrow distributions. Figure 3 (a) shows the recurrence interval of GS 1826−238 by observations of HETE-2. It is reported using the data by WFC on Beppo-SAX from 1996 to 1998 that the bursts from GS 1826−238 recurred with an interval of 5.76 hours [3]. The WXM/HETE-2 observations suggest a shorter recurrence interval of 4.20, 3.57 and 3.73 hours in 2001, 2002 and 2003 respecively. The recurrence intervals are shorter than that of Beppos-SAX until 2002 and slightly increasing in 2003. Figure 3 (b) shows the fluctuation of persistent flux. It is overlaped the result of Ginga, Beppo-SAX, RXTE/PCA, ASCA and RXTE/ASM [4][5][6][7]. All results are consistent with each other. Figure 3 (c) shows the relationship between persistent flux and recurrence interval. The symbols of plus and square show the result of HETE-2 and RXTE respectively. It shows that the recurrence interval is decreasing correspond to the increasing of the persistent flux.

TABLE 3. The results of the recurrence intervals compared with other satellite.

date	recurence interval (hour)	satellite	reference
1997.Apl.06	6.4	Beppo-SAX/NFI	[8]
1998.Mar.31	5.4	ASCA	[5]
1996 ~ 1998	5.76	Beppo-SAX/WFC	[3]
2001	4.20	HETE-2	this paper
2002	3.57	HETE-2	this paper
2003	3.73	HETE-2	this paper

CONCLUSION

We have analyzed the X-ray bursts observed by HETE-2 for the period from 1 January 2001 through 31 July 2003. We mainly studied about GS 1826−238 and found the spectra of GS 1826−238 were well fitted by black body model. The kT and burst luminosity distribution peaked at about 2 keV and 0.13×10^{39} erg s^{-1} respectively. The results of the recurrence intervals compared with other satellite are summarized in Table 3. We found that the recurrence interval of GS 1826−328 were decreasing until 2002 and slightly increasing in 2003. Figure 3 (b) provides the relationship that the increasing of the persistent flux comes up with decreasing of the recurrence interval. This is consistent with previous reported result obtained by Beppo-SAX/WFC [4].

REFERENCES

1. Ricker, G. R., and HETE Science Team, AAS **198**, #35.04 (2001).
2. Shirasaki, Y., et al., Publ. Astron. Soc. Japan **55**, 1033S (2003).
3. Ubertini, P., et al., ApJ **514** 27U (1999).
4. Cornelisse, R., et al., A&A **405**, 1033C (2003).
5. Kong, A.K.H., et al., MNRAS **311**, 405 (2000).
6. Galloway, D.K., et al., astro-ph/0304500 (2003).
7. Tanaka, Y., et al., in "Proc. 23rd ESLAB Symposium on Two Topics in X-ray Astronomy", eds.J.Hunt & B.Battrick, ESA SP-296, p.1 (2003).
8. in 't Zand, J.J.M., et al., A&A **347**, 89 (1999).

Magnetic Field Generation in Relativistic Shocks

Jorrit Wiersma* and A. Achterberg*

Sterrenkundig Instituut, Universiteit Utrecht, Netherlands

Abstract. We present an analytical estimate for the magnetic field strength generated by the Weibel instability in ultra-relativistic shocks in a hydrogen plasma. We find that the Weibel instability is, by itself, not capable of converting the kinetic energy of protons penetrating the shock front into magnetic field energy. Other (nonlinear) processes must determine the magnetic field strength in the wake of the shock.

INTRODUCTION

The fireball model for Gamma-ray Bursts (GRBs) explains GRB afterglows as synchrotron radiation coming from the external shocks that are formed when a relativistically expanding fireball (or jet) interacts with surrounding gas [1]. The spectra and luminosity of the observed afterglows indicate a magnetic field strength in the radiating material of about 10% of the equipartition field strength [2]: $B^2/8\pi \sim \varepsilon$ with ε the total post-shock energy density. Such a magnetic field is much stronger than what is expected from simple shock physics: for instance, passive compression of a dynamically unimportant magnetic field in a relativistic shock yields a post-shock field satisfying (for example, see [3]) $B^2/8\pi\varepsilon \sim (v_A/c)^2$, with v_A the Alfvén speed *ahead* of the shock, which usually satisfies $v_A \ll c$.

It may be possible that much stronger magnetic fields are generated in the shock transition itself through a Weibel-like instability. This low-frequency electromagnetic beam-instability can develop at the point where the shocked and unshocked plasma penetrate each other. It converts the kinetic energy of the penetrating particle beams into thermal motions and unordered magnetic fields [4]. For external shocks propagating into a hydrogen plasma incoming protons carry most of the kinetic energy in the shock frame. The instability must convert a significant fraction of this energy into magnetic energy in order to reach the required field strength [5]. Here we investigate if this is indeed possible.

SIMPLE COLLISIONLESS SHOCK MODEL

Within the shock transition, where the fireball material encounters the surrounding plasma, the mixing of the shocked (relativistically hot) and the unshocked (cold) plasma produces a plasma with a very anisotropic velocity distribution. As in non-relativistic collisionless shocks, it is quite likely that a significant fraction of the incoming ions is reflected at the shock transition by a large-scale electrostatic or magnetic field. The resulting situation is known to be unstable [4]: in the so-called Weibel instability the penetrating and reflected particles will bunch together, causing electric currents that induce a magnetic field in the plasma, which in turn causes the particles to bunch even more. If the generated magnetic fields become sufficiently strong they will trap the beam particles, eventually saturating the instability.

Recent numerical simulations [6], [7], [8] show that the electric currents generated by the Weibel instability merge with each other in the wake of the unstable region. This separate process has a longer time-scale, and forms structures on a larger length-scale than the Weibel instability, which occurs mostly on a scale of the order of the plasma skin depth c/ω_{pe}, with ω_{pe} the electron plasma frequency (see below).

In this paper we will concentrate on the proton-driven Weibel instability.

THE WEIBEL INSTABILITY AND ITS SATURATION

Because of the small electron mass, the electron-driven Weibel instability evolves very rapidly [5], with the magnetic field strength growing as $e^{\sigma t}$ with $\sigma \approx \omega_{pb}$ where $\omega_{pb} = \sqrt{4\pi e^2 n_{0b}/m_e}$ is the beam particle plasma frequency based on the proper beam density n_{0b}. When the electron-instability has saturated, the electron velocity distribution will be close to isotropic. The electrons then form a relativistically hot background plasma in which the much slower proton-driven Weibel instability develops because the proton velocity distribution is still very anisotropic (see, for example, Figure 6 of [7]). We will investigate the behavior of the protons in this situation.

The main features of the resulting instability can be reproduced by investigating a water-bag proton velocity distribution. We assume two counter-streaming proton beams moving along the x-direction, with a small velocity spread in the z-direction to model thermal motions. The proton momentum $\vec{p}$ is then distributed as:

$$F(\vec{p}) = \frac{n_p}{4p_{z0}}[\delta(p_x - p_{x0}) + \delta(p_x + p_{x0})]\delta(p_y)[\Theta(p_z - p_{z0}) - \Theta(p_z + p_{z0})], \quad (1)$$

where n_p is the total proton density, p_{x0} is the beam momentum, p_{z0} is the maximum momentum in the perpendicular direction, $\delta(x)$ is the Dirac delta function and $\Theta(x)$ is the unit step function. We consider the evolution of a wave perturbation with wave vector $\vec{k} = k\vec{e}_z$ and frequency ω. The growth rate of the proton instability is calculated in the usual manner by looking for wave solutions of the linearized equations of motion and Maxwell's equations (for example [9]). The resulting dispersion relation links the complex wave frequency ω to the wave number k. The Weibel instability obeys a dispersion equation of the form (using the notation of [9]):

$$\omega^2 - k^2 c^2 + C_{xxz} = 0, \quad (2)$$

where C_{xxz} contains contributions from both the electrons and the protons. Since the electrons are relativistically hot, their contribution is $C_{xxz,e} = -\omega_{pe}^2$, with ω_{pe} the effective electron plasma frequency which equals

$$\omega_{pe} = \sqrt{4\pi e^2 n_e / m_e h}, \quad (3)$$

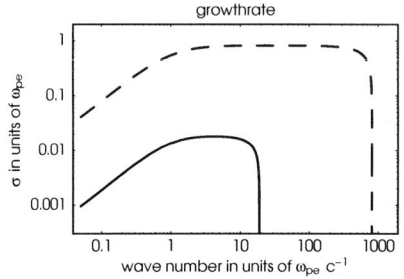
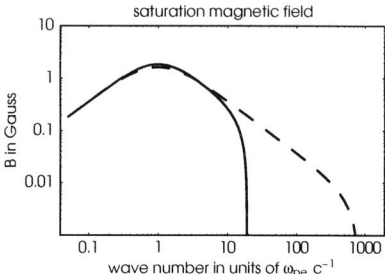

FIGURE 1. Left: growth rate as a function of wave number for a shock with Lorentz factor 1000 and thermal velocity spread $v_{z0} = 0.001$. The solid line is the result for an electron-proton plasma. The dashed line is the result for pure pair plasmas. Right: magnetic field strength as a function of wave number for a shock with the same parameters and $n_p = 2\,\text{cm}^{-3}$.

where $h \equiv (\varepsilon + P)_e / n_e m_e c^2$ is the electron enthalpy divided by the rest mass energy density, which parameterizes the relativistic mass correction in a relativistically hot plasma.

The proton contribution is [9]:

$$C_{xxz,p} = \omega_{pi}^2 \left\{ -\left(\overline{\frac{1}{\gamma_b}}\right) + \frac{1}{\gamma_{b0}} \frac{u_{x0}^2}{1+u_{x0}^2} - \frac{1}{\gamma_{b0}} \frac{k^2 v_{x0}^2}{\omega^2 - k^2 v_{z0}^2} \right\}, \qquad (4)$$

where $\omega_{pi} = \sqrt{4\pi e^2 n_p / m_p}$ is the (non-relativistic) proton plasma frequency based on the lab-frame beam density, m_p is the proton rest mass, $u_i = p_i/(m_p c)$, $\gamma_{b0} = (1 + u_{x0}^2 + u_{z0}^2)^{1/2}$, $v_i = u_i c / \gamma_{b0}$ and

$$\left(\overline{\frac{1}{\gamma_b}}\right) = \int d\vec{p}\, \frac{F(\vec{p})}{\gamma} = \frac{1}{2u_{z0}} \ln\left(\frac{1+v_{z0}/c}{1-v_{z0}/c}\right). \qquad (5)$$

With these results the dispersion equation (2) becomes a biquadratic equation for ω that has one positive imaginary solution giving the growth rate $\sigma = \text{Im}(\omega)$ of the unstable mode (Figure 1, left).

One expects the instability to saturate when the quiver motion of the beam particles in the wave reaches an amplitude Δz such that $k\Delta z \approx 1$. An expression for the corresponding magnetic field strength is derived in [4] and reads in our notation:

$$B_{\text{sat}} = \frac{\gamma_{b0} m_p}{v_{x0} e} \frac{\sigma^2}{k}. \qquad (6)$$

Using the dispersion relation between σ and k we can then find B_{sat} as a function of k (figure 1, right).

A straightforward calculation shows that for small v_{z0} the peak of B_{sat} lies at wave number k_{peak} given by

$$k_{\text{peak}}^2 c^2 \simeq \omega_{pe}^2 + \omega_{pi}^2 / \gamma_{b0}^3, \qquad (7)$$

with a corresponding growth rate

$$\sigma(k_{\text{peak}}) \simeq \frac{\omega_{\text{pi}}}{\sqrt{2\gamma_{b0}}} \left(\frac{v_{x0}}{c}\right). \tag{8}$$

The proton-driven Weibel instability occurs for a relatively small range in wavelength compared to the electron-positron case (Figure 1): modes with wavelength longer than the electron skin depth ($k < c/\omega_{\text{pe}}$) are inhibited by the response of the background electrons to the proton perturbations. The location where the growth rate levels off corresponds with the location where the saturation magnetic field peaks ($k = k_{\text{peak}}$). If the background electrons had not been as responsive, this location might have been at much lower wave number, and since the wave number is in the denominator of expression (6) for B_{sat}, this would have resulted in a much higher magnetic field. However, we find that the electrons set the location of the peak ($k_{\text{peak}} \simeq \omega_{\text{pe}}/c$) and that the proton Weibel instability can only produce slightly stronger magnetic fields than the electron instability (Figure 1, right), despite the larger kinetic energy of the protons.

CONCLUSION

The Weibel instability as it was modeled here is not efficient at converting the kinetic energy of the protons into magnetic fields. However, numerical simulations of collisionless shocks in electron-proton plasmas ([6], [7], [8]) *do* show efficient production of magnetic fields. This can probably be attributed to merging of the electric currents produced by the Weibel instability, *after* the instability that was considered here has stopped. To verify this, the current merging processes will need to be investigated further.

The properties of the magnetic fields generated by these processes will be important for determining the radiation that may be produced in these shocks and that we see in the form of gamma-ray burst afterglows.

ACKNOWLEDGMENTS

This research is supported by the Netherlands Research School for Astronomy (NOVA).

REFERENCES

1. Rees, M. J., and Mészáros, P., *Mon. Not. Roy. Astron. Soc.*, **258**, 41P (1992).
2. Gruzinov, A., and Waxman, E., *ApJ*, **511**, 852–861 (1999).
3. Kennel, C. F., and Coroniti, F. V., *ApJ*, **283**, 694–709 (1984).
4. Yang, T.-Y. B., Arons, J., and Langdon, A. B., *Physics of Plasmas*, **1**, 3059–3077 (1994).
5. Medvedev, M. V., and Loeb, A., *ApJ*, **526**, 697–706 (1999).
6. Fonseca, R. A., Silva, L. O., Tonge, J. W., Mori, W. B., and Dawson, J. M., *Physics of Plasmas*, **10**, 1979–1984 (2003).
7. Frederiksen, J. T., Hededal, C. B., Haugbølle, T., and Nordlund, Å., Magnetic field generation in collisionless shocks; pattern growth and transport, astroph/0308104 (2003).
8. Haruki, T., and Sakai, J.-I., *Physics of Plasmas*, **10**, 392–397 (2003).
9. Silva, L. O., Fonseca, R. A., Tonge, J. W., Mori, W. B., and Dawson, J. M., *Physics of Plasmas*, **9**, 2458–2461 (2002).

Very Short Gamma Ray Bursts: New Physics?

David. B. Cline

Astrophysics Division, Department of Physics & Astronomy, University of California, Los Angeles

Abstract. We review our study of 46 very short gamma ray bursts with a mean time duration of 50ms. We claim that they are likely from galactic sources and we speculate on their possible origin.

INTRODUCTION

Over the past decade we have studied all of the available GRBs with time duration less than 100ms. There are about 50 such events recorded since the discovery of GRBs in the late 1960s. We believe these events constitute a separate class of GRB. We show a beautiful example of this in Figure 1a, 1b from the PHEBUS detector. The duration of this event is about 50ms (the reference for this event can be found in [1]). This work is partially adapted from Astroparticle Physics **18** (2003) 531 and references [1], [2], and [3].

PROPERTIES OF VERY SHORT GRB

The time distribution T90 for all GRBs from the BATSE detector up to May 26, 2000 shows two clear peaks and a small excess below 100 ms duration. We divide the GRBs into three classes in time duration: L ($\tau > 1$s) long; M ($1s > \tau > 0.1$ s) medium; and S ($\tau < 100$ ms) very short. The duration time of T90 is used for all of this analysis. In this paper, we confine the discussion to the M and S classes of GRBs. Since these events are adjacent in time, it is important to contrast the behavior.

We assume that the S GRBs constitute a separate class of GRBs, and we fit the time distribution with a three-population model. The fit is excellent but does not in itself give significant evidence for a three population model. We now turn to the angular distributions of the S and M GRBs. In Figure 2, we show this distribution for the very short bursts (46 events in total). We can see directly that this is not an isotropic distribution. To ascertain the significance of the anisotropy, we break up the Galactic map into eight equal probability regions. In Figure 2, we show the distribution of events in the eight bins; clearly one bin has a large excess.

To contrast the distribution of the S GRBs and to test for possible errors in the analysis, we plot the same distributions for the M sample in Fig 3. As can be seen from Figure 3, this distribution is consistent with isotropy. Figure 3 shows the same analysis as Figure 2, indicating that there are no bins with a statistically significant deviation from the hypothesis of an isotropic distribution. Of the 46 very short S GRBs, there are 20 in the excess region. We have studied events from S and M samples and can find no real differences between the properties of the excess events and those outside of the excess region.

We have also calculated V/V_{max} for these samples $\langle V/V_{max} \rangle = 0.76 \pm 0.14$ that the S events are likely from Galactic, or possibly more local (solar neighborhood), sources. The V/V_{max} parameter is related to the volume of space that the source occupies divided by the maximum volume for such a source and is related to the geometry of the sources: $V/V_{max} = \frac{1}{2}$ for a Euclidean geometry. This is the first convincing evidence of some GRBs that are probably at non cosmological distances.

To further contrast the GRBs in S and M regions, we have calculated V/V_{max} for each event using C_p values (the event counts) from the BATSE data. As we have previously noted, this distribution is totally consistent with $\langle V/V_{max} \rangle = 0.5$ for a local distribution. In contrast, the same distribution for M events show in Figure 2 indicate a $\langle V/V_{max} \rangle$ much less than 0.5 consistent with the same mean values for the L (long) events, which is now widely interpreted as being due to the cosmological sources for those GRBs. It is probable that the M events are also from cosmological sources; however the S events appear to came from local sources. We note that the short bursts are strongly consistent with a Cp spectrum, indicating a Euclidean source distribution, as was shown previously. In the medium (from 100 ms to 1 s) time duration, the $\ln N - \ln S$ distribution seems to be non Euclidean. The $\langle V/V_{max} \rangle$ for the S, M, and L class of events is, respectively, 0.76 ± 0.14, 0.37 ± 0.03, three standard deviations different.

To determine the statistical probability for such deviation, we calculate the Poisson probability for eight bins with a total of 46 events (see Figure 2). The probability of observing 20 events in a single bin is 1.6×10^{-5}. We consider this as a very significant deviation from the isotropic distribution. We note that the M event distribution (Figure 3) is fully consistent with isotropy of the sources on contrast.

We perform also the likelihood analysis testing two hypothesis:

(h1) Poisson distribution with l = 5.75 = 46/8

In this case the logarithmic probability $\ln(p)$ calculated for the experimental sample is ~ -3.13, while the average $\ln(p)$ estimated from 105 randomly generated samples using the Poisson distribution with the same l and total number of events is ~ -17.76, with the standard deviation SD ~ 1.74. We conclude that the observed value is about 7.8 SD below the Poisson average. This corresponds to the $\sim 1.4 \times 10^{-6}$ chance to observe such a configuration and discards (h1).

(h2) Poisson distribution with extra source in "anomalous" angular bin.

Testing (h2), we first estimate the hypothesis that the underlying distribution for 7 angular bins except the anomalous one is indeed Poisson with l = 3.714 = 26/7. The $\ln(p)$ of the experimental sample is in this case ~ -15.29, which is less then one SD (~ 1.532) from the average for the random Poisson samples (~ -13.905). That confirms the Poisson hypothesis. One can get a rough estimate of the parameters characterizing (h2): $\lambda \sim 3.7, X \sim 20 - 3.7 = \sim 16$. The direct minimization of the likelihood for (h2) gives $\lambda \sim 3.78 \pm 0.7, X \sim 15.7 \pm 1.5$ in a good agreement with the previous estimate.

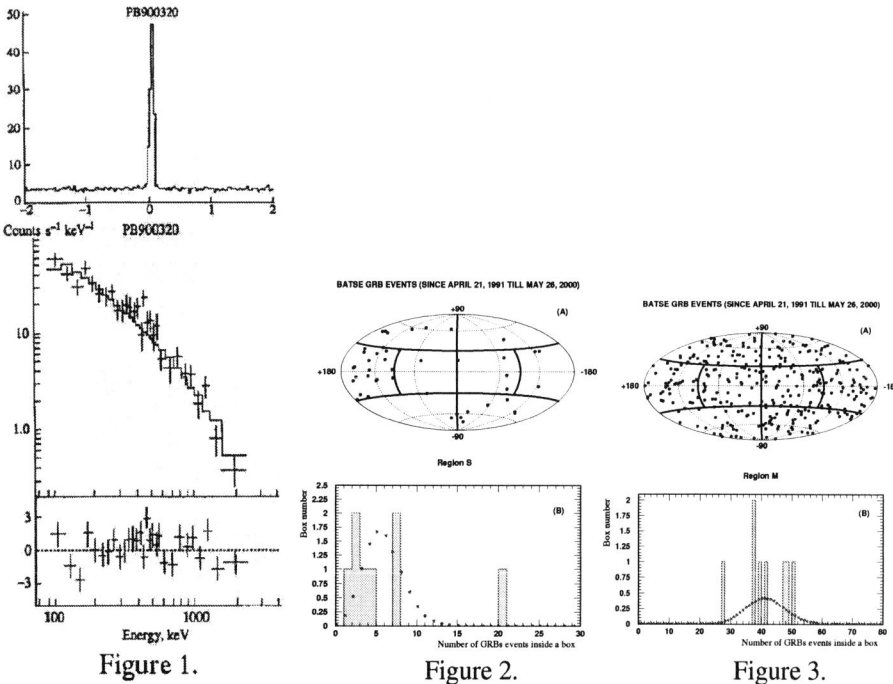

FIGURE 1. (Including sub-figures 1-3) – 1a: A very short GRB from the PHEBUS Detector (see Ref. 1); 1b: Energy distribution of the PHEBUS event in 1a. 2a/2b: The Galactic coordinate angular distribution of the GRB events (upper graph) with very short time duration (S). We define eight regions that correspond to equal angular surface. The distribution of the number of GRB events (lower graph) in each of the eight regions. Points correspond to prediction of the Poisson distribution for 42 events divided into eight equiprobable groups. Both distributions are normalized to the same surface. 3a/3b: The Galactic coordinate angular distribution of the GRB events (upper graph) with short time duration (M). Eight regions correspond to equal angular surface (like Figure 1a). The distribution of the number of GRB events (lower graph) in each of the eight regions. Points correspond to prediction of the Poisson (Gauss) distribution for 269 events divided into eight equiprobable groups. Both distributions are normalized to the same surface.

The likely minimum value is ~ -15.5 (which should be compared with 15.29).

We consider that this results are strong evidence for acceptance of (h2) and conclude that S GRBs are distributed isotropically with mean ~ 3.7, but there is an extra source which yields ~ 16 events in the anomalous angular region.

Independent of a direct association of the GRB events with specific galactic sources, it would be even harder to explain the distributions in Figure 3 as being due to extragalactic or even cosmological sources. The value of the $\langle V/V_{max} \rangle$ and the location asymmetry would seem to strongly support a Galactic origin of the sources for the S GRBs. It is also clear that this distinction would equally argue for a separate class of S GRBs from the M or L classes.

FUTURE STUDIES

We believe a future study of the S GRB population to a possible shorter time distribution will be fruitful. We may only have detected a fraction of the short bursts. The bulk of the very short bursts identified here all have time duration at or below the BATSE 64 ms integration time. We therefore believe that the BATSE trigger is likely an inefficient method of identifying such events; we also believe that many weak bursts may have been missed. There could be as many missed GRBs of 1 ms duration as the number that have been detected at 10 s.

Possible source contributors to the angular asymmetry of astrophysical sources:

1. It would seen unlikely for any extragalactic source to give such an asymmetric distribution.
2. We have studied various objects such as nearby stars and can find no angular excess in the same regions as Figure 2. Future tests should confirm this search.
3. If neutron stars were producing very short duration GRBs we would expect: (a) An excess at the galactic center (b) An excess at the galactic plane. Neither appears in Figure 2.
4. We conclude that the excess likely arises from some other source such as Primordial Black Holes and may cluster in clumps like CDM.

We thank Stan Otwinowski and Christina Matthey for collaboration on this study.

REFERENCES

1. Cline, D.B., Matthey, C., Otwinowski, S., ApJ **486**, 169 (1999).
2. Cline, D.B., Hong, W., ApJ **401**, L57 (1992).
3. Cline, D.B., Nucl. Phys. A **610**, 500c (1996).

Firework Model: Time Dependent Spectral Evolution of GRB

Guido Barbiellini*, Francesco Longo*, G. Ghirlanda[†], A. Celotti** and Z. Bosnjak**

Department of physics, University of Trieste and INFN, sezione di Trieste, Italy
[†]*Istituto di Astrofisica Spaziale e Fisica Cosmica, CNR, sezione di Milano, Italy*
***SISSA, Trieste, Italy*

Abstract. The energetics of the long duration GRB phenomenon is compared with models of a rotating BH in a strong magnetic field generated by an accreting torus. The GRB energy emission is attributed to magnetic field vacuum breakdown that gives origin to a $e^{\pm}$ fireball. Its subsequent evolution is hypothesized in analogy with the in-flight decay of an elementary particle. An anisotropy in the fireball propagation is thus naturally produced. The recent discovery in some GRB of an initial phase characterized by a thermal spectrum could be interpreted as the photon emission of the fireball photosphere when it becomes transparent. In particular, the temporal evolution of the emission can be explained as the effect of a radiative deceleration of the out-moving ejecta.

INTRODUCTION

At cosmological distances the observed GRB fluxes imply energies up to a solar rest-mass ($\sim 10^{54}$ erg), and as they vary on timescales of the order of milliseconds from causality and opacity arguments these must arise in regions whose size is of the order of few 10^{12} cm. This implies that an $e^{\pm}, \gamma$ fireball must form, which would expand relativistically. The fireball is energised and possibly collimated, mechanically or magnetically, close to the engine (for a review see e.g. [1]). Subsequently, it adiabatically expands and accelerates until the Thomson transparency is reached. The GRB phenomenology gives compelling reasons for the bulk motion of the emitting plasma to be highly relativistic with Lorentz factors of the order $\Gamma \sim 10^2 - 10^3$. The degree of isotropy/collimation of the ejected fireball is however still unclear. Recently a few GRB afterglows were observed at many wavelengths and suggest an axisymmetric jet-like structure for the fireball, thus strongly reducing the estimate of the energetics with respect to the isotropic case. The temporal decays of the emission at different frequencies, interpreted according to the fireball model, suggest jet beaming with opening angles $\theta \sim 3°$ [2]. The anisotropy of a collimated fireball alternatively can account for the observed phenomenology ([3],[4]). In our work [5] we focus on two aspects of the 'standard' scenario for the GRB event. The first concerns the extraction of energy from a rotating compact object and its conversion into a certain number of photon-$e^{\pm}$ fireball shells. Secondly, we suggest that the acceleration and collimation could occur in two phases: the first one consists in energising and collimating the shells, the second one is a radiation dominated expansion. This mechanism predicts that the observed Lorentz factor is determined by the product of the

Lorentz factor of the shell close to the engine and of the Lorentz factor achieved through the expansion, thus naturally giving rise to an anisotropic fireball [5].

Since their discovery the spectral analysis of Gamma Ray Bursts (GRBs) has been a crucial tool to investigate the nature of the emission process and the physics of their progenitors (see [6] for a recent review). In the standard picture the GRB spectrum is interpreted as the synchrotron emission by relativistic electrons ([7], [8]) which are accelerated at shocks produced by the encounters of the shells generated by the central engine ([9]). Recently, the analysis of some extremely hard BATSE GRB([10]) questioned synchroton as the leading emission process. Moreover, the time resolved spectra of the first 2 sec of these bursts were found to be consistent with a thermal (black body) model. The typical temperature of this "thermal phase" is $kT_{BB} \sim 100$ keV and, together with the luminosity L_{BB}, seems to evolve similarly with time in different bursts. A possible explanation of this observed temporal evolution of the thermal phase and of the fact that different bursts seems to show a similar spectral evolution is given[11].

GAMMA-RAY BURST PROGENITOR

In 1977 Blandford and Znajek proposed the interaction via magnetic fields between a rotating BH and an accretion disk to explain the energetics of Active Galactic Nuclei [12]. The same mechanism could be a good candidate for GRB engines as already pointed out (e.g. [13], [14]). In the following considerations it will be assumed that a dissipative interaction is at work between the BH and the torus surrounding it, due to an internal torque. In the approximation proposed in[5] considering the torus approximately at the last stable orbit, the loss of rotational ($\Delta E_{\rm rot}$) and gravitational energy ($\Delta E_{\rm g}$) can be derived as $\Delta E_{\rm rot} + \Delta E_{\rm g} \simeq 0.7 M_{\rm t} c^2$, ranging between $10^{53} - 10^{54}$ erg for a torus mass $M_{\rm t} = 0.1 - 1 M_\odot$.

A model for the generation of the GRB fireball is the vacuum breakdown in the volume close to the polar cap of the BH [15], where a value of $B_c \sim 4.5 \times 10^{13}$ G for the vacuum breakdown in the ergosphere is found. In the proximity of the BH is thus possible to generate $e^{\pm}$ pairs which could give origin to the GRB fireball, provided a sufficiently clean environment in order to avoid previous electric field discharge. After their creation by vacuum breakdown, the particles undergo three important processes, particle acceleration by an electric field of the order of 2×10^{15} V/cm, momenta randomisation by collision with the ambient photon density and single particle collimation in the direction of the magnetic field by synchrotron radiation. The momentum components perpendicular to field line for all the particles are damped. The result is the formation of a plasmoid made of a stream of particles with velocity parallel to the external field lines with bulk Lorentz factor $\Gamma_1 = \gamma \sim 30 \eta_{\rm acc}$ where we take into account also the acceleration efficiency.

COLLIMATION AND ACCELERATION: TWO PHASE EXPANSION

The subsequent jet evolution is composed by two distinct phases, the first one (phase-1), occurring close to the engine responsible of energising and collimating the burst. Phase-1 ends (at R_1) when the pre–existent collimating mechanism (by magnetic field) cannot balance the jet pressure any further. We could give a rough estimate of R_1 considering the distance when the collimation time scale becomes equal to the randomisation time scale. It then follows the second phase (phase-2), which consists of adiabatic expansion and corresponding acceleration of the relativistic particle fluid. This phase lasts for the radiation dominated phase and ends at the radius $R_{\text{pair}} \sim 100 R_0$ where the fireball becomes radiation dominated (e.g. [1]). Therefore, for a radiation dominated expansion [16]: $\frac{\Gamma'_2}{\Gamma'_1} \sim \frac{R_{\text{pair}}}{R_0} \sim 100$, where $\Gamma'_1 \sim 1$ is derived from the mean energy after the collimation phase measured in the comoving frame of a shell, Γ'_2 is the Lorentz factor at the end of phase 2 measured in the same reference frame.

At the moment of transparency the ejecta are thus moving in such a way that a particle accelerated during the radiation dominated expansion in the collimation direction will have $\Gamma_{\parallel} = 2\Gamma_1 \Gamma'_2$ where $\Gamma_{\parallel}$ is the bulk Lorentz factor in the axis direction being Γ_1 the Lorentz factor of the moving shell in the observer frame. The opening angle of the conical jet structure generated at the end of the two phases is determined by the particles accelerated perpendicularly to the collimation direction moving with Lorentz factor $\Gamma_{\perp} = \Gamma'_2$. The collimation angle θ_c is then: $\theta_c \sim \tan\theta_c = \frac{\Gamma_{\perp}}{\Gamma_1 \Gamma'_2} = \frac{1}{\Gamma_1}$. Assuming $\Gamma_{\parallel} \sim 10^3$ (e.g. [17]) and estimating $\Gamma'_2 \sim 100$ from the previous considerations, the value Γ_1 at the end of the collimation phase has to be of the order of $\Gamma_1 \sim 5$ and consequently $\theta_c \sim \frac{1}{\Gamma_1} \sim 2 \times 10^{-1}$. Furthermore, in the above scenario, the observed angular distribution of the expanding fireball is expected to be anisotropic, with a behaviour qualitatively similar to that postulated by [3] and [4] to account for the phenomenological findings by [2]. (See [5] for further details).

THERMAL DECELERATION

The recent observations of a thermal–like spectrum characterizing the initial $\sim$2 sec of 5 GRB[10] confirm a fundamental prediction of the standard fireball scenario. In particular, the temporal evolution of the temperature and luminosity at the beginning of these bursts could be interpreted as the effect of the mutual deceleration induced by the photons escaping from the shells while they are becoming transparent. In this case indeed the predicted decrease of the bulk Lorentz factor Γ with time reproduces the observed evolution of the temperature and luminosity of the spectrum.

The observed temperature of the thermal emission is $\Gamma k T$ where kT is the temperature of the photons emitted by the shells at the transparency transition and Γ is their Lorentz factor. Photons emitted by the expanding shells when their optical depth is $\tau \sim 1$ can decelerate following shells and produce the observed spectral evolution.

In order to estimate this effect, consider a group of 3 shells (labeled *i-1, i, i+1*) with relative position $R_{i-1} < R_i < R_{i+1}$. The momentum transfer due to the photons that they emit is

$$I_{i-1} - I_{i+1} \simeq m_i c \frac{d\Gamma_i}{\Gamma_i} \qquad (1)$$

where m_i is the mass of the central shell, $I_{i\pm1}$ is the momentum transferred due to photon interaction by the preceeding and following shells. The change of momentum induced by the *(i-1)* and *(i+1)* shells on shell *i* is: (see [11] for further details)

$$dI_i \propto R_i^2 \Gamma_i^2 \left(\frac{1}{R_{i-1}} - \frac{1}{R_{i+1}} \right) dt \qquad (2)$$

If the shells are produced by the central engine with typical separation Δ, the distance between the first and the last shells is $R_{i+1} = R_{i-1} + 2\Delta$ which gives $dI \propto 2\Delta \Gamma_i^2 dt$, and the conservation equation becomes

$$2\Delta \Gamma_i^2 dt \simeq \frac{d\Gamma}{\Gamma^3} \qquad (3)$$

This implies that the Lorentz factor of the *i*-th shell should decrease with time as $\Gamma \propto t^{-1/4}$. Assuming that the temperature of the thermal photons emitted by the different shells does not change much within the first 2 seconds the received photons have a temperature which is at least Γ times their source frame temperature. If so the observed temporal evolution[10] of the thermal peak $T_{BB} \propto t^{-1/4}$ is produced by the decrease of the bulk lorentz factor Γ. This result naturally accounts for the value (δ) of the temporal decay exponent for the temperature reported in [10]).

REFERENCES

1. Piran, T., Phys. Rep. **314**, 575 (1999).
2. Frail, D.A. et al.,, ApJ **562**, L55 (2001).
3. Zhang, B., Meszaros, P., ApJ **571**, 876 (2002).
4. Rossi, E., Lazzati, D., Rees M.J., MNRAS **332**, 945 (2002).
5. Barbiellini, G., Celotti, A. and Longo, F., MNRAS **339**, L17 (2003).
6. Hurley et al., astro-ph/0211620 (2002).
7. Katz J.I., ApJ **432**, L107 (1994).
8. Tavani, M., Physical Review Letters **76**, 3478 (1996).
9. Rees M.J. & Mészáros P., ApJ **430**, L93 (1994).
10. Ghirlanda, G., Celotti, A. and Ghisellini G., A&A **406**, 879 (2003).
11. Barbiellini, G. et al. in preparation (2004).
12. Blandford, R.D., Znajek, R.L., MNRAS **179**, 433 (1977).
13. Paczynski, B., ApJ **494**, L95 (1998).
14. Lee, H.K. et al., Phys. Rep. **325**, 83 (2000).
15. Heyl, J.S., Phys.Rev.D **63**, 064028 (2001).
16. Paczynski, B., ApJ **308**, L43 (1986).
17. Lithwick, Y., Sari, R., ApJ **555**, 540 (2001).

ANALYSIS AND OBSERVATION TECHNIQUES

The Gamma-Ray Burst ToolSHED is Open for Business

Timothy W. Giblin*, Jon Hakkila*, David J. Haglin[†] and Richard J. Roiger[†]

Department of Physics and Astronomy, The College of Charleston
[†]*Computer and Information Sciences Department, Minnesota State Universtiy, Mankato*

Abstract. The GRB ToolSHED, a Gamma-Ray Burst SHell for Expeditions in Data-Mining, is now online and available via a web browser to all in the scientific community. The ToolSHED is an online web utility that contains pre-processed burst attributes of the BATSE catalog and a suite of induction-based machine learning and statistical tools for classification and cluster analysis. Users create their own login account and study burst properties within user-defined multi-dimensional parameter spaces. Although new GRB attributes are periodically added to the database for user selection, the ToolSHED has a feature that allows users to upload their own burst attributes (e.g. spectral parameters, etc.) so that additional parameter spaces can be explored. A data visualization feature using GNUplot and web-based IDL has also been implemented to provide interactive plotting of user-selected session output. In an era in which GRB observations and attributes are becoming increasingly more complex, a utility such as the GRB ToolSHED may play an important role in deciphering GRB classes and understanding intrinsic burst properties.

INTRODUCTION

The GRB ToolSHED is designed to probe the BATSE GRB database in greater detail than ever before. Data mining tools can be valuable in the search for GRB classes and provide insight to the relationships between burst attributes (cf [1, 2, 3]). Note that the conclusions drawn from the learning model output depends strongly on the user's interpretation.

To facilitate analysis of GRB data it is desireable to provide the research community with: (i) widespread access to GRB data, (ii) easy-to-use data selection capabilities, and (iii) data analysis software tools. Examples of data analysis tools include data visualization and induction-based learning (frequently called *data mining* tools). These capabilities come together in the GRB ToolSHED, a web-based colocation of GRB data, selection and filtering capabilities, data visualization tools, and a range of data mining tools based on such strategies as statistical methods, decision trees, concept hierarchies, and neural networks. The ToolSHED is accessible from anywhere on the web by aiming a web browser to http://grb.mnsu.edu/grbts.[1]

---

[1] A mirror site is available at http://grb2.cofc.edu/grbts.

USING THE TOOLSHED

The ToolSHED must remember previous work performed by a user in order to reduce the set up time required to perform analysis. It is therefore necessary that all users register with the ToolSHED and login for each analysis session. Once registered, users can navigate through a menu of tasks including selecting data, defining new attributes directly computed from existing attributes, uploading attributes calculated outside the ToolSHED, selecting learning algorithms, building learned models, and visualizing the data. To be successful at using the repository of data and tools, the user needs to have a clear objective and understand that the output will not be an answer, it will only guide the user in the performance of scientific analysis.

Data Selection. Users must initially select which BATSE catalog to work with. The catalog is treated as a large table of data with many columns (attributes) and many rows (burst instances). Users may select some of the columns to work with and may provide a filter for selecting some of the rows. Any "selection" may be saved by the ToolSHED for later retrieval by the user.

New attributes definition. A classic example of defining new attributes directly computable from existing attributes is the value known as *HR32*, which is defined as the ratio of channel 3 fluence divided by channel 2 fluence. Other *binary* operations supported include multiplication, addition and subtraction. *Unary* operations include common and natural logarithms. Another supported computation is the notion of threshold. For example, a categorical attribute of "class" can be defined as *short* if $T90 \leq 2s$ and *long* if $T90 > 2s$. And finally, for users familiar with SQL, any generic SQL expression can be used to define a new attribute. All of the user-defined attributes can be used wherever a pre-defined attribute.

Attribute uploading. Users may wish to compute burst attributes outside of the ToolSHED but to include that information in their analysis. This can be done by uploading computed attributes to the ToolSHED. Understanding the importance of protecting access to this data, only the user who uploaded the data is allowed to see that information. As for user-defined attributes, all of the user-uploaded attributes can be used wherever a pre-defined attribute.

Learning algorithm selection. There are currently twelve machine learning algorithms supported in the ToolSHED (see Figure 1). It is expected that more algorithms will be added in the future. The existing algorithms represent a cross-section of machine learning strategies so that a variety of data analysis can be performed. The selection of the algorithm is critical to successful analysis.

Users should have an objective in mind and select the algorithm most likely able to meet that objective. For example, to find rules that "explain" membership in *long* and *short* classes, the user must define a categorical attribute for these two classes based on $T90$. The user then selects this "class" attribute along with several other attributes. To produce a set of rules, select the *PART* learning algorithm. The output will be rules based on the selected attributes that most effectively divide the bursts into the *long* and *short* classes.

There is an online manual available at the ToolSHED site to guide the user through the data mining process.

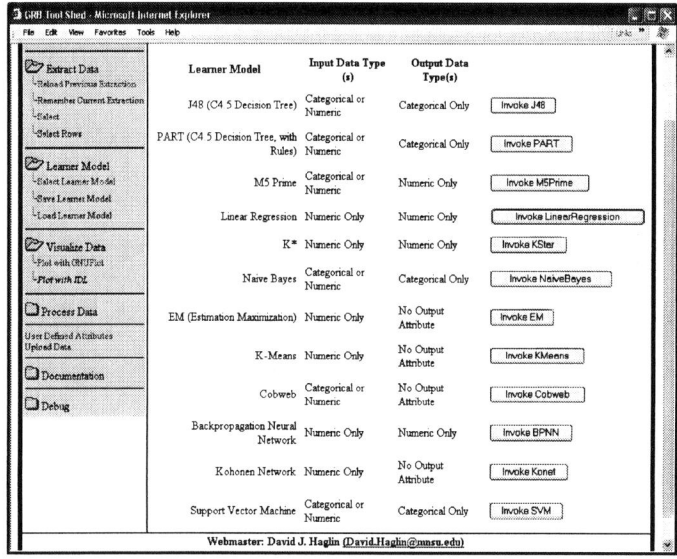

FIGURE 1. Supported learner algorithms

Learner Models. Once a learning algorithm is invoked, it uses the information selected and/or provided by the user for building an abstract representation of the information. This representation comes in many forms, depending upon the selected learning algorithm. There are two major types of learning algorithms supported in the ToolSHED[4].

The first type are the *supervised* classifiers that attempt to represent the essence of class membership. A popular example is the decision tree, which explains class membership as a series of decisions starting with the root of the tree and ending at a leaf node. Figure 2 shows a simple decision tree differentiating long and short bursts. Another commonly used algorithm is the *Back Propogation Neural Network* that builds a weighted network used to identify class membership; class descriptions based on these weights are unfortunately unobtainable.

The other type are the *unsupervised* clusterers that attempt to break a set of instances into a small number of partitions that maximize some criteria. These algorithms have no *a priori* knowledge of class membership. They work with the instances and some identified criteria (such as *distance* in n-space) to find instance clusters that make sense. One example of how this is done is finding a cutting plane in n-space, thereby defining two clusters).

Data visualization. Any of the selected data may be handed to either GNUplot or IDL to produce a plot of the data. The user is asked to choose which column of data — filtered by their row selection criteria — is to be used for the X-axis and which column is to be used for the Y-axis. The generated figure can be viewed as well as saved in EPS format on the users local machine.

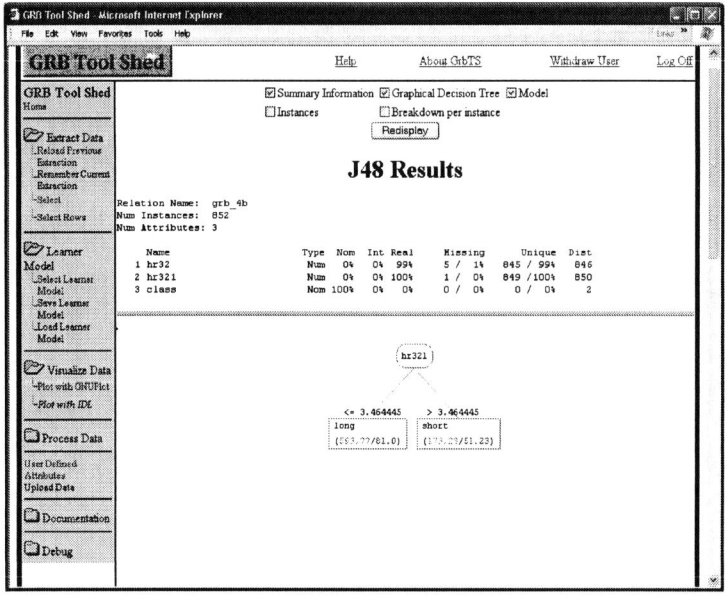

FIGURE 2. Sample decision tree output

CONCLUSIONS

We have successfully used the ToolSHED to guide our research leading to several significant results[1, 3]. We plan to continually add more GRB attributes as they become available to us. Examples include the BATSE spectral catalog[5], color color diagrams[6], and internal luminosity functions[7]. We also encourage the community to donate measured/calculated attributes to the ToolSHED repository.

ACKNOWLEDGMENTS

We gratefully acknowledge support from NASA grant NRA-98-OSS-03 (the Applied Information Systems Research program) and NSF grant AST-0098499 (Research in Undergraduate Institutions).

REFERENCES

1. Hakkila, J., et al., *ApJ*, **538**, 165 (2000).
2. Rajeniemi, and Mahonen, *ApJ*, **566**, 202 (2002).
3. Hakkila, J., et al., *ApJ*, **582**, 320 (2003).
4. Roiger, and Geatz, Addison-Wesley, 2003.
5. Preece, et al., *ApJS*, **126**, 19 (2000).
6. Giblin, T., et al., *ApJ*, **570**, 572 (2002).
7. Hakkila, J., et al., *These Proceedings* (2003).

Future Prospects for High Energy Polarimetry of Gamma-Ray Bursts

M. L. McConnell

Space Science Center, University of New Hampshire, Durham, NH 03824

Abstract. The recent detection of linear polarization from GRB120206 has piqued the interest of the community in this relatively unexplored avenue of research. Here, we review the current status and prospects for GRB polarimetry at hard X-ray and soft γ-ray energies. After reviewing the most recent results, we present a brief survey of current and planned experiments that are capable of making GRB polarization measurements in the energy range between 30 keV and 30 MeV.

INTRODUCTION

For many years, astronomers have generally been slow to accept the idea that polarimetry could be a useful tool for γ-ray astronomy. This has been both because of the experimental difficulty in making such a measurement and because the levels of polarization were expected to be quite low. Even at lower energies (1–10 keV), where source fluxes are considerably greater, the astronomical community has been slow to embrace the potential value of polarimetry. All this may have changed, however, with the recent detection of γ-ray polarization from GRB021206. This paper provides a brief overview of the experimental status and future prospects of GRB polarimetry at energies above ~ 30 keV.

POLARIMETRY TECHNIQUES

At these energies, there are three physical processes that can be exploited to measure linear polarization. These are: the photoelectric effect, Compton scattering (and its low-energy equivalent, Thomson scattering), and electron-positron pair production. In each case, the byproducts of the initial photon interaction (photoelectron, scattered photon, or electron-positron pair) have angular distributions that go as $\cos^2 \theta$. A measurement of the angular distribution of these secondaries provides a measure of not only the direction but also the magnitude of the linear polarization of the incident flux. The phase of the distribution is directly related to the direction of the incident polarization. The amplitude of the modulation in the angular distribution is directly related to the magnitude of the incident polarization. Much of the technical challenge for experimentalists arises from the difficulty in measuring these distributions.

RECENT RESULTS FROM RHESSI

Although originally designed as a hard X-ray solar imager, the Ramaty High Energy Solar Spectroscopic Imager [RHESSI; 1] has proven itself to be a valuable polarimeter. Two recent results demonstrate how RHESSI can do polarimetry utilizing two different techniques. Both techniques make use of RHESSI's 9-element Ge spectrometer array [2]. For polarization measurements at low energies (20 – 100 keV), a small block of passive Be (strategically located within the Ge array) is used to scatter photons into the rear segments of adjacent Ge detectors [3]. The polarization of a transient event (such as a solar flare) can be determined by a careful analysis of the counting rates in the Ge detectors that are closest to the Be scattering block. This mode is limited to a small FoV ($\sim 1°$) by the collimation of the Be scattering element through the front of the telescope assembly. At higher energies, scattering events between the Ge detectors within the spectrometer array can be used to measure polarization. The lack of significant amounts of shielding surrounding the Ge array means that this mode is sensitive to events over a much larger area of the sky. In both cases, the rotation of the RHESSI spacecraft (required for imaging with RHESSI's rotation modulation collimators) greatly facilitates effective polarization measurements by reducing systematic uncertainties and providing a more uniform sampling in the azimuthal direction.

A preliminary result from the low energy polarimetry mode has indicated a 20–40 keV polarization of $\sim 27\%$ from the solar flare of 23-July-2002 [4]. The high energy polarimetry mode of RHESSI has been dramatically demonstrated with a result from GRB 021206 that indicates polarization at a level of $80(\pm 20)\%$ in the 150 keV – 2 MeV energy range [5].

FUTURE PROSPECTS FOR GRB POLARIMETRY

One of the most important requirements of a GRB polarimeter is a large FoV. For most (if not all) other polarization studies, a large FoV can be detrimental in terms of increase background, etc. It may therefore be necessary, especially at lower energies, to consider dedicated GRB polarimeters.

Polarimetry in the 30–300 energy band requires low-Z scattering elements (coupled with high-Z photon absorbers) for achieving the best result. Unfortunately, instruments that operate in this energy band are usually not constructed using position sensitive *low-Z* material, but rather they are designed with *high-Z* materials to maximize photon absorption. Consequently, there is a need in this regime for dedicated instrumentation. One dedicated design, referred to as GRAPE, has been developed [6]. Based on Compton scattering from a low-Z plastic scintillator into a high-Z inorganic scintillator (CsI or LaBr$_3$), an early GRAPE design has been demonstrated in the laboratory (Figure 1a). Its very wide FoV also makes it ideal for studying the polarization of γ-ray bursts. A more recent version of the GRAPE concept (Figure 1b) permits tiled arrays, to serve as either a large area, wide FoV detector or as a detection plane for a coded mask imaging polarimeter.

At higher energies (300 keV – 10 MeV), Compton polarimeters based on the use

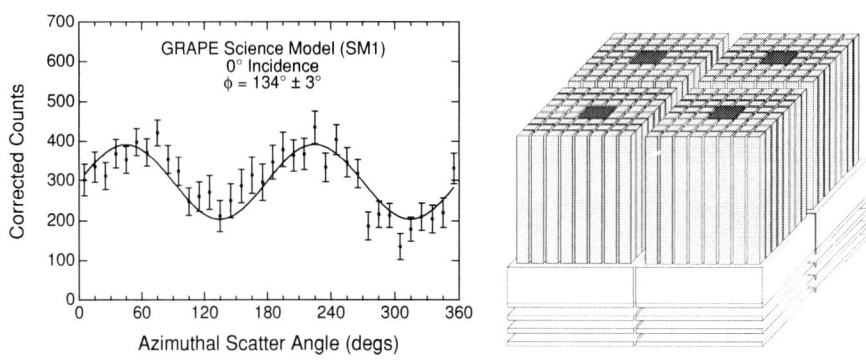

FIGURE 1. a) Laboratory results for a GRAPE prototype at 290 keV. b) A more recent GRAPE concept utilizes a flat-panel MAPMT for readout, shown here in a tiled configuration.

of high-Z scattering elements (coupled with high-Z absorbers) becomes viable. For example, the Ge double scatter approach used by RHESSI becomes most effective at energies above ~ 300 keV. Multiple scatter events in high-Z coded mask detection planes also offer possibilities for polarimetry. The use of a Ge strip detector has been demonstrated in this energy range [7]. An imaging polarimeter based on the use of CdTe is also being developed [8]. In principle, both the IBIS and SPI instruments on INTEGRAL are capable of polarimetry in this energy band [9]. Unfortunately, the lack of rotation makes the polarization analysis of these data difficult and telemetry limitations may limit the capabilities of IBIS. In principle, the CdZnTe detection plane of the Swift BAT instrument [10] might make for a good polarimeter, but the packaging design of the detectors and the design of the signal processing electronics results in a loss of the necessary multiple scatter event information.

This energy range is also the domain of Compton telescopes. A properly configured Compton telescope can serve as a very powerful polarimeter. The one Compton telescope that has flown in orbit, the COMPTEL instrument on CGRO [11], was very limited in its ability to do polarimetry. This was due both to its inability to precisely measure the interaction sites and also to a very poor Compton scattering geometry that required scatter angles $< 90°$. Although some efforts have been made to study polarization with COMPTEL data, no successful results have so far been obtained [12].

Compton telescope designs that are currently being studied offer a much more favorable geometry for polarization measurements. With the elimination of time-of-flight measurements, recent designs are much more compact. This results in significantly improved detection efficiency and a significantly larger FoV. It also provides a far more optimized well-type geometry for Compton polarimetry. The next generation of Compton telescopes will offer substantial improvements in polarization sensitivity. Recent Compton telescope designs can be characterized as those that attempt to track the scattered electron, such as TIGRE [13] and MEGA [14], and those that don't, such as LXeGRIT [15] and NCT [16]. One concept for the Advanced Compton Telescope (ACT) involves a large (1 m^2) stack of Si strip detectors that is used to track multiple Compton intertactions [17].

The potential utility of pair production for measuring polarization at energies above 2 MeV has been recognized for some time [e.g., 18]. Unfortunately, effective polarization measurements with pair production telescopes are limited by the effects of multiple coulomb scattering, which makes it difficult to define the plane of pair production. Efforts to measure polarization both with COS-B [19] and with CGRO/EGRET [20] have been unsuccessful, largely for this reason. It also appears that both GLAST and AGILE will suffer from similar difficulties, making polarization measurements with those instruments unlikely. One recent design for an effective pair production polarimeter involves the use of gas micro-well detectors for tracking the electron-positron pair with minimal scattering [21].

ACKNOWLEDGMENTS

This work has been supported by NASA grants NAG5-10203 (RHESSI) and NAG5-5324 (GRAPE).

REFERENCES

1. Lin, R.P., et al.*Solar Physics*, **210**, 3 (2002).
2. Smith, D.M., et al. *Solar Physics*, **210**, 33 (2002).
3. McConnell, M.L., et al. *Solar Physics*, **210**, 125 (2002).
4. McConnell, M.L., et al. *BAAS*, **35**, 616 (2003).
5. Coburn, W. and Boggs, S., *Nature*, **423**, 415 (2003).
6. McConnell, M.L., Ledoux, J., Macri, J., and Ryan, J. *SPIE Conf. Proc.*, **5165**, in press (2003).
7. Kroeger, R.A., Johnson, W.N., Kurfess, J.D., and Phlips, B.F., *Nucl. Instr. Methods*, **A436**, 165 (1999).
8. Caroli, E., et al., *SPIE Conf. Proc.*, **4140**, 573 (2000).
9. Lei, F, Dean, A.J., and Hills, G.L. *Sp. Sci. Rev.*, **82**, 309 (1997).
10. Barthelmy, S.D., *SPIE Conf. Proc.*, **4140**, 50 (2000).
11. Schönfelder, V., et al. *Ap. J. Supp.*, **86**, 657 (1993).
12. Lei, F, Hills, G.L., Dean, A.J., and Swinyard, B.M. *Astron. Astrophys. Supp.*, **C120**, 695 (1996).
13. O'Neill, T.J., et al. *Astron. Astrophys. Supp.*, **C120**, 661 (1996).
14. Kanbach, G., et al., *GAMMA 2001: Gamma-Ray Astrophysics*, AIP Conf. Proc. 587, eds.Ritz et al. (AIP, New York), 887 (2001).
15. Aprile, E., Bolotnikov, D., Chen, R., Mukerjee, R., and Xu, R., *Ap.J.Supp.*, **92**, 689 (1984).
16. Boggs, S.E., et al., *GAMMA 2001: Gamma-Ray Astrophysics*, AIP Conf. Proc. 587, eds.Ritz et al. (AIP, New York), 877 (2001).
17. Kurfess, J.D., and Kroeger, R.A., *GAMMA 2001: Gamma-Ray Astrophysics*, AIP Conf. Proc. 587, eds.Ritz et al. (AIP, New York), 867 (2001).
18. Maximon, L.C., and Olsen, H. *Phys. Rev.*, **126**, 310 (1962).
19. Mattox, J.R.., Mayer-Hasselwander, H.J., and Strong, A.W. *Ap. J.*, **363**, 270 (1990).
20. Mattox, J.R.. *Exp. Astron.*, **2**, 75 (1991).
21. Bloser, P.F., Hunter, S.D., Depaola, G.O., and Longo, F. (2003), astro-ph/0308331.

The KLENOT Telescope and GRBs

M. Tichý*, J. Tichá*, M. Kočer*, R. Hudec† and V. Šimon†

*Kleť Observatory & KLENOT Project, Zátkovo nábřeží 4, České Budějovice, Czech Republic
†Astronomical Institute, Academy of Sciences of the Czech Republic, 251 65 Ondřejov, Czech Republic

Abstract. The new 1.06-m KLENOT telescope equipped with a CCD camera is now in routine operation at the Kleť Observatory, with the primary goal to investigate the near-earth objects (NEOs) and other unusual objects. We describe this new telescope and the related systems and discuss the use of the telescope in GRB optical follow-up observations. The KLENOT Project website is http://klenot.klet.org.

THE KLENOT PROJECT

KLENOT is a project of the Kleť Observatory Near Earth and Other unusual objects observations Team (and Telescope) [1]. The KLENOT Project website is at the following URL: http://klenot.klet.org. The goals are as follows:

Confirmatory observations of newly discovered fainter NEO candidates: Some of new search facilities produce discoveries of near-earth objects (NEOs) fainter than $m_V = 20$ (for example the 1.8-m Spacewatch II, 1.2-m Palomar/NEAT) which need a larger telescope for the confirmation and early follow-up. A 1-m class telescope is also very suitable for the confirmation of very fast moving objects, and our larger FOV enables to search for NEO candidates having a larger ephemeris uncertainty.

Follow-up astrometry of poorly observed NEOs: It is necessary to observe the newly discovered NEOs in a longer arc during the discovery opposition before they get fainter. Special attention is given to "Virtual Impactors" and PHAs, target of future space missions or radar observations. On the other hand, it is necessary to find and use an optimal observing strategy to maximize the orbit improvement of each asteroid.

Recoveries of NEOs in the second opposition: For the determination of reliable orbits, it is required to observe asteroids in more then one opposition. If the observed arc in a discovery apparition is long enough, the chance for a recovery in the next apparition is good. If the observed arc at a single opposition is not so good, we plan to search along the line of variation. For this purpose, a larger field of view is an advantage.

Follow-up astrometry of other unusual objects: We plan to make follow-up astrometry of other unusual objects, that is, Centaurs and transneptunian objects, both in the discovery opposition and the next apparitions. To obtain positions of brighter transneptunians, we propose to use longer exposures with the magnitude limit $m_V \approx 22$. Considering the problem of securing adequate data for the orbit computation of these objects, follow-up astrometry, at least of some of them, will be useful.

FIGURE 1. The KLENOT telescope (left) and the Klet' Observatory (right).

Cometary features: The majority of new ground-based discoveries of comets come from large surveys devoted, predominantly, to Near Earth Asteroids. The first step in distinguishing these newly discovered members of the population of cometary bodies consists of confirmatory astrometric observations along with the detection of their cometary features. A timely recognition of the cometary features of a particular body having an unusual orbit can help in planning further observing campaigns.

Search for new asteroids: Our primary goal is the astrometric follow-up of NEOs and other unusual objects. Moreover, all CCD images are processed not only for the target objects, but also tested for the possible new object(s), because the effective field of view, observing time, limiting magnitude of $m_V \approx 22$ enable us to find new object(s). The obtained images will be processed with a special reference to fast moving objects and slow moving objects.

Follow-up of Gamma-ray burst (GRB) optical counterparts: A part of observing time is devoted to GRB optical follow-up observations as a target of opportunity.

PARAMETERS OF THE INSTRUMENT AND THE OBSERVING SITE

The KLENOT telescope characteristics: (a) 1.06-meter reflector; (b) four lenses primary focus corrector; (c) 1.06-m f/2.7 optical system.

CCD camera characteristics: CCD camera Photometrics Series 300; chip SITe 003B 1024×1024 pixels; pixel size 24 microns; liquid nitrogen cooling; FOV 33×33 arcmin; image scale 1.9 arcsec per pixel; Q.E.>80 per cent in the range 550–800 nm; Q.E.>60 per cent in the range 370–880 nm; limiting magnitude $m_V \approx 22$ for 180-sec exposure time in standard weather condition.

Software: A special software package has been developed for the KLENOT Project at Klet' using a combination of programs running on Windows and Linux platforms. The system consists of observation planning tools, data-acquisition, camera control and data processing tools. The code KAC (Klet' Atlas Coeli) (Fig. 2) shows stars and solar system objects with the line showing their daily motion in the sky in a selected region of the sky. The size of the region corresponds to the FOV of the telescope used so it is also used to check the telescope position during observation. As a source of the positions and

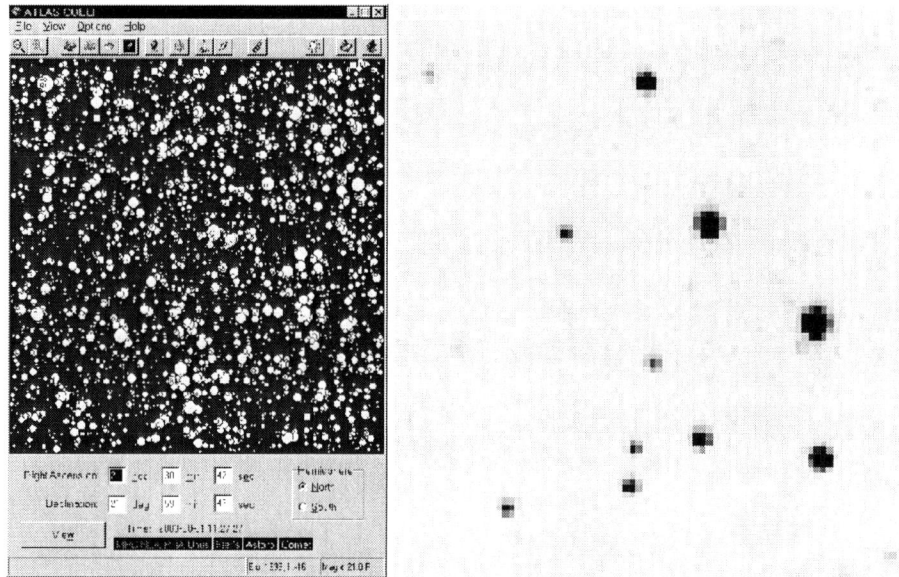

FIGURE 2. The code KAC - Klet' Atlas Coeli (left). The field of GRB030227 (FOV 2×2 arcmin, 60-sec exposure) observed by the KLENOT telescope (right).

magnitudes of the stars, the USNO-A2.0 star catalog is used.

Observing site: The Klet' Observatory is located in Czech Republic in Central Europe. The IAU/MPC observatory code is (246) Klet' Observatory-KLENOT. The geographical position of the observatory is $\lambda = +14°17'17"$ E, $\phi = +48°51'48"$ N, $h=1068$ m. There are about 140–200 clear nights per year, the best weather conditions are in February, March, August, September and October.

EXAMPLES OF GRB OBSERVATIONS WITH KLENOT

We have been engaged in GRB optical follow-up observations since 1992, first using wide-field photographic 0.63-m Maksutov telescope and then, since 1994, using 0.57-m telescope equipped with a SBIG CCD camera [2, 3, 4].

Since 2002, we can use the new 1.06-m KLENOT telescope equipped with a much more efficient CCD camera for these observations, which enables us to take larger field of view together with the magnitude limit up to magnitude $m_V \approx 22$. Several GRB fields were taken with the KLENOT Telescope up to now, as shown in the examples given below.

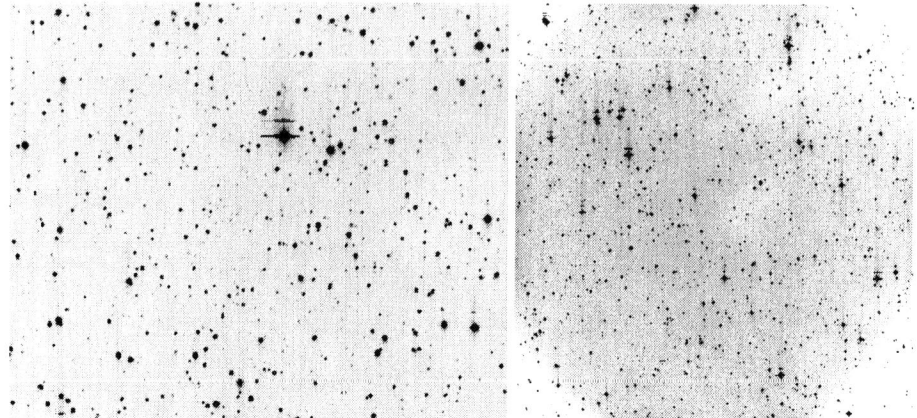

FIGURE 3. The fields of GRB030823 (FOV 10×10 arcmin, 60-sec exp.) (left) and GRB021212 (FOV 33×33 arcmin, 60-sec exp.) (right), observed by the KLENOT telescope.

ACKNOWLEDGMENTS

The GRB observations with KLENOT are supported by the grant A3003206 provided by the Grant Agency of the Academy of Sciences of the Czech Republic.

REFERENCES

1. Tichá, J., Tichý, M., and Kočer, M., "," in *Asteroids, Comets, and Meteors 2002*, edited by B. Warmbein, ESA SP 500, ESA/ESTEC, Noordwijk, the Netherlands, 2002, pp. 793–796.
2. Hudec, R., and et al., *Astrophysics and Space Science*, **231**, 335 (1995).
3. Hudec, R., and et al., *Experimental Astronomy*, **7**, 319 (1997).
4. Polcar, J., Hudec, R., Topinka, M., Tichá, J., Tichý, M., Masetti, N., and Pizzichini, G., "," in *Gamma Ray Burst and Afterglow Astronomy 2001*, AIP Cof. S. 662, AIP, USA, 2003, pp. 363–365.

Burst Populations and Detector Sensitivity

David L. Band

Code 661, NASA/Goddard Space Flight Center, Greenbelt, MD 20771

Abstract. The F_T (peak bolometric photon flux) vs. E_p (peak energy) plane is a powerful tool to compare the burst populations detected by different detectors. Detector sensitivity curves in this plane demonstrate which burst populations the detectors will detect. For example, future CZT-based detectors will show the largest increase in sensitivity for soft bursts, and will be particularly well-suited to study X-ray rich bursts and X-ray Flashes. Identical bursts at different redshifts describe a track in the F_T-E_p plane.

INTRODUCTION

Burst detectors are sensitive in different energy bands, and some bursts emit most of their energy at 10 keV while others emit at 1 MeV. Here I present a methodology to compare the burst populations that different detectors will detect. The issue is what bursts will cause a detector to trigger.

The standard rate trigger monitors a detector's count rate for a statistically significant increase[1]. The counts are binned over an energy range ΔE and time bin Δt, and the number of counts in this bin is compared to the expected number of background counts. For example, BATSE used ΔE=50–300 keV for most of the *CGRO* mission and Δt =0.064, 0.256 and 1.024 s. Usually the expected background is calculated from the counts accumulated over a period (e.g., 20 s long) before the bin being tested, and periodically (e.g., every 10 s) the background is recalculated. Most often the background is modelled as a constant in time, but polynomials can be fit.

A rate trigger tests the null hypothesis that only background counts are present. The test is whether the probability that the observed number of counts in a ΔE-Δt bin is a fluctuation is smaller than a threshold value. In the Gaussian limit the fluctuations are measured in units of σ, the square root of the expected number of counts, and the threshold fluctuation probability can be mapped into the number of σ. Thus the test is whether the observed number of counts exceeds the expected number by more than a threshold multiple of σ.

When more than one detector is present (e.g., BATSE's 8 detectors, the 12 NaI detectors of GLAST's GBM) the requirement may be that more than one detector trigger. Almost always a detector's sensitivity varies across the field-of-view (FOV). If two detectors must trigger, the spatial sensitivity depends on the angle to each detector.

For coded mask detectors such as HETE II's WXM and Swift's BAT the rate trigger is followed by imaging, and a new point source in the image must confirm the burst trigger.

A COMMON SENSITIVITY VOCABULARY

A detector's sensitivity is the minimum count rate that triggers the detector. This corresponds to the burst's peak count rate when integrated over ΔE and Δt. Because bursts are not constant for seconds and burst lightcurves differ at different energies, the sensitivity will differ for different sets of ΔE and Δt, and a smaller threshold count rate for one set of ΔE and Δt compared to another set may not indicate that fainter bursts will be detected. In addition, because of the detector's finite energy resolution, the count rate over ΔE is not the photon flux integrated over ΔE, and the count rates for two detectors in the same nominal ΔE cannot be compared directly. This complicates the comparison between different detectors and different triggers for the same detector. A common sensitivity measure is necessary; this quantity will be burst-dependent.

The sensitivity measure should be a measure of the burst's intensity that is independent of the detector or the trigger specifics[2]. I propose F_T, the photon flux integrated over 1–1000 keV and 1 s. This may not be the most physically interesting or relevant intensity measure, but it is closely related to the count rate that triggers the detector. Converting the count rate over ΔE to F_T requires the spectrum at the time of the peak flux, but this spectrum is required for the conversion from peak count rate to peak photon flux (e.g., BATSE's 50–300 keV peak flux). For a given detector trigger (i.e., with a choice of ΔE and Δt) F_T will depend on the particulars of the burst: the lightcurve around the time of the peak flux, and the spectrum at this time. Here I do not focus on the dependence on the lightcurve (i.e., on Δt), but instead on the spectrum.

Burst spectra can be described as $N_E \propto E^\alpha \exp[-E/E_0]$ at low energy and $N_E \propto E^\beta$ at high energy. The peak of $E^2 N_E \propto \nu f_\nu$ occurs at $E_p = (2+\alpha)E_0$, and thus E_p is a measure of spectral hardness. F_T is most strongly a function of E_p, although there is a residual dependence on α and β.

Therefore, detectors should be compared through sensitivity curves in the F_T-E_p plane. Bursts populate this plane: XRFs have small values of both F_T and E_p, while the burst hardness-intensity correlation means that bursts will populate a band from the lower left to the upper right of the F_T-E_p plane.

CALCULATION METHODOLOGY

While the sensitivity can be calculated in detail for each detector as a function of the source position in the FOV, taking into account variations in the background, here I describe simplified calculations[2].

Detector response—I assume that the detectors are "diagonal," the measured count energy is equal to the actual photon energy (i.e., infinite energy resolution). Since I integrate spectra and background over broad energy bands, this is a reasonable assumption, although at high energy a significant fraction of the photons are downscattered. In general I parameterize the energy dependence of the effective area as a series of power laws.

Background—In general I model the background as the sum of the aperture flux plus a constant (cts s^{-1} keV^{-1}); in some cases I use the background rates for a given detector quoted in the literature. The aperture flux is the product of the cosmic background flux

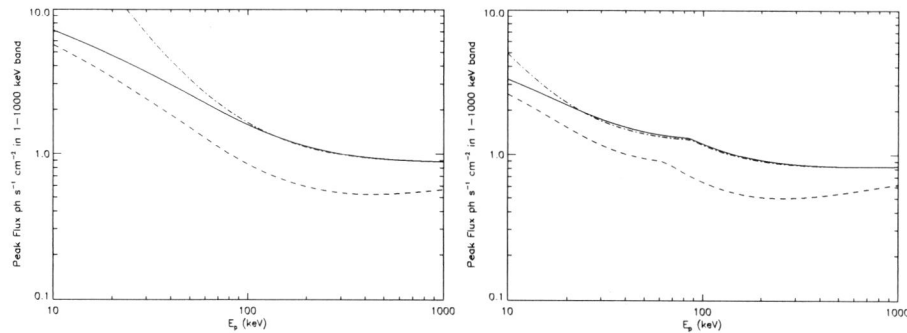

FIGURE 1. Sensitivity for BATSE (left) and Swift (right). Three curves are shown: solid line—$\alpha = -1$, $\beta = -2$; dashed line—$\alpha = -0.5$, $\beta = -2$; dot-dashed line—$\alpha = -1$, $\beta = -3$. In all cases I show the maximum sensitivity over the FOV, and assume $\Delta t = 1$ s.

and Ω, the average solid angle of the sky visible on the detector plane. Ω is calculated from formulae for rectangular and circular detectors.

Energy bands—I use the sets of ΔE applicable for each detector. For proposed detectors I use a set that optimizes the sensitivity.

Additional factors—I also consider the fraction of a coded mask that is open, the fraction of the detector plane that is active, and the angle to the second most sensitive detector when two detectors must trigger.

Factors not considered—I do not consider the high energy transparency of the detectors' side shields or coded masks (except when the mask transparency is included in the effective area). I also do not include scattering off the Earth or the spacecraft. The background calculation is crude, particularly at high energy.

APPLICATIONS

This methodology was originally developed to compare different detectors. Fig. 1 compares the sensitivity of BATSE, a set of NaI detectors, to Swift, a CZT detector. CZT is sensitive in the 10–150 keV band as opposed to NaI's 30–1000 keV sensitivity. Therefore, Swift's increase in sensitivity over BATSE will be greatest at low energy.

The scalloping in the Swift sensitivity curve results from triggering on more than one ΔE. The optimal set of ΔE can be found by comparing the sensitivities of different ΔE, as is shown by Fig. 2 from a trade study for GLAST's GBM NaI detectors[3]. In this case the second set of ΔE is as effective as the first, but does not include the $\Delta E = 50$–300 keV that was BATSE's primary trigger band.

Fig. 3 shows the position in the F_T-E_p plane that bursts at $z = 1$ with $F_T = 7.5$ ph cm^{-2} s^{-1} and $E_p = 30$, 100, 300 and 1000 keV would have if they originated at higher z. The calculation accounts for the narrowing of pulses at higher energy. The '+' are at half-integral z values up to $z = 10$. Also shown are the sensitivities for BATSE, Swift and EXIST for bursts with $\alpha = -1$ and $\beta = -2$.

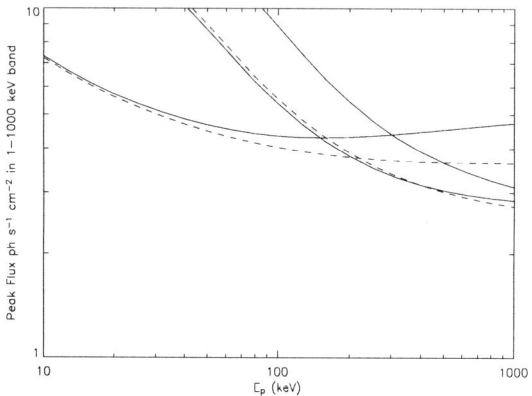

FIGURE 2. Sensitivity for the GBM NaI array on GLAST for $\Delta t = 1$ s and two sets of ΔE; $\alpha = -1$ and $\beta = -2$ are assumed. For the first set (solid curves, left to right) ΔE=5–100, 50–300, and 100–1000 keV while for the second set (dashed curves, left to right) ΔE=5–1000 and 50—1000 keV.

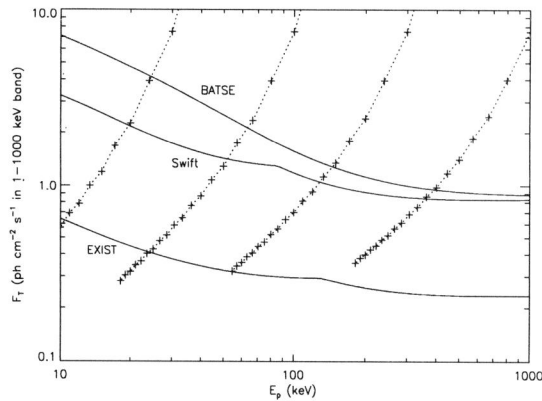

FIGURE 3. Tracks in the F_T-E_p plane for identical bursts at different z. Also shown are the sensitivities for BATSE, Swift and EXIST for bursts with $\alpha = -1$ and $\beta = -2$. The bursts would have F_T=7.5 ph cm^{-2} s^{-1} at $z = 1$.

REFERENCES

1. Band, D., ApJ, **578**, 806 (2002).
2. Band, D., ApJ, **588**, 945 (2003).
3. Band, D., et al., *GLAST's GBM Burst Trigger* in these proceedings (2004).

A Brief Comment on one of the "Amati Relationships" for Gamma-Ray Bursts

G. Pizzichini

IASF/CNR, Sezione di Bologna, via Gobetti 101, 40129 Bologna, Italy

Abstract. The measurement of redshift for a number of Gamma-Ray Bursts or their hosts now makes it possible to estimate their isotropic total radiation and its dependence on redshift, as it has been done by [1]. I discuss whether or not it is already possible, from the available data, to prove that there is a dependence of the properties of Gamma-Ray Bursts on redshift.

INTRODUCTION

The isotropic total radiated energy (hereafter E_{rad}) of Gamma-Ray Bursts (GRB) can be estimated, once their redshift is known. This has been done by [1] for nine events with firm redshift estimates, plus three more with less certain data. The same can be done for more GRBs discovered by HETE-2 [2] The fact that E_{rad} for detected GRBs substantially increases with their redshift has been interpreted as proof that GRBs, or their progenitors, must have had a strong cosmological evolution. I discuss here the possibility that instrumental selection effects and the luminosity function(s) of GRBs can either completely mimic or at least heavily mask a cosmological evolution and, at least with the data available at this time, prevent us from prooving that such an evolution took place.

DATA

Evidently there is no doubt, for example from Figure 2 in Amati and et al. [1], that E_{rad} for detected GRBs increases with their measured redshift. But, in order to prove that the increase is due to cosmological effects, we need to consider the probability that a GRB with that E_{rad} is actually detected at that redshift and that its redshift is measured.

If Gamma-Ray Bursts were both standard candles as well as isotropically and homogeneously distributed in space, and if reddening had no influence on the observed spectra, the number of events detected at the earth would be proportional to V_{max}, the volume corresponding to the maximum distance compatible with the instruments' detection limits. Thus the number would obey the well known Log N - Log P law, that is $(N > P) \propto P^{-3/2}$. But, even if we take into account the effect of redshift on detection limits for GRB instruments, which often trigger on E_{peak}, a quantity which in turn depends on energy, [3], we would expect the number of detected bursts to increase sensibly with redshift and then practically stop abruptly, a behaviour which is not observed. The

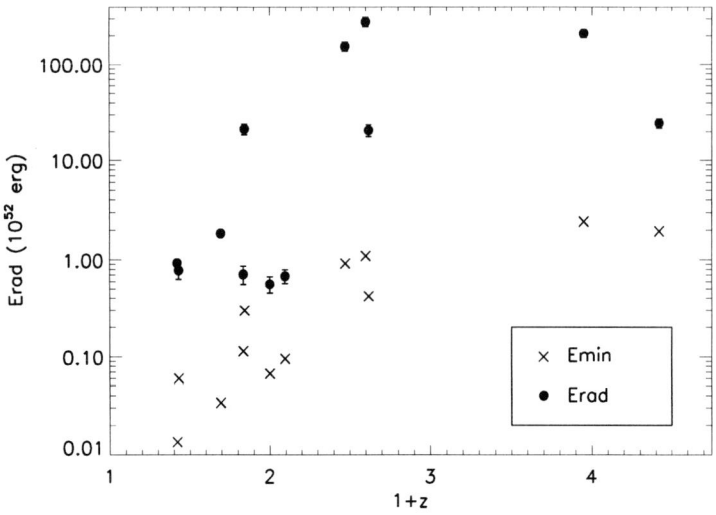

FIGURE 1. E_{rad} and its minimum detectable value versus z, from Amati and et al. [1]

probability of observing an event of a given E_{rad} at a given redshift is the product of the probability of E_{rad} times V_{max}. But, the event must be above the detection limits of GRB observatories, which take into account only intensities observed near earth. Thus, the minimum detected E_{rad} shall necessarily increase with redshift. Since the number of GRBs detected until now at different known redshifts, at least in the present limited sample, seems neither to increase nor to decrease very much, we must assume that the luminosity function, either dependent or not on redshift, decreases with burst intensity, be it E_{rad} or E_{peak}, roughly as much as V_{max} increases. Then, for any given detection limit the maximum detection probability shall not be too much above the detection limit. Such a situation corresponds well enough to what is shown in Figure 1, which shows the data in table 1. Data are taken from table 1 and figure 5 in Amati and et al. [1]. In order to derive E_{min}, I assume that E_{rad}/E_{min} is proportional to $S_{gamma}/S_{gamma,min}$ in that paper. The lack of events in the lower right hand corner is easily interpreted as due to the detection limit increasing with redshift, while the lack of events in the upper left hand corner can be attributed to the luminosity function, namely to the decrease in the probability to detect intense bursts which is not enough balanced by the increase in V_{max}, the volume available for events which can be detected. As long the number of events is so small it is only possible to state that we cannot expect all the events to be just above the detection limit, that situation would also correspond to a low probability, but we have a very small probability of detecting any of them well above it.

As long as we do not know if the luminosity function depends on redshift, we must first estimate the detection limit for each GRB, divide the events both into redshift and into E_{peak} intervals. Only then we shall be able to estimate if the luminosity functions at different redshifts differ or not.

TABLE 1. Data on 12 GRBs, derived from [1]

	$1+z$	E_{rad}	* E_{min}
GRB 970228	1.695	1.86 ± 0.14	0.034
GRB 970508	1.835	0.71 ± 0.15	0.114
GRB 971214	4.42	24.5 ± 2.8	1.95
GRB 980326	2.0	0.56 ± 0.11	0.067
GRB 980329	3.95	210.7 ± 20.3	2.4
GRB 980613	2.096	0.68 ± 0.11	0.095
GRB 990123	2.6	278.3 ± 31.5	1.10
GRB 990510	2.619	20.6 ± 2.9	0.42
GRB 990705	1.843	21.2 ± 2.7	0.30
GRB 990712	1.43	0.78 ± 0.15	0.06
GRB 000214	1.42	0.93 ± 0.03	0.0134
GRB 010222	2.473	154.2 ± 17.0	0.92

* E_{rad} and E_{min} in 10^{52} erg units

CONCLUSIONS

As long as we do not know if the luminosity function depends on redshift, we must first estimate the detection limit for each GRB, divide the events both into redshift and into E_{peak} intervals. Only then we shall be able to estimate if the luminosity functions at different redshifts differ. For the time being, there are not enough events to allow us to perform this exercise.

We must also keep in mind that, until now, most redshifts come from optical observations, and that estimating the detection limit for Optical Transients of GRBs is quite difficult, because it depends on the observing conditions in many observatories. Thus, at present, the fact that the redshift of a particular event was actually measured does not carry much significance and makes deriving conclusions from it less useful.

ACKNOWLEDGMENTS

I am indebted to Prof. Adalberto Piccioni and Dr. Patrizia Ferrero for their very useful help in preparing this manuscript.

REFERENCES

1. Amati, L., and et al., *A&A*, **390**, 81 (2002).
2. Lamb, D., and et al., *astro-ph/0310414* (2003).
3. Fenimore, E., and et al., *ApJ Letters*, **448**, L101 (1995).

PRESENT SATELLITES

Gamma-Ray Bursts Observed by INTEGRAL

S. Mereghetti

INAF, IASF, via E.Bassini 15, Milano I-20133, Italy

Abstract. During the first six months of operations, six Gamma Ray Bursts (GRBs) have been detected in the field of view of the INTEGRAL instruments and localized by the INTEGRAL Burst Alert System (IBAS): software for the automatic search of GRBs and the rapid distribution of their coordinates. I describe the current performances of IBAS and review the main results obtained so far. The coordinates of the latest burst localized by IBAS, GRB 031203, have been distributed within 20 s from the burst onset and with an uncertainty radius of only 2.7 arcmin.

INTRODUCTION

The INTEGRAL satellite, devoted to high-resolution imaging and spectroscopy in the hard X–ray / soft γ-ray energy range, has been successfully launched on October 17, 2002. The spacecraft carries two main instruments, SPI [1] and IBIS [2], optimized respectively for spectroscopy and imaging performances. Both instruments provide images of the γ-ray sky in the $\sim$15 keV – 10 MeV energy range, using the coded aperture technique. These two main instruments are complemented by an X-ray monitor (JEM-X [3]), covering the 4-35 keV range, and by an optical camera (OMC [4]) operating in the V band. All the INTEGRAL instruments are co-aligned and provide a broad energy coverage of the targets in the central part of the IBIS and SPI field of view.

The GRBs detected by IBIS can be localized with a precision of a few arc minutes in near real time, exploiting the continuous data link with the ground during INTEGRAL observations. This task is performed by the INTEGRAL Burst Alert System (IBAS, [5]) described in Section 2. The preliminary results on six GRB imaged by IBIS and SPI during the first year of the mission are presented in Section 3.

THE INTEGRAL BURST ALERT SYSTEM

Contrary to most other γ-ray astronomy satellites, no on-board GRB triggering system is present on INTEGRAL. Since the data are continuously transmitted to ground without important delays, the search for GRB is done at the INTEGRAL Science Data Centre (ISDC, [6]). This offers the advantages of a larger computing power and more flexibility for software and hardware upgrades, with respect to systems operating on board satellites.

IBIS is the most appropriate instrument on board INTEGRAL for GRB localization, thanks to its large field of view ($29° \times 29°$) and its capability to locate sources at the arcminute level. Two different methods to look for GRBs, using the data from the IBIS

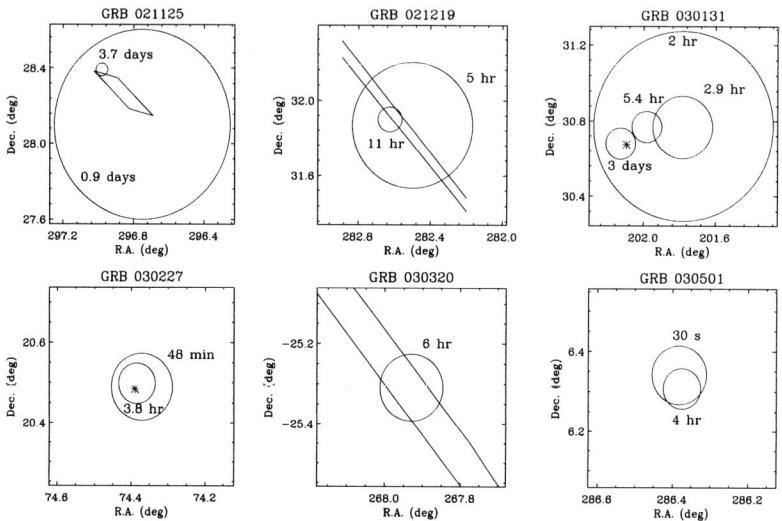

FIGURE 1. Error regions distributed for the six GRBs in the field of view of the INTEGRAL instruments, with the corresponding delays. The times refer to the public distribution of the GRB position. Note the different scale of the three upper (1° × 1°) and lower (0.5° × 0.5°) panels. The parallelogram and the straight lines indicate error regions independently derived with the IPN ([22, 23, 24]). The asterisks indicate the positions of the optical transients associated to GRB030131 ([25]) and GRB030227 ([26])

lower energy detector ISGRI [7], are used in parallel in IBAS.

In the first method the ISGRI counting rate is monitored to look for significant excesses with respect to a running average, in a way similar to traditional triggering algorithms used on-board previous satellites. Several different integration times, ranging from 2 ms to 5.12 s, are sampled in parallel. A rapid imaging analysis is performed only when a significant counting rate excess is detected. Imaging eliminates many false triggers caused, for example, by instrumental effects or background variations that do not produce a point source in the reconstructed sky images. The second method is entirely based on imaging. Images of the sky are continuously produced (integration times from 10 to 40 s) and compared with the previous ones to search for new sources.

An additional IBAS program is used to search for GRBs detected by the Anti Coincidence System (ACS) of the SPI instrument [8]. No directional or energetic information is available. The resulting light curves at 50 ms resolution are automatically posted on the ISDC WWW pages and are used for Inter-Planetary Network GRB localizations [9].

The GRB positions derived by IBAS are delivered via Internet to all the interested users. For the GRBs detected with high significance, this is done immediately by the software which sends *Alert Packets* using the UDP transport protocol. In case of events with lower statistical significance, the alerts are sent only to the members of the IBAS Localization Team, who perform further analysis and, if the GRB is confirmed, can distribute its position with an *Off-line Alert Packet*.

The time delay in the automatic distribution of coordinates results from the sum of several factors. There is a first delay of the data on board the satellite, which is variable

	Duration (s)	Distribution delay internal/public	Alert distribution	Peak Flux (ph cm^{-2} s^{-1})	Fluence (erg cm^{-2})	Ref.
021125	25	– / 0.9 days	OFF	22	7.4×10^{-6}	[10, 11]
021219	6	10 s / 5 hr	OFF	3.7	9×10^{-7}	[12, 13]
030131	150	21 s / 2 hr	ON	1.9	7×10^{-6}	[14, 15]
030227	20	35 s / 48 min	OFF	1.1	7.5×10^{-7}	[16, 17]
030320	50	12 s / 6 hr	ON	5.7	1.1×10^{-5}	[18, 19]
030501	40	30 s / 30 s	ON	2.7	3×10^{-6}	[20, 21]

and depends on the instrument. In the case of IBIS/ISGRI data the average delay is about 5 s, but it can be much longer for other instruments (e.g. approximately 20 s on average for the SPI ACS data). Signal propagation to the ground station is negligible (maximum ∼0.6 s), but some time is required before the data are received at the ISDC. This is on average 3 s when the ESA ground station in Redu (Belgium) is used, or 6 s when the NASA Goldstone ground station is used. The time to detect the GRB depends on the algorithm which triggers. The delay between the trigger time and the GRB onset is of course dependent on the intensity and time profile of the event. The IBAS simultaneous sampling in different timescales should ensure a minimum delay in most cases. Thus, for GRBs lasting a few tens of second, IBAS can in principle distribute the alerts with the position while the GRB is still ongoing.

The OMC covers only the central $5°\times5°$ of the IBIS and SPI field of view. During normal operations only the data from a number of small pre-selected windows around sources of interest are recorded and transmitted to the ground. The IBAS programs check whether the GRB position falls within the OMC field of view. In such a case, the appropriate telecommand with the definition of a new window centered on the interesting region is sent to the satellite. The OMC observation will consist of several frames with integration times of 60 s to permit variability studies and to increase the sensitivity for very intense but short outbursts. The expected limiting magnitude is of the order of V∼14 (60 s at high Galactic latitude).

RESULTS

IBAS has been running almost continuously since the launch of INTEGRAL. The first two months of operations were devoted to the tuning of the IBAS parameters. Some changes in the algorithms were also required to adapt them to the in-flight data characteristics. Delivery of the *Alert Packets* to the external clients started on January 17, 2003. Since then it has always been enabled, except for the period from February 12 to 28 (during calibration observations of the Crab Nebula), and for a short interruption (4 hours) on April 23 (for hardware maintenance reasons).

Six GRBs have been discovered in the field of view of IBIS during the first six months of operations. All of them were at relatively large off-axis angles, outside the fields of view of the OMC and JEM-X instruments. Two GRBs occurred during the initial performance and verification phase (GRB 021125 and GRB 021219), and at the time

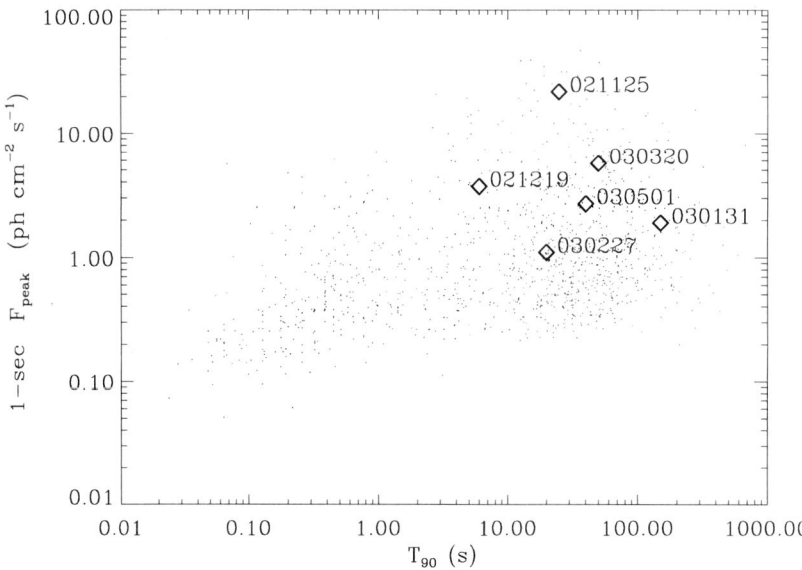

FIGURE 2. Peak flux versus duration for the six GRBs detected by INTEGRAL compared to the corresponding values of the BATSE catalog (dots).

of GRB 030227, as mentioned above, the automatic distribution of alerts to the external users was disabled. Of the remaining three bursts, only GRB 030501 was detected at a significance level high enough for automatic delivery of the coordinates.

The error regions derived for these six GRBs are shown in Figure 1, with the corresponding time delays in the public distribution of the coordinates. Note that at the beginning of the mission the in-flight instrument misalignment was not calibrated yet. Therefore, conservative error radii as large as 20$'$ or 30$'$ were given. The error regions obtained with the IPN, and the coordinates of the optical transients discovered for the two GRBs for which prompt observations could be done, are also shown in the figure. Their agreement with the INTEGRAL positions confirms that the IBAS localizations are correct.

The main properties of these six GRBs are summarized in Table 1. All of them where of the long duration class. Figure 2 shows their peak flux and T_{90} duration compared to the corresponding values of the bursts in the BATSE Catalog [27].

The first burst to be imaged by INTEGRAL, GRB 021125 is also the only one for which data from the high-energy IBIS/PICsIT [28] detector were obtained. In fact during this observation, PICsIT was operated in photon-by-photon mode (in the standard mode of operations PICsIT collects images integrated over a few thousand seconds and does not have enough time resolution for GRBs studies). Thus the spectrum of GRB 021125 could be studied up to $\sim$500 keV [11].

The second burst in the IBIS field of view, GRB 021219, was found and correctly localized by IBAS in real time [12]. The position derived by the IBAS software within $\sim$10 s had an accuracy of $\sim 20'$. As mentioned above, this error was largely dominated by systematic uncertainties present in the early phase of the mission. In less than four hours the error region could be significantly reduced, exploiting the presence of Cyg X-1 in the field of view of the same observation. Time resolved spectroscopy of GRB 021219 indicated a clear hard to soft evolution: the 15-200 keV spectrum was fit by a single power law with photon index evolving from 1.3 to 2.5 [13].

GRB 030131 lasted $\sim$150 s, but only the first 20 s were observed during a stable pointing, after which the satellite started a slew to the next pointing direction. Therefore, the signal to noise ratio of the trigger was not high enough for the automatic alert distribution. The moving satellite aspect complicated the analysis, resulting initially in a wrong localization (see Fig. 1). Nevertheless, a faint optical transient could be identified [25]. This burst had a fluence of 7×10^{-6} erg cm^{-2} (20-200 keV) and an average spectrum well described by a Band function with break energy E_0=70$\pm$20 keV, $\alpha = 1.4\pm0.2$ and β=3.0$\pm$1.0 [15]. Time resolved spectroscopy indicated also for this burst a hard-to-soft evolution.

The quick localization [16] obtained for GRB 030227 led to the discovery of both its X–ray [17] and optical afterglows [26]. This burst had a duration of about 20 s and a peak flux of $\sim$1.1 photons cm^{-2} s^{-1} in the 20-200 keV energy range. The spectrum was a power law with average photon index $\sim$2 and some evidence for a hard to soft evolution [17]. *XMM-Newton* started a Target of Opportunity Observation only 8 hours after the GRB. The X–ray afterglow was discovered with a 0.2-10 keV flux decreasing as t^{-1} from 1.3×10^{-12} to 5×10^{-13} erg cm^{-2} s^{-1}. The afterglow spectrum was well described by a power law with photon index 1.94$\pm$0.05. Interestingly, a significant absorption in addition to the Galactic value was required to fit the X–ray data [17]. The exact value of this intrinsic absorption depends on the (unknown) redshift, but is in any case of the order of a few times 10^{22} cm^{-2}. This supports the scenarios involving the occurrence of GRBs in regions of star formation. Some evidence for an emission line at 1.67 keV, which if attributed to Fe would imply a redshift $z \sim$3, was also found in the *XMM-Newton* spectrum [17]. Contrary to recent claims [29], we find that dividing the observation in short time intervals, all the spectra are well fit by the non-thermal power law spectrum, without the need of emission lines from light elements.

The next burst, GRB 030320 demonstrated the IBAS capability to discover and correctly locate GRBs even at very large off-axis angles [19]. The photons of this GRB, located more than 15° from the pointing direction, illuminated only a very small fraction of the IBIS/ISGRI detector and only three of the 19 SPI pixels. This GRB was coming from a direction very close to that of the Galactic Center and no optical observations of its error region were reported.

The Labour Day burst, GRB 030501, is the one with the best combination of speed and accuracy in localization. Its coordinates with an uncertainty of only 4.4$'$ reached all the IBAS users 30 s after the beginning of the event. Observations with robot telescopes started while the gamma-ray emission was still visible [30], but unfortunately GRB 030501 was at low Galactic latitude, in a region of very high interstellar absorption which hampered sensitive searches for counterparts.

Finally, thanks to the extension in the deadline for submission of these proceedings, I can add the results on the latest burst located by IBAS, GRB 031203 [31, 32]. It triggered with high significance on several timescales, setting a new record in localization speed and accuracy. Its position with an error of 2.7 arcmin was distributed by IBAS less than 20 s from the burst start time.

CONCLUSIONS

Thanks to the good imaging capabilities of the IBIS instrument and the continuous contact with the ground stations during the INTEGRAL observations, IBAS represents a step forward compared to previous GRB localization facilities. As demonstrated by GRB 030501 and GRB 031203, IBAS is currently able to provide small error regions ($\sim 3-4'$ radius) within a few tens of seconds from the GRB onset.

REFERENCES

1. Vedrenne G., Roques J.-P., Schönfelder V., et al., A&A **411**, L63 (2003).
2. Ubertini P., Lebrun F., Di Cocco G., et al., A&A **411**, L131 (2003).
3. Lund N., Budtz-Jorgensen C., Westergaard N.J., et al., A&A **411**, L231 (2003).
4. Mas-Hesse J.M., Giménez A., Culhane L., et al., A&A **411**, L261 (2003).
5. Mereghetti S., Götz D., Borkowski J., et al., A&A **411**, L291 (2003).
6. Courvoisier T.J.-L., Walter R., Beckmann V., et al., A&A **411**, L53 (2003).
7. Lebrun F., Leray J.P., Lavocat P., et al., A&A **411**,L141 (2003).
8. von Kienlin A., Beckmann V., Rau A., et al., A&A **411**, L299 (2003).
9. Hurley K., these proceedings (2004).
10. Bazzano, A. & Paizis A., GCN Circ. 1706 (2002).
11. Malaguti G., Bazzano A., Beckmann V., et al., A&A **411**, L307 (2003).
12. Mereghetti S., Götz D. & Borkowski J., GCN Circ. 1731 (2002).
13. Mereghetti S., Götz D., Beckmann V., et al., A&A **411**, L311 (2003).
14. Borkowski J., Götz D. & Mereghetti S., GCN Circ. 1836 (2003).
15. Götz D., Mereghetti S., Hurley K., et al., A&A **409**, 831 (2003).
16. Götz D., Borkowski J. & Mereghetti S., GCN Circ. 1895 (2003).
17. Mereghetti S., Götz D., Tiengo A., et al., ApJ **590**, L73 (2003).
18. Mereghetti S., Götz D. & Borkowski J., et al., GCN Circ. 1941 (2003).
19. von Kienlin A., Beckmann V., Covino S., et al., A&A **411**, L321 (2003).
20. Mereghetti S., Götz D., Borkowski J. & Courvoisier T. , GCN Circ. 2183 (2003).
21. Beckmann V., Borkowski J., Courvoisier T.J.-L., et al., A&A **411**, L327 (2003)
22. Hurley K., Mazets E., Golenetskii S., et al., GCN Circ. 1709 (2002).
23. Hurley K., Cline T., Götz D., et al., GCN Circ. 1772 (2002).
24. Hurley K., Cline T., Mitrofanov I., et al., GCN Circ. 1943 (2003).
25. Fox D.W., Price P.A., Heter T., et al., GCN Circ. 1857 (2003).
26. Castro-Tirado A.J., Gorosabel J., Guziy S., et al., A&A **411**, L315 (2003).
27. Paciesas W.S., Meegan C.A., Pendleton G.N., et al., ApJS **122**, 465 (1999).
28. Labanti C., Di Cocco G., Ferro G., et al., A&A **411**, L149 (2003).
29. Watson D., Reeves J.N., Hjorth J., et al., ApJ **595**, L29 (2003).
30. Boer M. & Klotz A., GCN Circ. 2188 (2003).
31. Götz D., Mereghetti S., Beck M., Borkowski J. & Mowlavi N., GCN Circ. 2459 (2003).
32. Mereghetti S. & Götz D., GCN Circ. 2460 (2003).
34. Ricker G.R., these proceedings (2004).
34. Ricker G.R., Atteia J.-L., Crew G.B., et al., in *Gamma-ray Burst and Afterglow Astronomy 2001*, eds. G.R. Ricker & R.K. Vanderspek, AIP Conference Proceeding 662, p. 3 (2003).

The Past, Present, and Future of the Third Interplanetary Network

K. Hurley* and T. Cline[†]

UC Berkeley, Space Sciences Laboratory, Berkeley, CA 94720-7450
[†]*NASA - Goddard Space Flight Center, Greenbelt, MD 20771*

Abstract. The Third Interplanetary Network for cosmic gamma-ray bursts began with the launch of the Ulysses spacecraft in October 1990, and has operated continuously since then. Over two dozen spacecraft or experiments have been incorporated into it over the years; today it comprises six spacecraft, which detect about 200 gamma-ray bursts per year. The composition of the network is reviewed, and some of the factors affecting the accuracy and rapidity of the localizations are explained. Future operations are discussed.

THE PAST

The Third Interplanetary Network (IPN) began operations with the launch of the Ulysses spacecraft in October 1990, and has now celebrated its 13th birthday. The experiments and missions which have played a role in it, besides Ulysses, are Pioneer Venus Orbiter, Ginga, the Defense Meteorological Satellites, the Compton Gamma-Ray Observatory, the WATCH, SIGMA, and PHEBUS experiments aboard the Granat spacecraft, Coronas, Yohkoh, the WATCH experiment aboard EURECA, the ill-fated Mars Observer, the Transient Gamma-Ray Spectrometer and Konus experiments aboard the Wind spacecraft, the Stretched Rohini Satellite Series-C, the Near Earth Asteroid Rendezvous Mission, BeppoSAX, the Rossi X-Ray Timing Explorer, Shenzou-2, the High Energy Transient Explorer (HETE-2), the HEND and GRS experiments on Mars Odyssey, the Ramaty High Energy Solar Spectroscopic Imager, and the IBIS and SPI-ACS experiments aboard INTEGRAL. This is twenty-five different experiments in total, not counting another ill-fated Mars mission, Mars-96, which had a small gamma-ray burst (GRB) detector, but failed to achieve orbit.

THE PRESENT

The IPN in a few numbers

To characterize the present IPN in a few numbers,

- It consists of **6** spacecraft, Ulysses, Wind, HETE, Mars Odyssey, RHESSI, and INTEGRAL,

- These spacecraft have a total of **20** detectors or experiments: two on Ulysses, two on Wind (Konus), four on HETE (FREGATE), two on Mars Odyssey (HEND and GRS), nine on RHESSI, and one on INTEGRAL (SPI-ACS),
- **58** people are involved in these experiments and missions,
- Over **1800** GRBs have been localized since 1990, and
- When all the permutations and combinations of the experiments and their operating modes are taken into account, there are over **2700** ways to detect and triangulate a burst.

The configuration of the IPN is shown in figure 1. The IPN website is http://ssl.berkeley.edu/ipn3/index.html.

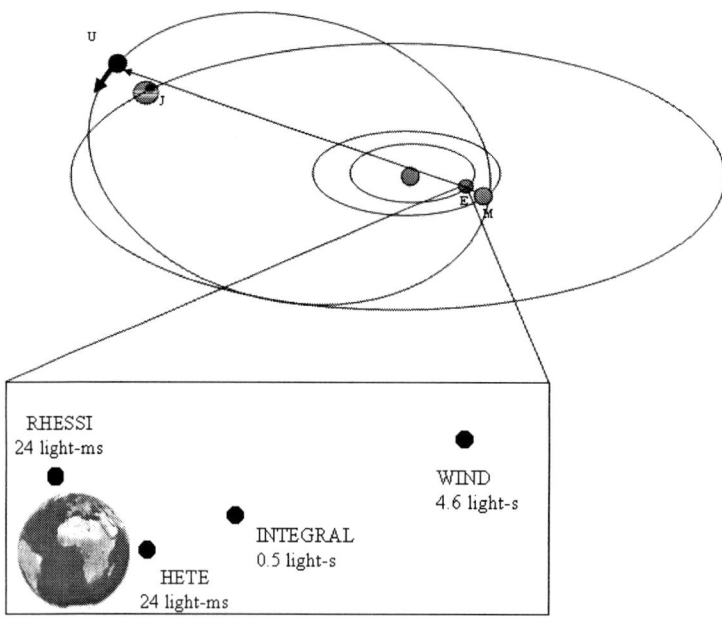

FIGURE 1. The configuration of the present IPN. This view is from a point north of the ecliptic, and shows the Sun and the orbits of the Earth, Mars, Jupiter, and Ulysses. The zoom in the lower left shows the near-Earth spacecraft, with their distances, or maximum distances from Earth center in light-seconds. GRB localizations to small error boxes are made possible by the fact that the network has three long baselines: Earth-Mars, Earth-Ulysses, and Ulysses-Mars.

Delays

The most important factor which determines the delay to get a GRB position from the IPN is usually the time between downlinks from the various spacecraft. This varies widely from one mission to another. A second factor which comes into play is often the question of whether data processing at the ground station can be done while the downlink is in progress, or whether the downlink must be complete to initiate it. Even light-travel time itself can also be important, since the most distant spacecraft in the IPN, Ulysses, can be over 3000 light-seconds from Earth.

HETE data on GRB light curves must be transmitted to one of the 3 primary ground stations, and the delay can be up to 60 minutes. Mars Odyssey telemetry passes go through NASA's Deep Space Network (DSN), and occur every 3-24 hours. However, the passes are long, and data can be processed during them, with the result that many Odyssey bursts can be found in near-real time. Ulysses passes similarly go through the DSN, and take place every 24 hours; but again, the data can be processed during the pass. RHESSI data are processed every 24 hours, but some quick-look data is available. Konus-Wind passes take place every 5-59 hours, and the downlink must be complete before data processing takes place. The INTEGRAL spacecraft is always within view of its ground station, and the data are downlinked continuously, but GRBs are found by processing the data at the INTEGRAL Science Data Center, and some manual intervention is required.

The net result is that positions are determined and sent out via the GCN with delays of 6 hours or more in most cases.

Error box sizes

Many factors determine the size of an IPN error box. Roughly in order of increasing importance, they are:

- The GRB intensity,
- The arrival direction of the GRB with respect to the vectors between the spacecraft; the width of an IPN annulus diverges when they coincide,
- The GRB time history; short or spiky time histories can be cross- correlated more accurately than featureless ones,
- The distance between spacecraft in the network; the greater, the better, and
- The alignment of the spacecraft; when 3 spacecraft lie in a line, or close to a line, the annuli coincide, causing the error boxes to become elongated.

Luck also plays a role. When only three spacecraft observe a burst, triangulation alone generally gives two possible error boxes. However, there are various ways to resolve the ambiguity. One is Earth-blocking for one of the spacecraft in low Earth orbit: one of the two IPN positions is clearly visible to a spacecraft which detects the burst, while the other is blocked. Another is Mars-blocking for the Odyssey spacecraft. A third is the fact that the Konus-Wind experiment can determine the ecliptic latitude of a burst to within about 10 degrees in the best cases; this often excludes one of the possible positions.

COMPARISON WITH OTHER MISSIONS

The strengths of the IPN may be summarized as follows. The network *as a whole* has roughly isotropic response, even though not all of its detectors do. It is sensitive to the short duration, hard spectrum class of GRBs; imaging instruments which localize bursts in the energy range below about 20 keV often have difficulties with these events, because they tend to be X-ray deficient. The IPN has good longevity: it is starting its fourteenth year. It is capable of sub-arcminute accuracy. The GRB detection rate by two or more spacecraft is roughly 200/year. It is modular; when one spacecraft finishes its mission, it can usually be replaced by another with little effort. And finally, it is inexpensive, since many of the experiments in it have been built for other purposes, and modifying them to detect bursts is relatively easy.

Missions such as HETE, INTEGRAL, and soon, Swift, have or will have capabilities that the IPN does not have, such as sensitivity to very weak bursts, and real-time localization capability. However, the instruments on these spacecraft localize bursts in only $\sim$ 13%, 4%, and 16% of the sky, respectively. Thus, assuming equal duty cycles, the IPN will detect intense bursts ($> 10^{-7} \mathrm{erg\,cm^{-2}}$) at $\sim$ 8, 25, and 6 times the rates of these missions. It can also detect bursts in regions of the sky which are inaccessible to them, due to Earth-limb, Sun, Moon, and bright star avoidance constraints. A good example is GRB021206, the "polarized burst" [1], which was only 18 degrees from the Sun. Although optical observations could not be carried out for several months, VLA observations did take place and a potential counterpart has been identified [2]. Another function that the IPN can carry out is to act as a monitor of soft gamma repeater activity, both from known SGRs and previously undiscovered ones.

OVERVIEW OF RESULTS

Over 80 bursts have been localized relatively rapidly and accurately by the IPN in recent years. Some were not followed up at optical or radio wavelengths, in many cases because they were not favorably located. However, of the ones which were, about 50% had detectable counterparts. This is about the same fraction as that of the BeppoSAX bursts [3], although there is a growing suspicion that many "dark" bursts would in fact have optical counterparts if they were observed quickly enough. Figure 2 compares the redshifts of IPN bursts with the distribution of redshifts for all GRBs. It is interesting to note that the two distributions do not appear to be significantly different, even though the IPN detects only the more intense bursts; indeed, the record GRB redshift, 4.5, is for an IPN burst [4].

FUTURE OF THE IPN

The Ulysses, Wind, and RHESSI missions have recently been approved for extensions by the Sun-Earth Connection Senior Review. Although Mars Odyssey's primary science mission ends in August 2004, it too may be extended. Swift [5] and Super-AGILE [6]

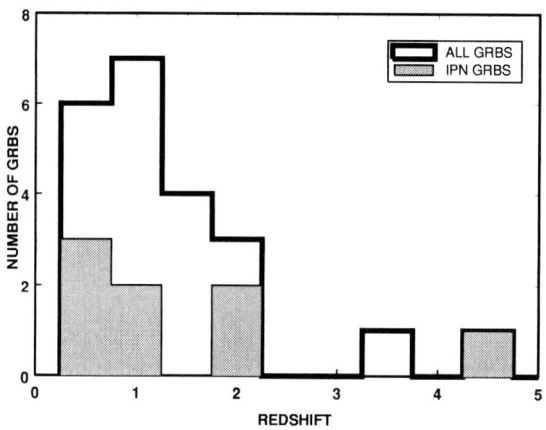

FIGURE 2. Redshift distributions for all GRBs with measured redshifts, and for IPN bursts.

will be added to the IPN in 2004 and 2005, respectively. Thus the IPN will be able to assist Swift in its checkout phase, as it has done for other missions in the past, and complement it when it commences routine operations.

ACKNOWLEDGMENTS

KH is grateful for Ulysses support under JPL Contract 958056, and for IPN support under NASA grants NAG5-11451, NAG5-12614, and NAG5-13080.

REFERENCES

1. Coburn, W., and Boggs, S., Nature **423**, pp. 415-417 (2003).
2. Frail., D., GCN Circ. 2280, 2003.
3. Piro, L., *Gamma-Ray Bursts in the Afterglow Era*, Springer, Berlin, pp. 97-105 (2000).
4. Andersen, M., et al., Astron. Astrophys. **364**, L54-L61 (2000).
5. Gehrels, N., *Gamma-Ray Burst and Afterglow Astronomy 2001*, American Institute of Physics, New York, pp. 465-468 (2003).
6. Feroci, M.,*Gamma-Ray Bursts in the Afterglow Era*, Springer, Berlin, pp. 368-370 (2000).

In-flight Calibration of the HETE-2 WXM Detector Response

T. Sakamoto[*†], Y. Nakagawa[**], K. Torii[†], Y. Shirasaki[‡], T. Tamagawa[†], N. Kawai[*], A. Yoshida[**], M. Matsuoka[§], E. E. Fenimore[¶], M. Galassi[¶], D. Q. Lamb[‖], C. Graziani[‖], T. Q. Donaghy[‖], J-L. Atteia[††], C. Barraud[††], M. Boer[‡‡], J-P. Dezalay[‡‡], J-F. Olive[‡‡] and HETE-2 team[§§]

[*]*Tokyo Institute of Technology*
[†]*The Institute of Physical and Chemical Research (RIKEN)*
[**]*Aoyama Gakuin University*
[‡]*National Astronomical Observatory*
[§]*Tsukuba Space Center, National Space Development Agency of Japan*
[¶]*Los Alamos National Laboratory*
[‖]*University of Chicago*
[††]*Observatoire Midi-Pyreneés*
[‡‡]*CNRS*
[§§]

Abstract. We present the in-flight calibration results of the detector response matrices of the X-ray instrument, Wide-field X-ray Monitor (WXM), on board the HETE-2 satellite. WXM consists of four one-dimensional position-sensitive proportional counters and two sets of one-dimensional coded aperture. Its energy range is 2–25 keV.

The detector response matrix (DRM) of WXM is calculated numerically using the ground measurements for each anode wire. It is then calibrated with the observation of the Crab nebula in December 2001 and January 2002. The spectral parameters and the normalization of the spectra are investigated at various incident angles, and compared with the known values in literature. With the current DRM, the derived spectral parameters are mostly consistent with the values in literature, expect for 10–20% uncertainty in normalization for individual anode wires, and at large incidence angles.

WIDE-FIELD X-RAY MONITOR (WXM)

The Wide-field X-ray Monitor (WXM) is located in the center of the spacecraft and the key instrument for localizing GRBs. WXM consists of one-dimensional position sensitive proportional counters (PSPC) and coded mask apertures. There are two sets of counters called the X-camera and the Y-camera. Each camera has two PSPCs. Each PSPC has three anode wires which are made by a carbon fiber of 10 μm in diameter. The counters are filled with xenon (97%) and carbon dioxide (3%) at 1.4 atm pressure at room temperature. The beryllium windows of 100 μm thickness is placed at the front of the detector. Details about WXM are described in the literature [1].

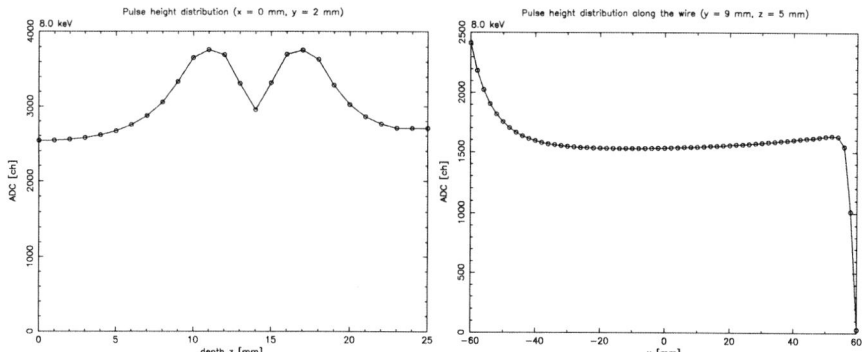

FIGURE 1. Modeling two major characteristics of the WXM detector. The pulse height as a function of the depth from the counter window (left) and the pulse height as a function of the position along the anode wire (right) for 8.0 keV monochromatic X-rays.

CALCULATION METHOD

The approach for calculating energy response function is an analytical method using the experimental equation based on the calibration data taken before launch. The difference of the gas gain at each position inside the detector is calculated as a function of the strength of the electric field which is based on the Garfield simulation (Figure 1).

Since the WXM counters are operating at the limited proportionality region to achieve the good position resolution, two major characteristics have to be taken into account for the DRM calculation. First, there is the highest gain region ~ 3 mm away from the anode wire (Figure 1, left). Although the physical reason for this phenomenon is still unclear, we know from the experiments that the extra gas gain (anomalous gain) is the source of this effect and this gain traces the strength of the electrical field [2]. The second effect is the gain variation at the both ends of the detector (Figure 1, right). This is due to the distortion of the electric field. Our DRM models these detector characteristics and successfully reproduce these phenomena.

THE WXM DETECTOR RESPONSE MATRIX

The input energy is calculated from 1.5 to 30.0 keV in 0.1 keV step. Since the photoelectric absorption is the dominant physical process in the WXM energy range, our analytical DRM calculation only takes into account this process. The DRMs are created in the Flexible Image Transport System (FITS) format to make it easy to handle in the standard softwares widely used in the high-energy astrophysical field (e.g. XSPEC). Each wire has its individual DRM.

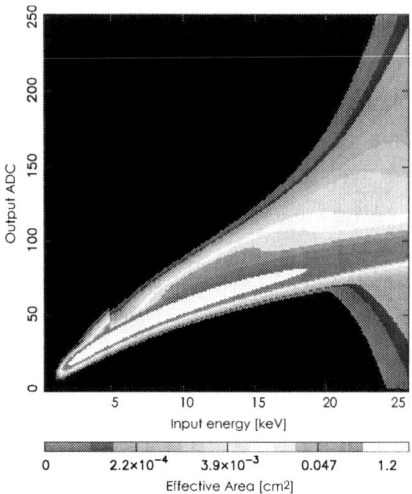

FIGURE 2. The WXM DRM of XA0 wire. The horizontal axis is the input energy and the vertical axis is the output ADC channels.

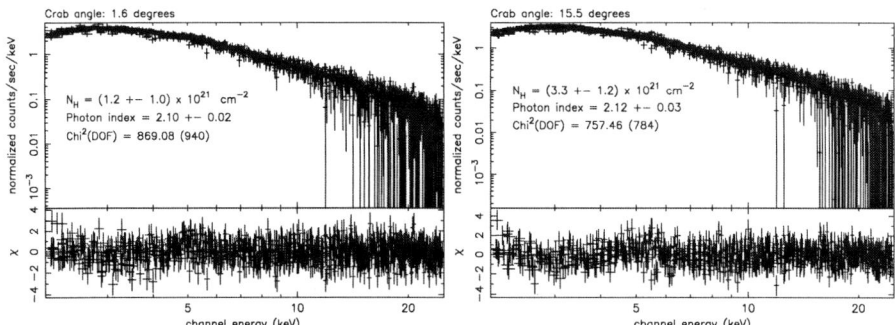

FIGURE 3. The spectra of the Crab nebula at the angle of 1.6 (left) and 15.5 (right) degrees. The spectral parameters of the Crab nebula in the literature are photon index of 2.1 and $N_H = 3.0 \times 10^{21}$ cm^{-2}.

CRAB CALIBRATION

The "RAW" data (the finest pulse height data) were collected for the Crab observation from December 2001 to January 2002. Since we need to check the DRMs with various incident angles, the Crab data with the different incident angles were recorded. The WXM Crab spectra at incident angle of 1.6 and 15.5 degrees are shown in Figure 3. The overall spectral parameters of the Crab are mostly consistent with the known values. However, there are several problems. The N_H values are relatively small. Also, variations of the normalization among the wires are seen. There is 10–20% uncertainty in the normalization for individual anode wires in the current DRM.

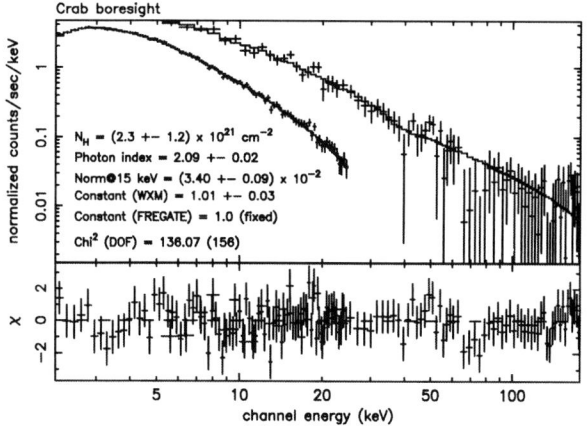

FIGURE 4. The joint WXM and FREGATE Crab spectrum at the boresight.

CROSS CALIBRATION BETWEEN WXM AND FREGATE

The joint WXM and FREGATE Crab spectrum at the boresight angle is shown in figure 4. The spectrum successfully represents the Crab spectral parameters.

SUMMARY

The WXM DRMs are well calibrated using the Crab nebula. Although further investigations of N_H and the normalization among wires are needed, The WXM DRMs can represent the Crab spectral parameters to a good precision. The cross calibration of WXM and FREGATE Crab data shows good agreement including the normalization between two instruments.

REFERENCES

1. Y. Shirasaki, et al., PASJ, **55**, 1033 (2003).
2. I. Sakurai, et al., SPIE **4140**, 511 (2000).

Gamma-Ray Bursts Observed with the Spectrometer SPI Onboard INTEGRAL

A. von Kienlin*, A. Rau*, V. Beckmann[†] and S. Deluit**

*Max-Planck-Institut für extraterrestrische Physik, Giessenbachstrasse, 85748 Garching, Germany
[†]NASA Goddard Space Flight Center, Greenbelt, Maryland 20771, USA
**INTEGRAL Science Data Centre, Chemin d' Écogia 16, 1290 Versoix, Switzerland

Abstract. The spectrometer SPI is one of the main detectors of ESA's INTEGRAL mission. The instrument offers two interesting and valuable capabilities for the detection of the prompt emission of Gamma-ray bursts. Within a field of view of 16 degrees, SPI is able to localize Gamma-ray bursts with an accuracy of 10 arcmin. The large anticoincidence shield, ACS, consisting of 512 kg of BGO crystals, detects Gamma-ray bursts quasi omnidirectionally above ~ 70 keV. Burst alerts from SPI/ACS are distributed to the interested community via the INTEGRAL Burst Alert System. The ACS data have been implemented into the 3rd Interplanetary Network and have proven valuable for the localization of bursts using the triangulation method. During the first 8 months of the mission approximately one Gamma-ray burst per month was localized within the field of fiew of SPI and 145 Gamma-ray burst candidates were detected by the ACS from which 40% have been confirmed by other instruments.

GRBS OBSERVED WITH THE CAMERA OF SPI

The aim of the spectrometer onboard INTEGRAL (SPI) [1] is to perform high-resolution spectroscopy of astrophysical sources in the energy range between 20 keV and 8 MeV. The imaging capability is good, but is exceeded by that of the imager IBIS which complements SPI by having higher imaging resolution, but lower spectroscopic resolving power. The detection and investigation of cosmic gamma-ray bursts (GRBs) is one of the important scientific topics of the INTEGRAL mission. The broad energy coverage of SPI is well suited to constrain the spectral shape, both below and above the energy at which the GRB power output is typically peaked (~ 250 keV) [2]. The long-standing controversy over the existence of short-lived spectral features in GRB spectra can be addressed by SPI's superb spectroscopic capabilities. In addition, the capability to cross-calibrate both spectra and images between the two experiments is extremely important, particularly in the case of such short-lived events as GRBs, which cannot be re-observed. Currently GRBs which occur inside SPI's field of view (FoV) are detected and analysed offline.

Since the start of the mission, six GRBs have been observed within the FoV of IBIS & SPI. The obtained scientific results are presented in Malaguti et al. [3] for GRB021125, Mereghetti et al. [4] for GRB 021219, Götz et al. [5] for GRB030131, Mereghetti et al. [6] for GRB0302227, von Kienlin et al. [7] for GRB030320, and Beckmann et al. [8] for GRB030501. In all cases the GRB alert was generated and distributed by IBIS, but SPI was also always able to detect the same event and to confirm in most cases the results obtained with IBIS. For the first three bursts the capabilites of SPI were weakened by the telemetry limitations at the beginning of the mission. The third event was the weak GRB030227 and the GRBs of March and May 2003 were observed at a large offset angle, thus only 15% to 25% of SPI's Ge-detectors were irradiated by the GRB. So currently the demonstration of SPI's full capabilities in the case of a strong event in the fully-coded FoV still has to take place. An overview on SPI's GRB detection capabilities, obtained for this first set of GRBs is given by von Kienlin et al. [9], which summarises the important quantities derived by SPI.

Most of the GRBs were detected by SPI with a S/N between 7 and 16. At this level the GRBs were located down to error radii of 20' – 30' (90% confidence) which are in most cases overlapping with the one of IBIS. Also the peak flux, fluence and photon indices are in agreement with the values derived by IBIS. For GRBs observed with SPI operating in full telemetry mode, spectra could be extracted. In one case (GRB030227) some evidence for a hard-to-soft spectral evolution was found in the data of both instruments, ISGRI and SPI [6].

GRBS OBSERVED WITH THE ANTICOINCIDENCE SHIELD ACS

Since December 2002 the anticoincidence shield (ACS) [10] has been added to the 3rd interplanetary network (IPN) of γ-ray detectors [11]. During the first year of the IN-TEGRAL mission the IPN consisted of Ulysses, Mars Odyssey 2001, Konus-WIND, HETE-2, RHESSI and INTEGRAL/SPI-ACS. The network had an excellent configuration, due to the large spacecraft separations between Earth, Mars and Ulysses, which is orbiting around the sun, out of the ecliptic plane. The analysis of GRBs detected with the SPI-ACS during the first eight months (November 2002 - June 2003) of the INTEGRAL mission is presented below. A more detailed description of these first results, obtained by SPI-ACS, can be found in von Kienlin et al. [9].

As SPI-ACS has no spatial resolution and thus cannot provide the position of the GRB [10], the GRB nature of a count rate increase observed by SPI-ACS can only be confirmed by the observation of the same event by another instrument (e.g. Ulysses, HETE-2). These bursts constitute only a subsample of the GRBs detected by SPI-ACS, as different instrumental properties of these missions (e.g. energy range, sensitivity) do not allow the simultaneous observations/detections of all GRBs seen by SPI-ACS. Obviously, it is worthwhile to study also bursts which are only visible in the SPI-ACS rates and not confirmed elsewhere. Probing the very high energies with the unprecedented sensitivity of the SPI-ACS [10] might open new insights into the burst populations and burst physics.

A sample of possible GRB events, based on the only property measurable by the SPI-ACS, the veto-countrate lightcurve, is selected for ACS by using only events which exceeds a predefined significance level above the background (the details of the selection procedure are described in von Kienlin et al. [9]). Each event is subsequently checked for solar or particle origin by comparing with events recorded by the X-ray monitor JEM-X and the radiation monitor IREM of INTEGRAL, or events noted on the GOES

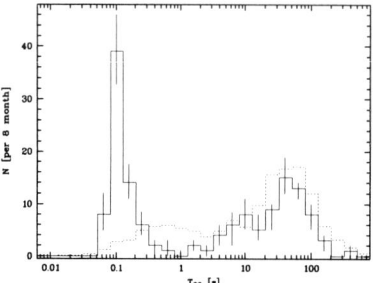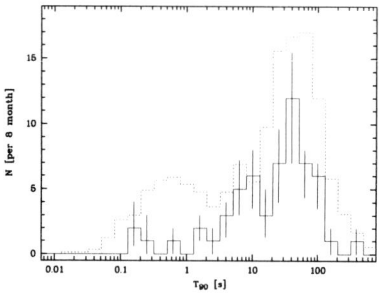

FIGURE 1. Left: Distribution of T_{90} for all GRB candidates (solid line) and for 1234 GRBs from the 4th BATSE GRB catalogue [13] (dotted). In order to compare with the SPI-ACS detections, the BATSE distribution is scaled to the elapsed INTEGRAL mission time (8 month). Note the very large fraction of short events compared to BATSE. Right: Same as left plot except that here only the confirmed SPI-ACS burst sample (solid line) is displayed. While most of the long duration bursts are confirmed for comparison an obvious lack of short GRBs can be noticed.

web page[1].

With the selection described above, a total of 145 GRB candidates were detected during the first 8 months of the mission. 58 of these have been confirmed by other instruments. Using the elapsed mission time, we find an approximate rate of GRBs detected by the SPI-ACS of $\sim$290 ($\sim$116 confirmed) per year which is in good agreement with the predictions given in [12] prior to the start of the mission. The total rate is comparable also to BATSE [13]. In addition to the number of events, the SPI-ACS overall rate provides the possibility of deriving the burst duration in the instrumental observer frame and the variability of the light curve. As no energy resolution exists, typical burst parameters such as fluence and peak flux cannot be derived. Only the total integrated counts and the counts in the burst maximum can be extracted from the light curve. Fig. 1 shows the distribution of the measure for the duration T_{90} (the time interval starting after 5% and ending after 95% of the background subtracted event counts have been observed) for the sample of SPI-ACS GRBs in comparison to the observed distribution of 1234 GRBs from the 4th BATSE GRB catalogue [13]. Despite the small sample, a bimodality in the distribution comparable to that found by BATSE is observed. But two main differences emerge: i) the SPI-ACS sample contains a significantly higher fraction of short burst candidates and ii) the maximum of the short distribution is offset towards shorter duration for the SPI-ACS sample.

The fraction of short ($<$1 s) duration GRBs is $\sim$0.48 (70/145) for the SPI-ACS sample compared to 0.20 for BATSE [13]. As BATSE was observing a softer energy band (50-320 keV) and was therefore more sensitive to X-ray rich (long) GRBs than SPI-ACS, a larger short/long rate was expected for the ACS sample. What is remarkable is the sharpness of the short distribution around 0.1 s. Due to the limited time resolution of 50 ms the short end cannot be sufficiently defined and resolved by our data. The offset

---

[1] http://www.sec.noaa.gov

of the maximum for the short events to smaller T_{90} might be due to the different energy bands of SPI-ACS and BATSE. An apparently shorter duration is measured as it would be if the bursts would have been observed by BATSE. As T_{90} depends strongly on the instrumental characteristics and as it is still unclear how this measure connects to the source frame quantity for a given burst, the discrepancies are neither surprising nor do they necessarily trace different burst populations. Still, the connection of the short events with real GRBs is not clear as this population is only marginally observed by other instruments. While a large fraction (73%; 55/75) of the long bursts are confirmed, less than 6% (4/70) of the short events were observed by other missions. This might be explained at least twofold. On the one hand we might observe a "real" short and very hard GRBs population, which could so far only be detected with SPI-ACS due to its high sensitivity at very high energies. As the current IPN members and HETE-2 are generally more sensitive at lower energies, the detection of a high fraction of unconfirmed short (and possible hard) events would not be surprising. These bursts should then have peak energies above 400 keV. On the other hand, a significant contribution to these short events from instrumental effects and/or cosmic ray events cannot be ruled out. A small contribution might also arise from soft gamma-ray repeaters (SGRs). Without localisation SGR bursts cannot be distinguished from short GRBs within SPI-ACS. The issue of origin of the short events is certainly of high interest and needs a more detailed investigation

ACKNOWLEDGMENTS

The SPI project has been completed under the responsibility and leadership of CNES. We are grateful to ASI, CEA, CNES, DLR, ESA, INTA, NASA and OSTC for support. The SPI/ACS project is supported by the German "Ministerium für Bildung und Forschung" through DLR grant 50.OG.9503.0.

REFERENCES

1. Vedrenne, G., Roques, J.-P., Schönfelder, V., et al., *A&A*, **411**, L63–L70 (2003).
2. Preece, R. D., Briggs, M. S., Mallozzi, R. S., et al., *ApJS*, **126**, 19–36 (2000).
3. Malaguti, G., Bazzano, A., Beckmann, V., et al., *A&A*, **411**, L307–L310 (2003).
4. Mereghetti, S., Götz, D., Beckmann, V., et al., *A&A*, **411**, L311–L314 (2003).
5. Götz, D., Mereghetti, S., Hurley, K., et al., *A&A*, **409**, 831–834 (2003).
6. Mereghetti, S., Götz, D., Tiengo, A., et al., *ApJ*, **590**, L73–L77 (2003).
7. von Kienlin, A., Beckmann, V., Covino, S., et al., *A&A*, **411**, L321–L325 (2003).
8. Beckmann, V., Borkowski, J., Courvoisier, T. J.-L., et al., *A&A*, **411**, L327–L330 (2003).
9. von Kienlin, A., Beckmann, V., Rau, A., et al., *A&A*, **411**, L299–L305 (2003).
10. von Kienlin, A., Arend, N., Lichti, G. G., et al., "Gamma-Ray Burst Detection with INTEGRAL/SPI," in *X-Ray and Gamma-Ray Telescopes and Instruments for Astronomy*, edited by J. E. Truemper and H. D. Tananbaum, 2003, vol. 4851 of *Proceedings of the SPIE*, pp. 1336–1346.
11. Hurley, K., "The 3rd Interplanetary Network," in *American Institute of Physics Conference Series*, 2004, p. this issue.
12. Lichti, G. G., Georgii, R., von Kienlin, A., et al., "The γ-Ray Burst-Detection System of SPI," in *American Institute of Physics Conference Series*, 2000, pp. 722–726.
13. Paciesas, W. S., Meegan, C. A., Pendleton, G. N., et al., *ApJS*, **122**, 465–495 (1999).

The Growing SXC Burst Catalog: A Transient for Each Detection

J. Villasenor*, J. G. Jernigan*, G. Crew*, R. Vanderspek*, A. Dullighan*, N. Butler*, G. Prigozhin*, J. Doty* and G. Ricker*

Center for Space Research, Massachusetts Institute of Technology

Abstract. The first year of full SXC operations has resulted in the detection of over a dozen GRBs and XRFs, and the SXC GRB catalog continues to grow at a rate of approximately 1/month. The small error circles, rapid position dissemination, and most importantly, aggressive follow up observations have led to a remarkable trend: nearly all SXC detections have an associated optical/IR/radio/X-ray counterpart. The systematic errors dominating the SXC astrometry are being reduced, and recent recalibration efforts will result in smaller error radii sent out to the GCN. Flight localizations calculated tens of seconds after the burst onset will also enable large telescopes with small fields of view to engage in rapid, deep searches. These two factors will aid in building a unique database of GRBs with associated transients.

A Revived Instrument

The HETE-2 Soft X-ray Camera (SXC) produced its first GRB localization with GRB020813, over a year and a half from the launch of the satellite. Early in the mission, atomic oxygen eroded the optical blocking filters protecting the SXC's focal plane CCDs, and excessive light leakage prevented this instrument's operation. Improvements in software and operational adjustments have dramatically revived the SXC. Better background light subtraction, and optimal event thresholding have doubled the photon counting efficiency. The adoption of head nodding, or the ability to point the spacecraft away from its normal anti-solar attitude, shifts the SXC field of view away from the moon during full moon, extending operations from 18 to 22 days for each lunar cycle.

The sensitivity of the SXC to bursts is now nearly comparable to that of the Wide-field X-ray Monitor (WXM), making it an excellent complement for vernier (fine) localization. SXC localizations now run an average of 1/month; since August of 2002, there have been 12 GRBs and 3 XRFs localized with the SXC. Aggressive follow-up observations in various bands (Table 1) have revealed a fading afterglow for nearly all of these bursts, making the case that most "dark" GRBs can be found with a combination of small, rapidly-disseminated error boxes and quick-reacting, deep-field telescopes.

Reducing the Error Boxes through Better Astrometry

Systematic variations are the dominant source of astrometric error. For bright bursts, statistical variations yield errors as low as 10-15" (2D), but thermal motion in the

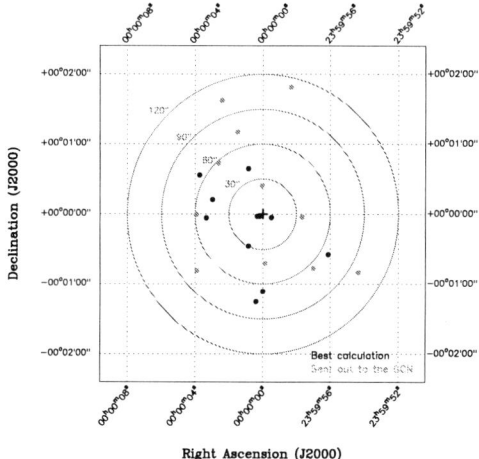

FIGURE 1. Bulls-eye plot comparing the accuracy of the GCN localizations compared with the positions adjusted with the latest thermal/deformation model.

reference optical cameras (due to the orbital day-night cycle) can be as large as an arcminute. Slight focal plane tilts/deformities are also a major source of error, adding arcminute corrections to off-axis detections. A recent astrometric calibration using the Crab now places the 2D, 90% confidence radius at 80", covering the whole field of view and valid at all phases of the orbit. The bulls-eye plot above (Fig.1) shows the improvement in the accuracy of the SXC GRB detections using these latest corrections.

SXC Flight Localizations

Due to a limited memory capacity, the onboard flight code is incapable of correcting for spacecraft motion, unlike ground code which uses the available aspect data. It also uses a fixed number of unfiltered photons, whereas the ground code removes particle and cosmic-ray induced events, as well as optimizes the foreground region for maximum S/N. Nevertheless, GRB021211 was the first demonstration of a SXC flight localization; the correct position was calculated on board to within 1.5', which was then broadcast via the real-time VHF along with the WXM position. The first GCN message disseminating the WXM position was sent out 22 seconds from burst onset, but since the SXC flight code was untested the SXC position was not sent out.

The ability of the SXC to obtain real time localizations for strong bursts had not been appreciated prior to this event. The improved SXC sensitivity, the adoption of optimized correlation parameters to dig out weak signals, and the stabler spacecraft attitude (due to better estimation parameters reducing spacecraft drift) have all been contributing factors. With continuing code improvement, we estimate that more than half of SXC detections may result in flight localizations, disseminated within a minute of the burst. The HETE/GCN interface now immediately sends out SXC positions of bright bursts.

TABLE 1. SXC Localizations from August 2002 to August 2003. Redshifts in brackets are pseudo-z estimates in lieu of actual measurements.

GRB	Loc. Axes	Distance to transient source	Quoted error radius	Error box size ($arcmin^2$)	Burst to GCN Delay time	Observed Afterglow	Redshift (z)	Burst Type
020813	2D	73.4", 25.6"	60", 50"	3.1, 2.2	184 m	V,IR,X,R	1.25	
020819	2D	68.7", 97.9"	130", 64"	14.7, 3.6	176 m	R	[1.5]	X-ray rich
020903	1D	70.6"*	120"	124	415 m **	V,R	0.25	XRF
021004	2D	33.1", 39.9"	120", 62"	12.6, 3.3	154 m	V,IR,X,R	2.3	X-ray rich
021211	2D	42.2"	120"	12.6	131 m	V,IR	1.01	X-ray rich
030115	2D	57.2"	120"	12.6	84 m	V,IR,R	[1.44]	X-ray rich
030226	2D	52.8"	120"	12.6	121 m **	V,IR,X	1.98	X-ray rich
030323	1D	15.4"*	120"	71.2	300 m	V	3.37	
030328	2D	103.8", 40.7"	120", 52"	12.6, 2.3	58 m	V,X	1.52	
030329	2D	24.1"	120"	12.6	73 m	V,IR,R,X	0.168	X-ray rich
030429	2D	73.2"	120"	12.6	112 m	V,IR	2.65	XRF
030528	2D	111.5", 59.3"	120", 150"	12.6, 19.6	107 m	IR,X	[0.36]	
030723	2D	61.2"	120"	12.6	430 m	V,X	[0.5]	XRF
030725	1D	126.4"*	120"	94.1	261 m	V	[0.7]	
030823	2D	??	120"	43.4	194 m	??	[0.64]	X-ray rich

* From center line ;   ** Untriggered bursts ;   V - visible   IR - Infrared   R - Radio   X - X-ray ;

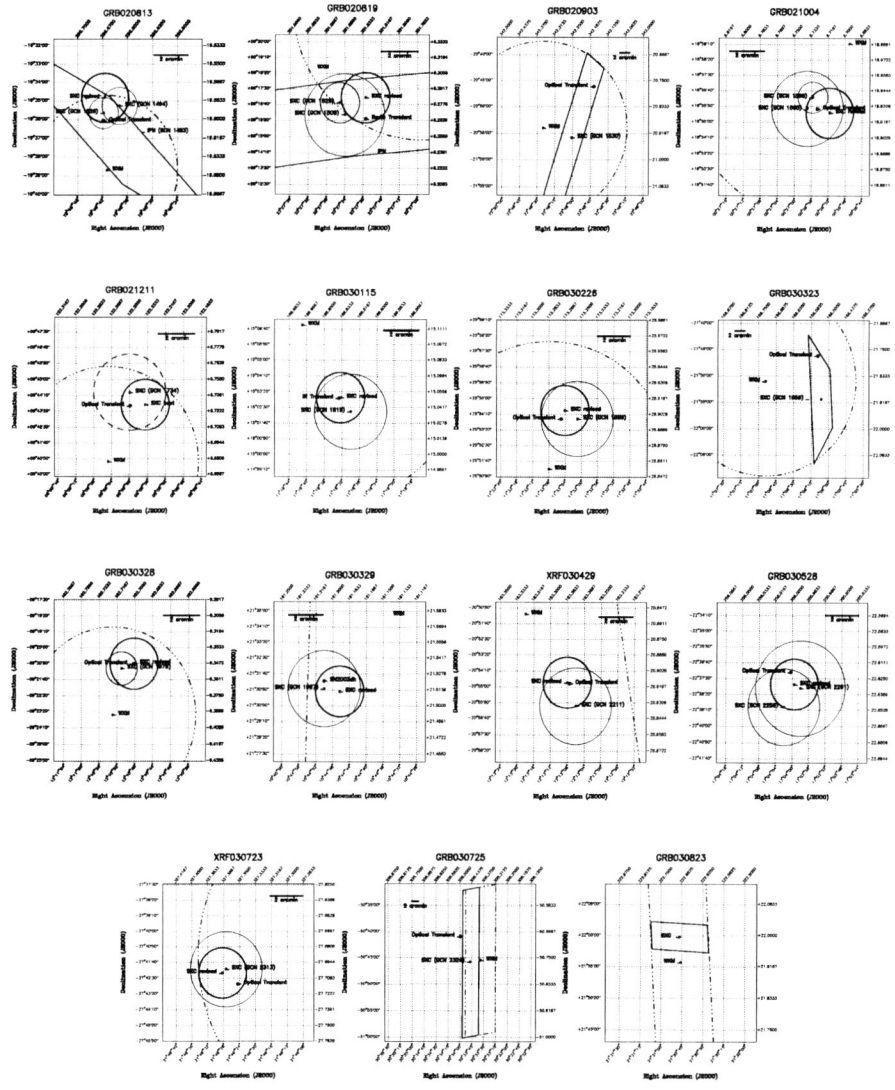

FIGURE 2. Skymaps of the 15 GRBs/XRFs localized by the SXC from Aug. 2002- Aug.2003.

Current Status of HETE-2 Operations

R. Vanderspek*, N. Butler*, G. B. Crew*, A. Dullighan*, G. Prigozhin*, J. P. Doty*, J. N. Villasenor*, G. R. Ricker*, T. Tamagawa†, K. Torii†, N. Kawai**†, T. Sakamoto**†, R. Sato**, M. Suzuki**, Y. Urata**‡, Y. Yamamoto**, A. Yoshida§†, Y. E. Nakagawa§, T. Yamazaki§, Y. Shirasaki¶†, C. Graziani‖, T. Donaghy‖, D. Q. Lamb‖, J. G. Jernigan††, K. Hurley††, J-L. Atteia‡‡, E. E. Fenimore§§ and M. Galassi§§

*Center for Space Research, Massachussetts Institute of Technology, Cambridge, MA 02139 USA
†RIKEN (The Institute of Physical and Chemical Research), 2-1 Hirosawa, Wako, Saitama 351-0198, Japan
**Tokyo Institute of Technology, 2-12-1 Ookayama, Meguro-ku, Tokyo 152-8551, Japan
‡RIKEN (The Institute of Physical and Chemical Research) 2-1 Hirosawa, Wako, Saitama 351-0198, Japan
§Department of Physics, College of Science and Engineering, Aoyama Gakuin University, 5-10-1 Fuchinobe, Sagamihara, Kanagawa 229-8558, Japan
¶National Astronomical Observatory of Japan, 2-21-1 Osawa, Mitaka, Tokyo 181-8588, Japan
‖Department of Astronomy & Astrophysics, University of Chicago, 5640 South Ellis Avenue, Chicago, IL 60637
††Space Sciences Laboratory, UC Berkeley, Berkeley, CA 94720-7450
‡‡Laboratoire d'Astrophysique, Observatoire Midi-Pyrénées, 14, Avenue E. Belin 31400 Toulouse France
§§NIS-2, Astrophysics and Machine Learning Team, MS B244, Los Alamos National Laboratory, Los Alamos, NM

Abstract. The High Energy Transient Explorer (HETE-2) has been in orbit for nearly three years. After a slow startup, the operation of the spacecraft and its instrumentation is now stable and efficient. GRBs are being localized at a rate of ~ 25 per year, and the Soft X-ray Camera (SXC) is determining burst positions with arcminute precision on a regular basis. As described elsewhere in this conference, HETE-2 has essentially solved the mystery of the "dark bursts" and helped confirm the connection between long GRBs and type Ic supernovae. Because of its excellent low-energy response, HETE has proven to be a capable detector of X-ray rich GRBs and X-ray flashes. In this paper, we give an update on the spacecraft and instruments and describe some of the more significant developments of the last 6-12 months. We also highlight issues which, although described in part on the HETE web page (http://space.mit.edu/HETE), may not be clear to many observers.

INSTRUMENT STATUS

The HETE spacecraft and instruments are working extremely well.

- The power system has shown only minor signs of aging, in that the power margin has dropped from $\sim 40\%$ to $\sim 30\%$.
- Fregate continues to operate well.
- The WXM YB detector was punctured by a micrometeorite in January, 2003, resulting in a 20% reduction in the Y camera's FOV (see Figure 1).

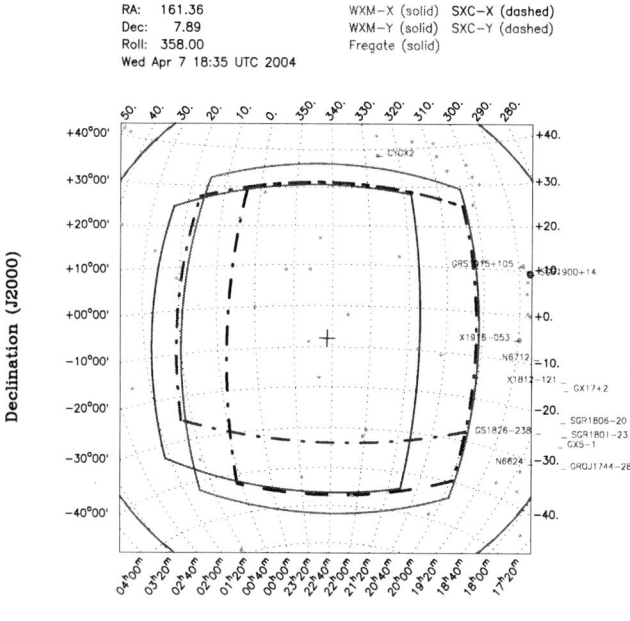

FIGURE 1. The extent of the FOVs of Fregate, the WXM, and the SXC as of September, 2003

- The performance of the SXC continues to improve as we improve analysis techniques to offset the loss of two CCDs, lost after the loss of the outer optical blocking filters (OBFs) in early 2001.

WHERE DOES HETE POINT?

The sensitivity of the SXC and optical systems on HETE to scattered light from the moon requires a modification to the nominal antisolar pointing strategy in the 10-12 days centered on full moon. As the full moon approaches the antisolar point ("waxing gibbous"), the spacecraft pointing is offset eastward from antisolar by up to 40 degrees; after full moon ("waning gibbous"), the offset is westward by up to 40 degrees. These offsets, or "nods", are applied in 5° steps; the precise timing of the changes in offset is a function of the lunar phase and the sensitivity of the SXC and optical cameras to scattered light. The predicted pointing of the HETE spacecraft is always available on the HETE web page: http://space.mit.edu/HETE/Latest_RADec.txt.

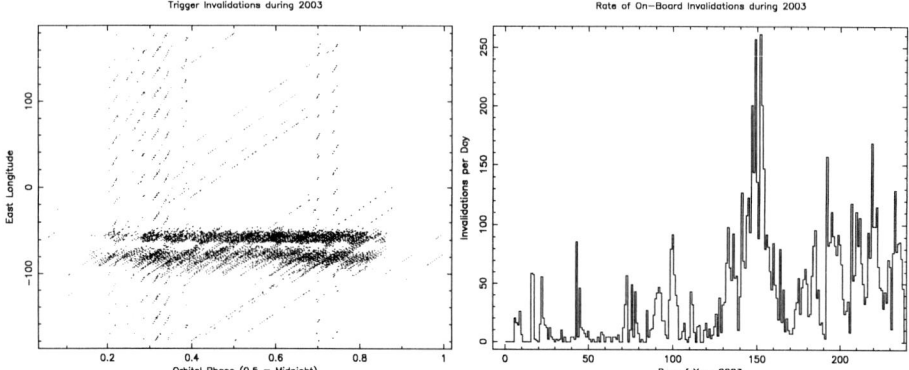

FIGURE 2. In the left panel are plotted the orbital phase vs the longitude of HETE at the time of all invalidations before August 27, 2003. The band of invalidations at longitudes between 50W and 100W are due to particle events in the Ecuador Anomaly. The slanted lines are invalidations of emersions and immersions of the Crab and Sco X-1. In the right panel, we plot the number of invalidations per day for each day of 2003. The strong peaks in May and June of 2003 are due largely to Sco X-1.

EFFECT OF TRAPPED PARTICLES ON TRIGGER EFFICIENCY

While HETE's equatorial orbit keeps it from passing through the heart of the South Atlantic Anomaly, we have detected a region of low-Earth orbit in which the impact of high-energy particles on the Fregate and WXM instruments can cause the on-board software to trigger.

During Solar Max, this region was active during periods of high geomagnetic activity, lasting typically 1–3 days. To suppress particle triggers during these periods, a "trigger restriction" mechanism was implemented on board: any trigger which occurs during a restricted time period (commanded from the ground) is immediately "invalidated" and all instruments are returned to Survey mode. Trigger restrictions are used to prevent particle triggers over Ecuador as well as triggers from X-ray sources rising over the Earth's limb.

Since Solar Max, the periods of high geomagnetic activity have been replaced by longer periods of high particle activity due to high-velocity solar wind streams; as a result, the sporadic episodes of particle restrictions have been replaced with an almost-continuous set of trigger restrictions over western South America.

While there is some loss in exposure to GRBs because of the trigger restrictions in place, the magnitude of the loss is small: the higher background rate in that region significantly reduces the sensitivity of Fregate and often causes the WXM high-count-rate safety mechanism to shuts off the WXM HV.

NEW REAL-TIME ALERTS TO THE GCN

Traditionally, burst localizations calculated on board HETE have been propagated to the GCN for triggers in the 7-80 keV and 30-400 keV band if

- the burst is localized by the WXM with an image S/N of >3.0 in both axes, in which case the error radius is $14'$.

or, if not

- the burst is localized by the WXM and the quadrature sum of the X and Y image S/N exceeds 3.7, in which case the error radius is $30'$.

However, recent modifications to flight and ground software now allow XRF localizations and small SXC localizations to be sent to the GCN in real time.

Real-time localizations of XRFs

HETE's low-energy response has allowed it to detect a significant number of XRFs: 1/3 to 1/2 of all GRBs detected by HETE are X-ray rich GRBs or XRFs. We have modified the algorithms controlling the propagation of burst localizations to the GCN to allow *real-time localizations of X-ray rich GRBs and X-ray flashes to be propagated to the GCN*. XRF030723 is the first example of such a localization.

Localizations of triggers in the 2-25 keV band will be propagated to the GCN if the image S/N exceeds 3.0 in both axes and the burst coordinates do not correspond to a known X-ray source. The error regions of such localizations will typically measure $14'$ in radius.

Real-time SXC localizations

Since the loss of the SXC outer OBFs, the HETE team has learned to compensate for the smaller effective area and higher background rate in the SXCs. As a result, ground analyses of downlinked burst data consistently result in a precise ($\sim 2'$) localization of bursts with sufficient soft X-ray fluence. We have ported a large fraction of this analysis process to the flight software, allowing $2'$ localizations of GRBs to be propagated to the GCN in real time. HETE trigger H2809, a trigger on the XRB X1812-121, is the first example of a real-time SXC localization.

Real-time SXC localizations will be propagated to the GCN if the WXM image S/N exceeds 2.5 in both axes *and* the best SXC localization within two degrees of the WXM position is within the WXM error circle. The error region of a prompt SXC localization will measure 2–2.5 arcminutes in radius.

SWIFT SATELLITE

The Swift Gamma-Ray Burst Mission

N. Gehrels

On behalf of the Swift team
Mail Code 661, Goddard Space Flight Center, Greenbelt, MD 20771

Abstract. Swift is a first-of-its-kind multiwavelength transient observatory for GRB astronomy. It has the optimum capabilities for the next breakthroughs in determining the origin of GRBs and their afterglows, as well as using bursts to probe the early Universe. Swift will also perform the first sensitive hard X-ray survey of the sky. The mission is being developed by an international collaboration and consists of three instruments, the Burst Alert Telescope (BAT), the X-ray Telescope (XRT), and the Ultraviolet and Optical Telescope (UVOT). The BAT, a wide-field gamma-ray detector, will detect >100 GRBs per year with a sensitivity >2 times that of BATSE. The sensitive narrow-field XRT and UVOT will be autonomously slewed to the burst location in 20 to 75 seconds to determine 0.3-5.0 arcsec positions and perform optical, UV, and X-ray spectrophotometry. Strong education/public outreach and follow-up programs will help to engage the public and astronomical community. The Swift launch is planned for mid 2004.

INTRODUCTION

The discovery by BeppoSAX and ground observers [1, 2, 3] of afterglow from GRBs has shown that they are cosmological, involving the most powerful explosions known. These explosions are thought to create super-relativistic blast-waves resulting in afterglow that fades from gamma-rays to radio. However, important information on the afterglow is lost by the ~ 8 hour or longer delay between the initial burst and follow-up observations.

Swift is a multiwavelength observatory that exploits the afterglow characteristics of GRBs to make a comprehensive study of hundreds of bursts. It will determine the origin of GRBs, tell us how the blast wave interacts with its surroundings, and identify classes of bursts. Swift will also investigate how GRBs can be used to study the early Universe.

SWIFT INSTRUMENTS

The Swift instrumentation was carefully chosen for GRB discovery. It incorporates a wide-field GRB detector and two sensitive narrow-field telescopes (Fig. 1).

Burst Alert Telescope

The Burst Alert Telescope (BAT) covers the 15-150 keV energy band and will detect >100 bursts per year. It has a CdZnTe (CZT) detector array with an area of 5200 cm^2 and a coded aperture mask covering 1.4 sr of the sky. The mask is positioned one meter

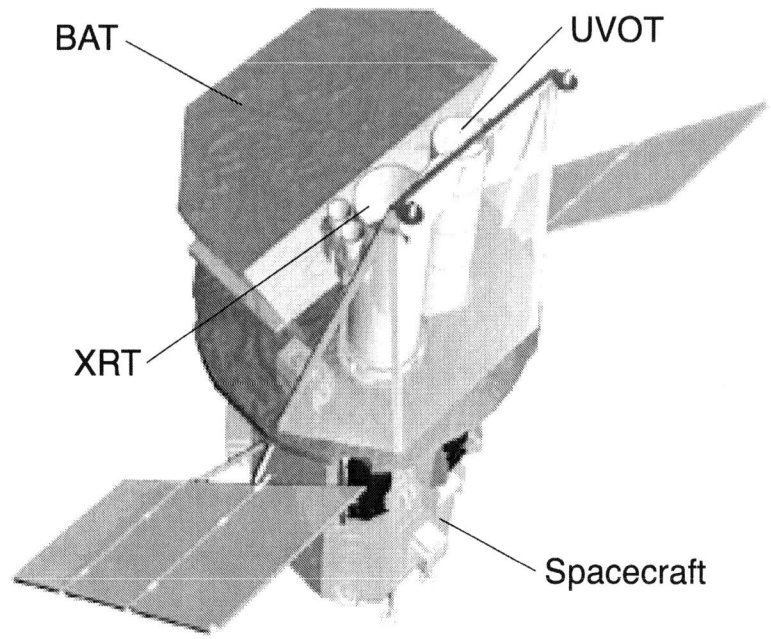

FIGURE 1. The Swift satellite.

TABLE 1. BAT parameters.

Energy Range	15-150 keV
Detecting Area	5200 cm^2
Detector Operation	Photon counting
Field of View	1.4 sr (half-coded)
Detector Element Size	$4 \times 4 \times 2$ mm^3
Telescope PSF	17 arcmin

away from the detectors and will provide positions of 1-4 arcmin accuracy. The large detector area and sophisticated triggering system will allow BAT to detect bursts of all durations to a sensitivity 2-5 times (depending on position in FOV) better than BATSE. The instrument development is led by Goddard Space Flight Center with the imaging flight software written at Los Alamos National Laboratory.

X-Ray Telescope

The X-Ray Telsecope (XRT) will locate bursts to 5 arcsec accuracy using flight-spare optics from the JET-X instrument on the Spectrum X-Gamma. The mirror has a 18 arcsec half-power diameter at 1.5 keV. The detector is a 600 square pixel CCD from

TABLE 2. XRT parameters.

Energy Range	0.2-10 keV
Telescope	JET-X Wolter 1
Effective Area	110 cm^3 @ 1.5 keV
Field of View	23.6 × 23.6 arcmin
Detection Elements	600 × 600 pixels
Telescope PSF	18 arcsec HPD @ 1.5 keV
Sensitivity	2×10^{-14} erg cm^{-2} s^{-1}

TABLE 3. UVOT parameters.

Wavelength Range	170 nm - 600 nm
Telescope	Modified Ritchey-Chrétien
Aperture	30 cm diameter
Detector	Intensified CCD
Detector Operation	Photon counting
Field of View	17 × 17 arcmin
Detection Elements	2048 × 2048 pixels
Telescope PSF	0.9 arcsec @ 350nm
Sensitivity	$B = 24$ in white light in 1000 s

the XMM/EPIC, giving a FOV of 24 arcmin square in an energy range of 0.2-10.0 keV. The instrument is being developed by Penn State University, University of Leicester, and Osservatorio Astronomico di Brera.

UV/Optical Telescope

The UV/Optical Telescope (UVOT) is a 30 cm diameter modified Ritchey-Crétien equipped with an image intensified CCD covering 170- 600 nm. It has a FOV 17 arcmin square and is based closely on the design of the XMM Optical Monitor (OM). The UVOT is able to reach $m_B = 24$ in 1000 s (open filter). A filter wheel provides 6 colors, two grisms and a 4× magnifier. The optical point spread function of the telescope is 0.9 arcsec allowing for excellent astrometry. By registering the field against foreground stars, the UVOT will provide < 0.3 arcsec positions. The instrument is being developed by Penn State University and Mullard Space Science Laboratory.

SWIFT MISSION

The strategy of the Swift mission is to slew to each new GRB and follow the afterglow as long as it is visible. To observe the earliest phase of the afterglow, new BAT positions will trigger an autonomous spacecraft slew followed by a programmed sequence of observations with the XRT and UVOT.

The initial GRB position is normally determined by the BAT, but positions can also be uploaded from other satellites through a real-time TDRSS uplink. Either case will trigger the spacecraft software to plan and execute an autonomous slew. All calculations of slew path and pointing constraints will be done on-board.

Each of the three Swift instruments rapidly produces alert messages after a GRB is detected. To ensure prompt delivery, these messages are sent through a real-time TDRSS downlink to the ground, and routed immediately to the GRB Coordinates Network (GCN) [4] for delivery to the community.

When Swift is not engaged in prompt observations of the most recent bursts, it will follow a schedule uploaded from the ground each working day and as needed. This schedule will provide for long term follow-up of GRB afterglows and other science. The PSU Mission Operational Center (MOC) will be capable of generating a new schedule in < 2 hours.

TABLE 4. Swift mission characteristics.

Autonomous slew decision capability
Fast Slew - 50° in < 75 s
Low Earth Orbit, 22° Inclination
Launch Vehicle: Delta 7320 with 3 meter fairing
Mass: 1500 kg
Power: 1000 W

GROUND SYSTEM AND DATA ANALYSIS

A layered data analysis approach will be used to achieve rapid dissemination of Swift results and data to the community. The most urgently needed results, namely GRB positons, are produced on the spacecraft. Quicklook results, including optical finding charts and multiwavelength light curves, are produced in the Penn State Mission Operations Center (MOC) in near real-time and distributed using the GCN. Definitive standard products, including spectra, multi-band light curves, and images, will be made into production FITS files.

All the Swift data will be processed at the Swift data center at Goddard and will be made available to the general public through the HEASARC in the US and data centers in the UK and Italy. The end result will be easy access for the entire community to a broad range of timely information on GRBs.

ORGANIZATION

Swift is the result of an international collaboration. The responsibilities of the various institutions are listed in Table 5.

TABLE 5. Swift mission responsibilities and institutions.

Responsibility	Institution
Mission Management	GSFC
XRT	PSU, LU, OAB
UVOT	PSU, MSSL
BAT	GSFC, LANL
Ground System	GSFC
Mission Operations	PSU
Data Centers	GSFC, ASI, LU
EPO	SSU, PSU, GSFC
GRB Follow-up Coordination	UCB

GSFC = Goddard Space Flight Center
PSU = Penn State University
LU = Leicester University
OAB = Osservatorio Astronomico di Brera
MSSL = Mullard Space Science Laboratory
LANL = Los Alomos National Laboratory
ASI = Italian Space Agency
SSU = Sonoma State University
UCB = University of California, Berkeley

REFERENCES

1. Costa, E., et al., *Nature*, **387**, 783 (1997).
2. Van Paradijs, J., et al., *Nature*, **386**, 686 (1997).
3. Frail, D.A., et al., *Nature*, **389**, 261 (1997).
4. Barthelmy, S. et al., *Gamma-Ray Bursts: 4th Huntsville Symposium*, edited by C. Meegan, R. Preece, and T. Koshut, AIP Conference Proceedings 428, American Insitute of Physics, New York, 1998, p 139.

The X-ray Telescope for the SWIFT Gamma-Ray Burst Mission

A Wells*, D N. Burrows[†], J. E. Hill[†], J. A. Nousek[†], G. Chincarini**, A. F. Abbey*, A. Beardmore*, J. Bosworth[‡], H. W. Brauninger[§], W. Burkert[§], S. Campana**, M. Capalbi[¶], W. Chang[‡], O. Citterio**, M. J. Freyberg[§], P. Giommi[¶], G. D. Hartner[§], R. Killough[∥], B. Kittle[‡], R. Klar[∥], C. Mangels[∥], M. McMeekin[‡], B.J. Miles[†], A. Moretti**, K. Mori[†], D. C. Morris[†], K. Mukerjee*, J. P. Osborne*, G. Tagliaferri**, F. Tamburelli[¶], D. J. Watson*, R. Willingale* and M. Zugger[†]

*Space Research Centre, University of Leicester, Leicester, LE1 7RH, UK
[†]Pennsylvania State University, 525 Davey Lab, University Park, PA 16802, USA
**INAF-Osservatorio Astronomico di Brera, Via Brera 28, 20121 Milano, Italy
[‡]Swales Aerospace, Inc.
[§]Max-Planck-Institut fur Extraterrestrische Physik, Garching bei Munchen, Germany.
[¶]ASI Science Data Center,Frascati, Italy
[∥]South West Research Institute

Abstract. The X-ray Telescope (XRT) for the SWIFT mission, built by the international consortium from Pennsylvania State University (US), University of Leicester (UK) and Osservatorio Astronomico di Brera (Italy), is already installed on the SWIFT spacecraft. The XRT has two key functions on SWIFT; to determine locations of GRBs to better than 5 arc seconds within 100 seconds of initial detection of a burst and to measure spectra and light curves of the X-ray afterglow over around four orders of magnitude of decay in the afterglow intensity. This paper summarises the XRT performance, operating modes and sensitivity for the detection of prompt and extended X-ray afterglows from gamma-ray bursts. The performance characteristics have been determined from data taken during the ground calibration campaign at MPE's Panter facility in September 2002.

INTRODUCTION

The Swift Gamma Ray Burst Explorer[1], chosen in October 1999 as NASA's next MIDEX mission, is now scheduled for launch in mid-2004. It carries three instruments: the Burst Alert Telescope[2] (BAT), which identifies gamma-ray bursts (GRBs) and determines their location on the sky to within a few arc minutes; the Ultraviolet/Optical Telescope[3] (UVOT) with sensitivity down to 24th magnitude and sub-arc second positional accuracy, and the X-ray Telescope (XRT). The three instruments combine to make a powerful multi-wavelength observatory with the capability of rapid determination of GRB positions to arc second accuracy within 1-2 minutes of their discovery, and the ability to measure light curves and redshifts of the bursts and after-glows.

The Swift XRT is a sensitive X-ray imaging spectrometer designed for autonomous operation to measure fluxes, spectra, and light curves of GRBs and afterglows over a dynamic range covering 7 orders of magnitude in flux. The XRT will deliver accurate

TABLE 1. XRT instrument characteristics

Telescope	3.5m Wolter I
Telescope PSF	18 arc sec HPD @ 1.5 keV 22 arc sec HDP @ 8.1 keV
Detector	E2V CCD-22, developed for XMM EPIC
CCD Format	600 × 602 pixels
Telescope Readout Modes	Photon-counting, Imaging and Timing
Field of View	23.6 x 23.6 arc min
Pixel scale	2.36 arc sec/pixel
Energy Range	0.2-10 keV
Effective Area	135 cm^2 @ 1.5 keV
Sensitivity	2×10^{-14} ergs cm^{-2} s^{-1} in 10^4 s
GRB location Accuracy	2.5 arc sec

GRB positions to the ground within 5 seconds of target acquisition for typical bursts, allowing optical telescopes to begin immediate optical, radio or infrared observations of the afterglow from ground based facilities. In many cases, the XRT will begin observations before the initial outburst has subsided and will be able to follow the subsequent afterglow flux and X-ray spectra for hours and days. Observations during the first few hours of the afterglow are especially important since this epoch of afterglow decay was unobservable with BeppoSax or RXTE.

XRT DESCRIPTION

The design and operation of the XRT has been described in an earlier paper[4]. A grazing incidence Wolter I telescope[5] and a focal plane camera assembly, incorporating an X-ray sensitive CCD[6] and an optical light blocking filter, are mounted in a highly stable carbon fibre telescope structure with a closure door which protects the mirrors during launch. Thermal baffles and heaters provide a controlled temperature environment for the XRT and prevent thermal distortion of the mirrors. A thermal radiator mounted on the anti-Sun side of the spacecraft is coupled, via a heat pipe, to a Peltier cooler attached to the CCD to enable the detector to be operated at a stable -100° C. Control, signal processing and data handling electronics are housed in a separate unit mounted on the spacecraft structure. Co-alignment of the XRT with Swift's star trackers is monitored in orbit by a separate optical system (the Telescope Alignment Monitor (TAM)). Background electrons, encountered in the Swift orbit, are deflected from the XRT field of view by a ring of rare earth permanent magnets serving as an electron deflector.

Performance of the XRT has been measured through an intensive calibration programme, including end-to-end testing in September 2002 in the 180m long X-ray beam Panter facility at the Max Planck Institut fur Extraterristriches Physik. The main results from the calibration, and their implications for the scientific performance of XRT in orbit, are reported in the rest of this paper.

XRT PERFORMANCE

GRB Position Determination. The XRT is required to measure afterglow positions to better than 5 arc seconds within 100 s of a burst alert from the BAT. The BAT provides a GRB position accurate to about 4 arc minutes and the spacecraft will slew to the BAT position in 20-75 s, depending on the position of the GRB on the sky. The XRT mirror PSF has a 18 arc seconds Half-Energy Width (HEW) on-axis (at 1.5 keV). The centroid of a point source image can be determined to sub-arcsecond accuracy in detector coordinates, given sufficient photons. Automatic source location has been tested as part of the Panter calibration effort. The brightest sources can be located within a single 0.1 second CCD integration frame and processed on-board within a further 2.5 seconds. Fainter sources are found by taking a second image frame with the integration time increased to 2.5 seconds. The results show that source positions can be determined to better than 2.5 arc seconds, in detector co-ordinates, for source fluxes between 0.05 and 50 Crab[7]. Accuracy declines for fainter sources. The automatic changeover, from the 0.1 second frame time to the 2.5 second frame time, occurs at around 2 Crab. Source position accuracy in sky co-ordinates is slightly degraded, to between 3 and 5 arc seconds, because of possible systematic alignment errors between the SWIFT star trackers and the XRT. Some of these errors can be measured and corrected for by the TAM.

Light Curves and Spectroscopy. The XRT is designed to provide accurate photometry and light curves, aiming to provide at least 10 ms time resolution in the early stages of a burst. Three timing and spectroscopy modes have been implemented in on-board software, with automatic switching between modes[8]. Photo-diode mode provides the best time resolution (0.14 ms), but integrates the count rate over the entire CCD and provides no spatial information. Windowed timing mode provides 2 ms time resolution with 1-D spatial resolution within a window 8 arc minutes wide. Photon counting mode retains full imaging and spectroscopic resolution but time resolution is only 2.5 seconds.

Each mode has been exercised during Panter calibration including measurement of the spectroscopic resolution of the CCD detector, over a range of X-ray energies. Table 2 below summarises imaging[9], timing and spectroscopic properties[10] for each mode, together with the range of source flux (in Crabs) applicable to each mode.

Source Detection Sensitivity. The end-to-end calibration of XRT at Panter has verified the instrument point spread function, focus and effective area. These results are now incorporated in the XRT response matrix function (RMF). Photometric accuracy of 10 % is achieved for source fluxes ranging over 7 orders of magnitude from a sensitivity limit of 2×10^{-14} ergs cm^{-2} s^{-1} (in 10^4 seconds). A typical response from a GRB X-

TABLE 2. Performance of the XRT in each of its operating modes

	Flux (Crabs)	Imaging	Spectroscopy at 6 keV	Timing	Data
Imaging	0.025-37	2-D	none	0.1/2.5 s	Image, Centroid
Photodiode timing (low-rate)	0.05-60	none	140 eV	0.14 ms	Events above threshold
Window timing	0.00005-15	1-D	138 eV	2.2 ms	Events above threshold
Photon counting	< 0.0001	2-D	133 eV	2.5 s	3x3 pixel matrix above threshold

ray afterglow has been simulated by combining the instrument RMF with source flux predicted for a 100 second observation of a 150 mCrab-like source at z=1.0, assuming a power law spectrum with a red-shifted Fe line (E_{rest} =6.4 keV).

This result, in Figure 1 above, illustrates the XRT's capability to extract spectral line features that may be present in an afterglow continuum spectrum.

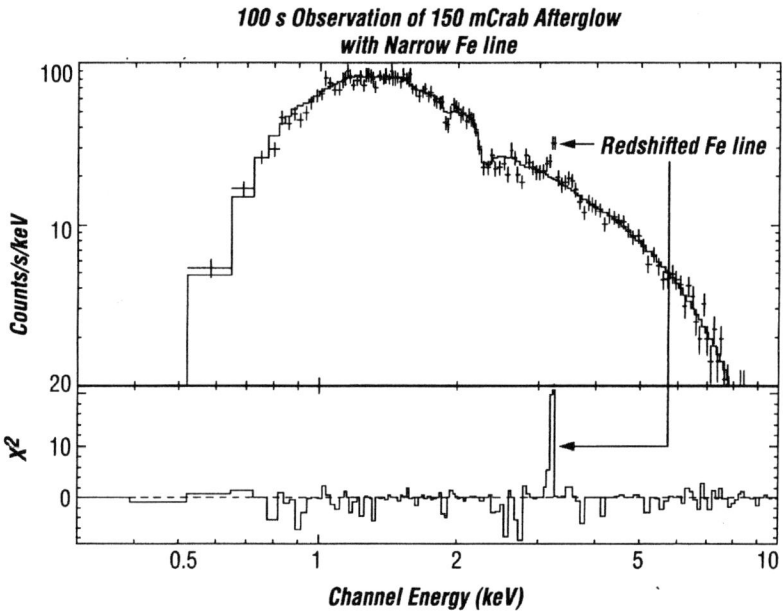

FIGURE 1. Simulated 150 mCrab XRT spectrum together with the residuals derived via an XSPEC fitting routine applied to the source count data

Automated Operation. Over the course of a typical GRB observation, the burst/afterglow flux will decrease by many orders of magnitude, yet must be observed over this wide dynamic range without detector saturation especially at early times when the burst and afterglow are bright. The XRT is designed for completely autonomous operation, switching between readout modes according to the instantaneous count rate in each CCD frame. For an initially acquired burst, the Imaging Mode determines the centroid of the source image. Its position is telemetered, together with a 2'×2' postage stamp image and a compressed image of the full CCD frame, through TDRSS and distributed immediately to the community through the Gamma-ray burst Coordinate Network (GCN). To cope with the rapidly changing source flux of GRBs, the XRT automatically determines the optimum readout mode appropriate for the instantaneous source intensity and adjust its readout and processing modes accordingly. Image, timing and spectroscopic data are stored on-board and transmitted to ground during SWIFT's ground passes over the Malindi ground station.

ACKNOWLEDGMENTS

This work is supported at Penn State by NASA contract NAS5-00136; at the University of Leicester by the Particle Physics and Astronomy Research Council; and at OAB by funding from ASI.

REFERENCES

1. Grehrels, N., "The SWIFT GRB Missions," in *in this proceeding*, AIP Conference Proceedings 000, American Institute of Physics, New York, 2004.
2. Bathelmy, S., and et al., "The Swift Burst Alert Telescope," in *Proc. SPIE*, edited by SPIE, SPIE Conference Proceedings 5165, SPIE, Bellingham, USA, 2004, p. in press.
3. Roming, P., and et al., "The Swift Ultra-violet Optical Telescope," in *Proc. SPIE*, edited by SPIE, SPIE Conference Proceedings 5165, SPIE, Wellingham, USA, 2004, p. in press.
4. Burrows, D., and et al., "The Swift X-ray Telescope," in *Proc. SPIE*, edited by SPIE, SPIE Conference Proceedings 5165, SPIE, Wellingham, USA, 2004, p. in press.
5. Citterio, O., and et al., "Characteristics of flight model optics for the JET-X telescope onboard the SPECTRUM X-gamma satellite," in *Proc. SPIE*, edited by SPIE, SPIE Conference Proceedings 2805, SPIE, Wellingham, USA, 2004, pp. 56–65.
6. Short, A. D. T., and et al., "Performance of the XMM EPIC MOS CCD Detector," in *Proc. SPIE*, edited by SPIE, SPIE Conference Proceedings 3445, SPIE, Bellingham, USA, 1998, pp. 13–27.
7. Moretti, A., and et al., "Swift XRT point spread function measured at Panter end-to-end tests," in *Proc. SPIE*, edited by SPIE, SPIE Conference Proceedings 5165, SPIE, Bellingham, USA, 2004, p. in press.
8. Hill, J., and et al., "Readout Modes and Readout Modes Automated Operation of the Swift X-ray Telescope," in *Proc. SPIE*, edited by SPIE, SPIE Conference Proceedings 5165, SPIE, Bellingham, USA, 2004, p. in press.
9. Tagliaferri, G., and et al., "Swift XRT effective area measured at Panter end-to-end tests," in *Proc. SPIE*, edited by SPIE, SPIE Conference Proceedings 5165, SPIE, Bellingham, USA, 2004, p. in press.
10. Mukerjee, K., and et al., "The spectroscopic performance of the Swift X-ray telescope for gamma-ray burst afterglow studies," in *Proc. SPIE*, edited by SPIE, SPIE Conference Proceedings 5165, SPIE, Bellingham, USA, 2004, p. in press.

Flight Calibration and Operations of the Swift X-ray Telescope (XRT)

D. N. Burrows*, J. E. Hill*, J. A. Nousek*, A. A. Wells[†], J. P. Osborne[†], K. Mukerjee[†], G. Chincarini**, G. Tagliaferri[‡] and S. Campana[‡]

*Pennsylvania State University, 525 Davey Lab, University Park, PA 16802, USA
[†]University of Leicester, Space Research Centre, University Road, Leicester LE1 7RH, UK
**INAF-Osservatorio Astronomico di Brera, Via E. Bianchi 46, Merate, 23807, Italy
[‡]Osservatorio Astronomico di Brera, Via E. Bianchi 46, Merate, 23807, Italy

Abstract. We present the current on-orbit calibration and operational plans for the *Swift* XRT. The XRT is a largely autonomous instrument and requires very little manual commanding for normal operations. A detailed calibration plan is being developed to verify the instrumental performance on-orbit, including effective area, point spread function, vignetting, spectroscopic performance, and timing accuracy. Operational plans include regular calibration measurements using on-board calibration sources as well as periodic calibration observations using celestial targets.

INTRODUCTION

The *Swift* Gamma-Ray Burst Observatory[1] is a NASA MIDEX-class Explorer mission designed to study Gamma-Ray Bursts and their afterglows. It incorporates a wide-field hard X-ray burst imager called the Burst Alert Telescope, as well as two narrow-field telescopes covering the UV/optical band (the UV/Optical Telescope, or UVOT) and X-ray band (the X-ray Telescope, or XRT). The spacecraft will perform rapid, autonomous slews to the burst positions provided by the BAT, allowing the narrow-field instruments to observe the burst/afterglow within about a minute of the initial burst trigger.

The XRT has three key science goals:

1. Determine GRB position with 5 arcsecond accuracy and transmit it to the ground within 100 s of the burst onset.
2. Measure the afterglow X-ray lightcurve.
3. Obtain X-ray spectroscopy of the afterglow.

These goals are met by a flexible, autonomous instrument design utilizing a Wolter I grazing incidence X-ray telescope[2] with an EPIC CCD[3] in the focal plane. Details of the XRT hardware design have been presented elsewhere[4]. The CCD is operated in one of four readout modes, depending on the source intensity: Image Mode (IM), which produces only an image; Photodiode (PD) mode, which produces spectral and rapid timing information but no image; Windowed Timing (WT) mode, which produces spectral and timing information with 1-D imaging; or Photon-Counting (PC) mode, which produces 2-D images with spectral information for the faintest sources. The XRT readout modes are described in more detail elsewhere[5].

FLIGHT CALIBRATION

The Point Spread Function (PSF) of the XRT has been measured at the Panter calibration facility operated by the Max-Planck-Institut für Extraterrestrische Physik (MPE) near Munich, Germany. While these measurements allowed us to characterize the PSF as a function of both energy and off-axis angle[6], they suffer from gravitational distortion of the mirrors and from a finite source distance. The PSF will be remeasured on orbit using isolated point sources of various brightnesses to allow us to accurately measure both the central core and the extended wings. The preliminary target list is given in Table 1.

TABLE 1. XRT PSF Calibration Targets

Source	Count Rate (cps)	Readout Mode	Exposure (ks)	Purpose
PKS 0537-441	0.45	PC	5	XRT/UVOT coalignment
NGC 2516	0.002–0.01	PC	50	Accurate boresight, plate scale
PKS 0312-770	0.08	PC	80	PSF core (hard source)
RX J0720.4-3125	0.16	PC	30	PSF core (soft source)
GX1+4	5.0	PC	20	PSF wings (hard source)
RXS J1708-4009	7.2	PC	20	PSF wings (soft source)

The effective area of the XRT was measured as a function of energy and off-axis angle at the Panter facility[7]. The on-axis effective area and vignetting will be verified on-orbit using the sources listed in Table 2.

TABLE 2. XRT Effective Area Calibration Targets

Source	Count Rate (cps)	Readout Mode	Exposure (ks)	Purpose
Cas A	21	PC, WT	150	Effective area (hard), vignetting, gain
2E0102-7212	4	PC, WT	126	Effective area (soft), vignetting, gain
Crab nebula	790	PD	10	Effective area, vignetting, gain

The energy resolution of the XRT detector was measured at the University of Leicester at 13 energies between 0.28 keV and 10.5 keV. The energy resolution of the full instrument was measured at Panter using five X-ray lines between 0.28 keV and 8.0 keV. The results of both calibrations have been used to create a model instrument response matrix [8]. In-flight measurements of the 5.9 keV and 6.5 keV lines from the on-board ^{55}Fe sources will be made on a regular basis (at least weekly) to monitor changes in the gain, charge-transfer efficiency, and energy resolution. We will also use several astrophysical sources with bright lines (Table 3) to monitor the instrument energy resolution after launch, with observations planned roughly every six months.

TABLE 3. XRT Energy Resolution Calibration Targets

Source	Count Rate (cps)	Readout Mode	Exposure (ks)	Purpose
AB Dor	3.9	WT	20	Energy resolution
HD 35850	0.9	PC	20	Energy resolution

The relative time accuracy of the XRT was tested at the Panter facility, but the measurements were limited by the stability of the chopper wheel, and we were not able to obtain an absolute time measurement. Both absolute and relative timing accuracy will be measured on-orbit using pulsars as celestial time references (Table 4).

TABLE 4. XRT Timing Calibration Targets

Source	Count Rate (cps)	Readout Mode	Exposure (ks)	Purpose
RXS J1708-4009	7.2	WT	10	Timing accuracy
Crab nebula	790	PD	10	Cross-calibration with RXTE

SCIENCE OPERATIONS

The *Swift* Observatory is a discovery-driven mission. Because GRBs occur randomly in both time and space, the observatory must be able to autonomously plan and execute its slews. *Swift* is the first astronomical observatory with this capability.

Swift is in a low Earth orbit (600 km) and has wide exclusion zones around the Sun, Moon, and Earth to avoid damaging the UVOT. Swift therefore has no continuous viewing zone, and each target will be interrupted by the Earth limb constraint during each orbit. Continuous observation periods (called "snapshots") are therefore limited to 20–30 minutes for typical targets. In order to maximize observing efficiency, *Swift* will observe between 4 and 6 targets during each orbit, slewing to a new target as the Earth's limb approaches the current target. This allows *Swift* to follow afterglows of up to a half-dozen GRBs at any one time, with each GRB observed for 2–3 weeks.

The XRT is designed to work autonomously within the overall *Swift* observatory operations[4]. When the observatory begins a slew, the XRT completes its current observation, closes the telemetry file, and begins accumulating bias information for the upcoming observation. If the new observation is a newly-discovered GRB, the XRT will measure its celestial position as soon as the slew ends, and will telemeter the position to the ground via the TDRSS satellite system, along with an image of the source. The position and image will be relayed via the GCN network to astronomers around the globe. A few minutes later an X-ray lightcurve and spectrum will also be transmitted via TDRSS and the GCN.

Following the initial position determination, the XRT enters its automated exposure mode, in which the readout mode is adjusted according to the measured count rate to provide the most information possible without detector saturation. (This is the same mode used for all other observations, e.g., followup of previously-discovered bursts, etc.) The count rates at which the XRT switches modes are selected so that detector saturation (or pile-up, in which individual photons can no longer be measured accurately because they fall on top of one another) is negligible for all but the very brightest sources. Sources brighter than about 3 Crabs ($\sim 10^{-7}$ photons cm^{-2} s^{-1}) will suffer from significant pile-up. Integrated lightcurves (in units of deposited charge per unit time) will still be valid in this case, but spectroscopic analysis will be challenging for severely piled-up data.

ACKNOWLEDGMENTS

The *Swift* XRT program is supported at Penn State University by NASA contract NAS5-00136; at the University of Leicester by the Particle Physics and Astronomy Research Council under grant PPA/G/S/00524; and at OAB by funding from ASI under grant number I/R/309/02. We gratefully acknowledge the contributions of dozens of members of the XRT team at PSU, UL, OAB, GSFC, and our subcontractors, who helped make this instrument possible.

REFERENCES

1. Nousek, J. A., "Swift GRB Mission," in *X-Ray and Gamma-Ray Instrumentation for Astronomy XIII*, edited by K. A. Flanagan and O. H. W. Siegmund, Proc. SPIE 5165, SPIE, Bellingham, Washington, 2003, in press.
2. Citterio, O., et al., "Characteristics of the Flight Model Optics for the JET-X Telescope onboard the SPECTRUM X-γ Satellite," in *Multilayer and Grazing Incidence X-ray/EUV Optics III*, edited by R. B. Hoover and A. B. C. Walker, Jr., Proc. SPIE 2805, SPIE, Bellingham, Washington, 1996, pp. 56–65.
3. Holland, A. D., et al., "MOS CCDs for the EPIC on XMM," in *EUV, X-Ray, and Gamma-Ray Instrumentation for Astronomy VII*, edited by O. H. W. Siegmund and M. A. Gummin, Proc. SPIE 2808, SPIE, Bellingham, Washington, 1996, pp. 414–420.
4. Burrows, D. N., et al., "The Swift X-Ray Telescope," in *X-Ray and Gamma-Ray Instrumentation for Astronomy XIII*, edited by K. A. Flanagan and O. H. W. Siegmund, Proc. SPIE 5165, SPIE, Bellingham, Washington, 2003, in press.
5. Hill, J. E., et al., "Readout Modes and Automated Operation of the Swift X-Ray Telescope," in *X-Ray and Gamma-Ray Instrumentation for Astronomy XIII*, edited by K. A. Flanagan and O. H. W. Siegmund, Proc. SPIE 5165, SPIE, Bellingham, Washington, 2003, in press.
6. Moretti, A., et al., "Swift XRT Point Spread Function Measured at the Panter End-to-End Tests," in *X-Ray and Gamma-Ray Instrumentation for Astronomy XIII*, edited by K. A. Flanagan and O. H. W. Siegmund, Proc. SPIE 5165, SPIE, Bellingham, Washington, 2003, in press.
7. Tagliaferri, G., et al., "Swift XRT Effective Area Measured at the Panter End-to-End Tests," in *X-Ray and Gamma-Ray Instrumentation for Astronomy XIII*, edited by K. A. Flanagan and O. H. W. Siegmund, Proc. SPIE 5165, SPIE, Bellingham, Washington, 2003, in press.
8. Mukerjee, K., et al., "The Spectroscopic Performance of the Swift X-Ray Telescope for Gamma-Ray Burst Afterglow Studies," in *X-Ray and Gamma-Ray Instrumentation for Astronomy XIII*, edited by K. A. Flanagan and O. H. W. Siegmund, Proc. SPIE 5165, SPIE, Bellingham, Washington, 2003, in press.

The Swift Ultra-Violet/Optical Telescope (UVOT)

Peter W.A. Roming[*], S.D. Hunsberger[*], John A. Nousek[*], Mariya Ivanushkina[*], Keith O. Mason[†] and Alice A. Breeveld[†]

[*]*Department of Astronomy & Astrophysics, Pennsylvania State University, 525 Davey Lab, University Park, PA 16802, USA*
[†]*Mullard Space Science Laboratory, University College London, Holmbury St. Mary, Dorking, Surrey RH5 6NT, UK*

Abstract. The Ultra-Violet/Optical Telescope (UVOT), one of three telescopes to fly on the Swift Gamma-ray Burst Observatory, is capable of detecting the early UV/optical photons and performing long-term UV/optical observations of GRB afterglows. The UVOT is a Ritchey-Chretien telescope with MCP intensified CCD detectors which operate in either a photon-timing or an imaging mode while providing sub-arcsecond resolution. A filter wheel accommodates broadband UV and visual filters for photometric studies including determination of photometric redshifts. UV and visual grisms for low-resolution spectroscopy are also housed in the filter wheel. We present a brief overview of the UVOT, calibration results, and science to be carried out.

OVERVIEW

The Swift Ultra-Violet/Optical Telescope (UVOT) consists of three basic units: the Telescope Module (TM) and two Digital Electronics Modules (DEMs). The TM houses the primary and secondary mirrors, the detectors, the filter wheels, and the corresponding electronics and power supplies. Figure 1 is an image of the UVOT TM just prior to delivery to NASA on December 16, 2002. The DEMs, one prime and the other

FIGURE 1. UVOT TM Just Prior to Delivery

redundant, each house an Instrument Control Unit (ICU) that operates the UVOT, a Digital Processing Unit (DPU) that handles the science data, and the corresponding electronics and power supplies. Figure 2 reveals the UVOT's placement on the Optical Bench (OB) with respect to the X-ray Telescope (XRT) and the Burst Alert Telescope (BAT). A more complete description of the UVOT can be found elsewhere [1].

FIGURE 2. UVOT Placement on OB

UVOT CALIBRATION

In November 2002 a comprehensive calibration program was executed on UVOT. During the course of the calibration the response of the telescope as a function of wavelength was determined. This response was compared to the UVOT's sister telescope, Optical Monitor (OM), and found to be significantly more sensitive in the UV. A measurement of the PSF in each filter was also characterized. The PSFs ranged from 3 to 1.7 arcsec FWHM for the UV to the V filters. The collimator PSF has not been deconvolved from these numbers; therefore, the true PSF in each filter is much smaller. Due to the fiber taper portion of the detector a significant amount of distortion is present. This was characterized during the testing and a distortion map was created that allows images to be corrected for the aberration. An end-to-end test of the system was also performed to demonstrate that a parameterized finding chart could be reproduced. Other parts of the program included determining the sensitivity, linearity, and dynamic range of the system, obtaining flat fields, and procuring grism data. A more detailed description of the UVOT calibration program can be found elsewhere [2].

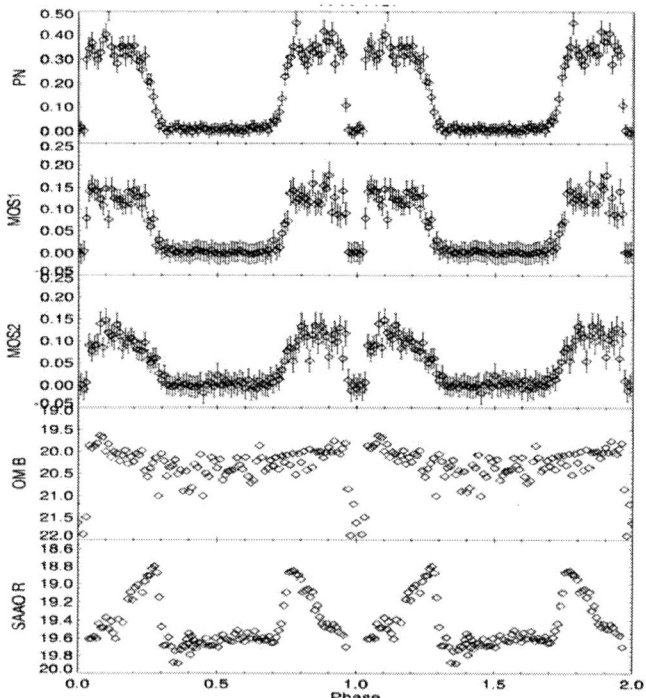

FIGURE 3. Sample UVOT Timing Information

UVOT SCIENCE

One of the primary science objectives of the UVOT is to send a finding chart to ground based astronomers so they can quickly follow-up GRB afterglows with larger telescopes. The BAT determines the GRB position to within an 8 arcminute (diameter) circle, the XRT then narrows that position to 5 arcseconds and the UVOT sends a finding chart centered on the XRT position to the ground. Once a finding chart is sent to the ground, an automated observing sequence using the different UVOT filters is begun. Sample sequences are listed elsewhere [3, 4]. The observing sequences are chosen so as to distinguish between the various GRB models. Unlike most UV/optical telescopes the UVOT is a photon counting device, therefore, timing information of the afterglow photons can also be obtained (see Figure 3 for a representative timing from OM).

ACKNOWLEDGMENTS

We would like to thank the members of the UVOT team from Pennsylvania State University, Mullard Space Science Lab, Southwest Research Institute, Starsys Research Corporation, Swales Aerospace, and NASA Goddard Space Flight Center for all their efforts. This work is sponsored at Penn State by NASA's Office of Space Science through contract NAS5-00136, and at MSSL by funding from PPARC.

REFERENCES

1. Roming, P. W. A., et al., "The Swift Ultra-Violet/Optical Telescope," in X-Ray and Gamma-Ray Instrumentation for Astronomy XIV, edited by K. A. Flanagan and O. H. W. Siegmund, SPIE Proceedings Series **5165**, accepted (2003).
2. Mason, K. O., et al., "Performance of the UV/optical Telescope (UVOT) on SWIFT," in X-Ray and Gamma-Ray Instrumentation for Astronomy XIV, edited by K. A. Flanagan and O. H. W. Siegmund, SPIE Proceedings Series **5165**, accepted (2003).
3. Roming, P. W. A., et al., "The Swift Ultra-Violet/Optical Telescope", in Gamma 2001: Gamma-Ray Astrophysics 2001, edited by S. Ritz, N. Gehrels, & C.R. Shrader, AIP Conference Proceedings **587**, 791-795 (2001).
4. Hunsberger, S.D., et al., "Afterglow Studies with the Swift UV/Optical Telescope", in Gamma-Ray Burst & Afterglow Astronomy 2001: A Workshop Celebrating the First Year of the HETE Mission, edited by G.R.Ricker & R.K.Vanderspek, AIP Conference proceedings, **662**, 497-499, (2003).

Swift Burst Alert Telescope Hard X-Ray Monitor and Survey

Hans A. Krimm*[†], Piotr Banat**, Scott D. Barthelmy*, Tomaso Belloni**, Jay R. Cummings*, Anthony Dean[‡], Edward E. Fenimore[§], Neil Gehrels*, Craig B. Markwardt*[¶], David M. Palmer[§], Ann M. Parsons*, Jack Tueller* and David Willis[‡]

*Code 661, NASA Goddard Space Flight Center, Greenbelt, MD 20771, USA
[†]Universities Space Research Assoc., 7501 Forbes Blvd., Suite 206, Seabrook, MD 20706, USA
**Osservatorio Astronomico di Brera, Via E. Bianchi 46, 23-807 Merate (LC), Italy
[‡]University of Southampton, Highfield, Southampton, United Kingdom SO17 1BJ
[§]Los Alamos National Laboratory, P.O. Box 1663, Los Alamos, NM 84545, USA
[¶]Department of Physics, University of Maryland, College Park, MD 20742, USA

Abstract. The Burst Alert Telescope (BAT) on the Swift gamma ray burst mission will perform the first new all sky hard X-ray survey since 1977. Swift will perform pointings covering >64% of the sky each day and achieve an integrated systematics limited sensitivity in three years of 0.6 milliCrabs for sources well off the galactic plane. This survey is expected to identify hundreds of new highly obscured AGN. BAT will also serve as a sensitive rapid response X-ray outburst and transient monitor.

THE BURST ALERT TELESCOPE ALL-SKY SURVEY

The Burst Alert Telescope[1, 2] on the Swift gamma-ray burst mission is a coded aperture telescope which provides the gamma-ray triggers for the mission. In addition to detecting and observing gamma-ray bursts, the BAT will provide a hard X-ray survey of unparalleled sensitivity.[3] The survey will be carried out during GRB pointings and daily sweeps of most of the sky.

The culmination of the survey will be an all-sky map containing flux measurements in four energy bands for every point on the sky, combined with a catalog of all BAT survey detections. The BAT data products produced for the All-Sky Map are described in [4]. More than 400 AGN will be detected in the survey. The paper [3] discusses the sensitivity for AGN detection and its relationship to models of the population of highly absorbed AGN we expect to detect in the BAT survey.

BAT will monitor an extensive catalog of known X-ray, optical and radio sources to determine if these sources have hard X-ray emission and if so, how the flux varies with time. The BAT science team will compile a list of known hard X-ray sources as well as likely hard X-ray candidates such as AGN, local black hole candidates, radio pulsars and hard spectra quasars. As the All Sky Map is developed and incremented, the BAT team will determine if there is a significant flux excess from each source location. This will increase the sensitivity of the survey and answer important questions about the hard

X-ray emission from a large number of important astrophysical sources.

TRANSIENT MONITOR

BAT will serve as a hard X-ray transient monitor in addition to its primary goal of discovering GRBs. For on-board imaging triggers on time scales > 90 seconds, transient alert and transient position messages will be sent out via TDRSS. Ground software will generate an email message (or a page if after hours) to a BAT on-call scientist who will examine the data and determine if the transient alert is genuine and worthy of being distributed through the GRB Coordinates Network (GCN)[5, 6]. This delay of up to an hour is required to assure data quality and reject false alerts.

BAT data will be examined on the ground for transients at a high sensitivity. But since this is part of the survey process, there will be a greater time delay, This is done by measuring the flux of sources in a BAT transient catalog on time scales from 5-45 minutes. This part of the survey is also sensitive to unknown transients. For every transient found and vetted by the BAT team, an alert will be sent out.

Timely BAT detection of transients will trigger numerous TOOs (target of opportunity) observations on other satellites. These observations will greatly increase the scientific impact of Swift transient detections. The Swift observatory has the capability for rapid reaction science, using the TDRSS uplink. This is further discussed in [3].

SURVEY SENSITIVITY

The BAT hard X-ray survey statistical sensitivity depends on the effective area of the telescope for imaging (Figure 1) convolved with the observing time. The statistical sensitivity is shown in Figure 2 and compared to earlier surveys such as HEAO A-4 in [3].

After about eight hours of observing time, systematic uncertainties in the BAT detector response[7] and effective area will dominate over systematics. See Figure 3 and its caption for an indication of the effect of systematics on survey sensitivity.

SIMULATIONS

The BAT team is making extensive use of two simulation packages for understanding the survey sensitivity. The photon tracking package *grmcflight*, developed by Ed Fenimore is described elsewhere.[3]

The Southampton University team has developed an extensive GEANT 4 model (Swift Mass Model or SwiMM) which it is using to model and study the particle background and its orbital variations.

SwiMM was developed in GEANT 4 with the (immediate) aims:

- Provide predictions of the small-scale pixel-to-pixel spatial variations, to reduce artifacts in a de-coded image.

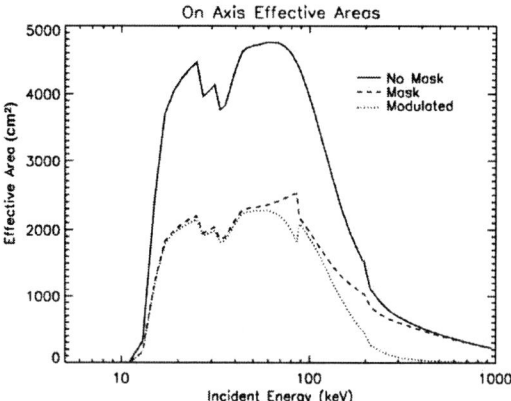

FIGURE 1. The BAT effective area. The dotted line shows the effective area for imaging. It is derived by subtracting the true effective area (dashed line) from the effective area that BAT would have without the mask (solid line). Imaging performance is greatly reduced above 200 keV, but the large size of the detector array gives a sizable unmodulated effective area up to above 1000 keV.

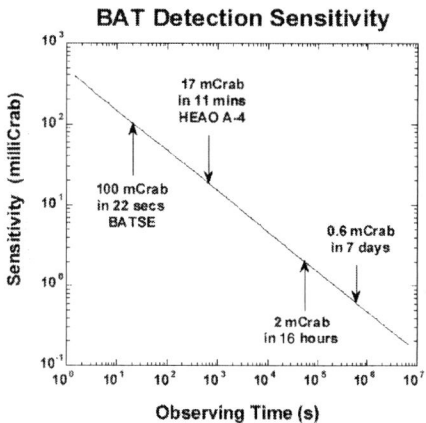

FIGURE 2. BAT 15 - 100 keV statistical detection sensitivity as a function of observing time based on a completed galactic pole simulation with 150 randomly distributed AGN in the field of view.

- Calibration Support (Block level). Provide a means of predicting/removing the residuals in the calibration fitting.
- Provide orbital background model (temporal/spatial variations including structural shadows) to predict the contributions of cosmic ray, cosmic diffuse X-ray, atmospheric albedo and induced radioactivity. Secondary step to include SAA decays.
- Incident GRB self-contamination spectral correction model.

Our current orbital background models are being refined. The effects of the various

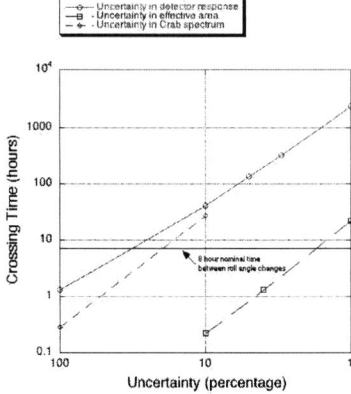

FIGURE 3. The effect of systematic uncertainty on the BAT survey sensitivity can be seen in this plot. The crossing time is the time at which systematic errors start to dominate statistics. Since the systematics can be smeared out by changing the satellite roll angle, it is essential to be able to reduce systematics so that they fall above the solid line (minimal time between roll angle changes allowed by S/C constraints). It can be seen that uncertainties in the detector effective area is the dominant source of systematics.

fits to the cosmic diffuse background in the literature can produce large variations in the levels of background in the BAT. It is imperative to get the proper input spectrum and so the modulations in the BATSE background are being used to provide an absolute spectrum to be used in SwiMM. The atmospheric model is also being developed further.

ACKNOWLEDGMENTS

This work was supported by NASA under the Swift MIDEX project.

REFERENCES

1. Barthelmy, S., "Burst Alert Telescope (BAT) on the Swift MIDEX Mission," in *X-Ray and Gamma-Ray Instrumentation for Astronomy XI, The Swift Mission*, edited by K. A. Flanagan and O. H. Siegmund, SPIE Proceedings 4140, 2000, pp. 50–63.
2. Barthelmy, S., "The Burst Alert Telescope (BAT) on the Swift MIDEX Mission," in [8].
3. Krimm, H. A., and et al., "Swift Burst Alert Telescope Hard X-Ray Monitor and Survey," in *Gamma-Ray Astrophysics 2001*, edited by S. Ritz, N. Gehrels, and C. Shrader, AIP Conference Proceedings 587, Baltimore, MD, 2001, pp. 796–800.
4. Krimm, H. A., and et al., "Swift Burst Alert Telescope Data Products and Analysis Software," in [8].
5. Barthelmy, S., "The GRB Coordinates Network (GCN): A Status Report," in [8].
6. Barthelmy, S., "The GRB Coordinates Network (GCN): A Status Report," in *Gamma-Ray Bursts 4th Huntsville Symposium*, edited by C. A. Meegan, R. D. Preece, and T. M. Koshut, AIP Conference Proceedings 428, Huntsville, AL, 1997, pp. 99–103.
7. Parsons, A., "Swift Burst Alert Telescope (BAT) Instrument Response," in [8].
8. These proceedings, AIP, Santa Fe, NM, 2003.

Swift Burst Alert Telescope Data Products and Analysis Software

Hans A. Krimm*†, Louis M. Barbier*, Scott D. Barthelmy*, Jay R. Cummings*, Edward E. Fenimore**, Neil Gehrels*, Derek D. Hullinger*‡, Craig B. Markwardt*‡, David M. Palmer**, Ann M. Parsons* and Jack Tueller*

*Code 661, NASA Goddard Space Flight Center, Greenbelt, MD 20771, USA
†Universities Space Research Assoc., 7501 Forbes Blvd., Suite 206, Seabrook, MD 20706, USA
**Los Alamos National Laboratory, P.O. Box 1663, Los Alamos, NM 84545, USA
‡Department of Physics, University of Maryland, College Park, MD 20742, USA

Abstract. The Burst Alert Telescope (BAT) on the Swift gamma-ray burst mission serves as the GRB trigger for Swift as well as a sensitive imaging telescope for the energy range of 15-150 keV. All BAT data products will be available to the astronomical community along with a complete set of analysis tools. Gamma-ray burst data products include rapid discovery messages delivered immediately via the GRB Coordinates Network, and event-by-event data from which light curves and spectra of the burst are generated. During nominal operations, the instrument provides accumulated survey histograms with 5-minute time sampling and appropriate energy resolution. These survey accumulations are analyzed in a pipeline to detect new sources and derive light curves of known sources. The 5-minute surveys will also be combined to produce the BAT all sky hard X-ray survey. In addition, the instrument accumulates high time resolution light curves of the brightest BAT sources in multiple energy bands, which are merged into a source light curve database on the ground. The BAT science data products and analysis tools will be described in this paper.

BAT GAMMA-RAY BURST DATA PRODUCTS

Immediately (~minutes):

BAT GRB messages are transmitted via TDRSS and distributed immediately to all interested observers by the GRB Coordinates Network (GCN)[1, 2] (http://gcn.gsfc.nasa.gov):

- GRB alert (includes trigger time and burst significance). These contain no position information or no guarantee that a position will be found.
- GRB position message. In addition to the sky location, this message includes trigger time, burst significance, peak intensity, burst and background fluence, and the burst and background time intervals used to produce burst location.
- TDRSS light curves. Four channel light curves from 24 seconds before to 186 seconds after the burst trigger (not background subtracted).
- Scaled maps. Used to produce a background subtracted sky image for the burst trigger interval.

All of these TDRSS messages are made available through the GCN in a variety of formats (e.g. email attachments, web access, etc.).

Swiftly (~couple hours):

- Background subtracted light curves derived from the event files in four energy channels. These will include the time covering the S/C slew to the burst. Time binning: uniform 64 ms and 1 s and Bayes blocks.
- Counts spectra on various time scales during and after the burst. Response matrix for each spectrum. Spectral fit parameters from XSPEC.
- Burst summary parameters:
  - Burst durations: (T50 and T90).
  - Energy fluences (in KeV bands: 10-25, 25-50, 50-100, 100-300, 2-2000).
  - Variability and time lags between fiducial energy bands.[1]
- Event files (10 minutes around the burst trigger time).
- Short time scale (1 minute) survey histograms until the next S/C slew.
- Images of the burst and surrounding sky before, during and after the burst.

A bit later (~days):

- Burst advocate generated products (possible examples):
  - Revised burst location.
  - Estimate of red shift from lag-variability analysis
  - Spectral fits for additional time intervals
  - Comparisons between BAT and Swift narrow field instrument analysis.
- Lists of follow-up observations and links to publicly available follow-up data.
- Gamma-ray burst catalog.

BAT SKY MONITOR PRODUCTS

- **Variable and Transient Alerts:** The observing community will be notified through the GCN when the BAT detects either a bright new source or a known source with a flux outside its normal limits. This happens on two timescales: ~minutes for transients detected on the spacecraft to ~hours for transients detected in the pipeline processing.
- **Detector Plane Histograms (DPH):** Eighty-channel spectral histograms for every BAT detector. The nominal accumulation time for a survey DPH is five minutes.

---

[1] These parameters may not be calculated in the pipeline.

The DPHs are also summed to cover a single spacecraft pointing (Swift snapshot of $\sim$20 minutes) and observing segment ($\sim$8 hours).
- **Sky images:** Sky images (both cleaned and with bright sources re-added) are archived for every $\sim$8 hour observation segment.
- **Monitored source light curves:** For each source that BAT monitors, a running light curve with five-minute time binning will be accumulated whenever the source is in the BAT field of view. For selected bright sources, 1.6 second time resolution background subtracted light curves are also available.
- **Source spectra and response matrices:** For each detected source, a counts spectrum and BAT response matrix is produced. These are produced in the pipeline for observation segment time scales, but can be generated on any desired time scale.
- **Catalogs:** Table summarizing monitored source properties for each observation segment. This is linked to the BAT master catalog of monitored and newly detected sources.

BAT HARD X-RAY SURVEY AND ALL-SKY MAP

The first delivery will be twelve months after launch and every six months thereafter. Deliveries will include six month and cumulative products.

- **Sky images:** Entire sky covered by 5 × 5 degree overlapping tiles with 10 arcmin resolution. Includes cleaned sky and sky with bright sources re-added; also exposure and noise maps. Sky maps will be produced in five energy bands.
- **Catalog:** Table of all sources detected on all survey time scales.
- **Flux calculation:** The user will be able to derive a flux or upper limit for each point in the All-sky map.

The BAT All-Sky Map will be more than a graphical representation of the sky or a list of detected sources – it will be developed so that a user can derive the flux and and background for any point in the sky on any time scale by accessing the archived survey data. The BAT Hard X-ray survey is described in more detail in[3, 4].

BAT DATA ANALYSIS TOOLS

- **batbinevt**: Multipurpose tool to create light curves and spectra from event files using uniform, constant signal/noise, or user supplied time bins. Produces background subtracted light curves if the event file has been mask weighted (see batmaskwtevt).
- **batcelldetect:** Searches sky images for point sources above user supplied threshold and derives flux estimates. User can also input a list of source locations for which flux measurements or upper limits are calculated.
- **batclean:** Cleans background model and source fluxes from detector plane; can be combined with batcelldetect for iterative cleaning.

- **batdrmgen:** Derives response matrix function (for use in XSPEC) for given location on the sky; uses parametrization of ground calibrations and flight data.
- **batfftimage:** Uses an inverse fast Fourier transform (FFT) to convolve the detector plane with the BAT coded mask to derive a sky image. Can also produce a map indicating the coded fraction as a function of sky location.
- **batmaskwtevt/img:** Uses forward projection ray tracing to calculate mask weighting factor for each BAT detector. Tools designed either for event-by-event or full detector plane analysis.
- **battblocks:** Derives optimum time intervals (GTIs) based on event or rate data, using the Bayesian Block algorithm. New Bayesian block intervals are chosen for each statistically significant change in the rate.

OTHER BAT PRODUCTS

Other public data products include:

- Calibration data base including the coded mask pattern, the Swift telemety definition file, spectral histogram energy bounds, and ground calibration summaries.
- Results of in-flight calibration (tagged radioactive sources and electronic pulser)
- Housekeeping and diagnostic files
- BAT rate files.
- Pulser folded data.

All BAT data (except for the hard X-ray survey) will be made publicly available within two hours of receipt at the Swift Science Data Center (SDC) Quicklook web site. After one week data will be available at the HEASARC (http://heasarc.gsfc.nasa.gov). All Swift products will also be available through European mirror sites.

ACKNOWLEDGMENTS

This work was supported by NASA under the Swift MIDEX project.

REFERENCES

1. Barthelmy, S., "The GRB Coordinates Network (GCN): A Status Report," in [5].
2. Barthelmy, S., "The GRB Coordinates Network (GCN): A Status Report," in *Gamma-Ray Bursts 4th Huntsville Symposium*, edited by C. A. Meegan, R. D. Preece, and T. M. Koshut, AIP Conference Proceedings 428, Huntsville, AL, 1997, pp. 99–103.
3. Krimm, H. A., and et al., "Swift Burst Alert Telescope Hard X-Ray Monitor and Survey," in *Gamma-Ray Astrophysics 2001*, edited by S. Ritz, N. Gehrels, and C. Shrader, AIP Conference Proceedings 587, Baltimore, MD, 2001, pp. 796–800.
4. Krimm, H. A., and et al., "Swift Burst Alert Telescope Hard X-Ray Monitor and Survey," in [5].
5. These proceedings, AIP, Santa Fe, NM, 2003.

The BAT-Swift Science Software

D. M. Palmer*, E. Fenimore*, M. Galassi*, K. Mclean*, T. Tavenner*, S. Barthelmy[†], M. Blau[†], J. Cummings[†], N. Gehrels[†], D. Hullinger[†], H. Krimm[†], C. Markwardt[†], R. Mason[†], J. Ong[†], J. Polk[†], A. Parsons[†], L. Shackelford[†], J. Tueller[†], S. Walling[†], Y. Okada**, H. Takahashi**, M. Toshiro**, M. Suzuki**, G. Sato**, T. Takahashi** and S. Watanabe**

Los Alamos National Laboratory
[†]*Goddard Space Flight Center, NASA*
**University of Tokyo, Saitama University and ISAS, Japan*

Abstract. The BAT instrument tells *Swift* where to point to make immediate follow-up observations of GRBs. The science software on board must efficiently process γ-ray events coming in at up to 34 kHz, identify rate increases that could be due to GRBs while disregarding those from known sources, and produce images to accurately and rapidly locate new Gamma-ray sources.

INTRODUCTION

BAT is a wide-field coded aperture gamma-ray telescope to be flown on the *Swift* Mission[1]. Its main purpose is to detect and accurately locate GRBs so that *Swift* may repoint itself to observe the GRBs with its X-ray and UV/Optical telescopes, sometimes while the GRB is still in its γ-ray flaring phase. It also provides the direct γ-ray observations of the GRB in its 15-150 keV energy range. While waiting for new GRBs, it provides secondary science such as a $\sim$mCrab all-sky survey, high time resolution observations of selected sources, and phase-resolved spectroscopy of pulsars.

BAT has 32,768 CdZnTe detectors, each $4 \times 4 \times 2$ mm^3 (5243 cm^2 total area) and a 2.6 m^2 mask with a 50%-filled random pattern of $5 \times 5 \times 1$ mm^2 lead tiles. The 5 mm mask scale divided by the 1 m mask-detector separation determines the 17 arcmin FWHM angular resolution. Detected γ-ray events and other housekeeping information are transmitted by 16 'Spacewire' (IEEE 1355) channels to the primary Image Processor Electronics (IPE) computer and to a redundant cold-spare IPE.

Each IPE includes a 25 MHz Rad6000 that serves as the main processor, and a rad-hard 25 MHz ADSP21020 digital signal processor (DSP) for image production and processing.

The Rad6000 software runs under the VxWorks operating system and is written in C++. Significant parts of the engineering code (*e.g.*, the Command and Data Handling Software Bus) were originally developed for the Triana spacecraft. The DSP software is written in C and is based on a simple purpose-built executive.

This paper briefly describes the instrument-unique science software in the IPE.

DATA INGEST

Each γ-ray detected is processed by the Data Ingest task. The digitized signal is converted from ADC units (~ 0.2keV resolution) to equivalent keV by gain and offset parameters continually determined for each detector by an electronic calibration system. Each event also includes a timestamp with $100\mu s$ resolution referenced to a 1 pulse-per-second hardline signal from the spacecraft.

The data is binned in time, energy, and detector region in multiple ways:

- On 5 minute time intervals, 80 channel energy histograms are made for each of the 32k detectors. These are processed on the ground to produce the all-sky survey.
- Accumulations in 4 energy bins in each of the 32k detectors over 8 seconds are generated and stored to be used in background subtraction when a burst is detected.
- The events are pulsar-folded into an 80 channel, 32 phase bin structure using a spin ephemeris based on a 13th order polynomial (generated on the ground using the spacecraft's predicted orbit for barycentering).
- Sets of 32k detector weights corresponding to the exposure through the mask allow individual 'mask-tagged' monitoring [2] of up to 3 selected sources.
- Accumulations in 4 energy bins in the 4 quadrants of the detector are generated every 64 ms and passed to the Trigger Algorithm.
- On 5 timescales from 4 ms to 64 ms, special code calculates the count rate in 4 energy ranges and 9 overlapping detector regions (from individual quadrants to the full detector plane) and tells the trigger system the maximum count rate achieved in each of the 180 time-energy-region combinations during a 320 ms time interval.

The large collecting area and FOV (1.4 sr to the half-coding point) will produce a nominal count rate that simulations predict to be ~ 12 kHz, although this will be greatly exceeded as the spacecraft passes near or through the South Atlantic Anomaly (SAA) in its $22°$ inclination orbit. The hardware and software can handle sustained operation at 34 kHz. Processing at this data rate, using a fraction of the 25 MHz processor, required programming for extreme efficiency (at the cost of clarity and maintainability).

At count rates above 34 kHz, or when the processor has episodic demands placed on it (*e.g.*, during data compression for telemetry) a backlog of data will accumulate in the input buffer. This data is processed in the order it is received so that processing falls behind briefly, to catch up when the CPU demand eases or the count rate declines. When the backlog exceeds 5×10^7 counts, it is assumed that we are in a high count rate situation such as the SAA, and the detector plane is placed in a mode where it merely counts events rather than reporting each one. The detector is returned to normal mode when the backlog has been processed and the count rate returns to a reasonable level. The SAA threshold is almost two orders of magnitude higher than the expected highest-fluence burst of the year (equivalent to ~ 200 kHz for 5 s), and so we do not expect to lose any data due to excessive flux from a GRB during the course of the mission.

TRIGGERING

The triggering code is more completely described elsewhere[3]. The binned count rates produced by Data Ingest allow triggering on timescales from 4 ms to many seconds. Rate increases are detected by comparing the count rate during a foreground interval to that extrapolated or interpolated from one, two or three background intervals. Mask-tagging to track the flux of individual sources allows us to ignore count rate increases when they are due to a known source (vetoing) or to subtract the source's count rate from the time series for triggering (canceling), which mitigates the effects of noisy sources such as Cyg X-1 and Sco X-1.

When the triggering system detects a rate increase, it also determines the optimal energy range and time intervals for foreground and background accumulations. These time intervals are used to produce a 'background-subtracted' detector map, where the pattern of net counts in the detectors as revealed by imaging should be dominated by flux from the varying source.

When no significant rate increase is detected, 'image triggering' is attempted by accumulating detector maps on 1 minute, 5 minute, and per-pointing intervals and passing them to the imaging system to detect weaker and slowly varying sources.

IMAGING

Imaging is handled by the DSP coprocessor, which is optimized for the FFT operations that dominate the execution time for imaging.

The detector map is processed to remove variation from systematic effects and known sources in the field of view. A sequence of balancing steps that subtract the average constant value from all or subclasses of detectors is combined with an optional CLEAN step.

CLEANing uses a linear regression of the detector patterns predicted for multiple bright known sources in the field of view to remove the effects of those sources. An image made of the residual map is free from the main peaks and systematic noise from those sources, allowing unknown sources to stand out. The linear regression jointly includes known background variation patterns, such as the tendency for the center of the detector plane to have more solid angle exposure to aperture flux than the periphery.

Because background subtraction eliminates many of the systematic effects, the pre-processing strategies can be separately specified for background-subtracted rate trigger maps and non-background-subtracted image trigger maps.

The pre-processed detector map is then convolved with the mask pattern, using an efficient FFT-based process, to produce what we call an FFT image. This step can also include Wiener filtering (which has the effect of approximating a numerically stable convolution-deconvolution hybrid) and high-pass filtering with no additional CPU processing time. The effect on sensitivity of these additional filters will be tested with on-orbit data.

The FFT image has a pixel scale set by the detector spacing (4.2 mm) divided by the 1 m detector-mask spacing, and is comparable to the size of the point spread function.

The signal from a source can thus be split among multiple adjacent pixels. For this reason the image is searched for peaks consisting of individual, pairs, and sets of 4 pixels, to a level that picks up some noise peaks in addition to true peaks from weak sources. The peak locations are compared to an on-board catalog to identify known sources, and to ignore them unless they exceed intensities at which they are considered interesting.

Unidentified candidate peaks from the FFT image are used as locations for back-projection (BP) images. These BP images of small regions of the image plane have fine pixels, allowing the precise locations and intensities of the peaks to be determined.

Roughly 6 seconds after the start of imaging, a list of possible sources in the field of view is available with accurate locations (<4 arcmin), intensity, and detection significance. If the significance of a candidate source exceeds a statistical threshold, a GRB is announced and targeted for follow-up. If no source exceeds threshold, the trigger system is consulted to see if an improved detector map is available, based on the RAD6000 processing that continued during the DSP's image processing.

FOLLOW-UP

If a GRB is detected and located, the *Swift* spacecraft evaluates it for quality and observability, then rapidly slews to make X-ray, optical and UV observations beginning about a minute after the first rate increase detection. Simultaneously, it sends position and other information down the real-time TDRSS link to allow ground-based follow-up observations.

The BAT also continues its observations. Light curves in 4 energy bands and up to 128 ms resolution are generated and sent down TDRSS as segments are accumulated, along with the attitude information required to interpret the changing background during the slew. Event-by-event data, describing every photon detected by BAT from 5 minutes before the GRB to 5 minutes after, is spooled to the spacecraft solid state recorders, for later dumping on the next pass over the Malindi ground station. The cadence of survey accumulations is accelerated for better time resolution during the afterglow phase. The GRB is selected as a mask-tagged source for 1.6 s resolution source-specific flux measurements whenever the GRB is in the BAT's field of view.

Over the following days, automatic and ground-planned observations continue until the GRB has faded from sight.

REFERENCES

1. Gehrels, N. *et al.*, in *Gamma-ray Burst and Afterglow Astronomy 2001* edited by G.R. Ricker & R. K. Vanderspek, AIP Conference Proceedings 662, pp. 465-468.
2. Fenimore, E. E., Applied Optics **26** 2760-2769 (1987).
3. Fenimore, E. *et al.*, in *Gamma-ray Burst and Afterglow Astronomy 2001* edited by G.R. Ricker & R. K. Vanderspek, AIP Conference Proceedings 662, pp. 491-493.

Setting the Triggering Thresholds on Swift

Kassandra M. McLean*[†], E. E. Fenimore*, David Palmer*, S. Barthelmy**, N. Gehrels**, H. Krimm**, C. Markwardt** and A. Parsons**

Los Alamos National Laboratory
[†]*University of Texas at Dallas*
**Goddard Space Flight Center*

Abstract. The Burst Alert Telescope (BAT) on Swift has two main types of "rate" triggers: short and long. Short trigger time scales range from 4ms to 64ms, while long triggers are 64ms to ≈ 16 seconds. While both short and long trigger have criteria with one background sample (traditional "one-sided" triggers), the long triggers can also have criteria with two background samples ("bracketed" triggers) which remove trends in the background. Both long and short triggers can select energy ranges of 15-25, 15-50, 25-100 and 50-350 KeV. There are more than 180 short triggering criteria and approximately 500 long triggering criteria used to detect gamma ray bursts. To fully utilize these criteria, the thresholds must be set correctly. The optimum thresholds are determined by a tradeoff between avoiding false triggers and capturing as many bursts as possible. We use realistic simulated orbital variations, which are the prime cause of false triggers.

INTRODUCTION

Swift is a rapidly slewing satellite, that can quickly point the field of views (FOVs) of the X-Ray telescope (XRT) and the ultraviolet-optical telescope (UVOT) at the gamma-ray bursts (GRBs). In order to begin this process, the Burst Alert Telescope (BAT) must detect that a burst has occurred and locate its position. Swift's behavior in responding to GRBs depends crucially on BAT triggering, which means that it is vital to optimally set the thresholds for the various trigger criteria. The main drive behind the thresholds is one of balancing sensitivity against false triggers. This paper presents the procedure we used to select the thresholds for Swift's various trigger criteria.

BAT uses about 800 different criteria to detect GRBs, each defined by a large number of commandable parameters. Usually the critical parameter is the time scale of the sample being analyzed for a statistically significant increase. There are three triggering systems. One is called the "short triggers" and covers times ranging from 4ms to 64 ms. The short triggers have a fixed background sample duration of about 1 sec. The second system is the "long triggers" and covers time scales range from 64ms to as large as we dare command without trends in the background producing too many false triggers(≈ 16 sec currently). There is a third trigger system (which we will not discuss) that searches for new point sources in images that are reconstructed periodically (typically every 32 sec, 320 sec, and 1000 sec) from the detector plane observations. (See [1] for details on the imaging.)

Both short and long triggers can target the 15-25, 15-50, 25-100 and 50-350 KeV energy ranges. A criterion can also target a quadrant (or any combinations of quadrants) of the detector plane, primarily to be more sensitive to bursts at the edge of the FOV. Swift has the most comprehensive set of triggering criteria ever attempted.

We trigger on a burst when there is a statistically significant increase in the counts for any particular criteria. See [2] for more information on how the trigger is evaluated. Once the BAT has a rate trigger, it then images the detector plane to find any new point sources. If no new sources are found, the trigger is rejected as false. Thus, we can tolerate false triggers in orbit and be more sensitive to bursts.

To create a high fidelity simulation of BAT in its orbital conditions we simulate the steady, diffuse x-ray/gamma-ray sources, GRBs, and particle variations throughout orbit. Two software packages are used to accomplish this. The GRMCFLIGHT program simulates the gamma-ray transport through BAT. GRMCFLIGHT produces the photon energy, location, and time of incidence on the detector plane. GENERATE1355 is a program which simulates the BAT electronics, and produces the data stream which is fed to the flight code. The combination of GRMCFLIGHT and GENERATE1355 allows us to trace each photon from its origin (an astronomical source or the background) all the way through the BAT flight software.

The ability to inject BATSE burst profiles into GRMCFLIGHT allows us to go even further and test the high-level scientific features of the BAT flight software: namely triggering and imaging. This can then be used for system-wide BAT tests, such as slewing to a GRB location. The result of these tests tells us how many bursts we will trigger on, image, and slew to (see [3]).

To set the thresholds, we run simulations with no injected bursts under various background conditions. We typically run between 10 and 20 hours of background, and try to set the threshold such that there would be no false triggers over this period of time. All simulations have a diffuse x-ray background component of about 8 KHz plus an additional 4 kHz background due to particles. We add various bumps in the background to mimic the orbital variations of the particle background. These bumps were typically Gaussian in shape with a full width at half maximum of about 24 minutes. We studied, in particular, bumps that added 20 kHz and 4 MHz. The 4 MHz bump is what we expect in the SAA.

SETTING THE SHORT TRIGGER THRESHOLDS

We ran each of the 180 short trigger criteria separately through the three different background variations: flat 12 KHz, flat 12 KHz plus a 20 KHz bump, and flat 12 KHz plus a 4 MHz bump. Since the background for the short triggers is within 1 sec of the foreground sample, we find that the thresholds are approximately the same for all three background variations. In order to allow for statistical variations, we ran each short trigger (i. e., 180 criteria) through about 10 orbits (each with a 4 MHz SAA bump) to determine the maximum score. The short criteria tended to require about the same threshold for various permutations of areas of the detector plane and energy ranges. We set the thresholds for the short criteria to only depend on the time scale (i.e., 4, 8, 16, 32, and 64 msec). In Figure 1, the short criteria are denoted by filled squares and have thresholds of about 6σ to 7σ.

SETTING THE LONG ONE-SIDED TRIGGER THRESHOLDS

Swift uses two types of triggers in the > 64ms range: *one-sided* and *bracketed* triggers. One-sided triggers have only one background sample which is before the foreground sample, and they have been used in all past missions, such as BATSE. For the one-sided triggers, since they do not remove trends in the data, we set them to withstand orbital variations of about 20kHz, a rather large bump for Swift in a quiet, non-SAA orbits.

We have diagnostic reports to the telemetry of the maximum score for each criterion, each orbit. We plan to use this diagnostic output to set the thresholds on orbit once we have a few days of background data. To set the long thresholds on the ground prior to launch we simulate 14 orbits (about 20 hrs) of background with 20kHz bumps in each orbit. The telemetry provides a maximum score for each criteria in each orbit. Let $S_{\max,i}$ be the overall maximum score for the i-th criteria seen over the 14 orbits. Let σ_i be the RMS of the 14 maximum scores from each orbit. We set T_i, the threshold for the i-th criteria, to be

$$T_i = S_{\max,i} + 2\sigma_i \; . \tag{1}$$

The open squares in Fig.1 show the thresholds for the one-sided long triggers. The one-sided criteria have foreground durations of 0.128, 0.256, 0.512, and 1.024 sec. The one-sided criteria are very susceptible to trends in the background. A one percent trend over one second will produce a one sigma increase in the count rate. For that reason, we could not make one-sided criteria use foregrounds much longer than one second.

For each foreground duration, there can be 36 different criteria. There are four different energy ranges (15-25, 15-50, 25-100, 50-350 KeV) and nine combinations of sub-areas of the detector plane (the four quadrants, four halves, and the full detector plane). We do not necessarily use a set of criteria that include all 36 permutations.

SETTING LONG BRACKETED TRIGGER THRESHOLDS

Bracketed triggers have a background sample before and after the foreground sample, allowing a fit to trends in the background rate. To date these have only been used in the HETE flight trigger algorithm. The bracketed triggers efficiently remove background trends so we could set the thresholds to be much lower than the one-sided triggers and use much longer foreground time samples. They are occasionally sensitive to the *peak* of a bump in the background. We initially simulated bumps ranging from 200Hz to 2 MHz in steps of a factor of ten.

The opened circles in Fig. 1 give the thresholds for various permutations of energy and detector plane regions for foreground durations between 0.128 and 16 sec. We used 14 orbits of background with a flat background of 20 KHz. No bumps were used because the two-sided criteria removes trends and the size of the bump had little effect on the thresholds. Equation 1 was applied to the maximum scores reported by the diagnostic software to obtain the thresholds. By using a flat background, we were able to get scores in the range of 3σ to 5σ.

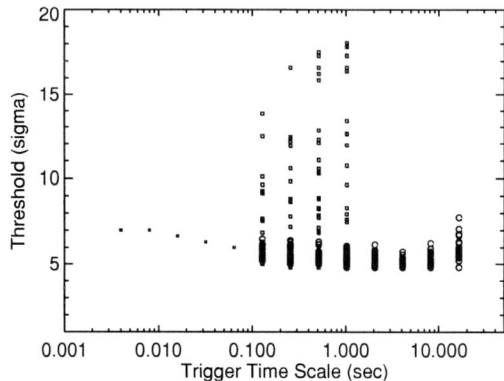

FIGURE 1. Thresholds for the BAT rate trigger criteria as a function of the foreground duration. The solid squares are for the short rate triggers. The open squares are for one-sided long criteria and the open circles are for two-sided long criteria. In contrast, the equivalent threshold for BATSE was 11 sigma and foreground time durations were only 64, 256, and 1024 ms.

CONCLUSION

Our strategy is to have the trigger thresholds as low as possible and allow approximately 2 to 3 false triggers per hour. The flight software will form an image at the time of the trigger and those false triggers will be rejected because there will not be a new point source in the image. Sources in a coded aperture image will usually have a smaller signal-to-noise than the signal-to-noise in the trigger. As a result, valid triggers near the threshold will not have significant new point sources in the image. Thus although Fig. 1 shows the thresholds required for a trigger, a successful image requires a stronger signal.

There is often the concern that the two-sided criteria introduce a delay in recognizing that a burst is occurring. This is true. Our strategy is that we will use one-sided criteria with durations as long as we can (which is ≈ 1 sec) and with threshold that avoid excessive false triggers. Those thresholds, unfortunately, can be as large as ≈ 15. The two-sided criteria indeed take longer, but they are only used when the one-sided were unable to detect the burst. It is better to get the burst later than not at all.

REFERENCES

1. Palmer, D., et al. these proceedings (2004).
2. Fenimore, E. E., et al., in *Gamma-ray Burst and Afterglow Astronomy 2001* edited by G.R. Ricker & R. K. Vanderspek, AIP 662, pp. 491-493 (2003).
3. Fenimore, E. E., et al., Baltic Astron. in press (2004).

Swift Burst Alert Telescope (BAT) Instrument Response

A. Parsons[*], S. Barthelmy[*], J. Cummings[*], N. Gehrels[*], D. Hullinger[*], H. Krimm[*], C. Markwardt[*], J. Tueller[*], E. Fenimore[†], D. Palmer[†], G. Sato[**], T. Takahashi[**], K. Nakazawa[**], Y. Okada[‡], H. Takahashi[‡], M. Suzuki[§] and M. Tashiro[§]

[*]*NASA/Goddard Space Flight Center, USA*
[†]*Los Alamos National Laboratory*
[**]*Institute of Space and Astronautical Science (ISAS), Japan*
[‡]*University of Tokyo, Japan*
[§]*Saitama University, Japan*

Abstract. The Burst Alert Telescope (BAT), a large coded aperture instrument with a wide field-of-view (FOV), provides the gamma-ray burst triggers and locations for the Swift Gamma-Ray Burst Explorer. In addition to providing this imaging information, BAT will perform a 15 keV – 150 keV all-sky hard x-ray survey based on the serendipitous pointings resulting from the study of gamma-ray bursts, and will also monitor the sky for transient hard x-ray sources. For BAT to provide spectral and photometric information for the gamma-ray bursts, the transient sources and the all-sky survey, the BAT instrument response must be determined to an increasingly greater accuracy. This paper describes the spectral models and the ground calibration experiments used to determine the BAT response to an accuracy suitable for gamma-ray burst studies.

INTRODUCTION

BAT instrument response information is needed to infer the incident 15 keV – 150 keV astronomical source photon spectrum from the BAT detector plane count rates. A telescope's spectral response is usually represented as an N x M matrix where N is the number of energy bins in the counts spectrum and M is the number of energy bins in the incident source photon spectrum. Each row of the response matrix gives the counts spectrum that would be produced by BAT if a 15 keV – 150 keV source flux of one photon/(cm^{2}s) were incident on the instrument. The response matrix units are cm^2 and thus the totals of each of the rows represent the effective area of the BAT as a function of incident photon energy.

For an ideal instrument, every incident photon would be recorded and would produce a count in a single bin corresponding to the incident photon's energy. A response matrix for such an instrument would therefore be diagonal. In reality, incident photons have a non-unity energy-dependent probability of being detected, and those photons that do interact in the detector usually produce a count in one of a range of different count bins. For the BAT detectors, the response matrix is somewhat triangular due to the charge trapping effects in the BAT CZT semiconductor detectors. These 4 mm x 4 mm x 2 mm CZT detectors each produce counts spectra with photopeaks that display a low energy "tail" that becomes especially prominent at energies greater than 60 keV. In this paper we present the current status of our efforts to determine the BAT response using ground calibration data to fit a parameterized instrument response function to a single spectrum representing the full array response.

BAT SPECTRAL RESPONSE MODELS

Defects within the CZT semiconductor material produce sites where drifting charge carriers (electrons and holes) are trapped and thus no longer contribute to the detector signal. Since the amount of charge collected is typically proportional to the incident photon energy, this charge loss will result in a measured "apparent" energy that is lower than the actual energy deposition in the detector. Interactions that occur deeper in the detector result in more charge trapping and therefore a lower apparent energy. The charge-trapping properties of a detector are quantified by the mobility-lifetime product of electrons ($\mu\tau_e$) and of holes ($\mu\tau_h$). Since the BAT detector plane contains 32768 individual CZT detectors, the most accurate BAT spectral analysis would require 32768 separate detector response matrices to separate mask modulation effects from individual detector efficiency differences. The accuracy requirements of the response needed for GRB spectroscopy analysis are fortunately much less severe and allow us to make an important approximation.

Since the GRB response generator has to be part of the quicklook burst data analysis pipeline and will be used to quickly fit burst spectra, the algorithm must be efficient and the parameter look-up tables must be modest in size. With these requirements, it is clear that generating 32768 individual response matrices for each burst position would be inefficient. The needs of the spectral response model for gamma ray bursts are such that we can approximate the full array response by averaging over the individual detector properties. If we determine the density of the mask shadow pattern on each detector from a point source at a given position in the BAT FOV, we can use this mask modulation information to create a weighted sum of all the individual detector counts spectra and produce a single spectrum that represents the measurement of only the photons coming directly from that source's position. We then can fit a parameterized spectral function to this full BAT spectrum. Our spectral model uses detector interaction depth distributions generated by Monte Carlo simulations to calculate the resulting output spectrum for a given source energy and angle. The average charge trapping effects of the detectors are captured as "effective" $\mu\tau_e$ and $\mu\tau_h$ fit parameters used in the calculation of the charge trapping fraction. Additional parameters such as overall intensity normalization, the number of detectors enabled, and electronics gain and noise are also needed to match the ground calibration data. For example, Figure 1 shows our model fit to a mask weighted summed spectrum from a Ba-133 source placed in the BAT FOV. The model fits the data fairly well with the biggest discrepancy occuring just below the large 31 keV line.

After finding the best-fit model parameter set for a large number of positions within the BAT FOV, we can use this large look-up table of model parameters to compute the weighted summed spectrum resulting from a point source anywhere within that FOV. For each incident photon energy, we compute the output spectrum and store it as a row

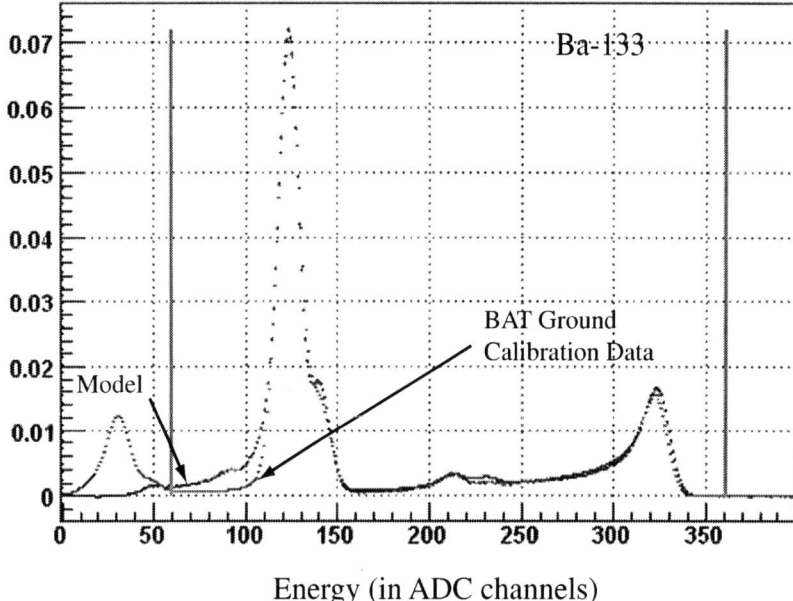

FIGURE 1. A fit of our preliminary BAT spectral response model to Ba-133 calibration data. The x-axis gives the energy bin number and the y-axis gives the output count rate for each bin.

in the response matrix. This process is repeated for each energy bin in the input photon spectrum, and the response matrix is built up, row by row.

The level of accuracy required for the BAT instrument response used for the hard x-ray survey is significantly higher than for bursts because this response must be used in the iterative clean algorithm for finding fainter sources. Because the bright sources add a lot of coding noise to the BAT sky image, fainter sources can be seen only after the counts due to the bright sources are removed. The iterative "clean" procedure of extracting counts due to bright sources requires a detailed knowledge of the BAT response since each source puts counts in each detector. For the hard x-ray survey data analysis, we have more computing time available and can thus treat detectors individually. Since the source detection sensitivity depends on the contrast between shadowed and illuminated areas, small scale mask non-uniformities and individual detector efficiencies are very important. The better we know the BAT response, the lower the noise in the cleaned spectrum and thus the more sensitive the survey.

BAT GROUND CALIBRATION TESTS

We have performed many different kinds of ground calibration measurements to determine the best fit model parameters and to verify that the spectral model is valid over

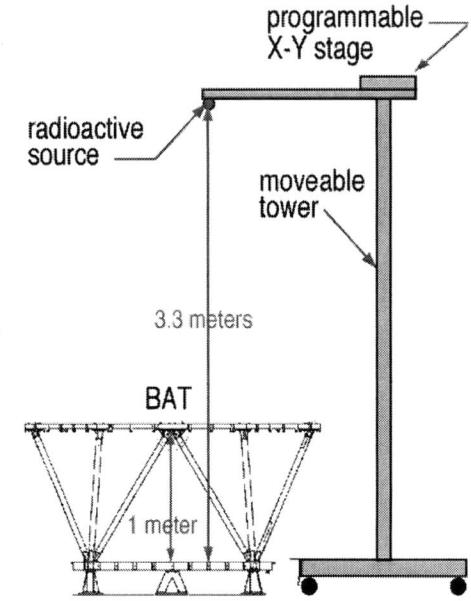

FIGURE 2. Schematic Diagram of the BAT Calibration Setup

the entire BAT FOV and energy range. We have used radioactive sources to directly illuminate individual BAT "Blocks" with known gamma-ray energies to study the CZT detector properties without the added complication of having the signal be modulated by the coded aperture mask. Key mobility and lifetime parameters $(\mu \tau_e)$ and $(\mu \tau_h)$ were determined for each of the 2048 detectors in a Block. These Block tests were performed for each of the 16 Blocks that make up the BAT array.

The next stage of testing occurred at the level of the fully assembled BAT instrument with all 16 Blocks installed and the coded aperture mask present. Radioactive sources such as Am-241, Co-57, Cd-109, and Ba-133 with line energies = 14.4, 17.8, 22.1, 25.0, 30.8, 35.0, 53.2, 59.5, 79.6, 81.0, 88.0, 122.1, and 136.5 keV were placed on a moveable fixture that suspended the source 3.3 meters above the BAT array. Figure 2 is a simple sketch of the experimental setup. The tower fixture shown in this figure was moved to locations all around the BAT, placing the source in many different positions all over the BAT FOV. Coarse grid measurements were taken at source positions about 50 cm apart and a programmable XY stage on top of the tower was used to move the source in a fine grid of positions with a separation of about a millimeter. We tried to cover as much of the BAT FOV as possible, given the physical constraints of the area around the BAT and we believe we have more than a sufficient amount of calibration data to define the GRB spectral response. We have also acquired over a terabyte of array calibration data that will be needed to map out the fine details required for an accurate response for use with the hard x-ray survey analysis.

FUTURE SATELLITES

Observing GRBs with EXIST

D. H. Hartmann*, J. Grindlay[†], J. Hong[†], A. Loeb[†], R. Blandford**, W. Craig[‡], J. Fishman[§], C. Kouveliotou[§], N. Gehrels[¶], D. Band[¶], F. Harrison[‖] and S. E. Woosley[††]

*Department of Physics & Astronomy, Clemson University, Clemson, SC
[†]Harvard-Smithsonian Center for Astrophysics, Cambridge, MA
**Stanford Linear Accelerator, Stanford University, Stanford, CA
[‡]Lawrence Livermore National Laboratory, Livermore, CA
[§]NASA, Marshal Space Flight Center, and NSSTC, Huntsville, AL
[¶]NASA, Goddard Space Flight Center, Greenbelt, MD
[‖]California Institute of Technology, Pasadena, CA
[††]Dept. of Astronomy & Astrophysics, The University of California, Santa Cruz, CA

Abstract. We describe the Energetic X-ray Imaging Survey Telescope EXIST, designed to carry out a sensitive all-sky survey in the 10 keV - 600 keV band. The primary goal of EXIST is to find black holes in the local and distant universe. EXIST also traces cosmic star formation via gamma-ray bursts and gamma-ray lines from radioactive elements ejected by supernovae and novae.

EXIST is proposed as a black hole finder probe in NASA's Beyond Einstein Program, and would operate in zenith pointing scanning mode (FoV of $180°\times75°$), to provide all-sky coverage every orbit. Variable sources on the sky would be monitored daily. Each source would be in the FoV for at least 20 min every 95 min orbit, and sampled with ms resolution. EXIST will monitor known variable sources, detect outbursts of known and new sources, and detect Gamma-Ray Bursts (GRBs) with a coverage probability of $P = FoV/4\pi \sim 0.5$. Technical details are presented at http://exist.gsfc.nasa.gov/. The "reference design" (Fig. 1) will be refined in a proposed mission concept study. EXIST surveys the sky from a low-inclination ($22°$), low-altitude (~ 500 km) orbit. With a mission life of five to ten years, EXIST could explore the hard X-ray universe in 2010 to 2020, a century after Albert Einstein developed special and general relativity.

EXIST employs 2.7-m^2 CZT for each of its three telescopes, and images the sky in the 10 keV - 600 keV range with Tungsten coded masks with angular resolution of a few arcmin, and 10-50 arcsec source localization. In the 10 - 200 keV band EXIST will have a continuum sensitivity of 2 mCrab (each orbit), and 0.05 mCrab for a one year survey. Above 200 keV the per-orbit sensitivity would be 20 mCrab, and 0.5 mCrab for a 1 year survey. EXIST would complement observations with GLAST, which will study sources in the 20 MeV - 300 GeV regime [1], and ground-based VHE/UHE instrumentation, such as HESS, MAGIC, and VERITAS. The development of new detector technology is launching gamma-ray astronomy into an exciting new era of breakthroughs in the MeV to TeV regime [2].

GRBs are associated with host galaxies at large redshifts [3, 4]. The red shift record is z = 4.5 for GRB000131 [5]. In at least two cases evidence exists for supernova

FIGURE 1. The current reference design of EXIST.

associations: GRB 980425 and SN1998bw at z = 0.008 and GRB030329-SN2003dh [6] at z = 0.169. The spectra of these two supernovae are very similar and characteristic of Type Ic hypernovae, such as SN1997ef [7, 8, 9]. The emerging paradigm is a link of massive stars with (long) GRBs, consistent with the predictions of the collapsar model [11]. In this scenario the collapse of a rotating, massive star to a black hole ultimately drives relativistic outflows through a collapsing envelope (e.g., [12] producing a GRB-jet after the breakout, and a subsequent supernova. Under some circumstances the formation of a stellar mass black hole is thus heralded by a burst of gamma-rays, followed by a decaying afterglow from X-rays to radio [3]. GRBs thus provide a unique opportunity to trace the process of star formation to large, cosmological distances. Theoretical studies [13, 14, 15] indicate that the first generation of stars may have formed at z > 10, re-ionizing the universe, perhaps with a bias towards more massive stars due to the low metallicity in the early universe. Observational support for the emergence of the first stars in the red shift range z∼10-30 (100 to 400 Myrs after the bigbang) comes from the polarization of the CMB [16]. Black holes from these population III stars may have played a significant role in seeding AGN cores. Even at such large distances it is easy to detect a red shifted GRB. A sensitivity of 1 mCrab in the hard X-ray band is sufficient to detect GRBs like GRB 990123 at essentially all realistic distances. Based on the assumption that the GRB rate traces the cosmic star formation rate, we estimate that 7% of all GRBs detected by EXIST originate at z > 10 (Fig. 2) To enable ground-based and space-based (e.g., NGST) studies of these bursts, EXIST provides 10"-50" locations in near real time. The total burst rate will be a few per day. EXIST would serve as the primary GRB observatory in the post-Swift era [17], during a time when sensitive neutrino detectors (IceCube) and gravity wave detectors (LISA, LIGO-II) may be able to detect the high-energy (> PeV) neutrino and gravity wave signature of GRBs.

Detection of the first generation of stars may require GRBs as a tracer, as galaxies beginning to form stars are faint. Spectroscopy of host galaxies will require observations with large aperture telescopes [15], and are likely to revolutionize our ability to trace

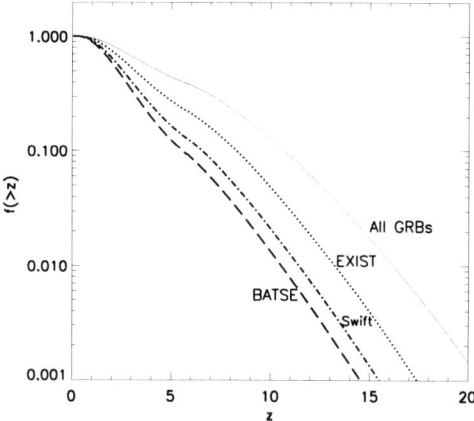

FIGURE 2. The estimated fraction of GRBs with red shift greater than z. Here, "all GRBs" is based on the assumption that GRBs trace the cosmic SFR. Estimates were made for EXIST, BATSE, and Swift.

and probe structure formation and galaxy evolution to the earliest times. GRBs provide signposts along "the road to galaxy formation" [18], and shed light on the crucial question of halo formation via a more or less smooth collapse or successive mergers of independently evolving proto-galactic fragments. The star formation rate during halo collapse is sensitive to stellar feedback (e.g., [19, 20]). Thus high-z GRBs could improve our understanding of a key ingredient in galaxy formation and evolution.

The star formation rate and GRB rate are related, but not one-to-one. As metallicity decreases, the IMF changes due to changes in the cooling function of the gas. After formation, mass loss from stellar winds is a sensitive function of the metal content in the stellar envelope, affecting subsequent evolution and probabilities for creating the right conditions for GRBs [21]. GRBs provide a new tool for studies of early galaxy evolution, and may lead to the elusive Population III stars and their feedback on proto-galaxies [22].

EXIST will also detect many X-ray rich GRBs (the so called X-ray Flashes, [23, 24]. This new burst phenomenon is in many ways similar to classical long duration GRBs, but, as the name indicates, characterized by a lower energy of the bulk of the emission. While the peak in the νF_ν spectrum of GRBs is clustered around 200 keV, the currently small sample of XRFs suggests a peak value of ~50 keV. The nature of XRFs is not yet clear. XRFs could be highly red shifted GRBs, but observations of the host galaxies of XRF 011030 and XRF 020427 [25] argue against that interpretation.

REFERENCES

1. Gehrels, N. & Michelson, (1999).
2. Schonfelder (2001).
3. van Paradijs, J., Kouveliotou, C., & Wijers, R. A. M. J., ARAA **38**, 379 (2002).
4. Meszaros, P., Ann. Rev. Astr. & Astrophys. **40**, 137 (2002).

5. Andersen, M. I. et al., A&A **364**, L54 (2000).
6. Hjorth, J., et al., Nature **423**, 847 (2003).
7. Woosley, S. E., Eastman, R., & Schmidt, B. P., ApJ **516**, 788 (1999).
8. Branch, D., in "Supernovae & Gamma-Ray Bursts" (Cambridge U. Press), p. 96 (2001).
9. Nomoto (2001).
10. Kawabato (2003).
11. Woosley, S. E., ApJ **405**, 273 (1993).
12. MacFadyen, A., & Woosley, S. E., ApJ **524**, 262 (1999).
13. Abel, T., Bryan, G. L., & Norman, M. L., Science **295**, 93 (2002).
14. Bromm, V., and Loeb, A., ApJ **575**, 111 (2002).
15. Loeb, A., astro-ph/0307231 (2003).
16. Kogut, A., New Astronomy Reviews astro-ph/0306048 (2003).
17. Grindlay, J. et al., astro-ph/0205323 (2002).
18. Keel, W. C., "The Road to Galaxy Formation""(Springer) (2002).
19. Hartmann, D. H., Myers, J., & The, L.-S., in "From Twilight to Highlight", eds. W. Hillebrandt & B. Leibundgut (Springer), p. 384 (2003).
20. Marri, S., & White, S. D. M., astro-ph/0207448 (2002).
21. Heger, A., et al., ApJ **591**, 288 (2003).
22. Barkana, R. & Loeb, A., Phys. Rep. **349**, 125 (2001).
23. Heise, J., in Gamma-Ray Burst and Afterglow Astronomy eds. G. R. Ricker & R. K. Vanderspek, AIP **662**, 229 (2003).
24. Heise, J., Nature, submitted (2003).
25. Bloom, J. S., et al., astro-ph/0303514 (2003).

GLAST and Gamma-Ray Bursts: Probing Photon Propagation over Cosmological Distances

Nicola Omodei*, J. Cohen-Tanugi[†] and Francesco Longo**

Department of physics, University of Siena and INFN, sezione di Pisa, Italy
[†]*INFN, sezione di Pisa, Italy*
**Department of physics, University of Trieste and INFN, sezione di Trieste, Italy*

Abstract. Theoretical models, especially within the framework of Quantum Gravity, allow for the possibility of a velocity dispersion effect for photons of different energies traveling cosmological distances. Due to their fine-scale time structure and their broad band emission, that is accurately modeled in this work, Gamma Ray Bursts could be good probes of such effect. GLAST will detect several GRBs per year at high energy where the effect could be detectable.

INTRODUCTION

GLAST is a next generation high-energy gamma ray observatory, onboard a satellite scheduled for launch in 2006. It consists of a Large Area Telescope (LAT, see e.g[1]) and a GLAST Burst Monitor (GBM, see e.g.[2]), and is designed for making observations of gamma-ray sources in the 10 keV to 300 GeV energy range.

The LAT is a pair conversion telescope, operating from 10 MeV to more than 300 GeV. It is composed of three subsystems: the Anti-Coincidence Detector (ACD) is responsible for vetoing the enormous charged cosmic-ray background and the Earth albedo secondary electrons and nuclei. The tracker (TKR) consists of a four-by-four array of tower modules, each of which being composed of 19 pairs of interleaved planes of silicon strip detectors and tungsten converter sheets. The silicon strip detectors track the electron positron pair created by the incident gamma. The calorimeter (CAL) is a segmented arrangement of CsI(Tl) bars, designed to give both longitudinal and transverse information on the energy deposition pattern.

The tracker will have the ability to determine the location of an object in the sky to within 0.5 to 5 arc minutes. The pair conversion signature is also used to help reject the background of charged cosmic rays. The GBM provides overlapping energy coverage in the range 10 keV to 25 MeV for bright transients such as bursts and solar flares. It is composed of 4 triplets of NaI(Tl) scintillators(low energy band) and 2 BGO scintillators (high energy band).

Due to its angular resolution (5 times better than EGRET) and its improved afterglow sensitivity, GLAST will extend the knowledge about GRB's. Its large field of view (about 2 sr) and its improved effective area (10 times EGRET's) will provide better burst statistics (up to 100 bursts/year with many more photons detected.)

GLAST AND GRB SIMULATIONS

The LAT team has set up a complete simulation chain, including generation of the incoming flux, full simulation of the detector response, reconstruction chain and analysis of the final trigger and alert. It is to be noted that the flux simulations include background (from albedo, cosmic and diffuse gamma ray events). Orbit, tilting and correct exposure are also taken into account.

An extension of this framework has been implemented for simulating transient sources such as Gamma-Ray Bursts (GRB). It can be used for studying the capability of GLAST for the observation of rapid transient fluxes in general.

The physics adopted is based on the fireball model of Gamma Ray Burst, for which a series of shells is injected in the circumburst medium with different Lorentz factors [5]. When a faster shell reaches a slower one, a shock occurs, and an accelerated electron distribution is obtained due to the shock acceleration process. Some of the energy dissipated during the shock is converted into a randomly oriented magnetic field. The electrons can loose their energy via synchrotron emission. The characteristic synchrotron spectrum is boosted (and beamed) thanks to the Lorentz factor of the emitting material. The higher energy part of a GRB spectrum can be obtained keeping into account the possibility of Compton scattering of the synchrotron photons against the electron accelerated by the shock (Inverse Compton Scattering) [3]. Figure 1 shows a light curve generated with this model.

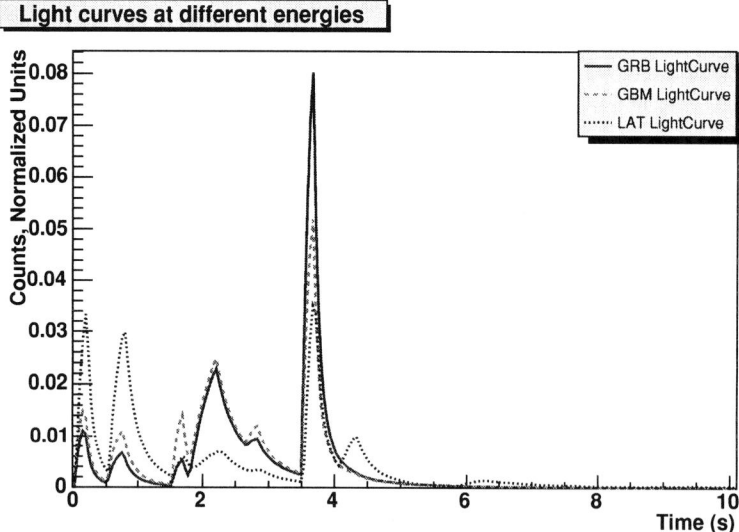

FIGURE 1. An example of GRB light curve from the GLAST simulator. Normalized light curves in the LAT and GBM energy bands are shown.

The physical model uses the BATSE bimodal distribution of GRB duration as input parameter and calculates the expected fluence[4]. In this way we have a calibration method to correlate the emission at high energy with the low energy observations.

GLAST AND QUANTUM GRAVITY

On the basis of recent quantum-gravity results, the possibility of space experiments to search for quantum properties of spacetime has been recently proposed[6].

In quantum-gravity phenomenology[6], one is looking for the small effects predicted by quantum-gravity theories, effects with magnitude set by the ratio between the energy of the particles involved and the huge Planck energy scale ($E_P \sim 10^{28}$) eV.

In several approaches to the quantum-gravity problem one finds some evidence of departures from ordinary Lorentz symmetry with the possible emergence of Planck-scale-modified dispersion relations:

$$E^2 = m^2 + \vec{p}^2 + f(\vec{p}^2, E, E_p), \tag{1}$$

Such a deformation in the dispersion relation leads to a small energy dependence of the speed of photons found by applying the relation $v = \frac{dE}{dp}$. An energy dependence of the speed of photons could be significant in the analysis of GRB coming from cosmological distance. Two photons with energy difference of order 100 MeV, over such a distance, will be detected with a relative time-delay that is of order of $t \sim 10^{-3}$ s Therefore such a quantum-gravity-induced time-of arrival delay could be detected by GLAST upon comparison of the structure of the signal in different energy channels[7].

To investigate such effect, a detailed simulation of the time structure of the GRB is necessary. The model presented in this paper is able to provide the light curve of GRB at different energy ranges. However, since the intrinsic time lag could mimic the expected phenomenon due to quantum gravity, the final analysis will need to correlate the measured delay between different energy bands with the measured source distance. Using a certain sample of GRBs observed by GLAST and with measured redshift, this effect could be tested. Some preliminary results of this analysis where the effects of quantum gravity are included in the photon propagation show that the LAT sensitivity permits to reconstruct the time structures of GRBs at high energies, allowing to measure the delay induced by the quantum space time structure[8].

REFERENCES

1. Bellazzini, R., et al., Nucl. Phys. Proc. Suppl. **113**, 303 (2002).
2. Kippen, R. M. et al., in AIP Conf. Proc., 587, GAMMA 2001, eds. S. Ritz, N. Gehrels, & C. R. Shrader, 801 (2001).
3. Omodei O. and Cohen-Tanugi J., http://www.pi.infn.it/~omodei/GRBSpectrum/
4. Cohen-Tanugi, J., et al., contribution for "Gamma Ray Bursts in the afterglow era - Third workshop, Rome, (2002) astro-ph/0301356.
5. Piran, T., Phys. Rep. **314**, 575 (1999).
6. Amelino-Camelia, G., To appear in a special issue "Fundamental physics on the International Space Station and in space" of General Relativity and Gravitation, (2003) astro-ph/0309174.
7. Norris, J.P, et al. (1999) astro-ph/9912136.
8. Cohen-Tanugi, J., Omodei, N., Longo, F., poster presented at the XXI Symposium on Relativistic Astrophysics (2002).

The GLAST Burst Monitor

P. N. Bhat[*], C. A. Meegan[†], G. G. Lichti[**], M. S. Briggs[*], V. Connaughton[*], R. Diehl[**], G. J. Fishman[†], J. Greiner[**], R. M. Kippen[‡], C. Kouveliotou[†§], W. S. Paciesas[*], R. D. Preece[*], V. Schönfelder[**], R. B. Wilson[†] and A. von Kienlin[**]

[*]*University of Alabama in Huntsville*
[†]*NASA Marshall Space Flight Center*
[**]*Max-Planck-Institut für extraterrestrische Physik*
[‡]*Los Alamos National Laboratory*
[§]*Universities Space Research Association*

Abstract. The Gamma Ray Large Area Space Telescope (GLAST) mission is a followup to the successful EGRET experiment onboard the Compton Gamma Ray Observatory (CGRO). It will provide a high-sensitivity survey of the sky in high-energy γ-rays, and will perform detailed observations of persistent and transient sources. There are two experiments onboard the GLAST - the Large Area Telescope (LAT) and the GLAST Burst Monitor (GBM).

The primary mission of the GBM instrument is to support the LAT in observing γ-ray bursts (GRBs) by providing low-energy measurements with high time resolution and rapid burst locations over a large field-of-view (≥ 8 sr). The GBM will complement the LAT measurements by observing GRBs in the energy range 10 keV to 30 MeV, the region of the spectral turnover in most GRBs. An important objective of the GBM is to compute the locations of GRB sources on-board the spacecraft and quickly communicate them to the LAT and to the ground to allow rapid followup observations. This information may be used to re-point the LAT towards particularly interesting burst sources that occurred outside its field-of-view.

The GBM consists of 14 uncollimated scintillation detectors coupled to phototubes to measure γ-ray energies and time profiles. Two types of detectors are used to obtain spectral information over a wide energy range: 12 NaI($T\ell$) detectors (10 keV to 1 MeV), and 2 BGO detectors (150 keV to 30 MeV). The detectors are distributed around the GLAST spacecraft to provide a large, unobstructed field of view. The 12 NaI($T\ell$) detectors are mounted with different orientations for use in locating GRB sources.

INTRODUCTION

Gamma Ray Bursts (GRBs), are unique in the sense that they release most of their energy as photons with energies in the range 30 keV to a few MeV [1]. Observations of GRB afterglows in the optical wavelengths have revealed the cosmological origin of these enigmatic explosions. The redshifts of about 35 GRB counterparts are now measured, confirming their enormous energy scale [2, 3, 4]. Models of the prompt emission powering these energetic events remain highly speculative. However, the optical afterglow observations of the recent GRB030329 have led to the conclusion that its origin is linked to a supernova explosion (SN2003dh), providing tantalizing observational clues pointing to supernovae (SNe) as a mechanism (through core collapse) for producing GRBs [5, 6, 7, 8, 9]. There are several emission models such as the relativistic shock model

[10] in which the prompt and afterglow emissions correspond to synchrotron radiation from shock accelerated electrons. In order to discriminate between various models for the prompt γ-ray emission more observational parameters are needed.

GRB energy spectra have been measured from $\sim$ 2 keV [11] to 18 GeV [12] with no evidence for spectral cutoff at high energies in many bursts. It is interesting to note that the GeV photons are delayed by as much as 1.5 hr with respect to the GRB trigger in GRB940217 [12]. There are bursts showing a distinct higher energy component (*e.g.* GRB941017 [13]) which could be related to relativistic electrons either through synchrotron emission or their interaction with the surrounding cloud. Similarly, there are a significant number of *X-ray rich* GRBs as well as *X-ray flashes* [14]. Even very high energy emission, in the range of GeV-TeV, is expected theoretically from inverse Compton scattering of electrons in external shocks [15] as well as from internal shocks in the prompt phase [16]. Hence there is an obvious need to study the GRB spectra over an extended energy range in order to understand the emission mechanisms.

The primary objective of the Glast Burst Monitor (GBM) instrument is to support the Large Area Telescope (LAT) in observing GRBs by providing low-energy measurements with high time resolution and by providing rapid burst locations over a large field-of-view ($\geq$ 8 sr). The GBM will complement the LAT (energy range: 10 MeV to > 100 GeV) measurements by observing GRBs in the energy range 10 keV to 30 MeV, the region of the prominent spectral turnover of most GRBs. Another important objective of GBM is to compute the locations of GRB sources on-board the spacecraft and quickly communicate them to the LAT and to the ground to allow rapid followup observations. This information may be used to re-point the spacecraft towards particularly interesting bursts that occurred outside the LAT field of view.

THE ROLE OF GBM

The goal of the GBM, which is functionally similar to its predecessor BATSE, is to enhance the science return of the GLAST mission in the study of γ-ray bursts by providing low energy context measurements with high time resolution (2 μs). The LAT will provide ground-breaking new GRB observations while the GBM will enable their evaluation in the context of prior observations and current knowledge.

FUNCTIONAL HIGHLIGHTS OF GBM

- GBM will provide spectra of bursts from 10 keV to 30 MeV, extending the LAT high energy (10 MeV to > 100 GeV) measurements with 2 more familiar energy domains.
- GBM will provide a wide sky coverage ($\sim$ 8 sr) and will generate a GRB trigger and locations enabling autonomous re-pointing for interesting bright bursts outside the LAT field-of-view for high energy afterglow studies.
- GBM with the LAT will provide an energy range spanning > 7 decades (10 keV to > 100 GeV) for the first time.

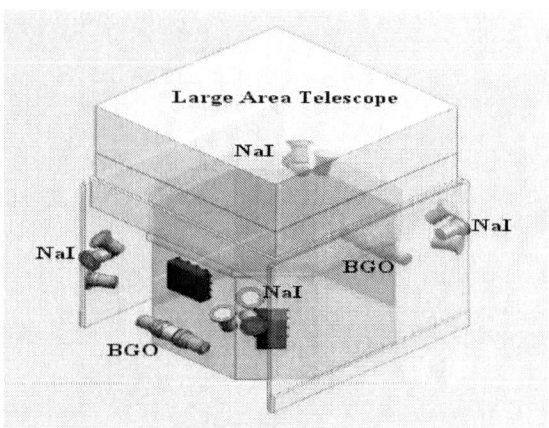

FIGURE 1. Placement of 12 NaI($T\ell$) and 2 BGO detectors onboard the satellite with respect to LAT, shown as the large structure above the spacecraft structure.

- GBM will provide rapid GRB locations for ground or space based quick followup observations and for further improvements in localization using the Swift or the inter-planetary network (IPN).

GBM DETECTOR DESIGN STRATEGY

GBM uses two types of uncollimated detectors: *viz.* NaI($T\ell$) to cover the energy range 10 keV to 1 MeV and BGO to cover 150 keV to 30 MeV. Each of the 12 NaI($T\ell$) detectors, having a thin Beryllium entrance window is 5″ diameter and 0.5″ thick; it is viewed by a 5″ phototube. Each of the two BGO detectors consists of a 5″ diameter and a 5″ thick BGO crystal viewed by two 5″ phototubes. The performance goals of these detectors are summarized in Table 1. The 12 NaI($T\ell$) detectors and 2 BGO detectors are placed as shown in Figure 1 with respect to the LAT.

PROSPECTS OF GBM ON THE GLAST MISSION

GRB Spectra: GBM has a very high time resolution (2 μs) and hence will provide time resolved spectra of GRBs in the energy range 10 keV - 30 MeV. This provides a unique advantage of studying the relation between the GRB spectra in keV - MeV - GeV energies. It will also enable us to explore whether those GRBs detected by the LAT form a separate class of bursts. Time resolved spectra will also allow us to study the temporal behavior & distribution of spectral parameters. Due to the large energy range available (10 keV to > 100 GeV) one can detect likely energy cut-offs in GRB spectra.

TABLE 1. GBM detector design capabilities

Parameter	Requirement	Goal	Current Capability
Energy Range	10 keV - 25 MeV	5 keV - 30 MeV	10 keV - 30 MeV
Energy Resolution	< 23.5% $FWHM$	<7%	~ 12% $FWHM$ 511 keV
On-board GRB locations	Within 2 s	15° within 1 s	< 15° within 1.8 s
Rapid ground GRB locations	5° accuracy (1 σ radius) within 5 s	3° within 1 s	TBD by analysis (scattering influenced)
Final GRB locations	3° accuracy (1 σ radius) within 1 day	No stated goal	TBD by analysis (scattering influenced)
GRB sensitivity (on ground)	0.5 photons cm^{-2} s^{-1} (peak flux 50-300 keV)	0.3 photons cm^{-2} s^{-1} (peak flux 50-300 keV)	0.35 photons cm^{-2} s^{-1} (peak flux 50-300 keV)
Field of View	8 sr	10 sr	~ 8.8 sr
Dead-time	< 10$\mu s/count$	< 3$\mu s/count$	~ 2.5$\mu s/count$

GRB Light-curves: The GBM can measure energy & time resolved light-curves to enable us to study possible spectral & temporal lags. It provides 8 channel spectra every 64 ms as well as 128 channel spectra every 2.048 s during a burst, for each type of detector. The time tagged event data (TTE) with a time resolution of 2 μs is available for > 300 s at a peak counting rate of 350 kHz during a burst. The TTE events also have a spectral resolution of 128 channels.

GRB Triggers: One would expect ~150 GRB triggers/year of which ~50-100 are expected to be observed by the LAT. Prior locations provided by the GBM will enable the LAT to detect weaker GRBs by limiting the search area, thus increasing its sensitivity to GRBs.

REFERENCES

1. van Paradijs, J., Kouveliotou, C. & Wijers, R. A. M. J., *ARA & A*, **38**, 379, 2000.
2. Metzger, M. R., *et al.*, *Nature*, **387**, 878, 1997.
3. K. Hurley, R. Sari and S. G. Djorgovski, *astro-ph/0211620*, 2003.
4. J. C. Greiner, http://www.mpe.mpg.de/~jcg/grbgen.html
5. Colgate, S. A., *Canadian Journal of Physics*, **46**, 476, 1968.
6. Woosley, S. E., *ApJ*, **405**, 273, 1993.
7. Woosley, S. E. and MacFadyen, *A & AS*, **138**, 499, 1999.
8. Hjorth, J., *et al.*, *Nature*, **423**, 847, 2003.
9. Stanek, K. Z., *et al.*, *ApJL*, **591**, L17, 2003.
10. Meszaros, P., *ARA & A*, **40**, 137, 2002
11. Frontera, F., *et al.*, *ApJS*, **127**, 59, 2000.
12. Hurley, K., *et al.*, *Nature*, **372**, 652, 1994.
13. Gonzalez, M. M. *et al.*,*Nature*, **424**, 749, 2003.
14. Lamb, D. Q. *et al.*, *astro-ph/0309462*, 2003.
15. Meszaros, P., Rees, M. J. and Papathanassiou, H., *ApJ*, **432**, 181, 1994.
16. Papathanassiou, H. and Meszaros, P., *ApJL*, **471**, L91, 1996.

GLAST's GBM Burst Trigger

D. Band*, M. Briggs†, V. Connaughton†, M. Kippen** and R. Preece†

Code 661, NASA/Goddard Space Flight Center, Greenbelt, MD 20771
†National Space Science and Technology Center, Huntsville, AL 35805
**NIS-2, Los Alamos National Laboratory, Los Alamos, NM 87545

Abstract. The GLAST Burst Monitor (GBM) will detect and localize bursts for the GLAST mission, and provide the spectral and temporal context in the traditional 10 keV to 25 MeV band for the high energy observations by the Large Area Telescope (LAT). The GBM will use traditional rate triggers in up to three energy bands, and on a variety of timescales between 16 ms and 16 s.

THE MISSION

The Gamma-ray Large Area Space Telescope (GLAST) is the next NASA general gamma-ray astrophysics mission, which is scheduled to be launched into low Earth orbit in February, 2007, for 5–10 years of operation. It will consist of two instruments: the Large Area Telescope (LAT) and the GLAST Burst Monitor (GBM). A product of a NASA/DOE/international collaboration, the LAT will be a pair conversion telescope covering the <20 MeV to >300 GeV energy band. The LAT will be ∼30 times more sensitive than *CGRO's* EGRET.

The GBM will detect and localize bursts, and extend GLAST's burst spectral sensitivity to the <10 keV to >25 MeV band. Consisting of 12 NaI(Tl) (10–1000 keV) and 2 BGO (0.15–25 MeV) detectors, the GBM will monitor >8 sr of the sky, including the LAT's field-of-view (FOV). Bursts will be localized to < 15° (1σ) by comparing the rates in different detectors.

During most of the mission GLAST will survey the sky by rocking ∼ 35° above and below the orbital plane around the zenith direction once per orbit. The first year will be devoted to a sky survey while the instrument teams calibrate their instruments. During subsequent years guest investigators may propose pointed observations, but continued survey mode is anticipated because it will usually be most efficient.

Both the GBM and the LAT will have burst triggers. When either instrument triggers, a notice with a preliminary localization will be sent to the ground through TDRSS and then disseminated by GCN. Additional data will be sent down through TDRSS for an improved rapid localization on the ground. "Final" positions will be calculated from the full downlinked data. All positions will be disseminated as GCN Notices, and additional information (e.g., fluences and durations) will be sent out as GCN Circulars.

Using its own and the GBM's observations, the LAT will decide autonomously whether to slew to the burst location for a 5 hour followup pointed observation. The threshold will be higher for GBM-detected bursts outside the LAT's FOV.

The GBM's NaI and BGO detectors will provide the number of counts detected in

8 energy bands every 16 ms; the GBM will trigger off these rates. Rate triggers test whether the increase in the number of counts in an energy band ΔE and time bin Δt is statistically significant; the expected number of non-burst photons in the ΔE–Δt bin is estimated by accumulating counts before the time bin being tested. Building on our experience with the BATSE trigger, we are performing trade studies to optimize the sensitivity of these triggers. Here we present the results of our studies focusing on the choice of ΔE and Δt.

CHOICE OF ΔT

We consider two Δt hierarchies—Δt spaced by factors of $\times 2$ (e.g., 16 ms, 32 ms, 64 ms...) or $\times 4$ (e.g., 16 ms, 64 ms, 256 ms...)—and three time bin registrations—non-overlapping bins (e.g., bins separated by 1024 ms for Δt=1024 ms bins), half-step bins (e.g., bins separated by 512 ms for Δt=1024 ms bins), and all possible bins (e.g., bins every 16 ms). Varying Δt can maximize the signal-to-noise ratio, while more time registrations permit the bin to be centered over the peak flux, also maximizing the signal-to-noise ratio. To test these 6 triggers we applied them to the 64 ms resolution lightcurves of the 25 brightest BATSE bursts; for each lightcurve we chose 10 starting times at random. Note that the GBM rates will have 16 ms resolution.

The most sensitive trigger would have Δt spaced by $\times 2$ and every possible time bin. The next most sensitive trigger would have Δt spaced by $\times 2$ and bins spaced every half step. These triggers would test different numbers of bins—11264 bins vs. 3070 bins tested every 16.384 s.

Besides the increased computational burden, the risk of a false trigger increases as the number of bins tested increases, but the false trigger probability is not proportional to the number of bins because the bins are not independent. Our simulations indicate this is a <5% effect—for the same false trigger rate the trigger threshold should be raised by a few percent as the number of time bins tested increases.

CHOICE OF ΔE

Triggering on the counts accumulated in different ΔE can tailor the detector sensitivity to hard or soft bursts. The GBM will be able to trigger on more than one ΔE, and therefore we seek the set that will maximize the sensitivity for both hard and soft bursts, although hard bursts are a priority since their spectra are more likely to extend into the LAT's energy band (of course one must be careful of one's theoretical prejudices). For the study of detector sensitivity to different types of bursts and for comparisons between detectors, the F_T-E_p plane is useful[1], where F_T is the peak photon flux in a fiducial energy band (here 1–1000 keV) and E_p is the energy of the peak of $E^2 N(E) \propto \nu f_\nu$. Burst spectra are also characterized by asymptotic low and high energy power laws with spectral indices α and β, respectively; the dependence of F_T on α and β is not as great as on E_p. For a given set of spectral indices the detector sensitivity (the threshold value of F_T at a give E_p) is a curve in the F_T-E_p plane.

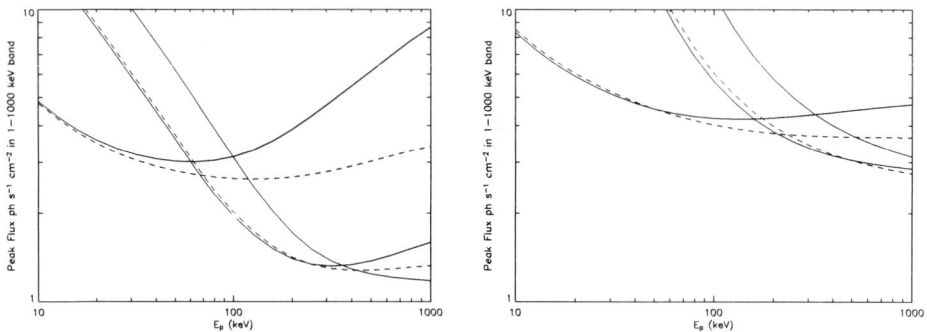

FIGURE 1. Sensitivity for two sets of ΔE for $\alpha=0$, $\beta=-2$ (left plot) and $\alpha=-1$, $\beta=-25$ (right plot). The solid curves are for (left to right within plot) ΔE=5–100, 50–300 and 100–1000 keV and the dashed for ΔE=5–1000 and 50–1000 keV. Lower curves are more sensitive.

To calculate these sensitivity curves we need both the number of counts a detector will detect in the nominal ΔE band for a given burst spectrum and the number of background counts in this ΔE. A code has been developed to calculate these numbers for each GBM detector for a burst in any direction relative to the spacecraft. Currently the code uses response matrices for the flux directly incident on the detectors (without scattering off the spacecraft or the Earth, but with obscuration by other parts of the observatory), and a model of the background on orbit. We used this code to calculate the sensitivity along the normal to the LAT for Δt=1.024 s assuming at least two detectors trigger at $\sigma_0 > 5.5$.

We calculated the sensitivity curves for a variety of ΔE. The extremes of our spectral index sets were α=0, $\beta=-2$ and $\alpha=-1$, $\beta=-25$. The first set is similar to the spectra sometimes observed early in a burst; its high energy tail would be more easily detected by the LAT. The second set is a spectrum without a high energy tail.

Figure 1 shows the sensitivity for two sets of ΔE. To compare the GBM and BATSE burst distributions we would like to include ΔE=50–300 keV, which was BATSE's primary trigger band. As can be seen, ΔE with a low energy cutoff of $\sim$50 keV is optimal for high energy sensitivity because it does not include the large low energy background. Conversely, ΔE should always extend to the highest energy possible because of the low background at high energy.

We find that triggering a single BGO detector with $\sigma_0 = 8$ increases the sensitivity for $E_p > 1$ MeV for α=0, $\beta=-2$. There is no increase in sensitivity for $\alpha = -1$.

Figure 2 compares the Δt=1 s sensitivity for the GBM (solid) and BATSE (dot-dashed) with the burst intensity (dashed) that when the spectrum is extrapolated to the LAT energy band would result in 25 detected photons per second. The burst is assumed to be on the LAT normal, $\alpha = -1$, $\beta = -2$, and the GBM triggers on ΔE=5–100 and 50–300 keV.

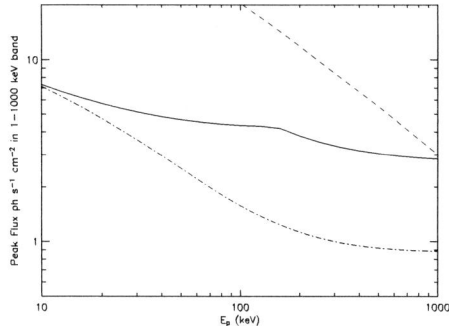

FIGURE 2. Comparison of GBM (solid curve) and BATSE (dot-dashed) sensitivities, and the flux necessary for a LAT detection of 25 photons (dashed).

CONCLUSIONS

The Δt calculations suggest that spacing Δt by factors of $\times 2$ (i.e., 16 ms, 32 ms, 64 ms...) and staggering the bins by half a timestep (e.g., the 1024 ms bins are accumulated every 512 ms) would be particularly efficient given the number of time bins that would be tested. Choosing two triggers with ΔE starting at 5 keV and 50 keV and extending to the detector's high energy cutoff would provide good high and low energy sensitivity. Using ΔE=50–300 keV would reproduce the BATSE trigger, but would reduce the $E_p >$500 keV sensitivity for the hardest bursts (which are more likely to have LAT flux); this can be mitigated by adding ΔE=100–1000 keV.

Future trade studies will focus on the methodology for measuring the background. GLAST will slew more frequently than *CGRO* did, and therefore the background rates in GLAST's different detectors will probably vary more rapidly. We will need to suppress spurious triggers resulting from variations in the background without a major reduction in the GBM's sensitivity.

Ultimately the trigger design will be constrained by the computational capabilities of the GBM's processor.

REFERENCES

1. Band, D., *Burst Populations and Detector Sensitivity*, in these proceedings (2004).

Analysis of Burst Observations by GLAST's LAT Detector

David L. Band*, Seth W. Digel†, the GLAST LAT Collaboration** and the GLAST Science Support Center**

Code 661, NASA/Goddard Space Flight Center, Greenbelt, MD 20771
†*Stanford University, Stanford, CA*
**GSSC*

Abstract. Analyzing data from GLAST's Large Area Telescope (LAT) will require sophisticated techniques. The PSF and effective area are functions of both photon energy and the position in the field-of-view. During most of the mission the observatory will survey the sky continuously, and thus, the LAT will detect each count from a source at a different detector orientation; each count requires its own response function! The likelihood as a function of celestial position and photon energy will be the foundation of the standard analysis techniques. However, the 20 MeV–300 GeV emission at the time of the $\sim$100 keV burst emission (timescale of $\sim$10 s) can be isolated and analyzed because essentially no non-burst counts are expected within a PSF radius of the burst location during the burst. Both binned and unbinned (in energy) spectral fitting will be possible. Longer timescale afterglow emission will require the likelihood analysis that will be used for persistent sources.

INTRODUCTION

The detection of the high energy emission from gamma-ray bursts is anticipated to be one of the spectacular observations by the Gamma-ray Large Area Space Telescope (GLAST), NASA's next general gamma-ray astrophysics mission. Scheduled to be launched into low Earth orbit in February, 2007, for 5–10 years of operation, GLAST will consist of two instruments: the Large Area Telescope (LAT) and the GLAST Burst Monitor (GBM).

A product of a NASA/DOE/international collaboration, the LAT builds on the success of *CGRO*'s EGRET. The LAT will be a pair conversion telescope: gamma rays will pair-produce in tungsten foils; silicon strip detectors will track the resulting pairs; the resulting particles will deposit energy in a CsI calorimeter; and an anticoincidence detector will veto charged particles. The anticoincidence detector will be segmented to limit the self-vetoing that plagued EGRET. The LAT will be 1.8 m×1.8 m×1m, and weigh $\sim$3000 kg.

The astrophysical photons will be only a small fraction of the total number of events detected by the LAT, most of which will result from charged particles. On board filtering of the events will reduce the $\sim$4 kHz trigger rate to the $\sim$30 Hz event rate that can be downlinked to the ground; ground processing will result in a $\sim$2 Hz photon rate.

The salient detector characteristics are: energy range of <20 MeV to >300 GeV; 1–10 GeV effective area of >8000 cm^2 with half maximum at 55°; angular resolution of < 3.5° at 100 MeV, < 0.15° at 10 GeV; field-of-view of >2 sr; deadtime $\sim$ 20μs per event (current, $< 100\mu$s required); and time resolution of $\sim 2\mu$s.

A descendant of CGRO's BATSE, the GBM will detect gamma-ray bursts and extend GLAST's burst spectral sensitivity to the <10 keV to >25 MeV band. Consisting of 12 NaI(Tl) (10–1000 keV) and 2 BGO (0.15–25 MeV) detectors, the GBM will monitor >8 sr of the sky, including the LAT's field-of-view (FOV). Bursts will be localized to $< 15°$ (1σ) by comparing the rates in different detectors.

Typically GLAST will survey the sky continuously. After a ~ 60 day checkout phase, GLAST will undertake a one year sky survey while the LAT team calibrates the instrument. In survey mode GLAST will rock $\sim 35°$ above and below the orbital plane about the zenith direction once per orbit. While pointed observations proposed by guest investigators will be feasible during subsequent years, continued survey mode operation will usually be most efficient, and is expected to predominate. Therefore most persistent sources will be observed at a variety of detector orientations; each count will be characterized by a different response function.

Both the GBM and the LAT will have burst triggers. The GBM will notify the LAT when it triggers. When either instrument triggers, a notice with a preliminary localization will be sent immediately to the ground through TDRSS and will then be disseminated by GCN. Additional data will be downlinked through TDRSS for an improved localization at the Mission Operations Center. Both Instrument Operations Centers will calculate "final" positions from the full downlinked data. GCN will disseminate all positions.

The LAT will determine whether the burst was intense enough for a followup 5 hour pointed observation at the burst location (interrupted by earth occultations). The threshold will be higher for bursts the GBM detected outside the LAT's FOV.

STANDARD SOURCE ANALYSIS

The LAT PSF is large ($\sim 3.5°$) at low energy (~ 100 MeV), small ($< 0.15°$) at high energy (~ 10 GeV). With the LAT's large effective area, many sources will be detected; their PSFs will merge at low energy. Therefore the analysis must be 3 dimensional—2 spatial and 1 spectral—and time is an additional dimension for variable sources. Diffuse emission underlies the point sources. For a typical analysis the source model must include: all point sources within a few PSF lengths of the region of interest; extended sources (e.g., supernova remnants); spatially variable diffuse Galactic emission (which must be modeled); and isotropic extragalactic emission. Sources are defined by positions, spectra, and perhaps time histories. Initial values may be extracted from the point source catalog the LAT team will compile. Consequently the source model will have many parameters. In an analysis some will be fitted, some will be fixed.

The instrument response (PSF, effective area, energy resolution) will at least be a function of energy and angle to the LAT normal; other parameters may be relevant such as the azimuthal angle around the LAT normal or the conversion layer (the front or back of the LAT). Since the LAT will usually survey the sky, a source will be observed at different instrument orientations. Each count will be characterized by many observables, and therefore a very large data space results. Even with 10^5 counts, this data space will be sparsely populated. Note that what high energy astrophysicists call a "count" is a "photon" to some particle physicists.

As was the case for EGRET[1] and earlier gamma-ray missions[2], likelihoods will be the foundation of our analyses (e.g., detecting sources, determining source intensities, fitting spectral parameters, setting upper limits). The likelihood is the probability of the data (the counts that were detected) given the model (the photon sources). The data consist of both the counts that were detected, and the regions of data space where counts were not observed. The calculation of the likelihood will be difficult because many counts will sparsely populate an enormous data space.

The likelihood will be calculated many times as the source model is changed (for example in fitting source parameters), and factors that are not model-dependent should be calculated once for a given analysis. Many of these quantities will have units of "exposure" (area×time).

BURST SPECTRAL ANALYSIS

The duration of the $\sim$100 keV burst emission is (relatively) short—at most 10's of seconds. Therefore, the LAT's pointing will not change significantly during the burst, and all the counts can be treated as having one response function. Within a PSF radius of the burst position less than one non-burst count per minute is expected: [$\sim$2 Hz cts over the FOV] / [2 sr FOV] × [$\pi(3.5°\pi/180)^2$ sr within PSF radius] = $\sim$0.01 Hz cts within a PSF radius or 0.7 cts/minute within a PSF radius. Therefore, we can treat all counts within 1–2 PSF radii as burst photons.

Since a) all the counts within a PSF radius of the burst originated in the burst, and b) all the counts have the same response function, multi-source spatial analysis is unnecessary for spectral analysis! Spatial analysis might be necessary for localizing the burst. All the counts within a PSF radius and within a time range can be binned into a count spectrum (apparent energy is the single dimension), and traditional spectral analysis can be applied to the resulting series of LAT count spectra. The GBM data (also a list of counts) can be binned with the same time binning, and then joint fits can be performed.

The afterglow will most likely produce a small number of counts accumulated over timescales of tens of minutes to hours. Thus, afterglow data must be analyzed with the general likelihood tool being developed for LAT data analysis.

The LAT team and the GLAST Science Support Center (GSSC) are developing a suite of tools to analyze both LAT and GBM data. These tools will use the HEAdas system supported on both Windows and LINUX platforms; most of the tools will be FTOOLS. Therefore the data will be in FITS files, and the tools will use IRAF-style parameter files. Here we describe the methodology for burst spectral analysis.

Extract LAT Counts: The user will extract the LAT counts from a specified time and region (here a circle around the burst position) from a GSSC database. The web-based extraction tool will return a FITS file with the requested counts and a second FITS file describing the instrument's pointing and livetime during the burst. Users will have a tool to perform further selections.

Extract GBM Counts: The GSSC will provide GBM counts in a FITS file. Users will fit polynomial (in time) background models to data before and after the burst.

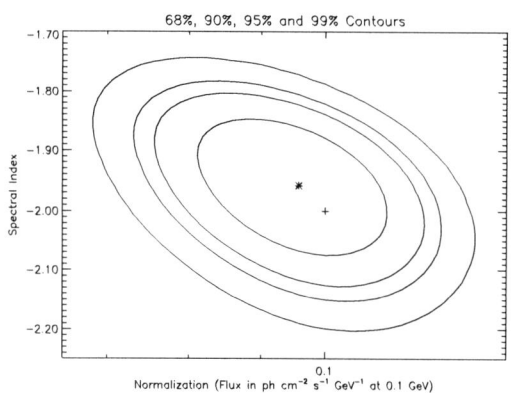

FIGURE 1. Confidence region for a sample power law LAT count spectrum with 115 counts. The asterisk marks the likelihood maximum and the cross the input parameters.

Bin Counts: An event binning tool will bin the LAT or GBM counts in both energy and time, resulting in a series of count spectra spanning the burst. The energy grid will be user specified. The time bins may be a) equally spaced, b) user specified, c) chosen to have equal signal-to-noise ratios in a user-specified energy band, or d) chosen by Bayesian Blocks[3]. The LAT counts are assumed to have no background, while the expected GBM background in each bin will be calculated from the background model.

DRM Generation: The LAT response function will be collapsed into a Detector Response Matrix (DRM), the product of the effective area at the position of the burst and the energy redistribution matrix. The effective area will account for the size of the region from which the counts were extracted. GBM DRMs will be supplied for each burst, and users will have a tool to calculate their own GBM DRMs.

Spectral Fitting—Binned Spectra: The spectra can be fit using XSPEC, with scripts automating the fitting of series of spectra. Spectra from the LAT, GBM and other missions (e.g., Swift) may be fit separately or jointly. Of course, the relative calibration of the different instruments will have to be understood. The XSPEC team is adding the capability of saving the results of the spectral fitting, along with the model spectra.

Spectral Fitting—Spectra Unbinned in Energy: For a burst with few counts a likelihood treatment using the standard likelihood tool may be more powerful; see Fig. 1. The likelihood function is for the apparent energy of LAT counts accumulated over a time period. Assuming the LAT spectrum is an extrapolation to higher energy of the GBM-observed spectrum, the GBM spectral fits can be used as priors on the LAT parameters.

REFERENCES

1. Mattox, J., et al., ApJ, **461**, 396, (1996).
2. Pollock, A., et al., A&A, **94**, 116, (1981).
3. Scargle, J., ApJ, **504**, 405, (1998).

SuperAGILE: The Hard X-ray Imager of AGILE

M. Feroci*, E. Costa*, L. Barbanera*, E. Del Monte*, G. Di Persio*, M. Frutti*, I. Lapshov*, F. Lazzarotto*, M. Mastropietro*†, E. Morelli**, L. Pacciani*, G. Porrovecchio*, B. Preger*, M. Rapisarda*, A. Rubini*, P. Soffitta*, M. Tavani*, A. Argan‡, G. Ghirlanda‡, S. Mereghetti‡, A. Pellizzoni‡, S. Vercellone‡, G. Barbiellini§, F. Longo§, M. Prest§ and E. Vallazza§

*Istituto di Astrofisica Spaziale e Fisica Cosmica - CNR-INAF - Roma
†Istituto di Metodologie Inorganiche e dei Plasmi - CNR - Montelibretti
**Istituto di Astrofisica Spaziale e Fisica Cosmica - CNR-INAF - Bologna
‡Istituto di Astrofisica Spaziale e Fisica Cosmica - CNR-INAF - Milano
§Istituto Nazionale di Fisica Nucleare - Sezione di Trieste

Abstract. SuperAGILE is the hard X-ray (10-40 keV) imager for the gamma-ray mission AGILE, currently scheduled for launch in mid-2005. It is based on 4 Si-microstrip detectors, with a total geometric area of 1444 cm^2 (max effective about 300 cm^2), equipped with one-dimensional coded masks. The 4 detectors are perpendicularly oriented, in order to provide pairs of orthogonal one-dimensional images of the X-ray sky. The field of view of each 1-D detector is 107°x68°, at zero response, with an overlap in the central 68°x68° area. The angular resolution on axis is 6 arcmin (pixel size). We present here the current status of the hardware development and the scientific potential for GRBs, for which an onboard trigger and imaging system will allow distributing locations through a fast communication telemetry link from AGILE to the ground.

INSTRUMENT PROPERTIES

SuperAGILE (SA) is the hard X-ray monitor for the gamma-ray space mission AGILE. It was designed with the constraints of using the same detectors and technology as the gamma-ray instrument, and being as light as possible to minimize the removal of high energy photons and the production of additional background for the gamma-ray imager.

The design of SA envisages a set of 4 Silicon microstrip detectors on a plane, oriented at 90 degrees in pairs, in such a way that two independent detectors continuously monitor the X-ray sky in two orthogonal directions. Each detector has a field of view limited to 107° x 68° by a collimator. The latter is composed of a carbon fiber (1 mm thick) structure, covered (for the purpose of X-ray shielding) with 100 micron thick Tungsten sheets. Each aperture of the collimator is covered with a 1-D coded mask, with the same dimension as the corresponding detector (190 mm x 190 mm). The coded mask, derived by a Hadamard sequence of 787 elements, is oriented in an antisymmetric configuration for the two homologous detectors in order to compensate for systematic effects related to the orientation of the individual detector.

The Silicon detectors are equipped with the SuperAGILE Front-End Electronics (SAFEE), that is based on the XAA1.2 ASIC chips. The SAFEE is hosted in part on

TABLE 1. Expected Instrumental Characteristics

Detector Type	410 μm thick Silicon Microstrip (1-D)
Detector Pitch	121 μm
Geometric Area	1444 cm^2
Energy Range	10-40 keV
On-axis Effective Area	$\sim$300 cm^2 (peak, @13 keV)
Energy Resolution	$\sim$5 keV FWHM
Timing Resolution	5 μs
Field of View (FWZR)	2 x (107° x 68°) crossed
Mask Transparency	50%
Focal Distance	140 mm
Nominal Mask Element	242 μm
On-axis Angular Resolution	6 arcmin (pixel size)
Source Location Accuracy	$\sim$2-3 arcmin for bright sources
Data Transmission	Photon by Photon

the detection plane and in part placed on the collimator wall through a flex multilayer printed circuit board. The SAFEE is powered, configured and read-out through the SuperAGILE Interface Electronics (SAIE), interfacing SA with the Payload Data Handling Unit and Power Supplies. The data transmission is photon by photon, with an expected telemetry load of about 20 kbit/s, mostly due to the contribution of the diffuse X-ray background, caused by the large field of view. A number of scientific ratemeters and housekeeping are also continuously transmitted into telemetry, with a time resolution going from 0.5 s to 16 s. The basic numbers of SA are summarized in Table 1.

STATUS OF HARDWARE

The AGILE mission is in its Phase C, approaching the Critical Design Review, planned by the end of 2003. A Simplified Engineering/Qualification Model (SEM) has been built for most of the Payload. For SuperAGILE it consists of a full-scale detection plane, composed of 2 real detectors (complete of their SAFEEs) and 2 dummy detectors. The mask-collimator assembly is instead complete, in a flight-like configuration. The SEM of the SAIE is mechanically representative but it drives only 1 SAFEE instead of 2. The SA SEM has been extensively electrically tested in the laboratory, and is currently undergoing thermal-vacuum and vibration tests. The XAA1.2 ASIC chips were successfully tested with respect to their properties in terms of radiation dose, single event upset and latch-up at the ions accelerator of INFN in Legnaro (Del Monte et al., [1]).

POTENTIAL FOR GAMMA RAY BURSTS

The SuperAGILE data will be fed in real time into the Payload Data Handling Unit, where they will be screened for counting excesses, based on an onboard GRB trigger algorithm. Due to the strong limitations in the onboard computing power, only a simple trigger algorithm will be applied: the counting rates in 2 energy bands and 4 detection

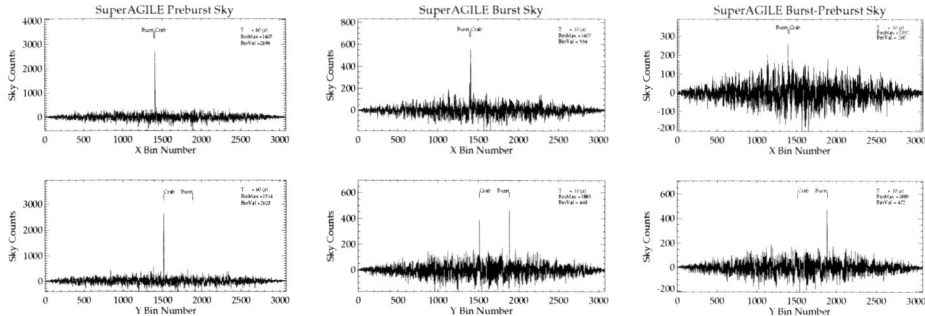

FIGURE 1. Simulation of a GRB in the Field of View of SuperAGILE shown in Figure 2. The three panels show the Sky Images, in the two orthogonal directions, *before* (left) and *during* (center) the GRB, and the the Net Sky Image (that is, *during* minus *before*, after exposure time normalization).

regions are simultaneously checked against background on timescales from 1 ms to 65 s. When an onboard trigger is declared, based on a configured combination of the analysis of the above counting rates, an onboard burst imaging procedure is started.

The onboard imaging procedure includes the Correction of the event position for the attitude drift, the integration of Background and Foreground Images (the latter accumulated over an optimal integration time, dynamically determined) for the 4 detectors, the Net Image determination (see Figure 1), the Correlation with the mask pattern, the sum of the 2 pairs of homologous Images, the determination of the peak position, the conversion into RA and Dec coordinates, the set-up of the Burst Telemetry Packet (including the basic GRB information, such as positions, trigger time scale, peak amplitude, etc.). The latter will then be sent down to ground, with a typical delay between few seconds and 2 minutes, through the ORBCOMM constellation of telecommunication satellites. After a prompt validation procedure on ground, the GRB position will be distributed via the internet to all the potential users, both ground and space based.

CONCLUSIONS

Thanks to its wide field of view, SuperAGILE will have a good chance of detecting and localizing serendipitous GRBs. The localization accuracy will depend on the event intensity, spectrum and position in the SA field of view, going from a 2-3 arcmin radius (bright events not too far off-axis) to several arcmin radius for weak and/or largely off-axis events. We computed the number of GRBs expected to be localized in both directions every year. They are shown in Figure 3 ("X and Y Trigger") together with the events localized in either one of the two directions ("X+Y Trigger"). They are approximately 1-2 events/months. This number, although similar of that of the Wide Field Cameras onboard BeppoSAX, will not be extraordinary in the Swift epoch. However, the SA events will have the unique property of being characterized also in the 100 MeV - 1 GeV range, thank to the combined use of all instruments on AGILE. In addition, the coordinates of the brightest events will be promptly distributed to the

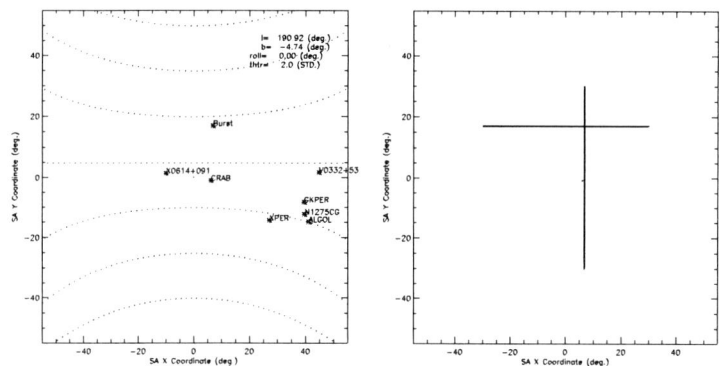

FIGURE 2. The sky location where the burst was simulated (left panel), together with the other sources included in the simulation, and the error box for the burst that could be derived (right panel). The highest-probability position reconstruction is given by the intersection of the two 1-D error boxes.

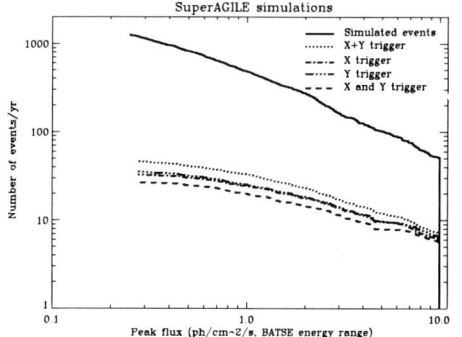

FIGURE 3. The expected number of GRB/year, with 1-D or 2x 1-D localizations.

community, including Swift itself, thus allowing a prompt observation of the X-ray afterglow.

We thank all the members of the AGILE collaboration and the the industrial partners cooperating in making AGILE a space mission for Astronomy. AGILE is a project funded by ASI, CNR/INAF, INFN and CIFS.

REFERENCES

1. Del Monte, E., et al., Nuclear Instruments and Methods A, *submitted* (2003).

The Test Equipment of the AGILE Minicalorimeter Prototype

M. Trifoglio*, A. Bulgarelli*, F. Gianotti*, E. Celesti*, G. Di Cocco*, C. Labanti*, A. Mauri*, M. Prest[†**], E. Vallazza** and T. Froysland[‡]

*Istituto di Astrofisica Spaziale e Fisica Cosmica sez. Bologna, CNR, Bologna, Italy
[†]Univ. della Insubria, Como, Italy
**Istituto Nazionale di Fisica Nucleare, sez. Trieste, Trieste, Italy
[‡]Istituto Nazionale di Fisica Nucleare, sez. Roma2, Roma, Italy

Abstract. AGILE is an ASI (Italian Space Agency) Small Space Mission for high energy astrophysics in the range 30 MeV - 50 GeV. The AGILE satellite is currently in the C phase and is planned to be launched in 2005. The Payload shall consist of a Tungsten-Silicon Tracker, a CsI Minicalorimeter, an anticoincidence system and a X-Ray detector sensitive in the 10-40 KeV range. The purpose of the Minicalorimeter (MCAL) is twofold. It shall work in conjunction with the Tracker in order to evaluate the energy of the interacting photons, and it shall operate autonomously in the energy range 250KeV-250 MeV for detection of transients and gamma ray burst events and for the measurement of gamma ray background fluctuations. We present the architecture of the Test Equipment we have designed and developed in order to test and verify the MCAL Simplified Electrical Model prototype which has been manufactured in order to validate the design of the MCAL Proto Flight Model.

INTRODUCTION

AGILE is a gamma-ray astrophysics Small Scientific Mission of the Italian Space Agency which is planned to be launched in 2005. The AGILE Payload [1] consists of three detectors: the Gamma-Ray Imaging Detector (GRID), based on the Silicon Tracker (ST), sensitive in the range 30 MeV - 50 GeV; the hard X-Ray imager named Super-Agile (SA) sensitive in the range 10 - 40 KeV, a non-imaging CsI(Tl) Minicalorimeter (MCAL) sensitive in the energy range 0.3 - 200 MeV; and an Anti-Coincidence auxiliary system (AC). The MCAL Simplified Electrical Model (SEM) consists of two orthogonal planes each composed of 4 CsI(Tl) Bars (out of 15 of the Proto Flight Model - PFM) with a photo-diode (PD) and a preamplifier placed at each end of each Bar, plus a Front End read out Electronics (FEE) whose design is representative of the actual PFM FEE. For each Preamplifier, the SEM FEE provides one Grid and one Burst electronic channels for the conditioning of the electrical signal and the generation of the Grid and Burst event data, respectively. The FEE interfaces the Payload Data Handling Unit (PDHU) through three differential LVDS serial bi-directional synchronous buses. Two are devoted to the Grid and Burst events download, the third is used for FEE configuration and Digital HK reading. These HK contain the rate of the signalS THAT ARE above the noise, as detected by the electronics chains which processes the single PD signal and the sum of the two PD signals. We present here the Test Equipment (TE) which has been developed for the functional and performance tests which have been carried out on the MCAL SEM at IASF (Bologna) and at a CERN Beam facility (Geneva) in the period June-August 2003.

TE REQUIREMENTS

Basically, the TE is required to simulate the PDHU power and data interfaces towards the FEE, and to monitor the FEE status and the HK parameters. Additional requirements come from the assumption that the quick look and on line analysis on the event data are assigned to another computer, the Science Console running the DISCoS software [2]. This approach is common to other AGILE TE, and in particular is the same applied to the Payload EGSE, the MCAL TE and the PDHU TE, where the TE for the flight model is provided by the detector manufacturer and is interfaced to the Science Console to be developed by the AGILE scientific collaboration [3].

The use of the DISCoS software requires that the TE be able to encapsulate into ESA Telemetry Standard packets the data acquired from the instrument, and, in near real time, to forward them to the Science Console where they are archived in NASA/FITS format and are analyzed during the measurement without interferring with the data acquisition.

Additional requirements come from the tests at the CERN Beam facility, where the MCAL SEM is integrated with the prototype of the ST detector and the AC subsystem. At CERN it is required that the TE interfaces the ST TE in order to synchronize the data taking and add a common time stamp to the acquired data. In addition, the TE acquires from the AC TE the signal which is used by the MCAL Burst chain to veto the event acquisition.

TE HARDWARE AND SOFTWARE ARCHITECTURE

The TE hardware consists mainly of some custom VME boards which interface the FEE and are controlled by a PCI-PC Host Computer (HC), through a Model 620 PCI-VMEbus Adapter with DMA (SBS Technologies, Inc.). The HC provides the operator console and forwards the acquired data to the Science Console.

In particular, three boards have been designed and customized for the TE. The HK/CONF board interfaces the FEE to configure the SEM Readout Electronics and to acquire the Digital HK data. The Burst board and the Grid board interface the FEE in order to acquire the Burst and the Grid data, respectively.

The HK and the Grid boards are based on a common layout which includes one Altera PLD for the data handling, the I/O connectors with the related LVDS/TTL converters, one memory unit for data storage, and one buffer for data transfer to/from the VME bus. The Burst board is equipped with an additional Altera for the generation of the time tag which is made available also to the Grid board, and it uses two FIFO memories for the storage of up to some tens of events.

A Junction panel provides the internal connections between the boards listed above and between these boards and the FEE. One I/O Adapter board handles the electrical signals generated by the ST TE in order to synchronize the measurement boundaries and to trigger the Burst chain. It also handles the Anticoincidence signal generated by the AC TE. One I/O Register board is used by the Host Computer in order to acquire the measurement synchronization signal in output from the Adapter board. One I/O 32 bit board generates the additional time tag that is used by both the MCAL TE and the ST TE in order to correlate the GRID data.

The application software running on the Host Computer (HC) has been written in C and C++ using the Linux KDE IDE, the Qt library [4] and the PacketLib library [5], under Mandrake Linux 9.1. A specific driver, installed as a kernel module, allows the application to access, by means of C-stream I/O statements (e.g. open, lseek, read), the configuration registers and the data memory area available on each VME board.

As shown in the block diagram sketched in Figure 1, the HC application software consists of one control process (MMC Console) with two threads and four processes, which are operated through a set of graphical widgets.

The MMC Console interacts directly with the VME HK board in order to send the commands which configure the FEE using the setting values defined by the operator either through a configuration file or by means of interactive graphical widgets. Hence, it acquires from the HK board and displays on its GUI the FEE status (e.g.: Operative mode, FIFO status) and the time profile of the FEE digital HK data. In addition, the MMC Console GUI provides the start/stop buttons which control the other tasks shown in Figure 1 and which automate the execution of the operations required to start/stop a single measurement. Namely, in order to start a measurement the MMC Console clears FEE FIFOs, changes the FEE operative mode from Idle to Observation, and sends to the Sender the ESA Telecommand (TC) Packet which notifies the start observation event to the Science Console. To stop the measurement, the MMC Console sends to the Sender a stop TC and sets the FEE operative mode to Idle.

The rest of the HC software is devoted to acquire and forward to the Science Console the event data and the HK data produced by the FEE during the observation period. As shown in Figure 1, the acquisition tasks are performed by the Grid and the Burst processes, and by the HK thread. They encapsulate the acquired data in the ESA packet format and, through a Unix shared queue, they communicate the resulting packets to the Sender process. Eventually, the Sender process reads the TM and TC packets from the shared queue, adds an incremental counter into their header, and writes each packet on the TCP/IP socket it has established with the Science Console at start-up. Optionally the Sender stores the packets into a local archive.

The purpose of the Monitor process (not implemented yet) is to gather and display summary information on the activity performed by the Grid, Burst and HK acquisiton tasks, e.g. the total number of packet sent to the SC.

The AcquisitionWatchDog Thread performs some automatic tasks associated with the occurrence of specific events. For example, during the CERN tests, in order to synchronize the data taking with the ST TE, this thread acquires the TTL signal generated by ST TE and handles via software the start/stop measurement button of the MMC Console.

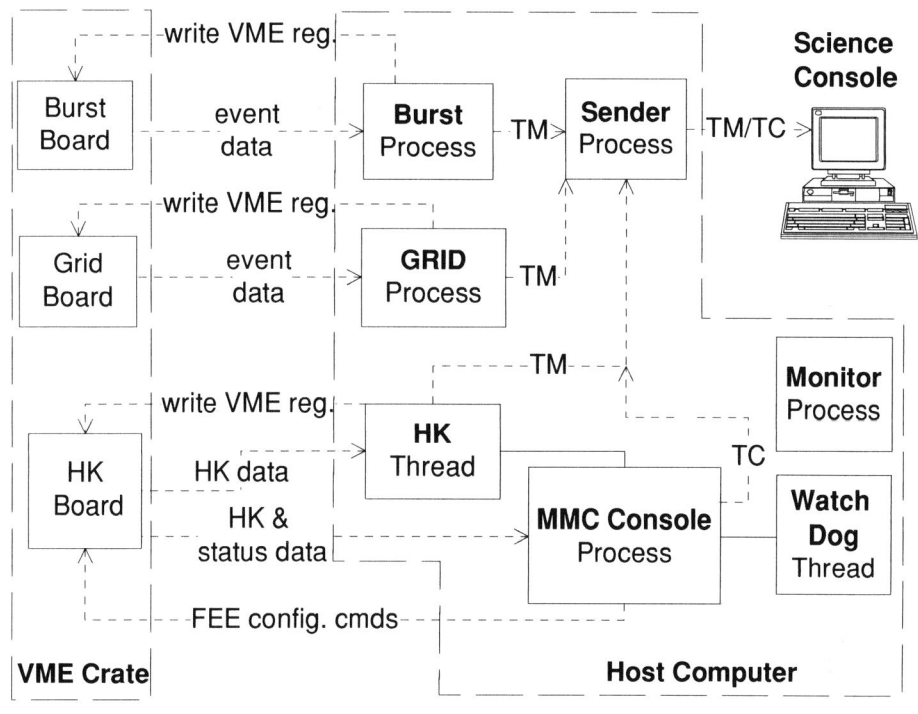

FIGURE 1. Block diagram of the Host Computer software architecture

REFERENCES

1. Tavani, M., and et al., "The AGILE Instrument," in *Astronomical Telescopes and Instrumentation, 22-28 August 2002*, SPIE Conference Proceedings, Waikoloa, Hawaii, USA, 2002.
2. Gianotti, F., and Trifoglio, M., "DISCoS - a detector-independent software for the on-ground testing and calibration of scientic payloads using the ESA Packet Telemetry and Telecommand Standards," in *ASP Conf. Ser. Vol. 238, Astronomical Data Analysis Software and System X*, ADASS Conference Proceedings, Boston, MA, USA, 2001.
3. Trifoglio, M., and et al., "The Ground Support Equipment for the Scientific tests and calibration of the AGILE instrument," in *Vol. 4140 pp 478-485*, SPIE Conference Proceedings, S.Diego, California, USA, 2000.
4. Trolltech, *The Qt Home Page* (2003), URL http://www.trolltech.com/products/qt/index.html.
5. Bulgarelli, A., Gianotti, F., and Trifoglio, M., "A C++ Library for Scientific Satellite Telemetry Applications," in *ASP Conf. Ser. Vol. 295, Astronomical Data Analysis Software and System XII*, ADASS Conference Proceedings, Baltimore, MD, USA, 2003.

AGILE Sensitivity and GRB Spectral Properties

G. Ghirlanda*, M. Galli[†], F. Longo**, B. Preger[‡], A. Argan*, G. Barbiellini**, S. Mereghetti*, A. Pellizzoni*, M. Tavani[‡] and S. Vercellone*

*IASF - Via Bassini 15, I-20133 Milano, Italy.
[†]ENEA - Via Don Fiammelli 2,I-40129 Bologna, Italy
**INFN - Via Valerio 2, I-34100 Trieste, Italy.
[‡]IASF - Via Fosso del Cavaliere, I-00133 Roma, Italy.

Abstract. Gamma Ray Bursts (GRBs) present different prompt and delayed emission properties which might suggest a possible taxonomy. Nonetheless, the existence of different classes could be the combined effect of different intrinsic and/or observational conditions. The latter, in particular, are related to the instrumental capabilities. For this reason, we studied the instrumental sensitivity of the AGILE satellite as a function of the spectral properties of GRBs. AGILE will be able to detect most of the bright–long bursts and its two detectors, with their independent trigger algorithms, compensate each other in detecting soft and high energy GRBs. Moreover, AGILE will possibly give a major contribution in the study of very energetic bursts with typical peak energies above ~ 1 MeV.

INTRODUCTION

More than 30 years of research in the field of GRBs revealed how major steps in the understanding of these intriguing astrophysical sources are intimately connected with the technological advances in designing instruments dedicated to the detection and study of transient gamma–ray sources. In particular, a key role has been played by the Compton Gamma Ray Observatory and by BeppoSAX which brought on two "revolutions" in the GRB field: the characterization of GRB global prompt emission properties and the identification of their afterglow emission, respectively. Although a general picture of the nature of GRBs has been progressively defined, still many open problems need to be solved (see [1] for a recent review).

The study of large samples of bursts in different energy bands revealed the possible existence of different "classes": (1) *short* and *long* bursts [2] characterized by ~ 0.5 and ~ 20 sec typical duration, respectively; (2) *dark bursts* with a normal X-ray (afterglow) emission but no optical counterpart ([3]); (3) GRBs associated with *Supernovae* like the cases of GRB980425-SN1998bw ([4]) and GRB030329-SN2003dh ([5]); (4) *high–energy* bursts (like GRB941017 - [6]) with MeV-GeV spectral emission evolving independently from the "classical" low energy component; (5) *soft* gamma ray burst (i.e. X ray flashes - XRF) with considerable emission concentrated below ~ 50 keV ([7]); (6) possible hard low energy spectral components and *thermal* like spectra ([8]) which question the leading GRB emission mechanism. Moreover, the prompt temporal and spectral properties of large samples of GRBs show a high degree of diversity (e.g. [9]) and no simple nor unique taxonomy is currently possible in the time-energy domain. Nevertheless, the instrumentation used to measure a burst might be responsible for the fact that these *appear* to be distinct GRB classes. For this reason we focus on the *sensitivity* of AGILE to GRBs with different *spectral properties*.

AGILE SENSITIVITY

The unpredictable appearance of GRBs in the sky requires, for their detection, a continuous monitoring of a large fraction of the sky in search of any significant increase of the flux over the monitored background. The appropriate choice of the search method (signal and background integration timescales and significance of their ratio) defines the instrumental *trigger method*. The algorithm that we use to compute the sensitivity (see also [10]) combines the instrumental parameters (e.g. effective area, energy integration of the background and source signal, etc.) and compares the background rate with the flux obtained assuming a parameterized function for the source spectrum. The result is the *sensitivity* defined as the *minimum peak flux* (expressed in phot/cm^2 s – integrated from 50-300 keV) which the burst should have in order to satisfy the trigger logic.

AGILE (see [11]) has two detectors for triggering transient γ–ray events: 1) Super-AGILE (SA) which is an X-ray coded mask detector (geometric area $\sim$ 1444 cm^2, FOV $\sim$ 0.8 sr) capable of triggering and accurately positioning GRBs in the energy range 10-40 keV; 2) Minicalorimeter (MCAL - geometric area $\sim$ 1472 cm^2, omni–directional) which will search for bursts in three energy ranges (0.3-1 MeV, 1-5 MeV, $\geq$5 MeV). The expected background rate of these two detectors strongly depends on the in–flight conditions. Currently, Monte Carlo simulations indicate a possible background rate of $\sim$1-0.7 kHz and $\sim$1-2 kHz for SA and MCAL, respectively, which are what we assume for the sensitivity computation. We also assume the Band model [9] which represents fairly well a typical burst spectrum, parametrized by a low and high energy powerlaw (with indices α and β, respectively) and by its EF$_E$ peak energy (E_{peak}). The significance parameter n of the burst signal with respect to the background fluctuation (i.e. $n = F_{source}/\sigma_{back}$) is assumed to be 4 (although it can be varied during the mission). Figure 1 represents the AGILE sensitivity as a function of the peak energy of the burst spectrum. We fixed α and β to their typical values -1 and -2.5, respectively, and varied E_{peak} between 10 keV and 2 MeV. The different curves for SA and MCAL refer to the energy ranges where these detectors will search for the burst. BATSE sensitivity is reported for comparison.

MCAL and SA complement each other due to their different sensitivity to bursts with peak energy in the keV or MeV energy range. Moreover, MCAL sensitivity improves for bursts with peak energy above 300 keV. Most of the well studied spectral samples of bursts ([9], [12] and [13] – open and filled circles in Figure 1) lie above the sensitivity of MCAL and SA. AGILE will be able to trigger events with spectral properties similar to normal bursts and peak fluxes above 2 phot/cm^2 s (integrated over 50-300 keV). Nonetheless, at lower flux levels SA and MCAL will independently trigger soft and hard bursts, i.e. with peak energies below and above $\sim$100 keV, respectively. From the instrumental sensitivity we might also estimate the expected GRB rate assuming (i) that most GRBs have a peak energy of $\sim$ 300 keV (which indeed is supported by the small dispersion of the distribution of this spectral parameter – e.g. [14]) and (ii) the BATSE

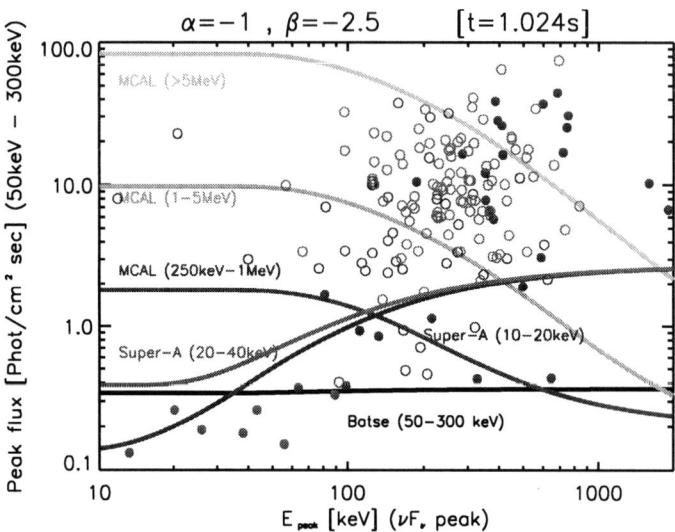

FIGURE 1. AGILE sensitivity vs. GRB peak energy. Curves represent the sensitivity expressed as the minimum peak flux (in the range 50-300 keV) which satisfies the trigger condition. Open circles are bright long bursts detected by BATSE from the Band catalog [9] and from the Preece catalog [12]. Filled points are the bursts detected by Hete II [13] and filled points in the low left corner are some XRF detected by BeppoSAX [7].

(1.024 sec) peak flux Log(N)-Log(S) (e.g. [15]). The expected rates (averaged over the realtive FOV) are $\sim$ 3-4 month^{-1} and $\sim$ 1-2 month^{-1} for MCAL and SA, respectively (see also [16]).

The dependence of the AGILE sensitivity on the other spectral parameters is reported in Figure 2: the peak energy is fixed to its central value (300 keV) and alternatively the low or high energy spectral slope are varied within their typical ranges. The sensitivity of MCAL is almost independent of the slope of the low and high energy spectral index. SA, instead, will trigger more efficiently on dim but relatively soft ($\alpha \leq -0.5$) events. We also tested the dependence of the sensitivity to the burst position with respect to the satellite axis and found that the sensitivity worsen by a factor $\sim$ 1.5 and $\sim$15 for MCAL and SA, respectively.

DISCUSSION

We have presented the expected AGILE sensitivity to GRBs with different spectral properties. The two detectors on board AGILE (MCAL and SA) should be able to trigger and study bright–long bursts with peak energies between few keV and some MeV. Moreover, MCAL has a good sensitivity for bursts with spectral energies above $\sim$ 0.5 MeV. This will represent a challenge for the investigation of high–energy events (indeed predicted by some theoretical models - e.g. [17]). Still SA has a very good sensitivity

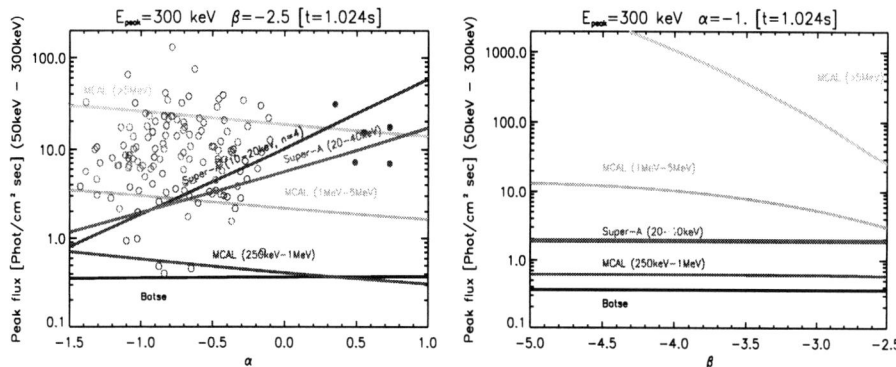

FIGURE 2. AGILE sensitivity vs. spectral slopes. *Left*: low energy spectral index α, with fixed $E_{peak} = 300$ keV and $\beta = -2.5$. *Right*: high energy spectral index β, with fixed $E_{peak} = 300$ keV and $\alpha = -1.0$.

at low peak energies: this will represent a unique opportunity to study the class of XRF and in particular to analyze simultaneously their broad band prompt emission.

ACKNOWLEDGMENTS

G. Ghirlanda and B. Preger would like to thank the organizing committee for the full grant support for the participation to this conference.

REFERENCES

1. Zhang, B., and Meszaros, P., *astro-ph/0311321* (2003).
2. Kouveliotou, C., *ApJL*, **413**, L101–L104 (1993).
3. De Pasquale, M. et al., *ApJ*, **592**, 1018–1024 (2003).
4. Galama, T. J. et al., *Nat*, **395**, 670–672 (1998).
5. Stanek, K. Z. et al., *ApJL*, **591**, L17–L20 (2003).
6. González, M. M. et al., *Nat*, **424**, 749–751 (2003).
7. Kippen, M. et al., "AIP Conf. Proc. 662," 2003, pp. 244–247.
8. Ghirlanda, G., Celotti, A., and Ghisellini, G., *A&A*, **406**, 879–892 (2003).
9. Band, D. et al., *ApJ*, **413**, 281–292 (1993).
10. Band, D. L., *ApJ*, **588**, 945–951 (2003).
11. Tavani, M. et al., "American Institute of Physics Conference Series," 2000, p. 746.
12. Preece, R. et al., *ApJS*, **126**, 19–36 (2000).
13. Barraud, C. et al., *A&A*, **400**, 1021–1030 (2003).
14. Zhang, B., and Mészáros, P., *ApJ*, **581**, 1236–1247 (2002).
15. Paciesas, W. S. et al., *ApJS*, **122**, 465–495 (1999).
16. Preger, B. et al., "AIP Conf. Proc. 526," 2000, p. 716.
17. Dermer, C. D., Böttcher, M., and Chiang, J., *ApJ*, **537**, 255–260 (2000).

Scaling and GRB Mission Optimization

John Doty

MIT Center for Space Research
Room 37-541
77 Massachusetts Avenue
Cambridge, Massachusetts, USA 02139

Abstract. Astrometry using wide field coded aperture cameras is the most effective way to obtain prompt locations of gamma ray bursts for observations by narrow field instruments. However, the rate of burst detections using this method has been disappointing. In an attempt to understand this problem, I have been investigating the scaling relationships between instrument characteristics and instrument performance. The ideas are very simple, but the results are sometimes counter-intuitive.

I will discuss the effects of field of view and detector area on detected burst rate. I will also discuss the relationship between instrument architecture and instrument volume, a major limitation and cost driver. I will show how attention to these relationships could lead to an inexpensive mission capable of locating bursts at a very high rate.

INTRODUCTION

Optimizing a GRB mission is a difficult applied physics problem. While simulations allow one to answer the "what if" questions, a broader mathematical context helps when considering which questions to ask. That is what I'm attempting to provide here.

My argument and results are essentially based on dimensional analysis, so I will omit factors of order unity and some cross terms for clarity.

RATE VERSUS DETECTOR AREA AND SOLID ANGLE COVERAGE

Assuming background limited sensitivity (a reasonable assumption for a burst of typical duration in a wide field system):

$$\text{sensitivity} \propto \sqrt{A/\Omega} \tag{1}$$

A is the total detector area and Ω is the solid angle coverage. To get a burst rate, we raise the sensitivity to the power s (the slope of the Log[N]/Log[S] curve) and multiply by Ω:

$$\text{detection rate} \propto A^{s/2} \Omega^{1-s/2} \tag{2}$$

For a homogeneous source distribution, s is 3/2, but for the burst population accessible to current instruments, s is closer to 2/3, so:

$$\text{detection rate} \propto A^{1/3} \Omega^{2/3} \tag{3}$$

The rate is proportional to $\Omega^{2/3}$: increasing the solid angle coverage raises the burst detection rate even though it decreases sensitivity. The real surprise is that the rate only increases as $A^{1/3}$: detector area is only a weak driver of the burst detection rate.

VOLUME AND RESOLUTION

For a detector positional resolution r and a mask-detector distance d, the angular resolution is:

$$\theta = r/d \tag{4}$$

If a is the mask area, the solid angle coverage is approximately:

$$\Omega = \frac{A+a}{d^2} \tag{5}$$

The volume of the camera is approximately:

$$V = d(A+a) \tag{6}$$

Given an available instrument volume, a required angular resolution, and a detector resolution, there is therefore a maximum possible solid angle coverage, (Ω_{max}):

$$\Omega_{max} = V \left(\frac{\theta}{r}\right)^3 \tag{7}$$

If $\Omega_{max} > 4\pi$ the field of view is unconstrained.

OBSERVING EFFICIENCY

The rate of burst detections is also affected by the operational efficiency of the mission. Occultation and radiation belt passages limit observing efficiency in low Earth orbit. Higher orbits may increase efficiency but greatly increase launch and operational costs (radiation exposure from solar flare protons is also a problem). If $\Omega_{max} \gg 4\pi$ a multiple satellite constellation in low Earth orbit appears to be the most reasonable approach to high observing efficiency.

If the total detector area is divided among n satellites, the satellites have uncorrelated observing schedules with efficiency (ε), the satellites detect bursts independently, and s is $2/3$, the burst rate is proportional to:

$$\frac{1-(1-\varepsilon)^n}{n^{1/3}} \tag{8}$$

The lower (dashed) curve Figure 1 shows this relative burst detection rate for an assumed ε of 0.4. For the brightest bursts the detection rate depends only on observing efficiency: sensitivity is not an issue. In this case the rate is proportional to $1-(1-\varepsilon)^n$. The upper

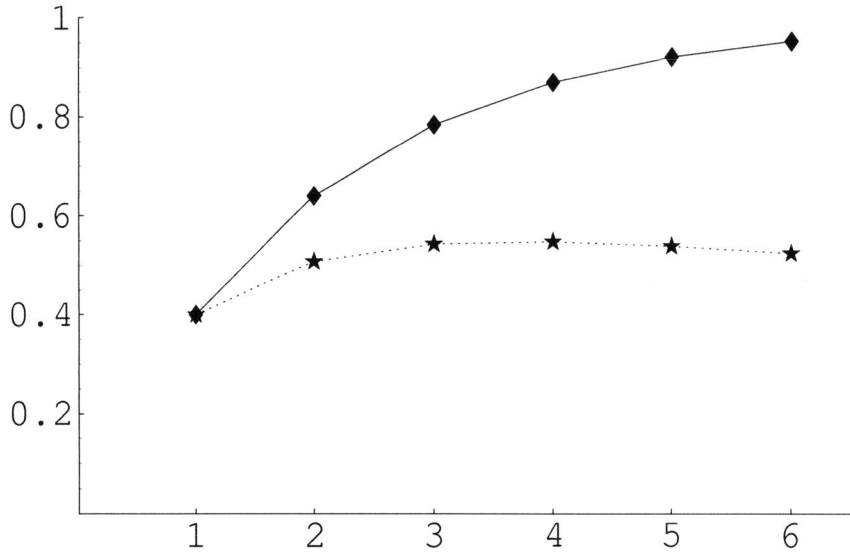

FIGURE 1. Relative burst detection rate versus number of satellites for a fixed detector area divided over several satellites in low Earth orbit. The dashed curve is for all bursts, the solid curve is for the brightest bursts.

(solid) curve in Figure 1 shows this. For faint bursts the loss in sensitivity of an individual satellite reduces the rate when the number of satellites is > 4. A constellation of 3-4 satellites is substantially more efficient than a larger single satellite, especially for bright bursts.

DISCUSSION

Collecting area is usually the first virtue we look for in an astronomical instrument. However, for these instruments, detector resolution versus available volume turns out to be much more important. High burst detection rates are more a product of solid angle coverage than of sensitivity. Detector resolution versus available volume has limited all missions of this type so far (BeppoSAX [1], HETE-2 [2], and Swift [3]) to fields of view of ~1 steradian. A really wide coverage system could be an order of magnitude more productive.

To get wide coverage we can ask for more volume (expensive), back off on angular resolution (undesirable), or improve detector spatial resolution. The last option seems most attractive.

In particular, silicon detectors can have resolutions two orders of magnitude better than gas detectors or exotic semiconductor detectors. This makes them very attractive for future high productivity localization missions. The HETE SXC operates only ~20% of the time and covers only ~1 steradian, so it detects only ~1.6% of bursts above its detection threshold. Nevertheless, it localizes $\sim$14 bursts/year using only 36 cm^2 of detector area. The scaling relations above suggest that an improved SXC with multiple modules to cover more solid angle and better optical blocking to improve operational efficiency could locate hundreds of bursts per year to arc minute accuracy using a few hundred cm^2 of silicon in an instrument with a volume of 0.01 m^3!

CONCLUSIONS

I have considered the productivity of a gamma ray burst localization mission using coded apertures.

- Detector area is only a minor factor in the localization rate. Small detectors can be very effective in an optimized system.
- For missions in low Earth orbit, small instruments on several satellites can detect more bursts than a single instrument with the same total volume and area.
- The detector virtue that enables the best optimization is spatial resolution.
- Silicon detectors are therefore the best technology for this job because they have much better spatial resolution than other technologies.

ACKNOWLEDGMENTS

These ideas grew from discussions of possible future missions with George Ricker, Roland Vanderspek, Geoff Crew, and Joel Villasenor. Matthew Wampler-Doty helped prepare this manuscript.

REFERENCES

1. Jager, R., et al., A&A Suppl, **125**, 557 (1997).
2. Kawai, N., et al., in *Gamma-Ray Burst and Afterglow Astronomy 2001: A Workshop Celebrating the First Year of the HETE Mission*, eds. R. Vanderspek and G. Ricker, AIP Conference Proceedings 662 (2002).
3. Band, David L., Ap. J., **588** (2003).

X–Ray Monitoring of GRBs with Lobster Eye Telescopes

L. Švéda*, R. Hudec[†], A. Inneman**, L. Pína* and G. Pizzichini[‡]

*Czech Technical University, Faculty of Nuclear Science, V Holešovičkách 2, CZ-180 00 Praha 8, Czech Republic
[†]Astronomical Institute of the Academy of Sciences of the Czech Republic, CZ-251 65 Ondřejov, Czech Republic
**Centre for Advanced X–ray Technologies, Reflex, Novodvorská 994, CZ-142 00 Praha 4, Czech Republic
[‡]IASF/CNR Sezione di Bologna, via P. Gobetti, 101, 40129 Bologna, Italy

Abstract. We present here the soft X–ray All-Sky Monitor (ASM). It is based on the current technological capabilities, sensitive in the $\sim 0.1 - 10.0$ keV range with angular resolution of $\sim 3 - 4$ arcmin, and has a limiting detectable flux $\sim 10^{-12}$ erg/s/cm^2 for daily scans in the mentioned energy range. The ASM will play a key role in studying transient X–ray sources like XRBs, GRBs, XRFs, X–ray novae, as well as in the study of the long term variability of X–ray sources like XRBs, AGN, or stellar X–ray flares.

LOBSTE EYE OPTICS

There exists several prototypes of the Schmidt design Lobster Eye (LE) modules today [1]. The optics consists of two perpendicular sets of metal coated plates (gold coated in the prototypes) arranged in a way similar to Figure 1.

The properties of the prototypes are displayed in Table 1. The variant of "mini 2" LE seems to be a good starting point for the development of the space based ASM module.

TABLE 1. LE Schmidt modules developed so far. Here plates have dimensions of $d \times l \times t$ and are arranged with spacing a. The modules have the focal length f.

Module	size d[mm]	thickness t[mm]	distance a[mm]	length l[mm]	foc. length f[mm]	resolution r[arcmin]
macro	300	0.75	10.80	300	6000	7
middle	80	0.3	2	80	400	20
mini 1	24	0.1	0.3	30	900	2
mini 2	24	0.1	0.3	30	250	6
micro	3	0.03	0.07	14	80	4

ASM MODULE PROPOSAL & SIMULATIONS

We have performed several simple ray–tracing simulations of the LE optics and have proposed the optics with the gold coated (microroughness $\sigma \sim 1$ nm) plates with di-

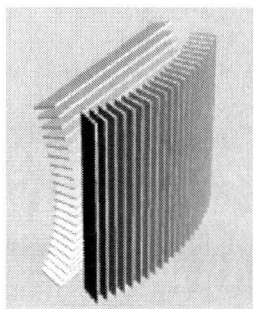

FIGURE 1. The plate arrangement in the Schmidt LE design.

mensions $78.0 \times 11.5 \times 0.1$ mm, 0.3 mm spacing, and the focal length $f = 375$ mm. This concept is based on current experience and technologies, and meets the ISS LE experiment requirements.

The front area of the proposed optics is $\sim 10\times$ larger than the front area of the "mini 2" prototype. The total length of the optics is $\sim 1/2$ of the prototype length. The weight of the prototype is ~ 0.1 kg. Hence we can estimate the weight of the proposed optics to $m \sim 0.5 - 1.0$ kg, depending on the exact plate assembly strategy.

The on–axis PSF FWHM of this optics ranges from 3.8 arcmin (420 μm) for 0.5 keV to 2.6 arcmin (280 μm) for 10.0 keV. The realistic FWHM will be smoothed and enlarged by various plate distortions, but we are near the required value. We have also simulated the point–to–point focusing LE system for 8 keV photons and compared it with the experiment. Even the simple ray–tracing model matched the experiment quite well.

The PSF consists of the central peak and the characteristic cross structure. The cross structure originates from photons reflected only once in the LE optics, while the photons gathered in the central peak are reflected twice (once from the horizontal plates, once from the vertical ones). The relative height of the central peak versus the cross structure decreases with the increasing photon energy. The shape of the PSF also changes with the off–axis distance, but the shape change is more significant at cross structure than in the case of the central peak.

The change of shape of the PSF is dominant in the central peak vs. cross structure position, while the shape of the central peak changes only slightly with the off–axis distance. The peak remains in the intersection of the two crossing lines (crossing structure), but the intersection is moved off the center of the line. In some extreme situations, when the lines are moved sidewise, there can be no intersection of the lines. The FOV can be estimated to $\sim 6 \times 6$ deg based on the PSF off–axis study.

The gain defined as the ratio of photons gathered inside the FWHM area with and without the LE optics is plotted in Figure 2. The gain will be also decreased by the effect of plate distortions, but the gain should reach up to one thousand below 2.5 keV.

We have performed the simulated scanning observation with the module mentioned above. We collected the data from the The ROSAT All-Sky Survey Bright Source Catalogue (1RXS) [2], and converted the count rates from the catalogue to fit the LE

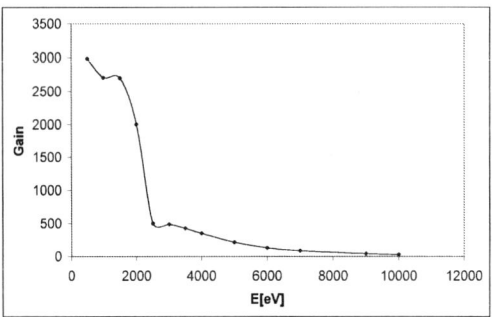

FIGURE 2. The gain computed from the ray–tracing simulation for the Schmidt LE module described in the text.

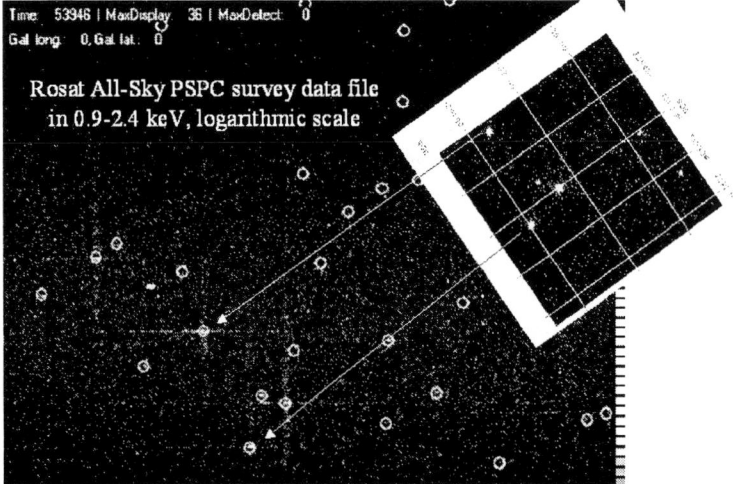

FIGURE 3. The LE module scanning observation simulation of the Galaxy center based on the ROSAT data [2] in the logarithmic scale. The effective exposure time is ∼ 1000 s, and the image has 512 × 325 pixels with 200 μm pixel size, what is equal to 15.0 × 9.5 deg in the sky. The Galactic longitude changes horizontally, the galactic latitude changes vertically. Each circle represents the position of 1RXS source. The simulation is compared with the ROSAT all–sky survey image of the Galaxy center in 0.9 – 2.4 keV band.

ASM module properties. We also added the background according to the model in [3].

We put the LE onto the 105 min orbit and performed Galactic plane scans. The result for the Galactic center after 8 revolutions is plotted in Figure 3.

GRBS & XRFS

Sources as faint as $\sim 10^{-12}$ erg/s/cm^2 for daily scanning observation in soft X-ray range are expected to be detected together with the moderate angular resolution (arcmins). Large FOV (6 × 180 deg for example) can be obtained thanks to the system modularity.

GRB X-ray prompt emission (about $\sim 10^{-8}$ erg/s/cm^2 in soft X-rays) is very well detectable by the LE telescopes, providing us with the flux measurement and relatively precise position measurement. Assuming the 6 × 180 deg FOV, 90 min revolutions, and the BATSE GRB detection rate (~ 300 yr^{-1}), approximately 20 GRBs per year will be observable during the GRB X-ray prompt emission phase. We can add some points into the GRB X-ray afterglow light curve for other GRBs (as the LE orbits the Earth, and before the flux falls below the LE detectable level).

The X-ray flashes, having very similar X-ray properties as the GRB prompt emission, will also be very well detectable by the LE ASM. Assuming the previous LE properties and XRF rate to be ~ 100 yr^{-1} [4] the LE detection rate will be ~ 8 yr^{-1}.

Depending on the X-ray detector, rough information about the energy of incoming photons can be obtained, hence a low energy GRB continuum spectra can be specified.

We expect that the LE ASM will be used in cooperation with other missions and/or projects. Large sensitive X-ray telescopes can be used to cover the later stages of the GRB afterglow for example. The LE will provide the GRB X-ray prompt emission characteristics and the early stage of the GRB afterglow, as well as the precise positioning of the events.

ACKNOWLEDGMENTS

We acknowledge the support provided by the Grant Agency of the Czech Republic, grant 102/99/1546. We also acknowledge the support from the Ministry of Industry and Trade of the Czech Republic, projects FB-C3/29/00 and FD-K3/052.

REFERENCES

1. Inneman A. et al.: *Progress in lobster eye X-ray optics development*, Proc. SPIE **4138**, p. 94–104 (2000).
2. Voges, W. et al.: *The ROSAT all-sky survey bright source catalogue*, Astronomy and Astrophysics, **349**, p. 389–405 (1999).
3. Priedhorsky, W. C. et al.: *An X-ray all-sky monitor with extraordinary sensitivity*, MNRAS **279**, p. 733–750 (1996).
4. Heise, J. et al.: *X-ray Flashes and X-ray rich Gamma Ray Bursts*, arXiv:astro-ph/0111246 (2001).

ROBOTIC OBSERVING SYSTEMS

Exploring the First Minute: New Technology for Measuring Color and Polarization Variations in Prompt Optical Emission

W. T. Vestrand*, D. J. Casperson*, C. Ho*, E. Raby[†], R. Shirey*, D. Thompson[†], R. R. White* and J. Wren*

*Los Alamos National Laboratory, ISR-2, MS B244, Los Alamos, NM 87545, USA
[†]Los Alamos National Laboratory, P-21, MS D454, Los Alamos, NM 87545, USA

Abstract. With the coming launch of the Swift satellite, there will be many new opportunities to study the physics of the prompt optical emission from gamma ray bursts with robotic ground-based telescopes. We discuss a new imaging system under development at Los Alamos National Laboratory that will provide simultaneous multicolor photometry of the rapidly evolving prompt optical emission in the first minutes after a burst trigger. This next generation system employs state-of-the-art photon-counting imaging technology at the focal plane of a rapidly slewing telescope. The imaging sensor is composed of an S-20 photocathode, stacked microchannel plates, and crossed delay line readout electronics that together are capable of measuring the time of arrival and positions for individual optical photons with 200 picosecond time resolution. The imager is coupled with electronically tunable liquid-crystal filters to provide essentially simultaneous linear polarization and multicolor photometric measurements of the prompt optical emission from a gamma ray bursts.

INTRODUCTION

To understand fully the evolution of prompt optical emission from gamma ray bursts (GRBs), one needs the ability to make simultaneous multi-color photometric observations during the first minute after the cataclysmic explosion. The extreme ultra-relativistic nature of the outflow means that the complex history of interactions and the deceleration of the bulk flow, which occurs on the timescale of a day in the plasma frame, is carried by emission that arrives at Earth within the span of a minute. During that rapid deceleration, there are important parameters that define the physics of the flow: the Lorentz factor, the ambient density, the fraction of the energy in mass of the particles, and the fraction of the energy in the magnetic field. By measuring simultaneously the spectral shape of the emission and how it evolves, each of the parameters can be determined [1]. Measurements of the spectral evolution during the first minute are therefore a powerful probe of the physics of the largest explosions since the Big Bang.

But studies of the spectral evolution of GRB emission at optical wavelengths must correct for spectral variations generated by material along the line of sight to the GRB. There is evidence that long duration GRBs are associated with star formation regions and that visual extinction along the line of sight is likely to be in the range of $A_V = 1 - 2$ magnitudes with values all the way up to $A_V = 5$ magnitudes or greater (e.g.[2, 3]). Further, the prompt optical-UV flux from a GRB explosion is sufficient to destroy the dust column within tens of seconds after the burst (e.g., [4]), producing strongly variable extinction and reddening. So to infer the intrinsic spectral and temporal evolution of a GRB explosion in its first minutes, the effects of the variable dust screen have to be measured. If the extinction law throughout the evaporating column has a fixed wavelength dependence, any shift in color-color space orthogonal to the reddening vector is intrinsic spectral evolution. So simultaneous photometry in three or more colors will enable the separation of color changes due to variable reddening from intrinsic spectral variability.

Another important diagnostic of the physical conditions in the ultra-relativistic burst plasma is the polarization of the optical emission. Since the prompt optical emission is believed to be synchrotron emission, it should be intrinsically linearly polarized. Polarization measurements have been made of the optical afterglow ten hours or more after the event and have shown linear polarizations of $\sim 1-3\%$ that can temporally vary in amplitude but show no variation in polarization position angle (e.g. [5]). However, new estimates indicate that the prompt emission could have polarization fraction as high as $\sim 60\%$, if the ordered magnetic field is advected into the relativistic plasma from the GRB progenitor[6]. Observations of polarization in the prompt optical phase and its variation both in amplitude and in position angle will therefore provide important clues about the geometry of the cataclysmic event and the interaction of burst plasma with the surrounding medium.

PHOTON-COUNTING MULTICOLOR PHOTOMETERY

At LANL we have constructed a new class of photon counting imagers with extremely high time resolution for astronomical applications. These imagers are composed of a photo-cathode, a stack of three micro-channel plates, and a crossed delay line (CDL) readout all hermetically sealed in a vacuum tube (see Figure 1a). When used for the detection of optical photons, we employ an S-20 photocathode, which has a quantum efficiency that allows effective detection of optical photons up to wavelengths of about 750 nanometers. Each photoelectron ejected by the photocathode generates a cascade in the micro-channel plate (MCP) stack that produces a gain of $\sim 10^7$. The emergent electron cloud spreads over about 10 wire pairs in the tightly wound CDL grid allowing centroiding of the event to better than 100 microns on the 40 mm diameter active area.

This sensor tube assembly is connected to ultra-fast electronics that measure individual photon arrival times with a time resolution of 200 pico-seconds and provide a throughput of more than 10^6 counts per second. This combination of imaging capability and unprecedented ability to detect individual optical photons with extremely high time resolution has already led to important applications of this technology for time-resolved imaging of cloud scattering and 3-D ranging and mapping of solid objects [7].

Like a conventional CCD imager operating at optical wavelengths, our photon counting imager employs a multicolor filter system to derive spectral information about the incident emission. But for studies of prompt GRB emission during the first minute, mechanical filter wheels are too slow. Instead, to obtain quasi-simultaneous multicolor photometry, a large aperture (35 mm square), electronically tunable, optical filter is located in front of the imager. The filter is a liquid-crystal tunable filter (LCTF) that is essentially a multistage Lyot-type polarization interference filter with an electronically controllable liquid-crystal waveplate in each stage to provide variable retardance. This LCTF is capable of switching between three broadband filters with no moving parts or vibration in under a millisecond. A controller is used to continuously cycle through the filter set at a rate of 1 Hz or faster. This filter/imager combination therefore time-tags and color-tags all detected photons yielding essentially simultaneous multi-color photometry.

Unlike a conventional CCD imager when used for ultra-high cadence imaging, our imager is essentially free of detector and read noise. The detector noise level for our photon counting imager is determined by thermal electron emission from the photocathode. At room temperature these electrons generate typically only a few thousand counts/s distributed randomly across the MCP face yielding a noise count rate of $\sim 10^{-2}$ per second per pixel. These rates are negligibly small compared to the available counting rate of 10^6 per second. Our imager has the ability to operate at room temperature with a sensitivity that is limited only by the natural sky background and the available photon statistics.

PHOTON-COUNTING POLARIMETRY

Another advantage of the LCTF technology is that it is intrinsically sensitive to the polarization properties of the incident radiation. When another electronically controllable stage is placed in front of the color filter, it is possible to select the color and linear polarization state of the photons that pass through to the imager. The rapid electronic switching possible with this system means that one can rapidly sample the spectral distribution and polarization properties of the prompt emission. The photon arrival time, position, color, and polarization are measured for all positions in the imager's field of view. As a result, one automatically gathers simultaneous "control" measurements of other objects in the field as well as insure that transients, whose locations are often only known with arcminute accuracy, are measured. The primary limitation of this approach is that the multiple stages of filters allow a very small fraction of the collected photons to be detected. As result, such detailed observations of polarization and spectral evolution during the first minute will only be possible for the very brightest optical transients.

APPLICATION TO THE FIRST MINUTE OF GRBS

Most models assume that the optical emission from GRBs is synchrotron radiation generated by energetic electrons in the burst outflow. Figure 1b shows the light-curve measured for GRB990123 [8] which, while it is only composed of four widely spaced points during the first few minutes of the burst, is still the most complete set of measurements available for prompt optical emission. By noting that the optical and gamma ray intensities were not correlated, many authors have argued that the emission in the two bands originated in different regions and suggested that the prompt optical outburst is generated as a reverse shock that traverses the relativistic outflow. Such a reverse shock can only occur once in a given GRB, and indeed, the ROTSE data seem to show a single peak followed by power-law decay. On the other hand, some models claim that rapid optical variations were present that should be correlated with the gamma ray fluctuations that were as short as tens of milliseconds in this event (e.g. [9]). Unfortunately, the lightcurves obtained by conventional CCD imaging will be too highly under-sampled to definitively distinguish between the models.

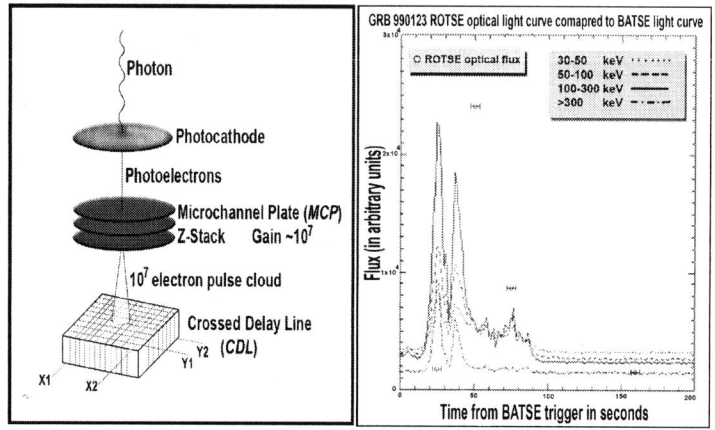

FIGURE 1. The panel on the left shows a schematic of the layout of the photon-counting imager. The panel on the right shows the gamma ray lightcurve and the four broadband optical lightcurve measurements of the optical flash from GRB990123.

The launch of the Swift satellite in the summer of 2004 will provide an exciting new opportunity to study the prompt optical emission from GRBs by providing accurate, real time, positions for roughly 100 GRBs per year. The Swift satellite will be able to reorient itself in a median time of just over 50 seconds and begin measurement of the prompt optical emission with a sensitive onboard UV/Optical Telescope. But to make measurements in the first minute, one will need a rapidly slewing ground-based telescope. One of our photon counting imagers will be deployed on a RAPTOR 0.4-meter telescope with a fully robotic rapidly slewing mount. The robotic mount is connected via socket to the GCN network for prompt response to GRB alerts and capable of pointing the telescope to any direction in the sky in only four seconds. When deployed on that telescope, the field of view for the imager is $12' \times 12'$. Operated without the LCTF, the system has a predicted 3-sigma sensitivity of $V \sim 19^{th}$ magnitude in 10 seconds. In an event as bright as GRB 990123 the system can measure fluctuations as short as a few milliseconds when operated without the tunable filter in the optical path. Such observations would be a powerful probe of the physics during the critical first minute.

REFERENCES

1. Wijers, R. and Galama, T., *ApJ* **525**, 177 (1999).
2. Paczynski, B., *ApJ* **494**, L45 (1999).
3. Lamb, D.Q., *Phys. Reports* **333**, 505 (2000).
4. Waxman, E. and Draine, B., *ApJ* **537**, 796 (2000).
5. Covino, S., et al., *A&A* **404**, L5 (2003).
6. Granot, J. and Konigl, A., *ApJ* **594**, L83 (2003).
7. Ho, C., et al., *Appl. Optics* **38**, 1833 (1999).
8. Akerlof, C.W. et al., *Nature* **398**, 400 (1999).
9. Liang, E. et al., *ApJ* **519**, L21 (1999).

The Search for Optical and Near-Infrared Counterparts of GRBs with the Super-LOTIS Telescope

G. G. Williams*, H. S. Park[†], S. D. Barthelmy**, D. H. Hartmann[‡], K. C. Hurley[§], P. A. Milne[¶], K. J. Lindsay[‡], M. Bradshaw[¶], R. E. Wurtz[†] and J. Wickersham[||]

*MMTO, 933 N Cherry Ave, Tucson, AZ 85721
[†]Lawrence Livermore National Laboratory, 7000 East Ave, Livermore, CA 94550
**NASA/Goddard Space Flight Center, Greenbelt, MD 20771
[‡]Department of Physics & Astronomy, Clemson University, Clemson, SC 29635-0978
[§]Space Sciences Laboratory, University of California, Berkeley, CA 94720
[¶]Steward Observatory, 933 N Cherry Ave, Tucson, AZ 85721
[||]UC Davis/LLNL, 7000 East Ave, Livermore, CA 94550

Abstract. The 0.6-m Super-LOTIS (Livermore Optical Transient Imaging System) telescope is a fully robotic system dedicated to the search for prompt optical emission from gamma-ray bursts. The telescope began routine operations from its Steward Observatory site atop Kitt Peak in April 2000. We summarize the current capabilities of the system and present some recent scientific results. A progress report is given on the upgrade of the system to allow for simultaneous near-infrared (NIR) and optical imaging. This upgrade will be completed to coincide with the launch of the Swift GRB explorer mission in mid-2004. Swift will have the capability of localizing very high redshift GRBs but absorption by the Ly-α forest prohibits optical detection of $z > 5$ bursts. NIR observations can detect GRBs out to $z \sim 10$. Although Swift is a multi-wavelength observatory capable of observing GRBs from the hard x-rays to the optical it has no NIR capability. The upgraded Super-LOTIS telescope will fill this NIR need.

CURRENT SYSTEM

Super-LOTIS is a fully automated robotic 0.6-m telescope. It is the third generation system designed specifically to search for and observe optical counterparts of gamma-ray bursts (GRBs). Its predecessors, GROCSE (Gamma-Ray Optical Counterpart Search Experiment) and LOTIS, were successful in establishing upper limits for the prompt optical flux from several gamma-ray bursts [1, 2, 3]. Each subsequent generation was designed and built with increased capabilities.

The configuration of the current Super-LOTIS telescope was conceived, constructed, and commissioned during the BATSE (Burst and Transient Source Experiment) era. Therefore, its design was largely driven by the characteristics of typical BATSE GRB triggers. The design requirements included a rapidly slewing telescope, a large field-of-view, and a faint limiting magnitude. The rapid slew rate was needed to observe a burst during the prompt gamma-ray emission phase; typically tens of seconds. The large field-of-view was necessary to cover the large error boxes provided by BATSE; typically a few

TABLE 1. Selected results from the current Super-LOTIS telescope.

Date	T_{obs}	R	Ref.
GRB 990308	28.2 m	> 15.3	[6], [7]
GRB 001025B	30.0 h	> 18.9	[8], [9]
GRB 010220	6.5 h	> 20.2	[10]
GRB 010222	23.7 h	20.0	–
GRB 010921	21.8 h	19.4	[11],[12]
GRB 020531	3.6 h	> 17.5	[13]
GRB 020813	106 m	18.4	[14]
GRB 021004	14.0 h	19.1	–
GRB 021211	143 s	15.2	[15]

degrees in diameter [4]. The sensitivity requirement was estimated from GROCSE and LOTIS upper limits as well as results from other experiments [5]. The current Super-LOTIS system consists of an automated f/3.5 0.6-m telescope equipped with a prime focus camera. This configuration provides a 51' x 51' field-of-view at a pixel scale of 1.5"/pixel. More details of the current system are given in Table 2.

RESULTS

Results from GROCSE and LOTIS are presented in Park et al. [1], Park et al. [2], and Williams et al. [3] and therefore will not be discussed further here. The Super-LOTIS telescope has observed more than 50 GRB error boxes since routine operations began in April 2000. Additional observations were performed during the engineering phase which began in March 1999. The upper limits and detections for some selected events are given in Table 1. More detailed information for these events can be found in the references listed in column four of Table 1.

The earliest detection of an optical counterpart by Super-LOTIS was during observations of GRB 021211. Super-LOTIS detected the counterpart at R = 15.2 just 143 s after the burst [13]. Figure 1 shows a 5.5' x 5.5' section of a Super-LOTIS image centered on the optical counterpart of GRB 021211.

UPGRADE DESIGN

The need for a wide-field instrument to search the historically large GRB error boxes was decreased with the de-orbit of the Compton Gamma-Ray Observatory and the subsequent launch of the HETE-2 satellite[1]. In addition, the Swift satellite, to be launched in mid-2004, will localize more than 100 GRBs per year and distribute prompt coordinates

---

[1] A wide-field is useful for covering some IPN error boxes but the IPN triggers are typically delayed by many hours and therefore do not allow for prompt follow-up.

FIGURE 1. A 5.5' x 5.5' section of the Super-LOTIS image centered on the optical counterpart of GRB 021211. Super-LOTIS detected the counterpart at R = 15.2 just 143 seconds after the burst [13].

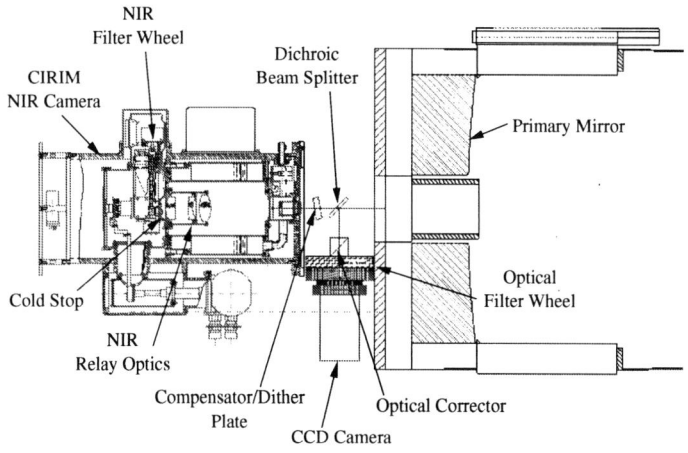

FIGURE 2. The design for the upgraded Super-LOTIS instrument package.

with typical error boxes of a few arc minutes radius [16, 17]. Swift is a multi-wavelength observatory with the capability to provide follow-up observations of GRBs with its on-board gamma-ray, x-ray and UV/optical telescopes. Swift is also capable of detecting high redshift GRBs [18] whose long wavelength counterparts are obscured in the optical by the Ly-α forest. NIR observations are required to detect counterparts for these most distant events. The Swift satellite has no on-board NIR instruments.

TABLE 2. Characteristics of the Super-LOTIS instruments.

	Current	Optical Upgrade	NIR Upgrade
Array	Loral 442A	EEV 42-40	HgCdTe NICMOS-3
Pixel Size (μm)	15	13.5	40
Pixels	2048x2048	2048x2048	256x256
Read Rate (kHz)	500	100-800	–
Read Noise (e^-)	50	5.59	50
Dark Current (e^-/s/pix)	1.0	0.07	2.5
Pixel Scale (arcsec)	1.5	0.5	1.5
Field-of-View	51' x 51'	17' x 17'	6.5' x 6.5'
Limiting Mag (60 sec)	R ~ 19.0	R ~ 19.0	J ~ 16.0

The upgraded Super-LOTIS telescope is designed to take full advantage of the precise localizations provided by the current and future GRB missions and to study wavelengths not covered by Swift's on-board telescopes. Figure 2 illustrates the design of the instrument package. The system will include the capability for simultaneous optical and NIR imaging. This will allow:

1. measurements of the intrinsic optical and NIR emission for both the prompt and early afterglow counterparts.
2. the detection and observation of GRBs which are optically obscured by material in their host star forming regions.
3. detection and observation of high redshift bursts which are optically obscured by the Ly-α forest. I-band drop-outs can be used as a indicator of potential high-z events.

The simultaneous multi-band imaging will be achieved by replacing the current prime focus camera with a 24.0-cm f/9 secondary. A dichroic beam splitter placed near the Cassegrain focus will split the light into NIR and optical beams. A compensator/dither plate is gimbaled to dither the NIR beam for background subtraction images. Details for the two cameras are provided in Table 2.

The optical camera is a commercial Spectral Instruments CCD camera. It contains a 2048 x 2048 pixel EEV 42-40 thinned CCD with 13.5 μm pixels. This configuration provides an optical field-of-view of 17' x 17'. The filter wheel will be populaed with B, V, R, and I filters.

The NIR camera, CIRIM (Cerro Tololo InfraRed IMager; see `http://www.ctio.noao.edu/instruments/ir_instruments/cirim/cirim.html`) is on loan from NOAO/CTIO. The NIR array is a 256 x 256 NICMOS-3 array. which provides a field-of-view of 6.5' x 6.5' with a pixel scale of 1.5"/pixel. The NIR filter wheel will house J, H, and K_s filters.

The upgraded system will be completed to coincide with the launch of the Swift satellite in mid-2004.

REFERENCES

1. Park, H. S., Ables, E., Band, D. L., Barthelmy, S. D., Bionta, R. M., Butterworth, P. S., Cline, T. L., Ferguson, D. H., Fishman, G. J., Gehrels, N., Hurley, K., Kouveliotou, C., Lee, B. C., Meegan, C. A., Ott, L. L., and Parker, E. L., *ApJ*, **490**, 99–+ (1997).
2. Park, H. S., Williams, G. G., Ables, E., Band, D. L., Barthelmy, S. D., Bionta, R., Butterworth, P. S., Cline, T. L., Ferguson, D. H., Fishman, G. J., Gehrels, N., Hartmann, D., Hurley, K., Kouveliotou, C., Meegan, C. A., Ott, L., Parker, E., and Wurtz, R., *ApJ*, **490**, L21+ (1997).
3. Williams, G. G., Park, H. S., Ables, E., Band, D. L., Barthelmy, S. D., Bionta, R., Butterworth, P. S., Cline, T. L., Ferguson, D. H., Fishman, G. J., Gehrels, N., Hartmann, D. H., Hurley, K., Kouveliotou, C., Meegan, C. A., Ott, L., Parker, E., and Porrata, R., *ApJ*, **519**, L25–L29 (1999).
4. Paciesas, W. S., Meegan, C. A., Pendleton, G. N., Briggs, M. S., Kouveliotou, C., Koshut, T. M., Lestrade, J. P., McCollough, M. L., Brainerd, J. J., Hakkila, J., Henze, W., Preece, R. D., Connaughton, V., Kippen, R. M., Mallozzi, R. S., Fishman, G. J., Richardson, G. A., and Sahi, M., *ApJS*, **122**, 465–495 (1999).
5. Akerlof, C., Balsano, R., Barthelmy, S., Bloch, J., Butterworth, P., Casperson, D., Cline, T., Fletcher, S., Frontera, F., Gisler, G., Heise, J., Hills, J., Hurley, K., Kehoe, R., Lee, B., Marshall, S., McKay, T., Pawl, A., Piro, L., Szymanski, J., and Wren, J., *ApJ*, **532**, L25–L28 (2000).
6. Williams, G. G., Park, H. S., and Porrata, R. A., GRB 990308 (Trig. 7457), optical observations (1999), GCN 272.
7. Schaefer, B. E., Snyder, J. A., Hernandez, J., Roscherr, B., Deng, M., Ellman, N., Bailyn, C., Rengstorf, A., Smith, D., Levine, A., Barthelmy, S., Butterworth, P., Hurley, K., Cline, T., Meegan, C., Kouveliotou, C., Kippen, R. M., Park, H.-S., Williams, G. G., Porrata, R., Bionta, R., Hartmann, D., Band, D., Frail, D., Kulkarni, S., Bloom, J., Djorgovski, S., Sadava, D., Chaffee, F., Harris, F., Abad, C., Adams, B., Andrews, P., Baltay, C., Bongiovanni, A., Briceno, C., Bruzual, G., Coppi, P., della Prugna, F., Dubuc, A., Emmet, W., Ferrin, I., Fuenmayor, F., Gebhard, M., Herrera, D., Honeycutt, K., Magris, G., Mateu, J., Muffson, S., Musser, J., Naranjo, O., Oemler, A., Pacheco, R., Paredes, G., Rengel, M., Romero, L., Rosenzweig, P., Sabbey, C., Sánchez, G., Sánchez, G., Schenner, H., Shin, J., Sinnott, J., Sofia, S., Stock, J., Suarez, J., Telléria, D., Vicente, B., Vieira, K., and Vivas, K., *ApJ*, **524**, L103–L106 (1999).
8. Park, H. S., Williams, G. G., Perez, D., Nemiroff, R., Barthelmy, S. D., Hartmann, D., Laver, C., and Hurley, K., GRB 001025B, optical observations (2000), GCN 873.
9. Hurley, K., Berger, E., Castro-Tirado, A., Castro Cerón, J. M., Cline, T., Feroci, M., Frail, D. A., Frontera, F., Masetti, N., Guidorzi, C., Montanari, E., Hartmann, D. H., Henden, A., Levine, S. E., Mazets, E., Golenetskii, S., Frederiks, D., Morrison, G., Oksanen, A., Moilanen, M., Park, H.-S., Price, P. A., Prochaska, J., Trombka, J., and Williams, G., *ApJ*, **567**, 447–453 (2002).
10. Williams, G. G., Uglesich, R., Bradshaw, M., Park, H. S., and Hartmann, D. H., GRB 010220, optical observations (2001), GCN 981.
11. Park, H. S., Williams, G. G., Barthelmy, S. D., Cline, T., Hartmann, D., Hurley, K., and Pereira, W., GRB 010921, optical observation (2001), GCN 1131.
12. Park, H. S., Williams, G. G., Hartmann, D. H., Lamb, D. Q., Lee, B. C., Tucker, D. L., Klose, S., Stecklum, B., Henden, A., Adelman, J., Barthelmy, S. D., Briggs, J. W., Brinkmann, J., Chen, B., Cline, T., Csabai, I., Gehrels, N., Harvanek, M., Hennessy, G. S., Hurley, K., Ivezić, Ž., Kent, S., Kleinman, S. J., Krzesinski, J., Lindsay, K., Long, D., Nemiroff, R., Neilsen, E. H., Nitta, A., Newberg, H. J., Newman, P. R., Perez, D., Periera, W., Schneider, D. P., Snedden, S. A., Stoughton, C., Vanden Berk, D. E., York, D., and Ziock, K., *ApJ*, **571**, L131–L135 (2002).
13. Park, H. S., Williams, G. G., and Lindsay, K., GRB 020531, optical observations (2002), GCN 1404.
14. Williams, G. G., Blake, C., and Hartmann, D., GRB 020813: Early Time Magnitude (2002), GCN 1492.
15. Park, H. S., Williams, G. G., and Barthelmy, S. D., GRB 021211 (HETE 2493), OT at 143 sec. (2002), GCN 1736.
16. Gehrels, N., "The Swift GRB Mission," 2004, these proceedings.
17. Barthelmy, S. D., "The Burst Alert Telescope (BAT) on the Swift MIDEX Mission," 2004, these proceedings.
18. Lamb, D. Q., and Reichart, D. E., *ApJ*, **536**, 1–18 (2000).

Mining the Sky for Explosive Optical Transients with Both Eyes Open

W. T. Vestrand*, K. Borozdin†, D. J. Casperson*, S. Davidoff†, H. Davis*, E. Fenimore*, M. Galassi*, K. McGowan†, D. Starr†, R. R. White*, P. Wozniak† and J. Wren*

*Los Alamos National Laboratory, ISR-2, MS B244, Los Alamos, NM 87545, USA
†Los Alamos National Laboratory, ISR-2, MS D436, Los Alamos, NM 87545, USA

Abstract. While it has been known for centuries that the optical sky is variable, monitoring the sky for optical transients with durations as short as a minute is an area of astronomical research that remains largely unexplored. Prompt follow-up observations of Gamma Ray Bursts have shown that bright, explosive, optical transients exist. However, there are many reasons to suspect the existence of explosive optical transients that cannot be located through sky monitoring by high-energy satellites. The RAPTOR sky monitoring system is an autonomous system of telescope arrays at Los Alamos National Laboratory that identifies fast optical transients as short as a minute and makes follow-up observations in real time. The core of the RAPTOR system is composed of two arrays of telescopes, separated by 38 kilometers, that stereoscopically monitor a field of about 1300 square degrees for transients down to about 12.5th magnitude in 30 seconds. Both arrays are coupled to real-time data analysis pipelines that are designed to identify transients on timescales of seconds. Each telescope array also contains a more sensitive higher resolution "fovea" telescope, capable of both measuring the light curve at a faster cadence and providing color information. In a manner analogous to human vision, each array is mounted on a rapidly slewing mount so that the "fovea" of the array can be rapidly directed for real-time follow-up observations of any interesting transient identified by the wide-field system. We discuss the first results from RAPTOR and show that stereoscopic imaging and the absence of measurable parallax is a powerful tool for distinguishing real celestial transients in the "forest" of false positives.

INTRODUCTION

Multi-wavelength observations have proven themselves as powerful probes of the physics of gamma ray bursts (GRBs). For example, optical observations taken many hours after the GRB have provided burst redshifts, identified host galaxies, and determined the long-term temporal behavior of burst afterglows (e.g.[1]). As a result, it is now believed that GRBs are rare cataclysmic endpoints of stellar evolution (e.g. [2]). Unfortunately, observations of afterglows taken hours later tell us little about the details of the cataclysmic events because variations that occur during the first day in the plasma frame arrive at Earth within the span of the first minute after the onset of the GRB. Observations taken many hours after the burst therefore only measure emission from regions located light years away from the progenitor. To fully unravel the puzzle of GRB origin, we need multi-wavelength observations taken during (*or starting even before*) the first minute of the event.

GRB PRECURSORS, SHORT DURATION GRBS, AND THE FIRST FEW SECONDS

Since it carries key information about the cataclysmic explosion, the goal for observers of prompt optical emission is to make measurements while the GRB is still emitting gamma rays. The distribution of GRB time durations measured at gamma ray energies is bimodal with a class of short GRBs having durations $T_{90} < 2$ seconds and a class of long GRBs with durations $T_{90} > 2$ seconds [3]. Observations of the prompt emission from the long duration event GRB 990123 showed an optical flash that started during the gamma ray emission, lasted about 80 seconds, and reached the astounding peak apparent magnitude of 9 —making it the most luminous optical source ever measured by man [4]. Further, while the prompt light curve for GRB 990123 was sparsely sampled, it had a shape consistent with that predicted for a reverse shock driven through the burst ejecta as it collides with ambient material. Afterglow observations have convincingly shown that long duration events, like GRB990123, are associated with supernovae and are probably located in star formation regions(e.g. [5]).

But the short duration GRBs may have quite different progenitors. One popular idea is that the short duration events are binary mergers of neutron stars or black holes (e.g. [6]). In that scenario one would expect the GRB explosion to occur in an environment with less dust and gas and therefore show less extinction of the optical emission and a very rapidly fading afterglow. To date, there have been no observations of prompt emission or afterglow emission from a short duration GRB. The runner-up for earliest detection of prompt optical emission from a GRB is our RAPTOR (RApid Telescopes for Optical Response) detection of GRB021211 that was $R_c = 14.0$ magnitudes at 65 seconds after the HETE trigger [7]. Extrapolation of the fading light curve measured during the first ten minutes predicts it had a magnitude of $R_c \sim 10$ at ten seconds after the trigger. Interestingly, GRB021211 had a gamma-ray duration of 2.5 seconds, which places it in the transition region between short-duration and long-duration events. Optical observations in the first few seconds of a burst could therefore be the key to understanding the different classes of GRBs. But rapidly slewing telescopes have little chance of observing the GRB location during the first few seconds.

There are also likely to be other explosive optical transients that cannot be found by rapidly slewing telescopes cued by high energy satellite monitors. Some GRB theories predict so-called optical "orphan" transients that are more common than GRBs (e.g.[8]). It has also been suggested that precursor flashes of optical emission could precede the GRB (cf. [9]). Further, our knowledge of the variability of optical sky is so incomplete that it is likely there are undiscovered classes of rapid optical transients that are completely unrelated to high-energy transients.

To search for these optical transients, as well as observe GRBs during their first few seconds, one needs to monitor a significant fraction of the sky and have a robust technique for rejecting the false positives generated by non-celestial transients. To study the physics of explosive transients, one also needs a monitoring system that can autonomously locate, in real time, the celestial transients and command follow-up observations in less than a minute. The RAPTOR program at Los Alamos National Laboratory is a pathfinding effort to construct such a fully automated sky monitoring system [10].

THE RAPTOR STEREOSCOPIC SKY MONITOR

As predators, we humans have evolved a highly sophisticated vision system for both imaging and change detection [11]. Human vision employs two spatially separated eyes viewing the same scene both to eliminate image faults like "floaters" and to extract distance information about objects in the scene. Each eye has a wide-field, low-resolution, imager (rod cells of the retina) as well as a narrow-field, high-resolution imager (cone cells of the fovea). Both eyes send image information to a powerful real-time processor, the brain, running "software" for the detection of interesting targets. If a target is identified, both eyes are rapidly slewed to place the target on the central fovea imager for detailed "follow-up" observations with color sensitivity and higher spatial resolution. During each step of the process, our brain is running powerful real-time software and comparing with an adaptive catalog—our memory—to study changes in the scene.

The stereoscopic, wide-field, sky monitoring component of the RAPTOR system is best understood as an analogue of human vision. To rapidly identify transients it employs two spatially separated telescope arrays (RAPTOR-A and RAPTOR-B). Each telescope array simultaneously images the same 1500 square-degree field with a wide-field imager and a central 4 square-degree with a narrow-field "fovea" imager. A sophisticated real-time software pipeline instantly analyzes images from RAPTOR A and B, identifies potential candidates, and the positions of any interesting transients without a measurable parallax are fed back to the mount controllers with instructions to point the fovea telescopes at the transient. The two fovea cameras then image the transient with higher spatial resolution and at a faster cadence to gather light curve information. Each fovea camera also images the transient through a different filter to provide color information. The RAPTOR A and B arrays therefore act as a binocular monitoring system employing closed loop feedback that autonomously identifies, generates alerts, and makes detailed follow-up observations of optical transients in real-time.

Previous attempts to search the sky for rapid optical transients from celestial objects have always been compromised by false triggers [12]. Those non-celestial false triggers are generated by a wide range of noise sources including—but not limited to—cosmic-ray hits, hot pixels, aircraft lights, image ghosts from bright stars, meteors, and glints from space debris as well as satellites. One observing strategy that has been used to reject many of the false positives is to require detection of the transient in consecutive images [13]. This persistence requirement filters out many false positives, but glints from slowly moving objects, like those in geostationary stationary orbit, can still be falsely identified as celestial. In addition, one loses all ability to find even bright flashes that have a duration shorter than the single image integration time.

To suppress false triggers, the RAPTOR system uses two wide-field arrays (RAPTOR-A and RAPTOR-B) to stereoscopically view the same scene. The RAPTOR-B telescope is located at the Los Alamos Neutron Science Center (LANSCE) in Los Alamos, New Mexico, and the RAPTOR-A telescope array is located almost 38 kilometers due west at our Fenton Hill Observatory site in the Jemez mountains. That 38 kilometer baseline yields a parallax shift of more than 220 arcseconds for non-celestial objects all the way out to the altitude of geostationary orbits at 36,000 kilometers. Our wide-field imagers have a single pixel resolution of 34 arcseconds so any transient generated at distance at least out to six times geostationary will have a detectable parallax. Figure 1 shows a

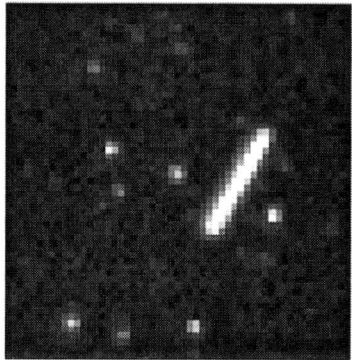 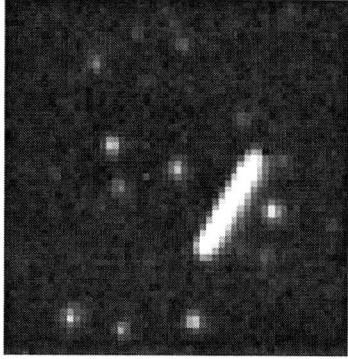

FIGURE 1. A small section trimmed from a stereoscopic pair of simultaneous images taken by the RAPTOR sky monitor. These 10-second images show the asteroid 2002 NY40 as it flew by earth at a distance of 1.3 times the distance to the moon. Notice the shift of the asteroid streak with respect to the backgrond stars.

demonstration of the stereoscopic system's ability to detect the parallax of an object at 1.3 times the distance to the moon.

The Wide-field imagers of the RAPTOR-A and RAPTOR-B arrays are each composed of four Canon 85mm f/1.2 lenses with CCD cameras at the focal planes. The cameras are thermo-electrically (TE) cooled Apogee AP-10 cameras, which employ a 2K×2K format Thomson 7899M CCD chip with 14-micron pixels. The camera electronics have been optimized to provide fast readout of the entire array in 5 seconds. Each camera of the array covers a 19.5° by 19.5° field and the four cameras are splayed so that they together simultaneously mosaic a field of approximately 1500 square-degrees. The 3-sigma limiting magnitude of this wide-field system is $\sim 12.5^{th}$ magnitude for a thirty-second exposure at the LANSCE site and $\sim 13^{th}$ magnitude at the darker Fenton Hill site. A nice feature of the mosaic approach is that each element of the array has its own dedicated computer for control and data acquisition. The signal processing task is therefore intrinsically parallel—which is a significant advantage when searching in real time for short duration transients.

In the center of each wide-field array is a fovea telescope. It is composed of a large 400mm focal length Canon telephoto lens with a 5.6-inch objective diameter and a Finger Lakes Instruments (FLI) MaxCam CM2-1 CCD camera. The camera uses a TE cooled, 1K×1K format, Marconi CCD-47 back-thinned chip with 13-micron pixels and a fast 4-second readout time. In this configuration the fovea cameras cover approximately a 2° by 2° FOV and have nearly five times the spatial resolution of the wide-field array. The limiting magnitude of these telescopes is about 16.5 magnitude for a 60-second exposure, making then well suited for faster cadence imaging of any identified transient.

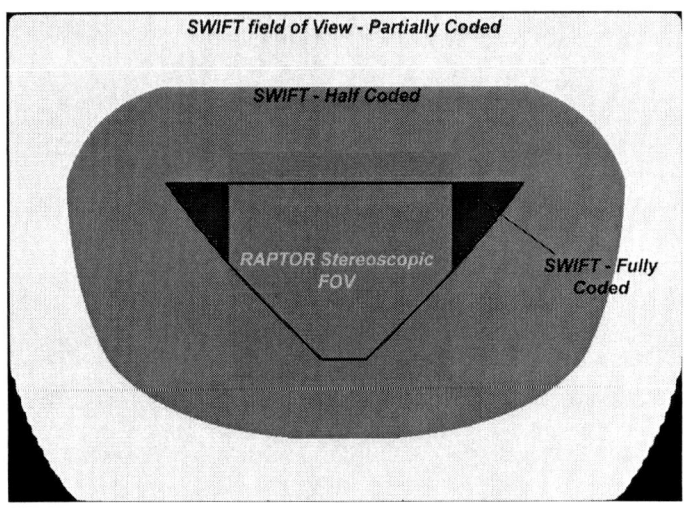

FIGURE 2. The RAPTOR stereoscopic sky monitor's field of view compared to the field monitored by the Swift Burst Alert Telescope.

CONCLUSIONS

We have described the RAPTOR stereoscopic sky monitor, now operating at LANL, that is designed to mine the optical sky for explosive transients. The system is capable of autonomously identifying transients with durations as short as a minute and commanding follow-up observations with more powerful telescopes in real time. By employing simultaneous stereoscopic imaging from sites separated by 38-km, the system is able to discriminate celestial transients from the non-celestial transients which are several orders of magnitude more common. The large field monitored by the system is comparable to the fully-coded FOV for Swift's Burst Alert Telescope—this makes it an effective system to search for optical emission from a broad range of explosive transients in the Swift era.

REFERENCES

1. Fruchter, A. et al. , *ApJ* **519**, L13 (1999).
2. Paczynski, B., *ApJ* **494**, L45 (1998).
3. Kouveliotou, C. et al , *ApJ* **413**, L101 (1993).
4. Akerlof, C.W. et al., *Nature* **398**, 400 (1999).
5. Stanek, C. et al., *ApJ* **591**, L17 (2003).
6. Meszaros, P. and Rees, M., *ApJ* **397**, 570 (1992).
7. Wozniak, P.R. et al.,*GCN Circ.*, 1757(2002).
8. Granot, J. et al., *ApJ* **570**, 61 (2002).
9. Paczynski, B., *astro-ph*/0108522 (2001).
10. Vestrand, W. T., et al.,*SPIE* **4845**, 126 (2002).
11. Hubel, D. H., *Eye, Brain and Vision*, W. H. Freeman and Co., New York, 1995, pp. 33-59.
12. Vanderspek, R., et al., in *Gamma-Ray Bursts*, AIP Conf. Proc. 307, New York, 2003, p. 438.
13. Kehoe, R. et al., *ApJ*, **577**, 845 (2002).

RAPTOR-scan: Identifying and Tracking Objects Through Thousands of Sky Images

Sherri Davidoff* and Przemyslaw Wozniak[†]

*alien@mit.edu
[†]wozniak@nis.lanl.gov

Abstract. The RAPTOR-scan system mines data for optical transients associated with gamma-ray bursts and is used to create a catalog for the RAPTOR telescope system. RAPTOR-scan can detect and track individual astronomical objects across data sets containing millions of observed points.

Accurately identifying a real object over many optical images (*clustering* the individual appearances) is necessary in order to analyze object light curves. To achieve this, RAPTOR telescope observations are sent in real time to a database. Each morning, a program based on the DBSCAN algorithm clusters the observations and labels each one with an object identifier. Once clustering is complete, the analysis program may be used to query the database and produce light curves, maps of the sky field, or other informative displays.

Although RAPTOR-scan was designed for the RAPTOR optical telescope system, it is a general tool designed to identify objects in a collection of astronomical data and facilitate quick data analysis. RAPTOR-scan will be released as free software under the GNU General Public License.

INTRODUCTION

Recognizing and tracking the behavior of real objects in a large collection of sky observations is very difficult due to environmental factors and inherent imprecision in telescope cameras and software. The RAPTOR telescope system, designed to detect optical transients associated with gamma-ray bursts, amasses an enormous quantity of sky data[1]. Every night, a single camera on the RAPTOR telescope system acquires approximately 200-400 optical images of the sky, where each image contains 15-30,000 observed objects. Over multiple exposures, the appearance of a real object will not occur at precisely the same coordinates each time. In addition, clouds and "hot pixels" affect the quality of an image, blocking some objects and producing false ones.

Real objects may be identified with high accuracy using the density of observed points in a given area. As shown in Figure 1, when multiple images of the same location in the sky are superimposed, static real objects appear as small, dense circular clusters of points, moving objects produce dense trail-shaped clusters, and false objects tend to be scattered points or sparse trails. The process of identifying these clusters and labeling each point with the appropriate cluster identification number is referred to as *clustering*.

RAPTOR-scan is designed to recognize real objects across a large series of potentially noisy sky observations, and to provide a means of analyzing their behavior over periods that range from hours to years. The system accomplishes this by storing all observations in a database. Objects are identified and labeled using a program based on the DBSCAN clustering algorithm[2]. RAPTOR-scan additionally provides a graphical analysis tool

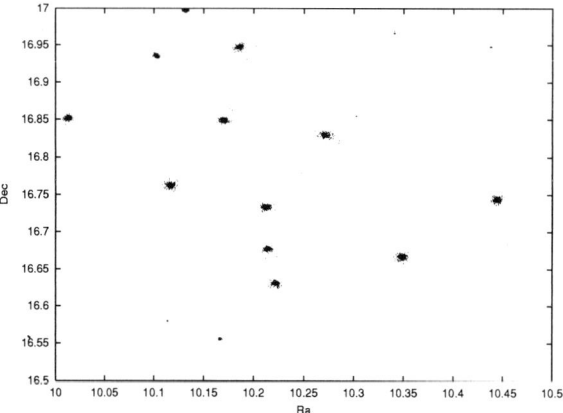

FIGURE 1. Spatial plot of objects found during one night in a small segment of the sky. Note how the points from individual exposures form clusters.

which produces visualizations of the data stored in the database. The RAPTOR-scan clustering system and graphical tool are currently being used to mine data for optical transients associated with gamma-ray bursts (GRBs) and to create a catalog for the RAPTOR telescope system.

DESIGN

Each night, a single camera on the RAPTOR telescope system produces hundreds of optical images. A list of objects is extracted from each image in real time and sent via internet to the RAPTOR-scan *database*. In the morning, when the previous night's data stream has completed, the *clustering program* automatically sorts through the database, adding a cluster identification number to each data point. After this process completes, the *analysis tool* may be used to produce various graphical analyses of the data. The clustering program may also be used to link data from many nights together to allow analyses over longer periods of time, or to produce catalogs.

Database

The RAPTOR-scan database contains six tables. The **starlist table** contains information extracted from the header of each optical image processed by the RAPTOR system. There is a separate starlist table for each night of data collection, and it stores one entry for each optical image. The starlist table contains environmental data (such as outside temperature and pressure); image statistics (such as mean, minimum and maximum pixel values); image information (including the date and time of the start of observation, the name of the source image, the telescope/camera identifier and the name of the dark and flat used to correct the image); and a unique image identification number.

The **object table** stores one entry for every object found in every image over the course of a night, typically six or seven million entries. For each object, the table contains: right ascension, declination, magnitude, standard deviation of the magnitude, the source image identification number, and a Hierarchical Triangle Mesh[3] (HTM) index which facilitates quick retrieval of objects from a section of the sky. This information may be used to analyze the spatial distribution of extracted objects, and (in combination with the cluster identification values) to analyze objects' magnitude and motion over time.

The cluster identification numbers and their corresponding object identifiers (OID from the object table) are stored in the **cluster_id table**. This information links the points in the object table to specific clusters. Storing the cluster identification numbers separately allows the clustering program to enter the data using the Postgres command "COPY", which is much quicker than updating a pre-existing table.

Detailed information about each cluster is stored in the **cluster table**. The cluster table contains the average right ascension and declination for each cluster (which are presumed to be the best estimate of the actual position for the associated real object), the average magnitude, the standard deviation of the magnitude, and the cluster's HTM identifier. In addition, the cluster table stores data to allow for re-calculation of the average values if new points are added to the cluster later.

Timing constraints limit the amount of data that can be clustered regularly to a few nights' worth at the most. Therefore, RAPTOR-scan clusters data every 24 hours. In order to track the behavior of a real object over a period of weeks, months, or years, there must be a way to recognize clusters over multiple cluster tables– in short, to "cluster the clusters". Fortunately, this problem is precisely the same as that for nightly object identification, but on a larger scale. After the desired period, the RAPTOR-scan clustering program may be run on the cluster table to produce clusters of clusters, or *superclusters*. The **supercluster_id** and **supercluster** tables then store meta-cluster information for analysis across multiple days.

Clustering

The clustering program uses the DBSCAN clustering algorithm to identify stars and transients, filter out noise and track moving objects over a series of images. This algorithm uses the information in the database to identify dense clusters of arbitrary shape which correspond to real objects. The RAPTOR-scan implementation of the DBSCAN clustering algorithm is summarized as follows:

1. Retrieve the entire list of points from the object table.
2. For each unclassified point:
   (a) Get the list of *neighboring points* that are within a radius Epsilon from the current point.
   (b) If the number of neighboring points is not greater than a specified minimum number, then the area is not dense enough to be the core of a cluster. The current point is labeled as *noise*.

(c) If the number of neighboring points is greater than the minimum number, then a cluster exists. The current point and all neighboring points that are currently unclassified or noise are labeled with the appropriate cluster identification number. Then, each neighboring point is recursively checked so that all unclassified or noise points within a radius of Epsilon from any neighboring point are also included in the current cluster.
3. If a cluster exists, store the corresponding information in memory and increment the current cluster identification number.
4. When all clusters have been found, store cluster identification numbers in the cluster_id table and cluster information in the cluster table.

Analysis Tool

The analysis tool is a command-line C program that can be used to produce the following types of graphical analyses:

- Header plots: graphs of selected header information over time. These plots are very useful for determining the quality of a night's data.
- Light curves for a specific object or all objects within an area.
- Magnitude versus standard deviation for all clusters in a specified table. This helps to determine the accuracy of the estimated cluster magnitudes.
- Spatial plots (right ascension versus declination) of objects, clusters, or both. The spatial plots allow researchers to visually examine object distribution and cluster assignments.

CONCLUSION

The RAPTOR-scan system accurately identifies and tracks stars and transients over thousands of optical images. It also provides an analysis tool which may be used to produce visualizations of the data stored in the database. RAPTOR-scan will be released as free software under the GNU general public license.

REFERENCES

1. Galassi, M., Starr, D., Wozniak, P., and Borozdin, K., "The Raptor Real-Time Processing Architecture," in *Astronomical Data Analysis Software and Systems XII*, edited by H. E. Payne, R. I. Jedrzejewski, and R. N. Hook, Astronomical Society of the Pacific, San Francisco, 2003, vol. 295 of *ASP Conference Series*, p. 225.
2. Sander, J., Ester, M., Kriegel, H.-P., and Xu, X., *Data Mining and Knowledge Discovery*, 2, 169–194 (1998).
3. Kunszt, P. Z., Szalay, A. S., Csabai, I., and Thakar, A. R., "The Indexing of the SDSS Science Archive," in *Astronomical Data Analysis Software and Systems IX*, edited by N. Manset, C. Veillet, and D. Crabtree, Astronomical Society of the Pacific, San Francisco, 2000, vol. 216 of *ASP Conference Series*, pp. 141–145.

A Rapid-Response Gamma-Ray Burst Afterglow Observing Program at Etelman Observatory in the US Virgin Islands

Timothy W. Giblin[*], James E. Neff[*], Jon Hakkila[*], Dieter Hartmann[†], Noretta Andresian-Thomas[**] and Donald M. Drost[**]

[*]*Department of Physics and Astronomy, The College of Charleston*
[†]*Department of Physics and Astronomy, Clemson University*
[**]*Division of Science and Math, University of the Virgin Islands*

Abstract. The College of Charleston (CofC) is one of three institutions that belong to a consortium led by the Division of Science and Mathematics at the University of the Virgin Islands (UVI) to maintain and operate a research grade telescope at Etelman Observatory on the island of St. Thomas (18 deg, 21′ N, 65 deg W at an elevation $\sim$ 1325 ft with $\leq 1''$ seeing). The location provides 80% sky coverage of the southern celestial hemisphere and, on average, $\sim$ 6 hours of clear sky per night, except during the peak of hurricane season. This makes the observatory an ideal facility for observing Gamma-Ray Bursts (GRBs). The observatory will serve a variety of needs to the consortium members that include research, teaching, and public outreach. The primary research function of this facility will be to perform rapid, automated followup observations of Gamma-Ray Bursts (GRBs) observed with NASA's Swift spacecraft, to be launched in June 2004, via the GCN. The newly renovated observatory houses a new robotic 0.5 m Cassegrain telescope with a back-illuminated Marconi 2024 × 2024 CCD42-40 imaging array and 12-position UBVRI filter wheel. Assuming $1.5''$ seeing and a S/N=5, we can obtain a limiting (unfiltered) magnitude of $\sim$ 19 with a 10 s integration time. The slew rate is ≥ 10 deg per second.

With the exceptional sky coverage at Etelman Observatory, we anticipate a detection rate of about 10-15% of the Swift detection rate, and we anticipate making a significant contribution to the global network of small telescopes dedicated to GRB observations.

INTRODUCTION

In the last few years, small observatories dedicated to Gamma Ray Burst (GRB) rapid-response have been playing a crucial role in deciphering the physics of Gamma-Ray Bursts. Observatories housing small telescopes (1 m or less) have the advantage of a rapid response to catch the early optical emission from the burst. A global network of small rapid response telescopes that cover a broad range of longitude is necessary to extend the time base of the counterpart and afterglow observations. Telescopes close to the equator provide broad sky coverage of both celestial hemispheres.

A consortium led by UVI has undertaken an effort to renovate the UVI-owned Etelman Observatory in St. Thomas. The observing site is strategically located at 65 deg W longitude, placing it as the most eastern GRB-dedicated observing site in the western hemisphere (see Figure 1). The consortium includes the University of the Virgin Islands, the College of Charleston, and South Carolina State University. CofC leads the primary

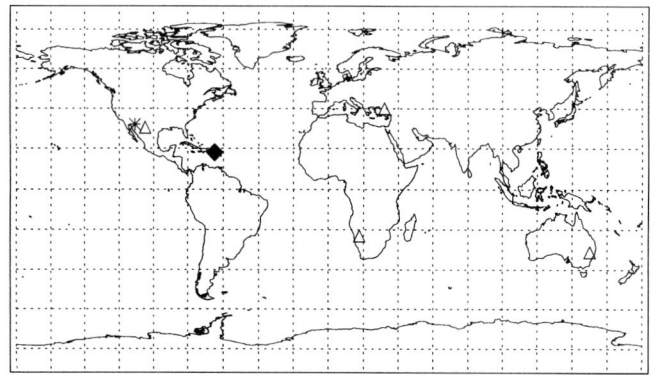

FIGURE 1. The UVI observing site (*filled diamond*). For comparison, the ROTSE-III nodes are also shown (*open triangles*), as well as S-LOTIS (*asterisk*).

research function of the observatory: a GRB rapid response observing program.

GRB OBSERVING PROGRAM OBJECTIVES

The rapid response capability of the telescope should allow observations of the prompt optical GRB emission for a few long bursts and place constraints on the reverse external shock physics. Observations of the early afterglow will further help fill in the gaps in the afterglow light curves and aid in the determination of break times, pre- and post-break slopes, and short-timescale variability. For brighter afterglows ($\leq$ 20th mag), colors in the BVRI Johnson-Cousins system will be measured to: (1) constrain the ordering of v_m and v_c and (2) aid in the search for orphan afterglows. Since the majority of the multi-wavelength afterglow data in hand suggests that GRBs are highly collimated, a significant number of orphan afterglows are expected [1].

OBSERVING SITE AND INSTRUMENTATION

The Observing Site. The site is situated at the crest of Crown Mountain on the island of St. Thomas in the US Virgin Islands. The facility is situated on a $\sim$ 2 acre site on a ridge at an elevation of $\sim$ 1325 ft, roughly 5 miles from UVI's St. Thomas campus. Two weather stations were recently installed at the site, both logging real-time conditions. A third weather station is located on the UVI campus, archiving hourly averages for the past several years. Meteorological records from the NOAA station at the airport and the USGS site in Puerto Rico indicate the sky brightness is very low and the atmospheric stability is extremely good on this island site in the trade-wind zone. Our visual observations indicate a seeing better than 1 arc-second. Local cloud patterns are very predictable, forming mainly near the east ridge of the mountain. To monitor this, we plan to install a 4th weather station about 0.5 miles east of the site. On average, we expect about 6 hours of clear sky per night except during hurricane season which typically lasts from July through October.

Observing Facility. The Etelman house and observatory, originally housing a 15-inch Fecker, was donated by the Etelman family to the University of the Virgin Islands in the mid-1960s. Thanks to an NSF grant to the Division of Science and Mathematics at UVI, a new 0.5 m robotic telescope, an automated dome and renovated control room and housing facility are now in place.

Telescope & Imaging System. The 0.5 meter, f/10 Cassegrain telescope was constructed by Optical Mechanics in Iowa City (see Figure 2). The linux-based "Talon" observatory control system enables fully-robotic operations, remote-control, or manual control of the telescope. The site is equipped with a dedicated high-speed internet link that will allow us to automatically transfer the data to CofC as well as a 100 GB file-server that will automatically archive the data. Observatory, telescope, and instrument control packages are fully integrated with each other and with various inputs (weather station, GPS, surveillance cameras), the data archiving system, and preliminary data analysis software. Identical software is running at CofC for student training, robotic-control algorithm development, and data analysis. The hardware specs are similar to ROTSE-III [2]. Hardware components and specifications are given below:
- 0.5 m Cassegrain f/10 w/ anodized 6061T6-511 Al Equatorial Fork Mount
- Parabolic primary with 0.76 wave RMS measured at 633 nm.
- Hyperbolic Secondary w/ open loop stepping focus
- GPS receiver, $< 10\,$deg/s slew time, $\pm 10''$ pointing accuracy
- 2048×2048, Marconi 42-40 Back-Illuminated CCD Imager, TE Cooled
- 13.5μ pixels, $19' \times 19'$ fov (same as *Swift* XRT)
- 12 position filter wheel w/ Johnson-Cousins UBVRI filters
- $20\mu s$/pixel readout
- Sensitivity: 19.7, S/N=5 unfiltered w/ 10 sec. integration time.

The telescope automation is governed by the Talon software, taking full advantage of the UNIX architecture. Long-running daemon processes are used to communicate with the hardware. Each daemon stores its current state in a common shared memory segment which processes may use to efficiently learn of a current system state and activity. These offer real-time control over the system and batch scheduling processes, as well as interrupts. The 0.5 m will respond to GRBs via the Gamma-ray burst Coordinate Network (GCN). The GCN will be configured for Etelman's latitude and longitude. Upon receipt of a GCN, a series of variables in the weather daemon log file will be scanned. A script then parses the GCN notice and forwards the GRB coordinates into a new scheduling file (`.sch`) which is given highest priority in a scan list file. The program `telrun`, the basic automated sequencer for the system, continually scans `telrun.sls` for new commands. Therefore a GCN trigger will take precedence over previously scheduled observations. In an automated response to a filtered GCN notice,

FIGURE 2. The 0.5 m automated telescope with filter wheel and CCD.

the telescope will rapidly slew to the center of the GRB error box and begin the series of observatons defined in the new scheduling file.

For at least the first six months of the GRB observing campaign, the telescope operators will be paged on new GCN burst triggers so that the presence of an OT can be identified manually by having the operator compare the image of the GRB error box with the POSS images. Later during the first year we expect to develop a technique to automate this procedure.

CURRENT STATUS & PLAN

The telescope was installed in early November 2003. We are currently engaged in an on-site calibration and testing phase. We are also testing remote operations and developing the robotic operations mode. The observatory will be ready in time for $Swift$, in late 2004. We anticipate ~ 15 GRB detections per year. A successful observing campaign with the 0.5 m telescope will lay the groundwork for a future plan to build a 1-2 meter class GRB dedicated telescope at the same site within 5-10 years.

ACKNOWLEDGMENTS

We gratefully acknowledge support for this project from NSF, NASA/EPSCoR, DoD, SC Space Grant, AAS, and The College of Charleston. We also thank Arne Henden for useful discussions.

REFERENCES

1. Rhoads, J., *ApJ*, **557**, 943 (2001).
2. Akerlof, C. W., et al., *PASP*, **115**, 132 (2003).

Watcher: A Telescope for Rapid Gamma-Ray Burst Follow-Up Observations

J. French[*], L. Hanlon[*], B. McBreen[*], S. McBreen[*], L. Moran[*], N. Smith[†], A. Giltinan[†], P. Meintjes[**] and M. Hoffman[**]

[*]*Department of Experimental Physics, University College Dublin, Dublin 4, Ireland.*
[†]*Department of Applied Physics and Instrumentation, Cork Institute of Technology, Bishopstown, Co. Cork, Ireland.*
[**]*Physics Department, University of the Free State, Bloemfontein, South Africa.*

Abstract. The Watcher telescope is planned to begin operation in Spring 2004 in South Africa. The system has been designed to respond primarily to very precise (arcminute) gamma-ray burst locations distributed via the internet by the GCN. Watcher will be fully automatic and the planned response time for GRBs is $\sim$ 30 seconds or better. In addition, the telescope will be used for blazar monitoring and the photometric detection of extra-solar planets when GRBs are not being observed.

INTRODUCTION

A significant number of robotic telescopes dedicated to the search for GRB optical counterparts and afterglows are now in operation or at various stages of development. These include BOOTES [1], BART [2], ROTSE [3], TAROT [4] and LOTIS [5]. So far, only one GRB has been detected optically during the burst by a robotic telescope. ROTSE detected GRB 990123 (Fig. 1(a)) just 22 seconds into the event [6]. More recently, very early observations of afterglows have been achieved by other robotic telescopes. GRB 021211 was detected 105 seconds after the initial trigger at an unfiltered magnitude of 14.6 [7]. The burst lasted just 2.5 seconds. The early afterglow of another burst, GRB 021004, was detected 193 seconds after the initial trigger by the RIKEN automated telescopes, at an unfiltered magnitude of 15.45 [8]. Watcher will respond to burst alerts within approximately 30 seconds or less, and will detect and monitor optical flash and afterglow emission down to a V-band magnitude of 19 with a photometric accuracy of $\sim$ 0.1 magnitudes (Fig. 1(b)).

Such early observations provide insights into the emission processes in GRBs. GRB 990123 provided the first evidence for the optical flash from the reverse shock wave predicted in the 'fireball' model [6]. The fast decline of the optical afterglow of GRB 021211 suggests that some of the so-called 'dark bursts' may not have been detected because observations began too late after the event [7]. Variability in the early afterglow of GRB 021004 may be evidence for continued activity of the central engine [8].

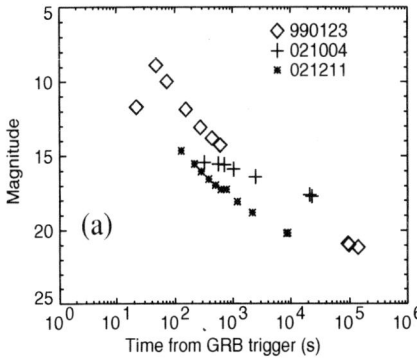 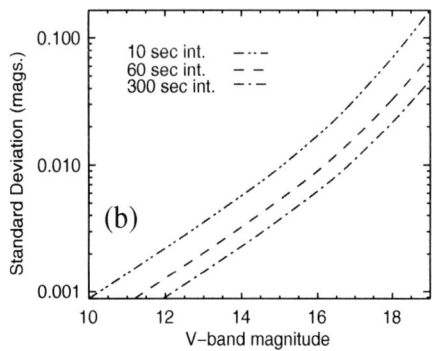

FIGURE 1. (a) Time profile of the optical counterparts and/or afterglow of GRB 990123 (V-Band)[6], GRB 021004 (unfiltered & R-Band)[7], and GRB 021211 (unfiltered)[8]. (b) Expected precision achievable with Watcher in terms of Standard Deviation (magnitudes.) vs. V-band magnitude

SYSTEM DESCRIPTION

The telescope chosen for the Watcher system is a 16″ Classical-Cassegrain telescope with an f-ratio of 14.25 and a thermally stable carbon fibre tube, available from Optical Guidance Systems. The telescope will be mounted on a Millennium Mount II robotic mount which can slew at various rates up to 7°/sec. An Apogee AP6E CCD camera will be mounted on the telescope. The image sensor in this camera is the KAF-1001E 1k x 1k array, with a pixel width of 24 μm. The camera has a fast readout time of $\sim$ 1 sec., an important consideration for early detection of gamma-ray bursts. Watcher's field of view will be 14.5′ x 14.5′, with a plate scale of 0.85″ per pixel. Observations will be made through a set of standard Bessell BVRI filters using an Optec IFW filter wheel system.

Observations, data acquisition, and the roof enclosure will be controlled by on-site computers. Roof control will be based on data from weather sensors which will monitor precipitation, temperature, wind speed and direction, cloud cover and air pressure. When not responding to burst alerts received from the GCN through an internet connection, Watcher will automatically perform routine scheduled observations. Data will be archived for subsequent analysis, with automated quick-look analysis being carried out for GRBs (Fig. 2). Watcher will be located at The Boyden Observatory (latitude 29°02′20″ South, longitude 26°24′20″ East), 24 km from the city of Bloemfontein. At an elevation of 1387m, the site has more than 300 clear nights per year. Among other telescopes, Boyden also has a 60″ reflector which may be used for follow-up observations.

HIGH-PRECISION PHOTOMETRY WITH WATCHER

The theoretical limit to the precision of the system is determined by the photon counting (Poisson) noise in the case of a bright source. Given the AP6's well-depth of 2×10^5 electrons, the maximum signal obtainable from a 9 pixel aperture would be 1.8×10^6

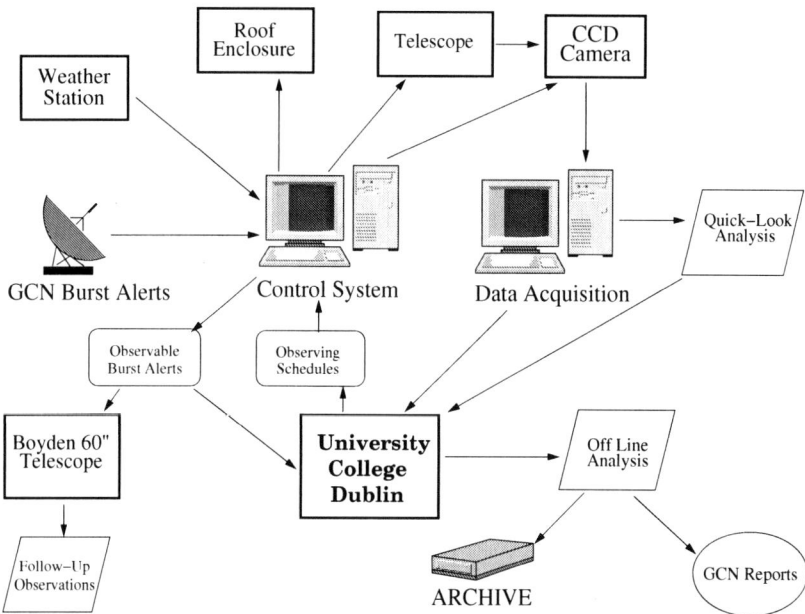

FIGURE 2. Schematic diagram of the system

electrons, yielding a photon limited precision of 8×10^{-4} magnitudes. When CCD and background noise terms are taken into account, the best possible precision that can realistically be obtained from a system is given by [9]:

$$\sigma_{mag} = \frac{1.0857\sqrt{N_* + n_{pix}(1 + n_{pix}/n_B)(N_S + N_D + N_{R^2})}}{N_*}$$

- N_* : no. of photons collected from object of interest
- n_{pix} : no. of pixels used for calculating S/N
- n_B : no. of background pixels used to estimate mean background level
- $1 + n_{pix}/n_B$: noise due to error in estimation of background level
- N_S : sky noise (no. of photons per pixel from background/sky)
- N_D : dark noise (no. of dark current electrons per pixel)
- N_{R^2} : no. of electrons per pixel due to read noise

In attempting to achieve photometry accurate to 1% or better for fainter sources (V-mag. > 14), previously minor sources of error become significant and must be examined [10]. Some examples are:

Pixel sampling Small changes in telescope pointing or atmospheric conditions can cause apparent variations in an undersampled signal. The combination of telescope f-ratio and CCD pixel size have been carefully chosen so that undersampling does not occur.

Reference star variability A chosen reference star may have intrinsic variability that will lead to inaccuracies in final magnitude measurements of the target star. A 'master' reference star can be formed by combining a number of reference stars.

Aperture photometry will be performed automatically on data frames using the IRAF package, and the resulting output used as input to an IDL program called QVAR. This program automatically searches for variability in a source by comparing it to the weighted average of a number of reference stars. QVAR can also address other sources of error, for example variations in the position of the image on the CCD array and in the width of the image, and variations in apparent magnitude of the source due to observations through different air masses. Noise artifacts due to cosmic ray hits may be rejected, and a variety of background determination methods may be applied to the data.

ACKNOWLEDGMENTS

We acknowledge support from the Irish Research Council for Science, Engineering and Technology (IRCSET).

REFERENCES

1. Castro Cerón, J. M., de Ugarte Postigo, A., García-Dabó, C. E., Mateo Sanguino, T. J., Páta, P., Bernas, M., Jelínek, M., Hudec, R., Berná, J. Á., Gorosabel, J., Más-Hesse, J. M., and Castro-Tirado, A. J., "The BOOTES experiment in support of the Gran Telescopio Canarias (GTC) in the study of the high energy Universe," in *Revista Mexicana de Astronomia y Astrofisica Conference Series*, 2003, pp. 77–80.
2. Jelínek, M., Hudec, R., Kubánek, P., Nekola, M., Soldán, J., Stoklasová, I., Topinka, M., Smída, R., Svéda, L., Hroch, F., Polcar, J., and Castro-Tirado, A. J., "BART - Recent Status," in *American Institute of Physics Conference Series*, 2003, pp. 520–522.
3. Akerlof, C. W., Kehoe, R. L., McKay, T. A., Rykoff, E. S., Smith, D. A., Casperson, D. E., McGowan, K. E., Vestrand, W. T., Wozniak, P. R., Wren, J. A., Ashley, M. C. B., Phillips, M. A., Marshall, S. L., Epps, H. W., and Schier, J. A., *PASP*, **115**, 132–140 (2003).
4. Boër, M., Bringer, M., Klotz, A., Moly, A. M., Toublanc, D., Calvet, G., Eysseric, J., Leroy, A., Meissonnier, M., Malina, R., Sanchez, P., Pollas, C., and Pedersen, H., *A&AS*, **138**, 579–580 (1999).
5. Park, H. S., Williams, G. G., Ables, E., Barthelmy, S. D., Cline, T., Gehrels, N., Hartmann, D., Hurley, K., Lindsay, K., Nemiroff, R., Pereira, W., and Perez-Ramirez, D., "Super-LOTIS and LOTIS for HETE2 GRB Triggers," in *American Institute of Physics Conference Series*, 2003, pp. 366–368.
6. Akerlof, C. W., Balsano, R., Barthelmy, S., Bloch, J., Butterworth, P., and Casperson, D., *Nature*, **398**, 400–402 (1999).
7. Li, W., Filippenko, A. V., Chornock, R., and Jha, S., *ApJ*, **586**, L9–L12 (2003).
8. Fox, D. W., *Nature*, **422**, 284–286 (2003).
9. Howell, S. B., *Handbook of CCD Photometry*, Cambridge University Press, 2000, pp. 53–58.
10. Howell, S. B., and Everett, M. E., "Ultra-High Precision CCD Photometry," in *Third Workshop on Photometry*, p. 1, 2001.

The University of Wyoming GRB Afterglow Follow-Up Program

S. L. Savage*, J. P. Norris[†], A. S. Kutyrev[†], M. Pierce* and R. Canterna*

University of Wyoming
[†]*NASA Goddard Space Flight Center*

Abstract. As the Swift era approaches, the University of Wyoming in Laramie has been preparing its two observatories for a robust GRB afterglow follow-up program. The 2.3-m Wyoming Infrared Observatory (WIRO) - first of its kind in collecting power and mid-infrared optimization - is located on Jelm Mt. (2944-m elevation) in a semi-arid atmosphere, 40 km southwest of Laramie. On dry, cold winter nights, our estimates show that WIRO's sensitivity in the K-band is comparable to that of a 4-m telescope at Mauna Kea observatory in Hawaii. Three instruments are currently in use at the observatory: WIRO-Prime, WIRO-Spec, and the Goddard IR camera. WIRO-Prime is a 2048^2 prime-focus camera with a 20 arcmin diameter FOV (f/2.1). Its sensitivity for a 300-s exposure will reach as faint as 24th (23rd) magnitude in V (R). WIRO-Spec is an integral field, holographic spectrometer which utilizes Volume-Phase-Holographic gratings with a 2048^2 detector. A bundle of 293 fiber optical cables (1 fiber $\sim$ 1 arcsec) connects the Cassegrain platform to the stationary spectrometer, optimizing the image by reduction from f/27 to f/9. At 20th magnitude, a 700-s exposure yields a S/N ratio of $\sim$ 10 at a resolution of $\sim$ 1 Angstrom, sufficient for resolving the MgII doublet [279.8 nm] in GRB host galaxies to determine a $0.4 < z < 2.5$ for an operational wavelength range of $\sim$ 400-1000 nm (WIRO-Prime and WIRO-Spec). The Goddard IR Camera is a 256^2 InSb camera (FOV $\sim$ 108 arcsec) mounted at Cassegrain and operated at 15K. Available filters for GRB observations include R, I, J, H, and K'. WIRO slew timescale ($\sim$ 120 s) is comparable to that of Swift. Red Buttes Observatory (RBO) is located 19 km south of Laramie in a dark site and houses a 0.6-m f/8 Cassegrain DFM reflector. RBO's Apogee AP8p 1024^2 camera (18 arcmin FOV, sufficiently large for BAT localizations) is available for use with filters U, B, V, R and I. We are in the final stages of implementing fully automated response to Swift BAT alerts at RBO, and expect an average acquisition timescale to random sky positions of $\sim$ 25 s. Thus, rapid GRB detections by RBO can be forwarded to WIRO even while Swift's pointed instruments are performing first integrations.

INTRODUCTION

To complement the GRB response from Swift, the combined efforts of the two research observatories at the University of Wyoming, WIRO and RBO, will enable rapid reaction to burst alerts, offer a large FOV for targeting bursts, and provide wavelength coverage ranging from optical through near-infrared. More detailed information on the observatories and their instruments can be found at http://physics.uwyo.edu/observatories/index.htm. RBO's automated response of $\sim$ 25 s from zenith to 2 air masses is comparable to Swift's slewing rate. Coupled with a large 18-arcmin FOV, it assures targeting a burst in its early afterglow stages and possibly detecting the elusive afterglow of short bursts – a feat yet to be accomplished. The large collecting power of the 2.3-m WIRO telescope and infrared optimization ensure deep imaging in wavelengths not accessible by Swift. WIRO and RBO will also be able to overlap with Swift's coverage in the optical. Because of the transient nature of bursts, GRB follow-up research must be performed on a target-of-opportunity basis. Both observatories, within an hour of Laramie, are owned by the university, making rapid and numerous prompt observations as well as long-term follow-up monitoring more feasible. Short bursts have yet to be associated with optical counterparts due to their predicted steep light curves and intrinsically low initial brightness [1]. Acquiring light curves for short bursts with the rapid reaction capabilities of RBO would be a significant contribution. In addition to imaging and photometry, spectroscopy is available for determining redshifts of high z burst sources – a necessary ingredient for modeling. GRBs are predicted to occur at considerable rates beyond $z \sim 5$, thus making them primary candidates as probes for cosmology in the study of the Lyman-α forest, the epoch of reionization, the evolution of metallicity, and large-scale, high-redshift structure [2, 3].

RED BUTTES OBSERVATORY

Red Buttes Observatory (RBO) is located 19 km south of Laramie on a dark site which, combined with the dry, thin atmosphere ($\sim$ 2200-m elevation), enables relatively deep imaging with its Apogee AP8p 1024^2 CCD camera mounted to a 0.6-m Cassegrain DFM reflector (f/8). Ten-minute exposures can yield typical limiting magnitudes of $\sim$ 19.5 (19) for V and R (B). The large 18-arcmin FOV is sufficient for rapidly acquiring and imaging Swift's BAT error regions ($\sim$ 4 arcmin radius) and relaying the locations to WIRO for deeper imaging in the IR and spectroscopy. To realize rapid response and to modernize the facility, RBO has been extensively renovated and upgraded. Among the upgrades are refurbishment of the telescope platform and facility, acquisition of faster computers, establishment of a microwave link, and installation of a GPS clock, flat-field lights/screen, a weather station, and an all-sky camera for local weather monitoring. All of the improvements were commensurate with the goal of realizing a completely automated GRB afterglow response. Current follow-up operations require human intervention; however, total automation is in its final stages.

WYOMING INFRARED OBSERVATORY

The Wyoming Infrared Observatory (WIRO), situated at 2944 m, 40 km west of Laramie in semi-arid conditions, is an optimal site for optical and infrared observations. The 2.3-m telescope is one of the largest and most readily available to afterglow follow-up research. WIRO's intermediate size allows fast acquisition strategies in comparison with larger telescopes. WIRO is currently undergoing renovations which include new and upgraded instruments as well as facility improvements. Recent additions are WIRO's three primary instruments: WIRO-Prime, WIRO-Spec, and the WIRO-Goddard IR camera. Beyond GRB targets of opportunity, WIRO is primarily dedicated to on-going observing programs by the faculty, new graduate students, and visiting astronomers with scientific emphasis on quasars, cataclysmic variables, globular clusters, mapping of the milky way, galaxy morphology, etc. In order for these instruments to operate, they must be vacuum-pumped and cooled with liquid nitrogen and/or liquid helium which can take the order of a day to reach reliable temperatures; consequently, for speed of acquisition the instrument in use at the time of a burst alert will be utilized to promptly pursue afterglows with photometry or spectroscopy.

WIRO INSTRUMENTATION

WIRO-PRIME

WIRO-Prime is an optical imaging instrument operating between $\sim$ 400 - 1000 nm with a 2048^2 13.5-μm pixel CCD. It is situated at the prime-focus mount (f/2.1) producing a plate-scale of 0.55 arcsec/pixel and a 20 arcmin diameter FOV. Under conditions of one arcsec seeing, a 5-minute exposure can yield images down to 24th V mag [4].

WIRO-SPEC

WIRO-Spec is a unique Volume-Phase-Holographic (VPH) optical spectrometer consisting of 293 fiber optic cables in a 15 x 20 array connected to a Cassegrain platform from a stationary freezer. One fiber corresponds to approximately one arcsec on the sky which allows for spectrographic sampling of a 15" x 20" FOV. Due to its high efficiency ($\sim$ 15% to $< \sim$ 40%), a 10-minute exposure typically yields a S/N $\sim$ 10 with a resolution of $\sim$ 5 Angstrom for a 20th magnitude point source considering the most optimistic values.

WIRO-GODDARD IR CAMERA

WIRO's near- to mid-infrared imaging camera, the WIRO-Goddard IR camera, was built in collaboration with NASA Goddard Space Flight Center. Its 256^2 InSb detector operates at $\sim$ 15 K necessitating the use of liquid nitrogen and liquid helium. At Cassegrain mount (f/27), the FOV of the detector is $\sim$ 2 arcmin. Available filters are R, I, J, H, and K'. Calibration and sensitivity measurements for this very recent WIRO instrumentation addition are in the process of being performed.

RESULTS AND EXPECTATIONS

Prior to the involvement of WIRO in the GRB program, RBO collected data on several GRBs in collaboration with the Follow-Up Network (FUN) GRB group in primary affiliation with Dan Reichart of the University of North Carolina, Chapel Hill. To date, eight GCN notices [5] have been archived from the University of Wyoming with detection or magnitude contributions. WIRO will continue the lightcurve analysis to deeper magnitudes than RBO, add infrared capability (J, H, and K' bands), and spectroscopy for determination of high redshift burst afterglows. Future projects are planned to include infrared star formation rate determination of GRB host galaxies.

ACKNOWLEDGMENTS

This research has been supported by NSF grant AST 00-97356, NASA EPSCoR grant NCC5-578, and NASA grant NAG5-11191.

REFERENCES

1. Panaitescu, A., Kumar, P., Narayan, R., ApJ, **561**, L171, (2001).
2. Bromm, V. & Loeb, A. ApJ, **575**, 111 (2002).
3. Lamb, D.Q. & Reichart, D.E., ApJ, **536**, 1, (2000).
4. Pierce, M. et al, in preparation.
5. GCNS: 1584, 1681, 1685, 1776, 2042, 2062, 2137, 2250

Scout or Cavalry? Optimal Discovery Strategies for GRBs

Robert J. Nemiroff

Michigan Technological University, Department of Physics, 1400 Townsend Drive, Houghton, MI 49931

Abstract. Many present and past gamma-ray burst (GRB) detectors try to be not only a "scout", discovering new GRBs, but also the "cavalry", simultaneously optimizing on-board science return. Recently, however, most GRB science return has moved out from the gamma-ray energy bands where discovery usually occurs. Therefore a future gamma-ray instrument that is *only* a scout might best optimize future GRB science. Such a scout would specialize solely in the initial discovery of GRBs, determining only those properties that would allow an unambiguous handoff to waiting cavalry instruments. Preliminary general principles of scout design and cadence are discussed. Scouts could implement observing algorithms optimized for finding GRBs with specific attributes of duration, location, or energy. Scout sky-scanning algorithms utilizing a return cadence near to desired durations of short GRBs are suggested as a method of discovering GRBs in the unexplored short duration part of the GRB duration distribution.

SCOUT OR CAVALRY

Detection algorithms for transients are rarely discussed in print. Recently Nemiroff [1] reviewed attributes common to many transient detection algorithms and attempted to draw general conclusions about useful mathematical optimizations. This paper is a preliminary attempt to apply and extend this analysis to gamma-ray bursts (GRBs) specifically.

"Scout" mode will be defined as a pointing algorithm optimized to *discover* transients. "Cavalry" mode will refer to an observational mode optimized to *study* transients. Many sky-monitoring programs try to combine "scout" and "cavalry" modes into a single instrument and pointing algorithm, including GRB searches. Additionally, a scout instrument may operate best in a particular energy band where, paradoxically, important science does not reside. Cavalry instruments optimized to mine key science attributes may therefore be inefficient scouts. Therefore, supposing sufficient resources, the GRB discovery rate (for example) is optimized by dissociating scout and cavalry modes, bandpasses, and/or instruments. A scout-specific instrument may be useful for future GRBs studies, particularly for very short duration GRBs.

SIGNAL OR NOISE

Here we will consider a canonical case with several attributes. First, the sky is considered isotropic. Once discovered by a scout, the cavalry attacks a transient. The effective apparent brightness distribution of transients $N(l)$ is assumed already known, and can be approximated at the dim limit by the power law l_{dim} as $N \sim l_{dim}^{\alpha}$.

A telescope will detect a source only if its signal peaks above the noise. At the limit of observation, one can quantify this as $l_{dim} \sim t_e^{\beta}$ for a given level of signal to noise, where t_e is exposure time. As discussed in Nemiroff [1], β approaches -1 when the signal is much greater than the background, but approaches only $-1/2$ when background noise is comparable to the transient signal.

TILE OR STARE

Combining the apparent brightness distribution with the exposure distribution, one can approximate the number of observed transients discovered during exposure time t_e as $N\, t_e^{\alpha\beta}$ [1].

Studying this simple equation can give significant insight into whether a scout should tile the sky, or stare at a single location. To calibrate intuition, let us consider the case of $\alpha\beta = 1$, so that the number of detected sources is just linear with the exposure time. An example case is when the background is low, β is about -1, and so α is about -1. Assuming it takes little time to slew to a new field, it then does not matter if one stares at the same location or tiles the sky: the same number of transients will be detected. Here the answer to "tile or stare" is a formal tie.

If the power-law index $\alpha\beta$ is less than unity, however, equation indicates that "tile" will detect more sources per unit exposure time than "stare." This is because when staring at a single field, new sources are appearing over the limiting brightness horizon at an increasingly slower rate. Higher rates are found by starting over on a new field. Given a total observing campaign time, the most sources will be found by dividing the time equally between all the available fields.

Similarly, if the power-law index $\alpha\beta$ is greater than unity, "stare" will detect more sources per unit exposure time than "tile." This is because when staring at a single field, new sources are pouring over the limiting brightness horizon at an increasingly fast rate. Lower rates would be found by starting over on a new field. Therefore, in general, given a total observing campaign time, the most sources will be found by staring at one field.

TILE OR STARE FOR GRB SCOUTS

Assume that GRBs are only detected when well above background, so that β is about -1.0. Bright GRBs in the local universe follow $N\, l_{dim}^{-1.5}$ so that $\alpha\beta$ is approximately 1.5 which is greater than unity. Application of the above analysis then indicates that "stare" mode is preferred. A GRB monitor then should just stare at one location to detect the most transients.

Below some limiting brightness, many GRBs will be detected that are outside our local universe. The brightness distribution of these GRBs will be affected by universe time and volume distortions convoluted with the co-moving luminosity function. For these "cosmological" GRBs, in the BATSE GRB detection bands, the apparent brightness distribution of GRBs roughly follows $N \sim l_{dim}^{-0.8}$, so that $\alpha \sim -0.8$. Given that even these are detected well above background ($\beta \sim -1$), then $\alpha\beta \sim 0.8$, which is less than unity. Therefore a "tile" mode will optimize transient discoveries.

It is interesting to note that most GRB detectors, past and present, work in "stare" mode and so are not optimized for GRB discovery.

OPTIMAL POINTING ALGORITHMS FOR GRB SCOUTS

Searching for GRB transients is conceptually similar to general transient search strategies discussed in Nemiroff [1]. In this particular paradigm, however, we will assume a GRB detector of area A operating in some optimal energy band for discovery. We will assume that this detector sees only an angular solid angle of the sky Ω and that the rest of the sky is shielded. One visualization of this is a non-imaging detector shaped like two consecutive meridian lines of longitude on the Earth, although other detector shapes are not excluded, including imaging detectors. This canonical detector spins with angular speed such that it returns under the same portion of the sky after a time t_{return}. The duration of a given GRB will be designated t_{dur}.

Two extreme paradigms occur, the first is when $t_{return} >> t_{dur}$. This corresponds to a relatively slow spin. Assuming a small Ω, many GRBs will be missed because the scout was not pointing toward them when they occurred. This is similar to a standard GRB "stare" detector. The usual fix is to try to make the detector point to a large Ω, suffering a trade-off where the background becomes increasingly dominant.

A second extreme paradigm is when $t_{return} << t_{dur}$. This corresponds to a relatively fast spin, indicating more of a "tile" paradigm than "stare." Here the scout is guaranteed to point toward a GRB during its emission phase, although actual detection likelihood depends on the instantaneous burst flux relative to the background.

The detection sensitivity for a tiling scout is greatest for GRBs between the two extremes, when $t_{return} \sim t_{dur}$ as discussed generally in Nemiroff [1]. Therefore, the spin rate of a GRB scout can be tuned to optimally detect GRBs of a given duration, including GRBs with a very short duration. Short duration GRBs might not be detectable any other way. *Such a capable scout might best explore the short duration part of the GRB duration distribution.*

In modern GRB studies, cavalry instruments working in other wavelength bands already obtain most GRB science. The effort to do as much burst science as possible with the scout itself now hinders the main usefulness of these instruments – to simply and optimally detect GRBs and hand them off.

Note that even if the scout is non-imaging, the scout's duty is to discover the GRB and report its location with enough accuracy and timeliness so that it can be handed off efficiently. In this light, two or three detectors spinning with non-parallel axes (complete orthogonality is one possibility) should be able to locate a GRB uniquely for handoff to a cavalry instrument.

Scouts can be tuned to optimize the discovery of GRBs with specific attributes such as duration, location, or energy. Besides duration, discussed above, the discovery energy band can itself be changed to optimize for GRBs of slightly different hardness, which may translate into an average GRB population with slightly different attributes or even a different distance class. Lastly, a scout can be tuned to look at specific locations where transients might occur, for example Soft Gamma Repeaters (SGRs) toward M31, or Terrestrial Gamma Flashes (TGFs) toward major thunderstorm regions on Earth.

Many of these same principles can be used to search for GRBs in other areas of the electromagnetic spectrum, including orphan afterglows in the optical. For GRBs, however, it appears relatively efficient to search in or near the soft gamma ray band where flux from a single GRB can more easily dominate the surrounding background.

ACKNOWLEDGMENTS

This research was supported, in part by grants from NASA and the NSF. I gratefully acknowledge useful conversations with Jerry Bonnell, Scott Gaudi, Jonathan Granot, Jay Norris, and Bohdan Paczynski. Part of this research was done while visiting the Institute for Advanced Study in Princeton, NJ.

REFERENCES

1. Nemiroff, R. J., AJ **125**, 2740 (2003).

RTS2 – Remote Telescope System, 2nd Version

Petr Kubánek*[†], Martin Jelínek[†], Martin Nekola[†], Martin Topinka*[†], Jan Štrobl*[†], René Hudec[†], Tomás de J. Mateo Sanguino**, Antonio de Ugarte Postigo[‡] and Alberto J. Castro-Tirado[‡]

*Charles University Prague, Faculty of Mathematics and Physics, Czech Republic
[†]Astronomical Institute of the Academy of Sciences of Czech Republic, Ondřejov, Czech Republic
**Centro de Experimentación de El Arenosillo (CEDEA-INTA), Mazagón, Huelva, Spain
[‡]Institute of Astrophysics of Andalusia, Granada, Spain

Abstract. BART is a small remote controlled robotic CCD telescope, devoted to rapid observation of prompt gamma ray burst transients. During its operation since early 2001, it had three prompt observations with world-competitive response time. The constraints to object magnitude were estimated and published in GCN circulars. Telescope is located in Astronomical Institute of the Czech Academy of Sciences in Ondřejov. This poster describes its new control system, named RTS2, which has been in service since February 2003.

INTRODUCTION

RTS2 is designed as a networked system for driving robotic telescopes. It is composed of several device servers, central server and various observational clients cooperating over a TCP network. RTS2 can control different types of mounts, domes and CCDs. Observation entries, requests and results are kept in database. Positions of GRBs are received from the Internet, and observed either immediately, or scheduled for later observation. When there is no request for GRB observations, the telescope monitors various active galaxies. The database is accessible through a WWW interface at http://lascaux.asu.cas.cz/bartdb.

RTS2 has been recently implemented for the cooperative Spanish-Czech BOOTES experiment. It is installed and performs well on stations BOOTES-1 and BOOTES-2 in southern Spain.

The system may be used on most Linux distributions. The vast majority of the code is written in the C programming language and uses the PostgreSQL database. The whole RTS2 package is available for download, with complete source code included.

The code is covered by the GNU General Public license, which enables anybody to modify it, if certain conditions are met. RTS2 web page is at http://lascaux.asu.cas.cz/rts2.

The system development is versioned using Concurrent Versions System - CVS. RTS2 uses GNU automake and autoconf tools for Makefile creation.

SYSTEM DESCRIPTION

The system consists of three kinds of programs:

central server
server/client programs for device control
client programs for running observations and monitoring of the system status

Central server holds a list of all connected devices and clients. Every device must be registered with the central server before it becomes accessible to clients. Every client must connect to the central server before it can monitor and control devices. Access to devices is priority-based – only the client with the highest priority can access the state-changing functions of a device. State-changing functions are functions which alter the physical state of device. For example, moving the mount is a state-changing function.

RTS2 is able to control the whole family of SBIG parallel port based CCD cameras. We are working on USB based SBIG, such as the Apogee and FLIcam CCDs. RTS2 can control the LX200 family of telescopes made by Meade, and we have an alpha level driver for the Paramount.

Implementing a driver-layer for nearly any device daemon should be relatively easy thanks to a well-planned design of the upper layers.

RTS2 has clients for scheduled observations, for gamma ray burst optical transient searches, for monitoring, and for camera focusing.

DATABASE STRUCTURE

The basic structure of the database can be seen in Figure 1. The most important are TARGETS, OBSERVATIONS and IMAGES tables.

All observation targets are kept in the TARGETS table. Each target can have an associated table of constraints which determine when and how it will be observed.

The OBSERVATIONS table forms a relation between TARGETS and IMAGES. It may be thought as a log of all light images acquired by the telescope. The IMAGES table stores only images with valid WCS headers.

IMAGE ANALYSIS

Acquired images can be processed with any image analysis software, which can run on a UNIX based operating system. Currently we use the modified Opera package, which was developed in Madrid. It adds sky coordinates (using standard World Coordinate System (WCS) headers) to processed images.

Coordinates of the processed images are sent to the mount driver. They are used for fine positioning.

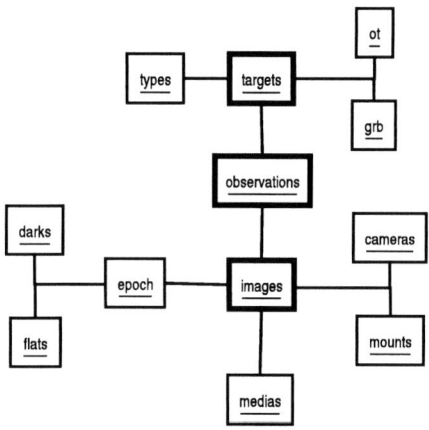

FIGURE 1. Database structure overview

SYSTEM OPERATION

The "scheduled observations client" obtains flat-field pictures during dusk and dawn. It also obtains dark-frames at scheduled times.

GRB coordinates are received primarily through the "gamma ray burst optical transient search client". This client is connected by Internet socket to the GCN server at NASA-GSFC. When it receives a GRB event which is above the local horizon, it asks the central server for priority, moves the mount, and after the telescope has been pointed at the GRB coordinates, it starts camera exposures. Pictures from the cameras are downloaded during readout through the network to a computer running the GRB client. The GRB cli9ent stores them, obtains WCS information, and inserts image information into the database.

If the socket connection fails, the backup system receives the more reliable e-mail alert. The GRB event is then observed from the main scheduler. The main scheduler also observes GRB error boxes which were bellow the horizon when a GRB event occurred, or which were received when the system was not observing due to unfavorable weather conditions.

SYSTEM STABILITY

RTS2 has been observing in Ondřejov for half a year. Most software errors have been tracked down. The system is running in semi-automatic mode, with staff available to check it 24 hours/day. The restriction is given by the lack of automatic roof control, thus the need of person to watch the weather nightly. This will change after complete roof automation in early 2004.

In Spain, at BOOTES-2, RTS2 has been installed recently, without any trouble. It performs well at the moment of this writing. It works there in fully automatic mode, without human intervention, controllable only through the Internet.

The RTS2 is based on experience gained from RTS1, which had been running on BART for two years.

CURRENT RESULTS

Between 15th and 18th August 2003, RTS2 responded to following GCN triggers:

Telescope	GCN #	RTS-2 #	Delay	first image	Notices
BART	2805	5740	1d 10:15	08-16 20:06:47	(1)
BOOTES-2	2805	67	09:18:26	08-15 19:10:05	(1)
BART	2808	5741	00:10:07	00:36:34	(2)
BOOTES-2	2808				(3)
BART	2809	5742	17:53:37	20:28:49	(2)(4)
BOOTES-2	2809				(3)
BART	2812	5743		no observation, bad weather	
BOOTES-2	2812	78	00:00:21	23:16:49	(5)

(**1**) Daytime GCN, burst localization observed from scheduler during night.
(**2**) Due to problems with rights on log file, GRB client wasn't running, all observations were from scheduler.
(**3**) GRB client crashed at 2003-08-16 around 16:30 UT, leaving no traces. Code review will follow.
(**4**) GRB was bellow horizon at receiving time.
(**5**) GRB was on horizon, images are sometimes disrupted with trees.

Exact timing of GRB-2812 processing on BOOTES-2

Time (UT)	Event
23:16:01	Burst detected on HETE
23:16:11	GCN message without localization
23:16:21	First GCN message with localization
23:16:22	Beginning of observation
23:16:22	GRB client asks for priority and gets it
23:16:22	GRB client starts moving telescope from 01:36:58, +15:49:12
23:16:22	scheduler request for common observation is ignored
23:16:48	Mount reach its position
23:16:49	Start of first 30 sec exposure
23:19:xx	First image with on-line astrometry

Limiting magnitude on narrow field was about 13 at the beginning of observation. No new object was found.

ACKNOWLEDGMENTS

We acknowledge the support provided by the grant A3003206 provided by the Grant Agency of the Czech Republic and by the ESA PRODEX, Project 14527.

Wide Field Optical Camera for Search and Investigation of Fast Cosmic Transients

A. Pozanenko*, G. Beskin[†], S. Bondar**, A. Biryukov [‡], K. Hurley [§], E. Ivanov**, S. Karpov[†], V. Loznikov*, V. Rumyantsev[¶] and Y. Zolotukhin [‡]

IKI RAS, Moscow, Russia
[†]*SAO RAS, Karachai-Cherkessia, Russia*
**Kosmoten Observatory, Karachai-Cherkessia, Russia*
[‡]*SAI MSU, Moscow, Russia*
[§]*UC Berkeley Space Sciences Laboratory*
[¶]*CrAO, Crimea, Ukraine*

Abstract. The primary purpose of the fast wide field optical camera (WFOC) is to perform continuous, alert-independent observations of optical transients and variable astrophysical sources simultaneously with space-born wide field X- and γ-ray telescopes. In particular the camera can detect possible optical precursors and early prompt emission from cosmic Gamma-Ray Bursts. The real-time source identification software generates alerts that also could be sent to global alert distribution networks such as the GCN. We estimate that in one year of continuous observation with the WFOC we will observe the following numbers of GRB error boxes simultaneously with space-borne telescopes: 1.6 (WXM/HETE-2), 0.5 (SPI/INTEGRAL), and 4 (BAT/SWIFT).

Even though GRBs were discovered as gamma-ray events, only optical afterglow observations placed them firmly at cosmological distances. While GRB science is getting more afterglow-related, only the study of the prompt emission can help unravel the emission mechanism. Indeed, the photon flux in the prompt optical emission can be as much as 10^2 - 10^3 more intense than the gamma-ray flux. The mystery of optical prompt emission seems to be ready for resolution after the observation of GRB990123 [1]. However, in the following 4 years no prompt emission has been observed. The problem is that the astronomical telescopes and their methods of observation are not fully suited to the problem.

To register prompt emission we need to look for celestial optical transients (OT) independent of the alert system. Although the time delay between a GRB trigger and an optical observation may be reduced, it will never vanish. Hence an alert-based observation cannot register neither early prompt emission nor possible optical precursors [2] or possible afterglows from short duration bursts [3].

To ensure that an OT is a counterpart to a GRB one needs to confirm the event in γ-rays. Simultaneous observations with space-borne GRB missions are therefore necessary. It is not necessary to correlate the optical and γ-ray data in real time; we need only assure that both telescopes observe the same field of view (FOV) of the sky simultaneously. The joint correlative analysis may be done later. Observation of only a specific part of the sky decreases the amount of data to be stored in comparison with all sky surveys [4, 5] and allows the time resolution of the survey to be improved.

The FOVs of the telescopes and time resolution of the detectors are crucial points in the search strategy. Astronomical telescopes have small FOVs compared to space-borne X- and γ-ray telescopes. Specialized telescopes may have a wide FOV (19.°5 × 19.°5 with 30 s time resolution, such as RAPTOR [6]) but do not posses a sufficient sensitivity with appropriate time resolution and vice versa (1.°85 × 1.°85 at 4 s, ROTSE-III [7]). To detect prompt emission efficiently the time resolution should be better than of nearly equal to the duration of the event. If the prompt emission consists of short duration optical flashes (e.g. [8, 9]) then a high time resolution detector should be used. Moreover, if the nature of the prompt optical emission is similar to that of the γ-ray emission, fast variability may be expected up to milliseconds. Fast optical variability or low on-off time ratio may explain the nondetection of optical prompt emission from GRB030329. The upper limit for prompt emission of $V = 5.1$ mag (32 s accumulation time) [10] may be converted to upper limit $V = 1.4$ mag if the prompt emission consists of only one flash with duration about 1 s. The nondetection of prompt emission cannot exclude the model of rapidly fading ($f \sim t^{-2}$) flare found in GRB990123 and GRB021211 [1, 11].

Obviously, the larger the FOV of the telescope, the larger the fraction of the GRB error-box that can be observed simultaneously, and a more sensitive detector has a greater chance to detect a faint OT from a GRB. On the other hand, for a fixed detector size (e.g. CCD-matrix), increasing the FOV decreases the sensitivity. One can show that number of detected GRB optical events per fixed amount of time and fixed size of the detector follows the formula $N_{OT detected} \sim \left(\frac{D}{\alpha}\right)^{3/2} \cdot FOV$, where D is the diameter of telescope aperture, α is angular resolution and FOV expressed in steradians; here we assume 3-D Euclidian space and uniform distribution of GRB sources $N(>S) \sim S^{-3/2}$ [12]. One can see that the telescope with larger FOV can detect more OTs, and the OT detection probability can be maximized while observing simultaneously with a given space-born telescope. Strategy of a search for very fast OT with field of view of about a few degrees with scanning telescopes was already discussed ([13, 14, 15]).

Taking into account this strategy we have developed a low cost optical camera for a wide field survey and an autonomous search for OTs. We use an image intensifier both to reduce the size of the image in the main objective focal plane to the small size of the TV-class CCD matrix, and also to amplify the light to compensate for the light lost in transmission through the optical system. Details of the camera can be found elsewhere [12].

The practical parameters of the instrument are a 17° × 20° FOV with spatial resolution $\sim$ 1' and a limiting magnitude (3 σ level) of 11.5 mag for a 0.13 s exposure. A limiting magnitude of 13 mag is reached for a 5.2 s exposure (co-added in PC memory). The maximum frame rate is 7.5 s^{-1}, and the spectral sensitivity is close to V. The system is mounted on an equitorial mount with a pointing and tracking accuracy of a few arc minutes and less than 1' per 8 hours, respectively. A photo of the WFOC and an example of a frame image are presented in Figures 1 and 2.

The relatively poor spatial resolution is the result of a compromise between cost and a wide field of view. Indeed the precise spatial resolution in fast observations is less important than the early detection: the precise localization can be done later by large aperture telescopes. Because of the high readout noise of the TV-CCD the sensitivity of the camera is restricted by the noise at minimum exposure time, and by sky background

FIGURE 1. Wide field camera at Special Astrophysical Observatory

at maximum exposures.

The software consists of the two independent elements. The storage system management accumulates all frames captured from the camera in a circular 100 Mb buffer and stores it in RAID mass memory. Simultaneously all frames are broadcast into a Gigabit Ethernet LAN. A separate PC which is included in the LAN captures the frames and compares successive frames with a reference frame (background frame) which is the sum of several preceding frames, acting for each pixel of the CCD matrix as a trigger scheme of the familiar GRB detector. Then each "triggered pixel" event is analyzed in space and time (at several successive frames after the event) and classified in accordance with predefined criteria.

The primary purpose of this WFOC is to perform continuous, alert-independent observations of optical transients and variable astrophysical sources simultaneously with space-borne X- and γ-ray telescopes. All the high time resolution data are stored in a data base for a limited amount of time (usually 2 days) and then can be used for detailed analysis. At the same time, real-time software performs on-the-fly identification of transient celestial sources estimating their brightness and location. For an event identified as a transient the location, brightness, and time are stored in a long term data base and can be used for correlative analysis using data in different wawelength. The real-time source identification software can generate alerts that could be sent to global alert distribution networks such as the GCN [16].

The camera has been operated in test mode since the end of May 2003. During the operation periods (May, 28 - Sep., 15; Nov.,17 - Dec., 1 of 2003) the total number of good nights was 39, total number of nights that the WFOC observed the FOV of the HETE-2 WXM is 27, and the number of meteors brighter than 9 mag was between 8 and 20 per hour. No synchronous GRB observation with HETE-2 was recorded, which is well within statistical expectations. Most autonomously generated triggers were due to active spacecraft and space debris.

This fully autonomous camera can detect and perform early photometry of prompt

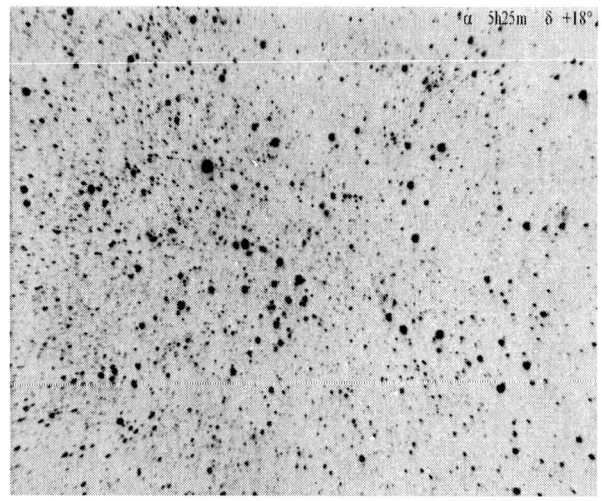

FIGURE 2. Center of HETE-2 FOV as seen with WFOC on Nov. 28 2003, 20:00 (UT), exp.=0.13 s

optical emission from both long and short duration GRBs, optical flashes preceding gamma-ray emission in GRBs, GRB afterglows not identified in γ-rays (orphan afterglows), optical outbursts related to Soft Gamma Repeaters, possible fast optical supernova precursors, optical flashes from LMXBs, and other compact X-ray transients, such as cataclysmic variables and related stars.

Acknowledgments. The project is supported by U.S. Civilian Research and Development Foundation (RP1-2394-MO-02). In SAO the work was supported by the Russian Foundation of Basic Researches (01-02-17857).

REFERENCES

1. Akerlof, C. et al., Nature **398**, 400 (1999).
2. Beloborodov, A. M., ApJ **565**, 808 (2001).
3. Hurley, K. et al., ApJ **567**, 447 (2002).
4. Paczynski, B., PASP **112**, 1281 (2000).
5. Nemiroff, R. J. Rafert, J. B., PASP **761**, 886 (1999).
6. Vestrand, W. T. et al., astro-ph/0209300 (2002).
7. Akerlof, C. et al., PASP **115**, 132 (2003).
8. Liang, E. et al., ApJ **519**, L21 (1999).
9. Eichler, D. and Beskin, G., Phys.Rev.L **85**, 2669 (2000).
10. Torii, K. et al., astro-ph/0309563 (2003).
11. Wozniak, P. et al., GCN Circ. 1757 (2003).
12. Pozanenko, A. et al., in "Proceedings of ADASS XII ASP Conference Series" **295**, 457 (2003).
13. Beskin, G. et al., AAS **138**, 589 (1999).
14. Piccioni, A. et al., in: "Proceeding of the 5th Huntsville Symposium" **526**, 756 (2000).
15. Beskin, G. et al., in: "Proceeding of the 2nd International workshop held in Rome", 387 (2001).
16. Barthelmy, S. D. et al., in: "Proceeding of the 2nd Huntsville Workshop" **307**, 643 (1994).

BOOTES: Technological Developments and Scientific Results by a Stereoscopic System with two Stations Spaced by 240 km

T. de J. Mateo Sanguino*[†], A. J. Castro-Tirado***, A.de Ugarte Postigo*, M.T. Fernández Palomo[†], J.M. Castro Cerón[‡], J.A. Berná Galiano[§], P. Páta[¶], J.Soldán[‖], M.Bernas[¶], R.Hudec[‖], M. Jelínek[‖], S. Vítek[¶], P. Kubánek[‖], S. McBreen[††], J. Gorosabel*, C.E. García Dabó[‡‡], T. Soria[§§], B.A. de la Morena Carretero[†] and J. Torres Riera[¶¶]

*IAA-CSIC, P.O.Box 03004, E-18080 Granada, Spain
[†]ESAt (INTA), E-21130 Mazagón, Huelva, Spain
**LAEFF-INTA, P.O. Box 50727, E-28080 Madrid, Spain
[‡]STScI, Baltimore MD 21218-2463, U.S.A.
[§]DFISTS, University of Alicante, Alicante, Spain
[¶]CTU-FEE, Dep. of Radioelectronics, 166 27 Praga, Czech Republic
[‖]Astronomical Institute, AV CR, 251 65 Ondřejov, Czech Republic
[††]Dep. of Experimental Physics, University College Dublin, Dublin, Irland
[‡‡]GRANTECAM, Dep. of Astrophysics, La Laguna, Tenerife, Spain
[§§]EELM-CSIC, Algarrobo Costa, Málaga, Spain
[¶¶]Div. of CC.EE. and TT.EE. (INTA), Torrejón de Ardoz, Madrid, Spain

Abstract. An overview of the technological developments at the Burst Observer and Optical Transient Exploring System (BOOTES) is given. The most important scientific results obtained so far are the detection of an OT in the GRB 000313 error box and the non detection of optical emission simultaneous for GRB 010220, GRB 030226 (long/soft events), GRB 020531 and GRB 021201 (short/hard events). With the recent instrumental and technical developments, BOOTES multiplies its science capabilities.

BOOTES: TWO STATIONS SPACED 240 KM

BOOTES is located in Spain and makes use of two sets of wide-field cameras, 240 km apart. These two stations taking simultaneous images will allow to distinguish near-Earth objects, closer to a distance of 1 million km, thus ruling out satellite glints, head-on meteors, etc in order to study the short duration optical transient phenomena that occur in the Universe.

RECENT SCIENTIFIC RESULTS

Besides the results obtained in 1998-2001 [1, 2], new results have been obtained in the period 2002-2003 [3].

TABLE 1. Current instrumentation at BOOTES (autumn 2003).

BOOTES-1A	BOOTES-1B	BOOTES-2
Schmidt-Cassegrain reflector telescope (0.3m,f/10) WFC-1:40°x28°FOV WFC-2:16°x11°FOV NFC:36'x36'FOV	Schmidt-Cassegrain reflector telescope (0.3m,f/10) Sky survey(16°x11°FOV)	Schmidt-Cassegrain reflector telescope (0.3m,f/10) NFC:36'x36'FOV WFC:42°x42°FOV

At the BOOTES-1 station, these are: 1) the detection of an OT (Optical Transient) related to GRB 030329 (R=16.4 the second night), and 2) the upper limits for simultaneous transient optical emission arising from the GRB 030115 [4] (with upper limit R=10) and GRB 030226 [5] (with upper limit R=11.5) error boxes. Both GRBs belong to the long/soft GRB class. For both GRB 020531 and 021201, belonging to the puzzling short/hard GRB class, upper optical limits of R=8 and R=10 respectively were imposed [6, 7].

The first event followed up with BOOTES-2 station is GRB 030913 obtaining co-added images with both wide-field and narrow-field cameras [8].

Regarding additional science, the BOOTES CCD cameras also detected the extraordinarily bright fireball of 27 January 2003, visible in Southern Spain and Northen Africa [9, 10].

LAST TECHNOLOGICAL DEVELOPMENTS

All these scientific results could not be obtained without the technological developments achieved at BOOTES by the Spanish-Czech collaboration [11, 12]. Thanks to automatic responses for HETE-2 alerts based on a new TCP/IP architecture installed in the observatories on June 2003, BOOTES reduces in three subsystems all the modules necessaries for astronomical observations and multiplies its science capabilities.

The first one is a new networked system based on client/servers applications devoted to drive robotic telescopes called RTS2 [13] - Remote Telescope System v.2- that communicates with others subsystems in order to get information about observatory status.

It is intended that RTS2 will be independent of different astronomical hardware, with access points for controlling mounts, domes and CCDs. Observation entries, requests and results are kept in a database with day-to-day reports sent by e-mail. Positions of GRBs are received from the Internet by socket communication and followed-up depending on weather conditions. Besides there is a scheduler for monitoring secondary science objectives and planning the remaining time.

The second subsystem is the automatic image analysis (IAM), a completely autonomous package that reduces the images from the different cameras calibrating with bias, dark frames and flat fields. Every time that detects objects in the field it calculates astrometry and correlates with catalogues and older BOOTES' images. When new ob-

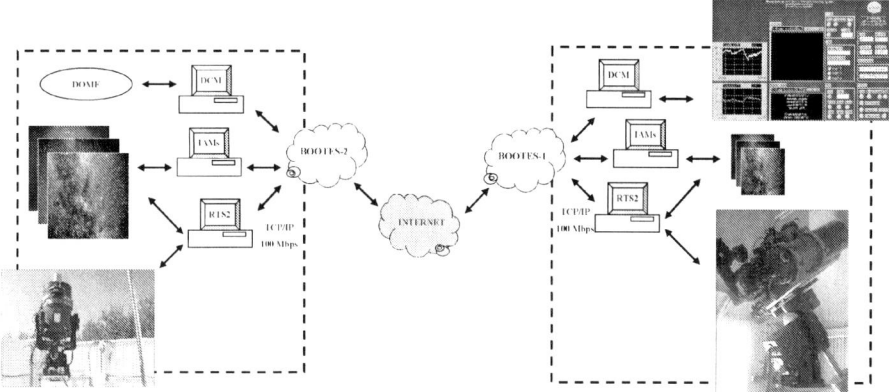

FIGURE 1. The BOOTES-1 and -2 modules showing hardware and software subsystems connections.

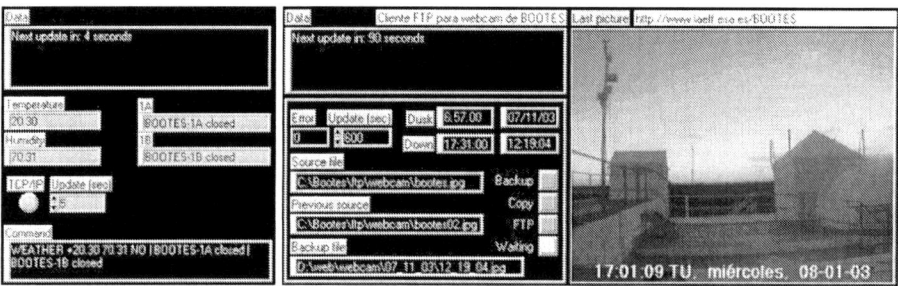

FIGURE 2. Socket communication and webcam programs from DCM module.

jects are found, IAM produces alerts that allow pointing of other instruments to observe these transient events [14].

During autumn 1999 we designed an instrument capable of obtaining GRB spectra for our rapid response system. When placed on the Cassegrain focus of the 0.3m telescope it is capable of acquiring slitless spectra of all the objects in a field of 43'x28'. This instrument allows us to have a quick pointing and fast switching from image to spectroscopic mode and saw first light on November 2000.

The third subsystem is the Dome Control Monitor (DCM) that controls the automatic opening system of the domes through a data acquisition board. It reads data in real time from weather sensors and takes decisions depending on environmental conditions [15]. Furthermore, DCM can be controlled by authorized users thanks to GSM calls and a Short Message System [16].

DCM send data sensors to RTS2 by socket communication in order to have BOOTES' status available. There is a module integrated with DCM that get these information and publish them in a private site in Internet, furthermore domes are monitorized by a webcam installed at the observatories [17]. More details at http://www.laeff.esa.es/BOOTES.

ACKNOWLEDGMENTS

We are grateful for the support given by Div. of CC.EE. and TT.EE. at INTA through the project IGE 4900506. This work is partially supported by Spain's MECD grant AP2002-0446 and by PNAYA 2002-0802 (including EU FEDER funds).

REFERENCES

1. Castro Cerón, J.M. et al., *Proc. of the Second GRB Workshop in the Afterglow Era*, Springer-Verlag, Berlín, eds. E.Costa, F.Frontera and J. Hjorth, ESO Astrophysics Symposia XIX, p.53 (2001).
2. Castro-Tirado, A.J. et al., A&A, **393** L55 (2002).
3. Castro-Tirado, A.J. et al., *Proc. of the Proceedings on Frontier Objects in Astrophysics and Particle Physics*, eds.: Giovanelli, F. & Mannochi, G., Italian Physical Society, in press (2004).
4. Castro-Tirado, A.J. et al., GCN Circ. 1826 (2003).
5. Castro-Tirado, A.J. et al., GCN Circ. 1887 (2003).
6. Castro-Tirado, A.J. et al., GCN Circ. 1430 (2002).
7. Castro-Tirado, A.J. et al., GCN Circ. 1720 (2002).
8. Castro-Tirado, A.J. et al., GCN Circ. 2389 (2003).
9. Gálvez Fernandez, F. et al., Tribuna de Astronomía y Universo, 51, 22 (2003).
10. Trigo-Rodríguez, J.M. et al., *Journal of the International Meteor Organization*, **31:2**, 49 (2003).
11. Mateo Sanguino, T.J. et al., *Proc. of the V Scientific Meeting of the Spanish Astronomical Society (SEA)*, Kluwer Academic Publishers, pag. 491 (2003).
12. Mateo Sanguino, T.J. et al., *Proc. of the Third GRB Workshop, Gamma-ray bursts in the Afterglow Era*, Astron. Soc. of the Pacific Conf. Series, in press (2003).
13. Kubánek, P. et al., *Gamma Ray Bursts: 30 Years of Discovery, Santa Fe (New México) USA*, AIP in press (2004).
14. Ugarte Postigo, A. de et al., *JENAM for 2003: 12th European Meeting for Astronomy and Astrophysics, Budapest (Hungary)*, Baltic Astronomy, vol.12, in press (2003).
15. Mateo Sanguino, T.J.,, Proyecto Fin de Carrera, University of Huelva (1998).
16. Mateo Sanguino, T.J.,, Proyecto Fin de Carrera , University of Granada (2001).
17. Mateo Sanguino, T.J.,, Trabajo de Investigación Tutelado, University of Granada (2003).

REM. Rapid Eye Mount

E. Molinari[*], S.D. Vergani[*], F.M. Zerbi[*], S. Covino[*] and G. Chincarini[†]

[*]INAF - OAB, via Bianchi 46, 23807 Merate, LC, Italy
[†]Università Milano Bicocca

Abstract. REM is a robotic fast moving telescope designed to immediately point and observe in optical and IR the GRBs detected by satellites. Its immediate data gathering capabilities and its accurate astrometry will issue early alerts for the VLT.

REM, THE TELESCOPE

From the very beginning the specifications for the REM telescope were very simple even if very demanding on both the hardware and software. The main science motivation, that is the propmt follow-up of newly exploded GRBs, demanded an instrumentation that could reveal near infrared radiation up to 2300 nm (K' band). This meant that the telescope had to slew on target immediately, without human intervention, after a satellite burst trigger, thus deciding automatically the observing strategy. A proper Figure of Merit (FoM) was then implemented in the control software, in order to give REM a limited intelligence necessary for the continuos update of the observing schedule. The quick look automatic software had to be capable of identifying right away, and measuring accurately, the position of the burst and its magnitude, so as to immediately alert the community and all the major telescopes, the ESO VLT in particular. Furthermore the combination of the two instruments described below, and the related software, had to allow the estimate of the redshift of the cosmic events via the Lyman-α line and the drop out technique. The specifications as dictated by the science needs suggested immediately the choice of an alt-azimuth mount in order to minimize the momentum of the instruments during slew.

After many interactions with the ESO staff, Cerro La Silla (Figure 1) was selected as the most convenient location for REM. The telescope uses a Ritchey-Chretien configuration with a 60 cm f/2.2 primary mirror and two Nasmyth f/8 focal stations. The telescope was manufactured by Teleskoptechnik Halfmann GmbH in Augsburg (Germany) and the optics by Carl Zeiss AG (Germany). To optimise the response in the near infrared the telescope optics were coated with silver and protected by a special overcoating. Accurate pointing, fast slewing and precise tracking are achieved using azimuth and elevation motors made by ETEL which allow a maximum speed of 12 deg/s and Heidenain encoders with 237 steps per arcsec. The instrumentation has been attached, together with the field de-rotator, in one of the Nasmyth foci (Figure 2). A beam splitter (dichroic) manufactured by ZAOT (Italy) according to our design leaves the Infrared beam (950 - 2300 nm) to continue along the optical axis where the IR Camera (REM-IR) is installed while it deflects the optical beam (450 - 950 nm) to an orthogonal axis where the optical

FIGURE 1. REM telescope installed in Notre Dome, La Silla, Chile.

instrument (ROSS) is installed (Figure 3).

THE CAMERAS

At present the infrared camera (REMIR) is working with 4 filters (Z, J, H and K') and one low dispersion grism. The camera optics allow us to have a 9.9x9.9 arcmin field of view on a 512x512 HgCdTe array produced by Rockwell. The IR array uses a Leach Controller with a read-out speed of 1.64 microseconds per pixel. The collimator and the camera (Silica - CaF2 and CaF2 - Silica) focus the image on the CCD after passing through the Cryostat window. The whole camera is mounted in a dewar manufactured by the Infrared Laboratories in Tucson (Arizona) so as to operate in a cold environment and is kept at a working temperature of about 80 K. The working temperature of the IR array is 80 K as well. The cryogenics are supported by a Stirling Cycle cryo-pump made by Leybold AG (Germany).

The visual camera ROSS consists of separated doublets made of ZKN7 - FPL53. The filter wheel accommodates the V-, R-, and I-filters and an Amici prism 66 mm long. The prism is made of Silica, BAF2 and CAF2 and the spectral range from 450 to 950 nm is displayed over 60 pixels. In order to match the optical thickness of the Amici prism and to avoid refocusing while passing from the imaging mode to the spectroscopy mode, the filters were glued on properly designed cylinders of optical glass. The detector head is a commercial Apogee AP47 camera hosting a Marconi 47-10 1Kx1K 13 μm pitch CCD.

OBSERVATIONAL OPERATIONS

REM is governed by a software system which includes by a central processing unit and a number of external sensors. These sensors continuously feed a local database

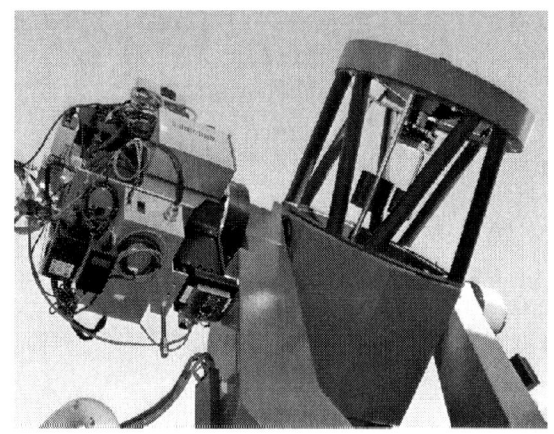

FIGURE 2. REM telescope with instrumentation at Nasmyth focus.

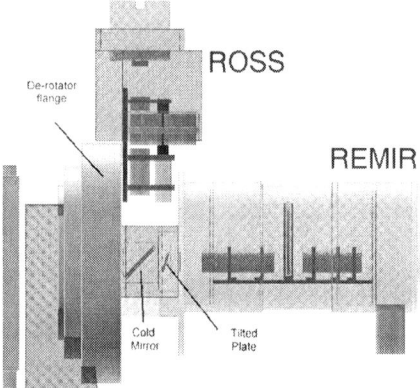

FIGURE 3. REMIR and ROSS cameras scheme at Nasmyth focus

providing crucial information for the decisional algorithm that is activated in case a new GRB alert is received. After a new alert, REM checks if the new target can be observed (astrometry, environmental conditions...) In case other observations are in progress REM also evaluates, following a continuously refined figure of merit, if these observations are to be aborted in order to start a new observational session devoted to the new target. As soon as data are available the REM software system automatically retrieves the optical and IR images, performs standard reduction and analysis, and delivers position and fluxes to the community. This information allows REM to refine the subsequent observational strategy and to automatically choose, in real-time, which large telescope observing program is the most suitable for the event under analysis.

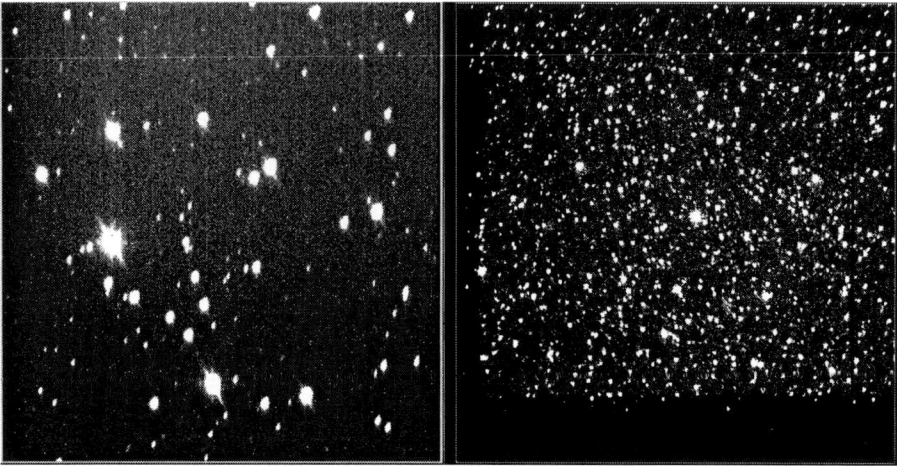

FIGURE 4. First images, showing the open star cluster M6, taken with the ROSS in the V-band (left) and with the REM in the K'-band (right). Note the striking difference between the images, showing the strong role played by the dust absorption along the galactic plane.

REM FIRST LIGHT: JUNE 2003

Soon after the transportation and in-dome installation all the equipment had been turned on and on June 22 we were able to point the telescope and get the first images. By Tuesday June 24 we had the first fairly good images (Figure 4) and standard stars both with the infrared camera (REMIR) and the ROSS instrument. Preliminary data reduction of the standard stars without correcting for flat fielding in the NIR, gives the following limiting magnitudes for 1 second integration and S/N=5: V=17.2, R=17.2, I=16.0, J=14.5, H=13.5, K'=13.0. These observations show that the sensitivity of the instrument already matches (or is even better in K') the sensitivity we expected as estimated in the original proposal. That also means that, by a proper reduction of the data and after fine-tuning of the software tools, the system will exceed the expectations.

ACKNOWLEDGMENTS

Thanks to all the REM team and the ESO-La Silla technical staff which made this dream possible.

GRB 2003 CONFERENCE PARTICIPANTS

- Aloy – Miguel-Angel, Max-Planck-Institut fur Astrophysik – maa@mpa-garching.mpg.de
- Antonelli – L. Angelo, INAF - Osservatorio Astronomico di Roma – angelo@mporzio.astro.it
- Anzenberg – Eitan, LANL NIS-2 – snake_eyez_313@yahoo.com
- Atteia – Jean-Luc, Laboratoire d'Astrophysique, OMP – atteia@ast.obs-mip.fr
- Bagoly – Zsolt, Eötvös University – zsolt.bagoly@elte.hu
- Band – David, NASA/GSFC – dband@lheapop.gsfc.nasa.gov
- Barnard – Vicki, Joint Astronomy Centre, Hawaii – vbarnard@jach.hawaii.edu
- Barraud – Celine, OMP toulouse – barraud@ast.obs-mip.fr
- Barthelmy – Scott, NASA-GSFC – scott@milkyway.gsfc.nasa.gov
- Belli – Bianca Maria, CNR – bianca@rm.iasf.cnr.it
- Beloborodov – Andrei, Columbia University – amb@phys.columbia.edu
- Berger – Edo, Caltech – ejb@astro.caltech.edu
- Bernardini – Maria Grazia, ICRA – maria.bernardini@icra.it
- Bhat – Narayana, University of Alabama in Huntsville – bhatn@uah.edu
- Bianco – Carlo Luciano, ICRA - Dipartimento di Fisica - Universita' "La Sapienza" – bianco@icra.it
- Bloom – Joshua, Harvard University/CfA – jbloom@cfa.harvard.edu
- Bloomina – Joshmina, Harvard University/CfA – jbloom@cfa.harvard.edu
- Boer – Michel, CESR/CNRS – Michel.Boer@cesr.fr
- Boggs – Steven, UC Berkeley – boggs@ssl.berkeley.edu
- Bonnell – Jerry, USRA/GSFC – bonnell@grossc.gsfc.nasa.gov
- Borozdin – Konstantin, Los Alamos National Laboratory – kbor@lanl.gov
- Bosnjak – Zeljka Marija, SISSA/ISAS, Trieste, Italy – bosnjak@sissa.it
- Braga – Joao, INPE – braga@das.inpe.br
- Brandt – Soren, Danish Space Research Institute – sb@dsri.dk
- Burbidge – Geoffrey, University of California, San Diego (UCSD) – gburbidge@ucsd.edu
- Burrows – David, Penn State – burrows@astro.psu.edu
- Butler – Nat, MIT CSR – nrbutler@space.mit.edu
- Canterna – Ron, University of Wyoming – canterna@uwyo.edu
- Castro-Tirado – Alberto J., IAA-CSIC, Granada – ajct@iaa.es
- Cenko – Brad, California Institute of Technology – cenko@srl.caltech.edu
- Chernenko – Anton, IKI – anton@cgrsmx.iki.rssi.ru
- Chevalier – Roger, University of Virginia – rac5x@virginia.edu
- Clark – Chris, Spectrum Astro, Inc. – chris.clark@specastro.com
- Cline – David, UCLA – dcline@physics.ucla.edu
- Cline – Thomas, NASA's Goddard Space Flight Center – Thomas.L.Cline@nasa.gov

- Clingempeel – Richard, AIIR/CCRF – oerlicon@uscyber.net
- Coburn – Wayne, UCB/SSL – wcoburn@ssl.berkeley.edu
- Cominsky – Lynn, Sonoma State University – lynnc@charmian.sonoma.edu
- Connaughton – Valerie, UAH – valerie@msfc.nasa.gov
- Conte – Dom, Spectrum Astro, Inc. – dom.conte@specastro.com
- Corsi – Alessandra, ICRA - Dipartimento di Fisica - Universita' "La Sapienza" – alessandra.corsi@icra.it
- Costa – Enrico, IASF-CNR, Roma – costa@rm.iasf.cnr.it
- Covino – Stefano, Brera Astronomical Observatory – covino@mi.astro.it
- Crew – Geoffrey, MIT Center for Space Research – gbc@space.mit.edu
- Crider – Anthony, Elon University – acrider@elon.edu
- Csatorday – Peter, MIT – csatorda@mit.edu
- Cusumano – Giancarlo, IASF/CNR – cusumano@pa.iasf.cnr.it
- Dai – Zigao, Department of Astronomy, Nanjing University – daizigao@public1.ptt.js.cn
- Daigne – Frédéric, Institut d'Astrophysique de Paris - Université Paris 6 – daigne@iap.fr
- Davidoff – Sherri, Los Alamos – alien@mit.edu
- Della Valle – Massimo, Arcetri Astrophysical Observatory (Florence) – massimo@arcetri.astro.it
- Dermer – Charles, Naval Research Laboratory – dermer@gamma.nrl.navy.mil
- Di Cocco – Guido, IASF-CNR – dicocco@bo.iasf.cnr.it
- Di Girolamo – Tristano, INFN, Naples (Italy) – tristano@na.infn.it
- Dingus – Brenda, Los Alamos National Laboratory – dingus@lanl.gov
- Djorgovski – George, Caltech – george@astro.caltech.edu
- Donaghy – Timothy, University of Chicago – quinn@abulafia.uchicago.edu
- Doty – John, MIT – jpd@space.mit.edu
- Drago – Alessandro, University of Ferrara – drago@fe.infn.it
- Dullighan – Allyn, MIT – allyn@space.mit.edu
- Fenimore – Ed, Los Alamos – efenimore@lanl.gov
- Feroci – Marco, IASF/CNR, Rome – feroci@rm.iasf.cnr.it
- Filippenko – Alex, University of California, Berkeley – alex@astro.berkeley.edu
- Fortini – Pierluigi, University of Ferrara – fortini@fe.infn.it
- Fox – Derek, Caltech – derekfox@astro.caltech.edu
- Frail – Dale, NRAO – dfrail@nrao.edu
- Fransson – Claes, Stockholm University – claes@astro.su.se
- Fraschetti – Federico, ICRA - Universita' di Trento – fraschetti@icra.it
- Friedman – Andy, Harvard University – afriedman@cfa.harvard.edu
- Fritzius – Bob, www.ShadeTreePhysics.com – RobertFritzius@msn.com
- Frontera – Filippo, University of Ferrara and IASF, CNR, Bologna, Italy – frontera@fe.infn.it

- Fruchter – Andrew, Space Telescope Science Institute – fruchter@stsci.edu
- Fryer – Chris, LANL – fryer@lanl.gov
- Fynbo – Johan, Department of Physics and Astronomy, University of Aarhus – jfynbo@phys.au.dk
- Galassi – Mark, Los Alamos – rosalia@galassi.org
- Gehrels – Neil, NASA-GSFC – gehrels@milkyway.gsfc.nasa.gov
- Ghirlanda – Giancarlo, IASF - CNR Miano – ghirland@mi.iasf.cnr.it
- Gianotti – Fulvio, CNR-IASF/BO – gianotti@bo.iasf.cnr.it
- Giblin – Timothy, The College of Charleston – giblint@cofc.edu
- Gomez – Enrique, University of Alabama – gomez002@bama.ua.edu
- Gonzalez – Magda, University of Wisconsin, LANL – magda@whopper.lanl.gov
- Gorosabel – Javier, IAA-CSIC/STScI – gorosabel@stsci.edu
- Gou – Lijun, Penn State University – lijun@astro.psu.edu
- Graziani – Carlo, University of Chicago – carlo@oddjob.uchicago.edu
- Greiner – Jochen, MPE Garching, Germany – jcg@mpe.mpg.de
- Gupta – Varsha, Department of Physics and Astrophysics, University of Delhi, Delhi-110007, India – varsha_gupta_98@yahoo.com
- Gursky – Herbert, Naval Research Laboratory – herbert.gursky@nrl.navy.mil
- Hakkila – Jon, College of Charleston – hakkilaj@cofc.edu
- Hardtke – Rellen, Cal Poly - Pomona – rellen@amanda.wisc.edu
- Hartmann – Dieter, Clemson University – hdieter@clemson.edu
- Heise – John, SRON Space Research org. Netherlands – j.heise@sron.nl
- Hill – Joe, Penn State University – jhill@astro.psu.edu
- Holland – Stephen, USRA/NASA/GSFC/SSC – sholland@milkyway.gsfc.nasa.gov
- Hoover – Andrew, Los Alamos National Laboratory – ahoover@lanl.gov
- Huang – Yong-Feng, Dept Astronomy, Nanjing University, P. R. China – hyf@nju.edu.cn
- Hudec – Rene, Astronomical Institute Ondrejov – rhudec@asu.cas.cz
- Hullinger – Derek, University of Maryland – derek@milkyway.gsfc.nasa.gov
- Hurley – Kevin, UC Berkeley – khurley@sunspot.ssl.berkeley.edu
- in 't Zand – Jean, Space Research Org. Netherlands – jeanz@sron.nl
- Inoue – Akio, Dept. of Phys., Kyoto Univ. – akinoue@scphys.kyoto-u.ac.jp
- Jarvis – Alexander, UCLA – ajarvis@physics.ucla.edu
- Jelínek – Martin, Observatory Ondrejov – mates@asu.cas.cz
- Jernigan – Garrett, Space Sciences Laboratory, UC Berkeley – jgj@ssl.berkeley.edu
- KALUZIENSKI – LOUIS, NASA – louis.j.kaluzienski@nasa.gov
- Kaneko – Yuki, University of Alabama in Huntsville – yuki.kaneko@msfc.nasa.gov
- Kawai – Nobu, Tokyo Tech – nkawai@phys.titech.ac.jp
- Kaye – Tom, Spectrashift.com – tkaye@airgun.com
- Kazanas – Demosthenes, NASA/GSFC – Demos.Kazanas-1@nasa.gov

- Kazanas – Demosthenes, NASA/GSFC – Demos.Kazanas@nasa.gov
- Kippen – R. Marc, Los Alamos National Laboratory – mkippen@lanl.gov
- Klebesadel – Ray, LANL – rklebesadel@juno.com
- Klose – Sylvio, Thuringer Landessternwarte – klose@tls-tautenburg.de
- Kobayashi – Shiho, Penn State – shiho@gravity.psu.edu
- Kocevski – Dan, Rice University – kocevski@rice.edu
- Kocharovsky – Vladimir, Institute of Applied Physics – kochar@appl.sci-nnov.ru
- Konigl – Arieh, University of Chicago – arieh@jets.uchicago.edu
- Krimm – Hans, USRA / NASA GSFC – krimm@milkyway.gsfc.nasa.gov
- Kulkarni – Shri, California Institute of Technology – srk@astro.caltech.edu
- Kumar – Pawan, UT Austin – pk@astro.as.utexas.edu
- Kuznetsov – Sergey, CEA-Saclay, France – kuznetz@discovery.saclay.cea.fr
- KWAK – KYUJIN, SUNY STONY BROOK – kkwak@grad.physics.sunysb.edu
- Lamb – Don, University of Chicago – d-lamb@uchicago.edu
- Lazzati – Davide, Institute of Astronomy – lazzati@ast.cam.ac.uk
- Lepore – Al, Spectrum Astro, Inc. – Albert.lepore@specastro.com
- Li – Hui, LANL – hli@lanl.gov
- Liang – Edison, Rice University – liang@spacsun.rice.edu
- Lingenfelter – Rich, Univ. of California, San Diego – rlingenfelter@ucsd.edu
- Lloyd-Ronning – Nicole, CITA – lloyd@cita.utoronto.ca
- Loeb – Abraham, Harvard University – aloeb@cfa.harvard.edu
- Longo – Francesco, University of Trieste and INFN, Trieste, Italy – francesco.longo@ts.infn.it
- Lustig – Harry, CUNY – lustig@aps.org
- Lyutikov – Maxim, McGill University – lyutikov@physics.mcgill.ca
- MacFadyen – Andrew, Caltech – andrew@tapir.caltech.edu
- Malesani – Daniele, SISSA/ISAS, Trieste, Italy – malesani@sissa.it
- Manchnada – Ravi, TIFR – ravi@tifr.res.in
- Mason – Keith, Mullard Space Science Lab – kom@mssl.ucl.ac.uk
- Matheson – Thomas, Harvard-Smithsonian Center for Astrophysics – tmatheson@cfa.harvard.edu
- McBreen – Sheila, University College Dublin – smcbreen@bermuda.ucd.ie
- McConnell – Mark, University of New Hampshire – mark.mcconnell@unh.edu
- McEnery – Julie, NASA/GSFC/UMBC – mcenery@milkyway.gsfc.nasa.gov
- McEnery – Julie, UMBC – rray@umbc.edu
- McLean – Kassandra, University of Texas at Dallas – kmclean@utdallas.edu
- McMahon – Erin, University of Texas, Austin – emcmahon@astro.as.utexas.edu
- Medvedev – Mikhail, University of Kansas – medvedev@ku.edu
- MELNICK – Jorge, ESO – jmelnick@eso.org
- Melott – Adrian, University of Kansas – melott@kusmos.phsx.ukans.edu

- Mereghetti – Sandro, IASF MILANO – sandro@mi.iasf.cnr.it
- Meszaros – Peter, Pennsylvania State University – pmeszaros@astro.psu.edu
- Michelson – Peter, Stanford University – peterm@leland.stanford.edu
- Mineo – Teresa, IASF-CNR Sezione di Palermo – teresa.mineo@pa.iasf.cnr.it
- Mirabal – Nestor, Columbia University – abulafia@astro.columbia.edu
- Mitrofanov – Igor, Institute for Space Research – imitrofa@space.ru
- MIZUMOTO – Yoshihiko, NAOJ – mizumoto.y@nao.ac.jp
- Mizuno – Yosuke, Department of Astronomy, Kyoto University – mizuno@kusastro.kyoto-u.ac.jp
- Mochkovitch – Robert, Institut d'Astrophysique de Paris – mochko@iap.fr
- Morales – Miguel, MIT Center for Space Research – mmorales@space.mit.edu
- Moran – Jane, University of North Carolina – moranj@physics.unc.edu
- Nakagawa – Yujin, Aoyama Gakuin University – yujin@phys.aoyama.ac.jp
- Nakamura – Takashi, Kyoto Univ. – takashi@tap.scphys.kyoto-u.ac.jp
- Nemiroff – Robert, Michigan Tech – nemiroff@mtu.edu
- Nicastro – Luciano, IASF-CNR, Sez. Palermo – nicastro@pa.iasf.cnr.it
- Nishikawa – Ken, NSSTC – ken-ichi.nishikawa@nsstc.nasa.gov
- Nishikawa – Ken, NSSTC – Ken-Ichi.Nishikawa@nsstc.nasa.gov
- Norris – Jay, NASA/GSFC – Jay.P.Norris@NASA.gov
- Nysewander – Melissa, UNC Chapel Hill – mnysewan@astro.unc.edu
- Osborne – Julian, University of Leicester – julo@star.le.ac.uk
- Palmer – David, Los Alamos National Laboratory – palmer@lanl.gov
- Panaitescu – Alin, University of Texas at Austin – adp@astro.as.utexas.edu
- Park – Hye-Sook, LLNL – park1@llnl.gov
- Parsons – Ann, NASA/ Goddard Space Flight Center – parsons@milkyway.gsfc.nasa.gov
- Petrosian – Vahe', Stanford University – vahe@astronomy.stanford.edu
- Piran – Tsvi, Hebrew University – tsvi@phys.huji.ac.il
- Piro – Luigi, IASF/INAF – piro@rm.iasf.cnr.it
- Pittori – Carlotta, Università di Roma "Tor Vergata" and INFN Roma2 – carlotta.pittori@roma2.infn.it
- Pizzichini – Graziella, IASF/CNR – pizzichini@bo.iasf.cnr.it
- Pozanenko – Alexei, IKI – apozanen@iki.rssi.ru
- Preger – Barbara, IASF/CNR – preger@rm.iasf.cnr.it
- Price – Paul, RSAA, ANU – pap@mso.anu.edu.au
- Proga – Daniel, JILA, University of Colorado – proga@colorado.edu
- Quimby – Robert, U. Texas / ROTSE / GRBlog – quimby@astro.as.utexas.edu
- Ramirez-Ruiz – Enrico, Institute of Astronomy, Cambridge University – enrico@ast.cam.ac.uk
- Rau – Arne, MPE – arau@mpe.mpg.de

- RAZAK – MOHAMMED, CONCERN YOUTH IN DEVELOPMENT – mazoo80gh@hotmail.com
- Reese – Adam, UNC – adam.reese@unc.edu
- Reeves – James, NASA Goddard Space Flight Center – jnr@milkyway.gsfc.nasa.gov
- Reichart – Dan, University of North Carolina at Chapel Hill – reichart@physics.unc.edu
- Richardson – Georgia, NRC/MSFC – georgia.richardson@msfc.nasa.gov
- Ricker – George, MIT – grr@space.mit.edu
- Roming – Pete, Penn State University – roming@astro.psu.edu
- Rossi – Elena, Institute of Astronomy, Cambridge, UK – emr@ast.cam.ac.uk
- Rosswog – Stephan, University of Leicester – sro@astro.le.ac.uk
- Ruffini – Remo, ICRA - Dipartimento di Fisica - Universita' "La Sapienza" – ruffini@icra.it
- Ryde – Felix, Stockholm Observatory – felix@astro.su.se
- Sahu – Sarira, INSTITUTO DE CIENCIAS NUCLEARES (ICN) – sarira@nuclecu.unam.mx
- Sakamoto – Takanori, Tokyo Institute of Technology – sakamoto@hp.phys.titech.ac.jp
- Salmonson – Jay D., Lawrence Livermore National Laboratory – salmonson@llnl.gov
- Sari – Re'em, caltech – sari@tapir.caltech.edu
- Sato – Rie, Tokyo Institute of Technology, Department of physics – rsato@tithp1.hp.phys.titech.ac.jp
- Savage – Sabrina, University of Wyoming – astrosabby@hotmail.com
- Schartel – Norbert, ESA, XMM-Newton SOC – nscharte@xmm.vilspa.esa.es
- Bhargavi – S. G., Indian Inst of Astrophysics, Bangalore – bhargavi_s@iiap.ernet.in
- Smirnov – Dmitry, Institute for Nuclear Research Russian Academy of Sciences – smirnovdm@yandex.ru
- Soderberg – Alicia, Caltech – ams@astro.caltech.edu
- Stamatikos – Michael, Department of Physics - University of Wisconsin, Madison – ms25@amanda.physics.wisc.edu
- Strong – IAN, LANL (Retired) – ianstrong@newmexico.com
- Suzuki – Motoko, Tokyo Institute of Technology, Department of physics – motoko@hp.phys.titech.ac.jp
- Tagliaferri – Gianpiero, Osservatorio Astronomico di Brera - INAF – tagliaferri@merate.mi.astro.it
- Tamagawa – Toru, RIKEN – tamagawa@crab.riken.go.jp
- Tavani – Marco, IASF - CNR - Rome – tavani@rm.iasf.cnr.it
- Tavenner – Tanya, NMSU – tanya@nmsu.edu
- Taylor – Greg, NRAO – gtaylor@nrao.edu
- Terrell – James, Los Alamos National Laboratory – jterrell@lanl.gov
- Topinka – Martin, Charles University of Prague – toast@lascaux.asu.cas.cz
- Trimble – Virginia, U. Maryland & UC Irvine – vtrimble@astro.umd.edu
- Uemura – Makoto, Kyoto University – uemura@kusastro.kyoto-u.ac.jp
- Urata – Yuji, RIKEN/Titech – urata@crab.riken.go.jp
- Vanderspek – Roland, MIT Center for Space Research – roland@space.mit.edu

- Vergani – Susanna, INAF-OAB – vergani@merate.mi.astro.it
- Vestrand – Tom, LANL – vestrand@lanl.gov
- Villasenor – Joel, MIT Center for Space Research – jsvilla@space.mit.edu
- Vitagliano – Luca, ICRA - Dipartimento di Fisica - Universita' "La Sapienza" – vitagliano@icra.it
- Vlahakis – Nektarios, University of Athens – vlahakis@phys.uoa.gr
- von Kienlin – Andreas, Max-Planck-Institut fuer extraterrestrische Physik – azk@mpe.mpg.de
- Vreeswijk – Paul, European Southern Observatory – pvreeswi@eso.org
- Wang – Lifan, Lawrence Berkeley National Laboratory – lwang@lbl.gov
- Wei – Daming, Purple Mountain Observatory – dmwei@pmo.ac.cn
- Wells – Alan, University of Leicester – aw@star.le.ac.uk
- Wick – Stuart, Naval Research Laboratory – wick@ssd5.nrl.navy.mil
- Wiersma – Jorrit, Sterrenkundig Instituut, Universiteit Utrecht – wiersma@astro.uu.nl
- Wijers – Ralph, University of Amsterdam – rwijers@science.uva.nl
- Williams – David A, UC Santa Cruz – daw@scipp.ucsc.edu
- Williams – Grant, MMTO – gwilliams@as.arizona.edu
- Woosley – Stan, Astronomy Dept., UCSC – woosley@ucolick.org
- Wozniak – Przemek, LANL – wozniak@lanl.gov
- Wren – James, LANL – jwren@lanl.gov
- Xue – she-sheng, ICRA, Physics Department, University of Rome, "La Sapienza" – xue@icra.it
- Yamazaki – Ryo, Kyoto University – yamazaki@tap.scphys.kyoto-u.ac.jp
- Yoshida – Atsu, Aoyama Gakuin University – ayoshida@phys.aoyama.ac.jp
- Zhang – Bing, Penn State University – bzhang@astro.psu.edu
- Zhang – Weiqun, UC SANTA CRUZ – zhang@ucolick.org

AUTHOR INDEX

A

Abbey, A. F., 642
Abel, T., 518
Achterberg, A., 570
Aloy, M.-A., 380
Andersen, M. I., 301
Andresian-Thomas, N., 737
Antonelli, L. A., 403
Argan, A., 696, 704
Armitage, P. J., 384
Atoyan, A., 141, 170, 557
Atteia, J.-L., 37, 42, 57, 81, 101, 106, 192, 217, 447, 618, 630

B

Bagoly, Z., 94
Balázs, L. G., 94
Banat, P., 655
Band, D. L., 65, 597, 677, 688, 692
Barbanera, L., 696
Barbiellini, G., 578, 696, 704
Barbier, L. M., 659
Barnard, V. E., 508
Bärnbantner, O., 269
Barraud, C., 37, 57, 81, 101, 106, 192, 217, 618
Barthelmy, S. D., 655, 659, 663, 667, 671, 723
Bartolini, C., 77, 471
Beardmore, A., 642
Beckmann, V., 225, 622
Begelman, M. C., 384
Belli, B. M., 90
Belloni, T., 655
Beloborodov, A. M., 187, 198
Benetti, S., 403
Bennett, K., 225, 244
Berger, E., 324
Berná Galiano, J. A., 761
Bernardini, M. G., 312, 424, 428
Bernas, M., 761
Bersier, D., 537
Beskin, G., 757
Bhargavi, S. G., 388

Bhat, P. N., 684
Bianco, C. L., 312, 424, 428
Biryukov, A., 757
Blain, A. W., 508
Blandford, R., 677
Blau, M., 663
Bloom, J. S., 497
Boer, M., 106, 192, 217, 447, 618
Boggs, S. E., 262
Bondar, S., 757
Bonnell, J. T., 65, 412
Boone, L. M., 166
Borozdin, K., 728
Bosnjak, Z., 578
Bosworth, J., 642
Bradshaw, M., 723
Bramel, D., 166
Brauninger, H. W., 642
Breeveld, A. A., 651
Briggs, M. S., 119, 203, 236, 244, 684, 688
Bulgarelli, A., 700
Bureau, M., 337
Burkert, W., 642
Burrows, D. N., 642, 647
Burud, I., 537
Butler, N., 57, 106, 111, 192, 217, 399, 435, 467, 626, 630

C

Campana, S., 642, 647
Canterna, R., 745
Capalbi, M., 642
Carson, J., 166
Casperson, D. J., 719, 728
Castro Cerón, J. M., 439, 491, 537, 761
Castro-Tirado, A. J., 301, 483, 491, 537, 753, 761
Celesti, E., 700
Celotti, A., 69, 578
Chang, W., 642
Chardonnet, P., 312, 424, 428
Cheng, K. S., 475
Chevalier, R. A., 355
Chincarini, G., 403, 443, 642, 647, 765

Chornock, R., 366
Citterio, O., 642
Cline, D. B., 574
Cline, T., 613
Coburn, W., 262
Cocozza, G., 403
Cohen-Tanugi, J., 681
Connaughton, V., 684, 688
Conway, M., 61
Cordell, D., 221
Corsi, A., 428
Costa, E., 696
Covault, C. E., 166
Covino, S., 403, 765
Craig, W., 677
Crew, G. B., 57, 106, 192, 217, 626, 630
Cummings, J. R., 655, 659, 663, 671

D

Dai, X., 208
Dai, Z. G., 463, 475
Daigne, F., 101, 328
D'Avanzo, P., 403
Davidoff, S., 728, 733
Davis, H., 728
Dean, A., 655
de Freitas Pacheco, J., 447
de J. Mateo Sanguino, T., 753, 761
de la Morena Carretero, B. A., 761
Della Valle, M., 403
Del Monte, E., 696
Deluit, S., 622
Dermer, C. D., 141, 170, 203, 557
de Ugarte Postigo, A., 491, 753
Dezalay, J.-P., 106, 192, 217, 618
Di Cocco, G., 700
Diehl, R., 684
Digel, S. W., 692
Di Girolamo, T., 150, 522
Dingus, B. L., 131, 203, 236, 244
Di Persio, G., 696
Di Sciascio, G., 150, 522
D'Olivo, J. C., 154
Donaghy, T. Q., 19, 37, 42, 47, 57, 106, 192, 217, 618, 630
Doty, J. P., 106, 192, 217, 626, 630, 708
Drago, A., 420
Drost, D. M., 737

Dullighan, A., 57, 111, 399, 435, 467, 626, 630
Dupuis, R., 371

E

Ellison, S., 453

F

Fathi, K., 337
Feldmeier, J., 388
Fenimore, E. E., 106, 192, 217, 566, 618, 630, 655, 659, 663, 667, 671, 728
Fernández Palomo, M. T., 761
Feroci, M., 696
Ferrero, P., 77, 471
Filippenko, A. V., 366
Fishman, G. J., 290, 684
Fishman, J., 677
Ford, P., 111, 399, 435
Fortin, P., 166
Fortini, P., 174
Frail, D., 324
Fraschetti, F., 312, 424, 428
French, J., 61, 225, 741
Freyberg, M. J., 642
Frontera, F., 31
Froysland, T., 700
Fruchter, A. S., 439, 453, 508, 537
Frutti, M., 696
Fryer, C. L., 371
Fugazza, D., 403
Fuller, S. P., 73
Fynbo, J. P. U., 301, 453, 537

G

Galassi, M., 106, 192, 217, 566, 618, 630, 663, 728
Galli, M., 704
Gao, W. H., 86
García Dabó, C. E., 761
Gehrels, N., 637, 655, 659, 663, 667, 671, 677
Georganopoulos, M., 294
Ghirlanda, G., 69, 316, 578, 696, 704

Ghisellini, G., 69, 316, 403
Gianotti, F., 700
Giblin, T. W., 52, 73, 585, 737
Gilmozzi, R., 403
Giltinan, A., 741
Gingrich, D. M., 166
Giommi, P., 642
Gómez, E. A., 278
González, M. M., 203, 236, 244
Gorosabel, J., 301, 439, 491, 537, 761
Gotthelf, E. V., 533
Gou, L., 518
Granot, J., 181
Grav, T., 269
Graziani, C., 37, 42, 47, 57, 106, 111, 192, 217, 566, 618, 630
Greiner, J., 269, 388, 439, 483, 684
Grindlay, J., 677
Grundahl, F., 301
Guarnieri, A., 77, 471
Guziy, S., 491

H

Haglin, D. J., 585
Hakkila, J., 52, 73, 585, 737
Halpern, J. P., 337, 366, 533
Hamuy, M., 403
Hanlon, L., 61, 225, 244, 741
Hanna, D., 166
Hardee, P. E., 278, 290
Hardtke, R., 158
Harrison, F., 677
Hartmann, D. H., 269, 333, 483, 677, 723, 737
Hartner, G. D., 642
Heger, A., 343, 376
Heise, J., 119
Henden, A. A., 269, 483
Hill, J. E., 642, 647
Hjorth, J., 269, 301, 453, 483, 537
Ho, C., 719
Hoffman, M., 741
Hong, J., 677
Horváth, I., 94
Howard, E., 333
Huang, Y. F., 475
Hudcová, V., 408

Hudec, R., 240, 339, 408, 479, 487, 593, 712, 753, 761
Hullinger, D. D., 659, 663, 671
Hunsberger, S. D., 651
Hurley, K., 57, 81, 106, 111, 192, 217, 399, 435, 613, 630, 723, 757

I

in't Zand, J. J. M., 119
Inneman, A., 712
Inoue, A. K., 514
Ioka, K., 115
Ishioka, R., 320
Ivanov, E., 757
Ivanushkina, M., 651

J

Janka, H.-T., 380
Jarvis, A., 166
Jehin, E., 269
Jelínek, M., 753, 761
Jensen, B. L., 301
Jernigan, J. G., 106, 111, 192, 217, 399, 435, 626, 630

K

Kaas, A. A., 269
Kaneko, Y., 203, 225, 236, 244
Karpov, S., 757
Kataoka, J., 307
Kato, T., 320
Kaufer, A., 269
Kawai, N., 81, 106, 192, 217, 307, 566, 618, 630
Kaye, T. G., 562
Kazanas, D., 294
Kildea, J., 166
Killough, R., 642
Kippen, R. M., 119, 225, 684, 688
Kittle, B., 642
Klar, R., 642
Klebesadel, R. W., 3, 541
Klose, S., 269, 333, 439, 483
Klotz, A., 447

Kobayashi, S., 125, 136, 208
Kočer, M., 593
Koide, S., 286, 392
Königl, A., 257, 282
Kouveliotou, C., 269, 508, 677, 684
Krimm, H. A., 655, 659, 663, 667, 671
Kubánek, P., 753, 761
Kulkarni, S., 324
Kumar, P., 458
Küpcü Yoldaş, A., 439
Kutyrev, A. S., 745

L

Labanti, C., 700
Lamb, D. Q., 19, 37, 42, 47, 57, 81, 106, 111, 192, 217, 399, 435, 566, 618, 630
Lapshov, I., 696
Lavagno, A., 420
Lazzarotto, F., 696
Lazzati, D., 251, 274, 403
Leake, M., 333
Ledoux, C., 453
Levan, A., 439, 537
Liang, E., 233
Lichti, G. G., 684
Lindner, T., 166
Lindsay, K. J., 333, 723
Lloyd-Ronning, N. M., 208
Loeb, A., 677
Longo, F., 578, 681, 696, 704
Loznikov, V., 757
Lu, T., 475
Lyutikov, M., 552

M

MacFadyen, A. I., 384
Malesani, D., 403
Malina, R., 447
Mangels, C., 642
Markwardt, C. B., 655, 659, 663, 667, 671
Masetti, N., 408, 483, 487, 491
Mason, E., 403
Mason, K. O., 651
Mason, R., 663
Mastichiadis, A., 294

Mastropietro, M., 696
Matsuoka, M., 106, 192, 217, 566, 618
Mauri, A., 700
Mazzali, P., 403
McBreen, B., 61, 225, 741
McBreen, S., 61, 225, 741, 761
McConnell, M. L., 589
McGowan, K., 728
McLean, K. M., 663, 667
McMahon, E., 529
McMeekin, M., 642
Medvedev, M. V., 25
Meegan, C. A., 684
Meintjes, P., 741
Mereghetti, S., 316, 491, 607, 696, 704
Mészáros, A., 94
Mészáros, P., 125, 208, 518
Miles, B. J., 642
Milne, P. A., 723
Mirabal, N., 337, 366, 533
Mizuno, Y., 392
Mochkovitch, R., 101, 328
Molinari, E., 443, 765
Møller, P., 301
Morales, M. F., 162
Moran, L., 61, 225, 741
Morelli, E., 696
Moretti, A., 642
Mori, K., 642
Morris, D. C., 642
Mueller, C., 166
Mukerjee, K., 642, 647
Mukherjee, R., 166, 533
Müller, E., 380
Murphy, J., 529

N

Nakagawa, Y. E., 106, 192, 566, 618, 630
Nakamura, T., 115, 416, 514
Nakar, E., 181
Nakazawa, K., 671
Natarajan, P., 508
Neff, J. E., 737
Nekola, M., 753
Nemiroff, R. J., 221, 749
Nieves, J. F., 154
Nishikawa, K.-I., 286, 290

Nogami, D., 320
Norris, J. P., 65, 412, 745
Nousek, J. A., 642, 647, 651
Nugent, P., 537

O

Okada, Y., 663, 671
Olive, J.-F., 81, 106, 192, 217, 618
Omodei, N., 681
Ong, J., 663
Ong, R. A., 166
Ortolan, A., 174
Osborne, J. P., 642, 647

P

Pacciani, L., 696
Paciesas, W. S., 684
Pagliara, G., 420
Palazzi, E., 269, 408
Palmer, D. M., 655, 659, 663, 667, 671
Panaitescu, A., 458
Pandey, S. B., 491
Park, H.-S., 269, 723
Parsons, A. M., 655, 659, 663, 667, 671
Páta, P., 761
Pedersen, H., 269, 447, 491
Pellizzoni, A., 696, 704
Perez-Ramirez, D., 221
Pérez Ramírez, M. D., 491
Perna, R., 388
Pian, E., 483
Piccioni, A., 77, 471
Pierce, M., 745
Pína, L., 712
Piran, T., 181
Pizzichini, G., 77, 106, 192, 339, 408, 471, 487, 601, 712
Polcar, J., 408
Polk, J., 663
Porrovecchio, G., 696
Pozanenko, A., 757
Preece, R. D., 119, 203, 225, 236, 244, 290, 684, 688
Preger, B., 696, 704
Prest, M., 696, 700
Price, P. A., 503

Prigozhin, G., 626, 630
Proga, D., 384

Q

Quimby, R., 529

R

Raby, E., 719
Ragan, K., 166
Ramirez-Ruiz, E., 349, 361, 508
Rapisarda, M., 696
Rau, A., 269, 439, 483, 622
Razzaque, S., 125
Rees, M. J., 198
Reimer, O., 269
Reinsch, K., 269
Reverte, D., 491
Rhoads, J., 388, 439, 537
Richardson, G., 286, 290
Ricker, G. R., 37, 57, 81, 106, 111, 192, 217, 399, 435, 467, 566, 626, 630
Ries, C., 269
Roiger, R. J., 585
Rol, E., 508
Roming, P. W. A., 651
Rossi, E. M., 198, 274, 316
Rosswog, S., 361
Rubini, A., 696
Ruffini, R., 312, 424, 428
Rumyantsev, V., 757

S

Sahu, S., 154
Sakamoto, T., 37, 42, 57, 81, 106, 192, 217, 566, 618, 630
Salmonson, J. D., 274, 547
Sari, R., 269
Sato, G., 663, 671
Sato, R., 192, 307, 566, 630
Savage, S. L., 745
Scalzo, R. A., 166
Schartel, N., 229, 316
Schmid, H. M., 269
Schmidt, B. P., 503

Schönfelder, V., 684
Shackelford, L., 663
Shaw, S., 333
Shibata, K., 286, 392
Shirasaki, Y., 57, 106, 192, 217, 566, 618, 630
Shirey, R., 719
Šimon, V., 339, 487, 593
Smith, I. A., 508
Smith, N., 741
Soffitta, P., 696
Sol, H., 290
Soldán, J., 761
Sollerman, J., 301
Soria, T., 761
Sprague, A. J., 73
Stallworth, A. D., 73
Stamatikos, M., 146
Starr, D., 728
Stecklum, B., 269, 333
Stella, L., 403
Straubmeier, C., 269
Štrobl, J., 753
Strong, I. B., 3
Suzuki, M., 57, 106, 192, 217, 307, 566, 630, 663, 671
Svéda, L., 712

T

Tagliaferri, G., 642, 647
Takagi, R., 307
Takahashi, D., 566
Takahashi, H., 663, 671
Takahashi, T., 663, 671
Tamagawa, T., 57, 106, 192, 217, 566, 618, 630
Tamburelli, F., 642
Tanvir, N. R., 439, 483, 491, 508
Tashiro, M., 671
Tavani, M., 696, 704
Tavenner, T., 663
Taylor, G., 324
Terndrup, D. M., 366
Terrell, J., 541
Testa, V., 403
Thiébaud, C., 447
Thompson, D., 719
Tichá, J., 593

Tichý, M., 593
Tiengo, A., 316, 491
Tilanus, R. P. J., 508
Topinka, M., 213, 408, 753
Torii, K., 106, 192, 217, 566, 618, 630
Torres Riera, J., 761
Toshiro, M., 663
Tovmassian, G., 269
Trifoglio, M., 700
Trimble, V., 7
Tueller, J., 655, 659, 663, 671

U

Uemura, M., 320
Urata, Y., 566, 630

V

Vallazza, E., 696, 700
van den Heuvel, E., 483
Vanderspek, R., 57, 81, 106, 111, 192, 217, 399, 435, 467, 626, 630
Vercellone, S., 696, 704
Vergani, S. D., 443, 765
Vernetto, S., 150
Vestrand, W. T., 719, 728
Vietri, M., 522
Villasenor, J. N., 57, 106, 192, 217, 626, 630
Vítek, S., 761
Vlahakis, N., 282
von Kienlin, A., 225, 622, 684
Vreeswijk, P., 301, 453

W

Walling, S., 663
Watanabe, S., 663
Watson, D. J., 642
Webb, J., 333
Wei, D. M., 86
Wells, A. A., 642, 647
White, R. R., 719, 728
Wick, S. D., 141
Wickersham, J., 723
Wiersma, J., 570

Wijers, R. A. M. J., 269, 453, 483, 508
Williams, D. A., 166
Williams, G. G., 269, 723
Williams, M., 333
Williams, O. R., 225, 244
Willingale, R., 642
Willis, D., 655
Wilson, R. B., 684
Winkler, C., 244
Woods, P. M., 119
Woosley, S. E., 106, 192, 217, 343, 376, 677
Wozniak, P., 728, 733
Wren, J., 719, 728
Wurtz, R. E., 723

X

Xue, S.-S., 312, 424, 428

Y

Yamada, S., 392
Yamamoto, Y., 566, 630
Yamaoka, H., 307, 320
Yamazaki, R., 115, 416, 514
Yamazaki, T., 57, 566, 630
Yamazaki, Y., 192
Yanagisawa, K., 307
Yatsu, Y., 307
Yonetoku, D., 416
Yoshida, A., 106, 192, 217, 566, 618, 630
Young, K. C., 73

Z

Zeh, A., 333, 483
Zerbi, F. M., 443, 765
Zhang, B., 125, 208, 518
Zhang, W., 343, 376
Zharikov, S., 269
Zolotukhin, Y., 757
Zugger, M., 642
Zweerink, J., 166